Stedman's
ANATOMY
& PHYSIOLOGY
WORDS

SECOND EDITION

Stedman's

ANATOMY & PHYSIOLOGY WORDS

SECOND EDITION

LIPPINCOTT
WILLIAMS
& WILKINS

Publisher: Rhonda M. Kumm, RN, MSN
Senior Manager: Julie K. Stegman
Senior Managing Editor: Nancy S. Wachter
Associate Managing Editor: Trista A. DiPaula
Associate Managing Editor: William A. Howard
Art Program Coordinator: Jennifer Clements
Assistant Production Manager: Kevin Iarossi
Typesetter: Peirce Graphic Services, Inc.
Printer & Binder: Malloy Litho

Printed in the United States of America

Second Edition, 2002

Library of Congress Cataloging-in-Publication Data

Stedman's anatomy & physiology words.—2nd ed.
   p. cm.
  ISBN 0-7817-3834-2 (alk. paper)
   1. Human anatomy—Nomenclature.   2. Human physiology—Nomenclature.
I. Title: Stedman's anatomy and physiology words. II. Title: Anatomy &
physiology words. III. Stedman, Thomas Lathrop, 1853–1938. IV. Lippincott
Williams & Wilkins.

QM81 S74 2002
612'.0014—dc21

                                                  2002066130

Contents

Acknowledgments

An important part of our editorial process is the involvement of medical transcriptionists—as advisors, reviewers, and/or editors.

We extend special thanks to A. Elaine Olson, CMT, and Patricia L. White, CMT, for editing the manuscript, helping resolve many difficult questions, and contributing material for the appendix sections. We are grateful to our Editorial Advisory Board members, including Pamela S. Harrmann, RN, BS, CCRN; Lin Harvell; Nancy Hill, MT; Carolyn Miles, CMT; Nancy Trueheart, CRI, CSR; and Pat Vargo, RHIT, who were instrumental in the development of this reference. They recommended sources and shared their valuable judgment, insight, and perspective.

We also extend thanks to Lin Harvell for working on the appendix. Additional thanks to Helen Littrell for performing the final prepublication review. Other important contributors to this edition include Marty Cantu, CMT; Shemah Fletcher; Sandy Kovacs, CMT; Robin Koza; and Tina Whitecotton, MT.

And, as always, Barb Ferretti played an integral role in the process by reviewing the content files for format, updating the database, and providing a final quality check.

As with all our *Stedman's* word references, this resource incorporates the suggestions and expertise of our many contacts in the medical transcriptionist community. Thanks to all of our advisory board participants, reviewers, and editors; AAMT meeting attendees; and others who have written us with requests and comments—keep talking, and we'll keep listening.

Editors' Preface

With co-editors living in Missouri and Texas, *Stedman's Anatomy & Physiology Words, Second Edition* book has traveled back and forth across the country, making its way back to the Lippincott Williams & Wilkins Baltimore office each time. We expect that this well-traveled title will prove to be a valuable reference for medical language professionals—from the seasoned professional to the new student.

A. Elaine Olson and Patricia White, both medical transcriptionists had to travel back to their student days of anatomy, physiology, and medical terminology courses as they edited this title. The combination of their backgrounds and experience, as well as valuable input from the book's Editorial Advisory Board and Stedman's customers have culminated in this new edition.

Medical terminology is the language that describes symptoms, diagnoses, treatments and procedures, anatomy, and instrumentation. They are the "windows" that medical transcriptionists see through as they document patient care. Lippincott Williams & Wilkins has always provided dependable, up-to-date references to support the MT. This new edition of *Stedman's Anatomy & Physiology Words* is no exception. As it is with most medical transcriptionists, I love reference books.

At my first American Association for Medical Transcription (AAMT) annual meeting in 1983, I went home with about 50 pounds of books! It was the beginning of a wonderful love affair that has never ended. I have always respected the authors and publishers of medical terminology reference materials, but having the honor of being a part of this team has only heightened my admiration. Thanks to those who have invested countless hours researching new terms, scanning medical publications, and surfing the Internet, all the while tracking down "leads" to make this book what it is.

A special thank you to Patty White, a wonderful editing partner, and LWW for the opportunity to participate in this project.

A. Elaine Olson, CMT

When I walked into Mrs. Miller's Anatomy, Physiology, and Medical Terminology class at Goldey-Beacom College more than 25 years ago, little did I know how important those words would become. At the time, I was a medical secretarial student, had yet to hear the word "transcriptionist," and certainly had no idea that I would ever BE one. I still have my original textbook, a 175-page overview of the structure and function of the human body. Looking back at it now, I realize just how simply it described the wonders of this mass of muscles, nerves, blood vessels, and organs we call the human body. When you consider that my textbook contained just 175 pages (and those pages contained all those extra words needed to make sentences), it's unbelievable to think that *Stedman's Anatomy and Physiology Words, Second Edition,* has over four times as many pages of anatomy and physiology terms . . . without any of those pesky "English" words.

From the largest organ down to the smallest cell, this new addition to the Stedman's library contains thousands of words presented in their trademark cross-referenced format and should serve as a resource for all of us. New MTs will find it invaluable as they embark on the path toward this great career. I can only wonder how much easier my early days of transcribing would have been had I had such a handy, easy-to-read reference book. . . . hmm, now that I think about it, ANY "reference book" would have been welcome back then. For those of us who have been at this for a while, we are well aware that our profession is a perpetual learning process, and just when we think we've got it mastered, a new doctor comes along and mumbles his way through the Latin version of a rarely-mentioned tendon or nerve, and we immediately groan (at the very least) and wish we had a comprehensive, easy-to-use anatomy reference.

Before starting this editing project, I had never seen the original edition of this word book. Now that I've had the chance to use it over these past few months and to help expand it into the book you're reading now, I realize just how much I've missed by not having it around. It is our hope that you will find *Stedman's Anatomy and Physiology Words, Second Edition,* to be a valuable addition to your Stedman's collection.

Special thanks go to my co-editor A. Elaine Olson; the many MT contributors and advisory board members; Barb Ferretti, database editor; and associate managing editor, Trista DiPaula. Many thanks, also, to the entire Stedman's team for always being one step ahead of us in developing new and improved reference materials that make us wonder how we ever made it this far without them.

Patricia L. White, CMT

Publisher's Preface

Stedman's Anatomy & Physiology Words, Second Edition, offers an authoritative assurance of quality and exactness to the wordsmiths of the healthcare professions—medical transcriptionists, medical editors and copyeditors, health information management personnel, court reporters, and the many other users and producers of medical documentation.

Over the years, customers have requested a word book that specifically covers anatomy and physiology. As a result, we decided to update, revise, and bring back the out-of-print *Stedman's Anatomy & Physiology Words,* originally published in 1992. The new edition is a comprehensive reference of anatomical and physiological terminology that the seasoned medical language specialist, as well as the student will find valuable.

In *Stedman's Anatomy & Physiology Words, Second Edition,* users will find thousands of words as they relate to the specialties of anatomy and physiology. Users will also find updated *Terminologia Anatomica* (TA) from *Stedman's Medical Dictionary, 27th Edition.* The appendix sections provide anatomical illustrations with useful captions and labels; tables of muscles, ligaments and tendons, nerves, and arteries; as well as anatomy words (English-Latin).

This compilation of more than 50,000 entries, fully cross-indexed for quick access, was built from a base vocabulary of approximately 30,000 medical words, phrases, abbreviations, and acronyms. The extensive A-Z list was developed from the database of *Stedman's Medical Dictionary, 27th Edition,* and supplemented by terminology found in current medical literature (see References on page xvi).

We at Lippincott Williams & Wilkins strive to provide you with the most up-to-date and accurate word references available. Your use of this word book will prompt new editions, which we will publish as often as updates and revisions justify. We welcome your suggestions for improvements, changes, corrections, and additions—whatever will make this *Stedman's* product more useful to you. Please complete the postpaid card at the back of this book, and send your recommendations care of "Stedman's" at Lippincott Williams & Wilkins.

Explanatory Notes

Medical transcription is an art as well as a science. Both approaches are needed to correctly interpret the dictation of a physician, whose language is a product of education, training, and experience. This variety in medical language means that there are several acceptable ways to express certain terms, including jargon. *Stedman's Anatomy & Physiology Words, Second Edition,* provides variant spellings and phrasings for many terms. These elements, in addition to complete cross-indexing, make *Stedman's Anatomy & Physiology Words, Second Edition,* a valuable resource for determining the validity of terms as they are encountered.

Alphabetical Organization

Alphabetization of main entries is letter by letter as spelled, ignoring punctuation, spaces, prefixed numbers, or other characters. For example:

O'Hara forceps
25(OH)D3
OHS

Terms beginning or ending with Greek letters show the Greek letters spelled out and listed alphabetically. For example:

alpha, α
    a. blockade
    estrogen receptor a. (ER alpha)
    a. gene

In subentry alphabetization, the abbreviated singular form or the spelled-out plural form of the noun main entry word is ignored.

Format and Style

All main entries are in **boldface** to expedite locating a sought-after term, to enhance distinction between main entries and subentries, and to relieve the textual density of the pages.

Irregular plurals and variant spellings are shown on the same line as the singular or preferred form of the word. For example:

bulla, pl. **bullae**
NCAM, N-CAM

Hyphenation

As a rule of style, multiple eponyms (e.g., Mears-Rubash approach) are hyphenated. Also, hyphens have been added between a manufacturer and one or more eponyms (e.g., Vital-Metzenbaum dissecting scissors). Please note that in many cases, hyphenation is a question of style, not of accuracy, and thus is a matter of choice.

Possessives

Possessive forms have been dropped in this reference for the sake of consistency and conformance with the guidelines of the American Association for Medical Transcription (AAMT) and other groups. Please note, however, that in many cases, retaining the possessive, like hyphenating, is a question of style, not of accuracy, and thus is a matter of choice. To form the possessive of a word, simply add the apostrophe or apostrophe "s" to the end of the word.

Cross-indexing

The word list is in an index-like main entry-subentry format that contains two combined alphabetical listings:

(1) A *noun* main entry-subentry organization, which is typical of the A-Z section of medical dictionaries like *Stedman's:*

epithelium
   ciliated e.
   e. lentis
   simple squamous e.

hyoid
   h. apparatus
   lesser horn of h.
   h. region

(2) An *adjective* main entry-subentry organization, which lists words and phrases as you hear them. The main entries are the adjectives or modifiers in a multiword term. The subentries are the nouns around which the terms are constructed and to which the adjectives or modifiers pertain:

distal
   d. colon
   d. esophagus
   d. interphalangeal joint
   d. surface of tooth

lower
   l. esophageal sphincter
   l. eyelid
   l. lobe of lung
   l. motor neuron

This format provides the user with more than one way to locate and identify a multiword term. For example:

Calot
    C. node

node
    Calot n.

It also allows the user to see together all terms that contain a particular descriptor, as well as all types, kinds, or variations of a noun entity. For example:

conjugate
    c. axis
    c. diameter of pelvic outlet
    false c.
    c. of outlet

Wherever possible, abbreviations are separately defined and cross-referenced. For example:

GFR
    glomerular filtration rate

glomerular
    g. filtration rate (GFR)

rate
    glomerular filtration r. (GFR)

References

In addition to the manufacturers' literature we gather at various medical meetings, scientific reports from hospitals, and the lists of our MT Editorial Advisory Board members (from their daily transcription work), we used the following sources for new terms in *Stedman's Anatomy & Physiology Words, Second Edition*.

Books

Drake E. Sloane's Medical Word Book, 4th Edition. Philadelphia: Saunders, 2001.

Greenfield LJ, Mulholland MW, Oldham KT, Zelenock GB, Lilemoe KD. Surgery: Scientific Principles and Practice, 3rd Edition. Philadelphia: Lippincott Williams & Wilkins, 2001.

Hiatt JL, Gartner LP. Textbook of Head and Neck Anatomy, 3rd Edition. Philadelphia: Lippincott Williams & Wilkins, 2000.

Hollinshead WH. Anatomy for Surgeons, The Head and Neck, 3rd Edition. Philadelphia: Lippincott Williams & Wilkins, 1982.

Lance LL. Quick Look Drug Book 2002. Baltimore: Lippincott Williams & Wilkins, 2002.

Loeser JD, Butler SH, Chapman CR, Turk DC. Bonica's Management of Pain, 3rd Edition. Philadelphia: Lippincott Williams & Wilkins, 2000.

Olson T. A.D.A.M. Student Atlas of Anatomy. Philadelphia: Lippincott Williams & Wilkins, 1996.

Stedman's Anatomy & Physiology Words. Baltimore: Lippincott Williams & Wilkins, 1992.

Stedman's Medical Dictionary, 27th Edition. Baltimore: Lippincott Williams & Wilkins, 2000.

Tessier C. The AAMT Book of Style. Modesto, CA: AAMT, 1995.

Tessier C. The Surgical Word Book, 2nd Edition. Philadelphia: Saunders, 1991.

Journals

Anesthesiology. Baltimore: Lippincott Williams & Wilkins, 2001.

Surgical Laparoscopy Endoscopy & Percutaneous Techniques. Baltimore: Lippincott Williams & Wilkins, 1999–2001.

Websites

http://www.hpisum.com

http://www.mtdaily.com

http://www.mtdesk.com

http://www.mtmonthly.com

A
> A bands
> A, B, C fiber
> A cells
> A disks

AA
> ascending aorta

AAL
> anterior axillary line

A1, A2 segment of anterior cerebral artery

a-aural

abampere

abapical pole

abaxial

abaxile

A, B, Cbile

ABD
> abdomen

abdomen (ABD)
>> anterior cutaneous nerve of a.
>> fatty layer of subcutaneous tissue of a.
>> membranous layer of subcutaneous tissue of a.
>> muscle of a.
>> rectus muscle of a.
>> subcutaneous vein of a.
>> transverse muscle of a.
>> upper a.
>> visceral lymph node of a.

abdominal
>> a. aorta
>> a. aortic plexus
>> aponeurosis of transverse a.
>> a. brain
>> a. canal
>> a. cavity
>> a. circumference
>> a. contents
>> a. esophagus
>> a. external oblique muscle
>> a. fat
>> a. fat pad
>> a. fissure
>> a. girth
>> a. heart
>> a. internal oblique muscle
>> a. lymph node
>> a. ostium of uterine tube
>> a. part
>> a. part of pectoralis major muscle
>> a. part of peripheral autonomic plexuses and ganglia
>> a. part of thoracic duct
>> a. part of ureter
>> a. pressure
>> a. reflex
>> a. region
>> a. respiration
>> a. ring
>> a. sac
>> a. viscera
>> a. wall
>> a. zone

abdominale
>> cerebrum a.

abdominalis
>> annulus a.
>> anulus a.
>> aorta a.
>> cavitas a.
>> globus a.
>> plexus nervosus aorticus a.
>> tunica a.

abdominis
>> aponeurosis of musculus transversus a.
>> cavum a.
>> fascia triangularis a.
>> intersectiones tendineae musculi recti a.
>> lamina anterior vagina musculi recti a.
>> lamina posterior vaginae musculi recti a.
>> linea arcuata vaginae musculi recti a.
>> membrana a.
>> musculi a.
>> musculus obliquus externus a.
>> musculus obliquus internus a.
>> musculus rectus a.
>> musculus transversalis a.
>> musculus transversus a.
>> nodi lymphoidei a.
>> obliquus externus a.
>> obliquus internus a.
>> panniculus adiposus telae subcutaneae a.
>> paracentesis a.
>> rectus a.
>> regio lateralis a.
>> regiones a.
>> subcutanea a.
>> tendinous intersections of rectus a.
>> transversus a.
>> vagina musculi recti a.
>> venae subcutaneae a.

abdominocardiac reflex

abdominocystic
abdominogenital
abdominopelvic
    a. cavity
    a. splanchnic nerve
abdominoperineal
abdominoscrotal
abdominothoracic arch
abdominovaginal
abdominovesical
abduce
abducens
    a. nerve
    nervus a.
    a. nucleus
    a. oculi
abducentis
    eminentia a.
    nervi a.
    nucleus a.
abducent nerve
abduct
abduction
    a. of first carpometacarpal joint
    a. of glenohumeral joint
    a. of metacarpophalangeal joint
    thumb a.
abductor
    a. digiti minimi muscle of foot
    a. digiti minimi muscle of hand
    a. digiti minimi quinti manus
    a. digiti minimi quinti pedis
    a. hallucis
    a. hallucis muscle
    a. indicis
    a. longus muscle
    a. magnus muscle
    a. muscle of great toe
    a. muscle of little finger
    a. muscle of little toe
    musculus a.
    a. pollicis brevis
    a. pollicis brevis muscle
    a. pollicis longus muscle
    a. pollicis longus tendon
abductus
    metatarsus a.
    pes a.
abembryonic
Abernethy fascia
aberrans
    ductulus a.
    ductus a.
    Haller vas a.
    Roth vas a.
    a. testis
    vas a.
aberrant
    a. bile duct

    a. bundles
    a. ductule
    a. ganglion
    a. obturator artery
aberrantes
    ductuli a.
    ductus a.
aberrantia
    Ferrein vasa a.
    vasa a.
aberration
abeyance
abfarad
abhenry
abiogenesis
abiogenetic
abiosis
abiotic
abiotrophy
ablastemic
ablepharia
abnerval
abneural
abnormal cleavage of cardiac valve
abohm
aborad
aboral
abort
abortion
abortive
abortus
above-elbow (AE)
abrachia
abrachiocephalia
abrachiocephaly
abscissa
absconsio
absolute
    atmosphere a.
    a. dehydration
    a. humidity
    a. terminal innervation ratio
    a. threshold
    a. viscosity
absorb
absorbancy
absorbent
    a. system
    a. vessel
absorption
    a. coefficient
    cutaneous a.
    external a.
    interstitial a.
    parenteral a.
    pathologic a.
absorptive cells of intestine
abterminal
abvolt

AC
    acromioclavicular
      AC joint
acantha
acanthion
acanthoid
acapnia
acapnial alkalosis
acarbia
acardia
acardiac
acardius
    a. acephalus
    a. amorphus
    a. anceps
acaryote
acatalasia
accelerans
accelerant
acceleration
    angular a.
    linear a.
    radial a.
accelerator
    a. fiber
    a. nerve
accelerometer
acces
    nucleus olivaris a.
accessoria, pl. **accessoriae**
    arteria obturatoria a.
    cartilagines nasales accessoriae
    cephalica a.
    glandulae lacrimales accessoriae
    glandulae suprarenales accessoriae
    glandulae thyroideae accessoriae
    glandula parotidea a.
    glandula parotis a.
    glandula thyroidea a.
    mamma a.
    organa oculi a.
    radix a.
    saphena a.
    structurae oculi accessoriae
    thyroidea a.
    vena cephalica a.
    vena hemiazygos a.
    vena saphena a.
    vena vertebralis a.
accessorii
    lymphonodi comitantes nervi a.
    nervi phrenici a.

    nodi lymphatici comitantes nervi a.
    nodi lymphoidei a.
    nucleus spinalis nervi a.
    pars spinalis nervi a.
    pars vagalis nervi a.
    radix cranialis nervi a.
    radix spinalis nervi a.
    rami musculares nervi a.
    ramus externus trunci nervi a.
    ramus internus trunci nervi a.
    truncus nervi a.
accessorium
    pancreas a.
    septum a.
accessorius
    ductus pancreaticus a.
    lien a.
    musculus flexor a.
    nervus a.
    nucleus cuneatus a.
    processus a.
    ramus meningeus a.
    spleen a.
    a. willisii
accessory
    a. adrenal gland
    a. arteriovenous connection
    a. auricle
    a. blood vessel
    a. breast
    a. cephalic vein
    a. cuneate nucleus
    a. flexor muscle of foot
    a. flocculus
    a. hemiazygos vein
    a. hepatic duct
    a. lacrimal gland
    a. maxillary hiatus
    a. meningeal branch
    a. meningeal branch of middle
      meningeal artery
    a. nasal cartilage
    a. navicular bone
    a. nerve lymph node
    a. nerve root
    a. nerve trunk
    a. obturator artery
    a. olivary nuclei
    a. organ
    a. organs of eye
    a. ovary
    a. pancreas

NOTES

accessory *(continued)*
  a. pancreatic duct
  a. parotid gland
  a. phrenic nerve
  a. placenta
  a. plantar ligament
  a. portion of spinal accessory nerve
  a. process
  a. process of lumbar vertebra
  a. quadrate cartilage
  a. root of tooth
  a. saphenous vein
  a. spleen
  a. suprarenal gland
  a. thyroid
  a. thyroid gland
  a. tubercle
  a. venous sinus of Verga
  a. vertebral vein
  a. visual apparatus
  a. visual structure
  a. volar ligament
acclimatization
acclimitation
accommodation
  histologic a.
  a. of nerve
accommodative
accompanying
  a. artery of sciatic nerve
  a. vein
  a. vein of hypoglossal nerve
accrementition
accreta
  placenta a.
accretionary growth
accretion line
acellular
acephalia
acephalism
acephalobrachia
acephalocardia
acephalocheiria, acephalochiria
acephalogaste
acephalogaster
acephalopodia
acephalorrhachia
acephalostomia
acephalothoracia
acephalous
acephalus
  acardius a.
  a. dibrachius
  a. dipus
  holoacardius a.
  a. monobrachius
  a. paracephalus
  a. sympus

acephaly
acervulus
acetabula (*pl. of* acetabulum)
acetabular
  a. artery
  a. branch
  a. fossa
  a. labrum
  a. lip
  a. margin
  a. notch
  a. rim
acetabulare
  labrum a.
acetabularis
  margo a.
  ramus a.
acetabuli
  arteria a.
  facies lunata a.
  fossa a.
  incisura a.
  ligamentum transversum a.
  limbus a.
acetabulum, pl. acetabula
  labrum of a.
  lunate surface of a.
  margin of a.
  notch of a.
  transverse ligament of a.
acetylcholine (ACH, Ach)
ACH, Ach
  acetylcholine
achalasia
acheales
acheilia
acheilous
acheiria
acheiropody
acheirous
achilia
achilles
  A. bursa
  tendo A.
  A. tendon
  A. tendon reflex
achillis
  bursa a.
achilous
achiria
achiropody
achirous
achondrogenesis
achondroplasia
  homozygous a.
achondroplasty
achroacyte
achromasia
achromatic apparatus

achromatin
achromatinic
achromatophil
achromatosis
achromia
achromophil
achromophilic
achromophilous
achromotrichia
acid
    amino a.
    apurinic a.
    a. cell
    gamma-aminobutyric a.
    a. gland
    separation of amino a.
    unesterified free fatty a. (UFA)
acid-base
    a.-b. balance
    a.-b. equilibrium
acidemia
    methylmalonic a.
acidocyte
acidophil
    a. cell
    a. granule
acidophile
acidophilic leukocyte
acidosis
    carbon dioxide a.
    compensated a.
    hyperchloremic a.
    metabolic a.
    respiratory a.
    uncompensated a.
acidotic
aciduria
    methylmalonic a.
acinar
    a. cell
    a. tissue
acini (*pl. of* acinus)
acinic
aciniform
acinose
acinotubular gland
acinous
    a. cell
    a. gland
acinus, pl. acini
    liver a.

ACL
    anterior cruciate ligament
aclasis
    diaphysial a.
acme
acnemia
acollis
    uterus a.
acolous
acorea
acormus
Acosta disease
acoustic
    a. area
    a. cell
    a. crest
    a. lemniscus
    a. meatus
    a. nerve
    a. papilla
    a. radiation
    a. spot
    a. striae
    a. tetanus
    a. tooth
    a. tubercle
    a. vesicle
acousticofacial
    a. crest
    a. ganglion
acquired hernia
acrania
acranial
Acrel ganglion
acribometer
acroblast
acrobrachycephaly
acrocephalia
acrocephalic
acrocephalopolysyndactyly
acrocephalosyndactyli
acrocephalosyndactylism
acrocephalosyndactyly
acrocephalous
acrocephaly
acrocinesia
acrocinesis
acrodolichomelia
acrodysostosis
acrodysplasia
acrofacial dysostosis
acrogeria

NOTES

acrokinesia
acromelia
acrometagenesis
acromial
    a. anastomosis of thoracoacromial artery
    a. angle
    a. arterial network
    a. articular facies of clavicle
    a. articular surface of clavicle
    a. bone
    a. branch of suprascapular artery
    a. branch of thoracoacromial artery
    a. end of clavicle
    a. extremity of clavicle
    a. facet of clavicle
    a. part of deltoid muscle
    a. plexus
    a. process
    a. reflex
acromiale
    os a.
    rete a.
acromialis
    angulus a.
    bursa subcutanea a.
acromii
    angulus a.
    facies articularis clavicularis a.
acromioclavicular (AC)
    a. articulation
    a. disk
    a. joint
    a. ligament
acromioclaviculare
    ligamentum a.
acromioclavicularis
    articulatio a.
    discus articularis a.
acromiocoracoid ligament
acromiohumeral
acromion
    articular surface of a.
    bursa of a.
    clavicular articular facet of a.
    a. process
    a. scapulae
acromioplasty
acromioscapular
acromiothoracic artery
acromphalus
acroosteolysis
acroosteolytica
    osteopetrosis a.
acropachyderma
acropetal
acrosomal
    a. cap

    a. granule
    a. vesicle
acrosome
acrosphenosyndactyly
acroteric
acrotic
actin filament
action
    calorigenic a.
    a. current
    a. potential
    salt a.
    sparing a.
    specific dynamic a. (SDA)
    thermogenic a.
activation
active
    a. congestion
    a. hyperemia
    a. length-tension curve
    a. movement
    a. transport
    a. vasoconstriction
    a. vasodilation
activity
    nonsuppressible insulinlike a. (NSILA)
    pulseless electrical a. (PEA)
    tertiary peristaltic a.
aculeate
aculeatum
    stratum a.
acuminate
acustic
    ramus meatus a.
acustica
    radiatio a.
    tuba a.
acusticae
    maculae a.
    teniae a.
acustici
    cartilago meatus a.
    dentes a.
    incisura cartilaginis meatus a.
    trigonum a.
acusticus
    nervus a.
    nucleus a.
acute
    a. angle
    a. margin of heart
acystia
AD
    right ear
adactylia
adactylism
adactylous
adactyly

adamantina
    membrana a.
    prismata a.
    substantia a.
adamantine membrane
Adamkiewicz
    arteries of A.
    A. artery
Adam's apple
adaptation syndrome of Selye
adaxial
Addison clinical plane
additive effect
adducent
adduct
adducta
    coxa a.
adduction of glenohumeral joint
adductor
    a. brevis
    a. brevis muscle
    a. canal
    a. compartment of thigh
    a. hallucis
    a. hallucis muscle
    a. hiatus
    a. longus muscle
    a. magnus
    a. magnus muscle
    a. magnus tendon
    a. minimus muscle
    a. muscle of great toe
    a. muscle of thumb
    musculus a.
    a. pollicis
    a. pollicis muscle
    a. pollicis obliquus
    a. reflex
    a. tubercle
    a. tubercle of femur
adductorium
    tuberculum a.
adductorius
    canalis a.
    hiatus a.
adductorum
    compartimentum femoris a.
adductovarus
    metatarsus a.
adductus
    metatarsus a.
    pes a.

adelomorphous
adendric
adendritic
adeniform
adenization
adenoblast
adenocyte
adenogenesis
adenogenous
adenohypophyseos
    pars intermedia a.
adenohypophysial
adenohypophysis
    intermediate part of a.
adenoidal
    a.-pharyngeal-conjunctival (A-P-C, APC)
adenoidea
    tonsilla a.
adenoid tissue
adenomere
adenose
adenous
adeps renis
adequal cleavage
adequate stimulus
adermia
adherens
    fascia a.
    macula a.
    zonula a.
adherent
    a. placenta
    a. zone
adhesio interthalamica
adhesion
    amniotic a.
    interthalamic a.
adhesiones
adiaphoresis
adiaphoretic
adiaphoria
adiathermancy
adipes
adipis
adipocellular
adipocyte
adipogenesis
adipogenic
adipogenous
adipoid
adipokinetic hormone

NOTES

adipokinin
adiposa
    hernia a.
adiposae
    plicae a.
adipose
    a. capsule of kidney
    a. cell
    a. folds of pleura
    a. fossae
    a. ligament
    a. renal capsule
    a. tissue
adiposum
    corpus a.
adiposus
    panniculus a.
aditus
    a. ad antrum
    a. ad antrum mastoideum
    a. ad saccum peritonei minorem
    a. glottidis inferior
    a. glottidis superior
    laryngeal a.
    a. laryngis
    a. of larynx
    a. to mastoid antrum
    a. orbitae
    a. pelvis
adix inferior
adjacent
    a. angle
    a. tissue
admaxillary gland
admedial
admedian
adminicula
adminiculum
    linea a.
    a. lineae albae
adnerval
adneural
adnexa, sing. adnexum
    ocular a.
    a. oculi
    a. uterus
adnexal
adrenal
    a. androgen-stimulating hormone
    a. body
    butterfly a.
    a. capsule
    a. cortex
    a. gland
    a. weight factor
adrenaline reversal
adrenergic
    a. fiber
    a. receptor

adrenic
adrenoceptive
adrenoceptor
adrenocortical hormone
adrenocorticomimetic
adrenocorticotrophic
adrenocorticotropic hormone
adrenolytic
adrenomimetic
adrenoreactive
adrenoreceptor
adrenotrophic
adrenotropic hormone
adrenotropin
adsorb
adsorption theory of narcosis
adsternal
adterminal
advehens, pl. advehentes
    vena a.
adventicia
adventitia
    membrana a.
    tunic a.
    tunica a.
adventitial
    a. bed
    a. cell
    a. sheath
    a. tissue
adventitious
    a. bursa
    a. membrane
adversive movement
AE
    above-elbow
    aryepiglottic
    AE fold
Aeby
    A. muscle
    A. plane
aequorin
aerate
aerated
aeration
aeroatelectasis
aerobic respiration
aerobiology
aerobiosis
aerobiotic
aerodynamics
aeroemphysema
aeropause
aeroplethysmograph
aerotonometer
affectomotor
afferens
    arteriola glomerularis a.
    vas lymphaticum a.

afferent
- a. fiber
- a. glomerular arteriole
- a. limb
- a. loop
- a. lymphatic
- a. nerve
- a. vein
- a. vessel

afferentia
- vasa a.

affinity

affixa
- lamina a.

afflux

affluxion

AFH
- anterior facial height

afibrillar cementum

afterbirth

aftercontraction, after-contraction

aftercurrent

afterdischarge, after-discharge

aftereffect

afterload
- ventricular a.

afterpotential

agamocytogeny

agamogenesis

agamogenetic

agamogony

agamous

agastric

age
- bone a.
- childbearing a.
- fetal a.
- gestational a.
- physiologic a.

agenesis
- gonadal a.
- renal a.
- thymic a.

agenitalism

agenosomia

agent
- ganglionic blocking a.

agger
- a. nasi
- a. nasi cell
- a. perpendicularis
- a. valvae venae

aggeres

agglomerate

agglomerated

agglomeration

aggregated
- a. lymphatic follicles
- a. lymphatic follicles of small intestine
- a. lymphatic follicles of vermiform appendix
- a. lymphatic nodule
- a. lymphoid nodule
- a. lymphoid nodules of small intestine

aggregate gland

aggregati
- folliculi lymphatici a.

aggregation

aging
- clonal a.

aglossia

aglossostomia

agmen peyerianum

agmina

agminated gland

agminate gland

agnathia

agnathous

agogue

agon

agonadal

agonist

agonistic muscle

agranular
- a. cortex
- a. endoplasmic reticulum

agranulocyte

agranuloplastic

agyria

ai
- thoracica suprema a.
- thyroidea ima a.

AICA
- anterior inferior communicating artery

aileron

air
- alveolar a.
- a. cell
- a. cell of auditory tube
- a. cell cast
- complemental a.
- complementary a.

NOTES

air (*continued*)
    functional residual a.
    minimal a.
    reserve a.
    residual a.
    a. sac
    a. saccule
    a. sinus
    a. space
    supplemental a.
    tidal a.
    a. vesicle
airway
    anatomic a.
    conducting a.
    esophageal a.
    lower a.
    nasal a.
    a. resistance
    respiratory a.
    upper a.
akanthion
akaryocyte
akaryote
ala, pl. **alae**
    a. auris
    a. cerebelli
    a. cinerea
    a. cristae galli
    a. of crista galli
    a. of ilium
    a. lobulis centralis
    a. major ossis sphenoidalis
    a. minor ossis sphenoidalis
    a. nasi
    a. of nose
    a. orbitalis
    a. ossis ilii
    sacral a.
    a. sacralis
    a. of sacrum
    a. temporalis
    a. of vomer
    a. vomeris
alacrima
alactic oxygen debt
alae (*pl. of* ala)
alar
    a. artery of nose
    a. bone
    a. branch of external maxillary artery
    a. cartilage
    a. chest
    a. fascia
    a. fold of intrapatellar synovial fold
    a. lamina of neural tube
    a. ligament

    a. part
    a. part of nasalis muscle
    a. plate of neural
    a. process
    a. scapula
    a. spine
alares
    plicae a.
alaria
    ligamenta a.
alaris
    lamina a.
alarm reaction
alba
    commissura ventralis a.
    linea a.
    substantia a.
albae
    adminiculum lineae a.
Albarran
    A. gland
    A. y Dominguez tubule
albi
    rami communicantes a.
albicans
    corpus a.
albicantia
albidus
Albini nodule
Albinus muscle
albocinereous
Albrecht bone
albuginea
    tunica a.
albugineous
albuminous
    a. cell
    a. gland
Alcock canal
alcohol diuresis
alecithal ovum
alemmal
aletocyte
algoid cell
alienia
aliform
alimentarium
    systema a.
alimentary
    a. apparatus
    a. canal
    a. system
    a. tract
alinasal
alinjection
alipotropic
alisphenoid cartilage
alkalosis
    acapnial a.

compensated a.
metabolic a.
respiratory a.
uncompensated a.
allantochorion
allantoenteric diverticulum
allantogenesis
allantoic
a. bladder
a. cavity
a. diverticulum
a. fluid
a. membrane
a. sac
a. stalk
a. vesicle
allantoidoangiopagus
allantois
allaxis
allelotaxis
allelotaxy
allied reflex
allocortex
allogamy
allogotrophia
allokinesis
allomeric function
allometron
allomorphism
allophenic
allophore
allosome
paired a.
allotherm
allotopia
allotrichia circumscripta
almond nucleus
alpha
a. cells of anterior lobe of hypophysis
a. cells of pancreas
a. fiber
a. granule
a. substance
a. unit
alpha-adrenergic receptors
alpha-fetoprotein
alta
patella a.
alteration
modal a.

qualitative a.
quantitative a.
alternation
altitude
a. chamber
a. erythremia
Altmann
A. granule
A. theory
alvei (*pl. of* alveus)
alveolar
a. air
a. angle
a. arch
a. arch of mandible
a. arch of maxilla
a. body
a. bone
a. border
a. branch of internal maxillary artery
a. canal of maxilla
a. canals
a. cell
a. crest
a. dead space
a. duct
a. foramina of maxilla
a. gas
a. gas equation
a. gland
a. index
a. macrophage
a. mucosa
a. nerve
a. part of mandible
a. periosteum
a. point
a. process
a. process of maxilla
rami a.
a. ridge
a. sac
a. septum
a. ventilation
a. yoke
alveolare
jugum a.
periosteum a.
alveolaria
foramina a.
alveolaris, pl. **alveolares**

NOTES

alveolaris *(continued)*
    canales alveolares
    ductulus a.
    a. inferior
    limbus a.
    processus a.
    sacculus a.
    squama a.
    alveolares superiores
    alveolares superiores anteriores
    a. superior posterior
alveolate
alveoli (*pl. of* alveolus)
alveolobuccal
    a. groove
    a. sulcus
alveolocapillary membrane
alveolodental
    a. canal
    a. ligament
    a. membrane
alveololabial
    a. groove
    a. sulcus
alveololabialis
alveololingual
    a. groove
    a. sulcus
alveolonasal line
alveolopalatal
alveoloschisis
alveolus, pl. **alveoli**
    dental alveoli
    a. dentalis
    pulmonary a.
    alveoli pulmonis
    tapetum alveoli
alveus, pl. **alvei**
    a. hippocampi
    a. urogenitalis
amacrine cell
Ambard law
Amberg lateral sinus line
ambiens
    cisterna a.
ambient cisterna
ambiguous nucleus
ambiguus
    nucleus a.
ambilateral
ambisexual
ambomalleal
ambos
amebaism
amebocyte
ameboid
    a. astrocyte
    a. cell
ameboididity

ameboidism
amelia
ameloblast
ameloblastic layer
amelodental junction
amelodentinal junction
amelogenesis
ametria
amicroscopic
amino acid
amis vermis
amitosis
amitotic
ammeter
Ammon
    A. fissure
    A. horn
    A. prominence
ammonis
    cornu a.
amni
amnii
    liquor a.
amniocardiac vesicle
amniochorionic
amnioembryonic junction
amniogenesis
amniogenic cells
amnionic
amnion ring
amniotic
    a. adhesion
    a. bands
    a. cavity
    a. corpuscle
    a. duct
    a. fluid
    a. fold
    a. hernia
    a. raphe
    a. sac
amorpha
    pars a.
amorphia
amorphism
amorphous
amorphus
    acardius a.
    holoacardius a.
amperage
ampere
ampheclexis
amphiarthrodial joint
amphiarthrosis
amphiarthrotic pubic symphysis
amphiaster
amphiblestrodes
amphicelous
amphicentric

amphicyte
amphidiarthrodial joint
amphidiarthrosis
amphikaryon
amphinucleolus
amphochromatophil, amphochromatophile
amphochromophil, amphochromophile
amphocyte
amphophil, amphophile
 a. granule
amphophilic
amphophilous
amplification
ampulla, pl. ampullae
 biliaropancreatic a.
 a. biliaropancreatica
 a. canaliculi lacrimalis
 a. chyli
 a. of ductus deferens
 a. ductus deferentis
 a. ductus lacrimalis
 duodenal a.
 a. duodeni
 a. of duodenum
 a. of gallbladder
 Henle a.
 hepatopancreatic a.
 a. hepatopancreatica
 a. of lacrimal canaliculus
 a. lactifera
 lactiferous a.
 a. of lactiferous duct
 a. membranacea
 membranous a.
 a. of milk duct
 musculus sphincter ampullae
 ampullae osseae canalium
  semicircularium
 osseous a.
 rectal a.
 a. recti
 a. of rectum
 sphincter of biliaropancreatic a.
 sphincter of hepatopancreatic a.
 a. tubae uterinae
 a. of uterine tube
 a. of vas deferens
 Vater a.
 a. of Vater
ampullar
ampullaria
 crura membranacea a.

ampullaris
 a. anterior
 crista a.
 cupula cristae a.
 a. lateralis
 neuroepithelium cristae a.
 plica a.
 a. posterior
 sulcus a.
ampullary
 a. crest
 a. crest of semicircular duct
 a. crura of semicircular duct
 a. fold of uterine tube
 a. groove
 a. limbs of semicircular ducts
 a. nerve
 a. sulcus
 a. type of renal pelvis
ampullula
amputation
 birth a.
 congenital a.
 intrauterine a.
 spontaneous a.
amu
 anatomic mass unit
amus
 a. digastricus
 a. iliacus
Amussat
 A. valve
 A. valvula
amyelencephalia
amyelencephalic
amyelencephalous
amyelia
amyelic
amyelinated
amyelination
amyelinic
amyeloic
amyelonic
amyelous
amygdala cerebelli
amygdalae
 nucleus a.
amygdaline
amygdaloid
 a. body
 a. complex
 a. fossa

NOTES

13

amygdaloid *(continued)*
    a. nucleus
    a. tubercle
amygdaloidei
    rami corporis a.
amygdaloideum
    corpus a.
amylaceous corpuscle
amylaceum, pl. **amylacea**
    corpus a.
amyloid
    a. bodies of prostate
    a. corpuscle
amyoplasia
amyotonia
anabolic
anabolism
anabolite
anacamptometer
anacatadidymus
anadidymu
anaerobic respiration
anagen
anagenesis
anagenetic
anal
    a. atresia
    a. canal
    a. canal artery
    a. cleft
    a. column
    a. crypt
    a. cushion
    a. ducts
    a. fascia
    a. gland
    a. membrane
    a. orifice
    a. papilla
    a. pecten
    a. pit
    a. plate
    a. reflex
    a. region
    a. sinuses
    a. sphincter
    a. transitional zone
    a. triangle
    a. valve
    a. verge
anales
    columnae a.
    sinus a.
    valvulae a.
analgesic
analgetic
analis
    canalis a.
    crena a.

    linea pectinata canalis a.
    musculi regionis a.
    pecten a.
    regio a.
    zona transitionalis a.
analog, analogue
analogous
analysis, pl. **analyses**
    bradykinetic a.
anamnionic
anamniotic
anamorphosis
anaplastic cell
anapophy
anapophysis
anastomose
anastomosing vessel
anastomosis, pl. **anastomoses**
    arteriolovenular a.
    a. arteriolovenularis
    a. arteriovenosa
    arteriovenous a.
    Béclard a.
    calcaneal a.
    Clado a.
    crucial a.
    cruciate a.
    cubital a.
    Galen a.
    genicular a.
    heterocladic a.
    homocladic a.
    Hoyer anastomoses
    Hyrtl a.
    intermesenteric arterial a.
    Jacobson a.
    patellar a.
    portacaval a.
    portal-systemic a.
    postcostal a.
    precapillary a.
    precostal a.
    ranine a.
    Riolan a.
    Sucquet a.
    Sucquet-Hoyer a.
anastomotic
    a. branch
    a. branch of middle meningeal
        artery with lacrimal artery
    a. fibers
    a. vein
    a. vessel
anastomotica
    a. inferior
    a. magna
    a. superior
anastomoticum
    vas a.

anastomoticus
> ramus a.

anatomic
> a. airway
> a. dead space
> a. mass unit (amu)
> a. position
> a. snuffbox
> a. sphincter

anatomica
> Basle Nomina A. (BNA)
> conjugata a.'s
> Nomina A. (NA)
> Terminologia A. (TA)

anatomical
> a. conjugate
> a. crown
> a. dead space
> a. element
> a. internal os of uterus
> a. neck of humerus
> a. position
> a. snuffbox
> a. sphincter

anatomically
anatomicomedical
anatomicopathologic
anatomicosurgical
anatomicum
> ostium a.

anatomique
> tabatière a.

anatomist
anatomy
> applied a.
> artificial a.
> artistic a.
> clastic a.
> clinical a.
> comparative a.
> dental a.
> descriptive a.
> developmental a.
> functional a.
> general a.
> gross a.
> living a.
> macroscopic a.
> medical a.
> microscopic a.
> physiologic a.
> physiological a.

> plastic a.
> practical a.
> radiologic a.
> radiological a.
> regional a.
> special a.
> surface a.
> surgical a.
> systematic a.
> systemic a.
> topographic a.
> transcendental a.
> ultrastructural a.

anaxon
anaxone
AnCC
> anodal closure contraction

anceps
> acardius a.

anchoring villus
ancillary
ancipital
ancipitate
ancipitous
ancon
anconad
anconal fossa
anconeal
anconeus
> a. muscle
> musculus a.

anconoid
ancyroid
Andernach ossicle
Andersch
> A. ganglion
> A. nerve

androgen binding protein
androgenesis
androgenic
> a. hormone
> a. zone

androgenous
androgynism
androgynoid
androgynous
androgyny
android pelvis
andromorphous
anelectroton
anelectrotonic
anelectrotonus

NOTES

anemic
    a. anoxia
    a. hypoxia
anemicus
    nevus a.
anemometer
anencephalia
anencephalic
anencephalous
anencephaly
    partial a.
anenzymia
anephric
aneroid manometer
anesthesia
    hydrate microcrystal theory of a.
aneurolemmic
angiectasia
angiectasis
angioarchitecture
angioblast
angioblastic cells
angiocyst
angiogenesis
angiogenic
angioid
angiokinesis
angiokinetic
angiologia
angiology
angiomyocardiac
angiopoiesis
angiopoietic
angiotonia
angiotonic
angiotrophic
angle
    acromial a.
    acute a.
    adjacent a.
    alveolar a.
    anorectal a.
    antegonial a.
    a. of anterior rib
    a. of antetorsion
    a. of anteversion
    axial a.
    basilar a.
    beta a.
    biorbital a.
    Broca basilar a.
    Broca facial a.
    cardiohepatic a.
    cardiophrenic a.
    carrying a.
    cavity line a.
    cephalic a.
    cephalomedullary a.
    cerebellar pontine a.

cerebellopontile a.
cerebellopontine a.
cervicoisthmic a.
conchal mastoid a.
costal a.
costophrenic a.
costovertebral a.
costoxiphoid a.
craniofacial a.
Daubenton a.
a. of declination
a. of depression
duodenojejunal a.
epigastric a.
ethmoid a.
facial a.
a. of femoral torsion
filtration a.
Frankfort-mandibular incisor a.
hepatic-renal a.
hepatorenal a.
a. of His
hypsiloid a.
impedance a.
a. of inclination
a. of inferior scapula
infrasternal a.
a. of iridocorneal
iridocorneal a.
a. of iris
Jacquart facial a.
a. of jaw
a. of lateral eye
left venous a.
a. of Louis
Louis a.
Ludwig a.
a. of Ludwig
lumbosacral a.
a. of mandible
medial a.
a. of medial eye
metafacial a.
a. of mouth
olfactory a.
ophryospinal a.
parietal a.
pectinate ligament of
    iridocorneal a.
pelvic-femoral a.
pelvivertebral a.
phrenopericardial a.
Pirogoff a.
pontine a.
a. of posterior rib
pubic a.
Quatrefages a.
Ranke a.
rib-vertebral a.

right venous a.
sacrohorizontal a.
Serres a.
spaces of iridocorneal a.
sphenoid a.
splenorenal a.
sternal a.
sternoclavicular a.
subcarinal a.
subcostal a.
subpubic a.
subscapular a.
substernal a.
sulcus a.
a. of superior scapula
sylvian a.
a. of Sylvius
Topinard facial a.
venous a.
Virchow a.
Virchow-Holder a.
Vogt a.
Weisbach a.
Welcker a.
xiphocostal a.
xiphoid a.

angular
   a. acceleration
   a. aperture
   a. artery
   a. convolution
   a. gyrus
   a. incisure
   a. notch
   a. sphincter
   a. spine
   a. tract of cervical fascia
   a. vein

angularis
   arteria gyri a.
   dens a.
   gyrus a.
   incisura a.
   sphincter a.
   spina a.
   sulcus a.
   vena a.

angulus, pl. **anguli**
   a. acromialis
   a. acromii
   a. costae
   a. frontalis ossis parietalis

a. inferior scapulae
a. infrasternalis
a. iridis
a. iridocornealis
a. lateralis scapulae
a. mandibulae
a. mastoideus ossis parietalis
a. occipitalis ossis parietalis
a. oculi lateralis
a. oculi medialis
a. oculi nasalis
a. oculi temporalis
a. oris
a. sphenoidalis ossis parietalis
a. sterni
a. subpubicus
a. superior scapulae

anhidrosis
anhidrotic
anhistic
anhistous
anhydration
ani (*pl. of* anus)
anial
anidean
anideus
   embryonic a.
anidous
anilinophil
anilinophile
anilinophilous
animal
   cold-blooded a.
   a. force
   a. pole
   warm-blooded a.
animation
anisodactylous
anisodactyly
anisogamy
anisognathous
anisomastia
anisomelia
anisospore
anisosthenic
anisotonic
anisotropic
   a. disks
   a. lipid
Anitschkow
   A. cell
   A. myocyte

NOTES

ankle
  a. bone
  central bone of a.
  a. clonus
  dorsiflexion of a.
  eversion of a.
  extensor retinaculum of a.
  fibular collateral ligament of a.
  inferior extensor retinaculum of a.
  inferior fibular retinaculum of a.
  inversion of a.
  a. jerk
  a. joint
  lateral collateral ligament of a.
  lateral joint of a.
  medial joint of a.
  a. mortise
  plantarflexion of a.
  a. reflex
  a. region
  superior extensor retinaculum of a.
  superior fibular retinaculum of a.
  talocrural joint of a.
ankylocolpos
ankylodactylia
ankylodactyly
ankyroid
anlage
anlagen
annectent gyrus
annexa
annexal
annular
  a. band
  a. cartilage
  a. ligament
  a. ligament of radius
  a. ligament of stapes
  a. ligaments of trachea
  a. pancreas
  a. placenta
  a. plexus
  a. pulley
  a. sphincter
annulare
  a. bulbi
  ligamentum a.
annularis
  digitus a.
  plexus a.
  rachitis fetalis a.
annulate lamellae
annuli (*pl. of* annulus)
annulospiral
  a. ending
  a. organ
annulu
annulus, pl. **annuli**
  a. abdominalis

  a. ciliaris
  a. conjunctiva
  a. femoralis
  a. fibrosus
  Haller a.
  a. hemorrhoidalis
  a. inguinalis profundus
  a. inguinalis superficialis
  a. lymphaticus cardiae
  a. ovalis
  superior a.
  a. tendineus communis
  a. of Zinn
AnOC
  anodal opening contraction
anococcygea
  raphe a.
anococcygeal
  a. body
  a. ligament
  a. nerve
anococcygei
  nervi a.
anococcygeum
  ligamentum a.
anococcygeus
  nervus a.
anocutanea
  linea a.
anocutaneous line
anodal
  a. closure contraction (AnCC)
  a. current
  a. opening contraction (AnOC, AOC)
anoderm
anodontia
  partial a.
anodontism
anogenital
  a. band
  a. raphe
anomalad
anomalous
  a. uterus
  a. viscosity
anomaly
  developmental a.
anonyma
  arteria a.
anonymous
  a. artery
  a. vein
anophthalmia
  consecutive a.
  primary a.
  secondary a.
anorchia
anorchism

anorectal
    a. angle
    a. flexure
    a. junction
    a. line
    a. lymph node
anorectalis, pl. **anorectales**
    flexura a.
    junctio a.
    linea a.
    lymphonodi anorectales
    nodi lymphoidei anorectales
anorectoperineales
    musculi a.
anorectoperineal muscle
anospinal center
anotia
anovesical
anovular ovarian follicle
anovulation
anovulatory cycle
anoxemia
anoxia
    anemic a.
    anoxic a.
    diffusion a.
    oxygen affinity a.
    stagnant a.
anoxic anoxia
ANP
    atrial natriuretic peptide
    ANP clearance receptor
Anrep effect
ANS
    anterior nasal spine
    autonomic nervous system
ansa, pl. **ansae**
    a. cervicalis
    a. cervicalis nerve
    a. cervicalis root
    Haller a.
    Henle a.
    a. hypoglossi
    lenticular a.
    a. lenticularis
    ansae nervorum spinalium
    peduncular a.
    a. peduncularis
    Reil a.
    a. sacralis
    a. subclavia

    a. subclavia nerve
    Vieussens a.
ansate
anserina
    bursa a.
anserine bursa
anserinus
    pes a.
ansiform
ansverse hermaphroditism
antagonism
antagonist
    associated a.
antagonistic
    a. muscle
    a. reflex
antalgic
ante
    artery to a.
    ductus semicircularis a.
    plexus venosus vertebralis
      internus a.
antebrachial
    a. fascia
    a. flexor retinaculum
    a. region
    a. vein
antebrachii
    chorda obliqua membranae
      interosseae a.
    facies anterior a.
    fascia a.
    margo lateralis a.
    margo medialis a.
    margo radialis a.
    margo ulnaris a.
    mediana a.
    membrana interossea a.
    vena cephalica a.
    vena intermedia a.
    vena mediana a.
antebrachium
antecardium
antecolic
antecubital
    a. fossa
    a. space
anteflex
antegonial
    a. angle
    a. notch
antehelical fold

NOTES

antenatal
anteprostate
anterior

a. abdominal cutaneous branch of intercostal nerve
a. abdominal wall
a. alveolar branch of maxillary nerve
ampullaris a.
a. ampullar nerve
a. ampullary nerve
a. angle of rib
a. annular ligament
a. antebrachial nerve
a. antebrachial region
a. arch of atlas
area intercondylaris a.
arteria alveolaris superior a.
arteria cecalis a.
arteria cerebelli inferior a.
arteria cerebri a.
arteria choroidea a.
arteria ciliaris a.
arteria circumflexa humeri a.
arteria communicans a.
arteria conjunctivalis a.
arteria ethmoidalis a.
arteria intercostalis a.
arteria interossea a.
arteria meningea a.
arteria parietales a.
arteria parietalis a.
arteria recurrens tibialis a.
arteria segmentalis a.
arteria segmentalis basalis a.
arteria spinalis a.
arteria temporalis a.
arteria tibialis a.
arteria tympanica a.
arteria vestibularis a.
a. articular surface of dens
a. atlantooccipital membrane
a. auricular branch of superficial temporal artery
a. auricular groove
auricularis a.
a. auricular muscle
a. auricular nerve
a. auricular vein
a. axillary fold
a. axillary line (AAL)
a. axillary lymph node
a. band of colon
a. basal branch
a. basal branch of superior basal vein of right and left inferior pulmonary vein
a. basal bronchopulmonary segment
a. basal segmental artery

a. basal vein
a. belly of digastric muscle
a. border
a. border of body of pancreas
a. border of eyelid
a. border of fibula
a. border of lung
a. border of radius
a. border of testis
a. border of tibia
a. border of ulna
a. brachial region
a. branch of the renal artery
a. bronchopulmonary segment
camera oculi a.
canales semicircularis a.
a. canaliculus of chorda tympani
a. cardiac vein
a. cardinal vein
a. carotid artery
a. carpal region
a. cecal artery
a. central convolution
a. central gyrus
cerebelli inferior a.
a. cerebral artery
a. cerebral vein
cerebri a.
a. cervical intertransversarii muscle
a. cervical intertransverse muscle
a. cervical lip
a. cervical region
a. chamber
a. chamber of eye
a. chamber of eyeball
a. choroidal artery
choroidea a.
a. ciliary artery
a. ciliary vein
circumflexa humeri a.
a. circumflex humeral artery
a. circumflex humeral vein
a. clinoid process
a. column
columna a.
a. column of medulla oblongata
a. column of spinal cord
commissura labiorum a.
a. commissure
a. communicating artery
a. compartment of arm
a. compartment of forearm
a. compartment of leg
a. compartment of thigh
a. component of force
a. condyloid canal of occipital bone
a. condyloid foramen
a. conjunctival artery

a. coronary periarterial plexus
a. corticospinal tract
a. costotransverse ligament
a. cranial base
a. cranial fossa
a. crest of stapes
crista lacrimalis a.
a. cruciate ligament (ACL)
a. crural nerve
a. crural region
a. crus of stapes
a. cubital region
a. cusp
cuspis a.
a. cusp of left atrioventricular valve
a. cusp of mitral valve
a. cusp of right atrioventricular valve
a. cusp of tricuspid valve
a. cutaneous branch of femoral nerve
a. cutaneous branch of iliohypogastric nerve
a. cutaneous branch of intercostal nerve
a. cutaneous nerve of abdomen
a. deep cervical lymph node
a. descending artery
diploica temporalis a.
a. divisions of brachial plexus
duplicitas a.
a. elastic layer
a. epithelium of cornea
a. ethmoidal air cell
a. ethmoidal branch of ophthalmic artery
ethmoidalis a.
a. ethmoidal nerve
a. ethmoid canal
a. external arcuate fiber
a. extremity
a. extremity of spleen
a. facial height (AFH)
a. facial vein
facies antebrachialis a.
facies brachialis a.
facies cruralis a.
facies cubitalis a.
facies femoralis a.
a. fascicle of palatopharyngeus muscle

fasciculus corticospinalis a.
fasciculus pyramidalis a.
a. femoral cutaneous nerve
a. fontanelle
fonticulus a.
fossa cranii a.
a. funiculus
funiculus a.
a. gastric branch of anterior vagal trunk
glandula lingualis a.
a. glandular branch of superior thyroid artery
a. gluteal line
a. gray commissure
a. great vessel
a. ground bundle
a. heart
a. heel
a. horn
a. humeral circumflex artery
incisura cerebelli a.
a. inferior cerebellar artery
a. inferior communicating artery (AICA)
a. inferior iliac spine
a. inferior renal segment
a. inferior segmental artery of kidney
a. intercondylar area
a. intercondylar area of tibia
a. intercostal branch of internal thoracic artery
a. intercostal vein
a. intermediate groove
a. intermuscular septum
interossea a.
interosseous a.
a. interosseous artery
a. interosseous nerve
a. interosseous vein
a. interventricular branch of left coronary artery
a. interventricular groove
a. interventricular sulcus
a. intestinal portal
a. intraoccipital joint
a. intraoccipital synchondrosis
jugularis a.
a. jugular lymph node
a. jugular vein
a. knee region

NOTES

21

anterior *(continued)*
a. labial branch of deep external pudendal artery
a. labial commissure
a. labial nerve
a. labial vein
a. lacrimal crest
lamina elastica a.
lamina limitans a.
a. lateral malleolar artery
a. lateral nasal branch of anterior ethmoidal artery
a. and lateral thoracic region
a. layer of rectus abdominis sheath
a. layer of thoracolumbar fascia
a. ligament of fibular head
a. ligament of head of fibula
a. ligament of Helmholtz
a. ligament of malleus
a. limb of internal capsule
a. limb of stapes
a. limiting lamina
a. limiting layer of cornea
a. limiting ring
linea axillaris a.
linea glutea a.
linea mediana a.
a. lingual gland
a. lip of external os of uterus
a. lip of uterine os
a. lobe of hypophysis
a. longitudinal ligament
a. lunate lobule
margo a.
a. medial malleolar artery
a. median fissure
a. median fissure of medulla oblongata
a. median fissure of spinal cord
a. median line
a. mediastinal artery
a. mediastinal lymph node
a. mediastinum
a. medullary velum
a. megalophthalmus
membrana atlantooccipitalis a.
meningea a.
a. meningeal branch of anterior ethmoidal artery
a. meniscofemoral ligament
musculus auricularis a.
musculus rectus capitis a.
musculus sacrococcygeus a.
musculus scalenus a.
musculus serratus a.
musculus tibialis a.
musculus tibiofascialis a.
a. naris
a. nasal meatus

a. nasal spine (ANS)
a. nasal spine of maxilla
a. nerve root
nervus ampullaris a.
nervus antebrachii a.
nervus ethmoidalis a.
nervus interosseus a.
nervus interosseus antebrachii a.
nodus tibialis a.
norma a.
a. notch of auricle
a. notch of cerebellum
a. notch of ear
a. nuclei of thalamus
a. oblique line of radius
a. ocular segment
a. palatine arch
a. palatine foramen
a. palatine nerve
a. palatine suture
a. palpebral margin
a. parietal artery
a. parolfactory sulcus
pars tibiotalaris a.
a. part of anterior commissure of brain
a. part of diaphragmatic surface of liver
a. part of fornix of vagina
a. part of pons
a. part of tongue
a. pectoral cutaneous branch of intercostal nerve
a. perforated substance
a. perforating artery
a. peroneal artery
a. pillar
a. pillar of fauces
a. pillar of fornix
a. piriform gyrus
a. pituitary gonadotropin
a. pituitary-like hormone
plexus venosus vertebralis externus a.
a. pole of eyeball
a. pole of lens
a. pontomesencephalic vein
a. portion of left medial segment IV of liver
a. and posterior (A&P)
a. and posterior radicular artery
a. and posterior superior pancreaticoduodenal artery
a. and posterior vestibular vein
a. primary division
a. process of malleus
processus clinoideus a.
a. pyramid
a. pyramidal tract

radix a.
a. rami of cervical nerve
rami hepatici trunci vagi a.
a. rami of lumbar nerve
rami musculares nervi interossei antebrachii a.
a. rami of sacral nerve
a. rami of thoracic nerve
ramus basalis a.
ramus interventricularis a.
a. ramus of lateral sulcus of cerebrum
a. ramus of spinal nerve
ramus temporalis a.
a. recess of tympanic membrane
recessus membranae tympani a.
rectus capitis a.
a. rectus fascia
a. rectus muscle of head
a. rectus sheath wall
recurrens tibialis a.
regio antebrachialis a.
regio antebrachii a.
regio brachialis a.
regio brachii a.
regio carpalis a.
regio cervicalis a.
regio cruris a.
regio cubitalis a.
regio femoris a.
regio genus a.
a. region of arm
a. region of elbow
a. region of forearm
a. region of knee
a. region of leg
a. region of neck
a. region of thigh
a. region of wrist
a. retinal orbital canal
a. root of spinal nerve
a. sacrococcygeal ligament
a. sacroiliac ligament
a. sacrosciatic ligament
a. sagittal pelvic inlet
a. scalene muscle
scalenus a.
a. scrotal branch of deep external pudendal artery
a. scrotal nerve
a. scrotal vein
a. segmental artery

a. semicircular canal
a. septal branch of anterior ethmoidal artery
serratus a.
a. serratus muscle
a. sinuses
a. spinal artery
spinalis a.
spina nasalis a.
a. spinocerebellar tract
a. spinothalamic tract
a. splenis muscle
a. sternoclavicular ligament
substantia perforata a.
subtendinous bursa of tibialis a.
sulcus auriculae a.
sulcus intermedius a.
sulcus interventricularis a.
sulcus lateralis a.
sulcus parolfactorius a.
a. superficial cervical lymph node
a. superior alveolar artery
a. superior alveolar branch of infraorbital nerve
a. superior dental artery
a. superior renal segment
a. superior segmental artery of kidney
a. supraclavicular nerve
a. surface
a. surface of arm
a. surface of cornea
a. surface of elbow
a. surface of eyelid
a. surface of forearm
a. surface of iris
a. surface of kidney
a. surface of leg
a. surface of lens
a. surface of lower limb
a. surface of maxilla
a. surface of patella
a. surface of petrous part of temporal bone
a. surface of prostate
a. surface of radius
a. surface of suprarenal gland
a. surface of thigh
a. surface of ulna
a. surface of uterus
synchondrosis intraoccipitalis a.

NOTES

anterior *(continued)*
- a. talar articular surface of calcaneus
- a. talocalcaneal ligament
- a. talofibular ligament
- a. talotibial ligament
- a. tarsal tendinous sheath
- a. temporal artery
- a. temporal branch
- a. temporal diploic vein
- a. tibial bursa
- tibialis a.
- a. tibialis tendon
- a. tibial lymph node
- a. tibial muscle
- a. tibial recurrent artery
- a. tibial tendon
- a. tibial vein
- a. tibiofibular ligament
- a. tibiotalar part
- a. tibiotalar part of deltoid ligament
- a. tibiotalar part of medial ligament of ankle joint
- a. tooth
- tractus corticospinalis a.
- tractus pyramidalis a.
- tractus spinocerebellaris a.
- tractus spinothalamicus a.
- a. triangle
- a. triangle of neck
- a. tubercle of atlas
- a. tubercle of cervi
- a. tubercle of cervical vertebrae
- a. tubercle of thalamus
- tympanica a.
- a. tympanic artery
- a. ulnar recurrent artery
- a. urethral valve
- a. vaginal fornix
- a. vaginal trunk
- a. vein of septum pellucidum
- vena auricularis a.
- vena basalis a.
- vena cerebri a.
- vena circumflexa humeri a.
- vena facialis a.
- vena jugularis a.
- vena pontomesencephalica a.
- vena septi pellucidi a.
- vena vertebralis a.
- a. vertebral vein
- a. vestibular artery
- a. wall of middle ear
- a. wall of stomach
- a. wall of tympanic cavity
- a. wall of vagina
- a. white commissure

anteriora
- ligamenta sacroiliaca a.

anterioris, pl. **anteriores**
- alveolares superiores anteriores
- arteriae alveolares superiores anteriores
- arteriae labiales anteriores
- arteriae mediastinales anteriores
- arteriae perforantes anteriores
- bursa subtendinea musculi tibialis a.
- cellulae anteriores
- cellulae ethmoidales anteriores
- ciliares anteriores
- conjunctivales anteriores
- cordis anteriores
- endothelium camerae a.
- intercostales anteriores
- labiales anteriores
- limbus palpebrales anteriores
- lymphonodi cervicales anteriores
- lymphonodi jugulares anteriores
- lymphonodi mediastinales anteriores
- nervi alveolares superiores anteriores
- nervi auriculares anteriores
- anteriores nervi infraorbitalis
- nervi labiales anteriores
- nervi scrotales anteriores
- nodi lymphoidei axillares anteriores
- nodi lymphoidei cervicales anteriores
- nodi lymphoidei jugulares anteriores
- nodi lymphoidei mediastinales anteriores
- pars anterior commissurae a.
- pars postcommunicalis arteriae cerebri a.
- pars posterior commissurae a.
- pars precommunicalis arteriae cerebri a.
- pars profunda compartimenti antebrachii a.
- pars superficialis compartimenti antebrachii a.
- plexus periarterialis arteriae cerebri a.
- rami auriculares anteriores
- rami duodenales arteriae pancreaticoduodenalis superioris a.
- rami gastrici anteriores trunci vagalis a.
- rami intercostales anteriores
- rami labiales anteriores
- rami nasales anteriores laterales arteriae ethmoidalis a.
- rami nasales externi nervi ethmoidalis a.

rami nasales interni nervi
ethmoidalis a.
rami nasales laterales nervi
ethmoidalis a.
rami nasales mediales nervi
ethmoidalis a.
rami parietooccipitales arteriae
cerebri anteriores
rami precuneales arteriae cerebri a.
rami scrotales anteriores
rami septales anteriores arteriae
ethmoidalis a.
rami temporales anteriores
ramus meningeus anterior arteriae
ethmoidalis a.
ramus palmaris nervi interossei
antebrachii a.
ramus perforans arteriae
interossei a.
ramus pyloricus trunci vagalis a.
scrotales anteriores
segmentum A1, A2 arteriae
cerebri a.
sinus anteriores
sinus ethmoidales anteriores
tibiales anteriores
trigonum musculare regionis
cervicalis a.
tuberculum musculi scaleni a.
tuberositas musculi serrati a.
vaginae tendinum tarsales anteriores
vagina tendinis musculi tibialis a.
venae cardiacae anteriores
venae ciliares anteriores
venae cordis anteriores
venae intercostales anteriores
venae labiales anteriores
venae scrotales anteriores
venae tibiales anteriores
**anterior/lateral/posterior glandular
branch of superior thyroid artery**
anterior-superior iliac spine (ASIS)
anterius
compartimentum antebrachii a.
compartimentum brachii a.
compartimentum cruris a.
compartimentum femoris a.
cornu a.
corpus quadrigeminum a.
foramen ethmoidale a.
foramen lacerum a.
labium a.

ligamentum auriculare a.
ligamentum capitis fibulae a.
ligamentum costotransversarium a.
ligamentum cruciatum a.
ligamentum longitudinale a.
ligamentum mallei a.
ligamentum meniscofemorale a.
ligamentum sacrococcygeum a.
ligamentum sternoclaviculare a.
ligamentum talofibulare a.
ligamentum talotibiale a.
ligamentum tibiofibulare a.
mediastinum a.
segmentum bronchopulmonale
basale a.
segmentum bronchopulmonale a. S
III
segmentum oculare a.
trigonum cervicale a.
trigonum colli a.
tuber a.
anteroapical
anterodistal border
anterodorsal
a. nucleus of thalamus
a. thalamic nucleus
anterodorsalis
nucleus a.
anteroexternal
anterofacial dysplasia
anterograde
anteroinferior
a. aspect
a. surface of pancreas
anterointernal
anterolateral
a. central artery
a. column of spinal cord
a. fontanelle
a. groove
a. neck
a. striate artery
a. sulcus
a. surface of arytenoid cartilage
a. surface of humerus
a. thalamostriate artery
anterolateralis, pl. **anterolaterales**
arteriae centrales anterolaterales
arteriae thalamostriatae
anterolaterales
fonticulus a.

NOTES

anterolateralium
    rami laterales arteriarum
      centralium a.
    rami mediales arteriarum
      centralium a.
anteromedia
anteromedial
    a. arm
    a. central artery
    a. frontal branch of callosomarginal
      artery
    a. intermuscular septum
    a. nucleus of thalamus
    a. surface of shaft of humerus
    a. thalamic nucleus
    a. thalamostriate artery
anteromedialis, pl. **anteromediales**
    arteriae centrales anteromediales
    arteriae thalamostriatae
      anteromediales
    nucleus a.
    rami centrales anteromediales
    ramus frontalis a.
anteromedian
anteroposterior
    a. diameter of pelvic inlet
    a. facial dysplasia
anteroseptal
anterosuperior
    arteria intercostalis a.
    a. surface of body of pancreas
anterotransverse diameter
anteroventral
    a. nucleus of thalamus
    a. thalamic nucleus
anteroventralis
    nucleus a.
antetorsion
    angle of a.
anteversion
    angle of a.
anthelicis
    crura a.
    crus a.
    fossa a.
anthelix
    fossa of a.
anthropoid pelvis
anthropology
    physical a.
anthropometer
anthropometric
anthropometry
anthroposcopy
anthroposomatology
antibrachial
antibrachium
anticlinal
anticnemion

anticus
    locus perforatus a.
    musculus scalenus a.
    musculus tibialis a.
    musculus tibiofascialis a.
    scalenus a.
antidromic
antienergic
antiG
antigravity muscle
antihelica
    fossa a.
antihelicis
    crura a.
antihelix
    crura of a.
    leg of a.
antiketogenesis
antiketogenic
antilobium
antiluteogenic
antimere
antimesenteric
    a. border
    a. fat pad
antimesocolic side of cecum
antimuscarinic
antinat
antinatriferic
antiniad
antinial
antinion
antinuclear
antiparastata
antipodal cone
antipode
antiport
antiporter
antiprostate
antiserum
    nerve growth factor a.
antithenar
antitragicus
    a. muscle
    musculus a.
antitragohelicina
    fissura a.
antitragohelicine fissure
antitragus
    muscle of a.
    a. muscle
antitrope
antitropic
antra (*pl. of* antrum)
antral
    a. floor
    a. fold
    a. mucosa

a. pouch
a. sphincter
antri
sphincter a.
antronasal
antropyloric
antrotonia
antrotympanic
antrum, pl. **antra**
aditus ad a.
aditus to mastoid a.
aperture of mastoid a.
a. auris
cardiac a.
a. cardiacum
distal a.
antra ethmoidale
antra ethmoidalia
falx of maxillary a.
follicular a.
frontal a.
a. of Highmore
mastoid a.
a. mastoideum
maxillary a.
pyloric a.
a. pyloricum
sphincter of gastric a.
tympanic a.
Valsalva a.
a. of Willis
anular
a. part of fibrous digital sheath of
digits of hand and foot
anulare
ligamentum a.
anularis
digitus a.
anules
anulus, pl. **anuli**
a. abdominalis
a. fibrocartilagineus membranae
tympani
a. fibrosus dexter/sinister cordis
a. fibrosus disci intervertebralis
a. fibrosus of intervertebral disk
a. of fibrous sheath
Haller a.
a. iridis
a. iridis major
a. iridis minor
a. lymphoideus pharyngis

a. tympanicus
a. umbilicalis
a. urethralis
Vieussens a.
anus, pl. **ani**
arcus tendineus musculi levatoris
ani
atresia ani
Bartholin a.
a. cerebri
corrugator cutis muscle of a.
crena ani
elevator muscle of a.
external sphincter muscle of a.
flexura perinealis canalis ani
imperforate a.
internal sphincter muscle of a.
levator ani
levator muscle of a.
musculus corrugator cutis ani
musculus levator ani
prolapsus ani
sphincter ani
tendinous arch of levator ani
vestibular a.
vulvovaginal a.
anvil
AOC
anodal opening contraction
aorta, pl. **aortae**
abdominal a.
a. abdominalis
appendicular a.
arch of a.
arcuate a.
arcus aortae
arteria a.
a. ascendens
ascending a. (AA)
bifurcatio aortae
bifurcation of a.
bronchial branch of thoracic a.
bulbus aortae
deep articular a.
a. descendens
descending bronchial branch of a.
descending groove for a.
descending groove of arch of a.
esophageal branch of thoracic a.
groove for abdominal a.
groove for arch of a.
groove for descending a.

NOTES

aorta *(continued)*
    isthmus aortae
    isthmus of a.
    mediastinal branch of thoracic a.
    ostium aortae
    pars abdominalis aortae
    pars ascendens aortae
    pars descendens aortae
    pars thoracica aortae
    pericardial branch of thoracic a.
    posterior articular a.
    primitive a.
    rami esophageales partis thoracicae aortae
    retroesophageal a.
    sinus aortae
    superior margin of a.
    terminal a.
    thoracic a.
    a. thoracica
    valva aortae
    valvula semilunaris dextra valvae aortae
    valvula semilunaris posterior valvae aortae
    valvula semilunaris sinistra valvae aortae
    ventral a.
    vestibulum aortae
aortal
aortic
    a. arch
    a. area
    a. area of auscultation
    a. atresia
    a. bifurcation
    a. body
    a. bulb
    a. commissure
    a. cusp
    a. foramen
    a. glomera
    a. hiatus
    a. impression of left lung
    a. isthmus
    a. knob
    a. knuckle
    a. lymphatic plexus
    a. lymph node
    a. nerve
    a. notch
    a. opening
    a. orifice
    a. ostium
    a. reflex
    a. ring
    a. root
    a. sac
    a. septal defect

    a. sinus
    a. spindle
    a. sulcus
    a. valve
    a. vestibule
    a. window
aortica
    glomera a.
aorticopulmonary septal defect
aorticorenal ganglion
aorticorenalia
    ganglia a.
aorticum
    corpus a.
    glomus a.
aorticus
    hiatus a.
    plexus a.
    sulcus a.
aortocoronary
aortoiliac
aortopulmonary septum
A&P
    anterior and posterior
apancreatic
aparathyroidism
A-P-C, APC
    adenoidal-pharyngeal-conjunctival
ape fissure
aperistalsis
aperta
    spina bifida a.
apertura, pl. **aperturae**
    a. canaliculi vestibuli
    a. externa aqueductus vestibuli
    a. externa canaliculi cochlea
    a. lateralis ventriculi quarti
    a. mediana vena
    a. pelvis inferior
    a. pelvis minoris
    a. pelvis superior
    a. piriformis
    a. sinus frontalis
    a. sinus sphenoidalis
    a. thoracis inferior
    a. thoracis superior
    a. tympanica canaliculi chordae tympani
aperture
    angular a.
    external acoustic a.
    frontal sinus a.
    inferior pelvic a.
    inferior thoracic a.
    laryngeal a.
    lateral cerebral a.
    margin of piriform a.
    a. of mastoid antrum
    medial a.

a. of orbit
pharyngeal a.
piriform a.
posterior nasal a.
sphenoidal sinus a.
superior pelvic a.
superior thoracic a.
apex, pl. **apices**
a. of arytenoid cartilage
a. of auricle
a. auriculae
a. capitis fibulae
cardiac a.
a. cartilaginis arytenoideae
a. cordis
a. cornus posterioris
a. cuspidis dentis
a. of cusp of tooth
a. of dens
a. of duodenal bulb
a. of external ring
a. of head of fibula
a. of heart
a. linguae
a. of lung
a. nasi
a. of nose
notch of cardiac a.
a. of orbit
orbital a.
a. ossis sacri
a. partis petrosae
a. partis petrosae ossis temporalis
a. of patella
petrous a.
a. of petrous part of temporal
  bone
a. prostatae
a. of prostate
pulmonary a.
a. pulmonis
a. radicis dentis
root a.
a. of sacrum
a. satyri
a. of tongue
a. of urinary bladder
a. vesicae
aphalangia
apheliotropism
apical
a. axillary lymph node

a. branch of inferior lobar branch
  of right pulmonary artery
a. branch of right superior
  pulmonary vein
a. bronchopulmonary segment
a. dendrite
a. dental foramen
a. foramen of tooth
a. gland
a. ligament of dens
a. process
a. segmental artery
a. segmental artery of superior
  lobar artery of right lung
a. vein
apicale
segmentum a.
segmentum bronchopulmonale a. S
  I
apicalis, pl. **apicales**
arteria segmentalis a.
nodi lymphoidei axillares apicales
ramus a.
vena a.
apices (*pl. of* apex)
apicis
apicoposterior
a. branch of left superior
  pulmonary vein
a. bronchopulmonary segment
ramus a.
a. vein
vena a.
apicoposterius
segmentum bronchopulmonale a.
apinealism
aplasia
aplastic
apleuria
apnea
deglutition a.
apneic oxygenation
apneumatosis
apneumia
apneusis
apneustic
apobiosis
apocrine gland
apodal
apodia
apodous
apody

NOTES

apogamia
apogamy
apolar cell
apomixia
aponeuro
aponeurology
aponeurosis, pl. **aponeuroses**
    a. of abdominal oblique muscle
    a. of biceps brachii
    bicipital a.
    a. bicipitalis
    Denonvilliers a.
    epicranial a.
    a. epicranialis
    extensor a.
    a. of external oblique muscle
    a. of iliocostalis
    a. of insertion
    a. of internal oblique muscle
    a. of investment
    levator a.
    a. linguae
    lingual a.
    a. musculi bicipitis brachii
    a. of musculus transversus
      abdominis
    a. of origin
    a. palatina
    palatine a.
    palmar a.
    a. palmaris
    Petit a.
    a. pharyngea
    plantar a.
    a. plantaris
    a. of plantar transverse fasciculi
    a. of posterior superior serratus
    Sibson a.
    a. of superior levator palpebra
    temporal a.
    thoracolumbar a.
    a. of transverse abdominal
    triangular a.
    a. of vastus muscle
    a. of velum
    a. of Zinn
    Zinn a.
aponeurotic
    a. falx
    a. fascia
aponeurotica
    falx a.
    galea a.
apophysary point
apophysial, apophyseal
    a. point
apophysis, pl. **apophyses**
    basilar a.
    a. conchae

    a. helicis
    lenticular a.
    temporal a.
aposome
aposthia
apparatus
    accessory visual a.
    achromatic a.
    alimentary a.
    attachment a.
    Barcroft-Warburg a.
    Beckmann a.
    Benedict-Roth a.
    branchial a.
    chromidial a.
    dental a.
    digestive a.
    a. digestorius
    genitourinary a.
    Golgi a.
    Haldane a.
    hyoid a.
    a. hyoideus
    juxtaglomerular a.
    lacrimal a.
    a. lacrimalis
    a. ligamentosus colli
    a. ligamentosus weitbrechti
    masticatory a.
    a. respiratorius
    respiratory a.
    Roughton-Scholander a.
    Scholander a.
    subneural a.
    a. suspensorius lentis
    urinary a.
    urogenital a.
    a. urogenitalis
    Van Slyke a.
    Warburg a.
apparent viscosity
appendage
    auricular a.
    cecal a.
    drumstick a.
    epiploic a.
    a.'s of eye
    a.'s of fetus
    left auricular a.
    omental a.
    right auricular a.
    a.'s of skin
    testicular a.
    uterine a.
    vermicular a.
    vermiform a.
    vesicular a.
appendical

appendiceal
- a. base
- a. mesentery
- a. tissue

appendices (*pl. of* appendix)

appendicis

appendicular
- a. aorta
- a. artery
- a. lymph node
- a. muscle
- a. skeleton
- a. vein

appendiculare
- skeleton a.

appendicularis, pl. **appendiculares**
- arteria a.
- lobus a.
- lymphonodi appendiculares
- nodi lymphoidei appendiculares
- vena a.

appendix, pl. **appendices**
- appendices adiposae coli
- aggregated lymphatic follicles of vermiform a.
- auricular a.
- cecal a.
- a. ceci
- a. cerebri
- ensiform a.
- a. epididymidis
- a. of epididymidis
- a. of epididymis
- epiploic a.
- a. epiploica
- a. fibrosa hepatis
- lumen of a.
- mesentery of a.
- Morgagni a.
- omental a.
- appendices omentales
- orifice of vermiform a.
- ostium of vermiform a.
- pelvic a.
- retrocecal a.
- a. testis
- valve of vermiform a.
- a. ventriculi laryngis
- vermiform a.
- a. vermiform
- a. vermiformis

- a. vesiculosa
- appendices vesiculosae

appestat

appetite juice

apple
- Adam's a.

applied anatomy

appositional growth

approximal surface of tooth

aproctia

aprosopia

apurinic acid

apus
- sympus a.

apyknomorphous

aqueduct
- a. of cerebrum
- cochlear a.
- Cotunnius a.
- external aperture of vestibular a.
- fallopian a.
- sylvian a.
- vein of cochlear a.
- vein of vestibular a.
- vestibular a.
- a. of vestibule

aqueductus
- a. cerebri
- a. cochlea
- a. cotunnii
- a. fallopii
- a. sylvii
- a. vestibuli

aqueous
- a. chambers
- a. humor
- a. vein

aquiparous

aquosus
- humor a.

arachnodactyly

arachnoid
- a. of brain
- cerebral part of a.
- a. cyst
- a. foramen
- a. granulation
- a. mater cranialis
- a. mater encephali
- a. mater and pia mater
- a. membrane
- a. of spinal cord

NOTES

arachnoid *(continued)*
    spinal part of a.
    a. villus
arachnoidal granulation
arachnoidea
    a. encephali
    a. mater
    a. mater spinalis
arachnoideales
    granulationes a.
arachnoides
arantii
    corpus a.
    ductus venosus a.
Arantius
    A. ligament
    A. nodule
    A. ventricle
arbor
    a. vitae
    a. vitae uterus
arbores
arborescent
arborization
arborize
arc
    auricular a.
    bregmatolambdoid a.
    crater a.
    flame a.
    interauricular a.
    lambdoid a.
    nasobregmatic a.
    nasooccipital a.
    Riolan a.
arcade
    arterial a.'s
    Flint a.
    intestinal arterial a.'s
    lower dental a.
    mandibular dental a.
    marginal a.
    maxillary dental a.
    pancreaticoduodenal arterial a.'s
    Riolan a.
    temporal a.
    upper dental a.
arcate
arch, pl. **arches**
    abdominothoracic a.
    alveolar a.
    anterior palatine a.
    a. of aorta
    aortic a.
    atlas a.
    axillary a.
    azygos vein a.
    branchial arches
    carpal arches

coracoacromial a.
Corti a.
costal a.
a. of cricoid cartilage
crural a.
deep crural a.
deep palmar arterial a.
deep palmar venous a.
deep plantar arterial a.
deep volar a.
dental a.
digital venous a.
dorsal carpal arterial a.
fallopian a.
faucial a.
femoral a.
arches of foot
glossopalatine a.
groove of aortic a.
Haller arches
hemal arches
hyoid a.
iliopectineal a.
inferior dental a.
inferior palpebral arterial a.
jugular venous a.
lamina of vertebral a.
Langer axillary a.
lateral longitudinal a.
lateral lumbocostal a.
lingual a.
longitudinal a.
lymph node of azygos a.
malar a.
mandibular a.
medial longitudinal a.
medial lumbocostal a.
nasal venous a.
neural a.
node of azygos a.
a. of palate
palatine a.
palatoglossal a.
palatopharyngeal a.
perforating branch of deep palmar a.
pharyngeal arches
pharyngopalatine a.
plantar arterial a.
plantar venous a.
popliteal a.
posterior palatine a.
postoral arches
prepancreatic a.
primitive costal arches
pubic a.
right aortic a.
Riolan a.
Salus a.

subcostal a.
superciliary a.
superficial femoral a.
superficial palmar arterial a.
superficial palmar venous a.
superior dental a.
superior palpebral arterial a.
supraorbital a.
tarsal a.
tendinous a.
a. of thoracic duct
transverse a.
Treitz a.
vertebral a.
visceral arches
zygomatic a.
arched crest
archenteric canal
archenteron
archeocyte
archeokinetic
arches (*pl. of* arch)
archetype
archicerebellum
archicortex
archipallium
architectonics
architecture
breast a.
hepatic a.
arch-loop-whorl system
arciform
a. artery
a. vein of kidney
arcual
arcuata
arteria a.
crista a.
eminentia a.
linea a.
zona a.
arcuate
a. aorta
a. artery
a. artery of kidney
a. crest
a. crest of arytenoid cartilage
a. eminence
a. fasciculus
a. fibers
a. line
a. line of ilium

a. line of rectus sheath
a. nucleus
a. popliteal ligament
a. pubic ligament
a. uterus
a. vein of kidney
a. zone
arcuati
nuclei a.
arcuation
arcuatum
ligamentum popliteum a.
arcuatus
nucleus a.
uterus a.
arcus
a. alveolaris mandibulae
a. alveolaris maxillae
a. anterior atlantis
a. aortae
a. cartilaginis cricoideae
a. costalis
a. costarum
a. dentalis inferior
a. dentalis mandibularis
a. dentalis maxillaris
a. dentalis superior
a. ductus thoracici
a. glossopalatinus
a. iliopectineus
a. inguinalis
a. lumbocostalis lateralis
a. lumbocostalis medialis
a. marginalis coli
a. palatini
a. palatoglossus
a. palatopharyngeus
a. palmaris profundus
a. palpebralis inferior
a. palpebralis superior
a. pedis longitudinalis
a. pedis longitudinalis pars lateralis
a. pedis longitudinalis pars medialis
a. pedis transversalis
a. plantaris profundus
a. posterior atlantis
a. pubis
a. raninus
a. superciliaris
a. tarseus
a. tendineus
a. tendineus fasciae pelvis

NOTES

arcus *(continued)*
a. tendineus musculi levatoris ani
a. tendineus musculi solei
a. tendineus of obturator fascia
a. tendineus of pelvic diaphragm
a. unguium
a. venosus dorsalis pedis
a. venosus juguli
a. venosus palmaris profundus
a. venosus palmaris superficialis
a. venosus plantaris
a. vertebrae
a. volaris profundus
a. volaris superficialis
a. zygomaticus

area, pl. areae
acoustic a.
anterior intercondylar a.
aortic a.
association areas
auditory a.
back, arm, neck, shoulder a.
BANS a.
Broca parolfactory a.
Brodmann areas
cardiac a.
Celsus a.
a. centralis
a. cochlea
cochlear a.
cortical a.
a. cribrosa papillae renalis
crural a.
dermatomic a.
embryonal a.
embryonic a.
entorhinal a.
a. for esophagus
excitable a.
facial nerve a.
a. of facial nerve
faucial a.
flank a.
Flechsig areas
frontal a.
frontoorbital a.
frontoparietal a.
gastric a.
a. gastrica
genital a.
germinal a.
a. germinativa
gyrous a.
Head a.
inferior vestibular a.
inguinal a.
insular a.
a. intercondylaris anterior
a. intercondylaris anterior tibiae

a. intercondylaris posterior
a. intercondylaris posterior tibiae
Kiesselbach a.
Killian-Jamieson a.
Laimer a.
a. of Laimer
Laimer-Haeckerman a.
lateral inferior hepatic a.
lateral superior hepatic a.
Little a.
malar a.
Martegiani a.
masseteric a.
midtemporal a.
motor a.
nasopharyngeal a.
a. nervi facialis
a. nuda hepatis
olfactory a.
parietal a.
a. parolfactoria
a. parolfactoria brocae
parolfactory a.
pear-shaped a.
perihilar a.
perineal a.
perispinal a.
peristriate a.
piriform a.
Pitres a.
postauricular a.
postcentral a.
posterior intercondylar a.
a. postrema
precentral a.
precommissural septal a.
prefrontal a.
premotor a.
preoptic a.
prestriate a.
pretectal a.
primary visual a.
Rolando a.
saddle a.
sagittal a.
scapular a.
scrotal a.
secondary visual a.
sensorial areas
sensorimotor a.
sensory areas
septal a.
Soemmering a.
somatosensory a.
somesthetic a.
striate a.
Stroud pectinated a.
a. subcallosa
subcallosal a.

subcostal a.
subglottic a.
submitral a.
subsegmental a.
superior vestibular a.
supraclavicular a.
supraorbital a.
suprarenal a.
sylvian a.
tailbone a.
temporal a.
vagus a.
vestibular a.
a. vestibularis inferior
a. vestibularis superior
visual a.
voluntary a.
Wernicke a.
areflexia
arenacea
corpora a.
areola, pl. **areolae**
a. of breast
glandulae areolares
a. mammae
a. of nipple
a. papillaris
tubercula areolae
areolar
a. gland
a. tissue
a. tubercle
a. venous plexus
areolaris
plexus venosus a.
argentaffin
a. cells
a. granule
argentaffine
argentation
argentophil, argentophile
argyrophil, argyrophile
argyrophilic
a. cells
a. fibers
arium
Arlt sinus
arm
anterior compartment of a.
anterior region of a.
anterior surface of a.
anteromedial a.

biceps muscle of a.
a. bone
deep artery of a.
deep fascia of a.
dynein a.
extensor compartment of a.
flexor compartment of a.
inferior lateral cutaneous nerve of a.
lateral surface of a.
lower lateral cutaneous nerve of a.
medial cutaneous nerve of a.
posterior compartment of a.
posterior cutaneous nerve of a.
posterior region of a.
posterior surface of a.
superior lateral cutaneous nerve of a.
triceps muscle of a.
upper lateral cutaneous nerve of a.
armpit
Arndt law
Arnold
A. bundle
A. canal
foramen of A.
A. ganglion
A. nerve
A. tract
Arnold-Chiari
A.-C. deformity
A.-C. malformation
arousal function
arrector
a. muscle of hair
a. pili muscle
arrectores pilorum
arrest
epiphysial a.
maturation a.
arrhinencephalia
arrhinencephaly
arrhinia
arrhythmia
arrhythmokinesis
artefact
arteria, pl. **arteriae**
a. acetabuli
arteriae alveolares superiores anteriores
a. alveolaris inferior
a. alveolaris superior anterior

NOTES

35

arteria *(continued)*

a. alveolaris superior posterior
a. anastomotica auricularis magna
a. anonyma
a. aorta
a. appendicularis
a. arcuata
arteriae arcuatae renis
a. arcuata pedis
a. articularis azygos
arteriae atriales
a. auditiva interna
a. auricularis posterior
a. auricularis profunda
a. axillaris
a. basilaris
a. brachialis
a. brachialis superficialis
a. buccalis
a. bulbi penis
a. bulbi urethrae
a. bulbi vaginae
a. bulbi vestibuli
a. calcarina
a. callosa mediana
a. callosomarginalis
a. canalis pterygoidei
arteriae caroticotympanicae
arteriae caroticotympanici
a. carotis communis
a. carotis externa
a. carotis interna
a. caudae pancreatis
a. cecalis anterior
a. cecalis posterior
a. celiaca
arteriae centrales anterolaterales
arteriae centrales anteromediales
arteriae centrales posterolaterales
arteriae centrales posteromediales
a. centralis brevis
a. centralis longa
a. centralis retinae
a. cerebelli inferior anterior
a. cerebelli inferior posterior
a. cerebri anterior
a. cerebri media
a. cerebri posterior
a. cervicalis ascendens
a. cervicalis profunda
a. cervicalis superficialis
a. cervicovaginalis
a. choroidea anterior
a. choroidea posterior
a. ciliaris anterior
a. ciliaris posterior brevis
a. ciliaris posterior longa
arteriae circumferentiales brevis
a. circumflexa femoris lateralis

a. circumflexa femoris medialis
a. circumflexa humeri anterior
a. circumflexa humeri posterior
a. circumflexa iliaca profunda
a. circumflexa iliaca superficialis
a. circumflexa ilium profunda
a. circumflexa ilium superficialis
a. circumflexa scapulae
a. cochlearis communis
a. cochlearis propria
a. colica
a. colica dextra
a. colica media
a. colica sinistra
a. collateralis media
a. collateralis radiali
a. collateralis radialis
a. collateralis ulnaris inferior
a. collateralis ulnaris superior
a. collicularis
a. comes nervi mediani
a. comitans nervi mediani
a. commissuralis mediana
a. communicans anterior
a. communicans posterior
a. conjunctivalis anterior
a. conjunctivalis posterior
a. coronaria dextra
a. coronaria sinistra
arteriae corticales radiatae
a. cremasterica
a. cystica
a. deferentialis
a. descendens genus
a. digitalis dorsalis
a. digitalis palmaris communis
a. digitalis palmaris propria
a. digitalis plantaris communis
a. digitalis plantaris propria
a. dorsalis clitoridis
a. dorsalis nasi
a. dorsalis pedis
a. dorsalis penis
a. dorsalis scapulae
a. ductus deferentis
arteriae encephali
a. epigastrica inferior
a. epigastrica superficialis
a. epigastrica superior
a. episcleralis
a. ethmoidalis anterior
a. ethmoidalis posterior
a. facialis
a. femoralis
a. fibularis
a. flexurae dextrae
a. frontobasalis lateralis
a. frontobasalis medialis
a. gastrica dextra

arteriae gastricae breves
a. gastrica posterior
a. gastrica sinistra
a. gastroduodenalis
a. gastroepiploica dextra
arteriae gastroepiploicae
a. gastroepiploica sinistra
arteriae gastro-omentales
a. gastroomentalis dextra
a. gastroomentalis sinistra
a. genus descendens
a. genus inferior lateralis
a. genus inferior medialis
a. genus media
a. genus superior lateralis
a. genus superior medialis
a. glutea inferior
a. glutea superior
a. gyri angularis
a. helicina
arteriae helicinae penis
arteriae helicinae uterus
a. hepatica communis
a. hepatica propria
a. hyaloidea
a. hypogastrica
a. hypophysialis inferior
a. hypophysialis superior
arteriae ileales
a. ileocolica
a. iliaca communis
a. iliaca externa
a. iliaca interna
a. iliolumbalis
a. inferior anterior cerebelli
a. inferior lateralis genus
a. inferior medialis genus
a. inferior posterior cerebelli
a. infraorbitalis
arteriae insulares
arteriae intercostales posteriores I
  et II
arteriae intercostales posteriores
  III–XI
arteriae intercostales posteriores
  prima et secunda
a. intercostalis anterior
a. intercostalis anterosuperior
a. intercostalis posterior
a. intercostalis suprema
arteriae interlobares
arteriae interlobulares

a. interlobulares hepatis
a. interlobulares renis
a. intermesenterica
a. interossea anterior
a. interossea communis
a. interossea posterior
a. interossea recurrens
a. interossea volaris
arteriae intestinales
arteriae intrarenales
a. ischiadica
a. ischiatica
arteriae jejunales
a. juxtacolica
arteriae labiales anteriores
a. labialis inferior
a. labialis superior
a. labyrinthi
a. lacrimalis
a. laryngea inferior
a. laryngea superior
a. lienalis
a. ligamenti teretis uterus
a. lingualis
a. lingularis
a. lingularis inferior
a. lingularis superior
arteriae lobares inferiores
arteriae lobares inferior et superior
arteriae lobares superiores
a. lobaris media
a. lobaris media pulmonis dextri
a. lobi caudati
a. lumbalis
a. lumbalis ima
a. lusoria
arteriae malleolares posteriores
  laterales
arteriae malleolares posteriores
  mediales
a. malleolaris anterior lateralis
a. malleolaris anterior medialis
a. mammaria interna
arteriae mammillares
a. marginalis coli
a. masseterica
a. maxillaris
a. maxillaris externa
a. media genus
arteriae mediastinales anteriores
arteriae medullares segmentales
arteriae membri inferioris

NOTES

arteria *(continued)*

arteriae membri superioris
a. meningea anterior
a. meningea media
a. meningea posterior
a. mentalis
a. mesenterica inferior
a. mesenterica superior
a. metacarpalis dorsalis
a. metacarpalis palmaris
a. metacarpea dorsalis
a. metacarpea palmaris
a. metatarsalis
a. metatarsalis dorsalis
a. metatarsalis plantaris
a. metatarsea dorsalis
a. metatarsea plantaris
arteriae musculares arteriae
  ophthalmicae
a. musculophrenica
arteriae nasales posteriores laterales
a. nasalis posterior septi
a. nasi externa
arteriae nervorum
a. nutriciae femoris
arteriae nutriciae humeri
a. nutricia tibiae
a. nutricia ulnae
a. nutriens femoris
a. nutriens fibulae
a. nutriens humeri
a. nutriens radii
a. nutriens tibiae
a. nutriens tibialis
a. nutriens ulnae
a. obturatoria
a. obturatoria accessoria
a. occipitalis
a. occipitalis lateralis
a. occipitalis medialis
a. ophthalmica
a. ovarica
a. palatina ascendens
a. palatina descendens
a. palatina major
a. palatina minor
arteriae palpebrales
arteriae palpebrales laterales et
  mediales
a. pancreatica dorsalis
a. pancreatica inferior
a. pancreatica magna
a. pancreaticoduodenalis
a. pancreaticoduodenalis inferior
a. pancreaticoduodenalis superior
a. paracentralis
arteriae parietales
a. parietales anterior

arteriae parietales laterales et
  mediales
a. parietales posterior
a. parietalis anterior
a. parietalis posterior
a. parietooccipitalis
arteriae perforantes
arteriae perforantes anteriores
arteriae perforantes arteriae
  profundae femoris
arteriae perforantes penis
arteriae perforantes radiatae renis
a. pericallosa
a. pericardiacophrenica
a. perinealis
a. peronea
a. pharyngea ascendens
a. phrenica inferior
a. phrenica superior
a. plantaris lateralis
a. plantaris medialis
a. plantaris profunda arteriae
  dorsalis pedis
a. plantaris profundus
a. polaris frontalis
a. polaris temporalis
arteriae pontis
a. poplitea
a. precunealis
a. prepancreatica
a. princeps pollicis
a. profunda brachii
a. profunda clitoridis
a. profunda femoris
a. profunda linguae
a. profunda penis
a. pterygomeningealis
arteriae pudendae externae
a. pudenda interna
a. pulmonalis
a. pulmonalis dextra
a. pulmonalis sinistra
a. quadrigeminalis
a. radialis indicis
arteriae radiculares anterior et
  posterior
a. radicularis magna
a. radii nutricia
ramus sinus cavernosi arteriae
  carotidis arteriae
a. ranina
a. rectalis inferior
a. rectalis media
a. rectalis superior
a. recurrens radialis
a. recurrens tibialis anterior
a. recurrens tibialis posterior
a. recurrens ulnaris
a. renalis

a. retinae centralis
a. retroduodenalis
a. sacralis lateralis
a. sacralis mediana
a. scapularis descendens
a. scapularis dorsalis
a. segmentalis anterior
a. segmentalis apicalis
a. segmentalis basalis anterior
a. segmentalis basalis lateralis
a. segmentalis basalis medialis
a. segmentalis lateralis
a. segmentalis posterior
a. segmentalis superior
a. segmenti
a. segmenti anterioris inferioris renis
a. segmenti anterioris superioris renis
arteriae segmenti hepaticae
a. segmenti inferioris renis
a. segmenti posterioris renis
a. segmenti superioris renis
arteriae sigmoideae
a. spermatica interna
a. sphenopalatina
a. spinalis anterior
a. spinalis posterior
a. splenica
a. striata medialis distalis
a. stylomastoidea
a. subclavia
a. subcostalis
a. sublingualis
a. submentalis
a. subscapularis
a. sulci centralis
a. sulci postcentralis
a. sulci precentralis
a. superior cerebelli
a. superior lateralis genus
a. superior medialis genus
a. suprachiasmatica
a. supraduodenalis
a. supraoptica
a. supraorbitalis
a. suprarenalis inferior
a. suprarenalis media
a. suprarenalis superior
a. suprascapularis
a. supratrochlearis
a. suralis

a. tarsea lateralis
a. tarsea medialis
a. temporalis anterior
a. temporalis intermedia
a. temporalis media
a. temporalis posterior
a. temporalis profunda
a. temporalis superficialis
a. testicularis
arteriae thalamostriatae anterolaterales
arteriae thalamostriatae anteromediales
a. thoracica interna
a. thoracica lateralis
a. thoracica superior
a. thoracoacromialis
a. thoracodorsalis
arteriae thymicae
a. thyroidea ima
a. thyroidea inferior
a. thyroidea superior
a. tibialis anterior
a. tibialis posterior
a. transversa cervicis
a. transversa colli
a. transversa faciei
a. tuberis cinerei
a. tympanica anterior
a. tympanica inferior
a. tympanica posterior
a. tympanica superior
a. umbilicalis
a. uncalis
a. urethralis
a. uterina
a. vaginalis
arteriae ventriculares
a. vertebralis
a. vesicalis inferior
a. vesicalis superior
a. vestibularis anterior
a. vestibuli
a. vestibulocochlearis
a. vitellina
a. volaris indicis radialis
a. zygomaticoorbitalis

arterial
a. arcades
a. arches of colon
a. arches of ileum
a. arches of jejunum

NOTES

arterial *(continued)*
 a. arch of lower eyelid
 a. arch of upper eyelid
 a. branch to dura mater
 a. bulb
 a. canal
 a. capillary
 a. cerebral circle
 a. circle of cerebrum
 a. circle of Willis
 a. cone
 a. duct
 a. grooves
 a. hyperemia
 a. ligament
 a. network
 a. plexus
 a. segments of kidney
 a. tension
 a. vein
arterialization
arteriarctia
arteriectasia
arteriectasis
arteries (*pl. of* artery)
arteriocapillary
arteriococcygeal gland
arteriola, pl. **arteriolae**
 a. glomerularis afferens
 a. glomerularis efferens
 a. maculae medius
 a. macularis inferior
 a. macularis superior
 a. medialis retinae
 a. nasalis retinae inferior
 a. nasalis retinae superior
 arteriolae rectae
 a. temporalis retinae inferior
 a. temporalis retinae superior
arteriolar network
arteriole
 afferent glomerular a.
 capillary a.
 efferent glomerular a.
 glomerular a.
 inferior macular a.
 inferior temporal retinal a.
 a. of kidney
 middle macular a.
 postglomerular a.
 spiral a.
 superior macular a.
 superior nasal retinal a.
 superior temporal retinal a.
arteriology
arteriolovenous
arteriolovenular
 a. anastomosis
 a. bridge

arteriolovenularis
 anastomosis a.
arteriomotor
arteriosa
 vena a.
arteriosi
 nodus lymphoideus ligamenti a.
 sulci a.
arteriosum
 ligamentum a.
 lymph node of ligamentum a.
 node of ligamentum a.
 ostium a.
 rete a.
arteriosus
 artery of circulus a.
 conus a.
 ductus a.
 patent ductus a.
 persistent truncus a.
 pseudotruncus a.
 truncus a.
arteriotony
arteriovenosa
 anastomosis a.
arteriovenous (AV)
 a. anastomosis
 a. carbon dioxide difference
 a. oxygen difference
artery, pl. **arteries**
 A1, A2 segment of anterior
  cerebral a.
 aberrant obturator a.
 accessory meningeal branch of
  middle meningeal a.
 accessory obturator a.
 acetabular a.
 acromial anastomosis of
  thoracoacromial a.
 acromial branch of suprascapular a.
 acromial branch of
  thoracoacromial a.
 acromiothoracic a.
 arteries of Adamkiewicz
 Adamkiewicz a.
 alar branch of external
  maxillary a.
 alveolar branch of internal
  maxillary a.
 anal canal a.
 anastomotic branch of middle
  meningeal artery with lacrimal a.
 angular a.
 a. of angular gyrus
 a. of angular nasal branch
 anonymous a.
 a. to ante
 anterior auricular branch of
  superficial temporal a.

anterior basal segmental a.
anterior branch of the renal a.
anterior carotid a.
anterior cecal a.
anterior cerebral a.
anterior choroidal a.
anterior ciliary a.
anterior circumflex humeral a.
anterior communicating a.
anterior conjunctival a.
anterior descending a.
anterior ethmoidal branch of ophthalmic a.
anterior glandular branch of superior thyroid a.
anterior humeral circumflex a.
anterior inferior cerebellar a.
anterior inferior communicating a. (AICA)
a. of anterior inferior segment of kidney
anterior intercostal branch of internal thoracic a.
anterior interosseous a.
anterior interventricular branch of left coronary a.
anterior labial branch of deep external pudendal a.
anterior lateral malleolar a.
anterior lateral nasal branch of anterior ethmoidal a.
anterior/lateral/posterior glandular branch of superior thyroid a.
anterior medial malleolar a.
anterior mediastinal a.
anterior meningeal branch of anterior ethmoidal a.
anterior parietal a.
anterior perforating a.
anterior peroneal a.
anterior and posterior radicular a.
anterior and posterior superior pancreaticoduodenal a.
anterior scrotal branch of deep external pudendal a.
anterior segmental a.
anterior septal branch of anterior ethmoidal a.
anterior spinal a.
anterior superior alveolar a.
anterior superior dental a.

a. of anterior superior segment of kidney
anterior temporal a.
anterior tibial recurrent a.
anterior tympanic a.
anterior ulnar recurrent a.
anterior vestibular a.
anterolateral central a.
anterolateral striate a.
anterolateral thalamostriate a.
anteromedial central a.
anteromedial frontal branch of callosomarginal a.
anteromedial thalamostriate a.
apical branch of inferior lobar branch of right pulmonary a.
apical segmental a.
appendicular a.
arciform a.
arcuate a.
ascending branch of the inferior mesenteric a.
ascending branch of superficial cervical a.
ascending cervical a.
ascending palatine a.
ascending pharyngeal a.
atlantic part of vertebral a.
atrial anastomotic branch of circumflex branch of left coronary a.
a. to atrioventricular node
auditory a.
auricular branch of occipital a.
auricular branch of posterior auricular a.
axillary a.
basal part of left and right inferior pulmonary a.
basal tentorial branch of internal carotid a.
basilar a.
bilateral internal mammary a. (BIMA)
brachial a.
brachiocephalic trunk a.
a. of brain
bronchial a.
bronchopulmonary segmental a. (BPSA)
buccal a.
buccinator a.

NOTES

41

artery *(continued)*

a. of bulb of penis
a. of bulb of vestibule
calcaneal a.
calcareous a.
calcarine branch of medial
  occipital a.
a. of calf
callosomarginal a.
capsular branch of intrarenal a.
capsular branch of renal a.
caroticotympanic a.
carotid a.
carpal a.
caudal pancreatic a.
a. of caudate lobe
cavernous branch of cavernous part
  of internal carotid a.
cavernous sinus branch of internal
  carotid a.
cecal a.
celiac a.
central retinal a.
central sulcal a.
a. of central sulcus
cerebellar a.
cerebral a.
a. of cerebral hemorrhage
cerebral part of internal carotid a.
cervical part of internal carotid a.
cervical part of vertebral a.
cervicovaginal a.
Charcot a.
a. of chiasmal region
choroidal a.
ciliary a.
cilioretinal a.
cingular branch of
  callosomarginal a.
a. of circulus arteriosus
circumferential pontine branch of
  pontine a.
circumflex branch of left
  coronary a.
circumflex branch of posterior
  tibial a.
circumflex femoral a.
circumflex fibular branch of
  posterior tibial a.
circumflex humeral a.
circumflex iliac a.
circumflex peroneal branch of
  posterior tibial a.
circumflex scapular a.
clavicular branch of
  thoracoacromial a.
clivus branch of cerebral part of
  internal carotid a.
coccygeal a.

cochlear branch of labyrinthine a.
cochlear branch of
  vestibulocochlear a.
colic branch of ileocolic a.
collateral digital a.
colli a.
collicular a.
common carotid a.
common cochlear a.
common hepatic a.
common iliac a.
common interosseous a.
common palmar digital a.
common plantar digital a.
communicating branch of fibular a.
communicating branch of
  peroneal a.
conjunctival a.
cord of umbilical a.
corkscrew a.
coronary a.
cortical part of middle cerebral a.
cortical radiate a.
costocervical a.
cremasteric a.
cricothyroid branch of superior
  thyroid a.
crural a.
cystic a.
deep auricular a.
deep brachial a.
deep branch of medial circumflex
  femoral a.
deep branch of medial plantar a.
deep branch of superior gluteal a.
deep branch of transverse
  cervical a.
deep circumflex iliac a.
deep circumflex inguinal a.
deep epigastric a.
deep external pudendal a.
deep femoral a.
deep lingual a.
deep middle cerebral a.
deep palmar branch of ulnar a.
deep plantar branch of dorsalis
  pedis a.
deep profunda brachial a.
deep temporal a.
deferential a.
dental a.
descending branch of lateral
  circumflex femoral a.
descending branch of medial
  circumflex femoral a.
descending branch of occipital a.
descending branch of superficial
  cervical a.
descending genicular a.

descending palatine a.
descending scapular a.
diaphragmatic a.
digital collateral a.
diploic a.
distal medial striate a.
distributing a.
dolichoectatic a.
dorsal branch of first and second
  posterior intercostal a.
dorsal branch of lumbar a.
dorsal branch of subcostal a.
dorsal branch of superior
  intercostal a.
dorsal carpal branch of radial a.
dorsal carpal branch of ulnar a.
dorsal digital a.
dorsal interosseous a.
dorsalis pedis a.
dorsal lingual branch of lingual a.
dorsal metacarpal a.
dorsal metatarsal a.
dorsal nasal branch of
  ophthalmic a.
dorsal pancreatic a.
dorsal scapular a.
dorsal thoracic a.
a. of Drummond
a. of ductus deferens
duodenal branch of anterior
  superior pancreaticoduodenal a.
duodenal branch of posterior
  superior pancreaticoduodenal a.
efferent a.
elastic a.
elastic laminae of arteries
elastic layers of arteries
end a.
epicardial coronary a.
epigastric a.
episcleral a.
esophageal branch of inferior
  thyroid a.
esophageal branch of left gastric a.
ethmoidal a.
external acoustic meatus a.
external carotid a.
external iliac a.
external mammary a.
external maxillary a.
external nasal a.
external pudendal a.

external spermatic a.
facial a.
fallopian a.
fascioscapulohumeral a.
femoral nutrient a.
fibular nutrient a.
first metacarpal a.
first and second posterior
  intercostal a.
a. forceps
frontal branch of middle
  meningeal a.
frontal branch of superficial
  temporal a.
frontopolar a.
funicular a.
ganglionic branch of internal
  carotid a.
gastric a.
gastroduodenal a.
gastroepiploic a.
gastroomental a.
genicular a.
gingival a.
glandular branch of facial a.
glandular branch of inferior
  thyroid a.
glaserian a.
gluteal a.
gonadal a.
great anastomotic a.
great anterior radicular a.
greater palatine a.
great pancreatic a.
great radicular a.
great segmental medullary a.
great superior pancreatic a.
groove of first rib for
  subclavian a.
groove of lung for subclavian a.
groove for middle temporal a.
groove for subclavian a.
groove for vertebral a.
helicine a.
hepatic a.
Heubner a.
a. of Heubner
highest intercostal a.
highest thoracic a.
Huebner recurrent a.
humeral nutrient a.
hyaloid a.

NOTES

artery *(continued)*
    hypogastric a.
    hypoglossal a.
    hypophyseal a.
    ileal a.
    ileocolic a.
    iliac branch of iliolumbar a.
    iliacus branch of iliolumbar a.
    ilioinguinal a.
    iliolumbar a.
    inferior alveolar a.
    inferior branch of superior gluteal a.
    inferior carotid a.
    inferior cerebellar a.
    inferior cerebral a.
    inferior dental a.
    inferior epigastric a.
    inferior hemorrhoidal a.
    inferior hypophysial a.
    inferior internal parietal a.
    inferior labial branch of facial a.
    inferior laryngeal a.
    inferior lateral genicular a.
    inferior lingular a.
    inferior lobar a.
    inferior medial genicular a.
    inferior mesenteric a. (IMA)
    inferior pancreatic a.
    inferior pancreaticoduodenal a.
    inferior phrenic a.
    inferior rectal a.
    a. of inferior segment of kidney
    inferior and superior lobar a.
    inferior suprarenal a.
    inferior temporal branch of retinal a.
    inferior thoracic a.
    inferior thyroid a.
    inferior tympanic a.
    inferior ulnar collateral a.
    inferior vesical a.
    infrahyoid branch of superior thyroid a.
    infraorbital branch of interior maxillary a.
    infrascapular a.
    inguinal branch of deep external pudendal a.
    innominate a.
    insular part of middle cerebral a.
    intercostal a.
    interior maxillary branch of external carotid a.
    interlobar a.
    interlobular a.
    intermediate atrial branch of left coronary a.
    intermediate atrial branch of right coronary a.
    intermediate temporal branch of lateral occipital a.
    intermediomedial frontal branch of callosomarginal a.
    internal auditory a.
    internal carotid a. (ICA)
    internal iliac a.
    internal mammary a. (IMA)
    internal maxillary a.
    internal pudendal a.
    internal rectal a.
    internal spermatic a.
    internal thoracic a.
    interosseous a.
    interventricular septal branch of left/right coronary a.
    intestinal a.
    intracranial part of vertebral a.
    intramural a.
    intrarenal a.
    a. island flap
    jejunal a.
    juxtacolic a.
    Kugel anastomotic a.
    labial a.
    a. of labyrinth
    labyrinthine a.
    lacrimal a.
    laryngeal a.
    lateral atrial branch of left coronary a.
    lateral atrial branch of right coronary a.
    lateral basal segmental a.
    lateral branch of pontine a.
    lateral circumflex femoral a.
    lateral costal branch of internal thoracic a.
    lateral frontobasal a.
    lateral inferior genicular a.
    lateral malleolar branch of fibular peroneal a.
    lateral mammary branch of lateral thoracic a.
    lateral and medial palpebral a.
    lateral and medial parietal a.
    lateral and medial posterior choroidal branch of posterior cerebral a.
    lateral medullary branch of intracranial part of vertebral a.
    lateral nasal branch of facial a.
    lateral occipital a.
    lateral orbitofrontal a.
    lateral palpebral a.
    lateral plantar a.
    lateral posterior nasal a.

lateral sacral branch of median
sacral a.
lateral segmental a.
lateral splanchnic arteries
lateral striate a.
lateral superior genicular a.
lateral superior geniculate a.
lateral tarsal a.
lateral thoracic a.
left anterior descending a. (LAD)
left circumflex a. (LCX, LCx)
left colic a.
left coronary a. (LCA)
left femoral a.
left gastric a.
left gastroomental a.
left hepatic a.
left marginal a.
left obturator a.
left ovarian a.
left perineal a.
left pulmonary a.
left subclavian a.
left testicular a.
left umbilical a.
left vertebral a.
lenticulostriate a.
lesser palatine a.
lienal a.
lingual a.
lingular a.
long central a.
long ciliary a.
long posterior ciliary a.
long thoracic a.
a. of lower limb
lowest lumbar a.
lowest thyroid a.
lumbar branch of iliolumbar a.
lumen of bronchial a.
macular a.
main pulmonary a.
malleolar a.
mammary a.
mammillary a.
mandibular a.
marginal atrial branch of right
coronary a.
marginal tentorial branch of
internal carotid a.
masseteric a.
mastoid branch of occipital a.

mastoid branch of posterior
auricular a.
mastoid branch of posterior
tympanic a.
maxillary a.
medial anterior malleolus a.
medial basal branch of
pulmonary a.
medial basal segmental a.
medial branch of pontine a.
medial circumflex femoral a.
medial collateral a.
medial commisural a.
medial cutaneous branch of dorsal
branch of posterior intercostal a.
medial femoral circumflex a.
medial frontobasal a.
medial inferior genicular a.
medial malleolar branch of
posterior tibial a.
medial medullary branch of
vertebral a.
medial occipital a.
medial orbitofrontal a.
medial palpebral a.
medial plantar a.
medial segmental a.
medial striate a.
medial superior genicular a.
medial tarsal a.
median callosal a.
median commissural a.
median sacral a.
mediastinal branch of internal
thoracic a.
medium a.
medullary spinal a.
meningeal branch of cavernous part
of internal carotid a.
meningeal branch of cerebral part
of internal carotid a.
meningeal branch of intracranial
part of vertebral a.
meningeal branch of occipital a.
mental branch of inferior
alveolar a.
mesenteric inferior a.
mesenteric superior a.
metatarsal a.
middle alveolar a.
middle cerebral a. (MCA)
middle colic a.

NOTES

artery *(continued)*

middle collateral a.
middle genicular a.
middle hemorrhoidal a.
middle lobar a.
middle meningeal a.
middle rectal a.
middle sacral a.
middle suprarenal a.
middle temporal branch of insular part of middle cerebral a.
middle temporal branch of lateral occipital a.
middle vesical a.
M2 segment of middle cerebral a.
muscular artery of ophthalmic a.
musculocutaneous a.
musculophrenic a.
mylohyoid branch of inferior alveolar a.
myometrial arcuate a.
myometrial radial a.
nasal septal branch of superior labial branch of facial a.
Neubauer a.
a. of nose
nutrient a.
obturator a.
occipital a.
omphalomesenteric a.
ophthalmic a.
orbital branch of middle meningeal a.
orbitofrontal a.
ovarian branch of uterine a.
palatine a.
palmar carpal branch of radial a.
palmar carpal branch of ulnar a.
palmar digital a.
palmar interosseous a.
palmar metacarpal a.
palpebral a.
paracentral branch of callosomarginal a.
paracentral branch of pericallosal a.
paramedian pontine branch of pontine a.
parathyroid a.
parent a.
parietal branch of medial occipital a.
parietal branch of middle meningeal a.
parietal branch of superficial temporal a.
parietooccipital branch of anterior cerebral a.
parietooccipital branch of posterior cerebral a.

patent part of umbilical a.
pectoral branch of thoracoacromial a.
a. of penis
perforating branch of anterior interosseous a.
perforating branch of fibular a.
perforating branch of internal thoracic a.
perforating branch of palmar metacarpal a.
perforating branch of peroneal a.
perforating branch of plantar metatarsal a.
perforating a. of deep femoral a.
perforating a. of internal thoracic a.
periarterial plexus of anterior cerebral a.
periarterial plexus of ascending pharyngeal a.
periarterial plexus of choroid a.
periarterial plexuses of coronary a.
periarterial plexus of facial a.
periarterial plexus of inferior phrenic a.
periarterial plexus of inferior thyroid a.
periarterial plexus of internal thoracic a.
periarterial plexus of lingual a.
periarterial plexus of maxillary a.
periarterial plexus of middle cerebral a.
periarterial plexus of occipital a.
periarterial plexus of ophthalmic a.
periarterial plexus of popliteal a.
periarterial plexus of posterior auricular a.
periarterial plexus of subclavian a.
periarterial plexus of superficial temporal a.
periarterial plexus of superior thyroid a.
periarterial plexus of testicular a.
periarterial plexus of thyroid a.
periarterial plexus of vertebral a.
pericallosal a.
pericardiacophrenic a.
perineal a.
peroneal a.
petrosal branch of middle meningeal a.
petrous part of internal carotid a.
pharyngeal branch of ascending pharyngeal a.
pharyngeal branch of descending palatine a.

pharyngeal branch of inferior
  thyroid a.
pharyngeal branch of internal
  mammary a.
phrenic a.
plantar digital a.
plantar metatarsal a.
plexus of anterior cerebral a.
plexus of choroid a.
plexus of middle cerebral a.
polar frontal a.
polar temporal a.
arteries of pons
pontine a.
popliteal a.
postcentral sulcal a.
a. of postcentral sulcus
postcommunicating part of anterior
  cerebral a.
postcommunicating part of posterior
  cerebral a.
posterior alveolar a.
posterior auricular a.
posterior auricular branch of
  external carotid a.
posterior branch of inferior
  pancreaticoduodenal a.
posterior branch of obturator a.
posterior branch of recurrent
  ulnar a.
posterior branch of renal a.
posterior branch of superior
  thyroid a.
posterior branch of ulnar
  recurrent a.
posterior cecal a.
posterior cerebral a.
posterior choroidal a.
posterior circumflex humeral a.
posterior communicating a.
posterior conjunctival a.
posterior dental a.
posterior descending a. (PDA)
posterior descending coronary a.
posterior ethmoidal branch of
  ophthalmic a.
posterior gastric a.
posterior glandular branch of
  superior thyroid a.
posterior humeral circumflex a.
posterior inferior cerebellar a.

posterior inferior communicating a.
  (PICA)
posterior intercostal a.
posterior interosseous a.
posterior interventricular branch of
  right coronary a.
posterior labial branch of internal
  perineal a.
posterior lateral nasal a.
posterior mediastinal a.
posterior meningeal a.
posterior pancreaticoduodenal a.
posterior parietal a.
posterior peroneal a.
posterior radicular a.
posterior scrotal branch of internal
  pudendal a.
posterior scrotal branch of
  perineal a.
posterior segmental a.
a. of posterior segment of kidney
posterior septal branch of
  sphenopalatine a.
posterior spinal a.
posterior superior alveolar a.
posterior temporal branch of
  middle cerebral a.
posterior tibial recurrent a.
posterior tympanic a.
posterior ulnar recurrent a.
posterior vestibular branch of
  vestibulocochlear a.
posterolateral central a.
posteromedial central a.
posteromedial frontal branch of
  callosomarginal a.
P1–P4 segment of posterior
  cerebral a.
precentral sulcal a.
a. of precentral sulcus
precommunical segment of anterior
  cerebral a.
precommunical segment of posterior
  cerebral a.
precommunicating part of anterior
  cerebral a.
precommunicating part of posterior
  cerebral a.
precuneal branch of anterior
  cerebral a.

NOTES

artery *(continued)*

prelaminar branch of spinal branch of dorsal branch of posterior intercostal a.
prepancreatic a.
preRolandic a.
preventricular a.
prevertebral part of vertebral a.
princeps cervicis a.
princeps pollicis a.
profunda brachii a.
profunda cervicalis a.
profunda femoris a.
profunda linguae a.
proper cochlear a.
proper hepatic a.
proper palmar digital a.
proper plantar digital a.
prostatic branch of inferior vesical a.
prostatic branch of middle rectal a.
proximal medial striate a.
pterygoid branch of maxillary a.
pterygoid branch of posterior deep temporal a.
a. of pterygoid canal
pterygomeningeal a.
pubic branch of inferior epigastric a.
pubic branch of obturator a.
pudendal a.
pulmonary a. (PA)
a. of pulp
pyloric a.
quadrigeminal a.
radial collateral a.
radial index a.
radialis indicis a.
radial recurrent a.
radicular a.
ranine a.
rectal a.
recurrent interosseous a.
recurrent radial a.
recurrent ulnar a.
renal a.
retinal a.
retroduodenal a.
retroesophageal a.
right atrial branch of right coronary a.
right colic a.
right femoral a.
right flexural a.
right gastric a.
right gastroepiploic a.
right gastroomental a.
right hepatic a.

right marginal branch of right coronary a.
right middle suprarenal a.
right obturator a.
right ovarian a.
right pulmonary a.
right subclavian a.
right testicular a.
rolandic sulcal a.
a. of round ligament of uterus
saphenous branch of descending genicular a.
scapular a.
sciatic a.
a. to sciatic nerve
screw arteries
scrotal a.
second lumbar a.
segmental medullary a.
septal a.
sheathed a.
short central a.
short ciliary a.
short circumferential a.
short gastric branch of lienal a.
short posterior ciliary a.
sigmoid a.
sigmoidal branch of inferior mesenteric a.
a. to the sinoatrial node
sinuatrial nodal branch of right coronary a.
sinuatrial node a.
small arteries
somatic arteries
spermatic a.
sphenoid part of middle cerebral a.
sphenopalatine branch of internal maxillary a.
spinal a.
spinoneural a.
spiral modiolar a.
splenic branch of splenic a.
stapedial branch of posterior tympanic a.
stapedial branch of stylomastoid a.
sternal branch of internal thoracic a.
sternocleidomastoid branch of occipital a.
sternocleidomastoid branch of superior thyroid a.
sternomastoid a.
straight a.
striate a.
stylomastoid a.
subclavian a.
subcostal a.
sublingual a.

submental a.
suboccipital part of vertebral a.
subscapular branch of axillary a.
sulcal a.
sulcus of meningeal a.
sulcus for middle temporal a.
sulcus of occipital a.
sulcus of subclavian a.
sulcus for vertebral a.
superficial brachial a.
superficial branch of medial
  circumflex femoral a.
superficial branch of medial
  plantar a.
superficial branch of superior
  gluteal a.
superficial branch of transverse
  cervical a.
superficial cervical a.
superficial circumflex iliac a.
superficial and deep external
  pudendal a.
superficial epigastric a.
superficial palmar branch of
  radial a.
superficial perineal a.
superficial temporal branch of
  external carotid a.
superficial temporalis a.
superficial temporary artery-middle
  cerebral a. (STA-MCA)
superficial volar a.
superior alveolar a.
superior branch of the superior
  gluteal a.
superior cerebellar a.
superior coronary a.
superior epigastric a.
superior gluteal a.
superior hemorrhoidal a.
superior hypophysial a.
superior intercostal a.
superior internal parietal a.
superior labial branch of facial a.
superior laryngeal a.
superior lateral genicular a.
superior lingular branch of lingular
  branch of superior lobar left
  pulmonary a.
superior medial genicular a.
superior mesenteric a. (SMA)
superior pancreaticoduodenal a.

superior phrenic a.
superior rectal a.
superior segmental a.
a. of superior segment of kidney
superior suprarenal a.
superior temporal branch of
  retinal a.
superior thoracic a.
superior thyroid a.
superior tympanic a.
superior ulnar collateral a.
superior vermian branch of
  superior cerebellar a.
superior vesical a.
suprachiasmatic a.
supraduodenal a.
suprahyoid branch of lingual a.
supraoptic a.
supraorbital a.
suprarenal a.
suprascapular a.
supratrochlear a.
supreme intercostal a.
sural a.
sylvian a.
a. to tail of pancreas
temporal a.
tentorial basal branch of internal
  carotid a.
tentorial marginal branch of
  cavernous part of internal
  carotid a.
terminal branch of middle
  cerebral a.
testicular a.
thoracoacromial a.
thoracodorsal a.
thymic branch of internal
  thoracic a.
thyrocervical a.
thyroid ima a.
tibial nutrient a.
tonsillar branch of facial a.
transversarial part of vertebral a.
transverse branch of lateral femoral
  circumflex a.
transverse cervical a.
transverse facial a.
transverse pancreatic a.
transverse scapular a.
triangle of lingual a.
triangle of vertebral a.

NOTES

artery *(continued)*
    tubal branch of ovarian a.
    tubal branch of uterine a.
    a. of tuber cinereum
    tympanic a.
    ulnar recurrent a.
    ultimobranchial a.
    umbilical a.
    uncal a.
    a. of upper limb
    ureteric branch of inferior
      suprarenal a.
    ureteric branch of ovarian a.
    ureteric branch of patent part of
      umbilical a.
    ureteric branch of renal a.
    ureteric branch of testicular a.
    urethral a.
    uterine a.
    vaginal a.
    a. to vas deferens
    venous a.
    ventral splanchnic arteries
    ventricular a.
    vertebral a. (VA)
    vesical a.
    vestibular branch of labyrinthine a.
    vestibulocochlear a.
    vidian a.
    vitelline a.
    volar interosseous a.
    Wilkie a.
    a. of Willis
    Zinn a.
    zygomaticofacial a.
    zygomaticoorbital a.
arthral
arthrodia
arthrodial
    a. articulation
    a. cartilage
    a. joint
arthrodysplasia
arthrogenous
arthrogryposis
arthrologia
arthrology
arthrosis
articular
    a. branch
    a. capsule
    a. cavity
    a. circumference of head of radius
    a. circumference of head of ulna
    a. corpuscles
    a. crescent
    a. crest
    a. disk
    a. disk of acromioclavicular joint

    a. disk of distal radioulnar joint
    a. disk of sternoclavicular joint
    a. disk of temporomandibular joint
    a. eminence
    a. eminence of temporal bone
    a. facet
    a. facet of head of fibula
    a. facet of head of rib
    a. facet of lateral malleolus
    a. facet of medial malleolus
    a. facet of radial head
    a. facet of tubercle of rib
    a. fossa of temporal bone
    a. labrum
    a. lamella
    a. lip
    a. margin
    a. meniscus
    a. muscle
    a. muscle of elbow
    a. muscle of knee
    a. pit of head of radius
    a. process
    a. recurrent nerve
    a. surface
    a. surface of acromion
    a. surface of arytenoid cartilage
    a. surface of head of fibula
    a. surface of head of rib
    a. surface of knee
    a. surface of mandibular fossa of
      temporal bone
    a. surface on calcaneus for cuboid
      bone
    a. surface of patella
    a. surface of talus
    a. surface of temporal bone
    a. surface of tubercle of rib
    a. tubercle
    a. tubercle of temporal bone
    a. vascular circle
    a. vascular network
    a. vascular network of elbow
    a. vascular network of knee
    a. vascular plexus
articulare
    cavitas a.
    cavum a.
    labrum a.
    rete vasculosum a.
    tuberculum a.
articularia
    corpuscula a.
articularis, pl. **articulares**
    capsula a.
    cartilago a.
    cavitas a.
    a. cubiti
    a. cubiti muscle

discus a.
eminentia a.
facies a.
a. genus
a. genus muscle
membrana fibrosa capsulae a.
meniscus a.
musculus a.
nervus a.
processus a.
rami articulares
stratum fibrosum capsulae a.
articulate
articulated skeleton
articulatio, pl. **articulationes**
  a. acromioclavicularis
  a. atlantoaxialis lateralis
  a. atlantoaxialis mediana
  a. atlanto-occipitalis
  a. bicondylaris
  a. calcaneocuboidea
  a. capitis costae
  a. carpi
  articulationes carpometacarpales
  a. carpometacarpalis pollicis
  articulationes carpometacarpeae
  a. carpometacarpea pollicis
  a. cartilaginis
  articulationes cinguli membri
  articulationes cinguli membri
   inferioris
  articulationes cinguli membri
   superioris
  articulationes cinguli pectoralis
  articulationes cinguli pelvici
  a. complexa
  a. composita
  a. condylaris
  a. costochondralis
  a. costotransversaria
  articulationes costovertebrales
  a. cotylica
  a. coxa
  a. coxofemoralis
  articulationes cranii
  a. cricoarytenoidea
  a. cricothyroidea
  a. cubiti
  a. cuneonavicularis
  a. cylindrica
  a. dentoalveolaris
  a. ellipsoidea

a. fibrosa
a. genus
a. glenohumeralis
a. humeri
a. humeroradialis
a. humeroulnaris
a. incudomallearis
a. incudostapedia
articulationes intercarpales
articulationes intercarpeae
articulationes interchondrales
articulationes intercuneiformes
articulationes intermetacarpales
articulationes intermetatarsales
articulationes intermetatarseae
articulationes interphalangeae
articulationes interphalangeae manus
articulationes interphalangeae pedis
articulationes intertarseae
a. lumbosacralis
a. mandibularis
a. mediocarpalis
articulationes membri inferioris
 liberi
articulationes membri superioris
 liberi
articulationes metacarpophalangeae
articulationes metatarsophalangeae
articulationes ossiculorum
 auditoriorum
articulationes ossiculorum auditus
a. ossis pisiformis
a. ovoidalis
articulationes pedis
a. plana
a. radiocarpalis
a. radiocarpea
a. radioulnaris distalis
a. radioulnaris proximalis
a. sacrococcygea
a. sacroiliaca
a. sellaris
a. simplex
a. spheroidea
a. sternoclavicularis
articulationes sternocostales
a. subtalaris
a. synovialis
a. talocalcanea
a. talocalcaneonavicularis
a. talocruralis
a. tarsi transversa

NOTES

articulatio *(continued)*
    articulationes tarsometatarsales
    articulationes tarsometatarseae
    a. temporomandibularis
    articulationes thoracis
    a. tibiofibularis
    a. trochoidea
    articulationes zygapophyseales
    articulationes zygapophysiales

articulation
    acromioclavicular a.
    arthrodial a.
    atlantoaxial a.
    atlantooccipital a.
    bicondylar a.
    brachiocarpal a.
    brachioulnar a.
    calcaneoastragaloid a.
    calcaneocuboid a.
    carpal a.
    carpometacarpal a.
    cartilaginous a.
    compound a.
    condylar a.
    costocentral a.
    costosternal a.
    costotransverse a.
    costovertebral a.
    cricoarytenoid a.
    cricothyroid a.
    cuneonavicular a.
    distal radioulnar a.
    ellipsoidal a.
    a.'s of foot
    glenohumeral a.
    a.'s of hand
    humeral a.
    humeroradial a.
    humeroulnar a.
    iliosacral a.
    incudomalleolar a.
    incudostapedial a.
    intercarpal a.
    interchondral a.
    intercostal a.
    intermetacarpal a.
    intermetatarsal a.
    interphalangeal a.
    intertarsal a.
    mandibular a.
    maxillary a.
    mediocarpal a.
    metacarpocarpal a.
    metacarpophalangeal a.
    metatarsophalangeal a.
    occipitoatlantal a.
    patellofemoral a.
    peg-and-socket a.
    phalangeal a.
    a. of pisiform bone
    proximal radioulnar a.
    radiocarpal a.
    radioulnar a.
    sacrococcygeal a.
    sacroiliac a.
    scapuloclavicular a.
    spheroid a.
    sternoclavicular a.
    sternocostal a.
    subtalar a.
    superior tibial a.
    talocalcaneonavicular a.
    talocrural a.
    talonavicular a.
    tarsometatarsal a.
    temporomandibular a.
    temporomaxillary a.
    tibiofibular a.
    tibiotarsal a.
    transverse tarsal a.
    trochoid a.
    trochoidal a.

articulationis
    recessus sacciformis a.

articulus

artifact

artifactual

artificial
    a. anatomy
    a. heart
    a. respiration

artistic anatomy

arycorniculata
    synchondrosis a.

arycorniculate synchondrosis

aryepiglottic (AE)
    a. fold
    a. muscle
    a. part of oblique arytenoid muscle

aryepiglottica
    plica a.

aryepiglotticus
    musculus a.

arytenoepiglottidean fold

arytenoid
    a. cartilage
    a. gland
    a. muscle
    a. process
    a. swelling

arytenoidal articular surface of cricoid

arytenoidea
    cartilago a.

arytenoideae
    apex cartilaginis a.
    basis cartilaginis a.
    colliculus cartilaginis a.
    crista arcuata cartilaginis a.

facies anterolateralis cartilaginis a.
facies articularis cartilaginis a.
facies medialis cartilaginis a.
facies posterior cartilaginis a.
fovea oblonga cartilaginis a.
fovea triangularis cartilaginis a.
processus muscularis cartilaginis a.
processus vocalis cartilaginis a.

arytenoideus
a. obliquus
a. transversus

aryvocalis
musculus a.

ascendens
aorta a.
arteria cervicalis a.
arteria palatina a.
arteria pharyngea a.
cervicalis a.
colon a.
lumbalis a.
mesocolon a.
musculus cervicalis a.
palatina a.
pharyngea a.
plexus pharyngeus a.
processus a.
ramus anterior a.
ramus posterior a.
vena lumbalis a.

ascendentis
plexus periarterialis arteriae
  pharyngeae a.
rami pharyngeales arteriae
  pharyngeae a.

ascending
a. aorta (AA)
a. branch
a. branch of the inferior
  mesenteric artery
a. branch of superficial cervical
  artery
a. cervical artery
a. colon
a. current
a. frontal convolution
a. frontal gyrus
a. lumbar vein
a. mesocolon
a. palatine artery
a. parietal convolution
a. parietal gyrus

a. part of duodenum
a. part of trapezius muscle
a. pharyngeal artery
a. pharyngeal plexus
a. process
a. ramus
a. ramus of lateral sulcus of
  cerebrum

Aschner
A. phenomenon
A. reflex

Aschner-Dagnini reflex
Aselli pancreas
asexual
a. generation
a. reproduction

ashen
a. tuber
a. tubercle
a. wing

Ashley phenomenon
ASIS
anterior-superior iliac spine

Askanazy cell
asoma
asomata
aspalasoma
aspect
anteroinferior a.
cephalad a.
facial a.
frontal a.
inferolateral a.
lateral a.
medial a.
occipital a.
outer a.
radial a.
rostral a.
superior a.
superolateral a.
ulnar a.
ventral a.
vertical a.
volar a.

aspera
lateral lip of linea a.
linea a.
medial lip of linea a.

asperae
labium laterale lineae a.
labium mediale lineae a.

NOTES

asphyxia
asphyxial
asphyxiant
asphyxiate
asphyxiating
asphyxiation
aspiration
asplenia
asplenic
Assézat triangle
assimilable
assimilation
    a. sacrum
assisted respiration
associated antagonist
association
    a. areas
    a. cortex
    a. fibers
    a. tract
aster
    sperm a.
asterion
asternal
asternia
asteroid
astomatous
astomerotomy
astomia
astomous
astragalar
astragalocalcanean
astragalofibular
astragalonavicular joint
astragaloscaphoid
astragalotibial
astragalus
astral fibers
astroblast
astrocele
astrocyte
    ameboid a.
    fibrillary a.
    fibrous a.
    gemistocytic a.
    protoplasmic a.
    reactive a.
astroglia cell
astroid
astrokinetic
astropyloric
astrosphere
asymmetric
asymmetros
    cephalothoracopagus a.
    syncephalus a.
asymmetrus
    janiceps a.
asymmetry

asynechia
asynergia
asynergic
asynergy
asystematic
atavic
atavicus
    metatarsus a.
atavism
atavistic epiphysis
atavus
atelectasis
atelectatic
atelia
ateliosis
aterales
athelia
athermancy
athermanous
athermosystaltic
athlete's heart
athrocytosis
athymia
athymism
athyrea
athyroidism
athyrosis
athyrotic
atlantad
atlantal
atlantic part of vertebral artery
atlantis
    arcus anterior a.
    arcus posterior a.
    facies articularis inferior a.
    facies articularis superior a.
    fasciculi longitudinales ligamenti
      cruciformis a.
    fovea articularis inferior a.
    fovea articularis superior a.
    fovea dentis a.
    ligamentum cruciatum a.
    ligamentum cruciforme a.
    ligamentum transversum a.
    massa lateralis a.
    tuberculum anterius a.
    tuberculum posterius a.
atlantoaxial
    a. articulation
    a. joint
atlantodidymus
atlantoepistrophic
atlantooccipital
    a. articulation
    a. joint
    a. ligament
    a. membrane
atlantooccipitalis
    articulatio a.

atlantoodontoid, atlanto-odontoid
atlas
    anterior arch of a.
    anterior tubercle of a.
    a. arch
    cruciate ligament of a.
    cruciform ligament of a.
    inferior articular facet of a.
    inferior articular pit of a.
    inferior articular surface of a.
    lateral mass of a.
    longitudinal bands of cruciform
      ligament of a.
    pit for dens of a.
    posterior arch of a.
    posterior tubercle of a.
    superior articular facet of a.
    superior articular pit of a.
    superior articular surface of a.
    transverse ligament of a.
atloaxoid
atlodidymus
atloid
atlooccipital
atm
    atmosphere
atmosphere (atm)
    a. absolute
    ICAO standard a.
    standard a.
atmospheric pressure
atmospherization
atonia
atonic
atonicity
atony
atrabiliaris
    glandula a.
atrabiliary capsule
atresia
    anal a.
    a. ani
    aortic a.
    biliary a.
    choanal a.
    esophageal a.
    a. folliculi
    intestinal a.
    a. iridis
    laryngeal a.
    pulmonary a.

    tricuspid a.
    vaginal a.
atresic teratosis
atretic ovarian follicle
atreticum
    corpus a.
atretoblepharia
atretocystia
atretogastria
atretopsia
atria (*pl. of* atrium)
atrial
    a. anastomotic branch of
      circumflex branch of left
      coronary artery
    a. auricle
    a. auricula
    a. branch
    a. natriuretic factor
    a. natriuretic peptide (ANP)
    a. septal defect
    a. transport function
atriales
    arteriae a.
    rami a.
atrichia
atrichosis
atrii
    auriculae a.
atriodigital dysplasia
atrio-His
    a.-H. pathway
    a.-H. tract
atrionector
atriopressor reflex
atriorum
    pars membranacea septi a.
atrioventricular (AV)
    a. band
    a. bundle
    a. canal
    a. canal cushions
    a. conduction system (AVCS)
    a. groove
    a. junction
    a. nodal branch
    a. node
    a. orifice
    a. ring
    a. septum
    a. sulcus

NOTES

atrioventricular *(continued)*
 a. trunk
 a. valve
atrioventriculare
 septum a.
atrioventricularis, pl. **atrioventriculares**
 crus dextrum fasciculi a.
 crus dextrum truncus a.
 crus sinistrum fasciculi a.
 crus sinistrum truncus a.
 cuspis septalis valvae a.
 fasciculus a.
 nodus a.
 rami subendocardiales fasciculi a.
 ramus nodi a.
 truncus fascicularis a.
atrium, pl. **atria**
 auricle of left a.
 auricle of right a.
 common a.
 a. cordis dextrum
 a. cordis sinistrum
 crista terminalis of right a.
 a. dextrum cordis
 a. glottidis
 a. of heart
 intervenous tubercle of right a.
 a. of lateral ventricle
 left a.
 a. meatus medii
 a. meatus medii nasalis
 a. of middle nasal meatus
 nasal a.
 oblique vein of left a.
 a. pulmonale
 right a.
 a. sinistrum cordis
 a. ventriculi lateralis
 wall of a.
atrophy
 vaginal a.
attachment
 a. apparatus
 epithelial a.
 muscle-tendon a.
 pericemental a.
attic
 tympanic a.
atticoantral
atticomastoid
attitude
 fetal a.
attitudinal reflex
attollens
 a. aurem
 a. auriculam
 a. oculi

attraction
 neurotropic a.
 a. sphere
attrahens
attrition
atymeric
atypia
atypism
AU, au
 auris uterque
 both ears
Aub-DuBois table
auditiva, pl. **auditivae**
 cartilago tubae auditivae
 cellulae pneumaticae tubae auditivae
 isthmus tubae auditivae
 lamina lateralis cartilaginis tubae
  auditivae
 lamina medialis cartilaginis tubae
  auditivae
 lamina membranacea cartilaginis
  tubae auditivae
 ostium pharyngeum tubae auditivae
 ostium tympanicum tubae auditivae
 pars cartilaginea tubae auditivae
 pars ossea tubae auditivae
 semicanalis tubae auditivae
 tuba a.
 tunica mucosa tubae auditivae
auditor
 pharyngeal opening of a.
auditoria, pl. **auditoriae**
 isthmus tubae auditoriae
 lamina lateralis cartilaginis tubae
  auditoriae
 lamina medialis cartilaginis tubae
  auditoriae
 lamina membranacea cartilaginis
  tubae auditoriae
 ostium pharyngeum tubae auditoriae
 pars cartilaginea tubae auditoriae
 pars ossea tubae auditoriae
 semicanalis tubae auditoriae
 sulcus tubae auditoriae
 tuba a.
 tunica mucosa tubae auditoriae
auditoriorum
 articulationes ossiculorum a.
 musculi ossiculorum a.
auditorium
 ligamenta ossiculorum a.
auditory
 a. area
 a. artery
 a. canal
 a. capsule
 a. cartilage
 a. cortex
 a. ganglion

A

a. hair
a. lemniscus
a. nucleus
a. ossicle
a. pit
a. placode
a. pore
a. process
a. receptor cells
a. striae
a. strings
a. tooth
a. tract
a. tube
a. tube nerve
a. vein
a. vesicle

auditus
articulationes ossiculorum a.
ligamenta ossiculorum a.
musculi ossiculorum a.
organum a.
ossicula a.

Auerbach
A. ganglia
A. mesenteric plexus

augmented lead

augmentor
a. fiber
a. nerve

augnathus

aural

aurem
attollens a.
musculus attollens a.
musculus attrahens a.
musculus retrahens a.
retrahens a.

aures

auricle
accessory a.
anterior notch of a.
apex of a.
atrial a.
concha of a.
eminence of triangular fossa of a.
isthmus of cartilaginous a.
left a.
a. of left atrium
ligament of a.
lobule of a.
oblique muscle of a.

pyramidal muscle of a.
right a.
a. of right atrium
terminal notch of a.
tip of a.
transverse muscle of a.
triangular fossa of a.

auricula, pl. **auriculae**
apex auriculae
atrial a.
auriculae atrii
a. atrii dextra
a. atrii sinistra
cartilago auriculae
concha auriculae
fossa navicularis auriculae
fossa triangularis auriculae
incisura anterior auriculae
lobulus auriculae
musculus obliquus auriculae
musculus pyramidalis auriculae
musculus transversus auriculae
obliquus auriculae
pyramidalis auriculae
sulcus posterior auriculae
transversus auriculae
tuberculum auriculae

auriculam
attollens a.
musculus attollens a.
musculus attrahens a.
musculus retrahens a.
retrahens a.

auricular
a. appendage
a. appendix
a. arc
a. branch of occipital artery
a. branch of posterior auricular artery
a. branch of vagus nerve
a. canaliculus
a. cartilage
a. fissure
a. ganglion
a. index
a. ligament
a. muscle
a. notch
a. point
a. surface of ilium
a. surface of sacrum

NOTES

auricular *(continued)*
    a. triangle
    a. tubercle
    a. tubercle of Darwin
    a. vein
auriculare
auricularia
    ligamenta a.
auricularis, pl. auriculares
    a. anterior
    a. anterior muscle
    digitus a.
    incisura terminalis a.
    isthmus cartilaginis a.
    a. magnus
    musculi auriculares
    a. posterior
    a. posterior muscle
    a. superior
    a. superior muscle
auriculocranial
auriculoinfraorbital plane
auriculomastoid line
auriculopressor reflex
auriculotemporali
    ramus communicans ganglii otici
       cum nervo a.
auriculotemporalis
    nervus a.
    rami membranae tympani nervi a.
    rami parotidei nervi a.
    rami temporales superficiales
       nervi a.
auriculotemporal nerve
auriculoventricular
    a. groove
    a. interval
    a. orifice
    a. valve
auriform
auris
    ala a.
    antrum a.
    a. externa
    fossa navicularis a.
    incisura anterior a.
    incisura terminalis a.
    a. interna
    isthmus cartilaginis a.
    a. media
    a. uterque (AU, au)
ausculatory triangle
auscultation
    aortic area of a.
    triangle of a.
auscultationis
    trigonum a.
auser ganglion
autoblast

autocoid
autocrine
autogamous
autogamy
autogenesis
autogenetic
autogenic
autokinesia
autokinesis
autokinetic
automatism
automatograph
automixis
autonomic
    a. division of nervous system
    a. ganglia
    a. motor neuron
    a. nerve
    a. nerve fiber
    a. nervous system (ANS)
    a. part
    a. part of peripheral nervous
      system
    a. plexus
autonomica, pl. autonomicae
    neurofibrae autonomicae
    pars a.
autonomici
    plexus gastrici systematis a.
    rami pulmonales systematis a.
autonomicorum
    ganglia plexuum a.
autonomicum
    systema nervosum a.
autonomicus
    nervus a.
autonomotropic
autonomous
autonomy
autophagia
autophagic vacuole
autophagolysosome
autophagy
autopod
autopodia
autopodium
autoregulation
    heterometric a.
    homeometric a.
autosite
autotemnous
autotomy
auxanology
auxesis
auxetic growth
auxiliary
auxilio
auxiliomotor
auxodrome

auxotonic
AV
   arteriovenous
   atrioventricular
     AV difference
     AV valve
avalanche conduction
avalvular
avascular
AVCS
   atrioventricular conduction system
avis
   calcar a.
   nidus a.
   unguis a.
axes (*pl. of* axis)
axial
   a. angle
   a. current
   a. filament
   a. muscle
   a. plate
   a. section
   a. skeleton
   a. surface
axiale
   skeleton a.
axialis
axifugal
axil
axile corpuscle
axilla, pl. **axillae**
   ligamentum suspensorium axillae
   suspensory ligament of a.
axillaris, pl. **axillares**
   arteria a.
   fascia a.
   fossa a.
   lymphonodi axillares
   nervus a.
   nodi lymphoidei axillares
   plexus lymphaticus a.
   plica a.
   rami musculares nervi a.
   rami subscapulares arteriae a.
   regio a.
   vena a.
axillary
   a. arch
   a. arch muscle
   a. artery
   a. cavity

   a. fascia
   a. fold
   a. foramen
   a. fossa
   a. hair
   a. line
   a. lymphatic plexus
   a. lymph node
   a. nerve
   a. region
   a. sheath
   a. space
   a. sweat gland
   a. triangle
   a. vein
axillobifemoral
axiodistal
axiodistocervical
axioincisal
axiolabial
axiomesial
axiomesiocervical
axiomesiodistal
axiomesioincisal
axion
axio-occlusal
axioplasm
axiopulpal
axipetal
axiramificate
axis, pl. **axes**
   basibregmatic a.
   basicranial a.
   basifacial a.
   biauricular a.
   bowel a.
   a. bulbi externus
   a. bulbi internus
   celiac a.
   cephalocaudal a.
   cerebrospinal a.
   ciliary a.
   condylar a.
   conjugate a.
   a. corpuscle
   craniofacial a.
   a. cylinder
   dens a.
   deutan a.
   embryonic a.
   encephalomyelonic a.
   a. externus bulbi oculi

NOTES

axis *(continued)*
    facial a.
    frontal a.
    hinge a.
    a. internus bulbi oculi
    a. of lens
    a. lentis
    a. ligament
    a. ligament of malleus
    longitudinal a.
    long posterior ciliary a.
    mandibular a.
    neural a.
    optic a.
    a. opticus
    orbital a.
    pelvic a.
    a. of pelvis
    pupillary a.
    sagittal a.
    short posterior ciliary a.
    supraopticohypophyseal a.
    thoracic a.
    thyroid a.
    transverse horizontal a.
    vertical a.
axoaxonic synapse
axodendritic synapse
axofugal
axograph
axolemma

axolysis
axon
    cervix of a.
    a. hillock
    a. terminal
axonal terminal boutons
axoneme
axonography
axopetal
axoplasm
axoplasmic transport
axosomatic synapse
Ayala
    A. index
    A. quotient
azurophile
azurophil granule
azygos, azygous
    arteria articularis a.
    a. artery of vagina
    a. lobe
    a. lobe of right lung
    lobus a.
    lymphonodus arcus vena a.
    nodus lymphoideus arcus venae a.
    a. vein
    a. vein arch
    a. vein principle
    vena a.
    a. venous line

B

B bile
B cell
B cell differentiating factor
B lymphocyte
baby tooth
baccate
bacciform
Bachmann bundle
bacillary layer
back
b., arm, neck, shoulder (BANS)
b., arm, neck, shoulder area
broadest muscle of b.
deep muscle of b.
low b.
muscle of b.
b. pressure
region of b.
b. tooth
true muscle of b.
backbone
back-knee
baculiform
Baer
B. law
B. vesicle
baffle
pericardial b.
bag
Douglas b.
nuclear b.
b. of waters
bahnung
Baillarger
B. bands
B. line
Bainbridge reflex
baja
patella b.
Baker cyst
balance
acid-base b.
Wilhelmy b.
balanic
b. hypospadias
balanopreputial
balanus
bale
ball
chondrin b.
ball-and-socket joint
band
A b.'s
amniotic b.'s

annular b.
anogenital b.
atrioventricular b.
Baillarger b.'s
Bechterew b.
Broca diagonal b.
b. cell
Clado b.
b.'s of colon
Essick cell b.'s
Gennari b.
b. of Giacomini
H b.
His b.
Hunter-Schreger b.'s
hymenal b.
I b.
iliotibial b. (ITB)
b. of Kaes-Bechterew
Lane b.
M b.
Maissiat b.
Meckel b.
moderator b.
b. neutrophil
Q b.'s
Reil b.
Simonart b.'s
Streeter b.'s
Z b.
zonular b.
BANS
back, arm, neck, shoulder
BANS area
bar
b. of bladder
Mercier b.
Passavant b.
sternal b.
terminal b.
barba
barbula hirci
Barcroft-Warburg apparatus
Bardinet ligament
bare
b. area diaphragm
b. area of liver
b. area of stomach
Barkow ligament
baroceptor
barograph
barometric pressure
barometrograph
baroreceptor nerve
baroscope

barostat
barotaxis
barotrauma
    otic b.
    sinus b.
barotropism
barrier
    blood-air b.
    blood-aqueous b.
    blood-brain b.
    blood-cerebrospinal fluid b.
    placental b.
Bartholin
    B. anus
    B. duct
    B., urethral, and Skene (BUS)
    B., urethral, and Skene gland
barye
bas
basad
basal
    b. cell
    b. cell layer
    b. corpuscle
    b. crest of cochlear duct
    b. diet
    b. ganglia
    b. granule
    b. lamina
    b. lamina of choroid
    b. lamina of ciliary body
    b. lamina of neural tube
    b. lamina of semicircular duct
    b. layer of choroid
    b. layer of ciliary body
    b. membrane of semicircular duct
    b. metabolic rate (BMR)
    b. metabolism
    b. part
    b. part of left and right inferior
      pulmonary artery
    b. part of occipital bone
    b. plate of neural tube
    b. ridge
    b. sphincter
    b. striation
    b. tentorial branch of internal
      carotid artery
    b. vein
    b. vein of Rosenthal
basale
    stratum b.
basalis
    cisterna b.
    decidua b.
    fibrocartilago b.
    lamina b.
    pars b.
    ramus tentorii b.

basaloid cell
base
    anterior cranial b.
    appendiceal b.
    b. of arytenoid cartilage
    b. of bladder
    cilia b.
    cranial b.
    b. deficit
    b. excess
    external surface of cranial b.
    b. of heart
    b. of hyoid bone
    internal surface of cranial b.
    b. line
    b. of lung
    b. of mandible
    b. of metacarpal
    b. of metatarsal
    b. of modiolus of cochlea
    b. of patella
    b. of phalanx
    b. of phalanx of foot
    b. of phalanx of hand
    b. of prostate
    b. of renal pyramid
    b. of sacrum
    b. of skull
    b. of stapes
    b. of tongue
    b. unit
    vitreous b.
basement
    b. lamina
    b. membrane
    b. tissue
bases (*pl. of* basis)
bas-fond
basialis
basialveolar
basibregmatic axis
basic electrical rhythm (BER)
basicranial
    b. axis
    b. flexure
basicranium
basifacial axis
basihyal bone
basihyoid
basilar
    b. angle
    b. apophysis
    b. artery
    Broca b.
    b. cartilage
    b. cell
    b. crest of cochlear duct
    b. fibrocartilage
    b. index

b. invagination
b. lamina
b. membrane
b. membrane of cochlear duct
b. part
b. part of occipital bone
b. part of pons
b. process
b. process of occipital bone
b. projection
b. sulcus
b. venous plexus
b. venous sinus
b. vertebra

basilare
os b.

basilaris
arteria b.
glandula b.
membrana b.
norma b.
pars b.
plexus venosus b.

basilateral
basilemma
basilic
b. hiatus
b. vein

basilica
mediana b.
vena b.
vena intermedia b.
vena mediana b.

basilicus
basinasal line
basioccipital bone
basiocciput
basioglossus
basion
basipetal
basipharyngeal canal
basis, pl. **bases**
b. cartilaginis arytenoideae
b. cordis
b. cranii
b. cranii externa
b. cranii interna
b. mandibulae
b. modioli cochlea
b. ossis metacarpalis
b. ossis metatarsalis
b. ossis sacri

b. patellae
b. phalangis
b. phalangis manus
b. phalangis pedis
b. pontis
b. prostatae
b. pulmonis
b. pyramidis renis
b. stapedis

basisphenoid bone
basitemporal
basium
ligamenta b.

basivertebralis, pl. **basivertebrales**
vena b.

basivertebral vein
basket
b. cell
fibrillar b.'s

Basle Nomina Anatomica (BNA)
basocyte
basolateral
basometachromophil
basometachromophile
basophil
b. cell of anterior lobe of hypophysis
b. granule
b. substance
tissue b.

basophile
basophilia
punctate b.
substantia b.

basophilic leukocyte
basophilocyte
basoplasm
bathmotropic
negatively b.
positively b.

Batson
B. plexus
B. vein complex

battledore placenta
Baudelocque diameter
Bauhin
B. gland
valve of B.
B. valve

Baumgarten
B. gland
B. vein

NOTES

bay
  celomic b.'s
  lacrimal b.
bayonet hair
beaded hair
beak
beaker cell
Beale cell
beard
bearing
Bechterew
  B. band
  layer of B.
  line of B.
  B. nucleus
Beckmann apparatus
Béclard
  B. anastomosis
  B. hernia
  B. triangle
bed
  adventitial b.
  b. of breast
  capillary b.
  crest of nail b.
  gallbladder b.
  hepatic b.
  liver b.
  nail b.
  parotid b.
  b. of parotid gland
  portal vascular b.
  scleral b.
  b. of stomach
  tissue b.
  vascular b.
behavior
  hookean b.
bel
belemnoid
Bell
  external respiratory nerve of B.
  long nerve of B.
  B. muscle
  B. palsy
  B. respiratory nerve
Bellini
  B. ducts
  B. ligament
belly, pl. **bellies**
  b. button
  b. of digastric muscle
  frontal b.
  muscle b.
  occipital b.
  b. of omohyoid muscle
bends
beneceptor

Benedict-Roth
  B.-R. apparatus
  B.-R. calorimeter
Bensley specific granule
benular nucleus
BER
  basic electrical rhythm
Béraud valve
Berger cells
Bergmann
  B. cords
  B. fibers
Bergmeister papilla
Bernard
  B. canal
  B. duct
Bernard-Cannon homeostasis
Bernhardt formula
Bernoulli
  B. effect
  B. law
  B. principle
  B. theorem
Berry ligament
Bertin
  B. bone
  B. columns
  B. ligament
  B. ossicle
beta
  b. angle
  b. cell of anterior lobe of hypophysis
  b. cell of pancreas
  b. fiber
  b. granule
beta-adrenergic receptors
beta-fetoprotein
beta-hypophamine
Betz cells
Bevan-Lewis cells
bevel
Bezold ganglion
Bianchi
  B. nodule
  B. valve
biarticular
biasterionic
biatriatum
  cor triloculare b.
biauricular
  b. axis
biaxial joint
bibulous
bicameral
bicanalicular sphincter
bicapsular
bicarbonate
  standard b.

bicellular
bicephalus
biceps
    b. brachii muscle
    b. femoris
    b. femoris muscle
    b. femoris reflex
    b. muscle of arm
    b. muscle of thigh
    b. tendon
Bichat
    B. canal
    B. fat-pad
    B. fissure
    B. foramen
    B. fossa
    B. ligament
    B. membrane
    B. protuberance
    B. tunic
bicipital
    b. aponeurosis
    b. fascia
    b. groove
    b. rib
    b. ridge
    b. tuberosity
bicipitalis
    aponeurosis b.
bicipitoradial bursa
bicipitoradialis
    bursa b.
Bickel ring
bicollis
    uterus bicornate b.
bicondylar
    b. articulation
    b. joint
bicondylaris
    articulatio b.
bicornate uterus
bicornis
    uterus b.
bicornous
bicornuate
bicuspid
    b. aortic valve
    b. tooth
bicuspidalis
    valvula b.
bicuspidi
    dentes b.

bicuspidus
    dens b.
bidactyly
bidiscoidal
    b. placenta
Biesiadecki fossa
bifascicular
bifid
    b. cranium
    b. penis
    b. rib
    b. tongue
    b. uterus
    b. uvula
bifida
    spina b.
bifidum
    cranium b.
bifidus
    uterus b.
biforate uterus
biforis
    uterus b.
bifurcate
    b. ligament
bifurcated ligament
bifurcatio
    b. aortae
    b. tracheae
    b. trunci pulmonalis
bifurcation
    b. of aorta
    aortic b.
    carotid b.
    b. lymph node
    b. of pulmonary trunk
    b. of trachea
    tracheal b.
    venous b.
bifurcatum
    ligamentum b.
Bigelow
    B. ligament
    B. septum
bigemina
    corpora b.
bigeminal
    b. bodies
    b. pregnancy
bigeminum
bigerminal
bilaminar blastoderm

NOTES

bilateral
    b. hermaphroditism
    b. internal mammary artery
      (BIMA)
    b. left-sidedness
bilateralism
bile
    A, B, C b.
    B b.
    C b.
    b. cyst
    b. duct
    b. duct lumen
    b. papilla
biliaris
    collum vesicae b.
    corpus vesicae b.
    ductus b.
    fossa vesicae b.
    fundus vesicae b.
    glandulae ductus b.
    infundibulum vesicae b.
    musculus sphincter ductus b.
    plicae mucosae vesicae b.
    rugae vesicae b.
    sphincter vesicae b.
    tunica mucosa vesicae b.
    tunica muscularis vesicae b.
    tunica serosa vesicae b.
    vesica b.
biliaropancreatica, pl. **biliaropancreaticae**
    ampulla b.
    musculus sphincter ampullae
      biliaropancreaticae
biliaropancreatic ampulla
biliary
    b. atresia
    b. canaliculus
    b. duct
    b. ductule
    b. gland
    b. plexus
    b. tract
    b. tree
bilifaction
biliferi
    ductuli b.
    ductus b.
    tubuli b.
biliferous
bilification
biligenesis
biligenic
biliosae
    glandulae mucosae b.
bilious
Billroth
    B. cords
    B. venae cavernosae

biloba
    placenta b.
bilobate
bilobed
bilobular
biloculare
    cor b.
bilocularis
    uterus b.
bilocular joint
biloculate
BIMA
    bilateral internal mammary artery
bimanual
bimastoid
bimodal
binary fission
binaural
binauricular a
Bingham
    B. flow
    B. plastic
binocular heterochromia
binotic
binovular
binuclear
binucleate
binucleolate
bioastronautics
bioavailability
biocenosis
bioclimatology
biodynamic
bioelectric potential
bioenergetics
biofeedback
biogenesis
biogenetic law
biogeochemistry
biogravics
bioinstrument
biokinetics
biologic
biological coefficient
biologist
biology
biolysis
biolytic
biome
biomechanics
biomedical
biometer
biometrician
biometry
bion
bionecrosis
bionomics
bionomy
biophagism

biophagous
biophagy
biophysics
bioplasm
bioplasmic
biorbital angle
biorheology
biorhythm
biosis
biosphere
biostatics
biota
biotaxis
Biot breathing
biotelemetry
biotic
    b. community
    b. factors
    b. potential
biotransformation
biovular
biparietal
    b. diameter (BPD)
    b. suture
bipartita
    placenta b.
bipartite
    b. uterus
    b. vagina
bipartitus
    uterus b.
bipennate muscle
bipennatus
    musculus b.
bipenniform
biperforate
bipolar
    b. cell
    b. neuron
bipotentiality
biramous
Birbeck granule
Birkett hernia
birth
    b. amputation
    b. canal
    b. fracture
    live b., livebirth
    b. weight
birthmark
bisacromial
bisaxillary

bisiliac
bistephanic
bistratal
bitemporal
biting strength
bitrochanteric
bitropic
bivalent
biventer
    b. cervicis
    lobulus b.
    b. mandibulae
    musculus b.
biventralis
    lobulus b.
biventral lobule
biventriculare
    cor triloculare b.
bizygomatic
Bizzozero
    B. corpuscle
    B. red cells
BL
    buccolingual
Black formula
blackout
bladder
    allantoic b.
    apex of urinary b.
    bar of b.
    base of b.
    body of b.
    circular layer of detrusor muscle
     of urinary b.
    b. dome
    dome of b.
    exstrophy of b.
    fundus of urinary b.
    gall b.
    ileocecal b.
    infundibulum of urinary b.
    lateral ligament of b.
    mucosa of urinary b.
    mucous membrane of urinary b.
    muscular coat of urinary b.
    muscular layer of urinary b.
    b. neck
    b. neck ridge
    neck of urinary b.
    b. pillar
    b. reflex
    serosa of urinary b.

NOTES

bladder *(continued)*
    sigmoid b.
    sphincter muscle of urinary b.
    trigone of b.
    uninhibited neurogenic b.
    urachal sinus of b.
    urinary b.
    uvula of b.
    venous plexus of b.
    b. wall

blade
    b. bone
    shoulder b.

Blandin gland
Blandin-Nuhn gland
Blasius duct
blast cell
blastema
    nephric b.

blastemic
blastocele
blastocelic
blastocoele
blastocoelic
blastocyst
blastocyte
blastoderm
    bilaminar b.
    embryonic b.
    extraembryonic b.
    trilaminar b.

blastoderma
blastodermal
blastodermic
    b. disk
    b. vesicle

blastodisk
blastogenesis
blastogenetic
blastogenic
blastolysis
blastolytic
blastomere
blastoneuropore
blastopore
blastoporic canal
blastula
blastular
blastulation
blennogenic
blennogenous
blennoid
blepharal
blepharocoloboma
blepharon
blepharoplast
blepharospasm
blighted ovum

blind
    b. foramen of frontal bone
    b. foramen of tongue
    b. gut
    b. spot

block
    exit b.
    palatine b.
    b. vertebrae

blocking
blood
    b. capillary
    b. cell
    cerebral b.
    b. circulation
    b. corpuscle
    b. disk
    b. flow ($\dot{Q}$)
    b. gas
    b. island
    b. islet
    b. mole
    b. plate
    b. pressure
    b. vessel

blood-air barrier
blood-aqueous barrier
blood-brain barrier
blood-cerebrospinal fluid barrier
bloodless decerebration
bloodstream
blood-vascular system
blue spot
Blumenau nucleus
Blumenbach clivus
BMR
    basal metabolic rate

BNA
    Basle Nomina Anatomica

Bochdalek
    flower basket of B.
    foramen of B.
    B. foramen
    B. ganglion
    B. gap
    B. hernia
    B. muscle
    B. valve

Bock
    B. ganglion
    B. nerve

body
    adrenal b.
    alveolar b.
    amygdaloid b.
    anococcygeal b.
    aortic b.
    basal lamina of ciliary b.
    basal layer of ciliary b.

bigeminal b.'s
b. of bladder
b. of breast
carotid b.
b. cavity
cell b.
chromaffin b.
ciliary b.
b. of clavicle
b. of clitoris
coccygeal b.
compressible cavernous b.'s
elementary b.'s
b. of epididymis
epithelial b.
fat b.
b. of femur
b. of fibula
b. of fornix
b. of gallbladder
geniculate b.
glomus b.
Hassall b.'s
Herring b.'s
Highmore b.
b. of humerus
hyaloid b.
b. of hyoid bone
b. of ilium
b. of incus
infrapatellar fat b.
intercarotid b.
intervertebral b.
iris ciliary b.
b. of ischium
juxtaglomerular b.
juxtarestiform b.
lateral geniculate b.
Lieutaud b.
long axis of b.
Luse b.'s
b. of Luys
Luys b.
malpighian b.
mamillary b.
mammary b.
b. of mammary gland
b. of mandible
b. of maxilla
b. mechanics
medial geniculate b.
b. of metacarpal

b. of metacarpal bone
b. of metatarsal
multilamellar b.
multivesicular b.'s
myelin b.
b. of nail
nerve cell b.
neuroepithelial b.
Nissl b.'s
nucleus of mamillary b.
nucleus of medial geniculate b.
Odland b.
olivary b.
orbital fat b.
osteocartilaginous b.
pacchionian b.'s
pampiniform b.
b. of pancreas
paraaortic b.
paranephric b.
paranuclear b.
paraterminal b.
parts of human b.
peduncle of mamillary b.
pedunculus of pineal b.
b. of penis
perineal b.
b. of phalanx
pigmented layer of ciliary b.
pineal b.
pituitary b.
polar b.
psammoma b.
pubic b.
b. of pubis
quadrigeminal b.'s
b. of radius
region of b.
restiform b.
b. of rib
b. righting reflex
Rosenmüller b.
Sandström b.
Savage perineal b.
Schmorl b.
b. of sesamoid bone
b. of sphenoid
b. of sphenoid bone
b. stalk
b. of sternum
b. of stomach
striate b.

NOTES

body *(continued)*

    suprarenal b.
    b. of sweat gland
    Symington anococcygeal b.
    b. of talus
    b. of thigh bone
    threshold b.
    thyroid b.
    b. of tibia
    tigroid b.'s
    b. of tongue
    trapezoid b.
    turbinated b.
    tympanic b.
    b. of ulna
    ultimobranchial b.
    b. of uterus
    b. of vagina
    ventral nucleus of trapezoid b.
    b. of vertebral
    vertebral b.
    Virchow-Hassall b.'s
    vitreous b.
    wall of b.
    Weibel-Palade b.'s
    wolffian b.
    yellow b.
    Zuckerkandl b.

body-weight ratio
Boerhaave syndrome
Bogros
    B. serous membrane
    B. space
Bohn nodule
Bohr equation
boiling point
Boll cells
bolometer
bombesin
bone

    accessory navicular b.
    acromial b.
    b. age
    alar b.
    Albrecht b.
    alveolar b.
    ankle b.
    anterior condyloid canal of occipital b.
    anterior surface of petrous part of temporal b.
    apex of petrous part of temporal b.
    arm b.
    articular eminence of temporal b.
    articular fossa of temporal b.
    articular surface of mandibular fossa of temporal b.

    articular surface on calcaneus for cuboid b.
    articular surface of temporal b.
    articular tubercle of temporal b.
    articulation of pisiform b.
    basal part of occipital b.
    base of hyoid b.
    basihyal b.
    basilar part of occipital b.
    basilar process of occipital b.
    basioccipital b.
    basisphenoid b.
    Bertin b.
    blade b.
    blind foramen of frontal b.
    body of hyoid b.
    body of metacarpal b.
    body of sesamoid b.
    body of sphenoid b.
    body of thigh b.
    breast b.
    Breschet b.
    bundle b.
    calcaneal process of cuboid b.
    calf b.
    b. canaliculus
    cancellated b.
    cancellous b.
    capitate b.
    carpal b.
    cartilage b.
    cecal foramen of frontal b.
    central b.
    cheek b.
    clavicle b.
    coccygeal b.
    coccyx b.
    collar b.
    compact b.
    conchal crest of palatine b.
    condyle b.
    convoluted b.
    cornu of hyoid b.
    b. corpuscle
    cortical b.
    coxal b.
    cranial b.
    crest of palatine b.
    crest of petrous part of temporal b.
    crest of petrous temporal b.
    cribriform plate of ethmoid b.
    cubital b.
    cuboid b.
    cuneiform b.
    b.'s of digits
    dorsal talonavicular b.
    ear b.
    elbow b.

endochondral b.
epactal b.
epipteric b.
episternal b.
ethmoid b.
ethmoidal crest of palatine b.
exoccipital b.
external surface of frontal b.
external surface of parietal b.
facial b.
femoral b.
fibular b.
first cuneiform b.
flank b.
flat b.
Flower b.
foot b.
b. of foot
foramen cecum of frontal b.
fourth turbinated b.
frontal angle of parietal b.
frontal border of parietal b.
frontal border of sphenoid b.
frontal process of zygomatic b.
funny b.
gladiolus b.
b. Gla protein
Goethe b.
greater horn of hyoid b.
greater multangular b.
greater wing of sphenoid b.
hamate b.
hamular process of lacrimal b.
hamular process of sphenoid b.
hamulus of hamate b.
haunch b.
head of metatarsal b.
head of thigh b.
heel b.
heterotopic b.
highest turbinated b.
hip b.
hollow b.
hooked b.
hook of hamate b.
horizontal plate of palatine b.
horn of greater hyoid b.
horn of lesser hyoid b.
horns of hyoid b.
humeral b.
hyoid b.
iliac b.

ilium b.
incarial b.
incisive b.
incus b.
inferior branch of pubic b.
b. of inferior limb
inferior surface of petrous part of temporal b.
inferior temporal line of parietal b.
inferior turbinated b.
inner table frontal b.
innominate b.
intermaxillary b.
intermediate cuneiform b.
internal surface of frontal b.
internal surface of parietal b.
interparietal b.
intrachondrial b.
irregular b.
ischial b.
jaw b.
joint of ear b.
jugal b.
jugular notch of occipital b.
jugular notch of petrous part of temporal b.
jugular process of occipital b.
jugular tubercle of occipital b.
knee b.
Krause b.
lacrimal b.
lambdoid border of occipital b.
lambdoid margin of occipital b.
lamella of b.
lamellar b.
lamina of palatine b.
lateral condyle b.
lateral cuneiform b.
lateral mass of ethmoid b.
lateral part of occipital b.
lateral surface of zygomatic b.
lenticular b.
lentiform b.
lesser multangular b.
lesser wing of sphenoid b.
limbus of sphenoid b.
lingual b.
long b.
b. of lower limb
lunate b.
malar b.
malleus b.

B

NOTES

bone *(continued)*
    mandibular b.
    marginal tubercle of zygomatic b.
    b. marrow
    mastoid angle of parietal b.
    mastoid border of occipital b.
    mastoid margin of occipital b.
    mastoid part of the temporal b.
    mastoid process of petrous part of
      temporal b.
    b. matrix
    maxillary surface of greater wing
      of sphenoid b.
    maxillary surface of palatine b.
    medial cuneiform b.
    membrane b.
    meniscal b.
    mesethmoid b.
    mesocuneiform b.
    metacarpal b.
    metatarsal b.
    middle cuneiform b.
    middle turbinated b.
    midshaft of b.
    multangular b.
    nasal border of frontal b.
    nasal crest of horizontal plate of
      palatine b.
    nasal margin of frontal b.
    nasal part of frontal b.
    nasal spine of frontal b.
    nasal surface of palatine b.
    navicular b.
    neck of thigh b.
    nonlamellar b.
    occipital angle of parietal b.
    occipital border of parietal b.
    occipital border of temporal b.
    occipital margin of temporal b.
    orbicular b.
    orbital eminence of zygomatic b.
    orbital lamina of ethmoid b.
    orbital layer of ethmoid b.
    orbital part of frontal b.
    orbital plate of ethmoid b.
    orbital process of palatine b.
    orbital tubercle of zygomatic b.
    outer table of frontal b.
    palatine crest of horizontal process
      of palatine b.
    palatine surface of horizontal plate
      of palatine b.
    parietal border of frontal b.
    parietal border of sphenoid b.
    parietal border of squamous part
      of temporal b.
    parietal margin of frontal b.
    part petrous of temporal b.
    patella b.

    pelvic b.
    perichondral b.
    periosteal b.
    periotic b.
    peroneal b.
    perpendicular plate of ethmoid b.
    perpendicular plate of palatine b.
    petrosal b.
    petrous part of temporal b.
    phalangeal b.
    pharyngeal tubercle of basilar part
      of occipital b.
    pipe b.
    Pirie b.
    pisiform b.
    pisohamate b.
    pneumatic b.
    pneumatized b.
    posterior border of petrous part of
      temporal b.
    posterior nasal spine of horizontal
      plate of palatine b.
    posterior surface of petrous part of
      temporal b.
    postsphenoid b.
    preinterparietal b.
    premaxillary b.
    presphenoid b.
    pterygoid process of sphenoid b.
    pterygoid ridge of sphenoid b.
    pubic b.
    pyramidal process of palatine b.
    radial b.
    ramus b.
    replacement b.
    b. resorption
    reticulated b.
    rider's b.
    Riolan b.
    rostrum of the sphenoid b.
    rudimentary b.
    sacral b.
    sacred b.
    sagittal border of parietal b.
    scaphoid fossa of sphenoid b.
    scroll b.
    second cuneiform b.
    semilunar b.
    septal b.
    sesamoid b.
    shaft of b.
    shank b.
    sheath process of sphenoid b.
    shin b.
    short b.
    sieve b.
    b. of skull
    sphenoid b.
    sphenoidal angle of parietal b.

sphenoidal border of temporal b.
sphenoidal margin of temporal b.
sphenoidal process of palatine b.
sphenoidal turbinated b.
sphenopalatine notch of palatine b.
spine of sphenoid b.
splint b.
spongy b.
squamosal border of parietal b.
squamous border of parietal b.
squamous border of sphenoid b.
squamous part of frontal b.
squamous part of occipital b.
squamous part of temporal b.
stapes b.
styloid process of temporal b.
styloid process of third
  metacarpal b.
subchondral b.
subperiosteal b.
superior border of petrous part of
  temporal b.
superior branch of the pubic b.
b. of superior limb
superior temporal line of
  parietal b.
superior turbinated b.
suprainterparietal b.
suprasternal b.
supreme turbinated b.
surface of orbital zygomatic b.
sutural b.
tail b.
tarsal b.
temporal line of frontal b.
temporal process of zygomatic b.
thigh b.
third cuneiform b.
three-cornered b.
b. tissue
toe b.
tongue b.
trabecular b.
trapezium b.
trapezoid b.
triangular b.
triquetral b.
triquetrum b.
tubercle of scaphoid b.
tubercle of trapezium b.
tuberculum of trapezium b.
tuberosity of cuboid b.

tuberosity of fifth metatarsal b.
tuberosity of first metatarsal b.
tuberosity of navicular b.
tuberosity of scaphoid b.
tuberosity of sesamoid b.
turbinated b.
tympanic part of temporal b.
tympanic plate of temporal b.
tympanohyal b.
ulnar styloid b.
unciform b.
uncinate process of ethmoid b.
upper jaw b.
b. of upper limb
vaginal process of sphenoid b.
vesalian b.
Vesalius b.
b. of visceral cranium
vomer b.
wedge b.
wormian b.
woven b.
wrist b.
yoke b.
zygomatic border of greater wing
  of sphenoid b.
zygomatic margin of greater wing
  of sphenoid b.
zygomatic process of frontal b.
zygomatic process of temporal b.

bonelet
Bonnet capsule
Bonwill triangle
bony
  b. ampullae of semicircular canal
  b. labyrinth
  b. limbs of semicircular canal
  b. nasal septum
  b. palate
  b. part of auditory tube
  b. part of external acoustic meatus
  b. part of pharyngotympanic tube
  b. part of skeletal system
border
  alveolar b.
  anterior b.
  anterodistal b.
  antimesenteric b.
  brush b.
  b. cells
  b.'s of eyelid
  free b.

NOTES

border *(continued)*
frontal b.
inferior b.
interosseous b.
b. of iris
lateral b.
medial b.
occipital b.
outer b.
parietal b.
b. of ramus
squamosal b.
squamous b.
sternal b.
striated b.
superior b.
b. of uterus
ventral b.
vermilion b.
Botallo
B. duct
B. foramen
B. ligament
both ears (AU, au)
botryoid
Böttcher
B. canal
B. cells
B. ganglion
B. space
Bouchard node
bouquet
Riolan b.
Bourgery ligament
bouton
axonal terminal b.'s
b.'s en passage
synaptic b.'s
terminal b.'s
b. terminaux
Bovero muscle
Bowditch effect
bowel
b. axis
large b.
loop of b.
b. loop
lumen of b.
b. lumen
small b.
b. wall
Bowman
B. capsule
B. disks
B. gland
B. membrane
B. muscle
B. space
B. theory

box
brain b.
Boyden
B. meal
B. sphincter
Boyer bursa
Boyle law
BPD
biparietal diameter
BPSA
bronchopulmonary segmental artery
brachia (*pl. of* brachium)
brachial
b. artery
b. autonomic plexus
b. fascia
b. gland
b. lymph node
b. muscle
b. plexus nerve
b. vein
brachialis, pl. **brachiales**
arteria b.
divisiones anteriores plexus b.
divisiones posteriores plexus b.
fasciculus lateralis plexus b.
fasciculus medialis plexus b.
fasciculus posterior plexus b.
lymphonodi brachiales
b. muscle
musculus b.
nodi lymphoidei brachiales
pars infraclavicularis plexus b.
pars supraclavicularis plexus b.
plexus autonomicus b.
rami musculares partis
   supraclavicularis plexus b.
b. superficialis
trunci plexus b.
truncus inferior plexus b.
truncus medius plexus b.
truncus superior plexus b.
venae brachiales
brachii
aponeurosis of biceps b.
aponeurosis musculi bicipitis b.
arteria profunda b.
bursa subtendinea musculi
   tricipitis b.
caput breve musculi bicipitis b.
caput laterale musculi tricipitis b.
caput longum musculi bicipitis b.
caput longum musculi tricipitis b.
caput mediale musculi tricipitis b.
facies anterior b.
facies lateralis b.
fascia b.
musculus biceps b.
musculus triceps b.

profunda b.
ramus deltoideus arteriae
  profundae b.
short head of biceps b.
subtendinous bursa of triceps b.
triceps b.
brachiocarpal articulation
brachiocephalic
  b. arterial trunk
  b. lymph node
  b. trunk artery
  b. vein
brachiocephalicae
  venae b.
brachiocephalicus, pl. **brachiocephalici**
  musculus b.
  nodi lymphoidei brachiocephalici
  truncus b.
brachiocrural
brachiocubital
brachioradial
  b. ligament
  b. muscle
brachioradialis
  b. muscle
  musculus b.
brachioulnar articulation
brachium, pl. **brachia**
  b. colliculi inferioris
  b. colliculi superioris
  b. conjunctivum cerebelli
  b. of inferior colliculus
  inferior quadrigeminal b.
  b. pontis
  b. quadrigeminum inferius
  b. quadrigeminum superius
  superior quadrigeminal b.
brachybasocamptodactyly
brachybasophalangia
brachycephalia
brachycephalic
brachycephalism
brachycephalous
brachycephaly
brachycheilia
brachychilia
brachycnemic
brachycranic
brachydactylia
brachydactylic
brachydactyly
brachyesophagus

brachyfacial
brachyglossal
brachygnathia
brachygnathous
brachykerkic
brachymelia
brachymesophalangia
brachymetacarpalia
brachymetacarpalism
brachymetacarpia
brachymetapody
brachymetatarsia
brachymorphic
brachyodont
brachypellic pelvis
brachypelvic
brachyphalangia
brachypodous
brachyprosopic
brachyrhinia
brachyrhynchus
brachyskelic
brachystaphyline
brachysyndactyly
brachytelephalangia
brachytype
brachyuranic
bradyesthesia
bradykinetic analysis
bradypnea
bradypragia
brain
  abdominal b.
  anterior part of anterior
    commissure of b.
  arachnoid of b.
  artery of b.
  b. box
  Broca motor speech area of b.
  commissure of anterior b.
  b. death
  dura mater of b.
  labyrinth of b.
  lateral fossa of b.
  lobe of b.
  b. mantle
  medullary arteries of b.
  membrane of b.
  posterior part of anterior
    commissure of b.
  b. region
  b. sand

B

NOTES

brain *(continued)*
 split b.
 sulci in b.
 visceral b.
 b. wave cycle
 Wernicke area of b.
braincase
brainstem
 reticular nuclei of b.
branch
 accessory meningeal b.
 acetabular b.
 anastomotic b.
 b. to angular gyrus
 anterior basal b.
 anterior temporal b.
 artery of angular nasal b.
 articular b.
 ascending b.
 atrial b.
 atrioventricular nodal b.
 b. to atrioventricular node
 b. of auriculotemporal nerve to
  tympanic membrane
 calcaneal b.
 carotid sinus b.
 circumflex b.
 cochlear b.
 communicating b.
 conus b.
 deep b.
 deltoid b.
 dental b.
 descending anterior b.
 descending posterior b.
 dorsal b.
 epiploic b.
 esophageal b.
 gastroepiploic b.
 genicular b.
 glandular b.
 b. of glossopharyngeal nerve to
  stylopharyngeus muscle
 inferior b.
 b. of internal carotid artery to
  trigeminal ganglion
 internal nasal b.
 joint b.
 lateral basal b.
 lateral cutaneous b.
 lateral mammary b.
 left b.
 lingual b.
 b. of lingual nerve to isthmus of
  fauces
 mammary b.
 medial mammary b.
 mediastinal b.
 meningeal b.

 muscular b.
 occipital b.
 b. of oculomotor nerve to ciliary
  ganglion
 omental b.
 pancreatic b.
 parietal b.
 parotid b.
 perforating b.
 peroneal communicating b.
 pharyngeal b.
 posterior basal b.
 right b.
 b. of segmental bronchi
 septal b.
 b. to sinuatrial node
 spinal b.
 superficial b.
 superior b.
 tracheal b.
 b. to trigeminal ganglion
 tubal b.
 ureteral b.
 ureteric b.
 ventral b.
branchial
 b. apparatus
 b. arches
 b. cartilage
 b. cells
 b. cleft cyst
 b. clefts
 b. duct
 b. efferent column
 b. fissure
 b. fistula
 b. groove
 b. mesoderm
 b. pouch
 b. sinus
branching type of renal pelvis
branchiogenic
branchiogenous
branchiomere
branchiomeric muscle
branchiomerism
branchiomotor nuclei
Braune
 B. muscle
 B. valve
break shock
breast
 accessory b.
 b. architecture
 areola of b.
 bed of b.
 body of b.
 b. bone
 b. implant

lateral margin of b.
male b.
papilla of b.
supernumerary b.
suspensory ligament of b.
suspensory retinaculum of b.
tail of axillary b.

breath
breathe
breath-holding test
breathing
Biot b.
glossopharyngeal b.
b. lung
mouth b.
positive-negative pressure b.
b. reserve
shallow b.

breech
bregma
bregmatic
b. fontanelle
bregmatolambdoid arc
Breschet
B. bone
B. canal
B. hiatus
B. sinus
B. vein
breve
caput b.
os b.
vinculum b.
brevi
musculus palmaris b.
brevia
vasa b.
brevicollis
brevis
abductor pollicis b.
adductor b.
arteria centralis b.
arteria ciliaris posterior b.
arteriae circumferentiales b.
arteriae gastricae b.'s
bursa musculi extensoris carpi
radialis b.
caput profundum musculi flexoris
pollicis b.
caput superficiale musculi flexoris
pollicis b.
ciliares posteriores b.'s

deep head of flexor pollicis b.
extensor carpi radialis b. (ECRB)
extensor digitorum b. (EDB)
extensor pollicis b. (EPB)
flexor digitorum b. (FDB)
flexor hallucis b. (FHB)
gastricae b.
b. muscle
musculi levatores costarum b.
musculus abductor pollicis b.
musculus adductor b.
musculus extensor carpi radialis b.
musculus extensor digitorum b.
musculus extensor hallucis b.
musculus extensor pollicis b.
musculus fibularis b.
musculus flexor digitorum b.
musculus flexor hallucis b.
musculus flexor pollicis b.
musculus palmaris b.
musculus peroneus b.
musculus supinator radii b.
nervi ciliares b.'s
nervus ciliaris b.
palmaris b.
peroneus b.
pollicis b.
processus b.
superficial head of flexor
pollicis b.
venae gastricae b.'s

bridge
arteriolovenular b.
cell b.'s
b. corpuscle
cytoplasmic b.'s
Gaskell b.
intercellular b.'s
mylohyoid b.
nasal b.
b. of nose
Wheatstone b.
bridle of clitoris
brim
pelvic b.
Brinell hardness number
bristle cell
broad
b. fascia
b. ligament of uterus
b. uterine ligament
broadest muscle of back

NOTES

Broca
- B. basilar
- B. basilar angle
- B. center
- B. diagonal band
- B. facial angle
- B. field
- B. fissure
- B. formula
- gyrus of B.
- B. motor speech area of brain
- B. parolfactory area
- B. pouch
- B. region
- B. space

brocae
- area parolfactoria b.

Brödel bloodless line

Brodie
- B. bursa
- B. fluid
- B. ligament

Brodmann areas

Broesike fossa

bronchi (*pl. of* bronchus)

bronchia

bronchial
- b. artery
- b. branch of thoracic aorta
- b. bud
- b. gland
- b. lumen
- b. mucosa
- b. tree
- b. tube
- b. vein

bronchiales
- glandulae b.
- rami b.
- venae b.

bronchic cell

bronchiogenic

bronchiole
- respiratory b.'s
- terminal b.

bronchioli respiratorii

bronchiolopulmonary

bronchiolus terminalis

bronchiomediastinalis
- truncus lymphaticus b.

bronchiorum
- tunica muscularis b.

bronchium

bronchoalveolar

bronchoaortic constriction

bronchoesophageal muscle

bronchoesophageus
- b. muscle
- musculus b.

bronchogenic

bronchomediastinal lymphatic trunk

bronchopulmonale
- lymphonodi b.
- nodi lymphoidei b.
- segmentum b.

bronchopulmonalis,
- pl. **bronchopulmonales**
- segmentum b.

bronchopulmonary
- b. lymph node
- b. segment
- b. segmental artery (BPSA)
- b. sequestration

bronchotracheal

bronchovesicular

bronchus, pl. **bronchi**
- branch of segmental bronchi
- eparterial b.
- fibromusculocartilagenous layer of bronchi
- intermediate b.
- b. intermedius
- intrasegmental bronchi
- bronchi intrasegmentales
- left main b.
- left upper lobe b.
- lingular b.
- bronchi lobares
- b. lobaris inferior
- b. lobaris medius
- b. lobaris superior
- main stem b.
- middle lobe b.
- mucosa of bronchi
- mucous membrane of b.
- muscular coat of bronchi
- muscular layer of bronchi
- primary b.
- b. principalis dexter
- b. principalis sinister
- right main b.
- segmental b.
- b. segmentalis
- stem b.
- tunica fibromusculocartilaginea bronchi
- tunica mucosa bronchi

brood cell

brow

brown
- b. fat
- b. layer
- b. striae

Browning vein

Bruch
- B. gland
- B. membrane

Brücke
- B. muscle
- B. tunic

Brunn
- B. membrane
- B. nest
- B. reaction

Brunner gland

brush border

bucca, pl. **buccae**
- corpus adiposum buccae

buccal
- b. artery
- b. branch of facial nerve
- b. cavity
- b. fat-pad
- b. gland
- b. groove
- b. lymph node
- b. mucosa
- b. region
- b. root of tooth
- b. sulcus
- b. surface
- b. vein
- b. vestibule

buccalis, pl. **buccales**
- arteria b.
- facies b.
- glandulae buccales
- nervus b.
- radix b.
- rami buccales
- regio b.

buccinator
- b. artery
- b. crest
- b. muscle
- musculus b.
- b. nerve
- b. node

buccinatoria
- crista b.

buccinatorius
- nodus lymphoideus b.

buccolingual (BL)

bucconasal membrane

bucconeural duct

buccopharyngea
- fascia b.

buccopharyngeal
- b. fascia

- b. membrane
- b. part
- b. part of superior pharyngeal constrictor

buccopharyngeus
- musculus b.

buccula

Buck fascia

bud
- bronchial b.
- end b.
- b. fission
- gustatory b.
- limb b.
- liver b.
- lung b.
- median tongue b.
- metanephric b.
- periosteal b.
- syncytial b.
- tail b.
- taste b.
- tooth b.
- ureteric b.
- vascular b.

budding

Budge center

Budin obstetrical joint

bulb
- aortic b.
- apex of duodenal b.
- arterial b.
- carotid b.
- commissure of vestibular b.
- conjunctival layer of b.
- b. of corpus spongiosum
- dental b.
- duodenal b.
- end b.
- b. of eye
- glomerular layer of olfactory b.
- b. of hair
- hair b.
- intermediate part of vestibular b.
- jugular b.
- b. of jugular vein
- Krause end b.'s
- b. of lateral ventricle
- molecular layers of olfactory b.
- olfactory b.
- b. of penis
- Rouget b.

NOTES

bulb *(continued)*
    taste b.
    b. of urethra
    vein of vestibular b.
    b. of vestibule
bulbar
    b. conjunctiva
    b. ridge
    b. septum
    b. sheath fascia
bulbi (*pl. of* bulbus)
bulbocavernosus
    b. muscle
    musculus b.
    b. reflex
bulbocavernous
    b. gland
    b. muscle
bulboid corpuscles
bulboidea
    corpuscula b.
bulbomembranous
bulbomimic reflex
bulbonuclear
bulbopontine
bulborum
    commissura b.
bulbosacral system
bulbospinal
bulbospongiosus
    b. muscle
    musculus b.
bulbourethral gland
bulbourethralis
    ductus glandulae b.
    glandula b.
bulboventricular
    b. fold
    b. loop
    b. ridge
bulbus, pl. **bulbi**
    annulare bulbi
    b. aortae
    camera anterior bulbi
    camerae bulbi
    camera posterior bulbi
    camera vitrea bulbi
    capsula bulbi
    b. cordis
    b. cornus posterioris
    b. duodeni
    bulbi fascia
    fascia bulbi
    fascia muscularis musculorum bulbi
    ligamentum anulare bulbi
    ligamentum suspensorium bulbi
    musculi bulbi
    obliquus inferior bulbi
    obliquus superior bulbi

    b. oculi
    b. olfactorius
    bulbi penis
    b. penis
    b. pili
    rectus inferior bulbi
    rectus lateralis bulbi
    rectus superior bulbi
    stratum pigmenti bulbi
    trochlea musculi obliqui superioris
      bulbi
    tunica conjunctiva bulbi
    tunica fibrosa bulbi
    tunica interna bulbi
    tunica vasculosa bulbi
    b. urethrae
    vagina bulbi
    b. venae jugularis
    b. vestibuli
    bulbi vestibuli
    bulbi vestibuli vaginae
    b. vestibuli vaginae
bulk modulus
bulla, pl. **bullae**
    ethmoid b.
    ethmoidal b.
    b. ethmoidalis
    intraepidermic b.
    subepidermic b.
bullosa
    concha b.
bundle
    aberrant b.'s
    anterior ground b.
    Arnold b.
    atrioventricular b.
    Bachmann b.
    b. bone
    commissural b.
    Flechsig ground b.'s
    forebrain b.
    Gantzer accessory b.
    Gierke respiratory b.
    ground b.'s
    Held b.
    Helie b.
    Helweg b.
    b. of His
    His b.
    His-Tawara b.
    Hoche b.
    Keith b.
    Kent b.
    Kent-His b.
    Killian b.
    Krause respiratory b.
    lateral ground b.
    left bundle of atrioventricular b.
    left crus of atrioventricular b.

Lissauer b.
Loewenthal b.
longitudinal medial b.
longitudinal pontine b.
medial forebrain b.
medial longitudinal b.
Meynert retroflex b.
Monakow b.
muscle b.
neurovascular b.
olfactory b.
olivocochlear b.
posterior longitudinal b.
precommissural b.
predorsal b.
Rathke b.
right bundle of atrioventricular b.
right crus of atrioventricular b.
Schütz b.
sinospiral muscle b.
solitary b.
subendocardial branch of
    atrioventricular b.'s
tendon b.
trunk of atrioventricular b.
Türck b.
Vicq d'Azyr b.
Bunsen solubility coefficient
Burdach
    B. column
    B. fasciculus
    B. nucleus
    B. tract
Burn and Rand theory
Burns
    B. falciform process
    B. ligament
    B. space
Burow vein
burr cell
burrow
bursa, pl. **bursae**
    Achilles b.
    b. achillis
    b. of acromion
    adventitious b.
    b. anserina
    anserine b.
    anterior tibial b.
    bicipitoradial b.
    b. bicipitoradialis
    Boyer b.

Brodie b.
b. of calcaneal tendon
Calori b.
coracobrachial b.
b. cubitalis interossea
deep infrapatellar b.
b. of extensor carp
b. of extensor carpi radialis brevis
    muscle
Fleischmann b.
foramen of superior recess of
    omental b.
bursae of gastrocnemius
gluteofemoral b.
gluteus medius b.
gluteus minimus b.
b. of great toe
b. of hyoid
iliac b.
b. iliopectinea
iliopectineal b.
iliopsoas b.
inferior recess of omental b.
infracardiac b.
infrahyoid b.
b. infrahyoidea
b. infrapatellaris profunda
infraspinatus b.
b. intermusculares musculorum
    gluteorum
intermuscular gluteal b.
interosseous cubital b.
intrapatellar b.
b. intratendinea olecrani
intratendinous olecranon b.
b. ischiadica musculi glutei maximi
b. ischiadica musculi obturatoris
    interni
ischial b.
laryngeal b.
lateral malleolar subcutaneous b.
lateral malleolus b.
Luschka b.
medial malleolar subcutaneous b.
Monro b.
b. of Monro
b. mucosa
b. musculi bicipitis femoris
    superior
b. musculi coracobrachialis
b. musculi extensoris carpi radialis
    brevis

NOTES

bursa *(continued)*
  b. musculi piriformis
  b. musculi semimembranosi
  b. musculi tensoris veli palatini
  bursae of obturator internus
  b. of obturator internus
  olecranon b.
  b. of olecranon
  omental b.
  b. omentalis
  ovarian b.
  b. ovarica
  b. pharyngea
  pharyngeal b.
  b. of piriformis
  b. of popliteus
  prepatellar b.
  b. quadrati femoris
  b. quadratus femoris
  quadratus femoris b.
  radial b.
  recess of omental b.
  retrocalcaneal b.
  retrohyoid b.
  b. retrohyoidea
  sac of lesser omental b.
  sartorius b.
  scapulothoracic b.
  b. of semimembranosus
  b. of semimembranosus muscle
  semimembranous b.
  splenic recess of omental b.
  subacromial b.
  b. subacromialis
  subcoracoid b.
  b. subcutanea acromialis
  b. subcutanea calcanea
  b. subcutanea infrapatellaris
  b. subcutanea malleoli lateralis
  b. subcutanea malleoli medialis
  b. subcutanea olecrani
  b. subcutanea prepatellaris
  b. subcutanea prominentiae
    laryngeae
  b. subcutanea trochanterica
  b. subcutanea tuberositas tibiae
  b. subcutanea tuberositatis tibiae
  subcutaneous acromial b.
  subcutaneous calcaneal b.
  subcutaneous infrapatellar b.
  subcutaneous olecranon b.
  subcutaneous prepatellar b.
  subdeltoid b.
  b. subdeltoidea
  b. subfascialis prepatellaris
  subfascial prepatellar b.
  subhyoid b.
  sublingual b.

  b. sublingualis
  subscapular b.
  bursae subtendineae musculi
    gastrocnemii
  bursae subtendineae musculi sartorii
  b. subtendinea iliaca
  b. subtendinea musculi bicipitis
    femoris inferior
  b. subtendinea musculi infraspinati
  b. subtendinea musculi latissimus
    dorsi
  b. subtendinea musculi obturatoris
    interni
  b. subtendinea musculi
    subscapularis
  b. subtendinea musculi teretis
    majoris
  b. subtendinea musculi tibialis
    anterioris
  b. subtendinea musculi trapezii
  b. subtendinea musculi tricipitis
    brachii
  b. subtendinea prepatellaris
  subtendinous iliac b.
  subtendinous prepatellar b.
  superior recess of omental b.
  suprapatellar b.
  b. suprapatellaris
  synovial b.
  b. synovialis
  synovial trochlear b.
  b. tendinis calcanei
  b. of tendo calcaneus
  b. of tensor veli palatine
  b. of tensor veli palatini muscle
  b. of teres major
  tibial intertendinous b.
  b. of trapezius
  triceps b.
  trochanteric b.
  b. trochanterica
  bursae trochantericae musculi glutei
    medii
  b. trochanterica musculi glutei
    maximi
  b. trochanterica musculi glutei
    minimi
  trochlear synovial b.
  ulnar b.
  vestibule of omental b.
  Voshell b.

bursal
burst
  respiratory b.
bursula testium
BUS
  Bartholin, urethral, and Skene
    BUS gland

butterfly
    b. adrenal
    b. vertebra
buttocks

button
      belly b.
Byzantine arch palate

NOTES

B

C

C bile
C cell

c

c muscle

Ca

calcium
cancer
carcinoma

CaCC

cathodal closure contraction

cacogenesis
cacomelia
cacoplastic
cacumen
cacumina
cacuminal
cadaver
cadaveric

c. rigidity

CaDTe

cathodal duration tetanus

caduca
caecum
cage

rib c.
thoracic c.

caisson disease
Cajal

C. cell
horizontal cell of C.

cake kidney
calamus scriptorius
calcanea

bursa subcutanea c.
regio c.

calcaneal

c. anastomosis
c. arterial network
c. artery
c. articular surface of talus
c. branch
c. facet
c. process of cuboid bone
c. region
c. sulcus
c. tendon
c. tubercle
c. tuberosity

calcanean tendon
calcanei

bursa tendinis c.
facies articularis cuboidea ossis c.
facies articularis talaris anterior c.

facies articularis talaris media c.
facies articularis talaris posterior c.
processus lateralis tuberis c.
processus medialis tuberis c.
rami c.
sulcus c.
trochlea fibularis c.
tuber c.
tuberculum c.

calcaneoastragaloid articulation
calcaneocavus
calcaneocuboid

c. articulation
c. joint
c. ligament
medial c.

calcaneocuboidea

articulatio c.

calcaneocuboideum

ligamentum c.

calcaneofibulare

ligamentum c.

calcaneofibular ligament
calcaneonaviculare

ligamentum c.

calcaneonavicular ligament
calcaneoplantar
calcaneoscaphoid
calcaneotibiale

ligamentum c.

calcaneotibial ligament
calcaneovalgocavus
calcaneovarus
calcaneum

rete c.

calcaneus

anterior talar articular surface of c.
bursa of tendo c.
cuboidal articular surface of c.
fibular trochlea of c.
interosseous groove of c.
middle talar articular surface of c.
peroneal trochlea of c.
posterior talar articular surface
of c.
talar articular surface of c.
tendo c.
tuberosity of anterior c.

calcar

c. avis
c. femorale
c. pedis
c. sclerae

calcareous artery

calcarina
> arteria c.
> fissura c.

calcarine
> c. branch of medial occipital
>   artery
> c. fasciculus
> c. fissure
> c. spur
> c. sulcus

calcarinus
> ramus c.
> sulcus c.

calces (*pl. of* calcis)
calcification lines of Retzius
calcified cartilage
calciostat
calcis, pl. **calces**
> os c.
> tuber c.

calcitonin gene-related peptide
calcium (Ca)
Caldani ligament
calf, pl. **calves**
> artery of c.
> c. bone
> lateral cutaneous nerve of c.
> triceps muscle of c.

calfbone, calf-bone
caliceal system
calices (*pl. of* calix)
caliciform cell
calicine
caliculus, pl. **caliculi**
> c. gustatorius
> c. ophthalmicus

calix, pl. **calices**
> major calices
> minor calices
> calices renales majores
> calices renales minores

Calleja
> islands of C.

callosal
> c. convolution
> c. gyrus
> c. sulcus

callosi
> corporis c.
> genu corporis c.
> pars frontalis corporis c.
> pars occipitalis corporis c.
> pedunculus corporis c.
> radiatio corporis c.
> raphe corporis c.
> rostrum corporis c.
> splenium corporis c.
> teniola corporis c.

> truncus corporis c.
> tuber corporis c.

callosomarginal
> c. artery
> c. fissure

callosomarginalis
> arteria c.
> rami paracentrales arteriae c.
> ramus cingularis arteriae c.
> ramus frontalis anteromedialis
>   arteriae c.
> ramus frontalis intermediomedialis
>   arteriae c.
> ramus frontalis posteromedialis
>   arteriae c.
> sulcus c.

callosum
> corpus c.
> dorsal vein of corpus c.
> genu of corpus c.
> peduncle of corpus c.
> radiation of corpus c.
> trunk of corpus c.

calmodulin
Calori bursa
caloric
calorie, calory
> gram c.
> kilogram c.
> large c.
> mean c.
> small c.

calorific
calorigenic action
calorimeter
> Benedict-Roth c.

caloritropic
calory (*var. of* calorie)
Calot
> C. node
> triangle of C.
> C. triangle

calvaria, pl. **calvariae**
> external table of c.
> internal table of c.
> lamina externa c.
> lamina interna c.

calvarial
calvarium
calves (*pl. of* calf)
calyceal
calyces (*pl. of* calyx)
calyciform ending
calycine
calycle
calyculus
calyx, pl. **calyces**
> superior pole of c.

cambium layer

cameloid cell
camera, pl. **camerae**
    c. anterior bulbi
    camerae bulbi
    c. oculi anterior
    c. oculi major
    c. oculi minor
    c. oculi posterior
    c. posterior bulbi
    c. postrema
    c. vitrea
    c. vitrea bulbi
    vitreous c.
Campbell ligament
Camper
    chiasm of C.
    C. chiasm
    C. fascia
    C. ligament
    C. line
campi foreli
camptodactylia
camptodactyly
camptomelia
campylodactyly
canal
    abdominal c.
    adductor c.
    Alcock c.
    alimentary c.
    alveolar c.'s
    alveolodental c.
    anal c.
    anterior ethmoid c.
    anterior retinal orbital c.
    anterior semicircular c.
    archenteric c.
    Arnold c.
    arterial c.
    artery of pterygoid c.
    atrioventricular c.
    auditory c.
    basipharyngeal c.
    Bernard c.
    Bichat c.
    birth c.
    blastoporic c.
    bony ampullae of semicircular c.
    bony limbs of semicircular c.
    Böttcher c.
    Breschet c.
    caroticoclinoid c.

    caroticotympanic c.
    carotid c.
    carpal c.
    caudal c.
    cavernous body of anal c.
    central c.
    cervical neural c.
    cervicoaxillary c.
    ciliary c.
    Civinini c.
    Cloquet c.
    cochlear c.
    condylar c.
    condyloid c.
    Corti c.
    Cotunnius c.
    cranial c.
    craniopharyngeal c.
    crura of bony semicircular c.
    crural c.
    deferent c.
    dental c.
    dentinal c.
    descending part of facial c.
    diploic c.
    Dorello c.
    Dupuytren c.
    ear c.
    endodermal c.
    endometrial c.
    eustachian c.
    external ear c.
    facial c.
    fallopian c.
    femoral c.
    Ferrein c.
    Fontana c.
    galactophorous c.
    Gartner c.
    gastric c.
    geniculum of facial c.
    genu of facial c.
    greater palatine c.
    c. of Guyon
    Hannover c.
    haversian c.'s
    Hensen c.
    c. of Hering
    Hering c.
    hiatus of facial c.
    Hirschfeld c.
    His c.

C

NOTES

canal *(continued)*
Holmgrén-Golgi c.'s
horizontal part of facial c.
Hoyer c.
Huguier c.
Hunter c.
Huschke c.
hyaloid c.
hypoglossal c.
incisive c.
incisor c.
inferior dental c.
infraorbital c.
inguinal c.
interdental c.
interfacial c.'s
intramedullary c.
introitus of facial c.
Jacobson c.
Kürsteiner c.'s
lacrimal c.
lateral crus of horizontal part of
  the facial c.
lateral semicircular c.
Lauth c.
Leeuwenhoek c.'s
c. for lesser palati
lesser palatine c.
c. for lesser palatine nerve
limb of bony semicircular c.
Löwenberg c.
mandibular c.
marrow c.
medial crus of the horizontal part
  of the facial c.
medullary c.
membranous c.
mental c.
musculotubal c.
nasolacrimal c.
nerve of pterygoid c.
neural c.
neurenteric c.
notochordal c.
c. of Nuck
nutrient c.
obturator c.
opening of carotid c.
optic c.
orbital c.
palatovaginal c.
palmate fold of cervical c.
part of optic nerve in c.
pelvic c.
pericardioperitoneal c.
perineal flexure of anal c.
persistent atrioventricular c.
Petit c.

pharyngeal branch of artery of
  pterygoid c.
c. for pharyngotympanic tube
plane of pelvic c.
pleuropericardial c.'s
pleuroperitoneal c.
portal c.'s
posterior internal orbital c.
posterior semicircular c.
prominence of facial c.
prominence of lateral
  semicircular c.
pterygoid c.
pterygopalatine c.
pudendal c.
pulp c.
pyloric c.
rectal c.
Reissner c.
Rivinus c.
Rosenthal c.
rufflec c.
sacral c.
Santorini c.
Schlemm c.
scleral c.
semicircular c.
septum of musculotubal c.
Sondermann c.
spinal c.
Stensen c.
Stilling c.
subsartorial c.
Sucquet c.
Sucquet-Hoyer c.
supraciliary c.
supraoptic c.
supraorbital c.
tarsal c.
temporal c.
tensor tympani c.
c. for tensor tympani muscle
Theile c.
threshold pads of anal c.
tubotympanic c.
tunnius c.
tympanic c.
uniting c.
urogenital c.
uterine c.
uterovaginal c.
van Horne c.
vein of pterygoid c.
Velpeau c.
venous plexus of hypoglossal c.
vertebral c. (VC)
vesicourethral c.
vestibular c.
vidian c.

Volkmann c.'s
vomerine c.
vomerobasilar c.
vomerorostral c.
vomerovaginal c.
Walther c.
white line of anal c.
Wirsung c.
canales (*pl. of* canalis)
canalicular
    c. duct
    c. sphincter
canaliculization
canaliculus, pl. **canaliculi**
    ampulla of lacrimal c.
    auricular c.
    biliary c.
    bone c.
    caroticotympanic canaliculi
    canaliculi caroticotympanici
    c. of chorda tympani
    c. of cochlea
    cochlear c.
    canaliculi dentales
    external aperture of cochlear c.
    external opening of cochlear c.
    inferior lacrimal c.
    intercellular c.
    intracellular c.
    lacrimal c.
    c. lacrimalis
    mastoid c.
    c. mastoideus
    opening of vestibular c.
    c. reuniens
    secretory c.
    tympanic c.
    c. tympanicus
    c. vein
    vein of cochlear c.
    vestibular c.
canalis, pl. **canales**
    c. adductorius
    canales alveolares
    canales alveolares corporis maxillae
    c. analis
    c. caroticus
    c. carpi
    c. centralis
    c. centralis medullae spinalis
    c. cervicis uterus
    c. condylaris

canales diploici
c. femoralis
c. gastricus
c. hyaloideus
c. hypoglossalis
canales incisivi
c. incisivus
c. infraorbitalis
c. inguinalis
introitus c.
canales longitudinales modioli
c. mandibulae
c. musculotubarius
c. nasolacrimalis
c. nervi facialis
c. nervi petrosi superficialis
    minoris
c. nutricius
c. obturatorius
c. opticus
canales palatini minores
c. palatinus major
c. palatovaginalis
c. pterygoidei
c. pterygoideus
c. pudendalis
c. pyloricus
c. radicis dentis
c. reuniens
c. sacralis
canales semicircularis anterior
canales semicircularis ossei
canales semicircularis posterior
c. spiralis cochlea
c. spiralis modioli
c. umbilicalis
c. umbilicus
c. ventriculi
c. vertebralis
c. vomerorostralis
c. vomerovaginalis
canalization
cancelled bone
cancellous
    c. bone
    c. tissue
cancellus, pl. **cancelli**
cancer (Ca)
candicans
canina
    fossa c.

NOTES

canine
    c. eminence
    c. fossa
    c. prominence
    c. tooth
canini
    dentes c.
caninus
    dens c.
    musculus c.
Cannon
    C. ring
    C. theory
cannula
    tracheal c.
canthal hypertelorism
canthomeatal plane
canthus, pl. **canthi**
    external c.
    inner c.
    internal c.
    lateral c.
    medial c.
    outer c.
    temporal c.
CaOC
    cathodal opening contraction
CaOCl
    cathodal opening clonus
cap
    acrosomal c.
    c. of ampullary crest
    duodenal c.
    fibrous c.
    head c.
    metanephric c.
    phrygian c.
    pyloric c.
capacitance
capacitation
capacitor
capacity
    cranial c.
    diffusing c.
    forced vital c.
    functional residual c. (FRC)
    inspiratory c.
    maximum breathing c. (MBC)
    oxygen c.
    residual c.
    respiratory c.
    total lung c.
    vital c.
capillar
capillare
    vas c.
capillariomotor
capillarity
capillaron

capillary
    arterial c.
    c. arteriole
    c. bed
    blood c.
    c. circulation
    continuous c.
    fenestrated c.
    c. lake
    c. lamina of choroid
    c. loop
    lymph c.
    c. muscle
    c. pericyte
    sinusoidal c.
    c. vein
    venous c.
    c. vessel
    c. wedge pressure
capillus, pl. **capilli**
capita (*pl. of* caput)
capital epiphysis
capitate bone
capitatum
    os c.
capitellum
capitis
    corona c.
    longissimus c.
    musculi c.
    musculus longissimus c.
    musculus longus c.
    musculus semispinalis c.
    musculus spinalis c.
    musculus splenius c.
    musculus transversalis c.
    regio frontalis c.
    regiones c.
    regio occipitalis c.
    regio parietalis c.
    regio temporalis c.
    semispinalis c.
    spinalis c.
    splenius c.
capitopedal
capitular
    c. joint
    c. process
capitulum, pl. **capitula**
    c. humeri
    c. of humerus
    c. of stapes
capnogram
capnograph
capsula, pl. **capsulae**
    c. adiposa perirenalis
    c. adiposa renis
    c. articularis
    c. articularis cricoarytenoidea

c. articularis cricothyroidea
c. bulbi
c. cordis
c. externa
c. extrema
c. fibrosa
c. fibrosa glandulae thyroideae
c. fibrosa perivascularis
c. fibrosa renis
c. glomeruli
c. interna
c. lienis
c. vasculosa lentis

capsular
c. branch of intrarenal artery
c. branch of renal artery
c. ligament
c. space

capsulare
ligamentum c.

capsularis, pl. **capsulares**
decidua c.
membrana c.
rami capsulares

capsulation
capsule
adipose renal c.
adrenal c.
anterior limb of internal c.
articular c.
atrabiliary c.
auditory c.
Bonnet c.
Bowman c.
cartilage c.
c. cell
c. of corpora
cricoarytenoid articular c.
c. of cricoarytenoid joint
cricothyroid articular c.
c. of cricothyroid joint
crystalline c.
external c.
extreme c.
eye c.
fatty renal c.
fibrous articular c.
fibrous layer of articular c.
fibrous layer of joint c.
fibrous membrane of joint c.
genu of internal c.
Gerota c.

Glisson c.
glomerular c.
c. of glomerulus
Heyman c.
internal c.
joint c.
c. of kidney
c. of knee
c. of lens
lens c.
lenticular c.
malpighian c.
medial c.
Müller c.
nasal c.
optic c.
otic c.
parotid c.
pelvioprostatic c.
perinephric c.
perirenal fat c.
perivascular fibrous c.
posterior limb of internal c.
prostatic c.
retrolenticular limb of internal c.
seminal c.
c. of spleen
splenic c.
sublenticular limb of internal c.
suprarenal c.
synovial c.
Tenon c.
c. of Tenon
thyroid c.

capsulolenticular
capsulopupillaris
membrana c.

Capuron point
caput, pl. **capita**
c. angulare quadrati labii
c. breve
c. breve musculi bicipitis brachii
c. breve musculi bicipitis femoris
c. costa
c. epididymidis
c. fibulae
c. gallinaginis
c. humerale
c. humerale musculi flexoris carpi ulnaris
c. humerale musculi pronatoris teretis

NOTES

caput (*continued*)

c. humeri
c. humeroulnare musculi flexoris digitorum superificialis
c. infraorbitale quadrati labii superioris
c. laterale
c. laterale musculi gastrocnemii
c. laterale musculi tricipitis brachii
c. longum
c. longum musculi bicipitis brachii
c. longum musculi bicipitis femoris
c. longum musculi tricipitis brachii
c. mallei
c. mandibulae
c. mediale
c. mediale musculi gastrocnemii
c. mediale musculi tricipitis brachii
c. medusae
c. obliquum
c. obliquum musculi adductoris hallucis
c. obliquum musculi adductoris pollicis
c. ossis femoris
c. ossis metacarpalis
c. ossis metatarsalis
c. pancreatis
c. phalangis manus et pedis
c. profundum musculi flexoris pollicis brevis
c. radii
c. stapedis
c. superficiale musculi flexoris pollicis brevis
c. tali
c. transversum
c. transversum musculi adductoris hallucis
c. transversum musculi adductoris pollicis
c. ulnae
c. ulnare
c. ulnare musculi flexoris carpi ulnaris
c. ulnare musculi pronatoris teretis
c. zygomaticum quadrati labii superioris

carbamide
carbohydrate metabolism
carbometry
carbon

c. dioxide acidosis
c. dioxide cycle
c. dioxide elimination

carbonmonoxy myoglobin
carbonometer
carbonometry
carcinoma (Ca)

card

plexus c.

cardia

gastric c.
lymphatic ring of c.
c. of stomach

cardiac

c. antrum
c. apex
c. area
c. competence
c. cycle
c. depressor reflex
c. fibrous skeleton
c. ganglia
c. gland
c. gland of esophagus
c. gland of stomach
c. histiocyte
c. hormone
c. impression of diaphragmatic surface of liver
c. impression of liver
c. impression of lung
c. index
c. jelly
c. lymphatic ring
c. mucosa
c. muscle
c. muscle tissue
c. nerve
c. notch
c. notch of lung
c. opening
c. orifice
c. output
c. part of stomach
c. plexus
c. prominence
c. segment
c. skeleton
c. tube
c. valve
c. valve leaflet
c. vein
c. volume (CV)

cardiaca

fovea c.
ganglia c.
incisura c.

cardiaci thoracici
cardiacum

antrum c.
ostium c.
segmentum c.

cardiacus

c. cervicalis inferior
c. cervicalis medius
c. cervicalis superior

plexus nervosus c.
ramus c.
cardiae
annulus lymphaticus c.
sphincter constrictor c.
cardial
c. notch
c. orifice
c. part of stomach
cardiatelia
cardiectopia
cardinal
c. ligament
c. vein
c. vessel
cardinale
ligamentum c.
cardioaortic
cardioarterial
cardiodynamics
cardioesophageal
c. junction
c. sphincter
cardiogenesis
cardiogenic plate
cardiohepatic
c. angle
c. sulcus
c. triangle
cardioid
cardiomegaly
cardiomuscular
cardionatrin
cardionector
cardionephric
cardioneural
cardiophrenic angle
cardiopulmonary splanchnic nerve
cardiopyloric
cardiorenal
cardiothoracic silhouette
cardiovasculare
systema c.
cardiovascular system
cardiovasculorenal
carina, pl. **carinae**
c. fornicis
tertiary c.
c. of trachea
c. tracheae
tracheal c.
c. urethralis vaginae

carinal lymph node
carinate
carinatum
pectus c.
carnea, pl. **carneae**
columnae carneae
trabeculae carneae
tunica c.
carnes (*pl. of* caro)
carnification
carnis
carnosa
membrana c.
carnosus
panniculus c.
caro, pl. **carnes**
c. quadrata sylvii
carotic
carotica
fossa c.
vagina c.
carotici
c. externi
ramus sinus c.
caroticoclinoid
c. canal
c. foramen
c. ligament
caroticotympanic
c. artery
c. canal
c. canaliculi
c. nerve
caroticotympanicae
arteriae c.
caroticotympanicus, pl. **caroticotympanici**
arteriae caroticotympanici
canaliculi caroticotympanici
nervi c.
rami caroticotympanici
caroticum
glomus c.
trigonum c.
tuberculum c.
caroticus
canalis c.
ductus c.
c. internus
nodulus c.
sinus c.
sulcus c.

NOTES

carotid
    c. artery
    c. bifurcation
    c. body
    c. branch of glossopharyngeal nerve
    c. bulb
    c. canal
    c. cavernous sinus
    c. duct
    c. foramen
    c. ganglion
    c. gland
    c. groove
    c. sheath
    c. sinus branch
    c. sinus nerve
    c. sinus reflex
    c. siphon
    c. sulcus
    c. triangle
    c. tubercle
    c. tubercle of vertebra
    c. vein
    c. venous plexus
    c. vessel
    c. wall of middle ear
    c. wall of tympanic cavity

carotis
    c. communis
    c. externa
    c. interna

carp
    bursa of extensor c.

carpal
    c. arches
    c. artery
    c. articular surface of radius
    c. articulation
    c. bone
    c. canal
    c. groove
    c. joint
    c. region
    c. tendinous sheath
    c. tunnel

carpalium
    vaginae tendinum c.

carpi (*pl. of* carpus)
carpocarpal
carpometacarpal (CMC)
    c. articulation
    c. joint
    c. joint of thumb
    c. ligament
    c. ligament dorsal and palmar

carpometacarpales
    articulationes c.

carpometacarpalia
    ligamenta c.

carpometacarpeae
    articulationes c.

carpopedal
carpophalangeal
carpus, pl. **carpi**
    articulatio carpi
    canalis carpi
    ossa carpi
    sulcus carpi

Carrel-Lindbergh pump
carrier cell
carrying angle
cart
    ensiform c.

cartilage
    accessory nasal c.
    accessory quadrate c.
    c. of acoustic meatus
    c. air cells
    alar c.
    alisphenoid c.
    annular c.
    anterolateral surface of arytenoid c.
    apex of arytenoid c.
    arch of cricoid c.
    arcuate crest of arytenoid c.
    arthrodial c.
    articular surface of arytenoid c.
    arytenoid c.
    auditory c.
    c. of auditory tube
    auricular c.
    base of arytenoid c.
    basilar c.
    c. bone
    branchial c.
    calcified c.
    c. capsule
    cellular c.
    ciliary c.
    circumferential c.
    colliculus of arytenoid c.
    conchal c.
    connecting c.
    corniculate c.
    cornu of thyroid c.
    costal c.
    cricoid c.
    cuneiform c.
    diarthrodial c.
    c. of ear
    ear c.
    elastic c.
    ensiform c.
    ensisternum c.
    epiglottic c.
    epiphysial c.

C

eustachian c.
facet c.
falciform c.
floating c.
foot pod of alar c.
greater alar c.
horn of inferior thyroid c.
horn of superior thyroid c.
Huschke c.
hyaline c.
hypsiloid c.
inferior horn of thyroid c.
innominate c.
interarticular c.
interosseous c.
interventricular c.
intervertebral c.
intraarticular c.
intrathyroid c.
investing c.
Jacobson c.
joint c.
c. lacuna
lamina of cricoid c.
lamina of thyroid c.
laryngeal c.
c. of larynx
lateral process of septal nasal c.
left plate of thyroid c.
lesser alar c.
Luschka laryngeal c.
major alar c.
mandibular c.
c. matrix
meatal c.
Meckel c.
medial surface of arytenoid c.
Meyer c.
minor alar c.
Morgagni c.
muscular process of arytenoid c.
nasal dome c.
nasal septal c.
c. of nasal septum
c. of nose
oblique line of thyroid c.
oblong fovea of arytenoid c.
oblong pit of arytenoid c.
orbitosphenoid c.
parachordal c.
paraseptal c.
patellar facet c.

periotic c.
permanent c.
c. of pharyngotympanic tube
posterior process of septal c.
posterior surface of arytenoid c.
precursory c.
primordial c.
quadrangular c.
quadrilateral c.
Reichert c.
reticular c.
retiform c.
right plate of thyroid c.
c. of Santorini
Santorini c.
secondary c.
Seiler c.
semilunar c.
septal nasal c.
sesamoid c.
c. space
sphenoid process of septal nasal c.
sternal c.
superior horn of thyroid c.
supraarytenoid c.
tarsal c.
temporary c.
tendon c.
thyroid articular surface of
  cricoid c.
tracheal c.
triangular fovea of arytenoid c.
triangular pit of arytenoid c.
triquetrous c.
triticeal c.
tubal c.
uniting c.
vocal process of arytenoid c.
vomerine c.
vomeronasal c.
Weitbrecht c.
Wrisberg c.
xiphoid c.
Y c.
yellow c.
Y-shaped c.
cartilaginea
  junctura c.
cartilagines (*pl. of* cartilago)
cartilagineus
  meatus acusticus externus c.
  vomer c.

NOTES

cartilaginis
    articulatio c.
    fovea triangularis c.
cartilaginoid
cartilaginous
    c. articulation
    c. joint
    c. neurocranium
    c. part of auditory tube
    c. part of external acoustic meatus
    c. part of pharyngotympanic tube
    c. part of skeletal system
    c. plate of auditory tube
    c. pyramid
    c. ring
    c. septum
    c. tissue
    c. vault
    c. viscerocranium
cartilago, pl. **cartilagines**
    c. alaris major
    c. articularis
    c. arytenoidea
    c. auriculae
    c. corniculata
    c. costalis
    c. cricoidea
    c. cuneiformis
    c. epiglottica
    c. epiphysialis
    cartilagines laryngis
    c. meatus acustici
    cartilagines nasales accessoriae
    c. nasi lateralis
    c. septi nasi
    c. sesamoidea laryngis
    c. sesamoidea ligamentum
      cricopharyngeum
    c. thyroidea
    cartilagines tracheales
    c. triticea
    c. tubae auditivae
    c. vomeronasalis
caruncle
    lacrimal c.
    Morgagni c.
    Santorini major c.
    Santorini minor c.
caruncula, pl. **carunculae**
    hymenal c.
    c. hymenalis
    c. lacrimalis
    c. myrtiform
    c. myrtiformis
    c. salivaris
    sublingual c.
    c. sublingualis
caryotheca, pl. **caryothecae**
    cisterna caryothecae

cascade
caseosa
    vernix c.
Casser
    Casser fontanelle
    C. perforated muscle
casserian muscle
cast
    air cell c.
castrate cells
castration cells
catabasial
catabiotic
catabolic
catabolism
catabolite
catachronobiology
catadidymus
catagen
catagenesis
cataplasia
cataplasis
cataract
    progeria with c.
catastalsis
catastaltic
catelectrotonus
caterpillar cell
cathodal
    c. closure contraction (CaCC)
    c. duration tetanus (CaDTe)
    c. opening clonus (CaOCl)
    c. opening contraction (CaOC)
cathode ray tube
cauda, pl. **caudae**
    c. epididymidis
    c. equina
    c. fascia
    c. helicis
    c. nuclei caudati
    c. pancreatis
    c. striati
caudad
caudal
    c. canal
    c. flexure
    c. helix
    c. ligament
    c. neuropore
    c. pancreatic artery
    c. pharyngeal
    c. pole
    c. retinaculum
    c. sac
    c. septum
    c. sheath
    c. transverse fissure
    c. vertebrae

caudale
    ligamentum c.
    retinaculum c.
caudalis
    nucleus tegmenti pontis c.
    pars c.
caudalward
caudate
    c. branch of left branch of portal vein
    c. lobe
    c. lobe of liver
    c. nucleus
    c. process
caudati
    arteria lobi c.
    cauda nuclei c.
    corpus nuclei c.
    rami c.
    rami caudae nuclei c.
    venae nuclei c.
caudatolenticular
caudatum
caudatus
    lobus c.
    nucleus c.
    processus c.
caudocephalad
caudolenticular
caul fat
cava, pl. **cavae**
    fold of left vena c.
    foramen of vena c.
    foramen venae cavae
    fossa venae cavae
    groove for inferior venae c.
    groove for superior venae c.
    inferior vena c. (IVC)
    ligament of left superior vena c.
    ligament of left vena c.
    opening of inferior vena c.
    opening of superior vena c.
    orifice of inferior vena c.
    orifice of superior vena c.
    sinus of the vena c.
    sulcus for vena c.
    sulcus venae cavae
    superior vena c.
    thoracic inferior vena c.
    valve of inferior vena c.
    vena c. (VC)

caval
    c. fold
    c. node
    c. opening of diaphragm
    c. valve
cavarum
    sinus venarum c.
cave
    Meckel c.
    trigeminal c.
cavea thoracis
caveola, pl. **caveolae**
cavern
    c.'s of corpora cavernosa
    c.'s of corpus spongiosum
caverna, pl. **cavernae**
    cavernae corporis spongiosi
    cavernae corporum cavernosorum
cavernosa, pl. **cavernosae**
    Billroth venae cavernosae
    cavernous space of corpora c.
    caverns of corpora c.
    cavity of corpora c.
    pars c.
    cavernosae penis
    trabeculae of corpora c.
    tunica albuginea of corpora c.
cavernosi
    c. clitoridis
    c. penis
    ramus sinus c.
cavernosorum
    cavernae corporum c.
    septum corporum c.
    trabeculae corporum c.
    tunica albuginea corporum c.
cavernosus
    plexus nervosus c.
    sinus c.
cavernous
    c. body of anal canal
    c. body of clitoris
    c. body of penis
    c. branch of cavernous part of internal carotid artery
    c. groove
    c. nerve of clitoris
    c. nerve of penis
    c. nervous plexus
    c. plexus of clitoris
    c. plexus of penis
    c. portion of urethra

C

NOTES

cavernous (*continued*)
 c. sinus branch of internal carotid artery
 c. space
 c. space of corpora cavernosa
 c. space of corporus spongiosum
 c. tissue
 c. vascular plexus of conchae
 c. vein of penis
 c. venous sinus
cavitary
cavitas
 c. abdominalis
 c. abdominis et pelvis
 c. articulare
 c. articularis
 c. conchae
 c. coronae
 c. coronalis
 c. cranii
 c. dentis
 c. glenoidalis
 c. glenoidalis scapulae
 c. infraglottica
 c. infraglotticum
 c. laryngis
 c. medullaris
 c. nasi
 c. oris
 c. oris propria
 c. pelvina
 c. pericardiaca
 c. pericardialis
 c. peritonealis
 c. pharyngis
 c. pleuralis
 c. pulparis
 c. thoracis
 c. tympanica
 c. uterus
cavitates
cavity
 abdominal c.
 abdominopelvic c.
 allantoic c.
 amniotic c.
 anterior wall of tympanic c.
 articular c.
 axillary c.
 body c.
 buccal c.
 carotid wall of tympanic c.
 cilia of nasal c.
 cleavage c.
 c. of concha
 c. of corpora cavernosa
 c. of corpus spongiosum
 cotyloid c.
 cranial c.

crown c.
ear c.
ectoplacental c.
ectotrophoblastic c.
endolymphatic c.
epamniotic c.
epidural c.
floor of tympanic c.
glenoid c.
greater peritoneal c.
groove of promontory of labyrinthine wall of tympanic c.
head c.
inferior laryngeal c.
inferior wall of tympanic c.
infraglottic c.
intermediate laryngeal c.
intracranial c.
labyrinthine wall of tympanic c.
laryngeal c.
c. of larynx
lateral wall of tympanic c.
lesser peritoneal c.
c. line angle
mastoid wall of tympanic c.
Meckel c.
medial wall of tympanic c.
medullary c.
membranous wall of tympanic c.
c. of middle ear
mucosa of tympanic c.
mucous membrane of tympanic c.
nasal c.
nephrotomic c.
olfactory groove of nasal c.
olfactory sulcus of nasal c.
optical c.
oral c.
orbital c.
pelvic peritoneal c.
pericardial c.
perilymphatic c.
peritoneal c.
perivisceral c.
pharyngonasal c.
c. of pharynx
pleural c.
pleuroperitoneal c.
posterior sinus of tympanic c.
posterior wall of tympanic c.
primitive perivisceral c.
promontory of tympanic c.
pulmonary c.
pulp c.
recesses of tympanic c.
respiratory region of mucosa of nasal c.
retroperitoneal c.
Retzius c.

roof of tympanic c.
segmentation c.
c. of septum pellucidum
somite c.
splanchnic c.
subarachnoid c.
subdural c.
subgerminal c.
subglottic c.
sulcus of promontory of
  tympanic c.
superior laryngeal c.
synovial c.
tegmental root of tympanic c.
tegmental wall of tympanic c.
thoracic c.
c. of tooth
trigeminal c.
tympanic c.
tympanomastoid c.
uterine c.
c. of uterus
visceral c.
vitreous c.

cavovalgus
cavovarus
cavum

c. abdominis
c. articulare
c. conchae
c. coronale
c. dentis
c. douglasi
c. epidurale
c. infraglotticum
c. laryngis
c. mediastinale
c. medullare
c. nasi
c. oris
c. pelvis
c. pericardii
c. peritonei
c. pharyngis
c. pleurae
c. psalterii
c. retzii
c. septi pellucidi
c. subarachnoidale
c. subarachnoidea
c. subarachnoideale
c. subdurale

c. thoracis
c. trigeminale
c. tympani
c. uterus
c. vergae
c. vesicouterinum

cavus

pes c.

C1–C5

segmenta cervicalia C.

C1–C7

vertebra C1–C7

C1–C8

segmenta medullae spinalis
  cervicalia C.

(C1–C8)
C1–Co

segmenta medullae spinalis C.

ceasmic teratosis
cebocephaly
ceca (*pl. of* cecum)
cecal

c. appendage
c. appendix
c. artery
c. fold
c. foramen of frontal bone
c. mesocolic lymph node
c. recess
c. vault

cecales

plicae c.

ceci

appendix c.

cecocolon
cecorectal
cecum, pl. **ceca**

antimesocolic side of c.
c. cupulare
habenula of c.
hepatic c.
intestinal c.
intestinum c.
mesentery of c.
vascular fold of c.
c. vestibulare

celenteron
celiac

c. arterial trunk
c. artery
c. axis
c. branch of posterior vagal trunk

NOTES

C

celiac *(continued)*
    c. branch of vagus nerve
    c. ganglia
    c. gland
    c. lymph node
    c. plexus
celiaca
    arteria c.
    ganglia c.
celiaci
    rami c.
celiacoduodenal part of suspensory muscle of duodenum
celiacus
    plexus nervosus c.
    truncus c.
cell
    A c.'s
    acid c.
    acidophil c.
    acinar c.
    acinous c.
    acoustic c.
    c. adhesion molecule
    adipose c.
    adventitial c.
    agger nasi c.
    air c.
    albuminous c.
    algoid c.
    alveolar c.
    amacrine c.
    ameboid c.
    amniogenic c.'s
    anaplastic c.
    angioblastic c.'s
    Anitschkow c.
    anterior ethmoidal air c.
    apolar c.
    argentaffin c.'s
    argyrophilic c.'s
    Askanazy c.
    astroglia c.
    auditory receptor c.'s
    B c.
    band c.
    basal c.
    basaloid c.
    basilar c.
    basket c.
    beaker c.
    Beale c.
    Berger c.'s
    Betz c.'s
    Bevan-Lewis c.'s
    bipolar c.
    Bizzozero red c.'s
    blast c.
    blood c.

    c. body
    Boll c.'s
    border c.'s
    Böttcher c.'s
    branchial c.'s
    c. bridges
    bristle c.
    bronchic c.
    brood c.
    burr c.
    C c.
    Cajal c.
    caliciform c.
    cameloid c.
    capsule c.
    carrier c.
    cartilage air c.'s
    castrate c.'s
    castration c.'s
    caterpillar c.
    c. center
    centroacinar c.
    chalice c.
    chromaffin c.
    Clara c.
    Claudius c.
    clear c.
    cleavage c.
    cleaved c.
    cochlear hair c.'s
    column c.'s
    columnar c.
    commissural c.
    compound granule c.
    cone c.
    connective tissue c.
    Corti c.'s
    crescent c.
    c. cycle
    cytomegalic c.'s
    cytotrophoblastic c.'s
    D c.
    dark c.'s
    daughter c.
    Davidoff c.'s
    decidual c.
    deep c.
    Deiters c.
    dendritic c.'s
    Dogiel c.'s
    dome c.
    Downey c.
    dust c.
    effector c.
    egg c.
    enamel c.
    endodermal c.'s
    endothelial c.
    enterochromaffin c.'s

enteroendocrine c.'s
entodermal c.'s
ependymal c.
epidermic c.
epithelial reticular c.
epithelioid c.
epitympanic c.
erythroid c.
ethmoid air c.
ethmoidal c.
external pillar c.'s
extremity air c.'s
exudation c.
Fananás c.
fasciculata c.
fat c.
fat-storing c.
floor c.
foam c.'s
follicular epithelial c.
follicular ovarian c.'s
foreign body giant c.
formative c.'s
fuchsinophil c.
c. fusion
G c.'s
ganglion c.
Gaucher c.'s
gemästete c.
gemistocytic c.
germ c.
germinal c.
ghost c.
Giannuzzi c.'s
giant c.
gitter c.
glial c.'s
glomerulosa c.
goblet c.
Golgi epithelial c.
Goormaghtigh c.'s
granule c.'s
granulosa lutein c.'s
great alveolar c.'s
guanine c.
gustatory c.'s
gyrochrome c.
hair c.'s
hairy c.'s
HeLa c.'s
Hensen c.
heteromeric c.

hilus c.'s
Hofbauer c.
Hortega c.'s
Hürthle c.
c. hybridization
hypotympanic c.
I c.
immunologically activated c.
immunologically competent c.
inclusion c.
c. inclusions
indifferent c.
infantilism air c.'s
intercapillary c.
internal pillar c.'s
interstitial c.'s
irritation c.
islet c.
Ito c.
juxtaglomerular c.'s
K c.
karyochrome c.
killer c.'s
Kirchner c.
Kulchitsky c.'s
Kupffer c.'s
lacis c.
Langerhans c.'s
Langhans c.'s
Langhans-type giant c.'s
LE c.
Leishman chrome c.'s
lepra c.'s
Leydig c.'s
lining c.
Lipschütz c.
littoral c.
Loevit c.'s
lupus erythematosus c.
luteal c.
lutein c.
lymph c.
lymphoid c.
macroglia c.
malpighian c.
Marchand wandering c.
marrow c.
Martinotti c.
mast c.
mastoid air c.
c. matrix
Mauthner c.

NOTES

cell *(continued)*

medullary c.
melanin-pigmented c.
c. membrane
Merkel tactile c.
mesangial c.
mesenchymal c.'s
mesoglial c.'s
mesothelial c.
Mexican hat c.
Meynert c.'s
microglia c.'s
microglial c.'s
middle ethmoidal air c.
midget bipolar c.'s
Mikulicz c.'s
mirror-image c.
mitral c.'s
monocytoid c.
mossy c.
mother c.
motor c.
mucoalbuminous c.'s
mucoserous c.'s
mucous neck c.
Müller radial c.'s
multipolar c.
mural c.
myeloid c.
myoepithelial c.
myoid c.'s
Nageotte c.'s
natural killer c.'s
nerve c.
Neumann c.'s
neurilemma c.'s
neuroendocrine transducer c.
neuroepithelial c.'s
neuroglia c.'s
neurolemma c.'s
neuromuscular c.
neurosecretory c.'s
nevus c.
Niemann-Pick c.
NK c.
noble c.'s
null c.'s
nurse c.'s
oat c.
olfactory receptor c.'s
oligodendroglia c.'s
Opalski c.
c. organelle
osseous c.
osteochondrogenic c.
osteogenic c.
osteoprogenitor c.
oxyntic c.
oxyphil c.'s

P c.
Paget c.'s
pagetoid c.'s
Paneth granular c.'s
parafollicular c.'s
paraganglionic c.'s
paraluteal c.
paranasal c.
parenchymal c.
parent c.
parietal c.
peptic c.
pericapillary c.
peripolar c.
perithelial c.
peritubular contractile c.'s
perivascular c.
pessary c.
petrous apex c.
petrous pyramid air c.
phalangeal c.'s
pharyngeal c.
pheochrome c.
photoreceptor c.'s
physaliphorous c.
Pick c.
pigment c.
pillar c.'s
pineal c.'s
plasma c.
pluripotent c.'s
polar c.
polychromatic c.
polychromatophil c.
posterior ethmoidal air c.
pregnancy c.'s
pregranulosa c.'s
prickle c.
primary embryonic c.
primitive reticular c.
primordial germ c.
prolactin c.
pseudounipolar c.
pseudoxanthoma c.
pulmonary epithelial c.
pulpar c.
Purkinje c.
pus c.
pyramidal c.'s
pyrrhol c.
Raji c.
reactive c.
red blood c.
Reed c.'s
Reed-Sternberg c.'s
Renshaw c.'s
resting wandering c.
restructured c.
reticular c.

reticularis c.
reticuloendothelial c.
rhagiocrine c.
Rieder c.'s
Rindfleisch c.'s
rod nuclear c.
Rolando c.'s
rosette-forming c.'s
Rouget c.
sarcogenic c.
scavenger c.
Schilling band c.
Schultze c.'s
Schwann c.'s
segmented c.
sensitized c.
septal c.
serous c.
Sertoli c.
sex c.
shadow c.'s
sickle c.
signet ring c.'s
silver c.
skein c.
small cleaved c.
smudge c.'s
somatic c.'s
sperm c.
spider c.
spindle c.
spine c.
splenic c.'s
squamous c.
squamous alveolar c.'s
stab c.
staff c.
stem c.
Sternberg c.'s
Sternberg-Reed c.'s
stichochrome c.
strap c.
supporting c.
sustentacular c.
sympathetic formative c.
sympathicotropic c.'s
sympathochromaffin c.
synovial c.
T c.
tactile c.
tanned red c.'s
target c.

tart c.
taste c.
tendon c.'s
thecal c.
theca lutein c.
tonsil air c.'s
totipotent c.
touch c.
Touton giant c.
transducer c.
transitional c.
tubal air c.'s
tufted c.
tunnel c.'s
Türk c.
tympanic air c.
type II c.'s
Tzanck c.'s
undifferentiated c.
unipolar c.
vasoformative c.
veil c.
vestibular hair c.'s
Virchow c.'s
virus-transformed c.
visual receptor c.'s
vitreous c.
c. wall
wandering c.
Warthin-Finkeldey c.'s
wasserhelle c.
white blood c.
wing c.
yolk c.'s
zymogenic c.

cella, pl. **cellae**
cellicolous
cellula, pl. **cellulae**
  cellulae anteriores
  cellulae coli
  cellulae ethmoidales
  cellulae ethmoidales anteriores
  cellulae ethmoidales mediae
  cellulae ethmoidales posteriores
  cellulae mastoideae
  cellulae pneumaticae tubae auditivae
  cellulae tympanicae
cellular
  c. cartilage
  c. tenacity
cellule
cellulifugal

NOTES

cellulipetal
cellulocutaneous
cellulosa
    vagina c.
celom
    extraembryonic c.
celoma
celomic
    c. bays
    c. metaplasia theory of
      endometriosis
celoschisis
celosomia
celothelium
Celsus area
cement
    c. corpuscle
    intercellular c.
    c. line
    tooth c.
cementoblast
cementoclast
cementocyte
cementodentinal junction
cementum
    afibrillar c.
    c. hyperplasia
    primary c.
    secondary c.
cenogenesis
cenosite
center
    anospinal c.
    Broca c.
    Budge c.
    cell c.
    chondrification c.
    ciliospinal c.
    dentary c.
    diaphysial c.
    ejaculation c.
    epiotic c.
    expiratory c.
    feeding c.
    genitospinal c.
    inspiratory c.
    Kerckring c.
    limbic c.
    medullary c.
    motor speech c.
    ossific c.
    c. of ossification
    ossification c.
    primary ossification c.
    reaction c.
    respiratory c.
    rotation c.
    c. of rotation
    secondary ossification c.

    semioval c.
    sensory speech c.
    speech c.
    sphenotic c.
    tendinous c.
    vital c.
    Wernicke c.
centimeter-gram-second (CGS)
    c.-g.-s. system
    c.-g.-s. unit (CGS)
centra (*pl. of* centrum)
centrad
central
    c. artery of retina
    c. axillary lymph node
    c. bone
    c. bone of ankle
    c. canal
    c. canal of cochlea
    c. canal of spinal cord
    c. canal of the vitreous
    c. gray substance
    c. gyri
    c. incisor
    c. incisor tooth
    c. inhibition
    c. lacteal
    c. lobule
    c. mesenteric lymph node
    c. nervous system
    c. palmar space
    c. perineum tendon
    c. pit
    c. placenta previa
    c. ramus of vertebra
    c. retinal artery
    c. retinal vein
    c. spindle
    c. sulcal artery
    c. sulcus
    c. superior mesenteric lymph node
    c. tegmental fasciculus
    c. tendon of diaphragm
    c. tendon of perineum
    c. transactional core
    c. vein of liver
    c. vein of retina
    c. vein of suprarenal gland
    c. venous pressure
centrale
    os c.
    systema nervosum c.
centralis, pl. **centrales**
    ala lobulis c.
    area c.
    arteria retinae c.
    arteria sulci c.
    canalis c.
    c. glandulae suprarenalis

Got a Good Word for STEDMAN'S?

Help us keep STEDMAN'S products fresh and up-to-date with new words and new ideas! How can we make your STEDMAN'S product the best medical word reference possible for you?

Do we need to add or revise any items? Is there a better way to organize the content?

Be specific! Fill in the lines below with your thoughts and recommendations and FAX the page to **ATTENTION STEDMANS, 410.528.4153**.

You are our most important contributor, and we want to know what's on your mind. Thanks!

Please tell us a little bit about yourself:

Name/Title: _____

Company: _____

Address: _____

City/State/Zip: _____

Day Telephone No.: (____) _____

E-mail Address: _____

TERMS YOU BELIEVE ARE INCORRECT:

Appears as: Suggested revision:

_____

_____

_____

NEW TERMS/WORDS YOU WOULD LIKE US TO ADD:

_____

_____

Other comments:

_____

_____

_____

May we quote you?   ☐ Yes   ☐ No

All done? Great, just FAX this page to the attention of STEDMAN'S at 410.528.4153 or MAIL the page to us at:

ATTN: STEDMAN'S
Lippincott Williams & Wilkins
P.O. Box 17344
Baltimore, MD 21298-9595

OR enter your information
ONLINE at **www.stedmans.com**

Thanks again!

A&P 738342

BUSINESS REPLY MAIL

FIRST CLASS   PERMIT NO. 724   BALTIMORE, MD

POSTAGE WILL BE PAID BY ADDRESSEE

LIPPINCOTT WILLIAMS & WILKINS
ATTN: STEDMAN'S MARKETING
351 WEST CAMDEN ST.
BALTIMORE MD. 21201-2436

centrales hepatis
lymphonodi superiores centrales
nodi lymphatici centrales
nodi lymphoidei centrales
nodi lymphoidei axillares centrales
nodi lymphoidei superiores centrales
placenta previa c.
c. retinae
substantia gelatinosa c.
substantia grisea c.
sulcus c.
tractus tegmentalis c.
centre médian de Luys
centrencephalic
centric
centriciput
centrifugal
    c. current
    c. nerve
centrilobular
centriole
    distal c.
    proximal c.
centripetal
    c. current
    c. nerve
centroacinar cell
centrocyte
centrokinesia
centrokinetic
centrolecithal ovum
centromedian nucleus
centromedianus
    nucleus c.
centromere
centroparietal region
centroplasm
centrosome
centrosphere
centrostaltic
centrum, pl. **centra**
    c. medianum
    c. medullare
    c. ossificationis
    c. ossificationis primarium
    c. ossificationis secundarium
    c. ovale
    c. semiovale
    c. tendineum diaphragmae
    c. tendineum diaphragmatis
    c. tendineum perinei
    vena centra

c. of vertebra
Vieussens c.
cephalad
    c. aspect
cephalic
    c. angle
    c. arterial rami
    c. flexure
    c. index
    c. pole
    c. triangle
    c. vein
    c. vein of forearm
cephalica
    c. accessoria
    mediana c.
    vena intermedia c.
    vena mediana c.
cephalization
cephalocaudad
cephalocaudal
    c. axis
cephalocele
cephalocercal
cephalochord
cephalodidymus
cephalodiprosopus
cephalogenesis
cephalogyric
cephalohematocele
cephalomedullary angle
cephalomegaly
cephalomelus
cephalometrics
cephalometry
cephalomotor
cephaloorbital index
cephalopagus
cephalopharyngeus
    musculus c.
cephalorrhachidian index
cephalothor
cephalothoracic
cephalothoracoiliopagus
cephalothoracopagus
    c. asymmetros
    c. disymmetros
    c. monosymmetros
ceptor
    chemical c.
    contact c.
    distance c.

C

NOTES

ceratocricoid
c. ligament
c. muscle
ceratocricoideum
ligamentum c.
ceratocricoideus
musculus c.
ceratoglossus
c. muscle
musculus c.
ceratohyal
ceratopharyngea
pars c.
ceratopharyngeal
c. muscle
c. part
c. part of middle pharyngeal
constrictor muscle of pharynx
ceratopharyngeus
musculus c.
cercus, pl. cerci
cerebella, pl. cerebellae
ramus tonsillae cerebellae
cerebellar
c. artery
c. cortex
c. falx
c. fissure
c. hemisphere
c. notch
c. peduncle
c. pontine angle
c. pyramid
c. sulci
c. tonsil
c. vein
cerebelli
ala c.
amygdala c.
arteria inferior anterior c.
arteria inferior posterior c.
arteria superior c.
brachium conjunctivum c.
corpus medullare c.
cortex c.
crura c.
falx c.
c. falx
fissura prima c.
fissura transversa c.
folia c.
frenulum c.
hemisphericum c.
hemispherium c.
horizontalis c.
c. inferior anterior
c. inferiores
c. inferior posterior
laminae albae c.

laminae medullares c.
lingua c.
lingula c.
lobulus centralis c.
nucleus dentatus c.
pons c.
stratum gangliosum c.
stratum granulosum c.
stratum moleculare c.
c. superior
c. superiores
tentorium c.
tonsilla c.
uvula c.
vallecula c.
venae superiores c.
vena precentralis c.
vincula lingulae c.
cerebellin
cerebellolental
cerebellomedullaris
cisterna c.
cerebellomedullary cistern
cerebello-olivary
cerebellopontile angle
cerebellopontine
c. angle
c. angle cistern
c. recess
cerebellorubralis
tractus c.
cerebellorubral tract
cerebellothalamic tract
cerebellothalamicus
tractus c.
cerebellum
anterior notch of c.
crescentic lobules of c.
dentate nucleus of c.
hemisphere of c.
horizontal fissure of c.
posterior notch of c.
primary fissure of c.
quadrangular lobule of c.
secondary fissure of c.
tentorium of c.
tongue of c.
transverse fissure of c.
vein of c.
cerebra (*pl. of* cerebrum)
cerebral
c. artery
c. blood
c. cortex
c. cranium
c. death
c. falx
c. fissure
c. flexure

c. fornix
c. fossa
c. hemisphere
c. index
c. layer of retina
c. localization
c. nerve
c. part of arachnoid
c. part of dura mater
c. part of internal carotid artery
c. peduncle
c. sinuses
c. sulcus
c. surface
c. vein
c. ventricle
c. vesicle

cerebrale
cranium c.
trigonum c.

cerebralia
juga c.

cerebralis
facies c.

cerebri
c. anterior
anus c.
appendix c.
aqueductus c.
circulus arteriosus c.
cisterna fossae lateralis c.
cisterna venae magnae c.
commissura posterior c.
communicans anterior c.
communicans posterior c.
cortex c.
crura c.
crus c.
epiphysis c.
c. falx
falx c.
fibrae arcuatae c.
fissura longitudinalis c.
fissura transversa c.
fornix c.
fossa lateralis c.
gyri profundi c.
gyri transitivi c.
hemispherium c.
hypophysis c.
c. inferiores
c. internae

lacuna c.
lamina terminalis c.
lobi c.
lobus frontalis c.
lobus occipitalis c.
lobus parietalis c.
c. magna
margo inferior c.
margo medialis c.
margo superior hemispherii c.
c. media
c. media profunda
c. media superficialis
membrana c.
pedunculus c.
polus frontalis c.
polus occipitalis c.
polus temporalis c.
c. posterior
ramus anterior sulci lateralis c.
ramus ascendens sulci lateralis c.
ramus posterior sulci lateralis c.
ramus posterior sulcus lateralis c.
riosus c.
sulci c.
sulcus lateralis c.
sulcus lunatus c.
sulcus medialis cruris c.
c. superiores
tomentum c.
tutamina c.
velamenta c.
venae anteriores c.
venae inferiores c.
venae superiores c.
vena magna c.

cerebriform
cerebroocular
cerebrophysiology
cerebrospinal
c. axis
c. fluid (CSF)
c. index
c. pressure
c. system

cerebrospinalis
liquor c.

cerebrovascular (CV)
cerebrum, pl. **cerebra**
c. abdominale
anterior ramus of lateral sulcus
of c.

NOTES

107

cerebrum *(continued)*
    aqueduct of c.
    arterial circle of c.
    ascending ramus of lateral sulcus
      of c.
    cistern of great vein of c.
    cistern of lateral fossa of c.
    gyri of c.
    lobes of c.
    longitudinal fissure of c.
    posterior ramus of lateral sulcus
      of c.
    transverse fissure of c.
    vein of c.
ceroplasty
cerulean
ceruleus
    locus c.
cerumen
ceruminal
ceruminosae
    glandulae c.
ceruminous gland
cervi
    anterior tubercle of c.
cervic
    superficial anterior c.
cervical
    c. branch of facial nerve
    c. diverticulum
    c. enlargement
    c. enlargement of spinal cord
    c. esophagus
    c. fascia
    c. flexure
    c. fringe
    c. gland
    c. gland of uterus
    c. heart
    c. iliocostal muscle
    c. interspinales muscle
    c. interspinal muscle
    c. ligament of uterus
    c. longissimus muscle
    c. loop
    c. lordosis
    c. margin of tooth
    c. musculature
    c. nerve (C1–C8)
    c. neural canal
    c. paratracheal lymph node
    pars c.
    c. part
    c. part of internal carotid artery
    c. part of thoracic duct
    c. part of vertebral artery
    c. patagium
    c. pleura
    c. plexus

    c. rib
    c. rotator muscle
    c. segments of spinal cord
    c. sinus
    c. spine
    c. splanchnic nerve
    c. triangle
    c. vein
    c. vertebra (CV)
    c. vesicle
    c. zone of tooth
cervicale
    trigonum c.
cervicales
    nervi c.
    regiones c.
    vertebrae c.
cervicalia
    segmentum medullae spinalis c.
cervicalis
    ansa c.
    c. ascendens
    costa c.
    descendens c.
    fascia c.
    inferior limb of ansa c.
    inferior root of ansa c.
    intumescentia c.
    lamina pretrachealis fasciae c.
    lamina prevertebralis fasciae c.
    lamina superficialis fasciae c.
    linea alba c.
    nervus transversus c.
    plexus c.
    processus uncinatus vertebrae c.
    c. profunda
    radix inferior ansae c.
    radix superior ansae c.
    ramus ascendens arteriae
      superficialis c.
    ramus thyrohyoideus ansae c.
    superior limb of ansa c.
    superior root of ansa c.
    thyrohyoid branch of ansa c.
cervicalium
    rami anteriores nervorum c.
    rami ventrales nervorum c.
    tuberculum anterius vertebrarum c.
    tuberculum posterius vertebrarum c.
cervices *(pl. of* cervix)
cervicis
    arteria transversa c.
    biventer c.
    ertransversarii posteriores c.
    iliocostalis c.
    ligamentum transversale c.
    longissimus c.
    lordosis c.
    musculi c.

musculi intertransversarii anteriores c.
musculi intertransversarii posteriores c.
musculi rotatores c.
musculus iliocostalis c.
musculus interspinalis c.
musculus longissimus c.
musculus semispinalis c.
musculus spinalis c.
musculus splenius c.
musculus transversalis c.
pars lateralis musculi intertransversarii posteriores c.
pars lateralis musculorum intertransversariorum posteriorum c.
pars medialis musculi intertransversarii posteriores c.
pars medialis musculorum intertransversariorum posteriorum c.
portio supravaginalis c.
portio vaginalis c.
princeps c.
rami profundi arteriae transversae c.
ramus descendens rami superficialis arteriae transversae c.
ramus superficialis arteriae transversae c.
semispinalis c.
splenius c.
venae transversae c.

cervicoaxillary canal
cervicobrachial
cervicofacial
cervicoisthmic angle
cervicooccipital
cervicothoracic
c. ganglion
c. transition
cervicothoracicum
ganglion c.
cervicotrochanteric
cervicovaginal artery
cervicovaginalis
arteria c.
cervicovesical
cervix, pl. **cervices**
c. of axon
c. columnae posterioris

c. dentis
lip of c.
supravaginal part of c.
c. of tooth
uterine c.
c. of uterus
vaginal c.
c. vesicae urinariae

CGS
centimeter-gram-second
centimeter-gram-second unit

chaeta
chain
c. ganglion
ganglionic c.
ossicular c.
c. reflex
superior cervical c.
superior ossicular c.
sympathetic c.

chalasia
chalasis
chalaza
chalice cell
chalone
chamber
altitude c.
anterior c.
aqueous c.'s
decompression c.
endothelium of anterior c.
c.'s of eyeball
heart c.
high altitude c.
hyperbaric c.
posterior c.
sinuatrial c.
vitreous c.

Chamberlain line
chamecephalic
chamecephalous
chameprosopic
channel
ion c.
ligand-gated c.
lymph c.
transnexus c.
vascular c.
voltage-gated c.

Charcot artery
Charcot-Böttcher crystalloids
Charles law

NOTES

Chassaignac
- C. axillary muscle
- C. space
- C. tubercle

Chaussier line

check
- c. ligament
- c. ligament of medial and lateral rectus muscle
- c. ligament of odontoid
- c. ligaments of eyeball

cheek
- c. bone
- fat body of c.
- c. muscle
- c. tooth

cheilion
cheilognathoglossoschisis
cheilognathopalatoschisis
cheilognathoprosoposchisis
cheilognathoschisis
cheilognathouranoschisis
cheiloschisis
cheiromegaly
chelidon
chemical ceptor
chemiotaxis
chemoceptor
chemodifferentiation
chemokinesis
chemokinetic
chemoreceptor
- medullary c.
- peripheral c.

chemoreflex
chemosmosis
chemotactic
chemotaxis
chemotropism
cherubism
chest
- alar c.
- flat c.
- foveated c.
- funnel c.
- c. index
- keeled c.
- phthinoid c.
- pterygoid c.
- region of c.
- c. wall

chewing force
Cheyne-Stokes respiration
chiasm
- Camper c.
- c. of Camper
- cistern of c.
- c. of digitus of hand
- c. of flexor sublimis

- c. of musculus flexor digitorum
- optic c.
- c. opticum
- tendinous c.

chiasma, pl. **chiasmata**
- c. opticum
- c. tendinum

chiasmatic
- c. cistern
- c. groove
- c. sulcus

chiasmatica
- cisterna c.

chiasmaticus
- ramus c.

chiasmatis
- cisterna c.

chief
- c. artery of thumb
- c. cell of corpus pineale
- c. cell of parathyroid gland
- c. cell of stomach

Chievitz
- C. layer
- C. organ

childbearing age
chilognathoglossoschisis
chilognathopalatoschisis
chilognathoprosoposchisis
chilognathoschisis
chilognathouranoschisis
chiloschisis
chimera
chimeric
chimerism
chin
- double c.
- c. jerk
- c. muscle
- c. reflex
- transverse muscle of c.

chiromegaly
chitoneure
chloride depletion
choana, pl. **choanae**
- c. narium
- primary c.
- primitive c.
- secondary c.
- vomerine crest of c.

choanal atresia
choanate
choanoid
cholangiole
cholanopoiesis
cholanopoietic
cholechromopoiesis
cholecyst
cholecystic

cholecystis
cholecystokinetic
choledocha
    fenestra c.
choledochal sphincter
choledoch duct
choledochi
    glandulae ductus c.
    musculus sphincter ductus c.
    sphincter ductus c.
choledochoduodenal junction
choledochous
choledochus
    ductus c.
cholepoiesis
cholepoietic
cholesterol cleft
cholesterologenesis
cholinergic
    c. fiber
    c. receptor
cholinoceptive
cholinolytic
cholinomimetic
cholinoreactive
cholinoreceptor
cholopoiesis
chondral
chondrification center
chondrin ball
chondroblast
chondroclast
chondrocostal
chondrocranium
chondrocyte
    isogenous c.
chondrodysplasia punctata
chondrodystrophia
    c. calcificans congenita
    c. congenita punctata
chondrodystrophy
chondroectodermal dysplasia
chondrogenesis
chondroglossus
    c. muscle
    musculus c.
chondroid tissue
chondrology
chondromalacia
chondromere
chondroosseous

chondropharyngea
    pars c.
chondropharyngeal
    c. part
    c. part of middle pharyngeal constrictor muscle of pharynx
chondropharyngeus
    musculus c.
chondroplast
chondroporosis
chondrosis
chondroskeleton
chondrosome
chondrosternal
chondrotrophic
chondroxiphoid ligament
chondrus
chonechondrosternon
Chopart joint
chorda, pl. **chordae**
    c. arteriae umbilicalis
    c. dorsa
    c. dorsalis
    c. dorsum
    c. gubernaculum
    c. magna
    c. obliqua
    c. obliqua membranae interosseae antebrachii
    c. saliva
    c. spermatica
    chordae tendineae
    c. tendineae cordis
    chordae tendineae falsae
    chordae tendineae of heart
    chordae tendineae spuriae
    c. tympani
    c. tympani nerve
    c. vertebralis
    chordae vocales
    c. vocalis
    chordae willisii
chordal
    c. process
    c. tissue
chorda-mesoderm
chordate
chordee
chordoskeleton
chorioallantoic
    c. membrane
    c. placenta

NOTES

chorioallantois
chorioamnionic p
choriocapillaris
    lamina c.
    membrana c.
choriocapillary layer
choriogonadotropin
choriomammotropin
chorion
    c. frondosum
    c. membrane
    previllous c.
    primitive c.
    shaggy c.
    smooth c.
chorionic
    c. gonadotrophic hormone
    c. gonadotropic hormone
    c. gonadotropin
    c. growth hormone-prolactin
    c. plate
    c. sac
    c. tissue
    c. vesicle
    c. villus
chorioretinal
choriovitelline placenta
chorista
choristoma
choroid
    basal lamina of c.
    basal layer of c.
    c. blood vessel
    capillary lamina of c.
    c. capillary layer
    c. fissure
    c. glomus
    c. plexus
    c. plexus of fourth ventricle
    c. plexus of lateral ventricle
    c. plexus of third ventricle
    c. skein
    c. tela of fourth ventricle
    c. tela of third ventricle
    vascular lamina of c.
    c. vein
    c. vein of eye
choroidal
    c. artery
    c. vessel
choroidea, pl. choroideae
    c. anterior
    fissura c.
    lamina c.
    lamina basalis choroideae
    lamina vasculosa choroideae
    plexus arteriae choroideae
    plexus periarterialis arteriae
      choroideae

    tenia c.
    vasa sanguinea choroideae
choroidei
    rami c.
choroideum
    glomus c.
choroideus
    plexus c.
choroidocapillaris
    lamina c.
chromaffin
    c. body
    c. cell
    c. system
    c. tissue
chromaphil
chromatic
    c. fiber
    c. granule
chromatin
    heteropyknotic c.
    c. network
    oxyphil c.
chromatinorrhexis
chromatokinesis
chromatophil
chromatophile
chromatophilia
chromatophilic
chromatophilous
chromatophorotropic hormone
chromatoplasm
chromidia (*pl. of* chromidium)
chromidial
    c. apparatus
    c. net
    c. substance
chromidiation
chromidiosis
chromidium, pl. chromidia
chromoblast
chromocyte
chromolipid
chromomere
chromonema
chromonemata
chromoph
chromophage
chromophil, chromophile
    c. granule
    c. substance
chromophilia
chromophilous
chromophobe
    c. cells of anterior lobe of
      hypophysis
    c. granule
chromophobic
chronaxia

chronaxie
chronaximeter
chronaximetry
chronaxis
chronaxy
chronic soroche
chronobiology
chronopharmacology
chronophotograph
chronotropic
chronotropism
    negative c.
    positive c.
chyle
    c. cistern
    c. vessel
chyli
    ampulla c.
    cisterna c.
    receptaculum c.
chylifaction
chylifactive
chylifera
    vasa c.
chyliferous
chylification
chylocyst
chylophoric
chylopoiesis
chylopoietic
chylorrhea
chylosis
chylous fluid
chyme
chymification
chymopoiesis
chymorrhea
chymous
Ciaccio gland
cilia (*pl. of* cilium)
ciliare
    corpus c.
    ganglion c.
    radix nervi oculomotorii ad
      ganglion c.
ciliari
    ramus communicans nervi
      nasociliaris cum ganglio c.
ciliaris, pl. **ciliares**
    annulus c.
    ciliares anteriores
    cornea c.

corona c.
corpus c.
fibrae meridionales muscularis c.
glandulae ciliares
lamina basalis corporis c.
lamina basilaris corporis c.
ciliares longi
musculus c.
orbicularis c.
orbiculus c.
plexus gangliosus c.
plicae ciliares
ciliares posteriores breves
ciliares posteriores longae
processus c.
radix brevis ganglii c.
radix longa ganglii c.
radix nasociliaris ganglii c.
radix oculomotoria ganglii c.
radix parasympathica ganglii c.
radix sensoria ganglii c.
radix sympathica ganglii c.
stratum pigmenti corporis c.
striae ciliares
venae ciliares
zona c.
zonula c.

ciliary
    c. artery
    c. axis
    c. body
    c. border of iris
    c. canal
    c. cartilage
    c. crown
    c. disk
    c. fold
    c. ganglion
    c. ganglionic plexus
    c. ganglion root
    c. gland
    c. ligament
    c. margin
    c. margin of iris
    c. muscle
    c. muscle of pupil
    c. nerve
    c. part of retina
    c. process
    c. reflex
    c. region
    c. ring

C

NOTES

ciliary *(continued)*
 c. vein
 c. vessel
 c. wreath
 c. zone
 c. zonule
ciliated epithelium
ciliogenesis
cilioretinal
 c. artery
 c. vein
cilioscleral
ciliospinal center
cilium, pl. **cilia**
 C. base
 cilia of eyelid
 cilia of nasal cavity
 cilia of paranasal sinus
cinematics
cinerea, pl. **cinereae**
 ala c.
 commissura c.
 fascia c.
 fasciola c.
 lamina c.
 nucleus alae cinereae
 substantia c.
cinereal
cinerei
 arteria tuberis c.
 rami laterales arteriarum tuberis c.
 rami mediales arteriarum tuberis c.
 ramituberis c.
cinereum
 artery of tuber c.
 lateral branch of artery of tuber c.
 medial branch of artery of
  tuber c.
 tuber c.
 tuberculum c.
cinereus
 locus c.
cineritious
cinetoplasm
cinetoplasma
cingular
 c. branch of callosomarginal artery
 isthmus of c.
cingularis
 ramus c.
cingulate
 c. convolution
 c. gyrus
 c. sulcus
cinguli
 gyrus c.
 c. gyrus
 isthmus gyri c.

 ramus marginalis sulci c.
 sulcus c.
cingulum, pl. **cingula**
 c. dentis
 c. membri inferioris
 c. membri superioris
 c. pectorale
 c. pelvici
 sulcus of c.
 c. of tooth
cinocentrum
circadian rhythm
circell
circellus venosus hypoglossi
circhoral
circinate
circle
 arterial cerebral c.
 articular vascular c.
 Haller c.
 Huguier c.
 c. of iris
 Ridley c.
 vascular c.
 c. of Willis
 Zinn vascular c.
circuit
 Papez c.
 reverberating c.
circular
 c. fibers
 c. folds
 c. fold of small intestine
 c. layer of detrusor muscle of
  urinary bladder
 c. layer of muscle coat of small
  intestine
 c. layer of muscular coat
 c. layers of muscular tunic
 c. layer of tympanic membrane
 c. pharyngeal muscle
 c. sulcus of Reil
 c. venous sinus
circularis, pl. **circulares**
 fibrae circulares
 plicae circulares
 sinus c.
circulation
 blood c.
 capillary c.
 collateral c.
 compensatory c.
 coronary c.
 cross c.
 embryonic c.
 enterohepatic c.
 fetal c.
 greater c.
 hypophysial portal c.

hypothalamohypophysial portal c.
intervillous c.
lesser c.
lymph c.
placental c.
portal hypophysial c.
pulmonary c.
Servetus c.
splanchnic c.
systemic c.
c. time
umbilical c.
vitelline c.
circulatory system
circulus, pl. **circuli**
  c. arteriosus cerebri
  c. arteriosus halleri
  c. arteriosus iridis major
  c. arteriosus iridis minor
  c. articularis vasculosus
  c. vasculosus nervi optici
  c. venosus halleri
  c. venosus ridleyi
  c. zinnii
circumanales
  glandulae c.
circumanal gland
circumarticular
circumaxillary
circumbulbar
circumcaval ureter
circumcorneal
circumduction
circumference
  abdominal c.
circumferentia
  c. articularis capitis radii
  c. articularis capitis ulnae
circumferential
  c. cartilage
  c. fibrocartilage
  c. lamella
  c. pontine branch of pontine artery
circumflex
  c. artery of scapula
  c. branch
  c. branch of left coronary artery
  c. branch of posterior tibial artery
  c. femoral artery
  c. femoral vein
  c. fibular branch of posterior tibial
   artery

c. humeral artery
c. iliac artery
c. iliac superficial fascia
c. iliac vein
c. nerve
c. peroneal branch of posterior
  tibial artery
c. scapular artery
c. scapular vein
c. vessel
circumflexa, pl. **circumflexae**
  c. femoris lateralis
  c. femoris medialis
  c. humeri anterior
  c. humeri posterior
  c. ilium profunda
  c. ilium superficialis
  c. scapulae
circumflexus
  ramus c.
circumgemmal
circumintestinal
circumlental
circumnuclear
circumocular
circumoral
circumorbital
circumrenal
circumscribed
circumscripta
  allotrichia c.
circumscriptum
  lymphangioma c.
circumscriptus
circumvallata
  placenta c.
circumvallate papilla
circumvascular
circumventricular organs
circumvolute
circus
  c. movement
  c. rhythm
cirrhonosus
cistern
  cerebellomedullary c.
  cerebellopontine angle c.
  c. of chiasm
  chiasmatic c.
  chyle c.
  c. of cytoplasmic reticulum
  great c.

C

NOTES

115

cistern *(continued)*
    c. of great vein of cerebrum
    interpeduncular c.
    lateral cerebellomedullary c.
    c. of lateral cerebral fossa
    c. of lateral fossa of cerebrum
    lumbar c.
    c. of nuclear envelope
    Pecquet c.
    pontine c.
    pontocerebellar c.
    posterior cerebellomedullary c.
    prepontine c.
    quadrigeminal c.
    subarachnoidal c.
cisterna, pl. **cisternae**
    c. ambiens
    ambient c.
    c. basalis
    c. caryothecae
    c. cerebellomedullaris
    c. cerebellomedullaris lateralis
    c. cerebellomedullaris posterior
    c. chiasmatica
    c. chiasmatis
    c. chyli
    c. cruralis
    c. fossae lateralis cerebri
    c. interpeduncularis
    c. lumbalis
    c. magna
    c. perilymphatica
    c. pontis
    c. pontocerebellaris
    c. quadrigeminalis
    cisternae subarachnoideae
    subsurface c.
    c. superior
    c. superioris
    suprasellar c.
    terminal cisternae
    c. venae magnae cerebri
cisternal
citric acid cycle
Civinini
    C. canal
    C. ligament
    C. process
Clado
    C. anastomosis
    C. band
    C. ligament
Clara cell
Clarke
    C. column
    C. nucleus
clasmatocyte
clasping reflex

clasp-knife
    c.-k. effect
    c.-k. rigidity
    c.-k. spasticity
clastic anatomy
clathrin
Claudius
    C. cell
    C. fossa
claustral layer
claustrum, pl. **claustra**
    c. gutturis
    c. oris
    c. virginale
clausura
clava
claval
clavate papilla
clavi (*pl. of* clavus)
clavicle
    acromial articular facies of c.
    acromial articular surface of c.
    acromial end of c.
    acromial extremity of c.
    acromial facet of c.
    body of c.
    c. bone
    conoid tubercle of c.
    shaft of c.
    sternal articular surface of c.
    sternal end of c.
    sternal extremity of c.
    sternal facet of c.
clavicula, pl. **claviculae**
    corpus claviculae
    extremitas acromialis claviculae
    extremitas sternalis claviculae
    facies articularis acromialis
      claviculae
    facies articularis sternalis claviculae
clavicular
    c. articular facet of acromion
    c. branch of thoracoacromial artery
    c. facet
    c. head of pectoralis major muscle
    c. notch
    c. notch of sternum
    c. part
    c. part of deltoid muscle
    c. part of pectoralis major muscle
    c. region
claviculare
    tuberculum conoideum c.
clavicularis
    incisura c.
    pars c.
    ramus c.
claviculus, pl. **claviculi**

clavipectoral
    c. fascia
    c. triangle
clavipectorale
    trigonum c.
clavipectoralis
    fascia c.
clavus, pl. **clavi**
    c. durum
clawfoot
clear
    c. cell
    c. layer of epidermis
clearance
    creatinine c.
    endogenous creatinine c.
    exogenous creatinine c.
    free water c.
    inulin c.
    maximum urea c.
    osmolal c.
    para-aminohippurate c.
    standard urea c.
    urea c.
cleavage
    adequal c.
    c. cavity
    c. cell
    complete c.
    determinate c.
    discoidal c.
    c. division
    equal c.
    equatorial c.
    holoblastic c.
    incomplete c.
    indeterminate c.
    c. line
    meridional c.
    meroblastic c.
    pudendal c.
    c. spindle
    subdural c.
    superficial c.
    unequal c.
    yolk c.
cleaved cell
cleft
    anal c.
    branchial c.'s
    cholesterol c.
    facial c.

    first visceral c.
    gill c.'s
    gluteal c.
    c. hand
    hyobranchial c.
    hyomandibular c.
    intergluteal c.
    interneuromeric c.'s
    c. lip
    natal c.
    c. nose
    oblique facial c.
    c. palate
    pudendal c.
    residual c.
    Schmidt-Lanterman c.'s
    Sondergaard c.
    c. spine
    subdural c.
    synaptic c.
    thenar c.
    c. tongue
    urogenital c.
    visceral c.
cleidal
cleidocostal
cleidocranial
    c. dysostosis
    c. dysplasia
cleidoepitrochlearis
    musculus c.
cleidomastoideus
    musculus c.
cleidooccipitalis
    musculus c.
Cleland cutaneous ligament
Clevenger fissure
clidocostal
clidocranial
    c. dysostosis
    c. dysplasia
clidoic
climbing fibers
clinica
    corona c.
    radix c.
clinical
    c. anatomy
    c. crown
    c. root
    c. root of tooth
clinocephalic

C

NOTES

clinocephalous
clinocephaly
clinoideus
    processus c.
clinoid process
clition
clitoridean
clitoridis
    arteria dorsalis c.
    arteria profunda c.
    cavernosi c.
    corpus cavernosum c.
    crus c.
    crus glandis c.
    dorsalis c.
    fascia c.
    frenulum preputii c.
    glans c.
    ligamentum fundiforme c.
    ligamentum suspensorium c.
    musculus erector c.
    nervi cavernosi c.
    nervus dorsalis c.
    os c.
    preputium c.
    profunda c.
    septum corporum cavernosorum c.
    smegma c.
    venae profundae c.
clitoris, pl. clitorides
    body of c.
    bridle of c.
    cavernous body of c.
    cavernous nerve of c.
    cavernous plexus of c.
    corpus cavernosum of c.
    crus of c.
    deep artery of c.
    deep dorsal vein of c.
    dorsal artery of c.
    dorsal nerve of c.
    dorsal vein of c.
    fascia of c.
    frenulum of c.
    fundiform ligament of c.
    glans of c.
    horn of c.
    os c.
    prepuce of c.
    septum of corpora cavernosa of c.
    superficial dorsal vein of c.
    suspensory ligament of c.
clival
clivus, pl. clivi
    Blumenbach c.
    c. branch of cerebral part of
      internal carotid artery
    lobulus clivi
    lobus clivi

    c. ocularis
    ramus clivi
cloaca, pl. cloacae
    ectopia cloacae
    endodermal c.
    exstrophy of c.
    odermal c.
cloacal
    c. membrane
    c. plate
clonal aging
clonic
    c. convulsion
clonicity
clonicotonic
clonism
clonograph
clonospasm
clonus
    ankle c.
    cathodal opening c. (CaOCl)
    wrist c.
C-loop of duodenum
Cloquet
    C. canal
    C. fascia
    C. ganglion
    C. ligament
    node of C.
    C. node
    C. pseudoganglion
    C. septum
    C. space
closed circuit method
closing
    c. contraction
    c. volume
cloverleaf skull
clubfoot
club hair
clubhand
cluneal nerve
clunes
clunium
    crena c.
    c. inferiores
    c. superiores
cluster
    egg c.
Clutton joints
CMC
    carpometacarpal
CO2-withdrawal seizure test
coalescence
coat
    circular layer of muscular c.
    longitudinal layer of muscular c.
    muscular c.
    sclerotic c.

seromuscular c.
serous c.
coccycephaly
coccygea, pl. **coccygeae**
    foveola c.
    vertebrae coccygeae
coccygeal
    c. artery
    c. body
    c. bone
    c. cornu
    c. dimple
    c. foveola
    c. ganglion
    c. gland
    c. horn
    c. joint
    c. ligament
    c. muscle
    c. nerve
    c. plexus
    c. segment of spinal cord
    c. sinus
    c. vertebrae
    c. whorl
coccygealia
coccygei
coccyges (*pl. of* coccyx)
coccygeum
    cornu c.
    corpus c.
    glomus c.
    segmentum medullae spinalis c.
coccygeus
    c. muscle
    musculus c.
    nervus c.
    plexus c.
    vortex c.
coccygis
    fovea c.
    musculus extensor c.
    os c.
coccyx, pl. **coccyges**
    c. bone
    muscle of c.
cochlea, pl. **cochleae**
    apertura externa canaliculi c.
    aqueductus c.
    area c.
    base of modiolus of c.
    basis modioli c.

    canaliculus of c.
    canalis spiralis c.
    central canal of c.
    columella c.
    crest of fenestrae c.
    crista fenestrae c.
    cupula of c.
    fenestra cochleae
    fenestra of c.
    fossula fenestrae cochleae
    ganglion spirale c.
    hamulus c.
    c. hamulus
    lamina basilaris c.
    lamina modioli c.
    lamina of modiolus of c.
    ligamentum spirale c.
    membranous c.
    spiral canal of c.
    spiral ganglion of c.
    spiral ligament of c.
    stria vascularis of c.
    tuber c.
    vena aqueductus c.
    vena canaliculi c.
    vena fenestrae c.
    vestibular fissure of c.
cochlear
    c. aqueduct
    c. area
    c. branch
    c. branch of labyrinthine artery
    c. branch of vestibulocochlear artery
    c. canal
    c. canaliculus
    c. cupula
    c. duct
    c. ganglion
    c. hair cells
    c. joint
    c. labyrinth
    c. nucleus
    c. part of vestibulocochlear nerve
    c. recess
    c. recess of vestibule
    c. root of vestibulocochlear nerve
    c. scalae
    c. window
cochleare
    ganglion c.
cochleares (*pl. of* cochlearis)

NOTES

cochleariformis
    processus c.
cochleariform process
cochlearis, pl. **cochleares**
    crista basilaris ductus c.
    crista spiralis ductus c.
    ductus c.
    labyrinthus c.
    lamina basilaris ductus c.
    ligamentum spirale ductus c.
    membrana tectoria ductus c.
    membrana vestibularis ductus c.
    nervus c.
    nuclei nervi c.
    paries externus ductus c.
    paries tympanicus ductus c.
    paries vestibularis ductus c.
    pars c.
    prominentia spiralis ductus c.
    radix c.
    ramus c.
    recessus c.
    stria vascularis ductus c.
    vas prominens ductus c.
    vestibularis ductus c.
cochleate
cochleo-orbicular reflex
cochleopalpebral reflex
cochleopapillary reflex
cochleostapedial reflex
cochleovestibular
cock's comb
codfish vertebrae
coefficient
    absorption c.
    biological c.
    Bunsen solubility c.
    creatinine c.
    extraction c.
    filtration c.
    Ostwald solubility c.
    oxygen utilization c.
    Poiseuille viscosity c.
    reflection c.
    respiratory c.
    ultrafiltration c.
    c. of viscosity
coeliaca
    ganglia c.
coeliaci
    lymphonodi c.
    nodi lymphoidei c.
coelom
cognitive laterality quotient
cogwheel
    c. phenomenon
    c. respiration
coiled artery of uterus
coil gland

coinosite
Coiter muscle
cold-blooded animal
cold-rigor point
coles
coli
    appendices adiposae c.
    arcus marginalis c.
    arteria marginalis c.
    cellulae c.
    haustra c.
    plica semilunaris c.
    stratum circulare tunicae
      muscularis c.
    stratum longitudinale tunicae
      muscularis c.
    tenia libera c.
    tunica mucosa c.
    tunica muscularis c.
    tunica serosa c.
colic
    c. branch of ileocolic artery
    c. flexure
    c. impression
    c. impression on liver
    c. impression of spleen
    c. lymph node
    c. omentum
    c. sphincter
    c. surface of spleen
    c. teniae
    c. valve
    c. vein
colica
    arteria c.
    c. dextra
    impressio c.
    c. media
    c. sinistra
colici
    nodi lymphatici c.
colla (*pl. of* collum)
collacin
collagen
    c. fiber
    c. fibrils
collagenation
collagenic
collagenization
collagenous fiber
collar
    c. bone
    renal c.
collarette
collastin
collasum
    sulcus of corpus c.
collateral
    c. branch of intercostal nerve

c. branch of posterior intercostal
c. circulation
c. digital artery
c. eminence
c. fissure
c. ligament
c. sulcus
c. trigone
c. vein
c. vessel

collaterale
ligamentum c.
trigonum c.
vas c.

collateralia
ligamenta c.

collateralis
eminentia c.
fissura c.
c. media
c. radialis
ramus c.
sulcus c.
c. ulnaris inferior
c. ulnaris superior

collateralization

collecting
c. structure
c. system
c. tube
c. tubule

Colles
C. fascia
C. ligament
C. space

colli
apparatus ligamentosus c.
arteria transversa c.
c. artery
fibromatosis c.
ligamentum transversalis c.
longissimus c.
lordosis c.
musculi c.
musculus longus c.
musculus semispinalis c.
musculus spinalis c.
musculus splenius c.
musculus subcutaneus c.
musculus transversalis c.
nervus transversus c.
nodi lymphoidei capitis et c.

pterygium c.
rami inferiores nervi transversi c.
rami inferiores nervi transversi
    cervicalis c.
rami superiores nervi transversi c.
ramus c.
ramus profundus arteriae
    transversae c.
ramus superficialis arteriae
    transversae c.
ramus superior nervi transversalis
    cervicalis c.
transversa c.
transversus c.
trigonum c.
venae transversae c.

collicular artery

collicularis
arteria c.

colliculus, pl. colliculi
c. of arytenoid cartilage
brachium of inferior c.
c. cartilaginis arytenoideae
facial c.
c. facialis
gray layer of superior c.
inferior c.
c. inferior
seminal c.
c. seminalis
c. superior
superior c.
c. urethralis

Collier tract

colloid
c. corpuscle
c. theory of narcosis
thyroid c.

colloidin

colloidoclasia

colloidoclasis

colloidoclastic

collum, pl. colla
c. anatomicum humeri
c. chirurgicum humeri
c. costae
c. dentis
c. fibulae
c. folliculi pili
c. glandis
c. glandis penis
c. mallei

C

NOTES

121

collum *(continued)*
    c. mandibulae
    c. ossis femoris
    c. pancreaticus
    c. radii
    c. scapulae
    c. tali
    c. vesicae
    c. vesicae biliaris
    c. vesicae felleae
colocutaneous
coloileal
colon
    anterior band of c.
    arterial arches of c.
    c. ascendens
    ascending c.
    bands of c.
    c. descendens
    descending c.
    distal c.
    fatty appendices of c.
    flexure of c.
    haustra of c.
    haustrations of c.
    left c.
    marginal artery of c.
    mesentery of sigmoid c.
    mesentery of transverse c.
    midsigmoid c.
    midtransverse c.
    mucosa of c.
    muscular coat of c.
    muscular layer of c.
    pelvic c.
    c. pelvinum
    plicae semilunares of c.
    proximal c.
    right c.
    sacculation of c.
    semilunar fold of c.
    serosa of c.
    sigmoid c.
    c. sigmoideum
    sphincter of hepatic flexure of c.
    splenic flexure of c.
    transverse c.
    c. transversum
    Waldeyer c.
colonic myenteric plexus
colonogram
colorectal mucosa
colostrum
colpatresia
columella, pl. **columellae**
    c. cochlea
    c. nasi
column
    anal c.

    anterior c.
    Bertin c.'s
    branchial efferent c.
    Burdach c.
    c. cells
    Clarke c.
    c. of fornix
    general somatic afferent c.
    general somatic efferent c.
    general visceral c.
    Goll c.
    Gowers c.
    gray c.'s
    lateral c.
    Morgagni c.'s
    primary curvature of vertebral c.
    rectal c.
    renal c.'s
    Rolando c.
    secondary curvatures of vertebral c.
    Sertoli c.'s
    special somatic afferent c.
    special visceral c.
    spinal c.
    splanchnic afferent c.
    splanchnic efferent c.
    Stilling c.
    Türck c.
    vaginal c.'s
    vein of vertebral c.
    ventral c. (VC)
    vertebral c.
columna, pl. **columnae**
    columnae anales
    c. anterior
    columnae carneae
    c. fornicis
    columnae griseae
    c. lateralis
    c. nasi
    c. posterior
    columnae renales
    columnae rugarum
    c. vertebralis
columnar
    c. cell
    c. epithelium
    c. layer
    c. mucosa
columnella, pl. **columnellae**
comb
    cock's c.
comb-growth test
comblike septum
comes, pl. **comites**
comfort zone
comitans, pl. **comitantes**
    c. nervi hypoglossi

c. vein
    vena c.
comitant artery of median nerve
comites (*pl. of* comes)
comma
    c. bundle of Schultze
    c. tract of Schultze
commissura, pl. **commissurae**
    c. anterior grisea
    c. bulborum
    c. cinerea
    c. fornicis
    c. habenularum
    c. hippocampi
    c. labiorum
    c. labiorum anterior
    c. labiorum posterior
    c. lateralis palpebrum
    c. medialis palpebrum
    c. palpebrarum lateralis
    c. palpebrarum medialis
    c. posterior cerebri
    c. posterior grisea
    commissurae supraopticae
    c. ventralis alba
commissural
    c. bundle
    c. cell
    c. fibers
    c. lip pit
    c. pulmonary valve
commissure
    anterior c.
    c. of anterior brain
    anterior gray c.
    anterior labial c.
    anterior white c.
    aortic c.
    c. of cerebral hemisphere
    c. of fornix
    Ganser c.'s
    great transverse c.
    c. of Gudden
    Gudden c.'s
    c. of habenulae
    habenular c.
    hippocampal c.
    interthalamic c.
    labial c.
    laryngeal c.
    lateral palpebral c.
    c. of lip

medial palpebral c.
Meynert c.'s
optic c.
palpebral c.
posterior cerebral c.
posterior labial c.
superior c.
supraoptic c.
transverse c.
c. of vestibular bulb
Wernekinck c.
white c.
common
    c. annular ring
    c. annular tendon
    c. anterior facial vein
    c. atrium
    c. basal vein
    c. bile duct
    c. cardinal vein
    c. carotid artery
    c. carotid nervous plexus
    c. cochlear artery
    c. crus of semicircular duct
    c. extensor
    c. extensor muscle of digit
    c. extensor tendon
    c. facial vein
    c. fibular nerve
    c. flexor sheath
    c. flexor sheath of hand
    c. hepatic artery
    c. hepatic duct
    c. iliac artery
    c. iliac lymph node
    c. iliac vein
    c. interosseous artery
    c. limb of membranous
      semicircular ducts
    c. modiolar vein
    c. palmar digital artery
    c. palmar digital nerve
    c. peroneal nerve
    c. peroneal tendon sheath
    c. plantar digital artery
    c. plantar digital nerve
    c. tendinous ring
    c. tendinous ring of extraocular
      muscle
commune
    integumentum c.
    mesenterium dorsale c.

NOTES

123

communes

digitales palmares c.
digitales plantares c.
lymphonodi iliaci c.
nervi digitales palmares c.
nervi digitales plantares c.
nodi lymphoidei iliaci c.

communicans, pl. **communicantes**

c. anterior cerebri
gray rami communicantes
macula c.
c. posterior cerebri
ramus c.
white rami communicantes

communicating

c. branch
c. branch of anterior interosseous nerve with ulnar nerve
c. branch of auriculotemporal nerve with facial nerve
c. branch of chorda tympani to lingual nerve
c. branch of facial nerve with glossopharyngeal nerve
c. branch of facial nerve with tympanic plexus
c. branch of fibular artery
c. branch of intermediate nerve with tympanic plexus
c. branch of internal laryngeal nerve with recurrent laryngeal nerve
c. branch of lacrimal nerve with zygomatic nerve
c. branch of lingual nerve with hypoglossal
c. branch of median nerve with ulnar nerve
c. branch of nasociliary nerve with ciliary ganglion
c. branch of otic ganglion to chorda tympani
c. branch of otic ganglion with medial pterygoid nerve
c. branch of otic ganglion with meningeal branch of mandibular nerve
c. branch of peroneal artery
c. branch of radial nerve with ulnar nerve
c. branch of spinal nerve
c. branch of superficial radial nerve with ulnar nerve
c. branch of sympathetic trunk
c. branch of tympanic plexus with auricular branch of vagus nerve
c. hydrocephalus
c. rami of sympathetic trunk

communis

annulus tendineus c.
arteria carotis c.
arteria cochlearis c.
arteria digitalis palmaris c.
arteria digitalis plantaris c.
arteria hepatica c.
arteria iliaca c.
arteria interossea c.
carotis c.
ductus hepaticus c.
hepatica c.
macula c.
musculus extensor digitorum c.
nervus fibularis c.
nervus peroneus c.
peroneus c.
plexus caroticus c.
plexus nervosus caroticus c.
ramus communicans fibularis nervi fibularis c.
ramus communicans peroneus nervi peronei c.
sacculus c.
c. tendon
truncus arteriosus c.
tunica vaginalis c.
vagina communis tendinum musculorum fibularium c.
vagina musculorum peroneorum c.
vagina tendinum musculorum fibularium c.
vagina tendinum musculorum peroneorum c.
vena basalis c.
vena facialis c.
vena iliaca c.
vena modioli c.

community

biotic c.

compact

c. bone
c. substance
c. tissue

compacta

substantia c.

compactum

stratum c.

compages thoracis

companion

c. artery to sciatic nerve
c. lymph node of accessory nerve
c. vein

comparative

c. anatomy
c. physiology

compartimentum

c. antebrachii anterius
c. antebrachii extensorum

c. antebrachii flexorum
c. antebrachii posterius
c. brachii anterius
c. brachii extensorum
c. brachii flexorum
c. brachii posterius
c. cruris
c. cruris anterius
c. cruris extensorum
c. cruris fibularium
c. cruris flexorum
c. cruris laterale peroneorum
c. cruris posterius
c. femoris adductorum
c. femoris anterius
c. femoris extensorum
c. femoris flexorum
c. femoris mediale
c. femoris posterius

compartment
medial c.
muscular space of retroinguinal c.
scrotal c.
c. of thigh for extensors of hip joint
c. of thigh for extensors of knee
c. of thigh for flexors of hip
c. of thigh for flexors of knee
tibial c.
vascular space of retroinguinal c.

compensated
c. acidosis
c. alkalosis

compensatory circulation
competence
cardiac c.

complemental air
complementary air
complete
c. cleavage
c. tetanus

complex
amygdaloid c.
Batson vein c.
Ghon c.
Golgi c.
internal hemorrhoidal c.
j-g c.
c. joint
junctional c.
juxtaglomerular c.
Ranke c.

Steidele c.
synaptinemal c.

complexa
articulatio c.

complexus
musculus c.
c. stimulans cordis

compliance
c. of heart
specific c.
static c.
thoracic c.
ventilatory c.

component
talar c.

composita
articulatio c.

composite joint
compound
c. articulation
c. gland
c. granule cell
c. joint
c. pregnancy

compressible cavernous bodies
compression
c. of tissue

compressor
c. muscle of lip
c. muscle of naris
c. naris muscle
c. urethrae
c. urethra muscle
c. venae dorsalis penis

conarium
concameration
concatenate
concave
concavity
concavoconcave
concavoconvex
concentration
plasma c.

concentric lamella
conception
conceptus, pl. **concepti**
concha, pl. **conchae**
apophysis conchae
c. of auricle
c. auriculae
c. bullosa

C

NOTES

concha *(continued)*
    cavernous vascular plexus of conchae
    cavitas conchae
    cavity of c.
    cavum conchae
    corpus cavernosum conchae
    cymba conchae
    c. of ear
    eminence of c.
    eminentia conchae
    ethmoidal process of inferior nasal c.
    highest c.
    inferior nasal c.
    lacrimal process of inferior nasal c.
    maxillary process of inferior nasal c.
    middle nasal c.
    Morgagni c.
    nasal c.
    c. nasalis
    c. nasalis inferior
    c. nasalis media
    c. nasalis superior
    c. nasalis suprema
    c. nasoturbinal
    plexus vascularis cavernosus conchae
    root of inferior nasal c.
    c. of Santorini
    Santorini c.
    sphenoidal c.
    conchae sphenoidales
    superior nasal c.
    supreme nasal c.
conchal
    c. cartilage
    c. crest
    c. crest of body of maxilla
    c. crest of palatine bone
    c. fossa
    c. mastoid angle
conchalis
    crista c.
concharum
    plexus cavernosi c.
conchoidal
concrement
concrescence
concretion
conditioned stimulus
conducting
    c. airway
    c. system of heart
conduction
    avalanche c.
    decremental c.

    nerve c.
    saltatory c.
    synaptic c.
    c. system
    c. system of heart
conductivity
    hydraulic c.
conductor
conduplicate
conduplicato corpore
condylar
    c. articulation
    c. axis
    c. canal
    c. emissary vein
    c. fossa
    c. joint
    c. process
    c. process of mandible
condylaris
    articulatio c.
    canalis c.
    fossa c.
    processus c.
    vena emissaria c.
condylarthrosis
condyle
    c. bone
    c. cord
    femoral c.
    c. of humerus
    lateral c.
    c. of mandible
    mandibular c.
    medial c.
    occipital c.
    tibial c.
condylion
condyloid
    c. canal
    c. fossa
    c. joint
    c. process
condyloideum
    emissarium c.
condylus
    c. humeri
    c. lateralis
    c. lateralis femoris
    c. lateralis tibiae
    c. medialis
    c. medialis femoris
    c. medialis tibiae
    c. occipitalis
cone
    antipodal c.
    arterial c.
    c. cell
    c. cell of retina

c. disks
elastic c.
fertilization c.
c. fiber
c. granule
Haller c.
implantation c.
layer of rods and c.'s
medullary c.
pulmonary c.
retinal c.
theca interna c.
twin c.
vascular c.

conexus intertendineus
confertus
configuration
confluence
c. of sinuses
confluens sinuum
congenerous muscle
congenita
chondrodystrophia calcificans c.
ectopia pupillae c.
fistula auris c.
fistula colli c.
congenital
c. amputation
c. diaphragmatic hernia
c. ectodermal defect
c. ectodermal dysplasia
c. hydrocele
c. hydrocephalus
c. nystagmus
c. torticollis
c. valve
congenitus
congestion
active c.
functional c.
passive c.
physiologic c.
congestive
conglobate
conglobation
conglomerate
coni (*pl. of* conus)
conicae
papillae c.
conical
c. lobules of epididymis
c. papilla

conic papilla
coniometer
coniophage
conjoined
c. asymmetrical twins
c. equal twins
c. nerve root
c. tendon
c. unequal twins
conjoint tendon
conjugata
c. anatomica
c. diagonalis
c. externa
c. recta
c. vera
conjugate
anatomical c.
c. axis
diagonal c.
c. diameter of pelvic inlet
c. diameter of pelvic outlet
effective c.
external c.
false c.
c. foramen
internal c.
median c.
obstetric c.
straight c.
true c.
conjugated
conjunctiva, pl. **conjunctivae**
annulus c.
bulbar c.
decussation of brachia c.
fornix of superior c.
palpebral c.
plica semilunaris of c.
saccus c.
sclera and c. (S&C)
tela c.
tunica c.
conjunctival
adenoidal-pharyngeal-c. (A-P-C, APC)
c. artery
c. cul-de-sac
c. fornix
c. gland
c. layer of bulb
c. layer of eyelid

NOTES

C

conjunctival *(continued)*
    c. limbus
    c. ring
    c. sac
    c. vein
conjunctivalis, pl. conjunctivales
    conjunctivales anteriores
    glandulae conjunctivales
    conjunctivales posteriores
    saccus c.
    venae conjunctivales
conjunctive
conjunctivi
    decussatio brachii c.
conjunctivus
    tendo c.
connecting
    c. cartilage
    c. stalk
    c. tubule
connectins
connection
    accessory arteriovenous c.
    intertendineus c.
connective
    c. tissue
    c. tissue cell
    c. tissue group
    c. tissue sheath
    c. tissue stalk
connexon
connexus intertendinei musculi extensoris digitorum
conniventes
    valvulae c.
conoid
    c. ligament
    c. process
    c. tubercle
    c. tubercle of clavicle
conoideum
    ligamentum c.
    tuberculum c.
conomyoidin
consecutive anophthalmia
conservation of energy
conservatrix
    vis c.
constant field equation
constitution
constitutive heterochromatin
constrictio
    c. bronchoaortica esophagea
    c. diaphragmatica esophagea
    c. partis thoracicae esophagea
    c. pharyngoesophagealis
    c. phrenica esophagea
constriction
    bronchoaortic c.

    inferior esophageal c.
    middle esophageal c.
    pharyngoesophageal c.
    primary c.
    pyloric c.
    c. ring
    secondary c.
    upper esophageal c.
    c.'s of ureter
constrictor
    buccopharyngeal part of superior pharyngeal c.
    glossopharyngeal part of superior pharyngeal c.
    c. muscle
    c. muscle of pharynx
    mylopharyngeal part of superior pharyngeal c.
    c. pharyngis inferior
    c. pharyngis medius
consumption
    oxygen c.
contact
    c. ceptor
    c. inhibition
    c. surface of tooth
contents
    abdominal c.
contiguity
contiguous
continence
continent
continuity
continuous capillary
contortus
    tubulus renalis c.
    tubulus seminiferus c.
contour lines of Owen
contract
contracted
    c. foot
    c. pelvis
contractile
contractility
contraction
    anodal closure c. (AnCC)
    anodal opening c. (AnOC, AOC)
    cathodal closure c. (CaCC)
    cathodal opening c. (CaOC)
    closing c.
    fibrillary c.'s
    front-tap c.
    Gowers c.
    hourglass c.
    hunger c.'s
    idiomuscular c.
    isometric c.
    isotonic c.
    myotatic c.

opening c.
paradoxical c.
postural c.
tetanic c.
tonic c.

contracture
Dupuytren c.
Volkmann c.

contralateral
c. partner

control
idiodynamic c.
reflex c.
synergic c.
tonic c.
vestibuloequilibratory c.

controlled respiration
conular
conus, pl. **coni**
c. arteriosus
c. branch
c. elasticus
c. elasticus laryngis
coni epididymidis
lateral c.
c. medullaris
pulmonary c.
coni vasculosi

convective heat
convergence nucleus of Perlia
conversion
conversive heat
convex
convexity
cortical c.
outer c.

convolute
convoluted
c. bone
c. gland
c. part of kidney lobule
c. seminiferous tubule
c. tubule of kidney

convolution
angular c.
anterior central c.
ascending frontal c.
ascending parietal c.
callosal c.
cingulate c.
first temporal c.
hippocampal c.

inferior frontal c.
inferior temporal c.
middle frontal c.
middle temporal c.
posterior central c.
second temporal c.
superior frontal c.
superior temporal c.
supramarginal c.
third temporal c.
transitional c.
transverse temporal c.'s
Zuckerkandl c.

convulsion
clonic c.
immediate posttraumatic c.
tetanic c.
tonic c.

Cooper
C. fascia
C. ligament
suspensory ligament of C.

coordinate
coordinated reflex
coordination
coossification
coossify
copula
His c.
c. linguae

cor
c. biloculare
c. triatriatum
c. triloculare
c. triloculare biatriatum
c. triloculare biventriculare

coracoacromial
c. arch
c. ligament

coracoacromiale
ligamentum c.

coracobrachial
c. bursa
c. muscle

coracobrachialis
bursa musculi c.
c. muscle
musculus c.

coracoclaviculare
ligamentum c.

coracoclavicularis
tuberositas ligamenti c.

C

NOTES

coracoclavicular ligament
coracohumerale
    ligamentum c.
coracohumeral ligament
coracoid
    c. notch
    c. process
    c. process of scapula
    c. tuberosity
coracoidea
    tuberositas c.
coracoideus
    processus c.
cord
    anterior column of spinal c.
    anterior median fissure of spinal c.
    anterolateral column of spinal c.
    arachnoid of spinal c.
    Bergmann c.'s
    Billroth c.'s
    central canal of spinal c.
    cervical enlargement of spinal c.
    cervical segments of spinal c.
    coccygeal segment of spinal c.
    condyle c.
    cornu of spinal c.
    covering of spermatic c.
    dental c.
    dorsal column of spinal c.
    dura mater of spinal c.
    false knots of umbilical c.
    false tendinous c.'s
    false vocal c.
    Ferrein c.
    gangliated c.
    genital c.
    germinal c.'s
    gonadal c.'s
    heel c.
    hepatic c.'s
    horn of lateral spinal c.
    intermediolateral cell column of
      spinal c.
    lateral column of spinal c.
    lateral funiculus of spinal c.
    lumbar enlargement of spinal c.
    lumbar segments L1–L5 of
      spinal c.
    lumbosacral enlargement of
      spinal c.
    medullary c.'s
    nephrogenic c.
    oblique c.
    posterior column of spinal c.
    posterior median fissure of
      spinal c.
    posterior median sulcus of
      spinal c.
    psalterial c.

    red pulp c.'s
    rete c.'s
    sacral part of spinal c.
    segments of spinal c.
    sex c.'s
    spermatic c.
    spinal c.
    splenic c.'s
    tendinous c.
    testicular c.
    testis c.'s
    thoracic spinal c.
    true knot of umbilical c.
    true vocal c.
    c. of tympanum
    umbilical c.
    c. of umbilical artery
    vein of spinal c.
    ventral column of spinal c.
    vitelline c.
    vocal c. (VC)
    Weitbrecht c.
    Wilde c.'s
    Willis c.
cordate pelvis
cordiform
    c. pelvis
    c. tendon of diaphragm
    c. uterus
cordiformis
    uterus c.
cordis
    c. anteriores
    anulus fibrosus dexter/sinister c.
    apex c.
    atrium dextrum c.
    atrium sinistrum c.
    basis c.
    bulbus c.
    capsula c.
    chorda tendineae c.
    complexus stimulans c.
    crena c.
    ectopia c.
    facies pulmonalis c.
    facies sternocostalis c.
    foramen ovale c.
    foramina venarum minimarum c.
    incisura apicis c.
    lacertus c.
    c. magna
    margo dexter c.
    c. media
    membrana c.
    c. minimae
    mucro c.
    ostium venosum c.
    c. para

pars muscularis septi
  interventricularis c.
plexus coronarii c.
plexus coronarius c.
posterior ventriculi sinistri c.
scrobiculus c.
sulcus interventricularis c.
sulcus terminalis c.
systema conducens c.
theca c.
trigona fibrosa c.
venae c.
ventriculus c.
vertex c.
vortex c.

core
central transactional c.

coria

Cori cycle

corii
papillae c.
rete cutaneum c.
stratum papillare c.
stratum reticulare c.
tunica propria c.

corium
papillae of c.
reticular layer of c.

corkscrew artery

cornea, pl. corneae
anterior epithelium of c.
anterior limiting layer of c.
anterior surface of c.
c. ciliaris
elastic layers of c.
endothelium posterius corneae
epithelium anterius corneae
facies anterior corneae
facies posterior corneae
lamina limitans anterior corneae
lamina limitans posterior corneae
limbus of c.
limbus corneae
limiting layers of c.
macula corneae
meridian of c.
posterior limiting lamina of c.
posterior limiting layer of c.
posterior surface of c.
substantia propria of c.
substantia propria corneae

vertex of c.
vertex corneae

corneal
c. corpuscles
c. layer of epidermis
c. limbus
c. margin
c. space
c. spot
c. vertex

corneocyte envelope

corneomandibular reflex

corneomental reflex

corneopterygoid reflex

corneosclera

corneoscleral
c. junction
c. part
c. part of trabecular tissue of
  sclera

corneoscleralis
pars c.

Corner-Allen test

corneum

corniculata
cartilago c.

corniculate
c. cartilage
c. process
c. tubercle

corniculatum
tuberculum c.

corniculopharyngeal ligament

corniculopharyngeum
ligamentum c.

corniculum laryngis

cornified

cornu, pl. cornua
c. ammonis
c. anterius
coccygeal c.
c. coccygeum
dorsal c.
c. of falciform margin of
  saphenous opening
c. of hyoid bone
c. inferius
c. inferius cartilaginis thyroideae
c. inferius hiatus saphenus
c. inferius marginis falciformis
  hiatus sapheni
c. inferius ventriculi lateralis

NOTES

cornu (*continued*)
  c. laterale
  c. of lateral ventricle
  c. majus
  c. majus ossis hyoidei
  c. minus
  c. minus ossis hyoidei
  c. of
  c. posterius
  c. posterius ventriculi lateralis
  sacral c.
  c. sacrale
  c. of spinal cord
  styloid c.
  c. superius
  c. superius cartilaginis thyroideae
  c. superius hiatus saphenus
  c. superius marginalis falciformis
  c. of thyroid cartilage
  c. uterus
cornual
  c. portion of uterus
cornuradicular zone
cornus
corona, pl. **coronae**
  c. capitis
  cavitas coronae
  c. ciliaris
  c. clinica
  cuspis coronae
  c. dentis
  c. glandis
  c. glandis penis
  c. of glans penis
  c. radiata
  tuberculum coronae
  c. vascularis
  Zinn c.
coronad
coronal
  c. plane
  c. pulp
  c. sulcus
  c. suture
coronale
  cavum c.
  pulpa c.
coronalia
  plana c.
coronalis
  cavitas c.
  pulpa c.
  sutura c.
coronaria
coronarii
  ostium sinus c.
  valvula sinus c.

coronarius
  sinus c.
  sulcus c.
coronary
  c. artery
  c. circulation
  c. cusp
  c. groove
  c. ligament
  c. ligament of knee
  c. ligament of liver
  c. node
  c. plexus
  c. reflex
  c. sinus
  c. sulcus
  c. tendon
  c. valve
  c. vein
coronion
coronoid
  c. fossa
  c. fossa of humerus
  c. process
  c. process of mandible
  c. process of ulna
coronoidea
  fossa c.
coronoideus
  processus c.
corpora (*pl. of* corpus)
corpore
  conduplicato c.
corporeal
corporis
  c. callosi
  regiones c.
corpse
corpus, pl. **corpora**
  c. adiposum
  c. adiposum buccae
  c. adiposum fossae ischioanalis
  c. adiposum fossae ischiorectalis
  c. adiposum infrapatellare
  c. adiposum orbitae
  c. albicans
  c. amygdaloideum
  c. amylaceum
  c. aorticum
  c. arantii
  corpora arenacea
  c. atreticum
  corpora bigemina
  c. callosum
  capsule of corpora
  corpora cavernosa recti
  c. cavernosum clitoridis
  c. cavernosum of clitoris
  c. cavernosum conchae

c. cavernosum penis
c. cavernosum urethrae
c. ciliare
c. ciliaris
c. claviculae
c. coccygeum
c. costa
c. dentatum
c. epididymidis
c. fibrosum
c. fibulae
c. fimbriatum
c. gastricum
c. geniculatum internum
c. geniculatum laterale
c. geniculatum mediale
c. glandulae sudoriferae
c. highmorianum
c. humeri
c. incudis
c. linguae
c. luteum
c. luteum hormone
c. luteum spurium
c. luteum verum
c. luysii
c. mamillare
c. mammae
c. mandibulae
c. maxillae
c. medullare cerebelli
c. metacarpale
c. metatarsale
c. nuclei caudati
c. olivare
c. ossis femoris
c. ossis hyoidei
c. ossis ilii
c. ossis ischii
c. ossis metacarpalis
c. ossis pubis
c. ossis sphenoidalis
c. pampiniforme
c. pancreatis
c. papillare
corpora para-aortica
c. paraterminale
c. phalangis
c. pineale
corpora quadrigemina
c. quadrigeminum anterius
c. quadrigeminum posterius

c. radii
c. restiforme
c. spongiosum penis
c. spongiosum urethrae muliebris
c. sterni
c. striatum
c. tali
c. tibiae
c. trapezoideum
c. triticeum
c. ulnae
c. unguis
uterine c.
c. of uterus
c. vertebrae
c. vesicae
c. vesicae biliaris
c. vesicae felleae
c. vitreum

corpuscle
amniotic c.
amylaceous c.
amyloid c.
articular c.'s
axile c.
axis c.
basal c.
Bizzozero c.
blood c.
bone c.
bridge c.
bulboid c.'s
cement c.
colloid c.
corneal c.'s
Dogiel c.
exudation c.
genital c.'s
Golgi-Mazzoni c.
Hassall concentric c.
inflammatory c.
lamellated c.'s
malpighian c.'s
Mazzoni c.
Meissner c.
Merkel c.
oval c.
pacchionian c.'s
pacinian c.'s
plastic c.
Purkinje c.'s
pus c.

NOTES

C

corpuscle *(continued)*
    renal c.
    reticulated c.
    Ruffini c.'s
    salivary c.
    Schwalbe c.
    splenic c.'s
    tactile c.
    taste c.
    terminal nerve c.'s
    third c.
    thymic c.
    touch c.
    Toynbee c.'s
    Tröltsch c.'s
    Valentin c.'s
    Vater c.'s
    Vater-Pacini c.'s
    Virchow c.'s
    white c.
    Zimmermann c.
corpuscular volume
corpusculum, pl. **corpuscula**
    corpuscula articularia
    corpuscula bulboidea
    corpuscula genitalia
    corpuscula lamellosa
    corpuscula nervosa terminalia
    c. renis
    c. tactus
correlative differentiation
corrugator
    c. cutis muscle of anus
    c. supercilia
    c. supercilii muscle
cortex, pl. **cortices**
    adrenal c.
    agranular c.
    association c.
    auditory c.
    cerebellar c.
    c. cerebelli
    cerebral c.
    c. cerebri
    deep c.
    dysgranular c.
    femoral c.
    fetal adrenal c.
    frontal c.
    fusiform cells of cerebral c.
    ganglionic layer of cerebellar c.
    ganglionic layer of cerebral c.
    c. glandulae suprarenalis
    granular layer of cerebellar c.
    granular layers of cerebral c.
    heterotypic c.
    homotypic c.
    insular c.
    c. of kidney

    laminated c.
    layers of cerebellar c.
    layers of cerebral c.
    c. of lens
    c. lentis
    c. of lymph node
    mastoid c.
    molecular layer of cerebellar c.
    molecular layer of cerebral c.
    motor c.
    c. nodi lymphatici
    olfactory c.
    orbitofrontal c.
    ovarian c.
    c. ovarii
    c. of ovary
    parastriate c.
    piriform c.
    plexiform layer of cerebral c.
    prefrontal c.
    premotor c.
    primary visual c.
    provisional c.
    c. reflex
    renal c.
    c. renalis
    c. renis
    rolandic c.
    secondary sensory c.
    secondary visual c.
    sensory c.
    somatic sensory c.
    somatosensory c.
    stellate cells of cerebral c.
    striate c.
    subcapsular c.
    supplementary motor c.
    suprarenal c.
    c. of suprarenal gland
    temporal c.
    tertiary c.
    c. thymi
    c. of thymus
    visual c. (VC)
Corti
    C. arch
    C. auditory tooth
    C. canal
    C. cells
    C. ganglion
    C. membrane
    organ of C.
    C. organ
    C. pillar
    C. rod
    C. tunnel
cortical
    c. arches of kidney
    c. area

c. bone
c. convexity
c. hormone
c. implantation
c. lobules of kidney
c. part
c. part of middle cerebral artery
c. radiate artery
c. substance
corticalis
pars c.
substantia c.
corticalization
cortices (*pl. of* cortex)
corticifugal
corticipetal
corticis
corticoafferent
corticobulbar
c. fibers
c. tract
corticobulbaris
tractus c.
corticocerebellum
corticoefferent
corticofugal
corticoliberin
corticomedial
corticonucleares
fibrae c.
corticonuclear fibers
corticopontinae
fibrae c.
corticopontine
c. fibers
c. tract
corticopontini
tractus c.
corticoreticulares
fibrae c.
corticoreticular fibers
corticospinal
c. fibers
c. tract
corticospinalis, pl. **corticospinales**
fibrae corticospinales
tractus c.
corticothalamic
corticotroph
corticotropic hormone
corticotropin

corticotropin-releasing
c.-r. factor
c.-r. hormone
cortisol
costa, pl. **costae**
angulus costae
articulatio capitis costae
caput c.
c. cervicalis
collum costae
corpus c.
crista capitis costae
crista colli costae
crista corporis c.
facies articularis capitis costae
facies articularis tuberculi costae
costae fluctuantes
costae fluitantes
ligamentum colli costae
ligamentum tuberculi costae
c. lumbalis
musculus levator costae
c. prima
costae spuriae
sulcus costae
tuberculum costae
costae verae
costal
c. angle
c. arch
c. cartilage
c. facet
c. groove
c. groove of rib
c. line of pleural reflection
c. margin
c. notch
c. notch of sternum
c. part of diaphragm
c. part of parietal pleura
c. pit of transverse process
c. portion of diaphragm
c. respiration
c. surface
c. surface of lung
c. surface of scapula
c. tuberosity
costale
os c.
costalis, pl. **costales**
arcus c.
cartilago c.

C

NOTES

costalis *(continued)*
    facies c.
    incisura c.
    pleura c.
    processus c.
    tuberositas c.
costarum
    arcus c.
    levatores c.
    musculi levatores c.
Costen syndrome
costicartilage
costicervical
costiform
costispinal
costoaxillary vein
costocentral articulation
costocervical
    c. arterial trunk
    c. artery
costocervicalis
    truncus c.
costochondral
    c. articulation of rib
    c. joint
    c. junction
    c. margin
costochondralis, pl. **costochondrales**
    articulatio c.
costoclavicular
    c. ligament
    c. line
costoclaviculare
    ligamentum c.
costoclavicularis
    impressio ligamenti c.
costocolic
    c. fold
    c. ligament
costocoracoid
costodiaphragmatic recess
costodiaphragmaticus
    recessus c.
costoinferior
costomediastinal
    c. recess
    c. sinus
costomediastinalis
    recessus c.
costophrenic
    c. angle
    c. sinus
    c. sulcus
costoscapular
costoscapularis
costosternal articulation
costosuperior
costotransversaria
    articulatio c.

costotransversarium
    foramen c.
    ligamentum c.
costotransverse
    c. articulation
    c. foramen
    c. joint
    c. ligament
costovertebral (CV)
    c. angle
    c. articulation
    c. joint
costovertebrales
    articulationes c.
costoxiphoid
    c. angle
    c. ligament
costoxiphoideum
    ligamentum c.
cotransport
cotunnii
    aqueductus c.
    liquor c.
Cotunnius
    C. aqueduct
    C. canal
    C. liquid
    C. space
cotyledon
    fetal c.
    maternal c.
cotyledonary placenta
cotylica
    articulatio c.
cotyloid
    c. cavity
    c. joint
    c. ligament
    c. notch
cotyloideum
    ligamentum c.
countercurrent
    c. exchanger
    c. mechanism
    c. multiplier
couple
covering
    outermost c.
    c.'s of spermatic cord
cow kidney
cowl muscle
Cowper
    C. gland
    C. ligament
coxa, pl. **coxae**
    c. adducta
    articulatio c.
    c. flexa
    c. magna

musculus triceps c.
os c.
c. plana
retinaculum capsulae articularis
coxae
c. valga
c. vara
c. vara luxans
zona orbicularis articulationis c.
coxal bone
coxale
punctum c.
coxofemoral
coxofemoralis
articulatio c.
Crabtree effect
Crampton
C. line
C. muscle
crania (*pl. of* cranium)
craniad
cranial
c. arachnoid mater
c. base
c. bone
c. canal
c. capacity
c. cavity
c. dura mater
c. extradural space
c. flexure
c. fossa
c. index
c. nerve (I–XII)
c. part of parasympathetic part of
autonomic division of nervous
system
c. reflex
c. root of accessory nerve
c. roots
c. segment
c. suture
c. synchondroses
c. synostosis
c. synovial joint
c. vault
c. venous sinus
c. vertebra
cranialis, pl. **craniales**
arachnoid mater c.
dura mater c.
nervi craniales

radices craniales
sulcus venae cavae c.
cranialium
nuclei nervorum c.
cranii
articulationes c.
basis c.
cavitas c.
fonticuli c.
foramen ovale basis c.
lamina externa c.
lamina interna ossium c.
ossa c.
periosteum c.
suturae c.
synchondroses c.
trabeculae c.
cranioaural
craniocarpotarsal
c. dysplasia
c. dystrophy
craniocele
craniocerebral
craniocervical part of peripheral
autonomic plexuses and ganglion
craniocleidodysostosis
craniodiaphysial dysplasia
craniodidymus
craniofacial
c. angle
c. axis
c. dysostosis
c. notch
craniofenestria
craniograph
craniography
craniolacunia
craniology
craniometaphysial dysplasia
craniometer
craniometric point
craniometry
craniopagus
c. occipitalis
c. parasiticus
craniopharyngeal
c. canal
c. duct
craniophore
craniorrhachidian
craniorrhachischisis

NOTES

C

craniosacral
    c. division of autonomic nervous
      system
cranioschisis
cranioscopy
craniospinal
craniostenosis
craniostosis
craniosynostosis
craniotympani
craniotympanic
cranium, pl. **crania**
    bifid c.
    c. bifidum
    bone of visceral c.
    cerebral c.
    c. cerebrale
    squamous suture of c.
    visceral c.
    c. viscerale
    wall of c.
crassi
    folliculi lymphatici solitarii
      intestini c.
    glandulae intestini c.
    tunica mucosa intestini c.
    tunica muscularis intestini c.
    tunica serosa intestini c.
crassum
    intestinum c.
crater arc
crateriform
crease
    digital flexion c.
    earlobe c.
    flexion c.
    inframammary c.
    nasolabial c.
    palmar c.
    simian c.
creatinine
    c. clearance
    c. coefficient
creep recovery
cremaster
    c. fascia
    c. muscle
    musculus c.
cremasteric
    c. artery
    c. fascia
    c. layer
    c. muscle
    c. reflex
    c. vein
cremasterica
    arteria c.
    fascia c.
crena, pl. **crenae**

    c. analis
    c. ani
    c. clunium
    c. cordis
    c. interglutealis
crescent
    articular c.
    c. cell
    Giannuzzi c.'s
    Heidenhain c.'s
    sublingual c.
crescentic lobules of cerebellum
crescograph
crest
    acoustic c.
    acousticofacial c.
    alveolar c.
    c. of alveolar ridge
    ampullary c.
    anterior lacrimal c.
    arched c.
    arcuate c.
    articular c.
    c. of body of rib
    buccinator c.
    cap of ampullary c.
    c. of cochlear opening
    conchal c.
    deltoid c.
    dental c.
    ethmoidal c.
    external lip of iliac c.
    external occipital c.
    falciform c.
    c. of fenestrae cochlea
    frontal c.
    ganglionic c.
    gluteal c.
    c. of greater tubercle
    c. of head of rib
    c. of humerus
    iliac c.
    ilium c.
    incisor c.
    infratemporal c.
    inguinal c.
    inner lip of iliac c.
    intermediate line of iliac c.
    intermediate sacral c.
    intermediate zone of iliac c.
    internal lip of iliac c.
    internal occipital c.
    interosseous c.
    intertrochanteric c.
    interureteric c.
    lacrimal c.
    lateral epicondylar c.
    lateral sacral c.
    lateral supracondylar c.

c. of lesser tubercle
marginal c.
maxillary c.
medial epicondylar c.
medial supracondylar c.
median sacral c.
c. of nail bed
c. of nail matrix
nasal c.
c. of neck of rib
neural c.
neuroepithelium of ampullary c.
obturator c.
outer lip of iliac c.
palatine c.
c. of palatine bone
c. of petrous part of temporal
  bone
c. of petrous temporal bone
posterior lacrimal c.
pubic c.
c. of round window
sacral c.
c. of scapular spine
sphenoid c.
sphenoidal c.
spiral c.
supinator c.
c. of supinator muscle
supramastoid c.
supraventricular c.
terminal c.
tibial c.
transverse c.
triangular c.
trigeminal c.
trochanteric c.
tubercle of iliac c.
turbinated c.
urethral c.
vestibular c.
c. of vestibule
cretin
cretinism
cretinistic
cretinoid
cretinous
crevice
gingival c.
crevicular
c. epithelium
c. fluid

cribra (*pl. of* cribrum)
cribrate
cribration
cribriform
c. area of the renal papilla
c. fascia
c. foramen
c. plate
c. plate of ethmoid bone
c. process
c. sinus
c. tissue
cribrosa, pl. **cribrosae**
ethmoidal lamina c.
fascia c.
foramina c.
lamina c.
macula c.
cribrous lamina
cribrum, pl. **cribra**
cricoarytenoid
c. articular capsule
c. articulation
c. joint
c. ligament
c. muscle
cricoarytenoidea
articulatio c.
capsula articularis c.
cricoarytenoideus
c. lateralis
c. posterior
cricoesophageal tendon
cricoesophageus
tendo c.
cricoid
arytenoidal articular surface of c.
c. cartilage
cricoidea, pl. **cricoideae**
arcus cartilaginis cricoideae
cartilago c.
facies articularis arytenoidea
  cricoideae
facies articularis thyroidea
  cricoideae
lamina cartilaginis cricoideae
cricopharyngea
pars c.
cricopharyngeal
c. ligament
c. muscle
c. part

C

NOTES

cricopharyngeal (*continued*)
    c. part of inferior constrictor
      muscle of pharynx
    c. sphincter
cricopharyngeum
    cartilago sesamoidea ligamentum c.
    ligamentum c.
cricopharyngeus
    c. muscle
    musculus c.
cricosantorinian ligament
cricothyroid
    c. articular capsule
    c. articulation
    c. arytenoid ligament
    c. branch of superior thyroid
      artery
    c. joint
    c. membrane
    c. muscle
cricothyroidea
    articulatio c.
    capsula articularis c.
    membrana c.
cricothyroidei
    pars obliqua musculi c.
    pars recta musculi c.
cricothyroideum
    ligamentum c.
cricothyroideus
    musculus c.
    ramus c.
cricotracheal
    c. ligament
    c. membrane
cricotracheale
    ligamentum c.
cricovocal membrane
crinis, pl. **crines**
crinogenic
crinophagy
crista, pl. **cristae**
    c. ampullaris
    c. ampullaris ductuum
      semicirculorum
    c. arcuata
    c. arcuata cartilaginis arytenoideae
    c. basilaris ductus cochlearis
    c. buccinatoria
    c. capitis costae
    c. choanalis vomeris
    c. colli costae
    c. conchalis
    c. conchalis corporis maxillae
    c. conchalis ossis palatini
    c. corporis costa
    cristae cutis
    c. dentalis
    c. dividens

c. ethmoidalis
c. ethmoidalis maxillae
c. ethmoidalis ossis palatini
c. fenestrae cochlea
c. frontalis
c. galli
c. glutea
c. helicis
c. iliaca
c. infratemporalis
c. infratemporalis alaris majoris
    ossis sphenoidalis
c. intertrochanterica
c. lacrimalis anterior
c. lacrimalis posterior
c. marginalis
c. marginalis dentis
cristae matricis unguis
c. medialis fibulae
cristae of mitochondria
cristae mitochondriales
c. musculi supinatoris ulnae
c. musculi supinatorius
c. nasalis
c. nasalis laminae horizontalis ossis
    palatini
c. nasalis processus palatini
    maxillae
c. obturatoria
c. occipitalis externa
c. occipitalis interna
c. palatina
c. palatina laminae horizontalis
    ossis palatini
c. phallica
c. pubica
c. quarta
cristae sacrales laterales
c. sacralis
c. sacralis intermedia
c. sacralis lateralis
c. sacralis medialis
c. sacralis mediana
c. of semicircular duct
c. spiralis
c. spiralis ductus cochlearis
c. supracondylaris lateralis
c. supracondylaris medialis
c. supraepicondylaris lateralis
c. supraepicondylaris medialis
c. supramastoidea
c. suprastyloidea radii
c. supraventricularis
c. temporalis mandibulae
c. terminalis
c. terminalis atrii dextri
c. terminalis of right atrium
c. transversa
c. transversalis

c. transversa meatus acustici interni
c. triangularis
c. tuberculi major
c. tuberculi majoris
c. tuberculi minoris
c. tympanica
c. urethralis
c. urethralis femininae
c. urethralis masculinae
c. verticalis meatus acustici interni
c. vestibuli

cross
c. circulation
hair c.'s
Ranvier c.'s
c. section

crossed
c. extension reflex
c. jerk
c. pyramidal tract

cross-section of heart
crossway
sensory c.

crotaph
crotaphion
crotaphyticobuccinatorius
porus c.

crown
anatomical c.
c. cavity
ciliary c.
clinical c.
c. of head
c. pulp
pulp cavity of c.
radiate c.
c. of tooth
c. tubercle

crown-heel length
crown-rump length
cruces (*pl. of* crux)
crucial
c. anastomosis
c. ligament

cruciate
c. anastomosis
c. eminence
c. ligament
c. ligament of atlas
c. ligament of leg
c. ligaments of knee

c. muscle
c. pulley

cruciatus
musculus c.

cruciform
c. eminence
c. ligament of atlas
c. part of fibrous digital sheath
c. pulley

cruciformis
eminentia c.

crura (*pl. of* crus)
crural
c. arch
c. area
c. artery
c. canal
c. fascia
c. fossa
c. interosseous nerve
c. ligament
c. region
c. ring
c. septum
c. sheath

cruralis
cisterna c.

crureus
cruris
compartimentum c.
facies anterior c.
facies lateralis c.
facies posterior c.
fascia c.
interosseous c.
ligamentum cruciatum c.
ligamentum transversum c.
membrana interossea c.
musculus biceps flexor c.
nervus interosseus c.
tegmen c.

crus, pl. **crura**
c. II
c. anterius capsulae internae
c. anterius stapedis
c. anthelicis
crura anthelicis
crura antihelicis
crura of antihelix
crura of bony semicircular canal
c. breve incudis
crura cerebelli

C

NOTES

crus *(continued)*
    c. cerebri
    crura cerebri
    c. clitoridis
    c. of clitoris
    c. corporis cavernosi penis
    c. dextrum diaphragmatis
    c. dextrum fasciculi
      atrioventricularis
    c. dextrum truncus atrioventricularis
    c. of diaphragm
    c. fornicis
    c. of fornix
    c. glandis clitoridis
    c. helicis
    c. of helix
    c. of incus
    crura of incus
    c. inferius marginis falciformis
      hiatus sapheni
    lateral c.
    c. laterale
    c. laterale anuli inguinalis
    c. laterale cartilaginis alaris majoris
    c. longum incudis
    medial c.
    c. mediale
    c. mediale anuli inguinalis
    c. mediale cartilaginis alaris
      majoris
    crura membranacea ampullaria
    crura ossea canales semicirculares
    c. of penis
    c. posterius capsulae internae
    c. posterius stapedis
    c. sinistrum diaphragmatis
    c. sinistrum fasciculi
      atrioventricularis
    c. sinistrum truncus
      atrioventricularis
    stapedial c.
    c. of stapes
    superior c.
    c. superius marginis falciformis
      hiatus sapheni
crusta, pl. **crustae**
Cruveilhier
    C. fascia
    C. fossa
    fossa navicularis C.
    C. joint
    C. ligament
    C. plexus
crux, pl. **cruces**
    c. of heart
    cruces pilorum
cryobiology
cryotolerant

crypt
    anal c.
    dental c.
    enamel c.
    c. of iris
    Lieberkühn c.'s
    c. of Lieberkühn
    c. of Lieberkühn of large intestine
    c. of Lieberkühn of small intestine
    lingual c.
    c.'s of Luschka
    Morgagni c.
    synovial c.
    tonsillar c.
crypta, pl. **cryptae**
    c. tonsillaris
cryptocrystalline
cryptodidymus
cryptophthalmia
cryptophthalmus
cryptorchidism
cryptorchid testis
cryptorchism
cryptotia
cryptozygous
crystal
    ear c.'s
crystalline
    c. capsule
    c. lens
crystalloid
    Charcot-Böttcher c.'s
    Reinke c.'s
CSF
    cerebrospinal fluid
cubital
    c. anastomosis
    c. bone
    c. fossa
    c. joint
    c. lymph node
    c. nerve
    c. region
    c. vein
cubitalis, pl. **cubitales**
    fossa c.
    lymphonodi cubitales
    nodi lymphoidei cubitales
cubitocarpal
cubitus, pl. **cubiti**
    articularis cubiti
    articulatio cubiti
    ligamentum collaterale radiale
      articulationis cubiti
    ligamentum collaterale ulnare
      articulationis cubiti
    mediana cubiti
    musculus articularis cubiti
    rete articulare cubiti

c. valgus
c. varus
vena intermedia cubiti
vena mediana cubiti

cuboid
c. bone

cuboidal
c. articular surface of calcaneus
c. epithelium

cuboidei
processus calcaneus ossis c.
tuberositas ossis c.

cuboideonavicular
c. joint
c. ligament

cuboideonaviculare
ligamenta c.
ligamentum c.

cuboideum
os c.

cuff
musculotendinous c.
suprahepatic caval c.
uterine c.
vaginal c.

cular ligament

cul-de-sac
conjunctival c.-d.-s.
c.-d.-s. of Douglas
Douglas c.-d.-s.
greater c.-d.-s.
Gruber c.-d.-s.
lesser c.-d.-s.
sacral c.-d.-s.
superior c.-d.-s.

culmen, pl. **culmina**

culminis
lobulus c.

cumulus, pl. **cumuli**
c. oophorus
c. ovaricus

cuneate
c. funiculus
c. nucleus

cuneati
nucleus funiculi c.
tuberculum nuclei c.

cuneatus
fasciculus c.
nucleus c.
tuberculum c.

cuneiform
c. bone
c. cartilage
c. joint
c. ligament
c. lobe
c. part of vomer
c. tubercle

cuneiforme
tuberculum c.

cuneiformis
cartilago c.
lobulus c.

cuneocerebellar tract

cuneocuboid
c. interosseous ligament
c. joint

cuneocuboideum
ligamentum c.

cuneometatarsal
c. interosseous ligament
c. joint

cuneonavicular
c. articulation
c. joint
c. ligament

cuneonavicularia
ligamenta c.

cuneonavicularis
articulatio c.

cuneoscaphoid

cuneus

cunnus

cup
Diogenes c.
ocular c.
optic c.
c. of palm

cupola

cupula, pl. **cupulae**
c. of cochlea
cochlear c.
c. cristae ampullaris
c. of pleura
c. pleurae
pleural c.
c. of semicircular duct

cupular
c. blind sac
c. cecum of cochlear duct
c. part
c. part of epitympanic recess

NOTES

143

cupulare
cecum c.
cupularis
pars c.
cupulate part
cupuliform
cupulogram
current
action c.
anodal c.
ascending c.
axial c.
centrifugal c.
centripetal c.
d'Arsonval c.
demarcation c.
descending c.
electrotonic c.
high-frequency c.
c. of injury
labile c.
Tesla c.
curvatura, pl. curvaturae
c. primaria columnae vertebralis
curvaturae secondariae columnae
vertebralis
c. ventriculi major
c. ventriculi minor
curvature
greater c.
lesser c.
curve
active length-tension c.
flow-volume c.
force-velocity c.
Frank-Starling c.
intracardiac pressure c.
isovolume pressure-flow c.
lordotic c.
lumbar lordotic c.
lumbosacral c.
muscle c.
passive length-tension c.
Starling c.
strength-duration c.
stress-strain c.
Traube-Hering c.
Cushing
C. effect
C. phenomenon
C. response
cushion
anal c.
atrioventricular canal c.'s
endocardial c.'s
c. of epiglottis
eustachian c.
hemorrhoidal c.
levator c.

Passavant c.
pharyngoesophageal c.'s
sucking c.
cusp
anterior c.
c. of anterior mitral valve
c. of anterior semilunar
aortic c.
c. of commissural mitral valve
coronary c.
c. of left aortic valve
c. of left atrioventricular valve
c. of left semilunar
posterior c.
c. of posterior aortic valve
c. of posterior mitral valve
c. of pulmonic valve
c. of right aortic valve
c. of right atrioventricular valve
c. of right semilunar
semilunar c.
septal c.
c. of septal
c. of tooth
cuspid
deciduous c.
cuspidate tooth
cuspidati
dentes c.
cuspidatus
dens c.
cuspis, pl. cuspides
c. anterior
c. anterior valvae atrioventricularis
dextrae
c. anterior valvae atrioventricularis
sinistrae
c. anterior valvae mitralis
c. anterior valvae tricuspidalis
c. coronae
c. dentis
c. posterior
c. posterior valvae atrioventricularis
dextrae
c. posterior valvae atrioventricularis
sinistrae
c. posterior valvae mitralis
c. posterior valvae tricuspidalis
c. septalis
c. septalis valvae atrioventricularis
c. septalis valvae tricuspidalis
cutanea
c. tarda
vena c.
**cutanei anterioris ramorum ventralium
nervorum thoracicorum**
cutaneomucosal
cutaneomucosus
musculus c.

cutaneomucous muscle
cutaneous
    c. absorption
    c. branch of anterior branch of
      obturator nerve
    c. branch of mixed nerve
    c. cervical nerve
    c. dorsalis medialis
    c. gland
    c. innervation
    lateral femoral c. (LFC)
    c. layer of tympanic membrane
    c. ligament of phalanx
    c. meningioma
    c. muscle
    c. vein
cutaneus
    c. antebrachii lateralis
    c. antebrachii medialis
    c. antebrachii posterior
    c. brachii lateralis inferior
    c. brachii lateralis superior
    c. brachii medialis
    c. brachii posterior
    c. dorsalis intermedius
    c. dorsalis lateralis
    c. femoris lateralis
    c. femoris posterior
    musculus c.
    nervus c.
    c. surae lateralis
    c. surae medialis
cuticle
    dental c.
    enamel c.
    c. of hair
    c. of nail
    Nasmyth c.
    c. of root sheath
cuticula, pl. **cuticulae**
    c. dentis
    c. vaginae folliculi pili
cuticular membrane
cutis
    cristae c.
    glandulae c.
    c. plate
    retinaculum c.
    stratum reticulare c.
    sulci c.
    c. vera
    c. verticis gyrata

cutization
cutting
    c. edge
Cuvier
    duct of C.
    C. ducts
CV
    cardiac volume
    cerebrovascular
    cervical vertebra
    costovertebral
cyanophil, cyanophile
cyanophilous
cybernetics
cybrid
cyclarthrodial
cyclarthrosis
cycle
    anovulatory c.
    brain wave c.
    carbon dioxide c.
    cardiac c.
    cell c.
    citric acid c.
    Cori c.
    dicarboxylic acid c.
    estrous c.
    fatty acid oxidation c.
    forced c.
    genesial c.
    glycine succinate c.
    glyoxylic acid c.
    hair c.
    Krebs-Henseleit c.
    Krebs ornithine c.
    Krebs urea c.
    life c.
    menstrual c.
    nitrogen c.
    ornithine c.
    ovarian c.
    reproductive c.
    restored c.
    returning c.
    succinic acid c.
    tricarboxylic acid c.
    urea c.
cyclencephalia
cyclencephaly
cyclocephalia
cyclocephaly
cyclopean eye

NOTES

cyclopia, cyclopea
cyclopian eye
cyclops
cylinder
    axis c.
cylindraxis
cylindrica
    articulatio c.
cylindrical
    c. epithelium
    c. joint
cylindricum
    stratum c.
cyllosoma
cyma
cymba conchae
cymbocephalic
cymbocephalous
cymbocephaly
cynocephaly
Cyon nerve
cyst
    arachnoid c.
    Baker c.
    bile c.
    branchial cleft c.
    duplication c.
    echinococcus c.
    Gartner c.
    c. of hypophysis
    involution c.
    laryngeal c.
    morgagnian c.
    preauricular c.
    sacrococcygeal c.
    sublingual c.
    teratomatous c.
    thyroglossal c.
    Tornwaldt c.
    wolffian c.
cystic
    c. artery
    c. duct
    c. duct lumen
    c. gall duct
    c. lymph node
    c. membrane
    c. mole
    c. plexus
    c. vein
cystica
    arteria c.
    pars c.
    spina bifida c.
    vena c.
cystic-choledochal junction
cystici
    plica spiralis ductus c.

cysticus
    ductus c.
    nodus lymphoideus c.
cystides
cystis
    c. fellea
    c. urinaria
cystoduodenal ligament
cystohepatic triangle
cystohepaticum
    trigonum c.
cystous
cytic series
cytoarchitectural
cytoarchitecture
cytobiotaxis
cytocentrum
cytochalasins
cytochemistry
cytochylema
cytocinesis
cytoclesis
cytocrine secretion
cytodieresis
cytogene
cytogenic reproduction
cytogenous
cytohyaloplasm
cytoid
cytokeratin filaments
cytokine network
cytokinesis
cytolemma
cytologic
cytology
cytolymph
cytolysosome
cytomatrix
cytomegalic cells
cytomembrane
cytometaplasia
cytomitome
cytomorphology
cytomorphosis
cyton
cytopempsis
cytophyletic
cytoplasm
cytoplasmic
    c. bridges
    c. matrix
cytoplast
cytopoiesis
cytosis
cytoskeleton
cytosol
cytosolic
cytosome
cytotrophoblast

cytotrophoblastic
    c. cells

c. shell

NOTES

C

D

D cell

DA

dopamine

dacryagogue

dacryocyst

dacryon

dacryostenosis

dactyl

dactyli (*pl. of* dactylus)

dactylia

dactylium

dactylomegaly

dactylus, pl. **dactyli**

Dale-Feldberg law

Dalton-Henry law

Dalton law

dammar

dark cells

Darkschewitsch

nucleus of D.

d'Arsonval

d. current

d. galvanometer

dartoic tissue

dartoid tissue

dartos

d. fascia

d. muliebris

d. muscle

tunica d.

Darwin

auricular tubercle of D.

D. tubercle

darwinian tubercle

datum plane

Daubenton

D. angle

D. line

D. plane

daughter cell

David

lyre of D.

davidis

lyra d.

Davidoff cells

DBP

diastolic blood pressure

deactivation

dead space

dearterialization

death

brain d.

cerebral d.

de Bordeau theory

Debove membrane

debt

alactic oxygen d.

lactacid oxygen d.

oxygen d.

decalvans

porrigo d.

decapacitation factor

decapitate

decapitation

decarbonization

deceleration

decentration

decerebrate

decerebration

bloodless d.

dechloridation

dechlorination

dechloruration

decidua

d. basalis

d. capsularis

membrana d.

d. menstrualis

d. parietalis

d. polyposa

d. reflexa

d. serotina

d. spongiosa

d. vera

decidual

d. cell

d. fissure

d. reaction

d. tissue

decidui

dentes d.

deciduous

d. cuspid

d. dentition

d. membrane

d. placenta

d. tissue

d. tooth

declination

angle of d.

declive

declivis

decompression

d. chamber

d. disease

explosive d.

rapid d.

decrement

decremental conduction

D

149

decurrent
decussate
    Forel d.
decussatio, pl. decussationes
    d. brachii conjunctivi
    d. fontinalis
    d. lemniscorum
    d. motoria
    d. nervorum trochlearium
    d. pedunculorum cerebellarium
     superiorum
    d. pyramidum
    d. sensoria
    decussationes tegmenti
decussation
    d. of brachia conjunctiva
    dorsal tegmental d.
    d. of fillet
    fountain d.
    Held d.
    d. of medial lemniscus
    Meynert d.
    motor d.
    optic d.
    pyramidal d.
    rubrospinal d.
    d. of superior cerebellar peduncle
    tectospinal d.
    tegmental d.
    d. of trochlear nerves
    ventral tegmental d.
dedifferentiation
deep
    d. anterior cervical lymph node
    d. anterior neck
    d. anterior wall
    d. artery of arm
    d. artery of clitoris
    d. artery of penis
    d. artery of thigh
    d. artery of tongue
    d. articular aorta
    d. auricular artery
    d. brachial artery
    d. branch
    d. branch of lateral plantar nerve
    d. branch of medial circumflex
     femoral artery
    d. branch of medial plantar artery
    d. branch of radial nerve
    d. branch of superior gluteal artery
    d. branch of transverse cervical
     artery
    d. branch of ulnar nerve
    d. cardiac plexus
    d. cell
    d. cerebellar nuclei
    d. cerebral vein
    d. cervical fascia

    d. cervical jugulodigastric node
    d. cervical vein
    d. circumflex iliac artery
    d. circumflex iliac vein
    d. circumflex inguinal artery
    d. cortex
    d. crural arch
    d. dorsal sacrococcygeal ligament
    d. dorsal vein
    d. dorsal vein of clitoris
    d. dorsal vein of penis
    d. epigastric artery
    d. epigastric vein
    d. epigastric vessel
    d. external pudendal artery
    d. facial vein
    d. fascia of abdominal wall
    d. fascia of arm
    d. fascia of forearm
    d. fascia of leg
    d. fascia of neck
    d. fascia of penis
    d. fascia of thigh
    d. femoral artery
    d. femoral vein
    d. fibular nerve
    d. flexor muscle
    d. flexor muscle of finger
    d. head
    d. head of flexor pollicis brevis
    d. infrapatellar bursa
    d. inguinal lymph node
    d. inguinal ring
    d. lamina
    d. lateral cervical lymph node
    d. layer
    d. layer of levator palpebrae
     superioris
    d. layer of temporal fascia
    d. lingual artery
    d. lingual vein
    d. lymphatic vessel
    d. lymph vessel
    d. middle cerebral artery
    d. middle cerebral vein
    d. muscle of back
    d. orbit
    d. origin
    d. palmar arterial arch
    d. palmar branch of ulnar artery
    d. palmar venous arch
    d. parotid lymph node
    d. part
    d. part of anterior compartment of
     forearm
    d. part of external anal sphincter
    d. part of flexor retinaculum
    d. part of masseter muscle

d. part of palpebral part of orbicularis oculi muscle
d. part of parotid gland
d. part of posterior flexor compartment of leg
d. perineal fascia
d. perineal pouch
d. perineal space
d. peroneal nerve
d. petrosal nerve
d. plantar arterial arch
d. plantar branch of dorsalis pedis artery
d. popliteal lymph node
d. posterior sacrococcygeal ligament
d. profunda brachial artery
d. reflex
d. sensibility
d. space compartment of hand
d. temporal artery
d. temporal nerve
d. temporal vein
d. tendon reflex (DTR)
d. transitional gyrus
d. transverse
d. transverse fiber
d. transverse metacarpal ligament
d. transverse metatarsal ligament
d. transverse muscle of perineum
d. transverse perineal muscle
d. vein of thigh
d. volar arch

defatigation
defecate
defecation
defect
aorticopulmonary septal d.
aortic septal d.
atrial septal d.
congenital ectodermal d.
endocardial cushion d.
fibrous cortical d.
ventricular septal d.
defective
defense reflex
deferens, pl. **deferentia**
ampulla of ductus d.
ampulla of vas d.
artery of ductus d.
artery to vas d.
diverticula of ampulla of ductus d.

funicular part of ductus d.
inguinal part of ductus d.
mucosa of ductus d.
mucous membrane of ductus d.
muscular coat of ductus d.
muscular layer of ductus d.
pelvic part of ductus d.
plexus of ductus d.
scrotal part of ductus d.
vas d.
vestige of ductus d.
deferent
d. canal
d. duct
deferentes
pars pelvica ductus d.
deferentia (*pl. of* deferens)
deferential
d. artery
d. plexus
deferentialis
arteria d.
plexus nervosus d.
deferentis
ampulla ductus d.
arteria ductus d.
diverticula ampullae ductus d.
pars funicularis ductus d.
pars inguinalis ductus d.
pars scrotalis ductus d.
tunica mucosa ductus d.
tunica muscularis ductus d.
deficit
base d.
oxygen d.
definitive lysosome
deflection
defluvium
defluxion
deformation
deforming
deformity
Arnold-Chiari d.
Erlenmeyer flask d.
lobster-claw d.
mermaid d.
parachute d.
pseudolobster-claw d.
reduction d.
Sprengel d.
deganglionate

D

NOTES

degeneration
    Nissl d.
    reaction of d.
deglutition
    d. apnea
    d. reflex
deglutitive
degranulation
degustation
dehydrate
dehydration
    absolute d.
    relative d.
    voluntary d.
dehydrocholate test
deiterospinal tract
Deiters
    D. cell
    D. nucleus
    D. process
    D. terminal frames
dejecta
dejection
delamination
delayed reflex
delomorphous
delta
    d. cell of anterior lobe of
      hypophysis
    d. cell of pancreas
    d. fiber
    d. fornicis
    Galton d.
    d. granule
    d. mesoscapulae
deltoid
    d. branch
    d. crest
    d. eminence
    d. fascia
    d. impression
    d. ligament
    d. muscle
    d. region
    d. tubercle of spine of scapula
    d. tuberosity
    d. tuberosity of humerus
deltoidea
    regio d.
    tuberositas d.
deltoidei
    pars acromialis musculi d.
    pars clavicularis musculi d.
    pars spinalis musculi d.
    pars tibiocalcanea ligamenti d.
deltoideopectoral
    d. triangle
    d. trigone

deltoideopectorale
    trigonum d.
deltoideum
    ligamentum d.
deltoideus
    musculus d.
    ramus d.
deltopectoral
    d. fascia
    d. groove
    d. sulcus
    d. triangle
deltopectorale
    trigonum d.
demarcation
    d. current
    d. potential
demilune
    Giannuzzi d.'s
    Heidenhain d.'s
    serous d.'s
demipenniform
Demoivre formula
den
    pulpa d.
dendraxon
dendriform
dendrite
    apical d.
dendritic
    d. cells
    d. depolarization
    d. process
    d. spine
    d. thorn
dendroid
dendron
denitrogenation
Denonvilliers
    D. aponeurosis
    D. ligament
dens, pl. **dentes**
    dentes acustici
    d. angularis
    anterior articular surface of d.
    apex of d.
    apical ligament of d.
    d. axis
    dentes bicuspidi
    d. bicuspidus
    dentes canini
    d. caninus
    dentes cuspidati
    d. cuspidatus
    dentes decidui
    d. in dente
    facet of atlas for d.
    dentes incisivi
    d. incisivus

d. lacteus
d. molaris
d. molaris tertius
d. permanens
pit of atlas for d.
posterior articular facet of d.
posterior articular surface of d.
dentes premolares
d. premolaris
d. sapientiae
d. serotinus
d. succedaneus

densa
lamina d.
macula d.

density
flux d.
vapor d.

dental
d. alveoli
d. anatomy
d. apparatus
d. arch
d. artery
d. branch
d. bulb
d. canal
d. cord
d. crest
d. crypt
d. cuticle
d. fibers
d. follicle
d. germ
d. groove
d. index
d. lamina
d. neck
d. nerve
d. papilla
plexus d.
d. process
d. pulp
d. rami
d. sac
d. shelf
d. tubercle
d. tubule

dentalis, pl. **dentales**
alveolus d.
canaliculi dentales
crista d.

rami dentales
tubuli dentales

dentary center
dentata, pl. **dentatae**
fissura d.
lamina d.
vertebra d.

dentate
d. fascia
d. fissure
d. gyrus
d. ligament
d. line
d. nucleus of cerebellum
d. suture

dentati
hilum nuclei d.

dentatothalamic tract
dentatum
corpus d.

dentatus
gyrus d.

dente
dens in d.

dentes (*pl. of* dens)
denticle
denticular hymen
denticulated
denticulate ligament
denticulatum
ligamentum d.

dentilabial
dentilingual
dentinal
d. canal
d. fibers
d. papilla
d. pulp
d. sheath
d. tubule

dentine
dentin globule
dentinocemental junction
dentinoenamel junction
dentinum
dentis
apex cuspidis d.
apex radicis d.
canalis radicis d.
cavitas d.
cavum d.
cervix d.

D

NOTES

dentis *(continued)*
    cingulum d.
    collum d.
    corona d.
    crista marginalis d.
    cuspis d.
    cuticula d.
    ebur d.
    facies approximalis d.
    facies articularis anterior d.
    facies articularis posterior d.
    facies contactus d.
    facies distalis d.
    facies facialis d.
    facies lingualis d.
    facies mesialis d.
    facies occlusalis d.
    facies vestibularis d.
    foramen apicis d.
    gubernaculum d.
    ligamentum apicis d.
    papilla d.
    pulpa d.
    radix d.
    radix clinica d.
    substantia ossea d.
    tuberculum d.
dentitio difficilis
dentition
    deciduous d.
    first d.
    mandibular d.
    maxillary d.
    primary d.
    secondary d.
    succedaneous d.
dentium
dentoalveolar
    d. joint
dentoalveolaris
    articulatio d.
dentogingival lamina
dentoliva
dentulous
Denucé ligament
denucleated
deozonize
depletion
    chloride d.
    d. response
    salt d.
    water d.
depolarization
    dendritic d.
depolarize
depression
    angle of d.
    d. of optic disk
    otic d.

    pacchionian d.'s
    pterygoid d.
    spreading d.
    sternal d.
    supratrochlear d.
depressomotor
depressor
    d. anguli oris
    d. anguli oris muscle
    d. fiber
    d. labii inferioris
    d. labii inferioris muscle
    d. muscle of epiglottis
    d. muscle of eyebrow
    d. muscle of lower lip
    d. muscle of septum
    d. nerve
    d. nerve of Ludwig
    d. reflex
    d. septi nasi
    d. septi nasi muscle
    d. supercilii
    d. supercilii muscle
deprivation
depuration
deradelphus
deranencephalia
deranencephaly
derencephalia
derencephalocele
derencephaly
derivation
derivative
dermad
dermal
    d. papilla
    d. ridge
    d. sinus
    d. system
    d. tissue
dermatic
dermatocele
dermatoglyphics
dermatoid
dermatome
    d. of female pelvis and perineum
    d. of lower limb
    d. of male pelvis and perineum
    d. of upper limb
dermatomegaly
dermatomere
dermatomic area
dermatoskeleton
dermic
dermis
    papilla of d.
    papillae d.
dermoblast
dermoepidermal interface

dermoid
>   inclusion d.
>   d. system

dermoskeleton

dermovascular

derodidymus

derotation

desaturate

desaturation

Descemet
>   membrane of D.
>   D. membrane

descendens
>   aorta d.
>   arteria genus d.
>   arteria palatina d.
>   arteria scapularis d.
>   d. cervicalis
>   colon d.
>   genus d.
>   d. hypoglossi
>   mesocolon d.
>   palatina d.
>   ramus anterior d.
>   ramus posterior d.
>   ramus profundus arteria
>   scapularis d.

descendentis
>   ramus pharyngeus arteriae
>   palatinae d.
>   ramus profundus arteriae
>   scapularis d.

descending
>   d. anterior branch
>   d. artery of knee
>   d. brain stem
>   d. branch of anterior segmental
>   artery of left and right lung
>   d. branch of hypoglossal nerve
>   d. branch of lateral circumflex
>   femoral artery
>   d. branch of medial circumflex
>   femoral artery
>   d. branch of occipital artery
>   d. branch of posterior segmental
>   artery of left and right lung
>   d. branch of superficial cervical
>   artery
>   d. bronchial branch of aorta
>   d. colon
>   d. current
>   d. genicular artery

>   d. genicular vein
>   d. groove for aorta
>   d. groove of arch of aorta
>   d. mesocolon
>   d. nucleus of trigeminus
>   d. palatine artery
>   d. part of duodenum
>   d. part of facial canal
>   d. part of iliofemoral ligament
>   d. part of trapezius muscle
>   d. posterior branch
>   d. scapular artery
>   d. tract of trigeminal nerve

descensus
>   d. paradoxus testis

descent

descriptive
>   d. anatomy
>   d. myology

desiccant

desiccate

desiccation

desiccative

desmocranium

desmodentium

desmodontium

desmogenous

desmography

desmoid

desmology

desmosome

desynchronous

determinate cleavage

detrusor
>   d. muscle
>   d. pressure
>   d. sphincter dyssynergia
>   d. urinae

deutan axis

deutencephalon

deuteroplasm

deuterosome

deuterotocia

deuterotoky

deutogenic

deutoplasm

deutoplasmic

deutoplasmigenon

deutoplasmolysis

development

developmental
>   d. anatomy

NOTES

developmental *(continued)*
    d. anomaly
    d. physiology
Deventer pelvis
deviant
deviation
devitalization
devitalize
devitalized
devolution
dexiocardia
dexter
    bronchus principalis d.
    ductus hepaticus d.
    ductus lobi caudati d.
    ductus lymphaticus d.
    ductus thoracicus d.
    lobus hepatis d.
    pulmo d.
    ramus d.
    ventriculus d.
dexter/sinister
    ventriculus cordis d.
dextra
    arteria colica d.
    arteria coronaria d.
    arteria gastrica d.
    arteria gastroepiploica d.
    arteria gastroomentalis d.
    arteria pulmonalis d.
    auricula atrii d.
    colica d.
    flexura coli d.
    gastrica d.
    gastroepiploica d.
    hemicardia d.
    intercostalis superior d.
    ovarica d.
    pars hepatis d.
    pulmonalis inferior d.
    pulmonalis superior d.
    testicularis d.
    valva atrioventricularis d.
    vena colica d.
    vena gastrica d.
    vena gastroomentalis d.
    vena intercostalis superior d.
    vena ovarica d.
    vena pulmonalis inferior d.
    vena pulmonalis superior d.
    vena suprarenalis d.
    vena testicularis d.
dextrad
dextrae
    arteria flexurae d.
    cuspis anterior valvae
      atrioventricularis d.
    cuspis posterior valvae
      atrioventricularis d.

    pars infralobaris venae posterioris
      venae pulmonalis superioris d.
    ramus apicalis lobi inferioris
      arteriae pulmonalis d.
    ramus atrialis intermedius arteriae
      coronariae d.
    ramus interventricularis posterior
      arteriae coronariae d.
    ramus lateralis rami lobaris medii
      arteriae pulmonalis d.
    ramus lobi medii arteriae
      pulmonalis d.
    ramus marginalis dexter arteriae
      coronariae d.
    ramus medialis rami lobaris medii
      arteriae pulmonalis d.
    ramus nodi sinuatrialis arteriae
      coronariae d.
    venae hepaticae d.
dextra/sinistra
    facies pulmonales cordis d.
dextri
    arteria lobaris media pulmonis d.
    crista terminalis atrii d.
    fissura horizontalis pulmonis d.
    lobus azygos pulmonis d.
    lobus medius pulmonis d.
    lymphonodi colici d.
    lymphonodi gastrici d.
    lymphonodi gastroomentales d.
    lymphonodi lumbales d.
    nodi lymphoidei colici d.
    nodi lymphoidei gastrici d.
    nodi lymphoidei gastroomentales d.
    nodi lymphoidei lumbales d.
    pars basalis arteriarum lobarium
      inferiorum pulmonis sinistri et d.
    pars intralobaris intersegmentalis
      venae posterioris lobi superioris
      pulmonis d.
    ramus posterior ductus hepatici d.
    sulcus terminalis atrii d.
    tuberculum intervenosum atrii d.
dextrocardia
    isolated d.
    type 1, 2 d.
    d. with situs inversus
dextrogastria
dextrogyration
dextroposition
dextrosinistral
dextrotropic
dextrum
    atrium cordis d.
    ligamentum triangulare d.
    ostium atrioventriculare d.
    segmentum hepatis anterius
      laterale d.

segmentum hepatis anterius
  mediale d.
segmentum hepatis posterius
  laterale d.
segmentum hepatis posterius
  mediale d.
trigonum fibrosum d.
diacele
diacrinous
diadochokinesia, diadochocinesia
diadochokinesis
diadochokinetic
diagonal
  d. conjugate
  d. conjugate diameter
  d. section
diagonalis
  conjugata d.
diakinesis
dial manometer
dialysance
dialysis
diamelia
diameter
  anterotransverse d.
  Baudelocque d.
  biparietal d. (BPD)
  diagonal conjugate d.
  external conjugate d.
  frontomental d.
  frontooccipital d.
  d. mediana
  d. medianus
  d. obliqua
  oblique d.
  obstetric conjugate d.
  occipitofrontal d.
  occipitomental d.
  suboccipitobregmatic d.
  trachelobregmatic d.
  d. transversa
  transverse d.
  zygomatic d.
diandria
diandry
diapause
  embryonic d.
diapedesis
diaphoresis
diaphoretic
diaphragm
  arcus tendineus of pelvic d.

bare area d.
caval opening of d.
central tendon of d.
cordiform tendon of d.
costal part of d.
costal portion of d.
crus of d.
dome of d.
fascia d.
foramen of sellar d.
inferior fascia of pelvic d.
inferior fascia of urogenital d.
intermediate tendon of d.
laryngeal d.
left crus of d.
lumbar part of d.
lumbocostal triangle of d.
d. of mouth
pelvic d.
d. of pelvis
pillar of d.
right crus of d.
right dome of d.
d. of sella
sellar d.
d. of sella turcica
sternal part of d.
sternocostal triangle of d.
superior fascia of pelvic d.
superior fascia of urogenital d.
urogenital d.
vertebral part of d.
diaphragma, pl. **diaphragmae**
  centrum tendineum diaphragmae
  musculus d.
  d. oris
  d. pelvis
  d. sellae
  d. urogenitale
diaphragmata
diaphragmatic
  d. artery
  d. constriction of esophagus
  d. hiatus
  d. ligament of mesonephros
  d. muscle
  d. nerve
  d. node
  d. part of parietal pleura
  d. silhouette
  d. surface

D

NOTES

diaphragmatic *(continued)*
    d. surface of heart
    d. surface of lung
diaphragmatica
    facies d.
    pleura d.
diaphragmatis
    centrum tendineum d.
    crus dextrum d.
    crus sinistrum d.
    pars costalis d.
    pars lumbalis d.
    pars sternalis d.
    trigonum lumbocostale d.
    trigonum sternocostale d.
diaphysial, diaphyseal
    d. aclasis
    d. center
    d. juxtaepiphysial exostosis
diaphysis, pl. **diaphyses**
diapiresis
diaplacental
diaplexus
diapophysis
diarthric
diarthrodial
    d. cartilage
    d. joint
diarthrosis, pl. **diarthroses**
diarticular
diastalsis
diastaltic
diastema
diastemata
diastematocrania
diastematomyelia
diaster
diastole
diastolic blood pressure (DBP)
diastrophism
diatela
diathermal
diathermancy
diathermanous
diathermic
diathermy
diathesis, pl. **diatheses**
diathetic
diauchenos
    dicephalus d.
dibrachius
    acephalus d.
    dicephalus dipus d.
dicarboxylic acid cycle
dicelous
dicephalous
dicephalus
    d. diauchenos
    d. dipus dibrachius

    d. dipus tetrabrachius
    d. dipus tribrachius
    d. dipygus
    d. monauchenos
dicheilia
dicheiria
dichilia
dichiria
dichorial twins
dichorionic
    d. diamniotic placenta
dichotic
dichotomous
dichotomy
dichromophil
dichromophile
Dickens shunt
dictyotene
didactylism
didelphic
didelphys
    uterus d.
didymus
diecious
dielectrography
diencephala
diencephalohypophysial
diencephalon
    ventricle of d.
diestrous
diestrus
diet
    basal d.
    high-fiber d.
    Kempner d.
    Ornish prevention d.
    Ornish reversal d.
    rice d.
difference
    arteriovenous carbon dioxide d.
    arteriovenous oxygen d.
    AV d.
differential
    d. blood pressure
    d. growth
    d. manometer
    threshold d.
    d. threshold
differentiated
differentiation
    correlative d.
    invisible d.
difficilis
    dentitio d.
diffusa
    placenta d.
diffused reflex
diffusible stimulant
diffusing capacity

diffusion
- d. anoxia
- d. hypoxia

digastric
- d. branch of facial nerve
- d. fossa
- d. groove
- d. muscle
- d. notch
- d. triangle

digastrica
- fossa d.

digastrici
- venter anterior musculi d.
- venter posterior musculi d.

digastricus
- amus d.
- musculus d.

digest

digestion
- gastric d.
- intestinal d.
- pancreatic d.
- peptic d.
- primary d.
- secondary d.

digestive
- d. apparatus
- d. system
- d. tract
- d. tube

digestorium
- systema d.

digestorius
- apparatus d.
- tubus d.

digit
- bones of d.'s
- common extensor muscle of d.
- d.'s of foot
- ventral surface of d.

digital
- d. collateral artery
- d. extensor tendon
- d. flexion crease
- d. fossa
- d. furrow
- d. joint
- d. pulp
- d. pulp of hand
- d. retinacular ligament

- d. vein
- d. venous arch

digitales
- d. dorsales hallucis lateralis et digiti secunda medialis
- d. dorsales manus
- d. dorsales nervi radialis
- d. dorsales nervi ulnaris
- d. dorsales pedis
- d. palmares
- d. palmares communes
- d. palmares communes nervi mediani
- d. palmares communes nervi ulnares
- d. palmares propriae
- d. palmares proprii nervi mediani
- d. palmares proprii nervi ulnaris
- d. plantares
- d. plantares communes
- d. plantares communes nervi plantaris lateralis
- d. plantares communes nervi plantaris medialis
- d. plantares propriae
- d. plantares proprii nervi plantaris lateralis
- d. plantares proprii nervi plantaris medialis

digitatae
- impressiones d.

digitate impression

digitation

digitationes hippocampi

digitorum
- chiasm of musculus flexor d.
- connexus intertendinei musculi extensoris d.
- facies palmares d.
- intertendinous connections of extensor d.
- ligamenta cruciata d.
- ligamentum anulare d.
- musculus extensor brevis d.
- musculus extensor longus d.
- musculus flexor brevis d.
- musculus flexor longus d.
- ossa d.

digitus, pl. **digiti**
- d. annularis
- d. anularis
- d. auricularis

NOTES

digitus *(continued)*
    digiti manus
    d. manus
    d. manus medius
    d. manus minimus
    d. manus primus
    d. manus quartus
    d. manus quintus
    d. manus secundus
    d. manus tertius
    musculus extensor minimi digiti
    musculus opponens minimi digiti
    digiti pedis
    d. pedis foot
    d. pedis minimus
    d. pedis primus
    d. pedis quartus
    d. pedis quintus
    d. pedis secundus
    d. pedis tertius
diglossia
dignathus
digynia
digyny
diiodotyrosine
dilatancy
dilatation
dilatator
    musculus d.
dilate
dilation
dilator
    d. iridis
    d. muscle of ileocecal sphincter
    d. muscle of nose
    d. muscle of pupil
    d. muscle of pylorus
    musculus d.
    d. naris
    d. of pupil
    d. pupillae
    d. pupillae muscle
    d. tubae
dimelia
dimension
    pelvic plane of greatest d.
    pelvic plane of least d.
    plane of least pelvic d.
dimerous
dimetria
dimidiata
    placenta d.
dimidiate hermaphroditism
dimple
    coccygeal d.
    postanal d.
Diogenes cup
diogenis
    poculum d.

diovular twins
diovulatory
DIP
    distal interphalangeal
      DIP joint
diphallus
diphasic
diphyodont
diploblastic
diplocardia
diplocephalus
diplocheiria
diplochiria
diploë
diploic
    d. artery
    d. canal
    d. vein
diploica
    d. frontalis
    d. occipitalis
    d. temporalis anterior
    d. temporalis posterior
    vena d.
diploici
    canales d.
diploid nucleus
diplokaryon
diplomyelia
diplonema
diploneural
diplopagus
diplopodia
diplosome
diplosomia
diplotene
dipodia
dip phenomenon
diprosopus
dipus
    acephalus d.
    sympus d.
dipygus
    dicephalus d.
direct
    d. nuclear division
    d. ovular transmigration
    d. pyramidal tract
direction
    pelvic d.
dirigation
dirigomotor
disaggregatio
disaggregation
disassimilation
disc *(var. of* disk)
discharge
discharging tubule
disci *(pl. of* discus)

disciform
discoblastic
discoblastula
discogastrula
discogenic
discoid
discoidal cleavage
discoplacenta
discus, pl. **disci**
    d. articularis
    d. articularis acromioclavicularis
    d. articularis radioulnaris
    d. articularis radioulnaris distalis
    d. articularis sternoclavicularis
    d. articularis temporomandibularis
    excavatio disci
    d. interpubicus
    d. intervertebralis
    d. lentiformis
    d. nervi optici
    d. proligerus
disdiaclast
disease
    Acosta d.
    caisson d.
    decompression d.
    Meniere d.
    Monge d.
dish face
disjunction
disjunctum
    stratum d.
disk, disc
    A d.'s
    acromioclavicular d.
    anisotropic d.'s
    anulus fibrosus of intervertebral d.
    articular d.
    blastodermic d.
    blood d.
    Bowman d.'s
    ciliary d.
    cone d.'s
    depression of optic d.
    embryonic d.
    excavation of optic d.
    fibrocartilaginous d.
    fibrous ring of intervertebral d.
    germ d.
    germinal d.
    H d.
    hair d.

    Hensen d.
    I d.
    interarticular d.
    intercalated d.
    intermediate d.
    interpubic d.
    intervertebral d.
    isotropic d.
    d. kidney
    mandibular d.
    Merkel tactile d.
    optic d.
    proligerous d.
    Q d.'s
    radioulnar articular d.
    Ranvier d.'s
    rod d.'s
    sacrococcygeal d.
    d. space
    sternoclavicular articular d.
    tactile d.
    temporomandibular articular d.
    transverse d.
    vertebral d.
    Z d.
disomic
disomy
disparate
disparity
dispermia
dispermy
disperse placenta
dispersion
    temporal d.
dispireme
displaceability
    tissue d.
displacement
    tissue d.
dissection
disseminated
dissepiment
Disse space
dissimilation
dissymmetry
distad
distal
    d. antrum
    d. bile duct
    d. centriole
    d. colon
    d. end

NOTES

D

distal *(continued)*
    d. end of radius
    d. esophagus
    d. femoral tensor
    d. fibula
    d. ileum
    d. interphalangeal (DIP)
    d. interphalangeal joint
    d. medial striate artery
    d. part of anterior lobe of
      hypophysis
    d. part of prostate
    d. part of prostatic urethra
    d. phalanx
    d. phalanx of foot
    d. phalanx of hand
    d. phalanx of thumb
    d. radioulnar articulation
    d. radioulnar joint
    d. spiral septum
    d. surface of tooth
    d. tibiofibular joint
    d. ulna
    d. ulnar collateral ligament
distalis
    arteria striata medialis d.
    articulatio radioulnaris d.
    discus articularis radioulnaris d.
    pars d.
    phalanx d.
    recessus sacciformis articulationis
      radioulnaris d.
    tuberositas phalangis d.
distance ceptor
distensibility
distension
distention
distichia
distichiasis
distomolar
distractionis
    lineae d.
distributing artery
distribution
    epidemiological d.
    exponential d.
    nerve d.
    d. volume
disymmetros
    cephalothoracopagus d.
diuresis
    alcohol d.
    osmotic d.
    water d.
diuretic
diurnal
    d. rhythm
divergence
divergent

diver's paralysis
diverticular hernia
diverticulum, pl. **diverticula**
    allantoenteric d.
    allantoic d.
    diverticula of ampulla of ductus
      deferens
    diverticula ampullae ductus
      deferentis
    cervical d.
    Heister d.
    Meckel d.
    metanephric d.
    Nuck d.
    pancreatic diverticula
    pituitary d.
    Rathke d.
    thyroglossal d.
    thyroid d.
    urethral d.
    ventricular d.
dividens
    crista d.
diving reflex
divisio
    d. autonomica systematis nervosi
      peripherici
    d. lateralis dextra hepatis
    d. lateralis sinistra
    d. lateralis sinistra hepatis
    d. medialis dextra hepatis
    d. medialis sinistra hepatis
division
    anterior primary d.
    cleavage d.
    direct nuclear d.
    ear d.
    equation d.
    indirect nuclear d.
    meiotic d.
    mitotic d.
    posterior primary d.
    reduction d.
divisiones
    d. anteriores plexus brachialis
    d. posteriores plexus brachialis
divisum
    pancreas d.
dizygotic twins
dizygous
doctrine
    Monro d.
    Monro-Kellie d.
Dogiel
    D. cells
    D. corpuscle
dolichocephalic
dolichocephalism
dolichocephalous

dolichocephaly
dolichocolon
dolichocranial
dolichoectatic artery
dolichofacial
dolichopellic pelvis
dolichopelvic
dolichoprosopic
dolichoprosopous
dolichostenomelia
dome
    bladder d.
    d. of bladder
    d. cell
    d. of diaphragm
    patellar d.
    d. of pleura
dominant hemisphere
Donders
    D. pressure
    space of D.
dopamine (DA)
    d. receptor
dopaminergic
Dorello canal
dorsa (*pl. of* dorsum)
dorsabdominal
dorsad
dorsal
    d. accessory olivary nucleus
    d. aponeurotic fascia
    d. artery of clitoris
    d. artery of foot
    d. artery of nose
    d. artery of penis
    d. branch
    d. branch of first and second
     posterior intercostal artery
    d. branch of lumbar artery
    d. branch of subcostal artery
    d. branch of superior intercostal
     artery
    d. branch of ulnar nerve
    d. calcaneocuboid ligament
    d. callosal vein
    d. carpal arcuate ligament
    d. carpal arterial arch
    d. carpal branch of radial artery
    d. carpal branch of ulnar artery
    d. carpal network
    d. carpal tendinous sheath
    d. carpometacarpal ligament

    d. column of spinal cord
    d. cornu
    d. cuboideonavicular ligament
    d. cuneocuboid ligament
    d. cuneonavicular ligament
    d. digital artery
    d. digital nerve of deep fibular
    nerve
    d. digital nerve of foot
    d. digital nerve of hand
    d. digital nerve of superficial
    fibular nerve
    d. digital nerve of ulnar nerve
    d. digital vein of foot
    d. digital vein of toe
    d. expansion of hand
    d. fascia of foot
    d. fascia of hand
    d. flex
    d. flexure
    d. funiculus
    d. hood
    d. horn
    d. intercarpal ligament
    d. intercuneiform ligament
    d. interossei of foot
    d. interossei of hand
    d. interosseous artery
    d. interosseous fascia
    d. interosseous muscle
    d. interosseous muscle of foot
    d. interosseous muscle of hand
    d. interosseous nerve
    d. lateral cutaneous branch of
    ramus
    d. lateral cutaneous nerve
    d. lateral muscular branch of
    ramus
    d. lingual branch of lingual artery
    d. lingual vein
    d. longitudinal fasciculus
    d. medial cutaneous branch of
    ramus
    d. medial cutaneous nerve
    d. mesocardium
    d. mesogastrium
    d. metacarpal artery
    d. metacarpal ligament
    d. metacarpal vein
    d. metatarsal artery
    d. metatarsal ligament
    d. metatarsal vein

D

NOTES

dorsal *(continued)*
- d. motor nucleus of vagus
- d. nasal branch of ophthalmic artery
- d. nerve of clitoris
- d. nerve of penis
- d. nerve root
- d. nerve of scapula
- d. nerve of toe
- d. nucleus
- d. pancreas
- d. pancreatic artery
- d. part of intertransversarii laterales lumborum muscle
- d. part of pons
- d. plate of neural tube
- d. primary ramus of spinal nerve
- d. radiocarpal ligament
- d. radius tubercle
- d. rami nerve
- d. rami of vertebra
- d. reflex
- d. region
- d. retinaculum
- d. root ganglion
- d. root of spinal nerve
- d. sacrococcygeal muscle
- d. sacrococcygeus muscle
- d. sacroiliac ligament
- d. scapular artery
- d. scapular nerve
- d. scapular vein
- d. sensory branch of ulnar nerve
- d. spine
- d. spinocerebellar tract
- d. surface
- d. surface of digit of hand or foot
- d. surface of sacrum
- d. surface of scapula
- d. talonavicular bone
- d. talonavicular ligament
- d. tarsal ligament
- d. tarsometatarsal ligament
- d. tegmental decussation
- d. thoracic artery
- d. tubercle of radius
- d. vein of clitoris
- d. vein of corpus callosum
- d. vein of penis
- d. vein of tongue
- d. venous arch of foot
- d. venous network of foot
- d. venous network of hand
- d. venous rete of hand
- d. vertebra

dorsale
- ligamentum carpi d.
- ligamentum cuboideonaviculare d.
- ligamentum cuneocuboideum d.
- ligamentum radiocarpale d.
- rete carpale d.
- rete carpi d.
- tuber d.
- tuberculum d.

dorsales (*pl. of* dorsalis)

dorsalia
- foramina sacralia d.
- ligamenta carpometacarpalia d.
- ligamenta cuneonavicularia d.
- ligamenta intercuneiformia d.
- ligamenta metacarpalia d.
- ligamenta metatarsalia d.
- ligamenta tarsi d.
- ligamenta tarsometatarsalia d.
- ligamentum carpometacarpalia d.
- ligamentum cuneonavicularia d.
- ligamentum intercarpalia d.
- ligamentum intercuneiformia d.
- ligamentum metacarpalia d.
- ligamentum metatarsalia d.
- ligamentum tarsi d.

dorsalia/palmaria
- ligamenta carpometacarpalia d.

dorsalis, pl. **dorsales**
- arteria digitalis d.
- arteria metacarpalis d.
- arteria metacarpea d.
- arteria metatarsalis d.
- arteria metatarsea d.
- arteria pancreatica d.
- arteria scapularis d.
- chorda d.
- d. clitoridis
- d. clitoridis profunda
- dorsales clitoridis superficiales
- facies d.
- fasciculus longitudinalis d.
- funiculus d.
- interosseous d.
- lamina d.
- dorsales linguae
- metacarpeae dorsales
- musculus sacrococcygeus d.
- d. nasi
- nervi digitales dorsales
- nervus interosseus d.
- nucleus d.
- d. pedis
- d. pedis artery
- d. pedis pulse
- d. pedis vein
- d. penis profunda
- d. penis superficiales
- radix d.
- ramus corporis callosi d.
- regiones dorsales
- sacrococcygeus d.

d. scapulae
sorius d.
spina d.
vena corporis callosi d.
venae metacarpeae dorsales
venae metatarseae dorsales
vena scapularis d.

dorsalium
vaginae tendinum carpalium d.

dorsi
bursa subtendinea musculi
latissimus d.
latissimus d.
musculi d.
musculus iliocostalis d.
musculus latissimus d.
musculus longissimus d.
musculus semispinalis d.
musculus spinalis d.
regiones d.
subtendinous bursa of latissimus d.

dorsiduct

dorsiflexion
d. of ankle
d. of foot

dorsiflexor compartment of leg

dorsiscapular

dorsispinal vein

dorsocephalad

dorsocuboidal reflex

dorsolateral
d. fasciculus
d. tract

dorsolateralis
fasciculus d.
tractus d.

dorsolumbar

dorsomedial
d. hypothalamic nucleus

dorsopancreaticus
ductus d.

dorsoventrad

dorsum, pl. **dorsa**
chorda dorsa
d. ephipii
d. of foot
d. of hand
d. linguae
d. manus
d. nasi
d. of nose
d. pedis

d. penis
d. rotundum
d. scapulae
d. sella
d. sellae
d. of tongue

double
d. aortic stenosis
d. chin
d. compartment hydrocephalus
d. penis

double-mouthed uterus

Douglas
D. bag
cul-de-sac of D.
D. cul-de-sac
D. fold
D. line
pouch of D.
D. pouch

douglasi
cavum d.

douloureux
tic d.

down
malignant d.

Downey cell

downy hair

Doyère eminence

drag
solvent d.

Dräger respirometer

drepanocyte

Dreyer formula

dromic

dromotropic
negatively d.
positively d.

drumhead

drum membrane

Drummond
artery of D.

drumstick appendage

dry vomiting

DTR
deep tendon reflex

DuBois formula

Du Bois-Reymond law

duct
abdominal part of thoracic d.
aberrant bile d.
accessory hepatic d.

D

NOTES

duct *(continued)*

accessory pancreatic d.
alveolar d.
amniotic d.
ampulla of lactiferous d.
ampulla of milk d.
ampullary crest of semicircular d.
ampullary crura of semicircular d.
ampullary limbs of
  semicircular d.'s
anal d.'s
arch of thoracic d.
arterial d.
Bartholin d.
basal crest of cochlear d.
basal lamina of semicircular d.
basal membrane of semicircular d.
basilar crest of cochlear d.
basilar membrane of cochlear d.
Bellini d.'s
Bernard d.
bile d.
biliary d.
Blasius d.
Botallo d.
branchial d.
bucconeural d.
d. of bulbourethral gland
canalicular d.
carotid d.
cervical part of thoracic d.
choledoch d.
cochlear d.
common bile d.
common crus of semicircular d.
common hepatic d.
common limb of membranous
  semicircular d.'s
craniopharyngeal d.
crista of semicircular d.
cupular cecum of cochlear d.
cupula of semicircular d.
Cuvier d.'s
d. of Cuvier
cystic d.
cystic gall d.
deferent d.
distal bile d.
eccrine d.
efferent d.
ejaculation d.
ejaculatory d.
endolymphatic d.
d. of epididymis
epithelium of semicircular d.
excretory d.
external surface of cochlear d.
external wall of cochlear d.
extrahepatic bile d.

frontonasal d.
galactophorous d.
gall d.
Gartner d.
genital d.
gland of common bile d.
guttural d.
hemithoracic d.
Hensen d.
hepatic d.
hepatocystic d.
Hoffmann d.
hypophysial d.
incisive d.
inferior lacrimal d.
intercalated d.'s
interlobar d.
interlobular d.
intrahepatic biliary d.
intralobular d.
jugular d.
lacrimal d.
lacrimonasal d.
lactiferous d.
left hepatic d.
d. lumen
Luschka d.'s
lymphatic d.
main pancreatic d. (MPD)
major sublingual d.
mamillary d.
mammary d.
membrana propria of
  semicircular d.
membranous ampullae of the
  semicircular d.
membranous limb of
  semicircular d.
mesonephric d.
metanephric d.
milk d.
minor sublingual d.
Müller d.
müllerian d.
nasal d.
nasofrontal d.
nasolacrimal d.
nephric d.
omphalomesenteric d.
opening of left parotid d.
opening of papillary d.
ovarian d.
pancreatic d.
papillary d.
paramesonephric d.
paraurethral d.
parotid d.
Pecquet d.
perilymphatic d.

periotic d.
pharyngobranchial d.'s
posterior branch of right hepatic d.
prepapillary bile d.
pronephric d.
proper membrane of semicircular d.
prostatic d.
reuniens d.
right hepatic d.
right lymphatic d.
Rivinus d.
saccular d.
salivary d.
Santorini d.
Schüller d.
secretory d.
semicircular d.
seminal d.
simple crus of semicircular d.
Skene d.
d. of Skene gland
spermatic d.
sphincter muscle of common
  bile d.
sphincter muscle of pancreatic d.
spiral crest of cochlear d.
spiral fold of cystic d.
spiral ligament of cochlear d.
spiral prominence of cochlear d.
spiral valve of cystic d.
Steno d.
Stensen d.
striated d.
stria vascularis of cochlear d.
subclavian d.
sublingual d.
submandibular d.
submaxillary d.
subvesical d.
sudoriferous d.
superior lacrimal d.
sweat d.
d. of sweat gland
tectorial membrane of cochlear d.
terminal bile d.
testicular d.
thoracic part of thoracic d.
thymic d.
thyroglossal d.
thyrolingual d.
tympanic surface of cochlear d.
tympanic wall of cochlear d.

uniting d.
utricular d.
utriculosaccular d.
vein of semicircular d.
vestibular cecum of the
  cochlear d.
vestibular surface of cochlear d.
vestibular wall of cochlear d.
vitelline d.
vitellointestinal d.
Walther d.
Wharton d.
d. of Wirsung
Wirsung pancreatic d.
wolffian d.

ductal sinus
ductibus
  glandulae sine d.
duction
ductless gland
ductular
ductule
  aberrant d.
  biliary d.
  inferior aberrant d.
  interlobular d.
  prostatic d.
  superior aberrant d.
ductulus, pl. **ductuli**
  d. aberrans
  d. aberrans inferior
  d. aberrans inferius
  d. aberrans superior
  d. aberrans superius
  ductuli aberrantes
  d. alveolaris
  ductuli biliferi
  d. efferens testis
  ductuli efferentes testis
  ductuli excretorii glandulae
  ductuli excretorii glandulae
    lacrimalis
  ductuli interlobulares
  ductuli paroophori
  ductuli prostatici
  ductuli transversi epoöphori
ductus, pl. **ductus**
  d. aberrans
  d. aberrantes
  d. arteriosus
  d. biliaris
  d. biliferi

D

NOTES

ductus *(continued)*
 d. caroticus
 d. choledochus
 d. cochlearis
 d. cysticus
 d. deferens vestigialis
 d. dorsopancreaticus
 d. ejaculatorius
 d. endolymphaticus
 d. epididymidis
 d. excretorius
 d. excretorius glandulae vesiculosae
 d. excretorius vesiculae seminalis
 d. glandulae bulbourethralis
 d. hemithoracicus
 d. hepaticus communis
 d. hepaticus dexter
 d. hepaticus sinister
 d. incisivus
 d. lacrimalis
 d. lactiferi
 d. lingualis
 d. lobi caudati dexter
 d. lobi caudati dexter hepatis
 d. lobi caudati sinister
 d. lobi caudati sinister hepatis
 d. longitudinalis epoöphori
 d. lymphaticus dexter
 d. mesonephricus
 d. nasolacrimalis
 d. omphalomesentericus
 d. pancreaticus
 d. pancreaticus accessorius
 d. paramesonephricus
 d. paraurethrales
 d. parotideus
 d. perilymphatici
 d. perilymphaticus
 d. pharyngobranchialis III, IV
 d. prostatici
 d. reuniens
 d. saccularis
 d. semicirculares
 d. semicircularis ante
 d. semicircularis lateralis
 d. semicircularis posterior
 d. sublinguales minores
 d. sublingualis major
 d. submandibularis
 d. submaxillaris
 d. sudoriferus
 d. thoracicus
 d. thoracicus dexter
 d. thyroglossus
 d. utricularis
 d. utriculosaccularis
 d. venosus
 d. venosus arantii
ductuum semicircularium

Duddell membrane
dumping stomach
Duncan
 D. folds
 D. ventricle
duodena (*pl. of* duodenum)
duodenal
 d. ampulla
 d. branch of anterior superior pancreaticoduodenal artery
 d. branch of posterior superior pancreaticoduodenal artery
 d. bulb
 d. cap
 d. fold
 d. fossae
 d. gland
 d. impression
 d. impression on liver
 d. lumen
 d. mucosa
 d. opening
 d. sphincter
 d. vein
 d. villus
duodenalis, pl. **duodenales**
 glandulae duodenales
 impressio duodenales
 rami duodenales
duodeni
 ampulla d.
 bulbus d.
 ligamentum suspensorium d.
 musculus suspensorius d.
 pars coeliacoduodenalis musculi ligamenti suspensorii d.
 pars descendens d.
 pars horizontalis d.
 pars inferior d.
 pars prima d.
 pars secundum d.
 pars superior d.
 pars tecta d.
 plica longitudinalis d.
duodenocolic ligament
duodenojejunal
 d. angle
 d. flexure
 d. fold
 d. fossa
 d. junction
 d. recess
 d. sphincter
duodenojejunalis
 flexura d.
 plica d.
duodenomesocolica
 plica d.
duodenomesocolic fold

duodenorenale
    ligamentum d.
duodenorenal ligament
duodenum, pl. **duodena**
    ampulla of d.
    ascending part of d.
    celiacoduodenal part of suspensory
      muscle of d.
    C-loop of d.
    descending part of d.
    first part of d.
    flexure of d.
    hidden part of d.
    horizontal part of d.
    inferior flexure of d.
    inferior part of d.
    longitudinal fold of d.
    pars ascendens dudeni
    phrenicoceliac part of suspensory
      muscle ligament of d.
    retroperitoneal part of d.
    second part of d.
    sphincter of third portion of d.
    superior flexure of d.
    superior part of d.
    suspensory ligament of d.
    suspensory muscle of d.
    third part of d.
duplex
    ileum d.
    placenta d.
    d. transmission
    uterus d.
    d. uterus
duplication cyst
duplicitas anterior
Dupré muscle
Dupuytren
    D. canal
    D. contracture
    D. fascia
dura
    lamina d.
    leaves of d.
    d. mater
    d. mater of brain
    d. mater cranialis
    d. mater encephali
    d. mater of spinal cord
    d. mater spinalis
dural
    d. part of filum terminale

    d. sac
    d. sheath
    d. sheath of optic nerve
    d. sinuses
duramatral
durum
    clavus d.
    palatine d.
    palatum d.
dust cell
Duverney
    D. fissure
    D. foramen
    D. gland
    D. muscle
dwarf pelvis
dynamic
    d. compliance of lung
    d. force
    d. friction
    d. viscosity
dynamogenesis
dynamogenic
dynamogeny
dynamograph
dynamometer
dynamoscope
dynamoscopy
dynatherm
dyne
dynein arm
dysbarism
dyscephalia
    d. mandibulo-oculofacialis
dyscephaly
dyschondrogenesis
dyschondroplasia
    d. with hemangiomas
dysdiadochokinesia, dysdiadochocinesia
dysembryoplasia
dysencephalia splanchnocystica
dysergia
dysfunction
dysgenesis
    gonadal d.
    iridocorneal mesodermal d.
dysgranular cortex
dyskaryosis
dyskaryotic
dyskinesia
    extrapyramidal d.'s
dysmature

NOTES

dysmelia
dysmorphia
    mandibulo-oculofacial d.
dysmorphism
dysmorphogenesis
dysmorphology
dysontogenesis
dysontogenetic
dysosteogenesis
dysostosis
    acrofacial d.
    cleidocranial d.
    clidocranial d.
    craniofacial d.
    mandibuloacral d.
    mandibulofacial d.
    metaphysial d.
    orodigitofacial d.
    otomandibular d.
    peripheral d.
dyspallia
dysplasia
    anterofacial d.
    anteroposterior facial d.
    atriodigital d.
    chondroectodermal d.
    cleidocranial d.
    clidocranial d.
    congenital ectodermal d.
    craniocarpotarsal d.
    craniodiaphysial d.
    craniometaphysial d.
    ectodermal d.

    encephalo-ophthalmic d.
    d. epiphysialis hemimelia
    d. epiphysialis multiplex
    d. epiphysialis punctata
    hidrotic ectodermal d.
    mandibulofacial d.
    metaphysial d.
    multiple epiphysial d.
    oculoauriculovertebral d.
    oculodentodigital d.
    oculovertebral d.
    pseudoachondroplastic
      spondyloepiphysial d.
    ventriculoradial d.
dysplastic
dysraphia
dysraphicus
    status d.
dysraphism
dyssomnia
dysspondylism
dyssynergia
    detrusor sphincter d.
dystonia
dystonic
dystopia
    d. transversa externa testis
    d. transversa interna testis
dystopic
dystrophy
    craniocarpotarsal d.
dysversion

ear

anterior notch of e.
anterior wall of middle e.
e. bone
both e.'s (AU, au)
e. canal
carotid wall of middle e.
cartilage of e.
e. cartilage
e. cavity
cavity of middle e.
concha of e.
e. crystals
e. division
external e.
inner e.
internal e.
isthmus of cartilage of e.
jugular wall of middle e.
labyrinthine wall of middle e.
lateral wall of middle e.
e. lobe
lop e.
mastoid wall of middle e.
medial wall of middle e.
membranous wall of middle e.
middle e.
e.'s, nose, and throat (ENT)
e. pinna
posterior wall of middle e.
right e. (AD)
saccule of e.
tegmental wall of middle e.
tip of e.
vessel of internal e.
vestibule of inner e.

eardrum
earlobe crease
earwax
Eberth perithelium
Ebner

E. gland
incremental lines of von E.
E. reticulum

eboris

membrana e.

ebullism
ebur dentis
eburnea

substantia e.

eburneous
eccentric
eccentropiesis

ecchondrosis

e. physaliformis
e. physaliphora

eccrine

e. duct
e. gland

eccrinology
eccrisis
eccritic
echinococcus cyst
echinocyte
Ecker fissure
ecological system
economy
ecosystem
ECRB

extensor carpi radialis brevis

ECRL

extensor carpi radialis longus

ecstrophe
ectad
ectal origin
ectental
ectethmoid
ectiris
ectoblast
ectocardia
ectocardiac
ectocardial
ectocervical
ectochoroidea
ectocornea
ectocrine
ectoderm

epithelial e.
superficial e.

ectodermal dysplasia
ectodermic
ectoentad
ectoental
ectoethmoid
ectogenic teratosis
ectogenous
ectomeninx
ectomere
ectomesenchyme
ectomorph
ectomorphic
ectopagus
ectopia

e. cloacae
e. cordis
e. lentis
e. pupillae congenita
e. renis

E

ectopia *(continued)*
    e. testis
    e. vesicae
ectopic
    e. teratosis
    e. testis
ectoplacental cavity
ectoplasm
ectoplasmatic
ectopy
ectoretina
ectosteal
ectotrophoblastic cavity
ectrocheiry
ectrochiry
ectrodactylia
ectrodactylism
ectrodactyly
ectrogenic
ectrogeny
ectromelia
ectromelic
ectropody
ectrosyndactyly
ectype
ECU
    extensor carpi ulnaris
ecuresis
ecussation
EDB
    extensor digitorum brevis
Eden-Lange procedure
edge
    cutting e.
    incisal e.
    ligament reflecting e.
    ligament shelving e.
    liver e.
    ossified e.
    shearing e.
Edinger-Westphal nucleus
EDL
    extensor digitorum longus
EDQ
    extensor digiti quinti
EDRF
    endothelium-derived relaxing factor
effect
    additive e.
    Anrep e.
    Bernoulli e.
    Bowditch e.
    clasp-knife e.
    Crabtree e.
    Cushing e.
    Fahraeus-Lindqvist e.
    Fenn e.
    Haldane e.
    Orbeli e.

    Pasteur e.
    second gas e.
    sigma e.
    Venturi e.
    Wedensky e.
effective
    e. conjugate
    e. osmotic pressure
    e. refractory period
    e. renal blood flow
    e. renal plasma flow
    e. temperature index
effector cell
efferens
    arteriola glomerularis e.
    vas lymphaticum e.
efferent
    e. artery
    e. duct
    e. ductule of testis
    gamma e.
    e. glomerular arteriole
    e. limb
    e. loop
    e. lymphatic
    e. nerve
    e. neuron
    e. vessel
efferentia
    vasa e.
efficiency
egg
    e. cell
    e. cluster
    e. membrane
Eglis gland
EHL
    extensor hallucis longus
Ehrenritter ganglion
eighth cranial nerve
EIP
    extensor indicis proprius
eisodic
ejaculate
ejaculatio
ejaculation
    e. center
    e. duct
ejaculatorius
    ductus e.
ejaculatory duct
ejecta
ejection
    e. fraction
    e. fraction systolic
ejector
ektoplasmic
ektoplastic
elastance

elastic
  e. artery
  e. cartilage
  e. cone
  e. fibers
  e. lamella
  e. laminae of arteries
  e. layers of arteries
  e. layers of cornea
  e. limit
  e. membrane
  e. tissue
elastica
  tela e.
  tunica e.
elasticity
  modulus of volume e.
elasticus
  conus e.
elastometer
Elaut triangle
elbow
  anterior region of e.
  anterior surface of e.
  articular muscle of e.
  articular vascular network of e.
  e. bone
  interosseous bursa of e.
  intratendinous bursa of e.
  e. jerk
  e. joint
  lateral ligament of e.
  lymph node of e.
  medial collateral ligament of e.
  medial ligament of e.
  point of e.
  posterior region of e.
  posterior surface of e.
  e. reflex
  e. structure
  tip of e.
  transverse ligament of e.
  triangle of e.
elbowed
electrical formula
electric irritability
electroaxonography
electrobioscopy
electrocontractility
electroconvulsive
electrodermal
electrogastrograph

electrogram
electrohysterograph
electrolyte metabolism
electromagnetic
  e. flowmeter
  e. unit
electromotive force
electromuscular sensibility
electromyogram
electromyograph
electromyography (EMG)
electron
  e. interferometer
  e. interferometry
electroneurography
electroneuromyography
electroolfactogram
electrophysiology
electrospinogram
electrospinography
electrostatic unit (esu)
electrostenolysis
electrotaxis
  negative e.
  positive e.
electrotonic
  e. current
  e. junction
  e. synapse
electrotonus
electrotropism
element
  anatomical e.
  extrachromosomal genetic e.
  labile e.'s
  morphologic e.
elementary
  e. bodies
  e. particle
elevata
  scapula e.
elevation
  frontonasal e.
  lateral nasal e.
  e. of levator palati
  medial nasal e.
  tactile e.'s
elevator
  e. muscle of anus
  e. muscle of prostate
  e. muscle of rib
  e. muscle of scapula

E

NOTES

elevator (*continued*)
    e. muscle of soft palate
    e. muscle of thyroid gland
    e. muscle of upper eyelid
    e. muscle of upper lip
    e. muscle of upper lip and wing
      of nose
eleventh cranial nerve
elimination
    carbon dioxide e.
Elliott law
ellipsoidal
    e. articulation
    e. joint
ellipsoidea
    articulatio e.
ellipsoid joint
elliptica
    fovea e.
elliptical
    e. recess
    e. recess of bony labyrinth
ellipticus
    recessus e.
elliptocyte
emanation
emancipation
emarginate
emargination
Embden-Meyerhof-Parnas pathway
Embden-Meyerhof pathway
embed
embole
emboli (*pl. of* embolus)
embolia
emboliformis
    nucleus e.
emboliform nucleus
embolus, pl. **emboli**
emboly
embryo
    heterogametic e.
    homogametic e.
    maxillary process of e.
    presomite e.
    previllous e.
    e. transfer
embryoblast
embryogenesis
embryogenetic
embryogenic
embryogeny
embryoid
embryologist
embryology
embryomorphous
embryonal area
embryonate

embryonic
    e. anideus
    e. area
    e. axis
    e. blastoderm
    e. circulation
    e. diapause
    e. disk
    e. shield
embryoniform
embryonization
embryonoid
embryony
embryopathy
embryoplastic
embryotoxicity
embryotoxon
embryotroph
embryotrophic
embryotrophy
emeiocytosis
emergency theory
emergent
EMG
    electromyography
emiction
eminence
    arcuate e.
    articular e.
    canine e.
    collateral e.
    e. of concha
    cruciate e.
    cruciform e.
    deltoid e.
    Doyère e.
    facial e.
    forebrain e.
    frontal e.
    genital e.
    hypobranchial e.
    hypoglossal e.
    hypothenar e.
    ileocecal e.
    iliopectineal e.
    iliopubic e.
    intercondylar e.
    intercondyloid e.
    intertubercular e.
    malar e.
    maxillary e.
    medial e.
    median e.
    nasal e.
    olivary e.
    orbital e.
    parietal e.
    pyramidal e.
    restiform e.

round e.
e. of scapha
thenar e.
thyroid e.
e. of triangular fossa
e. of triangular fossa of auricle
eminentia, pl. **eminentiae**
    e. abducentis
    e. arcuata
    e. articularis
    e. articularis ossis temporalis
    e. carpi radialis
    e. carpi ulnaris
    e. collateralis
    e. conchae
    e. cruciformis
    e. facialis
    e. fossae triangularis
    e. frontalis
    e. hypoglossi
    e. hypothenaris
    e. iliopubica
    e. intercondylaris
    e. intercondyloidea
    e. maxillae
    e. medialis
    e. mediana
    e. orbitalis
    e. orbitalis ossis zygomatici
    e. parietalis
    e. pyramidalis
    e. restiformis
    e. scaphae
    e. symphysis
    e. teres
    e. thenaris
emiocytosis
emissaria
    e. mastoidea
    e. occipitalis
    e. parietalis
    vena e.
emissariae
    venae e.
emissarium
    e. condyloideum
    e. mastoideum
    e. occipitale
    e. parietale
emissary
    e. sphenoidal foramen
    e. vein

emission
    nocturnal e.
emotiovascular
emulgent
emuresis
enamel
    e. cell
    e. crypt
    e. cuticle
    e. fibers
    e. germ
    e. hypocalcification
    e. layer
    e. membrane
    e. niche
    e. organ
    e. prisms
    e. rod
    e. rod sheath
enamelogenesis
enameloma
enamelum
enantiomorphism
enarthrodial joint
enarthrosis
encapsulated
encapsulation
encapsuled
encephala
encephali (*pl. of* encephalus)
encephalic vesicle
encephalization
encephalocele
encephaloclastic microcephaly
encephalodysplasia
encephalology
encephalomeningocele
encephalomere
encephalometer
encephalomyelocele
encephalomyelonic axis
encephalon
encephalo-ophthalmic dysplasia
encephalorrhachidian
encephaloschisis
encephalospinal
encephalus, pl. **encephali**
    arachnoidea encephali
    arachnoid mater encephali
    arteriae encephali
    dura mater encephali
enchondral

E

NOTES

enchondromatosis
enclave
encranial
encranius
encysted hernia
end
 e. artery
 e. bud
 e. bulb
 distal e.
 fixed e.
 mobile e.
 e. organ
 e. piece
 e. plate
endaural
endbrain
end-brush
end-feet
endgut
ending
 annulospiral e.
 calyciform e.
 epilemmal e.
 flower-spray e.
 free nerve e.'s
 grape e.'s
 hederiform e.
 nerve e.
 sole-plate e.
 synaptic e.'s
endoabdominal fascia
endoabdominalis
 fascia e.
endobasion
endoblast
endobronchial
endocardia
endocardiac
endocardial
 e. cushion defect
 e. cushions
 e. fibroelastosis
 e. sclerosis
endocardium
endoceliac
endocervical
 e. gland
 e. mucosa
 e. os
endocervix
endocholedochal
endochondral
 e. bone
 e. ossification
endocranial
endocranium
endocrinae
 glandulae e.

endocrine
 e. gland
 e. part of pancreas
 e. system
endocyma
endocytosis
endoderm
endodermal
 e. canal
 e. cells
 e. cloaca
 e. pouch
 e. sinus
endogamy
endogastric
endogenic
endogenous
 e. creatinine clearance
 e. fibers
endognathion
endolaryngeal
endolymph
endolympha
endolymphatic
 e. cavity
 e. duct
 e. fluid
 e. hydrops
 e. sac
 e. space
endolymphaticum
 spatium e.
endolymphaticus
 ductus e.
 sacculus e.
 saccus e.
endolymphic
endomeninx
endometria (*pl. of* endometrium)
endometrial
 e. canal
 e. tissue
endometrioid
endometriosis
 celomic metaplasia theory of e.
endometrium, pl. endometria
endometropic
endomorph
endomorphic
endomyocardial fibroelastosis
endomysium
endoneurium
endonucleolus
endopelvicae
 lamina retrorectalis fasciae e.
endopelvic fascia
endopelvina
 fascia e.
endopericardiac

endoplasm
endoplasmic reticulum
endoplastic
endorphinergic
endorrhachis
endoskeleton
endosmosis
endosteal
endosteum
endotendineum
endothelia (*pl. of* endothelium)
endothelial
    e. cell
    e. lining
    e. tissue
endothelialization
endothelin
endotheliochorial placenta
endothelio-endothelial placenta
endothelioid
endothelium, pl. endothelia
    e. of anterior chamber
    e. camerae anterioris
    e. posterius corneae
endothelium-derived relaxing factor
  (EDRF)
endothoracica
    fascia e.
endothoracic fascia
endotracheal
endovenosum
    septum e.
endovenous septum
endpiece
endplate
    motor e.
    e. of vertebrae
    vertebral e.
end-tidal sample
endyma
energetics
energometer
energy
    conservation of e.
    free e.
    Gibbs free e.
    kinetic e.
    latent e.
    law of specific nerve e.
    nutritional e.
    e. of position
    potential e.

    total e.
    unit of e.
enervation
engastrius
engineering
Englisch sinus
engorged
engorgement
enhancement
enkephalinergic
enlargement
    cervical e.
    lumbosacral e.
enorganic
ensiform
    e. appendix
    e. cart
    e. cartilage
    e. process
ensisternum cartilage
ENT
    ears, nose, and throat
ental
    e. origin
    e. tract
enteric
    e. nervous plexus
entericus
    liquor e.
    plexus e.
    plexus nervosus e.
enterocele
enterochromaffin cells
enteroendocrine cells
enterogastric reflex
enterograph
enterography
enterohepatic circulation
enterohepatocele
enterokinesis
enterokinetic
enteromegalia
enteromegaly
enterorenal
enterotropic
enthalpy
enthetic
entoblast
entochoroidea
entocornea
entocranial
entocranium

E

NOTES

entoderm
entodermal cells
entoectad
entomion
entopic
entoplasm
entoretina
entorhinal area
entry zone
entypy
enucleation
envelope
    cistern of nuclear e.
    corneocyte e.
    nuclear e.
enzygotic twins
enzyme inhibition theory of narcosis
EOM
    extraocular motion
    extraocular movement
    extraocular muscle
eosinocyte
eosinopenic reaction
eosinophile
eosinophil granule
eosinophilic leukocyte
eosinotactic
eosinotaxis
epactal
    e. bone
    e. ossicle
epamniotic cavity
eparterial bronchus
epaxial
EPB
    extensor pollicis brevis
ependyma
ependymal
    e. cell
    e. layer
    e. zone
ependymoblast
ependymoblastoma
ependymocyte
ependymoma
ephapse
ephaptic
ephipii
    dorsum e.
epiblast
epiblastic
epiblepharon
epibole
epiboly
epibranchial
epibulbar
epicanthal fold
epicanthic fold
epicanthus

epicardia
epicardial
    e. coronary artery
    e. space
    e. surface
epicardium
epichordal
epicomus
epicondylar ridge
epicondyle
    lateral e.
    medial e.
epicondyli
epicondylian
epicondylic
epicondylus
    e. lateralis
    e. lateralis humeri
    e. lateralis ossis femoris
    e. medialis
    e. medialis humeri
    e. medialis ossis femoris
epicoracoid
epicranial
    e. aponeurosis
    e. aponeurosis of scalp
    e. muscle
epicranialis
    aponeurosis e.
epicranium
epicranius
    e. muscle
    musculus e.
epicritic sensorium
epidemiological distribution
epiderm
epiderma
epidermal
    e. growth factor
    e. ridge
epidermatic
epidermic cell
epidermidis
    stratum corneum e.
    stratum granulosum e.
    stratum spinosum e.
epidermis, pl. **epidermides**
    clear layer of e.
    corneal layer of e.
    granular layer of e.
    horny layer of e.
epidermization
epididymal
epididymidis
    appendix e.
    appendix of e.
    caput e.
    cauda e.
    coni e.

corpus e.
ductus e.
globus major e.
globus minor e.
ligamentum e.
lobuli e.
sinus e.
epididymis, pl. **epididymides**
appendix of e.
body of e.
conical lobules of e.
duct of e.
head of e.
inferior ligament of e.
ligament of e.
lobules of e.
sinus of e.
superior ligament of e.
tail of e.
epidural
e. cavity
e. space
epidurale
cavum e.
epigastric
e. angle
e. artery
e. fold
e. fossa
e. inferior vein
e. reflex
e. region
epigastrica
fossa e.
e. inferior
plica e.
regio e.
e. superficialis
e. superior
e. superiores
epigastrium
epigastrius
epigastrocele
epiglottic
e. cartilage
e. fold
e. tubercle
e. vallecula
epiglottica, pl. **epiglotticae**
cartilago e.
plica e.
vallecula e.

epiglotticum
tuberculum e.
epiglottidean
epiglottidis
frenulum e.
petiolus e.
epiglottis
cushion of e.
depressor muscle of e.
stalk of e.
tubercle of e.
epignathus
epihyal ligament
epihyoid
epilamellar
epilemma
epilemmal ending
epimandibular
epimere
epimeric muscle
epimorphosis
epinephrine reversal
epinephros
epineural
epineurium
epionychium
epiotic center
epipapillary membrane
epipericardial ridge
epipharynx
epiphrenal
epiphrenic
epiphyseos
synchondrosis e.
epiphyses (*pl. of* epiphysis)
epiphysial, epiphyseal
e. arrest
e. cartilage
e. eye
e. line
e. plate
epiphysialis
cartilago e.
lamina e.
linea e.
epiphysiodesis
epiphysis, pl. **epiphyses**
atavistic e.
capital e.
e. cerebri
femoral e.
pressure e.

E

NOTES

epiphysis *(continued)*
    stippled e.
    traction e.
epipial
epiplocele
epiploic
    e. appendage
    e. appendix
    e. branch
    e. foramen
    e. foramen of Winslow
    e. tag
epiploica, pl. **epiploicae**
    appendix e.
    rami epiploicae
epiploici
    rami e.
epiploicum
    foramen e.
epiplomerocele
epiplomphalocele
epiploon
epiplosarcomphalocele
epiploscheocele
epipteric bone
epipygus
episclera
episcleral
    e. artery
    e. ganglion
    e. lamina
    e. layer of fibrous layer of eyeball
    e. space
    e. tissue
    e. vein
episclerale
    spatium e.
episcleralis, pl. **episclerales**
    arteria e.
    lamina e.
    venae episclerales
epispadia
epispadial
epispadias
epispinal
episternal bone
episternum
epistropheus
    odontoid process of e.
epitarsus
epitendineum
epitenon
epithalamus
epithelia *(pl. of* epithelium)
epithelial
    e. attachment
    e. body
    e. choroid layer

    e. ectoderm
    e. lamina
    e. lining
    e. plug
    e. reticular cell
    e. tissue
epithelialis
    lamina choroidea e.
epithelialization
epitheliochorial placenta
epitheliocyte
epitheliofibril
epithelioglandular
epithelioid cell
epithelium, pl. **epithelia**
    e. anterius corneae
    ciliated e.
    columnar e.
    crevicular e.
    cuboidal e.
    cylindrical e.
    e. ductus semicircularis
    external dental e.
    external enamel e.
    germinal e.
    gingival e.
    glandular e.
    inner dental e.
    inner enamel e.
    junctional e.
    laminated e.
    e. of lens
    e. lentis
    mesenchymal e.
    muscle e.
    olfactory e.
    pavement e.
    pigment e.
    pseudostratified e.
    reduced enamel e.
    respiratory e.
    e. of semicircular duct
    seminiferous e.
    simple squamous e.
    squamous e.
    stratified ciliated columnar e.
    stratified squamous e.
    sulcular e.
    suprarenal e.
    surface e.
    transitional e.
epithelization
epithesis, pl. **epitheses**
epitrichial layer
epitrichium
epitrochlea
epitrochlear node
epitrochleoanconaeus

epitrochleoanconeus
> e. muscle
> musculus e.

epitympanic
> e. cell
> e. recess
> e. space

epitympanici
> pars cupularis recessus e.

epitympanicum
> os e.

epitympanicus
> recessus e.

epitympanum

epivaginal connective tissue

EPL
> extensor pollicis longus

eponychial fold

eponychium

epoöphori
> ductuli transversi e.
> ductus longitudinalis e.
> tubuli e.

epoöphoron
> longitudinal duct of e.
> transverse ductule of e.
> vesicular appendages of e.

epulotic

equal cleavage

equation
> alveolar gas e.
> Bohr e.
> constant field e.
> e. division
> GHK e.
> Gibbs-Helmholtz e.
> Goldman e.
> Goldman-Hodgkin-Katz e.
> Nernst e.

equator
> e. bulbi oculi
> e. of eye
> e. of eyeball
> e. of lens
> e. lentis

equatorial cleavage

equiaxial

equicaloric

equilibration

equilibrium
> acid-base e.
> homeostatic e.

> nitrogenous e.
> nutritive e.
> physiologic e.

equina
> cauda e.

equinovalgus
> pes e.

equinovarus
> pes e.

equinus
> gastrocnemius e.

equivalent
> Joule e.
> metabolic e. (MET)
> nitrogen e.
> starch e.

erectile
> e. elements of the penis
> e. tissue

erection

erector
> e. muscle of hair
> e. muscle of spine
> e. spinae
> e. spinae muscle
> e. spinae tendon

ergastoplasm

ergodynamograph

ergoesthesiograph

ergogenic

ergograph
> Mosso e.

ergographic

ergometer

ergotropic

erigentes
> nervi e.

Erlenmeyer flask deformity

erratic
> mamma e.

erratica
> mamma e.

ertransversarii posteriores cervicis

erythremia
> altitude e.

erythroblast

erythrocyte

erythrocytic series

erythrocytoblast

erythrocytopoiesis

erythrogonia

erythrogonium

E

NOTES

erythroid cell
erythron
erythrophil
erythrophilic
erythropoiesis
erythropoietic
esodic nerve
esophagea
 constrictio bronchoaortica e.
 constrictio diaphragmatica e.
 constrictio partis thoracicae e.
 constrictio phrenica e.
 glandulae e.'s
 impressio e.
 venae e.'s
esophageal
 e. airway
 e. atresia
 e. branch
 e. branch of inferior thyroid artery
 e. branch of left gastric artery
 e. branch of recurrent laryngeal
  nerve
 e. branch of thoracic aorta
 e. branch of thoracic ganglion
 e. branch of vagus nerve
 e. gland
 e. groove
 e. hiatus
 e. impression
 e. impression on liver
 e. inlet
 e. introitus
 e. lumen
 e. mucosa
 e. nervous plexus
 e. opening
 e. sphincter
 e. vein
esophageales
 rami e.
esophageus, pl. **esophagei**
 hiatus e.
 plexus nervosus e.
 rami esophagei
esophagogastric
 e. mucosal junction
 e. orifice
 e. vestibule
esophagus, pl. **esophagi**
 abdominal e.
 area for e.
 cardiac gland of e.
 cervical e.
 diaphragmatic constriction of e.
 distal e.
 groove for e.
 impression of e.
 mucosa of e.

 mucous membrane of e.
 muscular coat of e.
 muscular layer of e.
 pars abdominalis esophagi
 pars cervicalis esophagi
 pars thoracica esophagi
 ridge formed by e.
 serosa of e.
 suspensory ligament of e.
 thoracic constriction of e.
 tunica mucosa esophagi
 tunica muscularis esophagi
 tunica serosa esophagi
 upper e.
 V-shaped area of e.
Essick cell bands
esthesia
esthesic
esthesiodic system
esthesiography
esthesiometer
esthesiometry
esthesiophysiology
estivation
estradiol
estrogenic hormone
estrous cycle
esu
 electrostatic unit
Et
 ethyl
ether test
ethmocranial
ethmofrontal
ethmoid
 e. air cell
 e. angle
 e. bone
 e. bulla
 e. infundibulum
 perpendicular plate of e.
ethmoidal
 e. artery
 e. bulla
 e. cell
 e. crest
 e. crest of maxilla
 e. crest of palatine bone
 e. fissure
 e. foramen
 e. fossa
 e. groove
 e. infundibulum
 e. labyrinth
 e. lamina cribrosa
 e. nerve
 e. notch
 e. process
 e. process of inferior nasal concha

e. sinus
e. vein
ethmoidale
antra e.
foramen e.
infundibulum e.
os e.
ethmoidalia
antra e.
ethmoidalis, pl. ethmoidales
e. anterior
bulla e.
cellulae ethmoidales
crista e.
fovea e.
hiatus e.
incisura e.
labyrinthus e.
lamina cribrosa ossis e.
lamina orbitalis ossis e.
lamina perpendicularis ossis e.
e. posterior
processus uncinatus ossis e.
sinus ethmoidales
sulcus e.
venae ethmoidales
ethmoidolacrimalis
sutura e.
ethmoidolacrimal suture
ethmoidomaxillaris
sutura e.
ethmoidomaxillary suture
ethmolacrimal
ethmomaxillary
ethmonasal
ethmopalatal
ethmosphenoid
ethmoturbinals
ethmovomerine plate
ethnology
ethyl (Et)
eubolism
eucapnia
eucaryote
eucaryotic
euchromatin
eucorticalism
eudiaphoresis
euhydration
eukaryote
eukaryotic
eukeratin

eukinesia
eumelanin
eumelanosome
eumetria
eumorphism
eunuchism
eunuchoidism
eupancreatism
eupepsia
eupeptic
euplasia
euplastic
eupnea
eupraxi
eupraxia
eurhythmia
eurycephalic
eurycephalous
eurygnathic
eurygnathism
eurygnathous
euryon
eurysomatic
euscope
eustachian
e. canal
e. cartilage
e. cushion
e. muscle
e. orifice
e. tonsil
e. tube
e. tuber
e. valve
eustachiana
tuba e.
eustachii
tuba e.
eusthenia
eusystole
eusystolic
euthermic
euthyroidism
eutonic
eutrophia
eutrophic
eutrophy
euvolia
evacuans
ostium urethrae internum e.
evacuant
evacuate

E

NOTES

evacuation
eversion of ankle
evocation
evocator
evoked
    e. potential
    e. response
ex
    ex vivo
excalation
excavatio
    e. disci
    e. papilla
    e. rectouterina
    e. rectovesicalis
    e. vesicouterina
excavation
    e. of optic disk
    physiologic e.
excavatum
    pectus e.
excentric
excess
    base e.
    e. lactate
    negative base e.
exchanger
    countercurrent e.
excitability
excitable area
excitant
excitation wave
excitatory
    e. junction potential
    e. postsynaptic potential
excitoglandular
excitometabolic
excitomotor
excitomuscular
excitoreflex nerve
excitor nerve
excitosecretory
excitovascular
exclave
excreta
excrete
excretion
excretorius
    ductus e.
excretory
    e. duct
    e. duct of lacrimal gland
    e. duct of seminal gland
    e. duct of seminal vesicle
    e. ductule of lacrimal gland
    e. system
exemia
exencephalia
exencephalic

exencephalous
exencephaly
exercise
    isometric e.
    isotonic e.
    e. test
exhalation
exhale
exhaustion
exit block
exitus
Exner plexus
exocardia
exoccipital bone
exocelomic membrane
exocervix
exocrine
    e. gland
    e. part of pancreas
exocytosis
exodic nerve
exogastrula
exogenetic
exogenous
    e. creatinine clearance
    e. fibers
exomphalos
exoplasm
exoskeleton
exosmosis
exostosis, pl. exostoses
    diaphysial juxtaepiphysial e.
    hereditary multiple exostoses
    multiple exostoses
exoteric
exothermic
expansion
    extensor digital e.
    hygroscopic e.
expectora
expectorant
expectorate
expectoration
expiration
expiratory
    e. center
    e. reserve volume
    e. resistance
expire
expired gas
explant
explantation
explosive decompression
exponential distribution
express
expression
    muscles of facial e.
expulsive
exsanguinate

exsanguination
exsanguine
exsiccant
exsiccate
exsiccation
exsomatize
exsorption
exstrophy
    e. of bladder
    e. of cloaca
extend
extension
    e. of carpometacarpal joint
    e. of elbow joint
    e. of hip
    e. of interphalangeal joint
    e. of knee
    e. of metacarpophalangeal joint
extensor
    e. aponeurosis
    e. carpi radialis brevis (ECRB)
    e. carpi radialis brevis muscle
    e. carpi radialis brevis tendon
    e. carpi radialis longus (ECRL)
    e. carpi radialis longus muscle
    e. carpi radialis longus tendon
    e. carpi ulnaris (ECU)
    e. carpi ulnaris muscle
    e. carpi ulnaris tendon
    common e.
    e. compartment of arm
    e. compartment of forearm
    e. compartment of leg
    e. compartment of thigh
    e. digital expansion
    e. digiti minimi muscle
    e. digiti minimi tendon
    e. digiti quinti (EDQ)
    e. digitorum brevis (EDB)
    e. digitorum brevis muscle
    e. digitorum brevis muscle of hand
    e. digitorum longus (EDL)
    e. digitorum longus muscle
    e. digitorum longus tendon
    e. hallucis brevis muscle
    e. hallucis longus (EHL)
    e. hallucis longus muscle
    e. hallucis longus tendon
    e. indicis muscle
    e. indicis proprius (EIP)
    e. indicis tendon
    e. muscle of little finger

    musculus e.
    e. pollicis brevis (EPB)
    e. pollicis brevis muscle
    e. pollicis brevis tendon
    e. pollicis longus (EPL)
    e. pollicis longus muscle
    e. pollicis longus tendon
    e. retinaculum
    e. retinaculum of ankle
    e. retinaculum of wrist
extensoris
    vagina tendinum musculorum
      abductoris longi et e.
    vagina tendinum musculorum
      extensoris digitorum et e.
extensorum
    compartimentum antebrachii e.
    compartimentum brachii e.
    compartimentum cruris e.
    compartimentum femoris e.
    pars lateralis compartimenti
      antebrachii posterioris e.
    retinaculum musculorum e.
exterior
exteriorize
externa
    arteria carotis e.
    arteria iliaca e.
    arteria maxillaris e.
    arteria nasi e.
    auris e.
    basis cranii e.
    capsula e.
    carotis e.
    conjugata e.
    crista occipitalis e.
    facies e.
    fascia spermatica e.
    iliac e.
    iliaca e.
    lamina e.
    membrana intercostalis e.
    organa genitalia feminina e.
    organa genitalia masculina e.
    organum genitalia feminina e.
    organum genitalia masculina e.
    palatina e.
    protuberantia occipitalis e.
    spermatica e.
    theca e.
    tunica e.
    vena iliaca e.

E

NOTES

externa *(continued)*
vena jugularis e.
vena palatina e.

externae
arteriae pudendae e.
nasales e.
partes genitales femininae e.
partes genitales masculinae e.
pudendae e.
venae nasales e.
venae pudendae e.

external
e. abdominal muscle
e. absorption
e. acoustic aperture
e. acoustic foramen
e. acoustic meatus
e. acoustic meatus artery
e. acoustic pore
e. anal sphincter
e. anal sphincter muscle
e. aperture of cochlear canaliculus
e. aperture of vestibular aqueduct
e. arcuate fibers
e. artery of nose
e. auditory foramen
e. auditory meatus
e. auditory pore
e. axis of eye
e. base of skull
e. branch of superior laryngeal nerve
e. branch of trunk of accessory nerve
e. calcaneoastragaloid ligament
e. canthus
e. capsule
e. carotid artery
e. carotid nerve
e. carotid nervous plexus
e. carotid vein
e. collateral ligament of wrist
e. conjugate
e. conjugate diameter
e. cruciate ligament
e. cuneate nucleus
e. dental epithelium
e. ear
e. ear canal
e. enamel epithelium
e. female genital organ
e. filum terminale of dura mater
e. filum terminale pia mater
e. genitalia
e. hydrocephalus
e. iliac artery
e. iliac lymphatic plexus
e. iliac lymph node
e. iliac vein

e. iliac vessel
e. inguinal ring
e. intercostal membrane
e. intercostal muscle
e. jugular vein
e. lateral ligament
e. ligament of malleus
e. lip of iliac crest
e. male genital organ
e. malleolus
e. mammary artery
e. mammary vein
e. mastoid plexus
e. maxillary artery
e. maxillary plexus
e. musculature
e. naris
e. nasal artery
e. nasal branch of infraorbital nerve
e. nasal vein
e. oblique fascia
e. oblique muscle
e. oblique ridge
e. obturator muscle
e. occipital crest
e. occipital protuberance
e. opening of cochlear canaliculus
e. opening of urethra
e. os
e. os of uterus
e. ovular transmigration
e. palatine vein
e. perineal fascia
e. pillar cells
e. popliteal nerve
e. pterygoid muscle
e. pterygoid nerve
e. pterygoid vein
e. pudendal artery
e. pudendal vein
e. pudic vein
e. radial vein
e. rectal sphincter
e. rectus muscle
e. rectus sheath
e. respiration
e. respiratory nerve of Bell
e. root sheath
e. rotation
e. salivary gland
e. saphenous nerve
e. semilunar fibrocartilage
e. sheath of optic nerve
e. spermatic artery
e. spermatic fascia
e. spermatic nerve
e. sphincter muscle of anus
e. spiral sulcus

e. surface
e. surface of cochlear duct
e. surface of cranial base
e. surface of frontal bone
e. surface of parietal bone
e. table of calvaria
e. urethral opening
e. urethral orifice
e. urethral sphincter
e. urethral sphincter of female
e. urethral sphincter of male
e. urinary meatus
e. wall of cochlear duct

externi
carotici e.
intercostales e.
isthmus meatus acustici e.
lymphonodi iliaci e.
meatus acustici e.
musculi intercostales e.
nervi carotici e.
nervus meatus acustici e.
nodi lymphoidei iliaci e.
pars profunda musculi sphincteri
  ani e.
pars subcutanea musculi sphincteri
  ani e.
pars superficialis musculi sphincteri
  ani e.
rami nasales e.

externum
filum terminale e.
ligamentum palpebrale e.
ligamentum tarsale e.
orificium urethrae e.
orpus geniculatum e.
ostium urethrae e.
ostium uteri e.
os uteri e.
perimysium e.
stratum plexiforme e.

externus
axis bulbi e.
meatus acusticus e.
musculus intercostalis e.
musculus obturator e.
musculus pterygoideus e.
musculus rectus e.
musculus sphincter ani e.
musculus sphincter urethrae e.
musculus thyroarytenoideus e.
musculus vastus e.

nasus e.
nervus spermaticus e.
plexus caroticus e.
plexus iliacus e.
plexus lymphaticus iliacus e.
plexus maxillaris e.
plexus nervosus caroticus e.
porus acusticus e.
ramus e.
sphincter ani e.
sphincter urethrae e.
sulcus spiralis e.

exteroceptive
exteroceptor
extima
tunica e.
extortor
extraarticular
extrabulbar
extracaliceal
extracapsularia
ligamenta e.
extracapsular ligament
extracarpal
extracellular
e. fluid
e. fluid volume
extrachorales
placenta e.
extrachromosomal genetic element
extracorporeal
extracorpuscular
extracostal muscle
extracranial
extracraniale
ganglion e.
extraction
e. coefficient
e. ratio
extradural
extradurale
spatium e.
extraembryonic
e. blastoderm
e. celom
e. mesoderm
extraepiphysial
extragenital
extraglomerular mesangium
extrahepatic bile duct
extraligamentous
extraluminal

E

NOTES

extramalleolus
extramedullary
extramural
extraneous
extranuclear
extraocular
    e. motion (EOM)
    e. movement (EOM)
    e. muscle (EOM)
    e. part of central retinal artery
     and vein
extraoral
extraosseous
extraovular
extrapapillary
extraparenchymal
extraperineal
extraperiosteal
extraperitoneal
    e. fascia
    e. space
    e. tissue
extraperitoneale
    spatium e.
extraperitonealis
    fascia e.
extraphysiologic
extraplacental
extrapleural fascia
extrapolar region
extraprostatic
extrapulmonary
extrapyramidal
    e. dyskinesias
    e. motor system
extrasaccular
extraserous
extrasomatic
extratarsal
extratracheal
extratubal
extrauterine
extravaginal
extravasate
extravasation
extravascular fluid
extraventricular
extrema
    capsula e.
extreme capsule
extremital
extremitas
    e. acromialis claviculae
    e. anterior splenica
    e. inferior
    e. inferior renis
    e. inferior testis
    e. posterior splenica
    e. sternalis claviculae

    e. superior
    e. superior renis
    e. superior testis
    e. tubaria ovarii
    e. uterina ovarii
extremity
    e. air cells
    anterior e.
    inferior e.
    lower e.
    posterior e.
    superior e.
    tubal e.
    upper e.
    uterine e.
extrinsic
    e. muscle
    e. muscle of eyeball
    e. sphincter
extrogastrulation
exudate
exudation
    e. cell
    e. corpuscle
exudative
exude
exumbilication
eye
    accessory organs of e.
    angle of lateral e.
    angle of medial e.
    anterior chamber of e.
    appendages of e.
    bulb of e.
    e. capsule
    choroid vein of e.
    cyclopean e.
    cyclopian e.
    epiphysial e.
    equator of e.
    external axis of e.
    fibrous tunic of e.
    globe of e.
    humor of e.
    internal axis of e.
    lateral angle of e.
    medial angle of e.
    orbicular muscle of e.
    parietal e.
    pineal e.
    plica semilunaris of e.
    posterior chamber of e.
    rudimentary e.
    e. socket
    substance of lens of e.
    suspensory ligament of e.
    e. tooth
    vascular layer of choroid coat
     of e.

vascular tunic of e.
vitreous chamber of e.
white of e.

eyeball
anterior chamber of e.
anterior pole of e.
chambers of e.
check ligaments of e.
episcleral layer of fibrous layer
  of e.
equator of e.
extrinsic muscle of e.
fascial sheath of e.
fibrous layer of e.
inner layer of e.
lateral check ligament of e.
medial check ligament of e.
meridians of e.
muscle of e.
nervous tunic of e.
posterior chamber of e.
posterior pole of e.
posterior segment of e.
postremal chamber of e.
sheath of e.
suspensory ligament of e.
vascular layer of e.

eyebrow
depressor muscle of e.
wrinkler muscle of e.

eye-ear plane
eyelash
eyelid
anterior border of e.
anterior surface of e.
arterial arch of lower e.
arterial arch of upper e.
borders of e.
cilia of e.
conjunctival layer of e.
elevator muscle of upper e.
free margin of e.
inferior e.
lower e.
margin of e.
orbital margin of e.
posterior border of e.
posterior surface of e.
skeleton of e.
superior e.
upper e.
vein of inferior e.
vein of superior e.

NOTES

E

fabella, pl. **fabellae**
fabellofibular ligament
Fabricius ship
face
    dish f.
    f. form
    motor nerve of f.
    f. region
    region of f.
    transverse artery of f.
    transverse vein of f.
facet, facette
    articular f.
    f. of atlas for dens
    calcaneal f.
    f. cartilage
    clavicular f.
    costal f.
    fibular f.
    inferior costal f.
    f. joint
    joint f.
    Lenoir f.
    f. on talus for calcaneonavicular
      part of bifurcate ligament
    f. on talus for plantar
      calcaneonavicular ligament
    superior articular f.
    superior costal f.
    transverse costal f.
facial
    f. angle
    f. artery
    f. aspect
    f. axis
    f. bone
    f. canal
    f. cleft
    f. colliculus
    f. eminence
    f. height
    f. hillock
    f. index
    f. lymph node
    f. motor nucleus
    f. muscle
    f. nerve
    f. nerve area
    f. nerve root
    f. plexus
    f. profile
    f. root
    f. skeleton
    f. surface of tooth
    transverse f.

    f. triangle
    f. vein
faciali
    rami communicantes nervi
      auriculotemporalis cum nervo f.
facialis, pl. **faciales**
    area nervi f.
    arteria f.
    canalis nervi f.
    colliculus f.
    eminentia f.
    geniculum canalis f.
    geniculum nervi f.
    geniculum nervus f.
    genu nervi f.
    hiatus canalis f.
    lymphonodi faciales
    musculi faciales
    nervus f.
    nodi lymphoidei faciales
    norma f.
    nucleus nervi f.
    plexus intraparotideus nervi f.
    plexus periarterialis arteriae f.
    prominentia canalis f.
    radix nervi f.
    rami buccales nervi f.
    rami glandulares arteriae f.
    rami parotidei venae f.
    rami temporales nervi f.
    rami zygomatici nervi f.
    ramus cervicalis nervi f.
    ramus colli nervi f.
    ramus digastricus nervi f.
    ramus labialis inferior arteriae f.
    ramus labialis superior arteriae f.
    ramus lateralis nasi arteriae f.
    ramus lingualis nervi f.
    ramus marginalis mandibulae
      nervi f.
    ramus stylohyoideus nervi f.
    ramus tonsillaris arteriae f.
    regio f.
    vena f.
faciei
    arteria transversa f.
    musculi f.
    ossa f.
    f. profunda
    transversa f.
    vena transversa f.
facies, pl. **facies**
    f. antebrachialis anterior
    f. antebrachialis posterior
    f. anterior antebrachii

F

facies *(continued)*

f. anterior brachii
f. anterior corneae
f. anterior corporis maxillae
f. anterior cruris
f. anterior glandulae suprarenalis
f. anterior iridis
f. anterior lateralis corporis humeri
f. anterior lentis
f. anterior medialis corporis humeri
f. anterior membri inferioris
f. anterior palpebrarum
f. anterior partis petrosae ossis temporalis
f. anterior patellae
f. anterior prostatae
f. anterior radii
f. anterior renis
f. anterior ulnae
f. anterior uterus
f. anteroinferior corporis pancreatis
f. anterolateralis cartilaginis arytenoideae
f. anterolateralis corporis humeri
f. anteromedialis corporis humeri
f. anterosuperioris corporis pancreatis
f. approximalis dentis
f. articularis
f. articularis acromialis claviculae
f. articularis anterior dentis
f. articularis arytenoidea cricoideae
f. articularis calcanea tali
f. articularis capitis costae
f. articularis capitis fibulae
f. articularis carpi radii
f. articularis cartilaginis arytenoideae
f. articularis clavicularis acromii
f. articularis cuboidea ossis calcanei
f. articularis fibularis tibiae
f. articularis fossae mandibularis ossis temporalis
f. articularis inferior atlantis
f. articularis inferior tibiae
f. articularis malleoli lateralis fibulae
f. articularis malleoli medialis tibiae
f. articularis malleoli tibiae
f. articularis navicularis tali
f. articularis partis calcaneonavicularis ligamenti bifurcati tali
f. articularis patellae
f. articularis posterior dentis
f. articularis sternalis claviculae
f. articularis superior atlantis
f. articularis superior tibiae
f. articularis talaris anterior calcanei
f. articularis talaris media calcanei
f. articularis talaris posterior calcanei
f. articularis thyroidea cricoideae
f. articularis tuberculi costae
f. auricularis ossis ilii
f. auricularis ossis sacri
f. brachialis anterior
f. brachialis posterior
f. buccalis
f. cerebralis
f. colica splenis
f. contactus dentis
f. costalis
f. costalis pulmonis
f. costalis scapulae
f. cruralis anterior
f. cruralis posterior
f. cubitalis anterior
f. cubitalis posterior
f. diaphragmatica
f. digitalis dorsalis manus et pedis
f. digitalis palmaris
f. digitalis plantaris
f. digitalis ventralis
f. distalis dentis
f. dorsalis
f. dorsalis ossis sacri
f. dorsalis scapulae
f. externa
f. externa ossis frontalis
f. externa ossis parietalis
f. facialis dentis
f. femoralis anterior
f. femoralis posterior
f. gastrica splenis
f. glutea ossis ilii
f. inferior linguae
f. inferior partis petrosae ossis temporalis
f. inferolateralis prostatae
f. infratemporalis alaris majoris ossis sphenoidalis
f. infratemporalis corporis maxillae
f. interlobares pulmonis
f. interna
f. interna ossis frontalis
f. interna ossis parietalis
f. intestinalis uterus
f. labialis
f. lateralis
f. lateralis brachii
f. lateralis cruris
f. lateralis digiti manus
f. lateralis digiti pedis
f. lateralis fibulae

f. lateralis membri inferioris
f. lateralis ossis zygomatici
f. lateralis ovarii
f. lateralis testis
f. lateralis tibiae
f. ligamenti calcaneonavicularis
  plantaris tali
f. lingualis dentis
f. lunata acetabuli
f. malleolaris lateralis tali
f. malleolaris medialis tali
f. masticatoria
f. maxillaris alaris majoris ossis
  sphenoidalis
f. maxillaris ossis palatini
f. medialis
f. medialis cartilaginis arytenoideae
f. medialis digiti pedis
f. medialis fibulae
f. medialis ovarii
f. medialis pulmonis
f. medialis testis
f. medialis tibiae
f. medialis ulnae
f. mediastinalis pulmonis
f. mesialis dentis
f. nasalis maxillae
f. nasalis ossis palatini
f. occlusalis dentis
f. orbitalis
f. palatina laminae horizontalis
  ossis palatini
f. palmares digitorum
f. pancreatica splenica
f. patellaris femoris
f. pelvica ossis sacri
f. poplitea femoris
f. posterior cartilaginis arytenoideae
f. posterior corneae
f. posterior corporis humeri
f. posterior cruris
f. posterior fibulae
f. posterior glandulae suprarenalis
f. posterior iridis
f. posterior lentis
f. posterior membri inferioris
f. posterior palpebrarum
f. posterior pancreatis
f. posterior partis petrosae ossis
  temporalis
f. posterior prostatae
f. posterior radii

f. posterior renis
f. posterior scapulae
f. posterior tibiae
f. posterior ulnae
f. pulmonales cordis dextra/sinistra
f. pulmonalis cordis
f. renalis glandulae suprarenalis
f. renalis lienis
f. renalis splenis
f. sacropelvina ossis ilii
f. scaphoidea
f. sternocostalis cordis
f. superior tali
f. symphysialis
f. urethralis penis
f. vesicalis uterus
f. vestibularis dentis
f. visceralis hepatis
f. visceralis lienis
f. visceralis splenis

facilitation
  Wedensky f.
factor
  f. 3
  adrenal weight f.
  atrial natriuretic f.
  B cell differentiating f.
  biotic f.'s
  corticotropin-releasing f.
  decapacitation f.
  endothelium-derived relaxing f.
    (EDRF)
  epidermal growth f.
  follicle-stimulating hormone-
    releasing f.
  galactagogue f.
  galactopoietic f.
  glycotropic f.
  granulocyte colony-stimulating f.
    (G-CSF)
  granulocyte-macrophage colony-
    stimulating f. (GM-CSF)
  growth f.
  growth hormone-releasing f.
    (GHRF)
  insulin-antagonizing f.
  lactogenic f.
  luteinizing hormone/follicle-
    stimulating hormone-releasing f.
  lymph node permeability f. (LNPF)
  macrophage colony-stimulating f.
    (M-CSF)

F

NOTES

factor (*continued*)
>   müllerian duct inhibitory f.
>   müllerian inhibiting f.
>   müllerian regression f.
>   multicolony-stimulating f.
>   nerve growth f. (NGF)
>   f. P
>   platelet-derived growth f. (PDGF)
>   prolactin-inhibiting f.
>   prolactin-releasing f.
>   relaxation f.
>   releasing f.
>   somatotropin release-inhibiting f.
>   somatotropin-releasing f.
>   sulfation f.
>   T-cell growth f.
>   thyroid-stimulating hormone-
>   releasing f.
>   thyrotropin-releasing f.
>   uterine relaxing f. (URF)

facultative heterochromatin
faculty
Fahraeus-Lindqvist effect
falcate
falces (*pl. of* falx)
falcial
falciform
>   f. cartilage
>   f. crest
>   f. fold of fascia lata
>   f. hymen
>   f. ligament of liver
>   f. lobe
>   f. lobule
>   f. margin
>   f. margin of saphenous opening
>   f. process
>   f. process of sacrotuberous
>   ligament

falciforme
>   ligamentum f.

falciformis
>   cornu superius marginalis f.
>   lobus f.
>   processus f.

falcine region
falcula
falcular
fallopian
>   f. aqueduct
>   f. arch
>   f. artery
>   f. canal
>   f. hiatus
>   f. ligament
>   f. tube

fallopii
>   aqueductus f.
>   tuba f.

falsa, pl. **falsae**
>   chordae tendineae falsae

false
>   f. chordae tendineae
>   f. conjugate
>   f. glottis
>   f. hermaphroditism
>   f. knots of umbilical cord
>   f. ligament
>   f. membrane
>   f. pelvis
>   f. rib
>   f. suture
>   f. tendinous cords
>   f. thirst
>   f. vertebrae
>   f. vocal cord

falx, pl. **falces**
>   aponeurotic f.
>   f. aponeurotica
>   cerebellar f.
>   f. cerebelli
>   cerebelli f.
>   cerebral f.
>   f. cerebri
>   cerebri f.
>   inguinal f.
>   f. inguinalis
>   ligamentous f.
>   f. of maxillary antrum
>   f. septi

Fananás cell
Farabeuf
>   triangle of F.
>   F. triangle

faradocontractility
faradomuscular
faradopalpation
Farre line
fascia, pl. **fasciae**
>   f. abdominalis parietalis
>   Abernethy f.
>   f. adherens
>   alar f.
>   anal f.
>   angular tract of cervical f.
>   antebrachial f.
>   f. antebrachii
>   anterior layer of thoracolumbar f.
>   anterior rectus f.
>   aponeurotic f.
>   arcus tendineus of obturator f.
>   f. axillaris
>   axillary f.
>   bicipital f.
>   brachial f.

f. brachii
broad f.
f. buccopharyngea
buccopharyngeal f.
Buck f.
bulbar sheath f.
f. bulbi
bulbi f.
Camper f.
cauda f.
cervical f.
f. cervicalis
f. cervicalis profunda
f. cinerea
circumflex iliac superficial f.
clavipectoral f.
f. clavipectoralis
f. clitoridis
f. of clitoris
Cloquet f.
Colles f.
Cooper f.
cremaster f.
cremasteric f.
f. cremasterica
cribriform f.
f. cribrosa
crural f.
f. cruris
Cruveilhier f.
dartos f.
deep cervical f.
deep layer of temporal f.
f. of deep penis
deep perineal f.
deltoid f.
deltopectoral f.
f. dentata hippocampi
dentate f.
f. diaphragm
f. diaphragmatis pelvis inferior
f. diaphragmatis urogenitalis inferior
f. diaphragmatis urogenitalis
 superior
dorsal aponeurotic f.
dorsal interosseous f.
f. dorsalis manus
f. dorsalis pedis
Dupuytren f.
endoabdominal f.
f. endoabdominalis
endopelvic f.

f. endopelvina
endothoracic f.
f. endothoracica
external oblique f.
external perineal f.
external spermatic f.
f. of extraocular muscle
extraperitoneal f.
f. extraperitonealis
extrapleural f.
fatty layer of superficial f.
f. of forearm
Gallaudet f.
geniohyoid f.
Gerota f.
gluteal f.
Godman f.
Hesselbach f.
hypothenar f.
iliac f.
f. iliaca
iliopectineal f.
inferior f.
f. infraspinata
infraspinatus f.
infraspinous f.
infundibuliform f.
f. of insertion
intercolumnar fasciae
internal oblique f.
internal spermatic f.
interosseous f.
f. investiens
f. investiens perinei superficialis
investing layer of cervical f.
lacrimal f.
f. lata
f. lata femoris
lateral oblique f.
f. of leg
lumbodorsal f.
masseteric f.
f. masseterica
membranous layer of superficial f.
middle cervical f.
muscular f.
f. muscularis musculorum bulbi
f. musculi quadrati lumborum
f. of neck
f. nuchae
nuchal f.
obturator f.

F

NOTES

fascia *(continued)*
  f. obturatoria
  f. of obturator internus
  f. of omohyoid muscle
  orbital fasciae
  fasciae orbitales
  palmar f.
  palpebral f.
  paraspinous f.
  parietal abdominal f.
  parietal pelvic f.
  parotid f.
  f. parotidea
  parotideomasseteric f.
  f. parotideomasseterica
  pectineal f.
  pectoral f.
  f. pectoralis
  pelvic f.
  f. pelvica
  f. pelvis parietalis
  f. pelvis profunda
  f. pelvis superficialis
  f. pelvis visceralis
  f. penis profunda
  f. penis superficialis
  perineal f.
  f. perinei
  perinephric f.
  perirenal f.
  perivesical f.
  pharyngobasilar f.
  f. pharyngobasilaris
  phrenicopleural f.
  f. phrenicopleuralis
  plantar f.
  popliteal f.
  Porter f.
  prececocolic f.
  f. prececocolica
  prepubic f.
  presacral f.
  f. presacralis
  pretracheal layer of cervical f.
  prevertebral layer of cervical f.
  f. prostatae
  f. of prostate
  psoas f.
  psoatic part of iliopsoas f.
  pubocervical f.
  pubovesicocervical f.
  quadratus lumborum f.
  rectal f.
  rectosacral f.
  f. rectosacralis
  rectovaginal f.
  rectovesical f.
  rectus f.
  renal f.
  f. renalis
  retrorectal lamina of endopelvic f.
  retrovisceral f.
  salpingopharyngeal f.
  scalene f.
  Scarpa f.
  semilunar f.
  Sibson f.
  spermatic f.
  f. spermatica externa
  f. spermatica interna
  subcutaneous f.
  subperitoneal f.
  f. subperitonealis
  subsartorial f.
  f. subscapularis
  subserous f.
  subvesical f.
  superficial abdominal f.
  superficial inguinal f.
  superficial layer of deep cervical f.
  superficial layer of temporal f.
  f. of superficial penis
  superficial perineal f.
  f. superior diaphragmatis pelvis
  temporal f.
  f. temporalis
  tendinous arch of pelvic f.
  thenar f.
  thoracic f.
  f. thoracolumbalis
  thoracolumbar f.
  thyrolaryngeal f.
  Toldt f.
  transversalis f.
  f. transversalis
  transverse f.
  Treitz f.
  triangular f.
  f. triangularis abdominis
  Tyrrell f.
  umbilical f.
  f. umbilicalis
  umbilical prevesical f.
  umbilicovesical f.
  underlying f.
  vastoadductor f.
  vesical f.
  vesicovaginal f.
  visceral pelvic f.
  Waldeyer f.
  Zuckerkandl f.

fascial
  f. layer
  f. sheath
  f. sheath of eyeball
  f. sheaths of extraocular muscle

fascicle
>muscle f.
>nerve f.

fascicular

fasciculata
>f. cell
>zona f.

fasciculate

fasciculated

fasciculation

fasciculus, pl. **fasciculi**
>f. anterior musculi palatopharyngei
>f. anterior proprius
>aponeurosis of plantar transverse fasciculi
>arcuate f.
>f. atrioventricularis
>Burdach f.
>calcarine f.
>central tegmental f.
>f. circumolivaris pyramidis
>f. corticospinalis anterior
>f. corticospinalis lateralis
>f. cuneatus
>dorsal longitudinal f.
>dorsolateral f.
>f. dorsolateralis
>Flechsig fasciculi
>Foville f.
>frontooccipital f.
>f. of Gowers
>f. gracilis
>hooked f.
>inate f.
>inferior longitudinal f.
>interfascicular f.
>f. interfascicularis
>intersegmental fasciculi
>f. lateralis plexus brachialis
>f. lateralis proprius
>f. lenticularis
>Lissauer f.
>fasciculi longitudinales ligamenti cruciformis atlantis
>fasciculi longitudinales pontis
>f. longitudinalis dorsalis
>f. longitudinalis inferior
>f. longitudinalis medialis
>f. longitudinalis superior
>f. macularis
>mamillotegmental f.
>f. mamillotegmentalis

>mamillothalamic f.
>f. mamillothalamicus
>f. marginalis
>f. medialis plexus brachialis
>medial longitudinal f.
>Meynert f.
>neate f.
>f. obliquus pontis
>occipitofrontal f.
>f. occipitofrontalis
>oval f.
>f. pedunculomamillaris
>perpendicular f.
>f. posterior musculi palatopharyngei
>f. posterior plexus brachialis
>proper fasciculi
>fasciculi proprii
>f. pyramidalis anterior
>f. pyramidalis lateralis
>retroflex f.
>f. retroflexus (FRF)
>f. of Rolando
>f. rotundus
>fasciculi rubroreticulares
>semilunar f.
>f. semilunaris
>septomarginal f.
>f. septomarginalis
>f. solitarius
>f. subcallosus
>superior longitudinal f.
>f. thalamicus
>f. thalamomamillaris
>transverse fasciculi
>fasciculi transversi
>unciform f.
>f. uncinatus
>wedge-shaped f.

fasciola, pl. **fasciolae**
>f. cinerea

fasciolar gyrus

fasciolaris
>gyrus f.

fascioscapulohumeral artery

fastigatum

fastigii
>nucleus f.

fastigiobulbaris
>tractus f.

fastigiobulbar tract

fastigium

F

NOTES

fat
>
> abdominal f.
> f. body
> f. body of cheek
> f. body of ischioanal fossa
> f. body of ischiorectal fossa
> f. body of orbit
> brown f.
> caul f.
> f. cell
> f. metabolism
> multilocular f.
> f. pad
> f. pad of ischioanal fossa
> paranephric f.
> preperitoneal f.
> properitoneal f.
> retrobulbar f.
> subcutaneous f.
> subcuticular f.
> subepicardial f.
> unilocular f.
> white f.

fatigability
fatigable
fatigue
fat-pad
>
> Bichat f.-p.
> buccal f.-p.
> Imlach f.-p.
> infrapatellar f.-p.
> ischiorectal f.-p.
> orbital f.-p.

fat-storing cell
fatty
>
> f. acid oxidation cycle
> f. appendices of colon
> f. capsule of kidney
> f. fold of pleura
> f. layer of subcutaneous tissue
> f. layer of subcutaneous tissue of abdomen
> f. layer of superficial fascia
> f. phanerosis
> f. renal capsule

fauces
>
> anterior pillar of f.
> branch of lingual nerve to isthmus of f.
> isthmus of f.
> pillars of f.
> posterior pillar of f.

faucial
>
> f. arch
> f. area
> f. branch of lingual nerve
> f. reflex
> f. tonsil

fauciales
>
> rami f.

faucium
>
> isthmus f.
> plica anterior f.
> plica posterior f.
> rami isthmi f.

faveolate
faveolus, pl. **faveoli**
FCR
>
> flexor carpi radialis

FCU
>
> flexor carpi ulnaris

FDB
>
> flexor digitorum brevis

FDL
>
> flexor digitorum longus

FDM
>
> flexor digiti quinti

FDS
>
> flexor digitorum sublimis
> flexor digitorum superficialis

features
fecal reservoir
Fechner-Weber law
fecund
fecundate
fecundation
fecundity
feedback
>
> f. inhibition
> negative f.
> positive f.
> f. system
> tubuloglomerular f.

feeding
>
> f. center
> fictitious f.
> sham f.

feet (*pl. of* foot)
Feiss line
fellea
>
> cystis f.
> vesica f.

felleae
>
> collum vesicae f.
> corpus vesicae f.
> fossa vesicae f.
> fundus vesicae f.
> infundibulum vesicae f.
> plicae tunicae mucosae vesicae f.
> sphincter vesicae f.
> tunica mucosa vesicae f.
> tunica muscularis vesicae f.
> tunica serosa vesicae f.

fellis
>
> vesicula f.

female
>
> f. external genitalia

external urethral sphincter of f.
f. gonad
f. hermaphroditism
f. internal genitalia
f. pronucleus
f. prostate
pubovesical ligament of f.
f. urethra
urethral crest of f.
urethral gland of f.
feminae
glandulae urethrales f.
femineus
penis f.
feminina, pl. **femininae**
crista urethralis femininae
glandulae urethrales femininae
musculus sphincter urethrae
externus femininae
tunica mucosa urethrae femininae
tunica muscularis urethrae
femininae
tunica spongiosa urethrae femininae
urethra f.
femininum
ligamentum pubovesicale f.
pudendum f.
femora (*pl. of* femur)
femoral
f. arch
f. bone
f. branch of genitofemoral nerve
f. canal
f. condyle
f. cortex
f. epiphysis
f. fossa
f. head
f. ligament
f. muscle
f. neck
f. nervous plexus
f. node
f. nutrient artery
f. opening
f. pulse
f. reflex
f. region
f. ring
f. septum
f. shaft
f. sheath

f. triangle
f. vein
femorale
calcar f.
septum f.
trigonum f.
femoralis
annulus f.
arteria f.
canalis f.
fovea f.
nervus f.
plexus nervosus f.
rami cutanei anteriores nervi f.
ramus f.
vena f.
femoris
arteriae perforantes arteriae
profundae f.
arteria nutriciae f.
arteria nutriens f.
arteria profunda f.
biceps f.
bursa quadrati f.
bursa quadratus f.
caput breve musculi bicipitis f.
caput longum musculi bicipitis f.
caput ossis f.
collum ossis f.
condylus lateralis f.
condylus medialis f.
corpus ossis f.
epicondylus lateralis ossis f.
epicondylus medialis ossis f.
facies patellaris f.
facies poplitea f.
fascia lata f.
fovea capitis ossis f.
inferior subtendinous bursa of
biceps f.
ligamentum capitis f.
ligamentum teres f.
linea intercondylaris f.
linea pectinea f.
musculus biceps f.
musculus quadratus f.
musculus quadriceps extensor f.
musculus rectus f.
musculus tensor fasciae f.
os f.
profunda f.
quadratus f.

F

NOTES

femoris *(continued)*
    quadriceps f.
    rectus f.
    regio f.
    short head of biceps f.
    superior bursa of biceps f.
    trigonum f.
    trochlea f.
    tuberculum adductorium f.
    vena profunda f.
femoroiliac
femoropatellar joint
femoropopliteal vein
femorotibial
femur, pl. **femora**
    adductor tubercle of f.
    body of f.
    fovea for ligament of head of f.
    greater trochanter of f.
    head of f.
    inferior articular surface of f.
    intercondylar line of f.
    intercondylar notch of f.
    lateral condyle of f.
    lateral epicondyle of f.
    lesser trochanter of f.
    ligament of head of f.
    medial condyle of f.
    medial epicondyle of f.
    neck of f.
    nutrient artery of f.
    patellar surface of f.
    pectineal line of f.
    pit of head of f.
    popliteal plane of f.
    popliteal surface of f.
    quadrate tubercle of f.
    quadriceps artery of f.
    round ligament of f.
    shaft of f.
fenestra, pl. **fenestrae**
    f. choledocha
    f. of cochlea
    f. cochleae
    f. ovalis
    f. rotunda
    f. vestibuli
fenestram
    fissula ante f.
fenestrata
    placenta f.
fenestrated
    f. capillary
    f. membrane
fenestration
Fenn effect
Ferrein
    F. canal
    F. cord

    F. foramen
    F. ligament
    F. pyramid
    F. tube
    F. vasa aberrantia
ferreini
    processus f.
ferrokinetics
ferruginea
    substantia f.
ferrugineus
    locus f.
fertile
fertility
fertilization
    f. cone
    f. membrane
    in vivo f.
fertilized ovum
fetal
    f. adrenal cortex
    f. age
    f. attitude
    f. circulation
    f. cotyledon
    f. habitus
    f. hydrops
    f. inclusion
    f. intrahepatic vein
    f. membrane
    f. movement
    f. ovoid
    f. placenta
    f. reticularis
    f. zone
fetalis
    hydrops f.
    placenta f.
    rachitis f.
fetoglobulins
fetography
fetology
fetopathy
fetoprotein
fetotoxicity
fetu
    fetus in f.
fetus, pl. **fetuses**
    appendages of f.
    f. in fetu
    f. papyraceus
FHB
    flexor hallucis brevis
fiber
    A f.
    A, B, C f.
    accelerator f.
    adrenergic f.
    afferent f.

alpha f.
anastomotic f.'s
anterior external arcuate f.
arcuate f.'s
argyrophilic f.'s
association f.'s
astral f.'s
augmentor f.
autonomic nerve f.
Bergmann f.'s
beta f.
cholinergic f.
chromatic f.
circular f.'s
climbing f.'s
collagen f.
collagenous f.
commissural f.'s
cone f.
corticobulbar f.'s
corticonuclear f.'s
corticopontine f.'s
corticoreticular f.'s
corticospinal f.'s
deep transverse f.
delta f.
dental f.'s
dentinal f.'s
depressor f.
elastic f.'s
enamel f.'s
endogenous f.'s
exogenous f.'s
external arcuate f.'s
gamma f.
Gerdy f.
Gratiolet f.'s
gray f.'s
inhibitory f.
inner cone f.
intercolumnar f.'s
intercrural f.'s
internal arcuate f.'s
intrafusal f.'s
intrinsic f.'s
James f.
Korff f.'s
Kühne f.
Mahaim f.
medullated nerve f.
meridional f.'s
mossy f.'s

motor f.
Müller f.'s
myelinated nerve f.
Nélaton f.
nerve f.
nonmedullated f.'s
nuclear bag f.
nuclear chain f.
osteocollagenous f.'s
osteogenetic f.
outer cone f.
pectinate f.
perforating f.'s
periodontal ligament f.
periventricular f.'s
pilomotor f.'s
postganglionic parasympathetic f.
postganglionic sympathetic f.
precollagenous f.'s
preganglionic parasympathetic f.
preganglionic sympathetic f.
pressor f.
projection f.'s
Prussak f.
pupillodilator f.
Purkinje f.
pyramidal f.'s
red f.'s
Reissner f.
Remak f.'s
reticular f.'s
rod f.
Rosenthal f.
Sappey f.
Sharpey f.
skeletal muscle f.'s
somatic nerve f.
spindle f.
sudomotor f.'s
superficial transverse f.
sympathetic nerve f.
tautomeric f.'s
Tomes f.'s
transseptal f.'s
transverse f.
unmyelinated f.'s
visceral motor f.
Weitbrecht f.
white f.
yellow f.'s
zonular f.

fibra, pl. **fibrae**

F

NOTES

fibra *(continued)*
    fibrae arcuatae cerebri
    fibrae arcuatae internae
    fibrae circulares
    fibrae corticonucleares
    fibrae corticopontinae
    fibrae corticoreticulares
    fibrae corticospinales
    fibrae intercrurales
    fibrae intercrurales anuli inguinalis
      superficialis
    fibrae lentis
    fibrae meridionales
    fibrae meridionales muscularis
      ciliaris
    fibrae obliquae tunicae muscularis
    fibrae obliquae ventriculi
    fibrae periventriculares
    fibrae pontis transversae
    fibrae pyramidales
    fibrae zonulares
fibril
    collagen f.'s
    muscular f.
    unit f.'s
fibrilla, pl. **fibrillae**
fibrillar baskets
fibrillary
    f. astrocyte
    f. contractions
    f. glia
fibrillate
fibrillated
fibrillation
fibrillogenesis
fibrinogenesis
fibrinogenic
fibrinogenous
fibrinous
    f. tissue
fibroadipose
fibroareolar
fibroblast
    human lung f.
fibroblastic
fibrocartilage
    basilar f.
    circumferential f.
    external semilunar f.
    interarticular f.
    internal f.
    interpubic f.
    interventricular f.
    semilunar f.
    stratiform f.
fibrocartilaginous
    f. disk
    f. joint
    f. material

    f. ring
    f. ring of tympanic membrane
fibrocartilago
    f. basalis
    f. interarticularis
    f. interpubica
    f. intervertebralis
fibrocellular
fibrocyte
fibrodysplasia
fibroelastic membrane of larynx
fibroelastosis
    endocardial f.
    endomyocardial f.
fibrofatty
fibrogenesis
fibrohyaline tissue
fibroid
fibroma
    nonosteogenic f.
fibromatosis colli
fibromuscular
    f. junction
    f. wall
fibromusculocartilagenous layer of bronchi
fibroplastic
fibroplate
fibroreticularis
    lamina f.
fibroreticulate
fibrosa, pl. **fibrosae**
    articulatio f.
    capsula f.
    junctura f.
    membrana f.
    meninx f.
    pars annularis vaginae fibrosae
    pars cruciformis vaginae fibrosae
    pericardium f.
    tunica f.
    vaginae f.
fibrose
fibroserous
fibrosum
    corpus f.
    pericardium f.
    stratum f.
fibrosus
    annulus f.
    lacertus f.
fibrous
    f. appendix of liver
    f. articular capsule
    f. astrocyte
    f. cap
    f. capsule of kidney
    f. capsule of liver
    f. capsule of parotid gland

f. capsule of spleen
f. capsule of thyroid gland
f. connective tissue
f. cortical defect
f. digital sheaths of foot
f. digital sheaths of hand
f. digital sheaths of toe
f. joint
f. layer
f. layer of articular capsule
f. layer of eyeball
f. layer of joint capsule
f. layer in or on deep aspect of fatty layer of subcutaneous tissue
f. membrane
f. membrane of joint capsule
f. nodule
f. pericardium
f. ring
f. ring of intervertebral disk
f. sac
f. sheaths of digits of hand
f. skeleton of heart
f. tendon sheath
f. trigones of heart
f. tunic of corpus spongiosum
f. tunic of eye

fibula

anterior border of f.
anterior ligament of head of f.
apex of head of f.
articular facet of head of f.
articular surface of head of f.
body of f.
distal f.
head of f.
interosseous border of f.
lateral surface of f.
ligaments of head of f.
malleolar articular surface of f.
medial crest of f.
medial surface of f.
neck of f.
nutrient artery of f.
posterior border of f.
posterior ligament of head of f.
posterior surface of f.
shaft of f.
styloid process of f.
upper extremity of f.

fibulae

apex capitis f.
arteria nutriens f.
caput f.
collum f.
corpus f.
crista medialis f.
facies articularis capitis f.
facies articularis malleoli lateralis f.
facies lateralis f.
facies medialis f.
facies posterior f.
fossa malleoli f.
ligamentum capitis f.
margo anterior f.
margo interosseus f.
margo posterior f.
ramus circumflexus f.

fibular

f. articular facet of tibia
f. articular surface of tibia
f. bone
f. collateral ligament
f. collateral ligament of ankle
f. compartment of leg
f. facet
f. lymph node
f. margin of foot
f. neck
f. nerve
f. notch
f. nutrient artery
f. peroneal border of foot
f. shaft
f. tarsal tendinous sheath
f. trochlea of calcaneus
f. vein

fibulare

ligamentum collaterale f.

fibularis, pl. fibulares

arteria f.
f. brevis muscle
incisura f.
f. longus muscle
f. longus tendon
nervus communicans f.
nodus lymphoideus f.
ramus communicans arteriae f.
ramus perforans arteriae f.
f. tertius muscle
f. tertius tendon

F

NOTES

fibularis (*continued*)
    vaginae tendinum tarsales fibulares
    venae fibulares
fibularium
    compartimentum cruris f.
    retinaculum musculorum f.
fibulocalcaneal
fibulotalar joint
Fick
    F. method
    F. principle
fictitious feeding
field
    Broca f.
    f.'s of Forel
    free f.
    H f.'s
    individuation f.
    magnetic f.
    nerve f.
    prerubral f.
    subdigastric lymph f.
    temporal f.
    Wernicke f.
Fielding membrane
fifth
    f. cranial nerve
    f. finger
    f. ventricle
figure
    myelin f.
fila (*pl. of* filum)
filaceous
filaggrin
filament
    actin f.
    axial f.
    cytokeratin f.'s
    intermediate f.'s
    keratin f.'s
    myosin f.
    pial f.
    root f.'s
    spermatic f.
    Z f.
filamentous
filamentum, pl. **filamenta**
filar substance
filiformis, pl. **filiformes**
    nucleus f.
    papillae filiformes
filiform papilla
fillet
    decussation of f.
    lateral f.
    f. layer
    medial f.
    triangle of f.
    trigone of f.

filling internal urethral orifice
filovaricosis
filtration
    f. angle
    f. coefficient
    f. fraction
    f. slit
    f. space
filtrum ventriculi
filum, pl. **fila**
    f. durae matris spinalis
    fila olfactoria
    fila radicularia
    f. of spinal dura mater
    terminal f.
    f. terminale
    f. terminale externum
    f. terminale internum
fimbria, pl. **fimbriae**
    f. hippocampi
    ovarian f.
    f. ovarica
    tenia fimbriae
    fimbriae tubae uterinae
    uterine tube f.
    fimbriae of uterine tube
fimbriata
    plica f.
fimbriate
fimbriated
    f. fold
    f. fold of inferior surface of
      tongue
    f. oviduct
fimbriatum
    corpus f.
fimbriodentate sulcus
fimbriodentatus
    sulcus f.
fine structure
finger
    abductor muscle of little f.
    deep flexor muscle of f.
    extensor muscle of little f.
    fifth f.
    first f.
    fourth f.
    index f.
    lateral surface of f.
    little f.
    middle f.
    opposer muscle of little f.
    palmar surfaces of f.
    pulp of f.
    ring f.
    second f.
    short flexor muscle of little f.
    spider f.

subcutaneous venous arch at root
of f.
superficial flexor muscle of f.
third f.
thumb f.
vinculum breve of f.
vinculum longum of f.
web of f.
finger-like villus
fingernail
fingers/toes
web of f.
finger-thumb reflex
fingertip
first
f. cervical vertebra
f. cranial nerve
f. cuneiform bone
f. dentition
f. duodenal sphincter
f. finger
f. metacarpal artery
f. parallel pelvic plane
f. part of duodenum
f. permanent molar
f. rib
f. and second posterior intercostal
artery
f. temporal convolution
f. visceral cleft
fission
binary f.
bud f.
multiple f.
simple f.
fissiparity
fissiparous
fissula ante fenestram
fissum
palatum f.
fissura, pl. **fissurae**
f. antitragohelicina
f. calcarina
f. cerebri lateralis
f. choroidea
f. collateralis
f. dentata
f. hippocampi
f. horizontalis pulmonis dextri
f. ligamenti teretis
f. ligamenti teretis hepatis
f. ligamenti venosi

f. longitudinalis cerebri
f. mediana anterior medullae
oblongatae
f. mediana anterior medullae
spinalis
f. obliqua
f. obliqua pulmonis
f. orbitalis inferior
f. orbitalis superior
f. parietooccipitalis
f. petrooccipitalis
f. petrosquamosa
f. petrotympanica
f. posterolateralis
f. prima cerebelli
f. pterygoidea
f. pterygoma
f. pterygomaxillaris
f. pterygopalatina
f. pudendi
f. sphenopetrosa
f. transversa cerebelli
f. transversa cerebri
f. tympanomastoidea
f. tympanosquamosa
fissural
fissuration
fissure
abdominal f.
Ammon f.
anterior median f.
antitragohelicine f.
ape f.
auricular f.
Bichat f.
branchial f.
Broca f.
calcarine f.
callosomarginal f.
caudal transverse f.
cerebellar f.
cerebral f.
choroid f.
Clevenger f.
collateral f.
decidual f.
dentate f.
Duverney f.
Ecker f.
ethmoidal f.
glaserian f.
great horizontal f.

NOTES

F

fissure (*continued*)
    great longitudinal f.
    Henle f.'s
    hippocampal f.
    inferior orbital f.
    interlobar f.
    lateral cerebral f.
    left sagittal f.
    f. for ligamentum teres
    f. for ligamentum venosum
    f.'s of liver
    longitudinal cerebral f.
    lunate f.
    f.'s of lung
    median f.
    oblique f.
    optic f.
    oral f.
    palpebral f.
    Pansch f.
    paracentral f.
    parietooccipital f.
    petrooccipital f.
    petrosphenoidal f.
    petrosquamous f.
    petrotympanic f.
    portal f.
    postcentral f.
    posterior median f.
    posterolateral f.
    posthippocampal f.
    postlingual f.
    postlunate f.
    postpyramidal f.
    postrhinal f.
    prenodular f.
    pterygoid f.
    pterygomaxillary f.
    rhinal f.
    right sagittal f.
    Rolando f.
    f. of Rolando
    f. of round ligament
    f. for round ligament of liver
    Santorini f.
    simian f.
    sphenoidal f.
    sphenomaxillary f.
    sphenopetrosal f.
    squamotympanic f.
    sternal f.
    superior orbital f.
    superior temporal f.
    sylvian f.
    f. of Sylvius
    tympanomastoid f.
    tympanosquamous f.
    umbilical f.

    f. of venous ligament
    zygal f.
fistula, pl. **fistulae**
    f. auris congenita
    branchial f.
    f. colli congenita
    lymphatic f.
    Mann-Bollman f.
    pharyngeal f.
    reverse Eck f.
    Thiry f.
    Thiry-Vella f.
    tracheal f.
    tracheobiliary f.
    tracheoesophageal f.
    Vella f.
    vitelline f.
fitness
    physical f.
fixa
    punctum f.
fixation
    nasomandibular f.
fixator
    f. muscle
    f. muscle of base of stapes
fixed
    f. end
    f. macrophage
fixing
fixus
    radius f.
flaccida
    membrana f.
flaccidity
flaccid part of tympanic membrane
Flack node
flagellum, pl. **flagella**
flame arc
flank
    f. area
    f. bone
flap
    artery island f.
    synovial f.
    thenar f.
flat
    f. bone
    f. chest
    f. foot
    f. muscle
    f. pelvis
flatfoot
flaval ligament
flavum, pl. **flava**
    ligamentum f.
    macula flava
    medulla ossium flava
flavus

Flechsig
> F. areas
> F. fasciculi
> F. ground bundles
> oval area of F.
> semilunar nucleus of F.
> F. tract

Fleischmann bursa
Fleisch pneumotachograph
Flemming
> intermediate body of F.
> minal center of F.

flesh
flex
> dorsal f.

flexa
> coxa f.

flexion
> f. crease
> f. of knee

flexor
> f. accessorius muscle
> f. carpi radialis (FCR)
> f. carpi radialis muscle
> f. carpi radialis tendon
> f. carpi ulnaris (FCU)
> f. carpi ulnaris muscle
> f. carpi ulnaris tendon
> f. compartment of arm
> f. compartment of forearm
> f. compartment of leg
> f. compartment of thigh
> f. digiti minimi brevis muscle of foot
> f. digiti minimi brevis muscle of hand
> f. digiti quinti (FDM)
> f. digitorum brevis (FDB)
> f. digitorum brevis muscle
> f. digitorum longus (FDL)
> f. digitorum longus muscle
> f. digitorum longus tendon
> f. digitorum profundus muscle
> f. digitorum profundus tendon
> f. digitorum sublimis (FDS)
> f. digitorum superficialis (FDS)
> f. digitorum superficialis muscle
> f. digitorum superficialis tendon
> f. hallucis brevis (FHB)
> f. hallucis brevis muscle
> f. hallucis brevis tendon
> f. hallucis longus muscle

> f. hallucis longus tendon
> musculus f.
> plantar f.
> f. pollicis brevis muscle
> f. pollicis longus muscle
> f. pollicis longus tendon
> f. reflex
> f. retinaculum
> f. retinaculum of forearm
> f. retinaculum of lower limb
> f. retinaculum of wrist
> f. sheath

flexorum
> compartimentum antebrachii f.
> compartimentum brachii f.
> compartimentum cruris f.
> compartimentum femoris f.
> retinaculum f.
> retinaculum musculorum f.
> vagina communis musculorum f.

flexura, pl. **flexurae**
> f. anorectalis
> f. colica splenica
> f. coli dextra
> f. coli hepatis
> f. coli sinistra
> f. duodeni inferior
> f. duodeni superior
> f. duodenojejunalis
> f. perinealis canalis ani
> f. perinealis recti
> f. sacralis recti
> f. sigmoidea

flexural
flexure
> anorectal f.
> basicranial f.
> caudal f.
> cephalic f.
> cerebral f.
> cervical f.
> colic f.
> f. of colon
> cranial f.
> dorsal f.
> duodenojejunal f.
> f. of duodenum
> hepatic f.
> iliac f.
> inferior duodenal f.
> left colic f.
> left colonic f.

F

NOTES

flexure *(continued)*
    f. line
    lumbar f.
    mesencephalic f.
    pontine f.
    f. of rectum
    right colic f.
    right colonic f.
    sacral f.
    sigmoid f.
    splenic f.
    superior duodenal f.
    telencephalic f.
    transverse rhombencephalic f.
flight or fight response
Flint arcade
floating
    f. cartilage
    f. organ
    f. rib
    f. villus
floccular fossa
floccule
flocculonodular lobe
flocculus, pl. **flocculi**
    accessory f.
    peduncle of f.
    pedunculus flocculi
Flood ligament
floor
    antral f.
    f. cell
    jugular f.
    f. of mouth
    f. of orbit
    orbital f.
    f. of orifice
    pelvic f.
    f. plate
    f. of tympanic cavity
flora
    intestinal f.
Florschütz formula
flow
    Bingham f.
    blood f. ($\dot{Q}$)
    effective renal blood f.
    effective renal plasma f.
    forced expiratory f.
    laminar f.
    newtonian f.
    peak expiratory f.
    shear f.
    venous f.
flower
    f. basket of Bochdalek
    F. bone
    F. dental index

flower-spray
    f.-s. ending
    f.-s. organ of Ruffini
flowmeter
    electromagnetic f.
flow-volume curve
fluctuance
fluctuantes
    costae f.
fluctuate
fluctuation
fluid
    allantoic f.
    amniotic f.
    Brodie f.
    cerebrospinal f. (CSF)
    chylous f.
    crevicular f.
    endolymphatic f.
    extracellular f.
    extravascular f.
    gingival f.
    interarterial f.
    interstitial f.
    intracellular f.
    intraocular f.
    newtonian f.
    non-newtonian f.
    perilymphatic f.
    pleural f.
    prostatic f.
    pseudoplastic f.
    Scarpa f.
    seminal f.
    spinal f.
    subdural f.
    subretinal f.
    sulcular f.
    synovial f.
    thixotropic f.
    tissue f.
    transcellular f.
    ventricular f.
    vitreous f.
fluidity
fluitantes
    costae f.
flumen
flumina pilorum
flux
    f. density
    net f.
    f. ratio
    unidirectional f.
fluxionary hyperemia
foam cells
fold
    AE f.

alar fold of intrapatellar
 synovial f.
amniotic f.
antehelical f.
anterior axillary f.
antral f.
aryepiglottic f.
arytenoepiglottidean f.
axillary f.
bulboventricular f.
caval f.
cecal f.
f. of chorda tympani
ciliary f.
circular f.'s
costocolic f.
Douglas f.
Duncan f.'s
duodenal f.
duodenojejunal f.
duodenomesocolic f.
epicanthal f.
epicanthic f.
epigastric f.
epiglottic f.
eponychial f.
fimbriated f.
gastric rugal f.
gastropancreatic f.
genital f.
glossoepiglottic f.
glossopalatine f.
gluteal f.
Guérin f.
Hasner f.
haustral f.
head f.
helical f.
hepatopancreatic f.
Houston f.
ileocecal f.
ileocolic f.
incudal f.
inferior duodenal f.
inferior rectal f.
inframammary f.
infrapatellar synovial f.
infraumbilical f.
inguinal aponeurotic f.
interdigital f.
interureteric f.
f. of iris

Kerckring f.
Kohlrausch f.
labial f.
labioscrotal f.'s
lacrimal f.
lateral glossoepiglottic f.
lateral nasal f.
lateral umbilical f.
left umbilical f.
f. of left vena cava
malar f.
mallear f.
malleolar f.
mammary f.
Marshall vestigial f.
medial canthic f.
medial nasal f.
medial umbilical f.
median glossoepiglottic f.
median umbilical f.
medullary f.
mesolateral f.
mesonephric f.
mesouterine f.
middle glossoepiglottic f.
middle transverse rectal f.
middle umbilical f.
mongolian f.
mucobuccal f.
mucosal f.
nail f.
nasojugal f.
nasolabial f.
Nélaton f.
neural f.
opercular f.
palatine f.
palatoglossal f.
palmate f.
palpebral f.
palpebronasal f.
pancreaticogastric f.
paraduodenal f.
parietocolic f.
parietoperitoneal f.
Passavant f.
patellar synovial f.
pericardiopleural f.
pharyngobasilar f.
pharyngoepiglottic f.
pleuroperitoneal f.
posterior axillary f.

NOTES

F

fold (*continued*)
    presplenic f.
    rectal f.
    rectouterine f.
    rectovaginal f.
    rectovesical f.
    retinal f.
    retrotarsal f.
    right umbilical f.
    Rindfleisch f.
    rugae f.
    rugal f.
    sacrogenital f.
    sacrouterine f.
    sacrovaginal f.
    sacrovesical f.
    salpingopalatine f.
    salpingopharyngeal f.
    Schultze f.
    semilunar conjunctival f.
    sentinel f.
    serosal f.
    sigmoid f.
    stapedial f.
    f. of stapes
    sublingual f.
    superior duodenal f.
    f. of superior laryngeal nerve
    superior rectal f.
    synovial f.
    systoletail f.
    tail f.
    tarsal f.
    tonsillar f.
    transverse palatine f.
    transverse rectal f.
    transverse vesical f.
    Treves f.
    triangular f.
    Tröltsch f.
    umbilical f.
    urachal f.
    ureteric f.
    urogenital f.
    urorectal f.
    uterosacral f.
    uterovesical f.
    vaginal f.
    Vater f.
    ventricular f.
    vestibular f.
    vestigial f.
    vocal f.
folding
    skin f.
folia (*pl. of* folium)
foliaceous
foliar

foliatae
    papillae f.
foliate papilla
folii
    lobulus f.
foliose
folium, pl. **folia**
    folia cerebelli
    folia linguae
    f. vermis
Folius muscle
follian process
folliberin
follicle
    aggregated lymphatic f.'s
    anovular ovarian f.
    atretic ovarian f.
    dental f.
    gastric lymphatic f.
    graafian f.
    granular layer of a vesicular
        ovarian f.
    growing ovarian f.
    hair f.
    intestinal f.'s
    Lieberkühn f.
    lingual f.'s
    lymph f.
    lymphatic f.
    mature ovarian f.
    Montgomery f.'s
    neck of hair f.
    ovarian f.
    polyovular ovarian f.
    primary ovarian f.
    primordial ovarian f.
    sebaceous f.'s
    solitary f.'s
    splenic lymph f.'s
    f.'s of thyroid gland
    vesicular ovarian f.
follicle-stimulating
    f.-s. hormone-releasing factor
    f.-s. hormone-releasing hormone
follicular
    f. antrum
    f. epithelial cell
    f. gland
    f. ovarian cells
folliculus, pl. **folliculi**
    atresia folliculi
    folliculi glandulae thyroideae
    folliculi linguales
    liquor folliculi
    folliculi lymphatici aggregati
    folliculi lymphatici aggregati
        appendicis vermiformis
    folliculi lymphatici laryngei
    folliculi lymphatici lienales

folliculi lymphatici recti
folliculi lymphatici solitarii
folliculi lymphatici solitarii intestini crassi
folliculi lymphatici solitarii intestini tenuis
f. lymphaticus
f. lymphaticus gastricus
f. ovaricus primarius
f. ovaricus vesiculosus
f. pili
theca folliculi
tunica externa thecae folliculi
tunica interna thecae folliculi

Folli process
Foltz valvule
Fontana
F. canal
space of F.
F. space
fontanelle, fontanel
anterior f.
anterolateral f.
bregmatic f.
Casser f.
frontal f.
Gerdy f.
mastoid f.
occipital f.
posterior f.
posterolateral f.
sagittal f.
sphenoidal f.
fonticulus, pl. fonticuli
f. anterior
f. anterolateralis
fonticuli cranii
f. mastoideus
f. posterior
f. posterolateralis
f. sphenoidalis
fontinalis
decussatio f.
foot, pl. feet
abductor digiti minimi muscle of f.
accessory flexor muscle of f.
anular part of fibrous digital sheath of digits of hand and f.
arches of f.
articulations of f.
base of phalanx of f.

bone of f.
f. bone
contracted f.
digits of f.
digitus pedis f.
distal phalanx of f.
dorsal artery of f.
dorsal digital nerve of f.
dorsal digital vein of f.
dorsal fascia of f.
dorsal interossei of f.
dorsal interosseous muscle of f.
dorsal surface of digit of hand or f.
dorsal venous arch of f.
dorsal venous network of f.
dorsiflexion of f.
dorsum of f.
fibrous digital sheaths of f.
fibular margin of f.
fibular peroneal border of f.
flat f.
flexor digiti minimi brevis muscle of f.
fundiform ligament of f.
head of phalanx of hand or f.
f. of hippocampus
inconstant arcuate artery of f.
intercapitular vein of f.
interphalangeal joint of f.
intrinsic muscle of f.
joint of f.
lateral border of f.
lateral dorsal cutaneous nerve of f.
lateral part of longitudinal arch of f.
ligament of f.
longitudinal arch of f.
lumbrical muscle of f.
lumbricals of f.
lumbricals (lumbrical muscles) of f.
medial border of f.
medial part of longitudinal arch of f.
Mendel dorsal reflex of f.
Morand f.
perforating artery of f.
peroneal border of f.
plantar aspect of f.
plantar flexion of f.
plantar ligament of interphalangeal joint of f.

F

NOTES

foot *(continued)*
    plantar surface of f.
    f. plate
    f. pod of alar cartilage
    primary digit of f.
    f. process
    pronation of f.
    proximal phalanx of f.
    reel f.
    root of f.
    skeleton of f.
    sole of f.
    supination of f.
    synovial sheaths of digits of f.
    tendinous sheath of extensor
      digitorum longus muscle of f.
    tendinous sheath of flexor
      digitorum longus muscle of f.
    tibial border of f.
    transverse arch of f.
    trochleae of phalanges of hand
      and f.
    tuberosity of distal phalanx of
      hand and f.
    vincula tendinea of digits of hand
      and f.
footplate
    stapes f.
foot-pound
foot-poundal
foot-pound-second (FPS)
    f.-p.-s. system
    f.-p.-s. unit
foramen, pl. **foramina**
    foramina alveolaria
    foramina alveolaria corporis
      maxillae
    anterior condyloid f.
    anterior palatine f.
    aortic f.
    apical dental f.
    f. apicis dentis
    arachnoid f.
    f. of Arnold
    axillary f.
    Bichat f.
    Bochdalek f.
    f. of Bochdalek
    Botallo f.
    f. bursae omentalis majoris
    caroticoclinoid f.
    carotid f.
    f. cecum of frontal bone
    f. cecum linguae
    f. cecum medullae oblongatae
    f. cecum ossis frontalis
    f. cecum posterius
    f. cecum of tongue
    conjugate f.

f. costotransversarium
costotransverse f.
cribriform f.
f. in cribriform plate
foramina cribrosa
f. diaphragmatis sellae
Duverney f.
emissary sphenoidal f.
epiploic f.
f. epiploicum
ethmoidal f.
f. ethmoidale
f. ethmoidale anterior et posterior
f. ethmoidale anterius
f. ethmoidale posterius
external acoustic f.
external auditory f.
Ferrein f.
f. of fourth ventricle
frontal f.
f. frontale
Galen f.
great f.
greater palatine f.
greater sciatic f.
Huschke f.
Hyrtl f.
incisive f.
f. incisivum
incisor f.
inferior dental f.
infraorbital f.
f. infraorbitale
infrapiriform f.
f. intermesocolica transversa
internal acoustic f.
internal auditory f.
internal neurocranial f.
interventricular f.
f. interventriculare
intervertebral f.
f. intervertebrale
f. ischiadicum
f. ischiadicum anterior et posterior
f. ischiadicum majus
f. ischiadicum majus et minor
f. ischiadicum minus
jugular f.
f. jugulare
f. of Key-Retzius
lacerated f.
f. lacerum
f. lacerum anterius
f. lacerum medium
f. lacerum posterius
Lannelongue f.
lateral f.
f. lateralis ventriculi quarti
lesser palatine f.

lesser sciatic f.
f. of Luschka
Luschka and Magendie f.
lymph node of anterior border of
  omental f.
Magendie f.
f. magnum
malar f.
f. mandibulae
mandibular f.
mastoid f.
f. mastoideum
medial f.
mental f.
f. mentale
f. of Monro
Monro f.
Morgagni f.
f. of Morgagni
nasal f.
foramina nervosa
neural f.
f. nutricium
nutrient f.
obturator f.
f. obturatum
occipital f.
oculomotor f.
olfactory f.
omental f.
f. omentale
optic f.
f. opticum
oval f.
f. ovale
f. ovale basis cranii
f. ovale cordis
f. ovale of heart
f. ovale ossis sphenoidalis
pacchionian f.
foramina palatina minora
palatine f.
f. palatinum majus
foramina papillaria renis
parietal f.
f. parietale
petrosal f.
f. petrosum
posterior condyloid f.
posterior palatine f.
postglenoid f.
primary interatrial f.

f. processus transversi
f. quadratum
f. recessus superioris bursae
  omentalis
Retzius f.
rivinian f.
root f.
f. rotundum
round f.
sacral f.
f. sacrale
foramina sacralia anterior et
  posterior
foramina sacralia dorsalia
foramina sacralia pelvina
foramina of Scarpa
Scarpa f.
Schwalbe f.
sciatic f.
f. of sclera
secondary interatrial f.
f. of sellar diaphragm
singular f.
f. singulare
foramina of smallest vein
foramina of the smallest vein of
  heart
Soemmering f.
solitary f.
sphenoidal emissary f.
sphenoid emissary f.
sphenopalatine f.
f. sphenopalatinum
sphenotic f.
f. spinosum
Stensen f.
stylomastoid f.
f. stylomastoideum
f. subseptale
f. of superior recess of omental
  bursa
supraorbital f.
f. supraorbitale
suprapiriform f.
thebesian f.
thyroid f.
f. thyroideum
f. transversarium
transverse f.
f. of transverse process
valve of oval f.
f. of vena cava

F

NOTES

foramen *(continued)*
    vena caval f.
    f. venae cavae
    foramina of the venae minimae
    foramina venarum minimarum
    foramina venarum minimarum
      cordis
    f. venosum
    venous f.
    vertebral f.
    f. vertebrale
    vertebroarterial f.
    f. vertebroarteriale
    f. vertebroarterialis
    Vesalius f.
    Vicq d'Azyr f.
    Vieussens f.
    Weitbrecht f.
    f. of Winslow
    Winslow f.
    zygomaticofacial f.
    f. zygomaticofaciale
    zygomaticoorbital f.
    f. zygomaticoorbitale
    zygomaticotemporal f.
    f. zygomaticotemporale
foraminalis
    nodus lymphoideus f.
foraminal lymph node
foraminiferous
foraminis
    nodus f.
foraminosus
    tractus spiralis f.
foraminula
foraminulosus
    tractus spiralis f.
foraminulum
force
    animal f.
    anterior component of f.
    chewing f.
    dynamic f.
    electromotive f.
    G f.
    f. of mastication
    masticatory f.
    nerve f.
    nervous f.
    occlusal f.
    reserve f.
    unit of f.
    van der Waals f.
    vital f.
forced
    f. cycle
    f. expiratory flow
    f. expiratory time
    f. expiratory volume

    f. respiration
    f. vital capacity
forceps
    artery f.
    f. major
    f. minor
    f. posterior
force-velocity curve
forcipate
forearm
    anterior compartment of f.
    anterior region of f.
    anterior surface of f.
    cephalic vein of f.
    deep fascia of f.
    deep part of anterior compartment
      of f.
    extensor compartment of f.
    fascia of f.
    flexor compartment of f.
    flexor retinaculum of f.
    intermediate vein of f.
    interosseous membrane of f.
    lateral border of f.
    lateral cutaneous nerve of f.
    lateral flexor muscle of f.
    lateral part of posterior extensor
      compartment of f.
    medial border of f.
    medial cutaneous nerve of f.
    medial flexor muscle of f.
    median vein of f.
    oblique cord of interosseous
      membrane of f.
    posterior branch of medial
      cutaneous nerve of f.
    posterior compartment of f.
    posterior cutaneous nerve of f.
    posterior region of f.
    posterior surface of f.
    pronation of f.
    radial border of f.
    radial part of posterior
      compartment of f.
    superficial part of anterior flexor
      compartment of f.
    supination of f.
    ulnar border of f.
    ulnar margin of f.
forebrain
    f. bundle
    f. eminence
    f. prominence
    f. vesicle
forefinger
forefoot
foregut
forehead
foreign body giant cell

forekidney
Forel
    F. decussate
    fields of F.
    fornix longus of F.
    space of F.
    tegmental fields of F.
foreli
    campi f.
foreskin
    frenulum of f.
    f. of penis
forestomach
form
    face f.
formatio, pl. **formationes**
    f. hippocampalis
    f. reticularis
formation
    reticular f.
formative cells
formula, pl. **formulae**
    Bernhardt f.
    Black f.
    Broca f.
    Demoivre f.
    Dreyer f.
    DuBois f.
    electrical f.
    Florschütz f.
    Gorlin f.
    Mall f.
    Meeh f.
    Meeh-Dubois f.
    Pignet f.
    Van Slyke f.
    vertebral f.
fornicate
fornicatus
    gyrus f.
    isthmus of gyrus f.
fornicis
    carina f.
    columna f.
    commissura f.
    crus f.
    delta f.
    stria f.
    tenia f.
fornix, pl. **fornices**
    anterior pillar of f.
    anterior vaginal f.

body of f.
cerebral f.
f. cerebri
column of f.
commissure of f.
f. conjunctivae inferior
f. conjunctivae superior
conjunctival f.
crus of f.
f. gastricus
f. of lacrimal sac
lateral part of vaginal f.
f. longus of Forel
pharyngeal f.
f. pharyngis
pillar of f.
posterior part of vaginal f.
posterior pillar of f.
posterior vaginal f.
f. sacci lacrimalis
f. of stomach
superior f.
f. of superior conjunctiva
tenia of f.
transverse f.
f. uterus
f. vaginae
vaginal f.
forskolin
fossa, pl. **fossae**
    acetabular f.
    f. acetabuli
    adipose fossae
    amygdaloid f.
    anconal f.
    antecubital f.
    anterior cranial f.
    f. anthelicis
    f. of anthelix
    f. antihelica
    f. axillaris
    axillary f.
    Bichat f.
    Biesiadecki f.
    Broesike f.
    f. canina
    canine f.
    f. carotica
    cerebral f.
    cistern of lateral cerebral f.
    Claudius f.
    conchal f.

NOTES

F

fossa (*continued*)
    condylar f.
    f. condylaris
    condyloid f.
    coronoid f.
    f. coronoidea
    f. coronoidea humeri
    cranial f.
    f. cranii anterior
    f. cranii media
    f. cranii posterior
    crural f.
    Cruveilhier f.
    cubital f.
    f. cubitalis
    digastric f.
    f. digastrica
    digital f.
    f. ductus venosi
    f. of ductus venosus
    duodenal fossae
    duodenojejunal f.
    eminence of triangular f.
    epigastric f.
    f. epigastrica
    ethmoidal f.
    fat body of ischioanal f.
    fat body of ischiorectal f.
    fat pad of ischioanal f.
    femoral f.
    floccular f.
    gallbladder f.
    f. for gallbladder
    Gerdy hyoid f.
    f. glandulae lacrimalis
    glenoid f.
    greater supraclavicular f.
    Gruber-Landzert f.
    f. of helix
    hepatic f.
    hyaloid f.
    f. hyaloidea
    hypophysial f.
    f. hypophysialis
    iliac f.
    f. iliaca
    iliacosubfascial f.
    f. iliacosubfascialis
    iliopectineal f.
    f. incisiva
    incisive f.
    incudal f.
    f. incudis
    f. of incus
    inferior duodenal f.
    infraclavicular f.
    f. infraclavicularis
    infraduodenal f.
    f. infraspinata

infraspinous f.
infratemporal f.
f. infratemporalis
inguinal f.
f. inguinalis lateralis
f. inguinalis medialis
f. innominata
innominate f.
intercondylar f.
f. intercondylaris
intercondylic f.
intercondyloid f.
f. intermesocolica transversa
interpeduncular f.
f. interpeduncularis
intrabulbar f.
ischioanal f.
f. ischioanalis
ischiorectal f.
f. ischiorectalis
Jobert de Lamballe f.
Jonnesco f.
jugular f.
f. jugularis
juxta-auricular f.
lacrimal f.
f. of lacrimal gland
f. of lacrimal sac
Landzert f.
lateral cerebral f.
lateral inguinal f.
f. lateralis cerebri
f. of lateral malleolus
lenticular f.
lesser supraclavicular f.
limiting sulcus of rhomboid f.
Malgaigne f.
f. malleoli fibulae
f. malleoli lateralis
mandibular f.
f. mandibularis
mastoid f.
medial inguinal f.
Merkel f.
mesentericoparietal f.
middle cranial f.
Mohrenheim f.
Morgagni f.
mylohyoid f.
navicular f.
f. navicularis auriculae
f. navicularis auris
f. navicularis Cruveilhier
f. navicularis urethrae
f. navicularis vestibulae vaginae
Ogura f.
f. olecrani
olecranon f.
omoclavicular f.

oval f.
f. ovalis
f. of oval window
ovarian f.
f. ovarica
paraduodenal f.
parajejunal f.
f. parajejunalis
pararectal f.
f. pararectalis
paravesical f.
f. paravesicalis
pelvic f.
peritoneal f.
petrosal f.
piriform f.
pituitary f.
f. poplitea
popliteal f.
posterior cranial f.
preauricular f.
prostatic f.
f. provesicalis
pterygoid f.
f. pterygoidea
pterygomaxillary f.
f. pterygopalatina
pterygopalatine f.
radial f.
f. radialis
f. radialis humeri
recess of anterior ischioanal f.
renal f.
retroduodenal f.
retromandibular f.
f. retromandibularis
retromolar f.
f. retromolaris
rhomboid f.
f. rhomboidea
Rosenmüller f.
f. of Rosenmüller
f. of round window
f. sacci lacrimalis
scaphoid f.
f. scaphoidea
f. scaphoidea ossis sphenoidalis
f. scarpae major
sigmoid f.
sphenomaxillary f.
f. subarcuata
subarcuate f.

subcecal f.
subinguinal f.
sublingual f.
submandibular f.
f. submandibularis
submaxillary f.
subpyramidal f.
subscapular f.
f. subscapularis
subsigmoid f.
superior duodenal f.
supraclavicular f.
f. supraclavicularis major
f. supraclavicularis minor
supramastoid f.
f. supraspinata
supraspinatus f.
supraspinous f.
suprasternal f.
supratonsillar f.
f. supratonsillaris
supravesical f.
f. supravesicalis
sylvian f.
f. of Sylvius
temporal f.
f. temporalis
f. terminalis urethrae
tibiofemoral f.
tonsillar f.
f. tonsillaris
transverse intermesocolic f.
Treitz f.
triangular f.
f. triangularis auriculae
trochanteric f.
f. trochanterica
trochlear f.
f. trochlearis
umbilical f.
urachal f.
valve of navicular f.
Velpeau f.
f. venae cavae
f. venae umbilicalis
f. venosa
vermian f.
f. vesicae biliaris
f. vesicae felleae
vestibular f.
f. of vestibule of vagina
f. vestibuli vaginae

F

NOTES

fossa (*continued*)
  Waldeyer fossae
  zygomatic f.
fossette
fossula, pl. **fossulae**
  f. fenestrae cochleae
  f. fenestrae vestibuli
  f. petrosa
  petrosal f.
  f. rotunda
  tonsillar fossulae
  fossulae tonsillarum palatini et
    pharyngealis
fossulate
fountain decussation
fourché
  main f.
fourchette
four-headed muscle
fourth
  f. cranial nerve
  f. finger
  f. lumbar nerve
  f. parallel pelvic plane
  f. toe
  f. turbinated bone
  f. ventricle
fovea, pl. **foveae**
  f. articularis capitis radii
  f. articularis inferior atlantis
  f. articularis superior atlantis
  f. capitis ossis femoris
  f. cardiaca
  f. centralis maculae luteae
  f. centralis retinae
  f. coccygis
  f. costalis inferior
  f. costalis processus transversi
  f. costalis superior
  f. dentis atlantis
  f. elliptica
  f. ethmoidalis
  f. of the femoral head
  f. femoralis
  f. hemielliptica
  f. hemispherica
  f. inguinalis interna
  f. for ligament of head of femur
  Morgagni f.
  f. oblonga cartilaginis arytenoideae
  pterygoid f.
  f. pterygoidea
  f. of radial head
  f. spherica
  f. sublingualis
  f. submandibularis
  f. submaxillaris
  superior f.
  f. supravesicalis

  f. triangularis cartilaginis
  f. triangularis cartilaginis
    arytenoideae
  trochlear f.
  f. trochlearis
foveal ligament
foveate
foveated chest
foveola, pl. **foveolae**
  f. coccygea
  coccygeal f.
  f. gastrica
  granular foveolae
  foveolae granulares
  f. ocularis
  radial f.
  f. of retina
  f. retinae
  f. suprameatalis
  f. suprameatica
foveolar
  f. cells of stomach
  f. gastric mucosa
  f. reflex
foveolate
Foville fasciculus
FPS
  foot-pound-second
fraction
  ejection f.
  filtration f.
fracture
  birth f.
frame
  Deiters terminal f.'s
framework
Frankenhäuser ganglion
Frankfort
  F. horizontal plane
  F. plane
Frankfort-mandibular incisor angle
Frank-Starling curve
fraternal twins
FRC
  functional residual capacity
free
  f. border
  f. border of nail
  f. border of ovary
  f. energy
  f. field
  f. macrophage
  f. margin
  f. margin of eyelid
  f. nerve endings
  f. part of lower limb
  f. part of upper limb
  f. tenia

f. villus
f. water clearance
frena (*pl. of* frenum)
frenal
frenulum, pl. **frenula**
    f. cerebelli
    f. of clitoris
    f. epiglottidis
    f. of foreskin
    f. of Giacomini
    f. of ileal orifice
    f. of ileocecal valve
    f. of labia minora
    f. labii inferioris
    f. labii superioris
    f. labiorum minorum
    f. labiorum pudendi
    f. linguae
    lingual f.
    f. of lower lip
    f. of M'Dowel
    f. of Morgagni
    f. ostii ilealis
    f. of penis
    f. of prepuce
    f. preputii
    f. preputii clitoridis
    f. of pudendal lip
    f. of superior medullary velum
    synovial frenula
    f. of tongue
    f. valvae ileocecalis
    f. veli medullaris superius
frenum, pl. **frena**
    labial f.
    lingual f.
    Morgagni f.
    synovial frena
frequency
    f. of micturition
    respiratory f.
fretum, pl. **freta**
Frey syndrome
FRF
    fasciculus retroflexus
friction
    dynamic f.
    starting f.
    static f.
fright reaction
frill
    iris f.

fringe
    cervical f.
    Richard f.
    synovial f.
frondosum
    chorion f.
frons
frontad
frontal
    f. angle of parietal bone
    f. antrum
    f. area
    f. aspect
    f. axis
    f. belly
    f. belly of occipitofrontalis muscle
    f. border
    f. border of parietal bone
    f. border of sphenoid bone
    f. branch of middle meningeal artery
    f. branch of superficial temporal artery
    f. cortex
    f. crest
    f. diploic vein
    f. eminence
    f. fontanelle
    f. foramen
    f. grooves
    f. gyrus
    f. horn
    f. lobe
    f. margin
    f. margin of sphenoid
    f. nerve
    f. notch
    f. paranasal sinus
    f. plane
    f. plate
    f. pole
    f. process
    f. process of maxilla
    f. process of zygomatic bone
    f. recess
    f. region
    f. region of head
    f. section
    f. segment
    f. sinus aperture
    f. squama
    f. suture

NOTES

F

frontal *(continued)*
    f. triangle
    f. tuber
    f. vein
    f. zygomatic suture line
frontale
    foramen f.
    os f.
    tuber f.
frontales
    venae f.
frontalia
    plana f.
frontalis
    apertura sinus f.
    arteria polaris f.
    crista f.
    diploica f.
    eminentia f.
    facies externa ossis f.
    facies interna ossis f.
    foramen cecum ossis f.
    gyrus f.
    incisura f.
    linea temporalis ossis f.
    margo nasalis ossis f.
    margo parietalis ossis f.
    f. muscle
    musculus f.
    nervus f.
    norma f.
    pars nasalis ossis f.
    pars orbitalis ossis f.
    processus zygomaticus ossis f.
    sinus f.
    spina nasalis ossis f.
    squama f.
    sutura f.
    torus f.
    venter f.
frontalium
    septum sinuum f.
fronte
    vis a f.
frontis
frontoethmoidalis
    sutura f.
frontoethmoidal suture
frontolacrimalis
    sutura f.
frontolacrimal suture
frontomalar suture
frontomarginalis
    sulcus f.
frontomaxillaris
    sutura f.
frontomaxillary suture
frontomental diameter

frontonasal
    f. duct
    f. elevation
    f. process
    f. suture
frontonasalis
    sutura f.
frontooccipital
    f.-o. diameter
    f.-o. fasciculus
frontooccipital
frontoorbital area
frontoparietal
    f. area
    f. suture
frontopolar artery
frontopontine tract
frontopontinus
    tractus f.
frontosphenoidal process
frontosphenoid suture
frontotemporale
frontotemporal tract
frontozygomatica
    sutura f.
frontozygomatic suture
front-tap contraction
Froriep ganglion
fuchsinophil
    f. cell
    f. granule
fuchsinophilia
fuchsinophilic
Fuchs stomas
fugacity
fulcrum, pl. **fulcra, fulcrums**
function
    allomeric f.
    arousal f.
    atrial transport f.
    isomeric f.
    sensory f.
functional
    f. anatomy
    f. congestion
    f. hypertrophy
    f. refractory period
    f. residual air
    f. residual capacity (FRC)
    f. sphincter
    f. terminal innervation ratio
functionale
    stratum f.
fundal
    f. gland
    f. portion of uterus
fundament
fundamentalis
    substantia f.

fundi (*pl. of* fundus)
fundic
    f. gland
    f. mucosa
fundic-antral junction
fundiform
    f. ligament of clitoris
    f. ligament of foot
    f. ligament of penis
fundus, pl. **fundi**
    gallbladder f.
    f. of gallbladder
    f. gastricus
    f. gland
    f. of internal acoustic meatus
    f. of internal auditory meatus
    f. meatus acustici interni
    ocular f.
    f. oculi
    f. of stomach
    f. tympani
    f. of urinary bladder
    uterine f.
    f. of uterus
    f. ventriculi
    f. vesicae biliaris
    f. vesicae felleae
    f. vesicae urinariae
fungiformes
    papillae f.
fungiform papilla
fungilliform
funic
funicle
funicular
    f. artery
    f. part of ductus deferens
    f. process
funiculus, pl. **funiculi**
    anterior f.
    f. anterior
    cuneate f.
    dorsal f.
    f. dorsalis
    f. gracilis
    f. lateralis
    funiculi medullae spinalis

    posterior f.
    f. posterior
    f. separans
    f. solitarius
    f. spermaticus
    f. teres
    f. umbilicalis
funiform
funis
funnel
    f. chest
    Martegiani f.
    pial f.
funnel-shaped pelvis
funny bone
fura-2
furcalis
    nervus f.
furcal nerve
furcula
furrow
    digital f.
    genital f.
    gluteal f.
    mentolabial f.
    palpebral f.
    primitive f.
    Schmorl f.
    skin f.
fusca
    f. lamina
    membrana f.
fuse
fused kidney
fusiform
    f. cells of cerebral cortex
    f. gyrus
    f. layer
    f. muscle
fusiformis
    gyrus f.
    lobulus f.
    musculus f.
fusimotor
fusion
    cell f.
fusocellular

NOTES

F

G

G cells
G force
gag reflex
galactagogue
g. factor
galactidrosis
galactobolic
galactophore
galactophori
tubuli g.
galactophorous
g. canal
g. duct
galactopoiesis
galactopoietic
g. factor
g. hormone
galactosis
Galant abdominal reflex
galea
g. aponeurotica
tendinous g.
Galeati gland
Galen
G. anastomosis
G. foramen
great vein of G.
G. nerve
vein of G.
G. vein
G. ventricle
gall
g. bladder
g. duct
Gallaudet fascia
gallbladder (GB)
ampulla of g.
g. bed
body of g.
fossa for g.
g. fossa
fundus of g.
g. fundus
infundibulum of g.
mucosa of g.
mucosal fold of g.
mucous membrane of g.
muscular coat of g.
muscular layer of g.
muscular tunic of g.
neck of g.
pelvis of g.
rugae of g.

serosa of g.
g. wall
galli
ala of crista g.
ala cristae g.
crista g.
wing of crista g.
gallinaginis
caput g.
Galton delta
galvanic
g. skin reaction
g. skin reflex
g. skin response (GSR)
g. threshold
galvanocontractility
galvanofaradization
galvanometer
d'Arsonval g.
galvanomuscular
galvanopalpation
galvanoscope
galvanotaxis
galvanotonus
galvanotropism
gametangium
gamete
gametocide
gametocyte
gametogenesis
gametogonia
gametogony
gametoid
gametokinetic hormone
gametophagia
gamic
gamma
g. cell of pancreas
g. efferent
g. fiber
g. loop
g. motor neuron
g. motor system
gamma-Abu
gamma-aminobutyric
g.-a. acid
gamma-fetoprotein
gamogenesis
gamogony
gamont
gamophagia
ganglia (*pl. of* ganglion)
ganglial
gangliate

G

gangliated
  g. cord
  g. nerve
gangliform
ganglioblast
gangliocyte
ganglioform
ganglioformis
  intumescentia g.
ganglion, pl. **ganglia**
  abdominal part of peripheral
    autonomic plexuses and ganglia
  aberrant g.
  acousticofacial g.
  Acrel g.
  Andersch g.
  aorticorenal g.
  ganglia aorticorenalia
  Arnold g.
  auditory g.
  Auerbach ganglia
  auricular g.
  auser g.
  autonomic ganglia
  ganglia of autonomic plexus
  basal ganglia
  Bezold g.
  Bochdalek g.
  Bock g.
  Böttcher g.
  branch of internal carotid artery to
    trigeminal g.
  branch of oculomotor nerve to
    ciliary g.
  branch to trigeminal g.
  cardiac ganglia
  ganglia cardiaca
  carotid g.
  celiac ganglia
  ganglia celiaca
  g. cell
  g. cervicale inferius
  g. cervicale medium
  g. cervicale superius
  cervicothoracic g.
  g. cervicothoracicum
  chain g.
  g. ciliare
  ciliary g.
  Cloquet g.
  coccygeal g.
  cochlear g.
  g. cochleare
  ganglia coeliaca
  communicating branch of
    nasociliary nerve with ciliary g.
  Corti g.
  craniocervical part of peripheral
    autonomic plexuses and g.

dorsal root g.
Ehrenritter g.
episcleral g.
esophageal branch of thoracic g.
g. extracraniale
g. of facial nerve
Frankenhäuser g.
Froriep g.
ganglionic branch of lingual nerve
  to sublingual g.
ganglionic branch of lingual nerve
  to submandibular g.
ganglionic branch of maxillary
  nerve to pterygopalatine g.
gasserian g.
geniculate g.
g. geniculatum
g. geniculi
glandular branch of
  submandibular g.
glossopharyngeal g.
Gudden g.
g. habenulae
hypogastric ganglia
g. impar
inferior cervical g.
inferior mesenteric g.
inferior part of vestibular g.
g. inferius nervi glossopharyngei
g. inferius nervi vagi
intercarotid g.
intercrural g.
ganglia intermedia
intermediate ganglia
g. of intermediate nerve
interpeduncular g.
interventricular g.
intervertebral g.
intracranial g.
g. isthmi
jugular g.
lacrimal g.
laryngopharyngeal branch of
  superior cervical g.
Laumonier g.
Lee g.
lenticular g.
lingual g.
Lobstein g.
long root of ciliary g.
Ludwig g.
ganglia lumbalia
lumbar ganglia
Meckel g.
Meissner g.
g. mesentericum inferius
g. mesentericum superius
middle cervical g.
motor root of ciliary g.

nasal g.
nasociliary root of ciliary g.
nephrolumbar g.
nerve g.
g. of nervus intermedius
neural g.
nodose g.
oculomotor root of ciliary g.
optic g.
orbital branch of pterygopalatine g.
otic g.
g. oticum
parasympathetic ganglia
ganglia parasympathetica
parasympathetic root of ciliary g.
parasympathetic root of otic g.
parasympathetic root of pelvic g.
parasympathetic root of
  pterygopalatine g.
parasympathetic root of
  submandibular g.
paravertebral ganglia
pelvic ganglia
ganglia pelvica
pelvic part of peripheral autonomic
  plexuses and g.
ganglia pelvina
petrosal g.
petrous g.
pharyngeal branch of
  pterygopalatine g.
phrenic ganglia
ganglia phrenica
ganglia plexuum autonomicorum
posterior superior lateral nasal
  branch of pterygopalatine g.
posterior superior medial nasal
  branch of pterygopalatine g.
prevertebral ganglia
prostatic g.
pterygopalatine g.
g. pterygopalatinum
Remak ganglia
renal ganglia
ganglia renalia
Ribes g.
g. ridge
root of ciliary g.
ganglia sacralia
Scarpa g.
Schacher g.
semilunar g.

g. sensorium nervi spinalis
sensory root of ciliary g.
sensory root of pterygopalatine g.
sensory root of sublingual g.
sensory root of submandibular g.
short root of ciliary g.
Soemmering g.
solar ganglia
sphenomaxillary g.
sphenopalatine g.
spinal g.
g. spinale
spiral g.
g. spirale cochlea
splanchnic g.
g. splanchnicum
stellate g.
g. stellatum
sublingual g.
g. sublinguale
submandibular g.
g. submandibulare
submaxillary g.
superior carotid g.
superior cervical g.
superior mesenteric g.
superior part of vestibular g.
superior vagal g.
g. superius nervi glossopharyngei
g. superius nervi vagi
suprarenal g.
sympathetic ganglia
sympathetic branch to
  submandibular g.
sympathetic root of ciliary g.
sympathetic root of otic g.
sympathetic root of
  pterygopalatine g.
sympathetic root of sublingual g.
sympathetic root of
  submandibular g.
g. of sympathetic trunk
synovial g.
terminal g.
g. terminale
thoracic ganglia
ganglia thoracica
thoracic cardiac branch of
  thoracic g.
thoracic part of peripheral
  autonomic plexuses and g.

NOTES

G

ganglion (*continued*)
 thoracic pulmonary branch of
  thoracic g.
 thoracic splanchnic g.
 g. thoracicum splanchnicum
 trigeminal g.
 g. trigeminale
 Troisier g.
 ganglia trunci sympathici
 g. of trunk of vagus
 tympanic g.
 g. tympanicum
 vagal g.
 Valentin g.
 ventricular g.
 vertebral g.
 g. vertebrale
 vestibular g.
 g. vestibulare
 Vieussens g.
 Walther g.
 Wrisberg ganglia
ganglionares
 rami g.
ganglionated
ganglionectomy
 sphenopalatine g.
ganglionic
 g. blocking agent
 g. branch of internal carotid artery
 g. branch of lingual nerve
 g. branch of lingual nerve to
  sublingual ganglion
 g. branch of lingual nerve to
  submandibular ganglion
 g. branch of maxillary nerve
 g. branch of maxillary nerve to
  pterygopalatine ganglion
 g. chain
 g. crest
 g. layer of cerebellar cortex
 g. layer of cerebral cortex
 g. layer of optic nerve
 g. layer of retina
 g. motor neuron
 g. saliva
Ganser
 G. commissures
 nucleus basalis of G.
Gantzer
 G. accessory bundle
 G. muscle
gap
 g. 1, 2
 Bochdalek g.
 g. junction
Gartner
 G. canal

 G. cyst
 G. duct
gas
 alveolar g.
 blood g.
 expired g.
 ideal alveolar g.
 inspired g.
 mixed expired g.
 serial blood g.
Gaskell
 G. bridge
 G. nerves
gasometer
gasometric
gasometry
gasserian ganglion
gaster
gastral mesoderm
gastric
 g. area
 g. artery
 g. branch of anterior vagal trunk
 g. branch of posterior vagal trunk
 g. canal
 g. cardia
 g. digestion
 g. gland
 g. impression
 g. impression on liver
 g. impression on spleen
 g. juice
 g. junction
 g. lymphatic follicle
 g. lymphoid nodule
 g. mucosa
 g. nervous plexus
 g. notch
 g. omentum
 g. outlet
 g. pit
 g. plexuses of autonomic system
 g. pouch
 g. pylorus
 g. ruga
 g. rugal fold
 g. secretion
 g. surface of spleen
 g. vein
gastrica, pl. **gastricae**
 area g.
 gastricae brevis
 g. dextra
 foveola g.
 glandulae gastricae
 impressio g.
 pars cardiaca gastricae
 plicae gastricae
 ruga gastricae

ruga g.
g. sinistra
tunica mucosa g.
tunica muscularis g.
tunica serosa gastricae
gastrici
gastricorum
plexus nervorum g.
gastricum
corpus g.
gastricus
canalis g.
folliculus lymphaticus g.
fornix g.
fundus g.
gastris
paries anterior g.
paries posterior g.
pars pylorica g.
gastroacephalus
gastroamorphus
gastrocardiac
gastrocele
gastrocnemius, pl. **gastrocnemii**
bursae of g.
bursae subtendineae musculi
gastrocnemii
caput laterale musculi gastrocnemii
caput mediale musculi gastrocnemii
g. equinus
g. muscle
musculus g.
g. reflex
gastrocolic
g. ligament
g. omentum
g. reflex
gastrocolicum
ligamentum g.
gastrocs
gastrodialysis
gastrodiaphragmatic ligament
gastroduodenal
g. artery
g. lumen
g. lymph node
g. mucosa
g. orifice
gastroduodenalis
arteria g.
musculus dilator pylori g.

gastroenteric
gastroepiploic
g. artery
g. branch
g. lymph node
g. vein
g. vessel
gastroepiploica, pl. **gastroepiploicae**
arteriae gastroepiploicae
g. dextra
g. sinistra
gastroesophageal
g. junction
g. sphincter
g. vestibule
gastrograph
gastrohepatic
g. ligament
g. omentum
gastroileac reflex
gastroileal reflex
gastrointestinal (GI)
g. hormone
g. mucosa
g. tract
gastrojejunal
gastrojejunocolic
gastrokinesograph
gastrolienale
ligamentum g.
gastrolienal ligament
gastromegaly
gastromelus
gastroomental artery
gastro-omentales
arteriae g.-o.
gastropagus
gastropancreatic
g. fold
g. ligament
g. reflex
gastropancreaticae
plicae g.
gastroparasitus
gastrophrenic ligament
gastrophrenicum
ligamentum g.
gastropneumonic
gastropulmonary
gastropyloric
gastroschisis

G

NOTES

227

gastrosplenic
 g. ligament
 g. omentum
gastrosplenicum
 ligamentum g.
gastrothoracopagus
gastrula
gastrulation
gate-control
 g.-c. hypothesis
 g.-c. theory
gating mechanism
Gaucher cells
gauge
 g. pressure
 strain g.
Gault cochleopalpebral reflex
gauss
Gavard muscle
Gay gland
GB
 gallbladder
G-CSF
 granulocyte colony-stimulating factor
gelatinoid
gelatinosa
 substantia g.
gelatinosus
 nucleus g.
gelatinous
 g. bone marrow
 g. substance
 g. tissue
gemästete cell
gemel
gemellology
gemellus
 g. inferior
 g. muscle
 g. superior
geminate
gemination
geminous
gemistocyte
gemistocytic
 g. astrocyte
 g. cell
gemma
gemmation
gemmule
 Hoboken g.
gena
genal gland
genera (*pl. of* genus)
general
 g. anatomy
 g. physiology
 g. somatic afferent column
 g. somatic efferent column

 g. stimulant
 g. visceral column
generales
 termini g.
generate
generation
 asexual g.
 nonsexual g.
 spontaneous g.
 virgin g.
generative
generator potential
genesial cycle
genesiology
genesis
genial tubercle
genian
genicula (*pl. of* geniculum)
genicula
genicular
 g. anastomosis
 g. artery
 g. branch
 g. vein
genicularis, pl. geniculares
 rami articulares arteriae
  descendentis g.
 ramus saphenus arteriae
  descendentis g.
 venae geniculares
geniculate
 g. body
 g. ganglion
geniculated
geniculatum
 ganglion g.
geniculi
 ganglion g.
 gyrus g.
geniculocalcarine
 g. radiation
 g. tract
geniculum, pl. genicula
 g. canalis facialis
 g. of facial canal
 g. of facial nerve
 g. nervi facialis
 g. nervus facialis
genioglossal muscle
genioglossus
 g. muscle
 musculus g.
geniohyoglossus
 musculus g.
geniohyoid
 g. fascia
 g. muscle
geniohyoideus
 musculus g.

genion
genital
    g. area
    g. branch of genitofemoral nerve
    g. branch of iliohypogastric nerve
    g. cord
    g. corpuscles
    g. duct
    g. eminence
    g. fold
    g. furrow
    g. gland
    g. ligament
    g. organs
    g. ridge
    g. swelling
    g. system
    g. tract
    g. tubercle
genitalia
    corpuscula g.
    external g.
    female external g.
    female internal g.
    male external g.
    male internal g.
    organa g.
    systema g.
genitalis
    ramus g.
genitocrural nerve
genitofemoralis
    nervus g.
    ramus femoralis nervi g.
    ramus genitalis nervi g.
genitofemoral nerve
genitoinguinale
    ligamentum g.
genitoinguinal ligament
genitospinal center
genitourinary
    g. apparatus
    g. region
    g. system
    g. tract
Gennari
    G. band
    line of G.
    G. stria
    stripe of G.
genodermatosis
gentianophil

gentianophile
gentianophilous
gentianophobic
genu, pl. **genua**
    g. capsulae internae
    g. corporis callosi
    g. of corpus callosum
    g. of facial canal
    g. of facial nerve
    g. impressum
    g. of internal capsule
    g. nervi facialis
    g. recurvatum
genual
genus, pl. **genera**
    arteria descendens g.
    arteria inferior lateralis g.
    arteria inferior medialis g.
    arteria media g.
    arteria superior lateralis g.
    arteria superior medialis g.
    articularis g.
    articulatio g.
    g. descendens
    g. inferior lateralis
    g. inferior medialis
    ligamenta cruciata g.
    ligamentum cruciatum tertium g.
    ligamentum transversum g.
    g. media
    musculus articularis g.
    rete articulare g.
    g. superior lateralis
    g. superior medialis
    venae g.
genyantrum
geotaxis
geotropism
Gerdy
    G. fiber
    G. fontanelle
    G. hyoid fossa
    G. interatrial loop
    G. knee tubercle
    G. ligament
Gerhardt triangle
Gerlach
    G. annular tendon
    G. tonsil
    G. valve
    G. valvula

G

NOTES

germ
g. cell
dental g.
g. disk
enamel g.
g. layer
g. layer theory
g. line
g. membrane
tooth g.
germinal
g. area
g. cell
g. cords
g. disk
g. epithelium
g. localization
g. membrane
g. pole
g. spot
g. vesicle
germinativa
area g.
macula g.
membrana g.
germinative
g. layer
g. layer of nail
germinativum
stratum g.
Gerota
G. capsule
G. fascia
gestational age
GFR
glomerular filtration rate
GHK
Goldman-Hodgkin-Katz
GHK equation
Ghon
G. complex
G. tubercle
ghost cell
GHRF
growth hormone-releasing factor
GHRH
growth hormone-releasing hormone
GI
gastrointestinal
Giacomini
band of G.
frenulum of G.
uncus band of G.
Giannuzzi
G. cells
G. crescents
G. demilunes
giant cell
Gibbs free energy

Gibbs-Helmholtz equation
Gierke respiratory bundle
gilbert
gill
g. arch skeleton
g. clefts
Gillette suspensory ligament
Gimbernat ligament
gingiva, pl. **gingivae**
gingival
g. artery
g. crevice
g. epithelium
g. fluid
g. groove
g. line
g. margin
g. mucosa
g. papilla
g. septum
g. space
g. sulcus
g. tissue
g. trough
g. zone
gingivalis
papilla g.
sulcus g.
gingivobuccal
g. groove
g. sulcus
gingivodental ligament
gingivolabial
g. groove
g. sulcus
gingivolingual
g. groove
g. sulcus
ginglyform
ginglymoarthrodial
ginglymoid joint
ginglymus
helicoid g.
lateral g.
girdle
joint of inferior limb g.
joint of pectoral g.
joint of pelvic g.
joint of superior limb g.
limb g.
pectoral g.
pelvic g.
shoulder g.
thoracic g.
girth
abdominal g.
gitter cell
gitterzelle
glabella

glabellad
glabelloalveolar line
gladiate
gladiolus bone
glairy mucus
gland

accessory adrenal g.
accessory lacrimal g.
accessory parotid g.
accessory suprarenal g.
accessory thyroid g.
acid g.
acinotubular g.
acinous g.
admaxillary g.
adrenal g.
aggregate g.
agminate g.
agminated g.
Albarran g.
albuminous g.
alveolar g.
anal g.
anterior lingual g.
anterior surface of suprarenal g.
apical g.
apocrine g.
areolar g.
arteriococcygeal g.
arytenoid g.
g. of auditory tube
axillary sweat g.
Bartholin, urethral, and Skene gland
Bauhin g.
Baumgarten g.
bed of parotid g.
biliary g.
g.'s of biliary mucosa
Blandin g.
Blandin-Nuhn g.
body of mammary g.
body of sweat g.
Bowman g.
brachial g.
bronchial g.
Bruch g.
Brunner g.
buccal g.
bulbocavernous g.
bulbourethral g.
BUS g.

cardiac g.
carotid g.
celiac g.
central vein of suprarenal g.
ceruminous g.
cervical g.
chief cell of parathyroid g.
Ciaccio g.
ciliary g.
circumanal g.
coccygeal g.
coil g.
g. of common bile duct
compound g.
conjunctival g.
convoluted g.
cortex of suprarenal g.
Cowper g.
cutaneous g.
deep part of parotid g.
duct of bulbourethral g.
ductless g.
duct of Skene g.
duct of sweat g.
duodenal g.
Duverney g.
Ebner g.
eccrine g.
Eglis g.
elevator muscle of thyroid g.
endocervical g.
endocrine g.
esophageal g.
g. of eustachian tube
excretory duct of lacrimal g.
excretory duct of seminal g.
excretory ductule of lacrimal g.
exocrine g.
external salivary g.
g. of female urethra
fibrous capsule of parotid g.
fibrous capsule of thyroid g.
follicles of thyroid g.
follicular g.
fossa of lacrimal g.
fundal g.
fundic g.
fundus g.
Galeati g.
gastric g.
Gay g.
genal g.

G

NOTES

gland *(continued)*
  genital g.
  Gley g.
  glomiform g.
  glossopalatine g.
  greater vestibular g.
  Guérin g.
  Havers g.
  hemal g.
  hematopoietic g.
  hemolymph g.
  Henle g.
  hibernating g.
  holocrine g.
  inferior parathyroid g.
  inferior thyroid g.
  infundibulum of pituitary g.
  inguinal g.
  internal salivary g.
  g. of internal secretion
  interscapular g.
  interstitial g.
  intestinal g.
  intraepithelial g.
  isthmus of thyroid g.
  jugular g.
  Knoll g.
  Krause g.
  labial g.
  lacrimal g.
  lactiferous g.
  g. of large intestine
  laryngeal g.
  lesser vestibular g.
  levator muscle of thyroid g.
  Lieberkühn g.
  lingual g.
  Littré g.
  lobe of mammary g.
  lobe of pituitary g.
  lobe of thyroid g.
  lobules of mammary g.
  lobules of thyroid g.
  Luschka cystic g.
  lymph g.
  major salivary g.
  g. of male urethra
  malpighian g.
  mammary g.
  marrow-lymph g.
  master g.
  maxillary g.
  medial border of suprarenal g.
  medulla of adrenal g.
  medulla of suprarenal g.
  Meibom g.
  meibomian g.
  merocrine g.
  Méry g.

  mesenteric g.
  milk g.
  minor salivary g.
  mixed g.
  molar g.
  Moll g.
  Montgomery g.
  g. of mouth
  mucilaginous g.
  muciparous g.
  mucosa of seminal g.
  mucous g.
  mucus-secreting g.
  muscular layer of seminal g.
  nasal g.
  Nuhn g.
  odoriferous g.
  oil g.
  olfactory g.
  orbital part of lacrimal g.
  oxyntic g.
  pacchionian g.
  palatine g.
  palpebral part of lacrimal g.
  papilla of parotid g.
  parathyroid g.
  paraurethral g.
  parenchyma of thyroid g.
  parotid g.
  pectoral g.
  peptic g.
  perspiratory g.
  Peyer g.
  pharyngeal g.
  pileous g.
  pineal g.
  pituitary g.
  Poirier g.
  posterior surface of suprarenal g.
  prehyoid g.
  preputial g.
  prostate g.
  pyloric lymph g.
  pyramidal lobe of thyroid g.
  renal surface of suprarenal g.
  retromandibular process of
    parotid g.
  retrosternal g.
  Rivinus g.
  Rosenmüller g.
  saccular g.
  salivary g.
  sebaceous g.
  seminal g.
  sentinel g.
  seromucous g.
  serous g.
  Serres g.
  sexual g.

sheath of thyroid g.
Skene g.
g. of small intestine
solitary g.
splenolymph g.
stroma of thyroid g.
sublingual g.
submandibular salivary g.
submaxillary g.
submental g.
substernal g.
sudoriferous g.
sudoriparous g.
superficial part of parotid g.
superior border of suprarenal g.
superior lacrimal g.
superior parathyroid g.
superior thyroid g.
suprahyoid g.
suprarenal g.
suspensory ligament of thyroid g.
Suzanne g.
sweat g.
synovial g.
target g.
tarsal g.
Terson g.
Theile g.
thymus g.
thyroid g.
Tiedemann g.
tracheal g.
trachoma g.
trigeminal g.
tubular g.
tubuloacinar g.
tubuloalveolar g.
tympanic g.
Tyson g.
unicellular g.
urethral g.
uterine g.
vaginal g.
vascular g.
venous circle of mammary g.
vesical g.
vestibular g.
Von Ebner g.
vulvovaginal g.
Waldeyer g.
Wasmann g.
Weber g.

Wepfer g.
Wölfler g.
Wolfring g.
Zeis g.
glandes (*pl. of* glans)
glandilemma
glandis
collum g.
corona g.
septum g.
glandula, pl. **glandulae**
glandulae areolares
g. atrabiliaris
g. basilaris
glandulae bronchiales
glandulae buccales
g. bulbourethralis
glandulae ceruminosae
glandulae cervicales uterus
glandulae ciliares
glandulae circumanales
glandulae conjunctivales
glandulae cutis
ductuli excretorii glandulae
glandulae ductus biliaris
glandulae ductus choledochi
glandulae duodenales
glandulae endocrinae
glandulae esophageae
glandulae gastricae
glandulae glomiformes
glandulae intestinales
glandulae intestini crassi
glandulae intestini tenuis
glandulae labiales
glandulae lacrimales accessoriae
g. lacrimalis
glandulae laryngeae
g. lingualis anterior
g. mammaria
glandulae molares
g. mucosa
glandulae mucosae biliosae
glandulae nasales
glandulae olfactoriae
glandulae oris
glandulae palatinae
g. parathyroidea
g. parotidea
g. parotidea accessoria
g. parotis
g. parotis accessoria

NOTES

G

233

glandula *(continued)*
    glandulae pharyngeae
    glandulae pharyngeales
    g. pituitaria
    glandulae preputiales
    glandulae propriae
    g. prostatica
    glandulae pyloricae
    g. salivaria
    glandulae salivariae majores
    glandulae salivariae minores
    glandulae sebaceae
    g. semi
    g. seminalis
    g. seromucosa
    g. serosa
    glandulae sine ductibus
    g. sublingualis
    g. submandibularis
    glandulae sudoriferae
    glandulae suprarenales accessoriae
    g. suprarenalis
    glandulae tarsales
    g. thyroidea
    g. thyroidea accessoria
    glandulae thyroideae accessoriae
    glandulae tracheales
    glandulae tubariae
    glandulae urethrales
    glandulae urethrales feminae
    glandulae urethrales femininae
    glandulae urethrales masculinae
    glandulae uterinae
    g. vesiculosa
    glandulae vestibulares minores
    g. vestibularis major
glandular
    g. branch
    g. branch of facial artery
    g. branch of inferior thyroid artery
    g. branch of submandibular
      ganglion
    g. epithelium
    g. substance of prostate
    g. system
    g. tissue
glandulares
    rami g.
glandule
glandulopreputial lamella
glandulous
glans, pl. **glandes**
    g. clitoridis
    g. of clitoris
    neck of g.
    g. penis
    septum of g.

glaserian
    g. artery
    g. fissure
glassy membrane
glenohumeral
    g. articulation
    g. joint
    g. ligament
glenohumeralia
    ligamenta g.
glenohumeralis
    articulatio g.
glenoid
    g. cavity
    g. cavity of scapula
    g. fossa
    g. labrum
    g. labrum of scapula
    g. ligament
    g. process
    g. surface
glenoidale
    labrum g.
    ligamentum g.
glenoidalis
    cavitas g.
glenoidal lip
Gley gland
gliacyte
gliae
    membrana limitans g.
glial
    g. cells
    fibrillary g.
    g. membrane
    g. sheath
gliding joint
glioblast
Glisson
    G. capsule
    G. sphincter
globe
    g. of eye
    ocular g.
    pale g.
globi *(pl. of globus)*
globosus
    nucleus g.
globular
    g. leukocyte
    g. process
globule
    dentin g.
    polar g.
globulin
    gonadal steroid-binding g.
    sex steroid-binding g.
globulus
globus, pl. **globi**

g. abdominalis
g. of heel
g. major
g. major epididymidis
g. minor
g. minor epididymidis
g. pallidus

glomal
glome
glomera
aortic g.
g. aortica

glomerular
g. arteriole
g. basement membrane
g. capsule
g. filtration rate (GFR)
g. layer of olfactory bulb

glomerule
glomeruli (*pl. of* glomerulus)
glomerulosa
g. cell
zona g.

glomerulose
glomerulus, pl. **glomeruli**
capsula glomeruli
capsule of g.
malpighian g.
g. of mesonephros
olfactory g.
g. of pronephros

glomiformes
glandulae g.

glomiform gland
glomus, pl. **glomera**
g. aorticum
g. body
g. caroticum
choroid g.
g. choroideum
g. coccygeum
g. intravagale
g. jugulare
g. pulmonale

glossa
glossal
glossi
glossodynamometer
glossoepiglottic
g. fold
g. ligament

glossoepiglottidean
glossohyal
glossopalatine
g. arch
g. fold
g. gland
g. muscle

glossopalatinus
arcus g.
musculus g.

glossopharyngea
pars g.

glossopharyngeal
g. breathing
g. ganglion
g. muscle
g. nerve
g. nerve root
g. part of superior pharyngeal
constrictor

glossopharyngeo
ramus communicans nervi facialis
cum nervo g.

glossopharyngeus, pl. **glossopharyngei**
ganglion inferius nervi
glossopharyngei
ganglion superius nervi
glossopharyngei
musculus g.
nervus g.
rami linguales nervi
glossopharyngei
rami pharyngei nervi
glossopharyngei
rami tonsillares nervi
glossopharyngei
ramus musculi stylopharyngei nervi
glossopharyngei
ramus sinus carotici nervi
glossopharyngei

glossoptosia
glossoptosis
glottal
glottic
glottidis
atrium g.
intercartilaginea rimae g.
intercartilaginous part of rima g.
intermembranous part of rima g.
pars intercartilaginea rimae g.
pars intermembranacea rimae g.

G

NOTES

235

glottidis *(continued)*
    rima g.
    g. rima
glottis, pl. **glottides**
    false g.
    g. respiratoria
    g. spuria
    true g.
    g. vera
    g. vocalis
glucocorticoid
glucocorticotrophic
glucokinetic
glucolysis
glucose transport maximum
glutea, pl. **gluteae**
    crista g.
    g. inferior
    linea g.
    g. superior
    tuberositas g.
gluteal
    g. artery
    g. cleft
    g. crest
    g. fascia
    g. fold
    g. furrow
    g. line
    g. lymph node
    g. muscle
    g. nerve
    g. reflex
    g. region
    g. ridge
    g. surface of ilium
    g. tuberosity
    g. vein
    g. vessel
glutealis, pl. **gluteales**
    lymphonodi gluteales
    nodi lymphoidei gluteales
    regio g.
gluteofemoral bursa
gluteoinguinal
gluteorum
    bursa intermusculares
      musculorum g.
gluteus
    g. inferior
    g. maximus
    g. maximus muscle
    g. medius
    g. medius bursa
    g. medius muscle
    g. minimus
    g. minimus bursa
    g. minimus muscle

    sulcus g.
    g. superior
glycine succinate cycle
glycocalyx
glycocorticoid
glycogen granule
glycolysis
glycolytic
glycosecretory
glycostatic
glycotropic factor
glyoxylic acid cycle
GM-CSF
    granulocyte-macrophage colony-
      stimulating factor
gnathic index
gnathion
gnathocephalus
gnathoschisis
GnRH
    gonadotropin-releasing hormone
goblet cell
Godman fascia
Goethe bone
Goldman equation
Goldman-Hodgkin-Katz (GHK)
    G.-H.-K. equation
Golgi
    G. apparatus
    G. complex
    G. epithelial cell
    G. internal reticulum
    G. tendon organ
    G. type I, II neuron
    G. zone
Golgi-Mazzoni corpuscle
Goll
    G. column
    nucleus of G.
    tract of G.
Gombault triangle
gomitoli
Gompertz hypothesis
gompholic joint
gomphosis
gonad
    female g.
    indifferent g.
    male g.
    suspensory ligament of g.
gonadal
    g. agenesis
    g. artery
    g. cords
    g. dysgenesis
    g. ridge
    g. steroid-binding globulin
    g. vein
gonadotroph

gonadotrophic
gonadotrophin
gonadotropic hormone
gonadotropin
    anterior pituitary g.
    chorionic g.
    human menopausal g.
gonadotropin-releasing hormone (GnRH)
gonaduct
Gonda reflex
gonecyst
gonecystis
gonia (*pl. of* gonion)
goniocraniometry
goniodysgenesis
gonion, pl. **gonia**
gonochorism
gonochorismus
gonocyte
gonophore
gonophorus
Goormaghtigh cells
Gorlin formula
Gowers
    G. column
    G. contraction
    fasciculus of G.
    G. tract
G-protein
graafian
    g. follicle
    g. ovule
    g. vesicle
gracile tubercle
gracilis
    fasciculus g.
    funiculus g.
    lobulus g.
    g. muscle
    musculus tibialis g.
    nucleus fasciculi g.
    nucleus funiculi g.
    processus g.
    g. tendon
    tubercle of nucleus g.
    tuberculum nuclei g.
gradient
gradus
    stratum helicoidale brevis g.
    stratum helicoidale longi g.
Graham law
gram calorie

gram-centimeter
gram-meter
grana
Granit loop
granular
    g. endoplasmic reticulum
    g. foveolae
    g. layer of cerebellar cortex
    g. layer of epidermis
    g. layers of cerebral cortex
    g. layers of retina
    g. layer of a vesicular ovarian
      follicle
    g. pit
    g. pneumonocyte
granulares
    foveolae g.
granulation
    arachnoid g.
    arachnoidal g.'s
    pacchionian g.'s
granulationes arachnoideales
granule
    acidophil g.
    acrosomal g.
    alpha g.
    Altmann g.
    amphophil g.
    argentaffin g.'s
    azurophil g.
    basal g.
    basophil g.
    Bensley specific g.'s
    beta g.
    Birbeck g.
    g. cells
    chromatic g.
    chromophil g.
    chromophobe g.'s
    cone g.
    delta g.
    eosinophil g.
    fuchsinophil g.
    glycogen g.
    juxtaglomerular g.'s
    kappa g.
    keratohyalin g.'s
    lamellar g.
    Langerhans g.
    Langley g.'s
    membrane-coating g.
    mucinogen g.'s

NOTES

G

granule *(continued)*
    neutrophil g.
    Nissl g.'s
    oxyphil g.
    proacrosomal g.'s
    prosecretion g.'s
    rod g.
    secretory g.
    seminal g.
    Zimmermann g.
granuloblast
granulocyte colony-stimulating factor (G-CSF)
granulocyte-macrophage colony-stimulating factor (GM-CSF)
granulomere
granuloplastic
granulosa
    g. lutein cells
    membrana g.
    pars g.
granum
grape
    g. endings
    g. mole
graphomotor
graphospasm
Grassi
    nerve of G.
Gratiolet
    G. fibers
    G. radiation
Gräupner method
grave wax
gravireceptor
gravitation
    law of g.
    newtonian constant of g.
gravitational unit
gravity
    zero g.
gray
    g. columns
    g. fibers
    g. layer of superior colliculus
    g. matter
    g. rami communicantes
    g. substance
    g. tuber
    g. tubercle
    g. wing
great
    g. adductor muscle
    g. alveolar cells
    g. anastomotic artery
    g. anterior radicular artery
    g. auricular nerve
    g. cardiac vein
    g. cerebral vein

    g. cistern
    g. foramen
    g. horizontal fissure
    g. longitudinal fissure
    g. pancreatic artery
    g. radicular artery
    g. saphenous vein
    g. sciatic nerve
    g. segmental medullary artery
    g. superior pancreatic artery
    g. toe
    g. transverse commissure
    g. vein of Galen
    g. vessel
greater
    g. alar cartilage
    g. arterial circle of iris
    g. circulation
    g. cul-de-sac
    g. curvature
    g. curvature of stomach
    g. horn
    g. horn of hyoid bone
    g. humerus tubercle
    g. ischiatic notch
    g. multangular bone
    g. occipital nerve
    g. omentum
    g. palatine artery
    g. palatine canal
    g. palatine foramen
    g. palatine groove
    g. palatine nerve
    g. pectoral muscle
    g. pelvis
    g. peritoneal cavity
    g. peritoneal sac
    g. petrosal nerve
    g. petrosal nerve hiatus
    g. posterior rectus muscle of head
    g. psoas muscle
    g. rhomboid muscle
    g. ring of iris
    g. sacrosciatic notch
    g. saphenous vein
    g. sciatic foramen
    g. sciatic notch
    g. splanchnic nerve
    g. superficial petrosal nerve
    g. supraclavicular fossa
    g. trochanter
    g. trochanter of femur
    g. trochanter muscle
    g. tubercle
    g. tubercle of humerus
    g. tuberosity of humerus
    g. tympanic spine
    g. vestibular gland
    g. wing of sphenoid bone

g. zygomati
g. zygomatic muscle
greatest gluteal muscle
grenz zone
grinding surface
grisea, pl. **griseae**
    columnae griseae
    commissura anterior g.
    commissura posterior g.
    substantia g.
grisei
    rami communicantes g.
griseum
    indusium g.
    stratum g.
griseus
groin
groove
    g. for abdominal aorta
    alveolobuccal g.
    alveololabial g.
    alveololingual g.
    ampullary g.
    anterior auricular g.
    anterior intermediate g.
    anterior interventricular g.
    anterolateral g.
    g. of aortic arch
    g. for arch of aorta
    arterial g.'s
    atrioventricular g.
    g. for auditory tube
    auriculoventricular g.
    bicipital g.
    branchial g.
    buccal g.
    carotid g.
    carpal g.
    cavernous g.
    chiasmatic g.
    coronary g.
    costal g.
    g. of crus of helix
    deltopectoral g.
    dental g.
    g. for descending aorta
    digastric g.
    esophageal g.
    g. for esophagus
    ethmoidal g.
    g. of first rib for subclavian artery
    frontal g.'s

gingival g.
gingivobuccal g.
gingivolabial g.
gingivolingual g.
greater palatine g.
g. for greater petrosal nerve
inferior petrosal g.
g. for inferior petrosal sinus
g. for inferior venae cava
infraorbital g.
interatrial g.
intercondylar g.
intercostal g.
intermuscular g.
interosseous g.
intertubercular g.
intertubular g.
interventricular g.
intraorbital g.
lacrimal g.
laryngotracheal g.
lateral bicipital g.
g. for left brachiocephalic vein
g. for lesser petrosal nerve
limbal g.
linear g.
linguogingival g.
Lucas g.
g. of lung for subclavian artery
malleolar g.
mastoid g.
medial bicipital g.
medullary g.
meningeal g.
g. of middle meningeal vessel
g. for middle temporal artery
musculospiral g.
mylohyoid g.
nasolabial g.
nasopalatine g.
nasopharyngeal g.
obturator g.
occipital g.
olfactory g.
optic g.
palatine g.
palatovaginal g.
paraglenoid g.
pectoral g.
pharyngeal g.'s
pharyngotympanic g.
pontomedullary g.

NOTES

G

groove *(continued)*
    popliteal g.
    g. for popliteus
    posterior auricular g.
    posterior intermediate g.
    posterior interventricular g.
    posterolateral g.
    preauricular g.
    primary labial g.
    primitive g.
    g. of promontory of labyrinthine wall of tympanic cavity
    g. of pterygoid hamulus
    pterygopalatine g.
    pulmonary g.
    radial g.
    g. for radial nerve
    rhombic g.
    g. for rib
    g. for right brachiocephalic vein
    sagittal g.
    Sibson g.
    sigmoid g.
    g. for sigmoid sinus
    skin g.'s
    Sondergaard g.
    g. for spinal nerve
    spiral g.
    subclavian g.
    g. for subclavian artery
    g. for subclavian vein
    g. for subclavius
    subcostal g.
    g. for superior petrosal sinus
    g. for superior sagittal sinus
    g. for superior vena cava
    supraacetabular g.
    g. for tendon of fibularis longus
    g. for tendon of flexor hallucis longus
    g. for tendon of long peroneal muscle
    g. for tendon of peroneus longus
    g. for tibialis posterior tendon
    tracheobronchial g.
    transverse anthelicine g.
    g. for transverse sinus
    tympanic g.
    ulnar g.
    g. for ulnar nerve
    uncinate g.
    urethral g.
    venous g.
    Verga lacrimal g.
    vertebral g.
    g. for vertebral artery
    vertical g.
    vomeral g.
    vomerine g.
    vomerovaginal g.

gross anatomy
ground
    g. bundles
    g. lamella
    g. substance
group
    connective tissue g.
growing ovarian follicle
growth
    accretionary g.
    appositional g.
    auxetic g.
    differential g.
    g. factor
    g. hormone-inhibiting hormone
    g. hormone-releasing factor (GHRF)
    g. hormone-releasing hormone (GHRH)
    interstitial g.
    intussusceptive g.
    multiplicative g.
    g. plate
    g. quotient
    g. rate
    Rubner laws of g.
Gruber
    G. cul-de-sac
    G. ligament
    petrospheno-occipital suture of G.
Gruber-Landzert fossa
Grynfeltt
    g. hernia
    g. triangle
gryochrome
GSR
    galvanic skin response
g-tolerance
guaiacin
guanine cell
gubernaculum
    chorda g.
    g. dentis
    Hunter g.
    g. testis
gubernatrix
    plica g.
Gudden
    commissure of G.
    G. commissures
    G. ganglion
Guérin
    G. fold
    G. gland
    G. sinus
    G. valve
gulae
    plexus g.

gullet
gum line
Gumprecht shadow
Günz ligament
gustation
gustatorium
  organum g.
gustatorius
  caliculus g.
  porus g.
gustatory
  g. bud
  g. cells
  g. lemniscus
  g. nucleus
  g. organ
  g. pore
gustus
  organum g.
gut
  blind g.
  lumen of g.
  postanal g.
  preoral g.
Guthrie muscle
gutter
  left g.
  paracolic g.
  paravertebral g.
  right g.
guttural duct
gutturis
  claustrum g.
Guyon
  canal of G.
  G. isthmus
gymnocyte
gynandrism
gynandroid
gynandromorphism
gynandromorphous
gynecoid pelvis
gynogenesis
gyrata
  cutis verticis g.
gyrate
gyration
gyre
gyrencephalic
gyri (*pl. of* gyrus)
gyrochrome cell

gyrorum
  impressiones g.
gyrose
gyrous area
gyrus, pl. **gyri**
  angular g.
  g. angularis
  annectent g.
  anterior central g.
  anterior piriform g.
  artery of angular g.
  ascending frontal g.
  ascending parietal g.
  branch to angular g.
  gyri breves insulae
  g. of Broca
  callosal g.
  central gyri
  gyri of cerebrum
  cingulate g.
  g. cinguli
  cinguli g.
  deep transitional g.
  dentate g.
  g. dentatus
  fasciolar g.
  g. fasciolaris
  g. fornicatus
  frontal g.
  g. frontalis
  g. frontalis inferior
  g. frontalis superior
  fusiform g.
  g. fusiformis
  g. geniculi
  Heschl gyri
  hippocampal g.
  g. hippocampi
  impression of cerebral gyri
  inferior frontal g.
  inferior occipital g.
  inferior parietal g.
  inferior temporal g.
  g. of insula
  interlocking gyri
  lateral occipitotemporal g.
  g. limbicus
  lingual g.
  g. lingualis
  g. longus insulae
  marginal g.
  g. marginalis

NOTES

G

gyrus *(continued)*
medial occipitotemporal g.
middle frontal g.
middle temporal g.
occipital gyri
g. occipitotemporalis lateralis
g. occipitotemporalis medialis
g. olfactorius lateralis of Retzius
g. olfactorius medialis of Retzius
orbital gyri
gyri orbitales
g. paracentralis
parahippocampal g.
g. parahippocampalis
paraterminal g.
g. paraterminalis
postcentral g.
g. postcentralis
posterior central g.
precentral g.
g. precentralis
prepiriform g.
gyri profundi cerebri
quadrate g.
g. rectus

Retzius g.
splenial g.
straight g.
subcallosal g.
g. subcallosus
subcollateral g.
superior frontal g.
superior occipital g.
superior parietal g.
superior temporal g.
supracallosal g.
g. supracallosus
supramarginal g.
g. supramarginalis
tail of dentate g.
temporal g.
gyri temporales transversi
g. temporalis inferior
g. temporalis medius
g. temporalis superior
transitional g.
gyri transitivi cerebri
transverse temporal gyri
uncinate g.
g. uncinatus

H

H band
H disk
H fields
H substance
habena, pl. **habenae**
habenal
habenar
habenula, pl. **habenulae**
h. of cecum
commissure of habenulae
ganglion habenulae
Haller h.
nucleus habenulae
habenulae perforata
pineal h.
Scarpa h.
trigone of h.
trigonum habenulae
h. urethralis
habenular
h. commissure
h. trigone
habenularum
commissura h.
habenulointerpeduncular tract
habenulopeduncularis
tractus h.
habitus
fetal h.
Haeckel law
hair
arrector muscle of h.
auditory h.
axillary h.
bayonet h.
beaded h.
h. bulb
bulb of h.
h. cells
club h.
h. crosses
cuticle of h.
h. cycle
h. disk
downy h.
erector muscle of h.
h. follicle
h.'s of head
lanugo h.
moniliform h.
h. papilla
primary h.
pubic h.
h. root

scalp h.
h. shaft
h. stream
taste h.'s
h.'s of tragus
h.'s of vestibule of nose
h. whorl
hairline
hairy cells
Hal
Haldane
H. apparatus
H. effect
H. transformation
H. tube
Haldane-Priestley sample
Hales piesimeter
half-moon
halitus
Haller
H. annulus
H. ansa
H. anulus
H. arches
H. circle
H. cone
H. habenula
H. insula
H. line
H. plexus
H. rete
H. tripod
H. tunica vasculosa
H. unguis
H. vas aberrans
H. vascular tissue
halleri
circulus arteriosus h.
circulus venosus h.
rete h.
hallex, pl. **hallices**
hallucal
halluces (*pl. of* hallux)
hallucis
abductor h.
adductor h.
caput obliquum musculi
adductoris h.
caput transversum musculi
adductoris h.
musculus abductor h.
musculus adductor h.
musculus flexor brevis h.
musculus flexor longus h.

H

hallus
hallux, pl. halluces
hamartomatous
hamate
    h. bone
    hook of h.
hamati
    hamulus ossis h.
hamatum
    os h.
Hamilton-Stewart method
hammer
hammock ligament
hamstring
    inner h.
    lateral h.
    medial h.
    h. muscle
    h. tendon
hamular
    h. notch
    h. process
    h. process of lacrimal bone
    h. process of sphenoid bone
hamulus, pl. hamuli
    h. cochlea
    cochlea h.
    groove of pterygoid h.
    h. of hamate bone
    lacrimal h.
    h. lacrimalis
    h. laminae spiralis
    medial h.
    h. ossis hamati
    pterygoid h.
    h. pterygoideus
    h. of spiral lamina
    sulcus of pterygoid h.
hand
    abductor digiti minimi muscle
      of h.
    articulations of h.
    base of phalanx of h.
    chiasm of digitus of h.
    cleft h.
    common flexor sheath of h.
    deep space compartment of h.
    digital pulp of h.
    distal phalanx of h.
    dorsal digital nerve of h.
    dorsal expansion of h.
    dorsal fascia of h.
    dorsal interossei of h.
    dorsal interosseous muscle of h.
    dorsal venous network of h.
    dorsal venous rete of h.
    dorsum of h.
    extensor digitorum brevis muscle
      of h.

fibrous digital sheaths of h.
fibrous sheaths of digits of h.
flexor digiti minimi brevis muscle
    of h.
intercapitular vein of h.
intermediate palm of h.
interphalangeal joint of h.
joint of h.
lumbrical muscle of h.
lumbricals of h.
lumbricals (lumbrical muscles)
    of h.
middle phalanges of foot and h.
navicular bone of h.
palm of h.
palmar ligament of interphalangeal
    joint of h.
perforating artery of h.
proximal phalanx of h.
h. ratio
split h.
superficial dorsum of h.
superficial palm of h.
synovial sheaths of digits of h.
trident h.
hanging septum
Hannover canal
Hapsburg
    H. jaw and lip
haptics
hard
    h. palate
    h. tissue
harelip
harmonic suture
Hartmann
    pouch of H.
    H. pouch
Hasner
    H. fold
    H. valve
Hassall
    H. bodies
    H. concentric corpuscle
Haubenfelder
haunch bone
haustoria
haustorium
haustra (pl. of haustrum)
haustral fold
haustration
    h.'s of colon
haustrum, pl. haustra
    haustra coli
    haustra of colon
Havers gland
haversian
    h. canals
    h. lamella

h. space
h. system
Hayem hematoblast
Hayflick limit
HCS
hourglass contraction of stomach
head
anterior ligament of fibular h.
anterior rectus muscle of h.
H. area
articular facet of radial h.
h. cap
h. cavity
crown of h.
deep h.
h. of epididymis
femoral h.
h. of femur
h. of fibula
h. fold
fovea of the femoral h.
fovea of radial h.
frontal region of h.
greater posterior rectus muscle
    of h.
hairs of h.
hourglass h.
humeral h.
humeroulnar h.
h. of humerus
h. of incus
inferior oblique muscle of h.
intraarticular ligament of costal h.
lateral rectus muscle of h.
long muscle of h.
h. of malleus
h. of mandible
medial h.
h. of metacarpal
metacarpal h.
h. of metatarsal
h. of metatarsal bone
muscle of h.
oblique h.
occipital region of h.
h. of pancreas
h. of phalanx
h. of phalanx of hand or foot
posterior ligament of fibular h.
h. process
radial h.
h. of radius

region of h.
h. of rib
saddle h.
semispinal muscle of h.
short h.
smaller posterior rectus muscle
    of h.
spinal muscle of h.
splenius muscle of h.
h. of stapes
superficial h.
superior oblique muscle of h.
h. of talus
talus h.
temporal region of h.
h. of thigh bone
transverse h.
h. of ulna
ulnar h.
headgut
hearing
organ of h.
heart
abdominal h.
acute margin of h.
anterior h.
apex of h.
artificial h.
athlete's h.
atrium of h.
base of h.
cervical h.
h. chamber
chordae tendineae of h.
compliance of h.
conducting system of h.
conduction system of h.
h. conduction system
cross-section of h.
crux of h.
diaphragmatic surface of h.
fibrous skeleton of h.
fibrous trigones of h.
foramen ovale of h.
foramina of the smallest vein
    of h.
h. hormone
inferior lobe of h.
infundibulum of h.
interventricular sulcus of h.
law of the h.
left atrium of h.

NOTES

H

245

heart *(continued)*
    left fibrous trigone of h.
    left margin of h.
    longitudinal sulcus of h.
    Mahaim bundle in h.
    middle lobe of h.
    h. muscle
    muscular part of interventricular
      septum of h.
    notch of apex of h.
    obtuse margin of h.
    h. orifice
    oval foramen of h.
    parchment h.
    posterior margin of h.
    pulmonary surface of h.
    Riedel lobe right and left fibrous
      rings of h.
    right atrium of h.
    right border of h.
    right fibrous trigone of h.
    right/left pulmonary surfaces of h.
    right/left ventricles of h.
    right margin of h.
    h. sac
    h. silhouette
    skeleton of h.
    skin h.
    sternocostal surface of h.
    subendocardial conducting system
      of h.
    sulcus of h.
    superior lobe of h.
    h. tissue
    transverse sulcus of h.
    h. valve
    vein of h.
    venous h.
    ventricles of h.
    vortex of h.
heartbeat
heart-lung preparation
heart-shaped
    h.-s. pelvis
    h.-s. uterus
heat
    convective h.
    conversive h.
    initial h.
    unit of h.
heat-rigor point
hecateromeric
hecatomeral
hecatomeric
hederiform ending
heel
    anterior h.
    h. bone
    h. cord

    globus of h.
    network of h.
    h. region
    h. tendon
Heidenhain
    H. crescents
    H. demilunes
    H. law
    H. pouch
height
    anterior facial h. (AFH)
    facial h.
    nasal h.
    orbital h.
height-length index
Heister
    H. diverticulum
    valve of H.
    H. valve
HeLa cells
Held
    H. bundle
    H. decussation
helical
    h. fold
    h. rim
helices *(pl. of* helix)
helicin
helicina
    arteria h.
helicine
    h. artery
    h. artery of penis
    h. artery of the uterus
helicis
    apophysis h.
    cauda h.
    crista h.
    crus h.
    incisurae h.
    h. major
    h. major muscle
    h. minor
    h. minor muscle
    musculus incisurae h.
    spina h.
    sulcus cruris h.
helicoid ginglymus
helicotrema
Helie bundle
heliotaxis
heliotropism
helix, pl. **helices**
    caudal h.
    crus of h.
    fossa of h.
    groove of crus of h.
    large muscle of h.
    limb of h.

muscle of notch of h.
smaller muscle of h.
spine of h.
tail of h.
Heller plexus
Helmholtz
anterior ligament of H.
H. axis ligament
Helmholtz-Gibbs theory
Helweg bundle
hemafacient
hemagogic
hemagogue
hemal
h. arches
h. gland
h. node
h. spine
hemangiectatic hypertrophy
hemangioblast
hemangiomas
dyschondroplasia with h.
hematachometer
hematein
hematherm
hemathermal
hemathermous
hematoblast
Hayem h.
hematocryal
hematocytoblast
hematogenesis
hematogenic
hematogenous
hematohistioblast
hematolymphangioma
hematoplastic
hematopoiesis
hematopoietic
h. gland
h. system
h. tissue
hematosis
hematothermal
hemiacardius
hemiacrosomia
hemianencephaly
hemiaplasia
hemiazygos
h. vein
vena h.

hemicardia
h. dextra
h. sinistra
hemicentrum
hemicephalia
hemicerebrum
hemidesmosomes
hemidiaphragm
hemidystrophy
hemiectromelia
hemielliptica
fovea h.
hemifacial
hemiglossal
hemihydranencephaly
hemilateral
hemilingual
hemimelia
dysplasia epiphysialis h.
hemipagus
hemisensory
hemiseptum
hemisphere
h. of bulb of penis
cerebellar h.
h. of cerebellum
cerebral h.
commissure of cerebral h.
dominant h.
inferior veins of cerebellar h.
inferolateral margin of cerebral h.
inferomedial margin of cerebral h.
superior margin of cerebral h.
superior veins of cerebellar h.
ventricle of cerebral h.
hemispherica
fovea h.
hemisphericum cerebelli
hemispherium
h. bulbi urethrae
h. cerebelli
h. cerebri
hemithoracic duct
hemithoracicus
ductus h.
hemithorax
hemivertebra
hemoblast
hemochorial placenta
hemocytoblast
hemodynamic
hemoendothelial placenta

NOTES

H

247

hemogenesis
hemogenic
hemoglobin
hemohistioblast
hemolamella
hemolymph
    h. gland
    h. node
hemophoresis
hemoplastic
hemoplasty
hemopoiesis
hemopoietic tissue
hemorheology
hemorrhage
    artery of cerebral h.
hemorrhoidal
    h. cushion
    h. nerve
    h. plexus
    h. vein
    h. zone
hemorrhoidalis
    annulus h.
    nervus h.
    zona h.
hemostasia
hemostasis
hemotachogram
hemotachometer
hemotroph
hemotrophe
Henke space
Henle
    H. ampulla
    H. ansa
    H. fenestrated elastic membrane
    H. fiber layer
    H. fissures
    H. gland
    H. loop
    H. nervous layer
    H. sphincter
    spine of H.
    H. spine
    H. triangle
    H. tubules
Henry
    H. law
    ligament of H.
Henry-Gauer response
Hensen
    H. canal
    H. cell
    H. disk
    H. duct
    H. knot
    H. line

    H. node
    H. stripe
Hensing ligament
hepatic
    h. architecture
    h. artery
    h. artery proper
    h. bed
    h. branch of anterior vagal trunk
    h. branch of vagus nerve
    h. cecum
    h. cords
    h. duct
    h. flexure
    h. fossa
    h. hilum
    h. laminae
    h. ligament
    h. lobe
    h. lobule
    h. lymph node
    h. nervous plexus
    h. outflow tract
    h. portal system
    h. portal vein
    h. prominence
    h. triad
    h. trinity
    h. vein
    h. venous segments
hepatica, pl. **hepaticae**
    arteriae segmenti hepaticae
    h. communis
    pars h.
    h. propria
    venae hepaticae
hepatici
    lymphonodi h.
    nodi lymphoidei h.
    rami h.
hepaticopulmonary
hepatic-renal angle
hepaticus
    plexus nervosus h.
hepatis
    appendix fibrosa h.
    area nuda h.
    arteria interlobulares h.
    centrales h.
    divisio lateralis dextra h.
    divisio lateralis sinistra h.
    divisio medialis dextra h.
    divisio medialis sinistra h.
    ductus lobi caudati dexter h.
    ductus lobi caudati sinister h.
    facies visceralis h.
    fissura ligamenti teretis h.
    flexura coli h.

impressio cardiaca faciei
  diaphragmaticae h.
impressio colica h.
impressio duodenalis h.
impressio esophagea h.
impressio gastrica h.
impressio renalis h.
impressio suprarenalis h.
incisura ligamenti teretis h.
interlobulares h.
ligamentum coronarium h.
ligamentum falciforme h.
ligamentum teres h.
ligamentum triangulare dextrum h.
ligamentum triangulare sinistrum h.
lobulus h.
margo inferior h.
pars anterior faciei
  diaphragmatis h.
pars dextra faciei
  diaphragmaticae h.
pars posterior faciei
  diaphragmatis h.
pars posterior facies
  diaphragmatis h.
pars quadrata h.
pars superior faciei
  diaphragmaticae h.
pars superior facies
  diaphragmatis h.
pars transversa rami sinistri venae
  portae h.
pars umbilicalis rami sinistri venae
  portae h.
pons h.
ponticulus h.
porta h.
processus papillaris lobi caudati h.
rami laterales rami sinistri venae
  portae h.
rami lobi caudati rami sinistri
  venae portae h.
rami mediales rami sinistri venae
  portae h.
ramus dexter venae portae h.
ramus posterior rami dextri venae
  portae h.
ramus sinister venae portae h.
segmenta h.
tuber omentale h.
tunica fibrosa h.
tunica serosa h.

vasa aberrantia h.
vas aberrans h.
vena afferens h.
venae centrales h.
venae interlobulares h.
vena portae h.
hepatobiliary
hepatocellular
hepatocolic ligament
hepatocolicum
  ligamentum h.
hepatocystic duct
hepatocystocolic ligament
hepatocyte
hepatoduodenale
  ligamentum h.
hepatoduodenal ligament
hepatoenteric recess
hepatoesophageal ligament
hepatoesophageum
  ligamentum h.
hepatogastric ligament
hepatogastricum
  ligamentum h.
hepatogastroduodenal ligament
hepatogenic
hepatojugular
hepatojugularometer
hepatonephric
hepatopancreatic
  h. ampulla
  h. fold
  h. sphincter
hepatopancreatica, pl. **hepatopancreaticae**
  ampulla h.
  musculus sphincter ampullae
    hepatopancreaticae
  sphincter ampullae
    hepatopancreaticae
hepatophrenic ligament
hepatopneumonic
hepatoportal
hepatopulmonary
hepatorenal
  h. angle
  h. ligament
  h. pouch
  h. recess
  h. recess of subhepatic space
hepatorenale
  ligamentum h.

NOTES

H

hepatorenalis
  recessus h.
hepatoumbilical ligament
hereditary multiple exostoses
heredofamilial
Hering
  canal of H.
  H. canal
  sinus nerve of H.
  H. sinus nerve
Hering-Breuer reflex
hermaphrodism
hermaphrodite
hermaphroditism
  ansverse h.
  bilateral h.
  dimidiate h.
  false h.
  female h.
  lateral h.
  male h.
  true h.
  unilateral h.
hernia
  acquired h.
  h. adiposa
  amniotic h.
  Béclard h.
  Birkett h.
  Bochdalek h.
  congenital diaphragmatic h.
  diverticular h.
  encysted h.
  Grynfeltt h.
  inguinoproperitoneal h.
  intermuscular h.
  interparietal h.
  linea alba h.
  Littré-Richter h.
  meningeal h.
  mucosal h.
  oblique h.
  omental h.
  ovarian h.
  parasternal h.
  paraumbilical h.
  pectineal h.
  pulsion h.
  rectal h.
  h. in recto
  retrosternal h.
  subpubic h.
  synovial h.
  thyroidal h.
  tonsillar h.
  uterine h.
  vaginal h.
  vaginolabial h.
hernial sac

herophili
  torcular h.
Herring bodies
hertz (Hz)
herz hormone
Heschl
  H. gyri
hesitancy
Hesselbach
  H. fascia
  H. ligament
  H. triangle
heteradelphus
heteralius
heteraxial
heteroblastic
heterocellular
heterocentric
heterocephalus
heterocheiral
heterochiral
heterochromatic
heterochromatin
  constitutive h.
  facultative h.
  satellite-rich h.
heterochromia
  binocular h.
heterochron
heterochronia
heterochronic
heterochronous
heterocladic
  h. anastomosis
heterocrine
heterocytotropic
heterodromous
heterodymus
heterogametic embryo
heterogenesis
heterogenetic
heterolateral
heterologous
  h. stimulus
  h. twins
heterology
heteromeral
heteromeric cell
heteromerous
heterometric autoregulation
heteromorphosis
heteromorphous
heteronomous
heteronomy
heteropagus
heterophagy
heteropyknotic chromatin
heterotaxia
heterotaxic

heterotaxis
heterotaxy
heterotherm
heterothermic
heterotonia
heterotopia
heterotopic bone
heterotype mitosis
heterotypic cortex
Heubner
    artery of H.
    H. artery
Heuser membrane
hexadactylism
hexadactyly
hexose monophosphate shunt
Hey ligament
Heyman capsule
hiatal
hiatus
    accessory maxillary h.
    adductor h.
    h. adductorius
    aortic h.
    h. aorticus
    basilic h.
    Breschet h.
    h. of canal for greater petrosal
      nerve
    h. canalis facialis
    h. canalis nervi petrosi majoris
    h. canalis nervi petrosi minoris
    h. of canal for lesser petrosal
      nerve
    diaphragmatic h.
    esophageal h.
    h. esophageus
    h. ethmoidalis
    h. of facial canal
    fallopian h.
    greater petrosal nerve h.
    lesser petrosal nerve h.
    h. maxillaris
    maxillary h.
    pleuropericardial h.
    pleuroperitoneal h.
    sacral h.
    saphenous h.
    h. saphenus
    scalene h.
    Scarpa h.
    semilunar h.

    h. semilunaris
    h. subarcuatus
    h. tendineus
    h. totalis sacralis
hibernating gland
hibernoma
    interscapular h.
hidden
    h. border of nail
    h. nail skin
    h. part
    h. part of duodenum
hidropoiesis
hidropoietic
hidroschesis
hidrosis
hidrotic ectodermal dysplasia
high
    h. altitude chamber
    h. endothelial postcapillary venule
highest
    h. concha
    h. intercostal artery
    h. intercostal vein
    h. nuchal line
    h. thoracic artery
    h. turbinated bone
high-fiber diet
high-frequency current
Highmore
    antrum of H.
    H. body
highmori
highmorianum
    corpus h.
hila (*pl. of* hilum)
hilar
    h. lymph node
    h. region
    h. shadow
    h. structure
hillock
    axon h.
    facial h.
    seminal h.
Hilton
    H. muscle
    H. sac
    H. white line
hilum, pl. **hila**
    h. of dentate nucleus
    hepatic h.

NOTES

H

hilum *(continued)*
    h. of kidney
    left h.
    h. lienis
    h. of lung
    h. of lymph node
    h. lymphonodi
    h. nodi lymphatici
    h. nuclei dentati
    h. nuclei olivaris
    h. of olivary nucleus
    h. ovarii
    h. of ovary
    h. pulmonis
    h. renalis
    right h.
    h. of spleen
    splenic h.
    h. splenicum

hilus
    h. cells
    kidney h.
    lung h.
    h. of ovary

hindbrain vesicle
hindgut
hind kidney
hinge
    h. axis
    h. joint

hip
    h. bone
    compartment of thigh for flexors of h.
    extension of h.
    h. joint
    retinaculum of articular capsule of h.
    triceps muscle of h.

hippocampal
    h. commissure
    h. convolution
    h. fissure
    h. gyrus

hippocampalis
    formatio h.

hippocampi
    alveus h.
    commissura h.
    digitationes h.
    fascia dentata h.
    fimbria h.
    fissura h.
    gyrus h.
    pes h.
    rudimentum h.
    sulcus h.
    tenia h.

hippocampus
    foot of h.
    h. major
    h. minor

hircus, pl. **hirci**
    barbula hirci

Hirschberg reflex
Hirschfeld canal
hirtellous
hirundinis
    nidus h.

His
    angle of H.
    H. band
    bundle of H.
    H. bundle
    H. canal
    H. copula
    isthmus of H.
    H. perivascular space
    H. spindle

His-Purkinje system
histangic
His-Tawara
    H.-T. bundle
    H.-T. node
    H.-T. system

histioblast
histiocyte
    cardiac h.
    sea-blue h.

histiogenic
histioid
histionic
histoangic
histoblast
histochemistry
histocyte
histodifferentiation
histogenesis
histogenetic
histogenous
histogeny
histogram
histoid
histologic accommodation
histological internal os of uterus
histologicum
    ostium h.

histologist
histology
histomorphom
histoneurology
histonomy
histopathogenesis
histophysiology
historrhexis
histotoxic
histotroph

histotrophic
hitchhiker thumbs
Hoboken
    H. gemmule
    H. nodule
    H. valve
Hoche
    H. bundle
    H. tract
hodoneuromere
Hofbauer cell
Hoffmann duct
Holden line
Holl ligament
hollow
    h. bone
    Sebileau h.
Holmgrén-Golgi canals
holoacardius
    h. acephalus
    h. amorphus
holoacrania
holoanenceph
holoblastic cleavage
holocephalic
holocrine gland
hologastroschisis
holomorphosis
holoprosencephaly
holotelencephaly
homalocephalous
homaluria
homaxial
Home lobe
homeocyte
homeometric autoregulation
homeomorphous
homeoplasia
homeoplastic
homeorrhesis
homeosis
homeostasis
    Bernard-Cannon h.
    ontogenic h.
    physiologic h.
    physiological h.
    waddingtonian h.
homeostatic
    h. equilibrium
homeotherm
homeothermal
homeothermic

homeotic
homergy
homigrade scale
hominal physiology
homoblastic
homocentric
homochronous
homocladic anastomosis
homocytotropic
homodromous
homogametic embryo
homogenesis
homogeny
homoioplasia
homoiothermal
homolateral
homolog, homologue
homologous stimulus
homology
homomorphic
homonomous
homonomy
homonymous
homoplastic
homorganic
Homo sapiens
homothermal
homotonic
homotopic
homotype
homotypical
homotypic cortex
homozygous achondroplasia
homunculus
hood
    dorsal h.
hook
    h. of hamate
    h. of hamate bone
    h. of spiral lamina
hookean behavior
hooked
    h. bone
    h. bundle of Russell
    h. fasciculus
Hooke law
horizontal
    h. cell of Cajal
    h. cells of retina
    h. fissure of cerebellum
    h. fissure of right lung
    h. part of duodenum

NOTES

H

horizontal *(continued)*
    h. part of facial canal
    h. plane
    h. plate of palatine bone
horizontalia
    plana h.
horizontalis cerebelli
hormion
hormonal
hormone
    adipokinetic h.
    adrenal androgen-stimulating h.
    adrenocortical h.
    adrenocorticotropic h.
    adrenotropic h.
    androgenic h.
    anterior pituitary-like h.
    cardiac h.
    chorionic gonadotrophic h.
    chorionic gonadotropic h.
    chromatophorotropic h.
    corpus luteum h.
    cortical h.
    corticotropic h.
    corticotropin-releasing h.
    estrogenic h.
    follicle-stimulating hormone-
      releasing h.
    galactopoietic h.
    gametokinetic h.
    gastrointestinal h.
    gonadotropic h.
    gonadotropin-releasing h. (GnRH)
    growth hormone-inhibiting h.
    growth hormone-releasing h.
      (GHRH)
    heart h.
    herz h.
    human chorionic
      somatomammotropic h.
    hypophysiotropic h.
    interstitial cell-stimulating h. (ICSH)
    lactogenic h.
    luteinizing hormone-releasing h.
    luteotropic h.
    mammotropic h.
    melanocyte-stimulating h.
    parathyroid h.
    pituitary gonadotropic h.
    pituitary growth h.
    placental growth h.
    progestational h.
    prolactin-inhibiting h.
    prolactin-releasing h.
    releasing h.
    salivary gland h.
    sex h.
    somatotropic h.
    thyroid-stimulating h. (TSH)

    thyrotropic h.
    thyrotropin-releasing h.
hormone-prolactin
    chorionic growth h.-p.
hormone-releasing
    luteinizing hormone/follicle-
      stimulating h.-r.
hormonogenesis
hormonogenic
hormonopoiesis
hormonopoietic
hormonoprivia
horn
    Ammon h.
    anterior h.
    h. of clitoris
    coccygeal h.
    dorsal h.
    frontal h.
    greater h.
    h. of greater hyoid bone
    h.'s of hyoid bone
    inferior h.
    h. of inferior thyroid cartilage
    h. of lateral spinal cord
    lateral ventricle h.
    h. of lateral ventricle
    lesser h.
    h. of lesser hyoid bone
    occipital h.
    posterior h.
    sacral h.
    h.'s of saphenous opening
    superior h.
    h. of superior thyroid cartilage
    temporal h.
    uterine h.
    h. of uterus
    vein of posterior h.
    ventral h.
Horner
    H. muscle
    H. syndrome
horny
    h. layer of epidermis
    h. layer of nail
horseshoe
    h. kidney
    h. placenta
Hortega cells
hourglass
    h. contraction
    h. contraction of stomach (HCS)
    h. head
    h. vertebrae
Houston
    H. fold
    H. muscle

valve of H.
H. valve
Howship lacunae
Hoyer
    H. anastomoses
    H. canal
Hubrecht protochordal knot
Huebner recurrent artery
Hueck ligament
Huguier
    H. canal
    H. circle
    H. sinus
human
    h. chorionic somatomammotropic
      hormone
    h. lung
    h. lung fibroblast
    h. menopausal gonadotropin
    h. subject
humani
    partes corporis h.
humeral
    h. articulation
    h. axillary lymph node
    h. bone
    h. head
    h. nutrient artery
    h. shaft
humerale
    caput h.
humerales
    nodi lymphoidei axillares h.'s
humeri (*pl. of* humerus)
humeroradial
    h. articulation
    h. joint
humeroradialis
    articulatio h.
humeroscapular
humeroulnar
    h. articulation
    h. head
    h. head of flexor digitorum
      superficialis muscle
    h. joint
humeroulnaris
    articulatio h.
humerus, pl. **humeri**
    anatomical neck of h.
    anterolateral surface of h.
    anteromedial surface of shaft of h.

arteriae nutriciae humeri
arteria nutriens humeri
articulatio humeri
body of h.
capitulum of h.
capitulum humeri
caput humeri
collum anatomicum humeri
collum chirurgicum humeri
condyle of h.
condylus humeri
coronoid fossa of h.
corpus humeri
crest of h.
deltoid tuberosity of h.
epicondylus lateralis humeri
epicondylus medialis humeri
facies anterior lateralis corporis
  humeri
facies anterior medialis corporis
  humeri
facies anterolateralis corporis
  humeri
facies anteromedialis corporis
  humeri
facies posterior corporis humeri
fossa coronoidea humeri
fossa radialis humeri
greater tubercle of h.
greater tuberosity of h.
head of h.
inferior greater tubercle of facet
  of h.
lateral border of h.
lateral condyle of h.
lateral epicondyle of h.
lateral supracondylar ridge of h.
lesser tubercle of h.
lesser tuberosity of h.
ligamentum transversum humeri
little head of h.
margo lateralis humeri
margo medialis humeri
medial border of h.
medial condyle of h.
medial epicondyle of h.
medial supracondylar ridge of h.
middle greater tubercle of facet
  of h.
neck of h.
nutriciae humeri
nutrient artery of h.

NOTES

H

humerus *(continued)*
    posterior surface of shaft of h.
    processus supraepicondylaris humeri
    pulley of h.
    radial fossa of h.
    shaft of h.
    supracondylar process of h.
    surgical neck of h.
    trochlea humeri
    trochlea of h.
    tuberculum majus humeri
    tuberculum minus humeri
    tuberositas deltoidea humeri
humidity
    absolute h.
    relative h.
humor
    aqueous h.
    h. aquosus
    h. of eye
    Morgagni h.
    ocular h.
    vitreous h.
    h. vitreus
humoral
humoris
Humphrey ligament
hunger
    h. contractions
    narcotic h.
Hunter
    H. canal
    H. gubernaculum
    H. ligament
    H. line
    H. membrane
hunterian perforator
Hunter-Schreger
    H.-S. bands
    H.-S. line
hunting
    h. phenomenon
    h. reaction
Hürthle cell
Huschke
    H. auditory tooth
    H. canal
    H. cartilage
    H. foramen
    H. ligament
    H. valve
Huxley
    H. layer
    H. membrane
    H. sheath
hyaline
    h. bodies of pituitary
    h. cartilage
    h. membrane

hyalinization
hyalocapsular ligament
hyalocyte
hyaloid
    h. artery
    h. body
    h. canal
    h. fossa
    h. membrane
    h. vessel
hyaloidea
    arteria h.
    fossa h.
    membrana h.
    stella lentis h.
hyaloideo-capsulare
    ligamentum h.-c.
hyaloideo-capsulario
    ligamentum h.-c.
hyaloideus
    canalis h.
hyalomere
hyaloplasm
    nuclear h.
hyaloplasma
hyalosome
hybridization
    cell h.
hydatid
    h. mole
    Morgagni h.
    nonpedunculated h.
    pedunculated h.
    sessile h.
    stalked h.
hydatidiform mole
hydatoid
hydranencephaly
hydrate microcrystal theory of anesthesia
hydration
hydraulic conductivity
hydrencephalocele
hydrencephalomeningocele
hydrencephalus
hydroadipsia
hydrocele
    congenital h.
    h. spinalis
hydrocephalocele
hydrocephaloid
hydrocephalus
    communicating h.
    congenital h.
    double compartment h.
    external h.
    h. ex vacuo
    internal h.
    noncommunicating h.

otitic h.
primary h.
hydrocephaly
obstructive h.
hydrocortisone
hydrodipsia
hydrodynamics
hydroencephalocele
hydrogen transport
hydrokinetic
hydrokinetics
hydrolability
hydromeningocele
hydromicrocephaly
hydromyelia
hydromyelocele
hydrophilia
hydroposia
hydrops
endolymphatic h.
fetal h.
h. fetalis
hydrorheostat
hydrostat
hydrostatic pressure
hydrotaxis
hydrothermal
hydrotomy
hydrotropism
hygrometer
hygrometry
hygroscopic expansion
hyla
hylic
hymen
denticular h.
falciform h.
hymenal
h. band
h. caruncula
h. orifice
h. ring
hymenalis, pl. **hymenales**
caruncula h.
hymenoid
hymenology
hyobranchial cleft
hyoepiglottic
h. ligament
hyoepiglotticum
ligamentum h.
hyoepiglottidean

hyoglossal
h. membrane
h. muscle
hyoglossus
h. muscle
musculus h.
hyoid
h. apparatus
h. arch
h. bone
bursa of h.
lesser horn of h.
h. muscle
h. region
hyoidei
cornu majus ossis h.
cornu minus ossis h.
corpus ossis h.
hyoideum
os h.
hyoideus
apparatus h.
hyomandibular cleft
hyopharyngeus
hyothyroid
hyothyroidea
membrana h.
hypanakinesia
hypanakinesis
hypaxial muscle
hypencephalon
hyper
hyperactivity
hyperanacinesia
hyperanacinesis
hyperanakinesia
hyperanakinesis
hyperbaric
h. chamber
h. oxygen
h. oxygenation
h. oxygen therapy
hyperbarism
hyperbrachycephaly
hypercapnia
hypercarbia
hypercementosis
hyperchloremic acidosis
hyperchromasia
hyperchromatic
hyperchromatism
hyperchromia

NOTES

H

hyperchromic
hypercinesia
hypercinesis
hypercyesia
hypercyesis
hypercytochromia
hyperdactylia
hyperdactylism
hyperdactyly
hyperdynamia
hyperdynamic
hyperemia
    active h.
    arterial h.
    fluxionary h.
    passive h.
    venous h.
hyperemic
hyperencephaly
hypereuryprosopic
hypergenesis
hypergenetic
hypergenic teratosis
hyperisotonic
hyperkinemia
hyperkinesia
hyperkinesis
hyperkinetic
hypermastia
hypermetabolism
hypermetropia
hypermorph
hypernephroid
hyperorchidism
hyperoxia
hyperpepsia
hyperphalangism
hyperplasia
    cementum h.
hyperpnea
hyperpolarization
hypersomia
hypersonic
hypersthenia
hypersthenic
hypertelorism
    canthal h.
    ocular h.
hypertensor
hyperthelia
hypertonia
hypertonic
hypertonicity
hypertrichosis
    h. lanuginosa
    nevoid h.
hypertrophic pyloric stenosis
hypertrophy
    functional h.

    hemangiectatic h.
    physiologic h.
    vicarious h.
hyperventilation
hypervolia
hypoadenia
hypobaria
hypobaric
hypobarism
hypobaropathy
hypoblast
hypoblastic
hypobranchial eminence
hypocalcification
    enamel h.
hypocapnia
hypocarbia
hypocelom
hypochondriaca
    regio h.
hypochondriac region
hypochondrium
hypochondroplasia
hypochordal
hypocinesia
hypocinesis
hypodactylia
hypodactylism
hypodactyly
hypoderm
hypodermis
hypodipsia
hypodontia
hypodynamia
hypodynamic
hypoeccrisis
hypoeccritic
hypofunction
hypogastric
    h. artery
    h. ganglia
    h. nerve
    h. region
    h. vein
hypogastrica
    arteria h.
    plica h.
    vena h.
hypogastricus
    nervus h.
hypogastrium
hypogastropagus
hypogastroschisis
hypogenesis
    polar h.
hypogenetic
hypogenitalism
hypoglossal
    h. artery

h. canal
h. canal venous plexus
communicating branch of lingual
  nerve with h.
h. eminence
h. fossa of sella turcica
h. nerve
h. nucleus
hypoglossalis
  canalis h.
hypoglossi
  ansa h.
  circellus venosus h.
  comitans nervi h.
  descendens h.
  eminentia h.
  nucleus nervi h.
  plexus venosus canalis h.
  rami linguales nervi h.
  rete canalis h.
  trigonum nervi h.
  tuberculum h.
  vena comitans nervi h.
hypoglossis
hypoglosso
  rami communicantes nervi lingualis
    cum nervo h.
hypoglossus
  nervus h.
hypoglottis
hypognathous
hypognathus
hypogonadism
  hypogonadotropic h.
  primary h.
  secondary h.
hypogonadotropic hypogonadism
hypohepatia
hypoisotonic
hypokinesia
hypokinesis
hypokinetic
hypomastia
hypomazia
hypomelanosis
hypomelia
hypomere
hypometabolism
hypomorph
hypomotility
hypomyelination
hypomyelinogenesis

hypomyotonia
hyponychial
hyponychium
hypopancreorrhea
hypoperistalsis
hypophalangism
hypopharyngeus
  musculus h.
hypopharynx
hypophyseal
  h. artery
  h. fossa of sella turcica
  h. pouch
hypophyseoportal vein
hypophyseos
  lobus anterior h.
  lobus glandularis h.
  lobus posterior h.
  pars nervosa h.
  pars pharyngea h.
hypophysial
  h. duct
  h. fossa
  h. portal circulation
  h. portal system
hypophysialis
  fossa h.
hypophysioportal system
hypophysiotropic hormone
hypophysis
  alpha cells of anterior lobe of h.
  anterior lobe of h.
  basophil cell of anterior lobe
    of h.
  beta cell of anterior lobe of h.
  h. cerebri
  chromophobe cells of anterior lobe
    of h.
  cyst of h.
  delta cell of anterior lobe of h.
  distal part of anterior lobe of h.
  infundibular part of anterior lobe
    of h.
  infundibulum of h.
  pharyngeal h.
  posterior lobe of h.
  tentorium of h.
hypoplasia
  optic nerve h.
  right ventricular h.
  thymic h.
hypoplastic

NOTES

H

hypopnea
hypoposia
hypopraxia
hyposmosis
hyposomia
hypospadiac
hypospadias
    balanic h.
    penoscrotal h.
    perineal h.
hypostomia
hypostosis
hypotension
hypotensive
hypothalami
    nucleus dorsomedialis h.
    nucleus posterior h.
    upraopticus h.
hypothalamic
    h. infundibulum
    h. sulcus
hypothalamicorum
    rami nucleorum h.
hypothalamicus
    ramus h.
    sulcus h.
hypothalamohypophysial
    h. portal circulation
    h. portal system
    h. tract
hypothalamus
    mamillary tubercle of h.
    ventromedial nucleus of h.
hypothenar
    h. eminence
    h. fascia
    h. muscle
    h. prominence
hypothenaris
    eminentia h.
hypothermal
hypothermia
hypothesis, pl. hypotheses
    gate-control h.
    Gompertz h.
    Makeham h.

    sliding filament h.
    Starling h.
hypotonia
hypotonic
hypotonicity
hypotonus
hypotony
hypotrichiasis
hypotympanic cell
hypotympanum
hypouresis
hypoventilation
hypovolia
hypoxemia
hypoxia
    anemic h.
    diffusion h.
    hypoxic h.
    ischemic h.
    oxygen affinity h.
    stagnant h.
    h. warning system
hypoxic hypoxia
hypsibrachycephalic
hypsicephalic
hypsicephaly
hypsiconchous
hypsiloid
    h. angle
    h. cartilage
    h. ligament
hypsistaphylia
hypsistenocephalic
hypsocephaly
hypsodont
Hyrtl
    H. anastomosis
    H. epitympanic recess
    H. foramen
    H. loop
    H. sphincter
hysteresis
    static h.
Hz
    hertz

I
> I band
> I cell
> I disk

iacus superficialis

ICA
> internal carotid artery

ICAO standard atmosphere

ichthyosis
> i. congenita neonatorum

icrocoria

ICSH
> interstitial cell-stimulating hormone

icter
> physiologic i.

icterus
> i. neonatorum

ideal alveolar gas

identical twins

ideokinetic

ideomotion

ideomotor

idiodynamic
> i. control

idiomuscular
> i. contraction

idioreflex

idiosome

idiotrophic

IJP
> internal jugular pressure

ileac

ileadelphus

ileal
> i. artery
> i. loop
> i. orifice
> i. papilla
> i. reservoir
> i. sphincter
> i. vein

ileale
> ostium i.

ilealis, pl. ileales
> arteriae ileales
> frenulum ostii i.
> musculus dilator pylori i.
> papilla i.

ilei
> jejunales et i.
> operculum i.
> pars terminalis i.
> venae jejunales et i.

ileocecal
> i. bladder

i. eminence
i. fold
i. junction
i. opening
i. orifice
i. valve

ileocecale
> ostium i.

ileocecalis
> frenulum valvae i.
> plica i.
> valva i.

ileocecocolic sphincter

ileocecum

ileocolic
> i. artery
> i. fold
> i. lymph node
> i. plexus
> i. valve
> i. vein

ileocolica, pl. ileocolicae
> arteria i.
> ramus colicus arteriae ileocolicae
> valvula i.
> vena i.

ileocolici
> lymphonodi i.
> nodi lymphoidei i.

ileocolonic

ileogastric reflex

ileorectal

ileosigmoid

ileum
> arterial arches of i.
> distal i.
> i. duplex
> intestinum i.
> sphincteroid tract of i.
> terminal i.

ileus
> meconium i.

ilia (*pl. of* ilium)

iliac
> i. bone
> i. branch of iliolumbar artery
> i. bursa
> i. crest
> i. externa
> i. fascia
> i. flexure
> i. fossa
> i. lymph node
> i. muscle
> i. nervous plexus

iliac *(continued)*
> i. region
> i. spine
> i. tubercle
> i. tuberosity
> i. vein
> i. vessel

iliaca, pl. iliacae
> bursa subtendinea i.
> crista i.
> i. externa
> fascia i.
> fossa i.
> i. interna
> labium externum cristae iliacae
> labium internum cristae iliacae
> linea intermedia cristae iliacae
> tuberositas i.

iliaci
> plexus i.

iliacosubfascial fossa
iliacosubfascialis
> fossa i.

iliacum
> os i.
> tuberculum i.

iliacus
> amus i.
> i. branch of iliolumbar artery
> i. minor muscle
> musculus i.
> plexus nervosus i.
> subtendinous bursa of i.

iliadelphus
ilii
> ala ossis i.
> corpus ossis i.
> facies auricularis ossis i.
> facies glutea ossis i.
> facies sacropelvina ossis i.
> linea arcuata ossis i.

iliocapsularis
> musculus i.

iliococcygeal
> i. muscle
> i. raphe

iliococcygeus
> i. muscle
> musculus i.
> raphe musculi i.

iliocostal
> i. muscle
> i. space

iliocostalis
> aponeurosis of i.
> i. cervicis
> i. cervicis muscle
> i. lumborum
> i. lumborum muscle

> musculus i.
> i. thoracis
> i. thoracis muscle

iliofemorale
> ligamentum i.

iliofemoralis
> pars descendens ligamenti i.
> pars transversa ligamenti i.

iliofemoral ligament
iliohypogastric
> i. muscle
> i. nerve

iliohypogastrici
> ramus cutaneus anterior nervi i.
> ramus cutaneus lateralis nervi i.

iliohypogastricus
> nervus i.

ilioinguinal
> i. artery
> i. nerve
> i. vein

ilioinguinalis
> nervus i.

iliolum
> ligamentum i.

iliolumbale
> ligamentum i.

iliolumbalis
> arteria i.
> ramus iliacus arteriae i.
> ramus lumbalis arteriae i.
> vena i.

iliolumbar
> i. artery
> i. ligament
> i. vein

iliolumbocos
iliopagus
iliopectinea
> bursa i.

iliopectineal
> i. arch
> i. bursa
> i. eminence
> i. fascia
> i. fossa
> i. ligament
> i. line

iliopectineale
> ligamentum i.

iliopectineus
> arcus i.

iliopelvic sphincter
iliopsoas
> i. bursa
> i. muscle
> musculus i.
> i. tendon

iliopsoaticae
    pars iliaca fasciae i.
    pars psoatica fasciae i.
iliopubic
    i. eminence
    i. tract
iliopubica
    eminentia i.
iliopubicus
    tractus i.
iliosacral articulation
iliosciatic notch
iliospinal
iliothoracopagus
iliotibi
    tractus i.
iliotibial
    i. band (ITB)
    i. tract
iliotibialis
    tractus i.
iliotrochanteric ligament
ilioxiphopagus
ilium, pl. **ilia**
    ala of i.
    arcuate line of i.
    auricular surface of i.
    body of i.
    i. bone
    i. crest
    gluteal surface of i.
    os i.
    sacropelvic surface of i.
    wing of i.
IMA
    inferior mesenteric artery
    internal mammary artery
ima, pl. **imae**
    arteria lumbalis i.
    arteria thyroidea i.
    lumbalis i.
    thyroidea i.
    vena thyroidea i.
imbalance
    sex chromosome i.
    sympathetic i.
Imlach
    I. fat-pad
    I. ring
immediate posttraumatic convulsion
immovable joint
immune response

immunoblast
immunoglobulin
    thyroid-stimulating i. (TSI)
immunohistochemistry
immunologically
    i. activated cell
    i. competent cell
imorphia
impar
    ganglion i.
    nervus i.
    plexus venosus thyroideus i.
    tuberculum i.
imparidigitate
impedance angle
imperforate anus
imperforation
impermeable
impersistence
    motor i.
impervious
implant
    breast i.
implantation
    i. cone
    cortical i.
    interstitial i.
impotence
impotency
impregnate
impregnation
impressio, pl. **impressiones**
    i. aortica pulmonis sinistri
    i. cardiaca faciei diaphragmaticae hepatis
    i. cardiaca pulmonis
    i. colica
    i. colica hepatis
    impressiones digitatae
    i. duodenalis
    i. duodenalis hepatis
    i. esophagea
    i. esophagea hepatis
    i. gastrica
    i. gastrica hepatis
    impressiones gyrorum
    i. ligamenti costoclavicularis
    i. petrosa pallii
    i. renalis
    i. renalis hepatis
    i. suprarenalis

NOTES

impressio *(continued)*
    i. suprarenalis hepatis
    i. trigeminalis
impression
    i. of cerebral gyri
    colic i.
    i. for costoclavicular ligament
    deltoid i.
    digitate i.
    duodenal i.
    esophageal i.
    i. of esophagus
    gastric i.
    renal i.
    rhomboid i.
    suprarenal i.
    trigeminal i.
impressum
    genu i.
impulse
    nerve i.
    neural i.
imus
    nervus splanchnicus i.
    splanchnicus i.
inaction
inadequate stimulus
inanimate
inassimilable
inate fasciculus
inattention
    sensory i.
    visual i.
inborn
incarcerated placenta
incarial bone
incarnatus
    unguis i.
incisal
    i. edge
    i. margin
    i. surface
    i. surface of tooth
incisalis
    margo i.
incision
    infraorbital i.
    lateral i.
incisiva
    fossa i.
    papilla i.
    semicrista i.
    sutura i.
incisive
    i. bone
    i. canal
    i. duct
    i. foramen
    i. fossa

    i. papilla
    i. suture
incisivi
    canales i.
    dentes i.
    i. labii inferioris
    i. labii superioris
incisivum
    foramen i.
    os i.
incisivus
    canalis i.
    dens i.
    ductus i.
incisor
    i. canal
    central i.
    i. crest
    i. foramen
    lateral i.
    second i.
    i. tooth
incisura, pl. incisurae
    i. acetabuli
    i. angularis
    i. angularis ventriculi
    i. anterior auriculae
    i. anterior auris
    i. apicis cordis
    i. cardiaca
    i. cardiaca pulmonis sinistri
    i. cardiaca ventriculi
    i. cartilaginis meatus acustici
    i. cerebelli anterior
    i. cerebelli posterior
    i. clavicularis
    i. costalis
    i. ethmoidalis
    i. fibularis
    i. frontalis
    incisurae helicis
    i. interarytenoidea
    i. intertragica
    i. ischiadica major
    i. ischiadica minor
    i. jugularis
    i. jugularis ossis occipitalis
    i. jugularis ossis temporalis
    i. jugularis sternalis
    i. lacrimalis
    i. ligamenti teretis hepatis
    i. mandibulae
    i. mastoidea
    i. nasalis
    i. pancreatis
    i. parietalis
    i. pterygoidea
    i. radialis
    i. rivini

i. Santorini
i. scapulae
i. semilunaris ulnae
i. sphenopalatina
i. supraorbitalis
i. tentorii
i. of tentorium
i. terminalis auricularis
i. terminalis auris
i. thyroidea inferior
i. thyroidea superior
i. tragica
i. trochlearis
i. tympanica
i. ulnaris
i. umbilicalis
i. vertebralis

incisure
angular i.
Lanterman i.
Rivinus i.
Santorini i.
Schmidt-Lanterman i.
tympanic i.

inclinatio, pl. **inclinationes**
i. pelvis

inclination
angle of i.
pelvic i.
i. of pelvis

inclusion
cell i.'s
i. cell
i. dermoid
fetal i.

incompatibility
physiologic i.
therapeutic i.

incompetence
incompetency
incomplete
i. cleavage
i. conjoined twins
i. tetanus

inconstant arcuate artery of foot
incontinence
incontinent
incontinentia
increment
incremental
i. line
i. lines of von Ebner

increta
placenta i.

incretion
incudal
i. fold
i. fossa

incudes (*pl. of* incus)
incudiformis
uterus i.

incudiform uterus
incudis
corpus i.
crus breve i.
crus longum i.
fossa i.
ligamentum i.
plica i.
processus lenticularis i.

incudomalleal
i. joint

incudomallearis
articulatio i.

incudomalleolar
i. articulation
i. joint

incudostapedia
articulatio i.

incudostapedial
i. articulation
i. joint

incurvation
incus, pl. **incudes**
body of i.
i. bone
crura of i.
crus of i.
fossa of i.
head of i.
lenticular process of i.
ligament of i.
long crus of i.
long limb of i.
long process of i.
posterior ligament of i.
short limb of i.
superior ligament of i.

indeciduate
indentation
indeterminate cleavage
index, pl. **indices**
alveolar i.
auricular i.

NOTES

index *(continued)*
Ayala i.
basilar i.
cardiac i.
cephalic i.
cephaloorbital i.
cephalorrhachidian i.
cerebral i.
cerebrospinal i.
chest i.
cranial i.
dental i.
effective temperature i.
i. extensor muscle
facial i.
i. finger
Flower dental i.
gnathic i.
height-length i.
length-breadth i.
length-height i.
metacarpal i.
mitotic i.
nasal i.
orbital i.
orbitonasal i.
palatal i.
palatine i.
palatomaxillary i.
pelvic i.
ponderal i.
pressure-volume i.
Röhrer i.
sacral i.
stroke work i.
superior facial i.
thoracic i.
tibiofemoral i.
total facial i.
transversovertical i.
uricolytic i.
vertical i.
zygomaticoauricular i.
indicator dilution method
indicis, pl. **indices**
abductor i.
arteria radialis i.
musculus extensor i.
i. proprius
radialis i.
vagina tendinum musculorum
extensoris digitorum et
extensoris i.
indifferent
i. cell
i. gonad
i. tissue

indirect
i. nuclear division
i. ovular transmigration
individuation field
induce
inductance
induction
spinal i.
inductor
inductorium
indulinophil
indulinophile
indusia
indusium griseum
inert
inertia
magnetic i.
i. time
infant
liveborn i.
post-term i.
preterm i.
stillborn i.
term i.
infantilism air cells
inferior
i. aberrant ductule
aditus glottidis i.
adix i.
i. alveolar artery
alveolaris i.
i. alveolar nerve
i. alveolar vein
i. anal nerve
anastomotica i.
i. anastomotic vein
i. angle of scapula
i. anterior segment of lung
apertura pelvis i.
apertura thoracis i.
arcus dentalis i.
arcus palpebralis i.
area vestibularis i.
arteria alveolaris i.
arteria collateralis ulnaris i.
arteria epigastrica i.
arteria glutea i.
arteria hypophysialis i.
arteria labialis i.
arteria laryngea i.
arteria lingularis i.
arteria mesenterica i.
arteria pancreatica i.
arteria pancreaticoduodenalis i.
arteria phrenica i.
arteria rectalis i.
arteria suprarenalis i.
arteria thyroidea i.
arteria tympanica i.

arteria vesicalis i.
arteriola macularis i.
arteriola nasalis retinae i.
arteriola temporalis retinae i.
i. articular facet of atlas
i. articular facet of rib
i. articular pit of atlas
i. articular process
i. articular process of vertebra
i. articular surface of atlas
i. articular surface of femur
i. articular surface of tibia
i. basal vein
i. belly of omohyoid muscle
i. border
i. border of body of pancreas
i. border of liver
i. border of lung
i. border of spleen
i. branch
i. branch of oculomotor nerve
i. branch of pubic bone
i. branch of superior gluteal artery
i. branch of transverse cervical
  nerve
bronchus lobaris i.
bursa subtendinea musculi bicipitis
  femoris i.
i. calcaneonavicular ligament
cardiacus cervicalis i.
i. cardiac vein
i. carotid artery
i. carotid triangle
i. cerebellar artery
i. cerebellar peduncle
i. cerebellar vein
i. cerebral artery
i. cerebral vein
i. cervical cardiac branch of vagus
  nerve
i. cervical facet of vertebra
i. cervical ganglion
i. choroid vein
i. cluneal nerve
collateralis ulnaris i.
colliculus i.
i. colliculus
i. commissure of lip
concha nasalis i.
i. constrictor muscle of pharynx
i. constrictor pharyngeal muscle
constrictor pharyngis i.

i. costal facet
i. costal pit
cutaneus brachii lateralis i.
i. dental arch
i. dental artery
i. dental branch of inferior dental
  plexus
i. dental canal
i. dental foramen
i. dental nerve
i. dental nervous plexus
i. dental rami
ductulus aberrans i.
i. duodenal flexure
i. duodenal fold
i. duodenal fossa
i. duodenal recess
epigastrica i.
i. epigastric artery
i. epigastric lymph node
i. epigastric vein
i. esophageal constriction
i. esophageal sphincter
i. extensor retinaculum
i. extensor retinaculum of ankle
extremitas i.
i. extremity
i. extremity of kidney
i. eyelid
i. fascia
fascia diaphragmatis pelvis i.
fascia diaphragmatis urogenitalis i.
i. fascia of pelvic diaphragm
i. fascia of urogenital diaphragm
fasciculus longitudinalis i.
i. fibular retinaculum
i. fibular retinaculum of ankle
fissura orbitalis i.
flexura duodeni i.
i. flexure of duodenum
fornix conjunctivae i.
fovea costalis i.
i. frontal convolution
i. frontal gyrus
i. frontal sulcus
i. ganglion of glossopharyngeal
  nerve
i. ganglion of vagus
i. ganglion of vagus nerve
gemellus i.
i. gemellus muscle

NOTES

inferior *(continued)*

i. gingival branch of inferior dental plexus
i. glenohumeral ligament
glutea i.
i. gluteal nerve
i. gluteal vein
gluteus i.
i. greater tubercle of facet of humerus
gyrus frontalis i.
gyrus temporalis i.
i. hemiazygos vein
i. hemorrhoidal artery
i. hemorrhoidal nerve
i. hemorrhoidal plexuses
i. hemorrhoidal vein
i. horn
i. horn of falciform margin of saphenous opening
i. horn of saphenous opening
i. horn of thyroid cartilage
i. hypogastric nervous plexus
i. hypophysial artery
i. ileocecal recess
incisura thyroidea i.
i. internal parietal artery
i. interosseous vein
i. labial branch of facial artery
i. labial branch of mental nerve
labialis i.
i. labial vein
i. lacrimal canaliculus
i. lacrimal duct
i. lacrimal papilla
i. lacrimal puncta
laryngea i.
i. laryngeal artery
i. laryngeal cavity
i. laryngeal nerve
i. laryngeal vein
laryngeus i.
i. lateral basal segment of lung
i. lateral brachial cutaneous nerve
i. lateral cutaneous nerve of arm
i. lateral genicular artery
i. lateral genicular vein
i. ligament of epididymis
i. limb
i. limb of ansa cervicalis
linea glutea i.
linea nuchae i.
linea temporalis i.
i. lingual muscle
i. lingular artery
i. lingular branch of lingular branch of left pulmonary
i. lingular bronchopulmonary segment

i. lobar artery
i. lobe of heart
i. lobe of left/right lung
lobulus parietalis i.
lobulus semilunaris i.
i. longitudinal fasciculus
i. longitudinal muscle of tongue
i. longitudinal sinus
i. lumbar facet of vertebra
i. lumbar triangle
macula cribrosa i.
i. macular arteriole
i. macular venule
i. margin
i. margin of liver
margo i.
i. maxilla
i. maxillary nerve
i. meatus
meatus nasi i.
i. medial basal segment of lung
i. medial genicular artery
i. medial genicular vein
i. mediastinum
i. medullary velum
i. member
mesenterica i.
i. mesenteric artery (IMA)
i. mesenteric ganglion
i. mesenteric lymph node
i. mesenteric nervous plexus
i. mesenteric vein
musculus constrictor pharyngis i.
musculus gemellus i.
musculus longitudinalis i.
musculus obliquus capitis i.
musculus rectus i.
musculus serratus posterior i.
musculus tarsalis i.
i. nasal arteriole of retina
i. nasal concha
i. nasal retinal venule
i. nasal venule of retina
nervus alveolaris i.
nervus cardiacus cervicalis i.
nervus cutaneus brachii lateralis i.
nervus gluteus i.
nervus laryngeus i.
norma i.
i. nuchal line
nucleus salivatorius i.
i. oblique (IO)
i. oblique muscle
i. oblique muscle of head
obliquus capitis i.
i. occipital gyrus
i. occipital triangle
oliva i.
i. olivary nucleus

i. olive
i. omental recess
i. ophthalmic vein
i. orbital fissure
palpebra i.
i. palpebral arterial arch
i. palpebral nerve
i. palpebral vein
i. pancreatic artery
i. pancreaticoduodenal artery
i. parathyroid gland
i. parietal gyrus
i. parietal lobe
i. parietal lobule
pars i.
i. part
i. part of duodenum
i. part of lingular vein of left
  superior pulmonary vein
i. part of trapezius muscle
i. part of vestibular ganglion
i. part of vestibulocochlear nerve
pedunculus cerebellaris i.
pedunculus thalami i.
i. pelvic aperture
i. peroneal retinaculum
i. petrosal groove
i. petrosal sinus
i. petrosal sulcus
i. phrenic artery
i. phrenic lymph node
i. phrenic vein
plexus dentalis i.
plexus hypogastricus i.
plexus mesentericus i.
plexus nervosus dentalis i.
plexus nervosus hypogastricus i.
plexus nervosus mesentericus i.
plexus thyroideus i.
plexus venosus vesicalis i.
plica duodenalis i.
i. pole
i. pole of kidney
i. pole of testis
i. pole of thyroid
polus i.
i. posterior serratus muscle
processus articularis i.
i. pubic ligament
i. pubic ramus
i. quadrigeminal brachium
i. radicular vein (IRV)

i. radioulnar joint
ramus lingularis i.
i. ramus of pubis
i. recess of omental bursa
recessus duodenalis i.
recessus ileocecalis i.
i. rectal artery
rectales i.
i. rectal fold
i. rectal nerve
i. rectal nervous plexus
i. rectal vein
rectus i.
i. rectus (IR)
i. rectus muscle
i. renal segment
i. retinaculum of extensor muscle
i. root
i. root of ansa cervicalis
i. root of cervical loop
i. root of vestibulocochlear nerve
i. sagittal sinus
i. salivary nucleus
i. segmental artery of kidney
i. semilunar lobule
serratus posterior i.
sinus petrosus i.
sinus sagittalis i.
spina iliaca anterior i.
spina iliaca posterior i.
i. strait
i. subtendinous bursa of biceps
  femoris
sulcus frontalis i.
sulcus temporalis i.
i. and superior lobar artery
i. suprarenal artery
suprarenalis i.
i. surface of petrous part of
  temporal bone
i. surface of tongue
tarsalis i.
i. tarsal muscle
tarsus i.
i. tarsus
tela choroidea i.
i. temporal arteriole of retina
i. temporal branch of retinal artery
i. temporal convolution
i. temporal gyrus
i. temporal line
i. temporal line of parietal bone

NOTES

inferior *(continued)*
i. temporal retinal arteriole
i. temporal retinal venule
i. temporal sulcus
i. temporal venule of retina
i. thalamic peduncle
i. thalamostriate vein
i. thoracic aperture
i. thoracic artery
i. thoracic facet of vertebra
i. thyroid artery
thyroidea i.
i. thyroid gland
i. thyroid notch
i. thyroid plexus
i. thyroid tubercle
i. thyroid vein
i. tibiofibular joint
i. tracheobronchial lymph node
i. transverse ligament
i. transverse scapular ligament
truncus i.
i. trunk
i. trunk of brachial plexus
i. turbinate
i. turbinated bone
tympanica i.
i. tympanic artery
i. ulnar collateral artery
i. veins of cerebellar hemisphere
vena anastomotica i.
vena azygos minor i.
vena basalis i.
i. vena cava (IVC)
vena cava i.
vena choroidea i.
vena epigastrica i.
vena labialis i.
vena laryngea i.
vena mesenterica i.
vena ophthalmica i.
vena phrenica i.
vena thyroidea i.
vena ventricularis i.
vena vermis i.
i. ventricular vein
venula macularis i.
venula nasalis retinae i.
venula temporalis retinae i.
i. vertebral notch
i. vesical artery
vesicalis i.
i. vesical venous plexus
i. vessel
i. vestibular area
i. vestibular nucleus
i. wall
i. wall of orbit

i. wall of tympanic cavity
zygapophysis i.
inferiore
ramus communicans cum nervo
  laryngeo i.
inferioris, pl. **inferiores**
arteriae lobares inferiores
arteriae membri i.
articulationes cinguli membri i.
brachium colliculi i.
cerebelli inferiores
cerebri inferiores
cingulum membri i.
clunium inferiores
depressor labii i.
facies anterior membri i.
facies lateralis membri i.
facies posterior membri i.
frenulum labii i.
incisivi labii i.
lymphonodi epigastrici inferiores
lymphonodi phrenici inferiores
lymphonodi tracheobronchiales
  inferiores
lymphonodus gluteales inferiores
lymphonodus inguinales inferiores
lymphonodus pancreatici inferiores
musculus depressor labii i.
musculus incisivus labii i.
musculus levator labii i.
musculus quadratus labii i.
musculus triangularis labii i.
nervi anales inferiores
nervi clunium inferiores
nervi rectales inferiores
nodi lymphatici mesenterici
  inferiores
nodi lymphoidei epigastrici
  inferiores
nodi lymphoidei membri i.
nodi lymphoidei mesenterici
  inferiores
nodi lymphoidei phrenici inferiores
nodi lymphoidei tracheobronchiales
  inferiores
nucleus colliculi i.
ossa membri i.
ostium venae cavae i.
palpebrales inferiores
pars cricopharyngea musculi
  constrictoris pharyngis i.
pars libera membri i.
pars thyropharyngea musculi
  constrictoris pharyngis i.
phrenicae inferiores
plexus nervosus rectalis inferiores
plexus periarterialis arteriae
  phrenicae i.
plexus rectales inferiores

processus ethmoidalis conchae
  nasalis i.
processus lacrimalis conchae
  nasalis i.
processus maxillaris conchae
  nasalis i.
rami cardiaci cervicales inferiores
rami dentales inferiores
rami dentales arteriae alveolaris i.
rami dentales inferiores plexus
  dentalis i.
rami esophageales arteriae
  thyroideae i.
rami gingivales inferiores
rami gingivales inferiores plexus
  dentalis i.
rami glandulares arteriae
  thyroideae i.
rami labiales inferiores
rami nasales posteriores inferiores
rami pharyngeales arteriae
  thyroideae i.
rami prostatici arteriae vesicalis i.
rami tracheales arteriae
  thyroideae i.
rami ureterici arteriae
  suprarenalis i.
ramus apicalis lobi i.
ramus mentalis arteriae alveolaris i.
ramus mylohyoideus arteriae
  alveolaris i.
ramus obturatorius arteriae
  epigastricae i.
ramus obturatorius rami pubici
  arteriae epigastricae i.
ramus posterior arteriae
  pancreaticoduodenalis i.
ramus pubicus arteriae
  epigastricae i.
ramus pubicus venae epigastricae i.
ramus superior lobi i.
ramus superior venae pulmonalis
  dextrae/sinistrae i.
rectales inferiores
regiones membri i.
retinaculum musculorum flexorum
  membri i.
sulcus sinus petrosi i.
valvula venae cavae i.
vellus olivae i.
venae cerebelli inferiores
venae cerebri inferiores

venae gluteae inferiores
venae hemispherii cerebelli
  inferiores
venae hemorrhoidales inferiores
venae membri i.
venae palpebrales inferiores
venae phrenicae inferiores
venae rectales inferiores
venae thalamostriatae inferiores

inferius
brachium quadrigeminum i.
cornu i.
ductulus aberrans i.
ganglion cervicale i.
ganglion mesentericum i.
ligamentum epididymidis i.
ligamentum pubicum i.
ligamentum transversum scapulae i.
mediastinum i.
membrum i.
retinaculum musculorum
  extensorum i.
retinaculum musculorum
  fibularium i.
retinaculum musculorum
  peroneorum i.
segmentum anterius i.
segmentum renale anterius i.
trigonum lumbale i.
tuberculum thyroideum i.
velum medullare i.

inferolateral
i. aspect
i. margin
i. margin of cerebral hemisphere
i. surface of prostate

inferomedial margin of cerebral
  hemisphere
inferomedian
inferonasal
inferoposterior
inferotemporal
infiltrate
infiltration
inflammatory corpuscle
infraauricular
i. deep parotid lymph node
i. subfascial parotid lymph node
infraauriculares
nodi lymphoidei parotidei
  profundi i.
infraaxillary

NOTES

infracardiac bursa
infracerebral
infraclavicular
    i. fossa
    i. part of brachial plexus
    i. region
    i. triangle
infraclavicularis
    fossa i.
    regio i.
infracortical
infracostalis, pl. infracostales
    musculi infracostales
    musculus i.
infracostal line
infracotyloid
infracristal
infradentale
infradian
infradiaphragmatic
infraduodenal fossa
infraglenoid
    i. tubercle of scapula
    i. tuberosity
infraglenoidale
    tuberculum i.
infraglottic
    i. cavity
    i. part of larynx
    i. space
infraglottica
    cavitas i.
infraglotticum
    cavitas i.
    cavum i.
infragranular layer
infrahepatic
infrahyoid
    i. branch of superior thyroid artery
    i. bursa
    i. lymph node
    i. muscle
infrahyoidea
    bursa i.
infrahyoidei
    musculi i.
infrahyoideus
    ramus i.
infralobar
    i. part
    i. part of posterior vein of right
      superior pulmonary vein
infralobaris
    pars i.
inframamillary
inframammaria
    regio i.
inframammary
    i. crease

    i. fold
    i. region
inframandibular
inframarginal
inframaxillary
infraoccipital nerve
infraorbital
    i. branch of interior maxillary
      artery
    i. branch of maxillary nerve
    i. canal
    i. foramen
    i. groove
    i. incision
    i. margin
    i. notch
    i. region
    i. suture
    i. vein
infraorbitale
    foramen i.
infraorbitalis
    anteriores nervi i.
    arteria i.
    canalis i.
    margo i.
    nervus i.
    rami labiales superiores nervi i.
    rami nasales externi nervi i.
    rami nasales interni nervi i.
    ramus alveolaris superior medius
     nervi i.
    regio i.
    sulcus i.
    sutura i.
infraorbitomeatal
    i. line
    i. plane
infrapalpebralis
    sulcus i.
infrapalpebral sulcus
infrapatellar
    i. branch of saphenous nerve
    i. fat body
    i. fat-pad
    i. synovial fold
infrapatellare
    corpus adiposum i.
infrapatellaris
    bursa subcutanea i.
    plicae alares plicae synovialis i.
    plica synovialis i.
    ramus i.
infrapiriform foramen
infrapsychic
infrarenal node
infrascapular
    i. artery
    i. region

infrascapularis
    regio i.
infrasegmental
    i. part
    i. vein
infrasegmentalis
    pars i.
infraspinata
    fascia i.
    fossa i.
infraspinati
    bursa subtendinea musculi i.
infraspinatus
    i. bursa
    i. fascia
    i. muscle
    musculus i.
    i. reflex
    subtendinous bursa of i.
    i. tendon
infraspinous
    i. fascia
    i. fossa
    i. muscle
infrasplenic
infrasternal
    i. angle
    i. angle of thorax
infrasternalis
    angulus i.
infratemporal
    i. crest
    i. crest of greater wing of
      sphenoid
    i. fossa
    i. region
    i. surface of body of maxilla
    i. surface of greater wing of
      sphenoid
infratemporalis
    crista i.
    fossa i.
infratentorial
infrathoracic
infratonsillar
infratrochlear
    i. branch of ophthalmic nerve
infratrochlearis
    nervus i.
    rami palpebrales nervi i.
infraumbilical fold
infundibula (*pl. of* infundibulum)

infundibular
    i. part of anterior lobe of
      hypophysis
    i. recess
    i. stalk
    i. stem
infundibularis
    pars i.
infundibuliform
    i. fascia
    i. sheath
infundibuliformis
    recessus i.
infundibuloovarian ligament
infundibulopelvic ligament
infundibulum, pl. **infundibula**
    ethmoid i.
    ethmoidal i.
    i. ethmoidale
    i. of fallopian tube
    i. of gallbladder
    i. of heart
    i. of hypophysis
    hypothalamic i.
    infundibula of kidney
    i. of lung
    i. of pituitary gland
    i. of right ventricle
    i. tubae uterinae
    i. of urinary bladder
    i. of uterine tube
    i. vesicae biliaris
    i. vesicae felleae
Ingrassia
    I. process
    I. wing
inguen
inguinal
    i. aponeurotic fold
    i. area
    i. branch of deep external
      pudendal artery
    i. canal
    i. crest
    i. falx
    i. fossa
    i. gland
    i. ligament
    i. ligament of kidney
    i. lymphatic plexus
    i. lymph node
    i. nerve

NOTES

inguinal *(continued)*
    i. part of ductus deferens
    i. reflex
    i. region
    i. ring
    i. sphincter
    i. triangle
    i. trigone
inguinale
    ligamentum i.
    trigonum i.
inguinalis, pl. inguinales
    arcus i.
    canalis i.
    crus laterale anuli i.
    crus mediale anuli i.
    falx i.
    plexus lymphaticus i.
    plica i.
    rami inguinales
    regio i.
inguinoabdominal
inguinocrural
inguinofemoral
inguinolabial
inguinoproperitoneal hernia
inguinoscrotal
inhalation
inhale
inherent
inherited
inhibit
inhibition
    central i.
    contact i.
    feedback i.
    reciprocal i.
    reflex i.
    Wedensky i.
inhibitory
    i. fiber
    i. junction potential
    i. nerve
    i. postsynaptic potential (IPSP)
iniencephaly
initial heat
injection mass
injury
    current of i.
    i. potential
inlet
    anterior sagittal pelvic i.
    anteroposterior diameter of pelvic i.
    conjugate diameter of pelvic i.
    esophageal i.
    laryngeal i.
    i. of larynx
    pelvic plane of i.
    plane of i.

    thoracic i.
    transverse pelvic i.
innate reflex
inner
    i. border of iris
    i. canthus
    i. cell mass
    i. cone fiber
    i. dental epithelium
    i. ear
    i. enamel epithelium
    i. hamstring
    i. layer of eyeball
    i. lip of iliac crest
    i. malleolus
    i. sheath of optic nerve
    i. spiral sulcus
    i. stripes of renal medulla
    i. table bone of skull
    i. table frontal bone
    i. zone of renal medulla
innermost
    i. intercostal muscle
    i. membrane of meninges
innervated muscle
innervation
    cutaneous i.
    i. of head and neck
    reciprocal i.
innominata
    fossa i.
    substantia i.
    vena i.
innominatal
innominate
    i. artery
    i. bone
    i. cardiac vein
    i. cartilage
    i. fossa
    i. substance
inosculate
inosculation
inotropic
    negatively i.
    positively i.
inscription
    tendinous i.
inscriptio tendinea
insemination
insensible
    i. perspiration
    i. thirst
insertion
    aponeurosis of i.
    fascia of i.
    parasol i.
    tendon i.
    velamentous i.

insheathed
insorption
inspiration
inspiratory
    i. capacity
    i. center
    i. reserve volume
inspire
inspired gas
inspirometer
instep
insufficiency
    mitral i. (MIS)
    respiratory i.
insula, pl. **insulae**
    gyri breves insulae
    gyrus of i.
    gyrus longus insulae
    Haller i.
    limen insulae
    long gyrus of i.
    short gyri of i.
    sulcus circularis insulae
insular
    i. area
    i. cortex
    i. part
    i. part of middle cerebral artery
insularis, pl. **insulares**
    arteriae insulares
    pars i.
    venae insulares
insulin-antagonizing factor
insulinogenesis
insulinogenic
insulogenic
integration
integument
integumentary system
integumentum commune
intensity
    unit of magnetic field i.
interacinar
interacinous
interalveolar
    i. pore
    i. septum
interalveolare
    septum i.
interalveolaria
    septa i.
interannular segment

interarterial fluid
interarticular
    i. cartilage
    i. disk
    i. fibrocartilage
    i. joint
interarticularis
    fibrocartilago i.
interarytenoid
    i. muscle
    i. notch
interarytenoidea
    incisura i.
interasteric
interatrial
    i. foramen primum
    i. foramen secundum
    i. groove
    i. septum
interatriale
    septum i.
interauricular arc
intercalary neuron
intercalated
    i. disk
    i. ducts
    i. nucleus
intercalatus
    nucleus i.
intercanalicular
intercapillary cell
intercapitales
    venae i.
intercapital vein
intercapitular
    i. vein
    i. vein of foot
    i. vein of hand
intercapitulares
    venae i.
intercarotic
intercarotid
    i. body
    i. ganglion
    i. nerve
intercarpal
    i. articulation
    i. joint
    i. ligament
intercarpales
    articulationes i.

NOTES

intercarpalia
ligamenta i.
intercarpeae
articulationes i.
intercartilaginea rimae glottidis
intercartilaginous
i. part of glottic opening
i. part of rima glottidis
i. rim
intercavernosi
sinus i.
intercavernosus septum
intercavernous
i. sinuses
i. venous sinus
intercellular
i. bridges
i. canaliculus
i. cement
i. junctions
i. space
intercentra
intercentral
intercentrum
intercerebral
interchondral
i. articulation
i. joint
i. ligament
interchondrales
articulationes i.
intercilium
interclavicular
i. ligament
i. notch
interclaviculare
ligamentum i.
interclinoid ligament
intercoccygeal
intercolumnar
i. fasciae
i. fibers
i. tubercle
intercondylar
i. eminence
i. fossa
i. groove
i. line
i. line of femur
i. notch
i. notch of femur
i. process
i. tubercle
intercondylare
tuberculum i.
intercondylaris
eminentia i.
fossa i.
linea i.

intercondylic fossa
intercondyloid
i. eminence
i. fossa
i. notch
intercondyloidea
eminentia i.
intercornual ligament
intercosta
intercostal
i. artery
i. articulation
collateral branch of posterior i.
i. groove
i. ligament
i. lymph node
i. margin
i. membranes
i. muscle
i. nerve
i. space
i. vein
i. vessel
intercostale
spatium i.
intercostalia
ligamenta i.
membranae i.
intercostalis, pl. intercostales
intercostales anteriores
intercostales externi
intercostales interni
lymphonodi intercostales
membranae intercostales
nervi intercostales
nodi lymphoidei intercostales
intercostales posteriores
ramus cutaneus anterior abdominalis
nervi i.
ramus cutaneus anterior pectoralis
nervi i.
i. superior dextra
i. superior sinistra
i. suprema
intercostalium
rami cutanei anteriores pectoralis et
abdominalis nervorum i.
rami mammarii laterales ramorum
cutaneorum lateralium nervorum i.
rami musculares nervorum i.
ramus collateralis nervorum i.
ramus cutaneus lateralis
abdominalis/pectoralis nervorum i.
ramus cutaneus lateralis ramorum
posteriorum arteriae i.
intercostobrachiales
nervi i.
intercostobrachial nerve
intercostohumeralis

intercostohumeral nerve
intercristal
intercrural
    i. fibers
    i. fibers of aponeurosis of abdominal oblique muscle
    i. fibers of superficial ring
    i. ganglion
intercrurales
    fibrae i.
intercuneiform
    i. joint
    i. ligament
intercuneiformes
    articulationes i.
intercuneiformia
    ligamenta i.
intercutaneomucous
interdeferential
interdental
    i. canal
    i. papilla
interdentalis
    papilla i.
interdigit
interdigital
    i. fold
    i. ligament
    i. space
interdigitalis
    plica i.
interdigitation
interface
    dermoepidermal i.
interfacial canals
interfasciale
    spatium i.
interfascial space
interfascicular fasciculus
interfascicularis
    fasciculus i.
interfemoral
interferometer
    electron i.
interferometry
    electron i.
interferon-beta2
interfibrillar
interfibrillary
interfibrous
interfilamentous

interfoveolare
    ligamentum i.
interfoveolar ligament
interfrontal
interganglionares
    rami i.
interganglionic branch of sympathetic trunk
intergemmal
interglobula
    spatium i.
interglobular
    i. space
    i. space of Owen
interglobularia
    spatia i.
intergluteal cleft
interglutealis
    crena i.
intergonial
intergyral
interhemicerebral
intérieur
    milieu i.
interiliaci
    lymphonodi i.
    nodi lymphoidei i.
interiliac lymph node
interior
    i. chest wall
    i. maxillary branch of external carotid artery
interischiadic
interkinesis
interlamellar
interlaminar jelly
interlobar
    i. artery
    i. artery of kidney
    i. duct
    i. fissure
    i. notch
    i. surfaces of lung
    i. vein
    i. vein of kidney
interlobares
    arteriae i.
    i. renis
interlobul
interlobular
    i. artery
    i. artery of kidney

NOTES

interlobular *(continued)*
    i. artery of liver
    i. duct
    i. ductule
    i. vein of kidney
    i. vein of liver
interlobulares
    arteriae i.
    ductuli i.
    i. hepatis
    i. hepatis venae
    i. renis
interlocking gyri
intermalleolar
intermammary
intermammillary
intermaxilla
intermaxillaris
    sutura i.
intermaxillary
    i. bone
    i. segment
    i. suture
intermedia, pl. **intermediae**
    arteria temporalis i.
    crista sacralis i.
    ganglia i.
    massa i.
    pars i.
    portio i.
    venae hepaticae intermediae
intermediary
    i. nerve
    i. system
intermediate
    i. antebrachial vein
    i. anterior wall
    i. atrial branch of left coronary artery
    i. atrial branch of right coronary artery
    i. basilic vein
    i. body of Flemming
    i. branch of hepatic artery proper
    i. bronchus
    i. cephalic vein
    i. cubital vein
    i. cuneiform bone
    i. digastric tendon
    i. disk
    i. dorsal cutaneous nerve
    i. filaments
    i. ganglia
    i. great muscle
    i. hepatic vein
    i. junction
    i. lacunar lymph node
    i. lamella
    i. laryngeal cavity

    i. layer
    i. line of iliac crest
    i. lumbar lymph node
    i. mesoderm
    i. omohyoid tendon
    i. palm of hand
    i. part
    i. part of adenohypophysis
    i. part of male urethra
    i. part of vestibular bulb
    i. sacral crest
    i. supraclavicular nerve
    i. temporal branch of lateral occipital artery
    i. tendon of diaphragm
    i. vastus muscle
    i. vein of forearm
    i. zone of iliac crest
intermedii
    lymphonodi lumbales i.
    lymphonodus iliaci communes i.
    lymphonodus iliaci externi i.
    nodi lymphoidei lumbales i.
    supraclaviculares i.
intermediolateral
    i. cell column of spinal cord
    i. nucleus
intermediolaterali
    nucleus i.
intermediomedial
    i. frontal branch of callosomarginal artery
    i. nucleus
intermediomedialis
    nucleus i.
intermedium
    os cuneiforme i.
    septum cervicale i.
intermedius
    bronchus i.
    cutaneus dorsalis i.
    ganglion of nervus i.
    musculus vastus i.
    nervus cutaneus dorsalis i.
    nervus supraclavicularis i.
    nodus lymphoideus lacunaris i.
    sphincter i.
    vastus i.
intermembranous
    i. part of glottic opening
    i. part of rima glottidis
intermeningeal
intermesenteric
    i. arterial anastomosis
    i. nervous plexus
intermesenterica
    arteria i.
intermesentericus
    plexus nervosus i.

intermetacarpal
   i. articulation
   i. joint
   i. ligament
intermetacarpales
   articulationes i.
intermetameric
intermetatarsal
   i. articulation
   i. joint
   i. ligament
intermetatarsales
   articulationes i.
intermetatarseae
   articulationes i.
intermetatarseum
   os i.
intermission
intermittent
intermuscular
   i. gluteal bursa
   i. groove
   i. hernia
   i. membrane
   i. septum
intermusculare
   septum i.
interna
   arteria auditiva i.
   arteria carotis i.
   arteria iliaca i.
   arteria mammaria i.
   arteria pudenda i.
   arteria spermatica i.
   arteria thoracica i.
   auris i.
   basis cranii i.
   capsula i.
   carotis i.
   crista occipitalis i.
   facies i.
   fascia spermatica i.
   fovea inguinalis i.
   iliaca i.
   jugularis i.
   lamina i.
   membrana intercostalis i.
   organa genitalia feminina i.
   organa genitalia masculina i.
   organum genitalia feminina i.
   organum genitalia masculina i.
   protuberantia occipitalis i.

   pudenda i.
   theca i.
   thoracica i.
   vena iliaca i.
   vena jugularis i.
   vena mammaria i.
   vena pudenda i.
   vena thoracica i.
internae
   cerebri i.
   crus anterius capsulae i.
   crus posterius capsulae i.
   fibrae arcuatae i.
   genu capsulae i.
   pars cavernosa arteriae carotidis i.
   pars cavernosa arteriae carotis i.
   pars cerebralis arteriae carotidis i.
   pars cerebralis arteriae carotis i.
   pars cervicalis arteriae carotidis i.
   pars cervicalis arteriae carotis i.
   pars petrosa arteriae carotidis i.
   pars petrosa arteriae carotis i.
   pars retrolentiformis capsulae i.
   pars sublentiformis capsulae i.
   plexus nervosus arteriae carotidis i.
   plexus periarterialis arteriae
     thoracicae i.
   rami capsulae i.
   rami clivales partis cerebralis
     arteriae carotidis i.
   rami intercostales anteriores arteriae
     thoracicae i.
   rami labiales posteriores arteriae
     pudendae i.
   rami mammarii mediales ramorum
     perforantium arteriae thoracicae i.
   rami mediastinales arteriae
     thoracicae i.
   rami perforantes arteriae
     thoracicae i.
   rami scrotales posteriores arteriae
     pudendae i.
   rami sternales arteriae thoracicae i.
   rami thymici arteriae thoracicae i.
   ramus basalis tentorii arteriae
     carotidis i.
   ramus costalis lateralis arteriae
     thoracicae i.
   ramus ganglionares trigeminales
     arteriae carotidis i.
   ramus marginalis tentorii arteriae
     carotidis i.

NOTES

internae *(continued)*
    ramus marginalis tentorii partis cavernosae arteriae carotidis i.
    ramus meningeus arteriae carotidis i.
    ramus meningeus arteriae carotis i.
    ramus meningeus partis cavernosae arteriae carotidis i.
    ramus meningeus partis cerebralis arteriae carotidis i.
    ramus sinus cavernosi arteriae carotidis i.
    ramus sinus cavernosi partis cavernosae arteriae carotidis i.
    thoracicae i.
    vasa auris i.
    vasa sanguinea auris i.
    venae cerebri i.
    venae thoracicae i.

internal
    i. abdominal muscle
    i. abdominal ring
    i. acoustic foramen
    i. acoustic meatus
    i. acoustic opening
    i. acoustic orifice
    i. acoustic pore
    i. anal sphincter
    i. arcuate fibers
    i. auditory artery
    i. auditory foramen
    i. auditory meatus
    i. auditory vein
    i. axis of eye
    i. base of skull
    i. branch of superior laryngeal nerve
    i. branch of trunk of accessory nerve
    i. calcaneoastragaloid ligament
    i. canthus
    i. capsule
    i. carotid artery (ICA)
    i. carotid nerve
    i. carotid nervous plexus
    i. carotid venous plexus
    i. cerebral vein
    i. collateral ligament of wrist
    i. conjugate
    i. ear
    i. female genital organ
    i. fibrocartilage
    i. filum terminale of pia mater
    i. hemorrhoidal complex
    i. hydrocephalus
    i. iliac artery
    i. iliac lymph node
    i. iliac vein
    i. inguinal ring

    i. intercostal membrane
    i. intercostal muscle
    i. jugular pressure (IJP)
    i. jugular vein
    i. lateral ligament
    i. limiting membrane
    i. lip of iliac crest
    i. male genital organ
    i. malleolus
    i. mammary artery (IMA)
    i. mammary plexus
    i. maxillary artery
    i. maxillary plexus
    i. maxillary vein
    i. naris
    i. nasal branch
    i. neurocranial foramen
    i. oblique fascia
    i. oblique line
    i. oblique muscle
    i. obturator muscle
    i. occipital crest
    i. occipital protuberance
    i. os
    i. ovular transmigration
    i. pillar cells
    i. pterygoid muscle
    i. pudendal artery
    i. pudendal vein
    i. ramus of accessory nerve
    i. rectal artery
    i. rectal nerve
    i. rectal sphincter
    i. respiration
    i. root sheath
    i. rotation
    i. salivary gland
    i. saphenous nerve
    i. semilunar fibrocartilage of knee joint
    i. sheath of optic nerve
    i. spermatic artery
    i. spermatic fascia
    i. spermatic vessel
    i. sphincter muscle of anus
    i. spiral sulcus
    i. surface
    i. surface of cranial base
    i. surface of frontal bone
    i. surface of parietal bone
    i. surface of trachea
    i. table of calvaria
    i. thoracic artery
    i. thoracic lymphatic plexus
    i. thoracic vein
    i. urethral opening
    i. urethral orifice
    i. urethral sphincter

internarial

internasalis
  sutura i.
internasal suture
interne
  milieu i.
interneuromeric clefts
interneurons
interni
  bursa ischiadica musculi
    obturatoris i.
  bursa subtendinea musculi
    obturatoris i.
  crista transversa meatus acustici i.
  crista verticalis meatus acustici i.
  fundus meatus acustici i.
  intercostales i.
  lymphonodi iliaci i.
  musculi intercostales i.
  nodi lymphoidei iliaci i.
  rami nasales i.
  ramus meatus acustici i.
interno
  ramus communicans nervi laryngei
    recurrentis cum ramo laryngeo i.
internodale
  segmentum i.
internodal segment
internode
internuclear
internum
  corpus geniculatum i.
  filum terminale i.
  ligamentum tarsale i.
  ostium urethrae i.
  ostium uteri i.
  os uteri i.
  perimysium i.
internuncial neuron
internus
  axis bulbi i.
  bursae of obturator i.
  bursa of obturator i.
  caroticus i.
  fascia of obturator i.
  meatus acusticus i.
  musculus intercostalis i.
  musculus obturator i.
  musculus pterygoideus i.
  musculus rectus i.
  musculus sphincter ani i.
  musculus sphincter urethrae i.
  musculus thyroarytenoideus i.

  musculus vastus i.
  nervus caroticus i.
  plexus caroticus i.
  plexus mammarius i.
  plexus maxillaris i.
  plexus venosus caroticus i.
  porus acusticus i.
  ramus i.
  sphincter ani i.
  sulcus spiralis i.
interoceptive
interoceptor
interolivary
interomedialis
  ramus frontalis i.
interorbital
interossea
  i. anterior
  bursa cubitalis i.
  ligamenta cuneometatarsalia i.
  ligamenta intercuneiformia i.
  ligamenta metacarpalia i.
  ligamenta metatarsalia i.
  ligamenta sacroiliaca i.
  ligamenta tarsi i.
  ligamentum intercarpalia i.
  ligamentum intercuneiformia i.
  ligamentum metacarpalia i.
  ligamentum metatarsalia i.
  i. posterior
  i. recurrens
interosseal
interossei (*pl. of* interosseus)
interosseous
  i. anterior
  i. anterior nerve
  i. artery
  i. border
  i. border of fibula
  i. border of radius
  i. border of tibia
  i. border of ulna
  i. bursa of elbow
  i. cartilage
  i. crest
  i. crest of radius
  i. crest of ulna
  i. cruris
  i. cruris nerve
  i. cubital bursa
  i. cuneocuboid ligament
  i. cuneometatarsal ligament

NOTES

interosseous *(continued)*
    i. dorsalis
    i. dorsalis nerve
    i. fascia
    i. groove
    i. groove of calcaneus
    i. groove of talus
    i. margin
    i. membrane
    i. membrane of forearm
    i. membrane of leg
    i. metacarpal ligament
    i. metacarpal space
    i. metatarsal ligament
    i. metatarsal space
    i. muscle
    i. nerve of leg
    i. posterior
    i. posterior nerve
    i. sacroiliac ligament
    i. talocalcaneal ligament
    i. tibiofibular ligament
    i. vein
interosseum
    ligamentum cuneocuboideum i.
    ligamentum talocalcaneare i.
interosseus, pl. **interossei**
    interossei dorsales manus
    interossei dorsales pedis
    margo i.
    musculi interossei
    interossei palmares
    interossei plantares
interpalpebral
    i. suture
    i. zone
interparieta
    os i.
interparietal
    i. bone
    i. hernia
    i. sulcus
    i. suture
interparietale
    os i.
interparietalis
    sutura i.
interpectorales
    lymphonodi i.
    nodi lymphoidei i.
interpectoral lymph node
interpediculate
interpeduncular
    i. cistern
    i. fossa
    i. ganglion
    i. joint space
    i. nucleus

interpeduncularis
    cisterna i.
    fossa i.
    nucleus i.
interphalangeae
    articulationes i.
interphalangeal
    i. articulation
    i. collateral ligament
    distal i. (DIP)
    i. joint (IPJ)
    i. joint of foot
    i. joint of hand
    proximal i. (PIP)
interphase
interplant
interplanting
interpleural space
interpositum
    velum i.
interpositus
    nucleus i.
interproximal
    i. papilla
    i. surface of tooth
interpubic
    i. disk
    i. fibrocartilage
interpubica
    fibrocartilago i.
    lamina fibrocartilaginea i.
interpubicus
    discus i.
interpulmonary septum
interpupillary
interradial
interradicular
    i. septa
    i. septa of maxilla and mandible
interradicularia
    septa i.
interrenal
interrupted respiration
interscalene triangle
interscapular
    i. gland
    i. hibernoma
    i. region
interscapulum
intersciatic
intersect
    tendinous i.
intersectio, pl. **intersectiones**
    i. tendinea
    intersectiones tendineae musculi
      recti abdominis
intersection
    tendinous i.

intersegmental
  i. fasciculi
  i. part
  i. part of pulmonary vein
intersegmentalis
  pars i.
interseptal
interseptovalvular space
interseptum
intersheath space of optic nerve
intersigmoideus
  recessus i.
intersigmoid recess
interspace
  vertebral i.
intersphincteric space
interspinal
  i. line
  i. muscle
  i. muscle of thorax
  i. plane
interspinale
  ligamentum i.
  planum i.
interspinalis, pl. **interspinales**
  interspinales cervicis muscle
  linea i.
  interspinales lumborum muscle
  musculi interspinales
  interspinales thoracis muscle
interspinous
  i. ligament
  i. plane
interspongioplastic substance
interstice
interstitial
  i. absorption
  i. cells
  i. cell-stimulating hormone (ICSH)
  i. fluid
  i. gland
  i. growth
  i. implantation
  i. lamella
  i. nucleus
  i. space
  i. tissue
interstitialis
  nucleus i.
interstitium

intertarsal
  i. articulation
  i. joint
intertarseae
  articulationes i.
intertendineus
  conexus i.
  i. connection
intertendinous connections of extensor digitorum
interthalamic
  i. adhesion
  i. commissure
interthalamica
  adhesio i.
intertragica
  incisura i.
intertragic notch
intertragicus
  musculus i.
intertransversalis
intertransversarii
  i. muscle
  musculi i.
intertransversarium
  ligamentum i.
intertransverse
  i. ligament
  i. muscle
intertrochanteric
  i. crest
  i. line
  i. ridge
intertrochanterica
  crista i.
  linea i.
intertubercular
  i. eminence
  i. groove
  i. line
  i. plane
  i. sulcus
  i. tendon sheath
intertuberculare
  planum i.
intertubercularis
  linea i.
  sulcus i.
  vagina tendinis i.
intertubular
  i. groove
  i. tissue

NOTES

interureteral
interureteric
    i. crest
    i. fold
    i. ridge
interureterica
    plica i.
intervaginal subarachnoid space of optic nerve
interval
    auriculoventricular i.
intervascular
intervenosum
    tuberculum i.
intervenous
    i. tubercle
    i. tubercle of right atrium
interventricular
    i. cartilage
    i. fibrocartilage
    i. foramen
    i. ganglion
    i. groove
    i. notch
    i. septal branch of left/right coronary artery
    i. septum
    i. space
    i. sulcus of heart
    i. vein
interventriculare
    foramen i.
    septum i.
interventricularis
    pars membranacea septi i.
    pars muscularis septi i.
intervertebral
    i. body
    i. cartilage
    i. disk
    i. foramen
    i. ganglion
    i. notch
    i. symphysis
    i. vein
intervertebrale
    foramen i.
intervertebralis
    anulus fibrosus disci i.
    discus i.
    fibrocartilago i.
    symphysis i.
    vena i.
intervillous
    i. circulation
    i. lacuna
    i. space
interzonal mesenchyme
intestina

intestinal
    i. arterial arcades
    i. artery
    i. atresia
    i. cecum
    i. digestion
    i. flora
    i. follicles
    i. gland
    i. juice
    i. loop
    i. lumen
    i. mucosa
    i. peritoneum
    i. rotation
    i. surface of uterus
    i. tract
    i. trunks
    i. vein
    i. villi
intestinalis, pl. **intestinales**
    arteriae intestinales
    glandulae intestinales
    tonsilla i.
    trunci lymphatici intestinales
    villi intestinales
intestine
    absorptive cells of i.
    aggregated lymphatic follicles of small i.
    aggregated lymphoid nodules of small i.
    circular fold of small i.
    circular layer of muscle coat of small i.
    crypt of Lieberkühn of large i.
    crypt of Lieberkühn of small i.
    gland of large i.
    gland of small i.
    large i.
    longitudinal layer of muscle coat of small i.
    mesenteric portion of small i.
    mucosa of large i.
    mucosa of small i.
    mucous membrane of large i.
    mucous membrane of small i.
    muscular coat of large i.
    muscular coat of small i.
    muscular layer of large i.
    muscular layer of small i.
    nonrotation of i.
    serosa of large i.
    serosa of small i.
    small i.
    solitary nodules of i.
intestinointestinal reflex
intestinum
    i. cecum

i. crassum
i. ileum
i. jejunum
i. rectum
i. tenue
i. tenue mesenteriale
intima
tunica i.
intimal layer
intimus
musculus intercostalis i.
intoe
intortor
intraabdominal viscera
intraacinous
intraadenoidal
intraalveolar septa
intraaortic
intraarterial
intraarticular
i. cartilage
i. ligament of costal head
i. ligament of head of rib
i. sternocostal ligament
intraarticulare
ligamentum capitis costae i.
ligamentum sternocostale i.
intraatrial
intraaural
intraauricular muscle
intrabronchial
intrabuccal
intrabulbar fossa
intracameral
intracanalicular
intracanicular part of optic nerve
intracapsularia
ligamenta i.
intracapsular ligament
intracardiac pressure curve
intracarpal
intracartilaginous
intracavernous plexus
intracavitary
intracelial
intracellular
i. canaliculus
i. fluid
i. water
intracerebellar
intracerebral
intracervical

intrachondrial bone
intracisternal
intracolic
intracondylar
intracordal
intracorporeal
intracorpuscular
intracostal
intracrania
intracranial
i. cavity
i. ganglion
i. part
i. part of optic nerve
i. part of vertebral artery
intracranialis
pars i.
intracutaneous
intracuticular
intracystic
intradermal
intradermic
intraduct
intraductal
intradural
intraembryonic mesoderm
intraepidermal
intraepidermic bulla
intraepiphysial
intraepithelial
i. gland
i. vessel
intrafascial space
intrafascicular
intrafilar
intrafusal fibers
intragastric
intragemmal
intraglandular deep parotid lymph node
intraglandulares
lymphonodi parotidei i.
nodi lymphoidei parotidei i.
intraglobular
intragracilis
sulcus i.
intragyra
intrahepatic biliary duct
intrahyoid
intrajugularis
processus i.
intrajugular process

NOTES

intralaminar
    i. nuclei of thalamus
    i. part of intralocular part of optic
    nerve
intralaryngeal
intraligamentous
intralobar
    i. part
    i. part of the posterior vein of
    the right superior pulmonary vein
intralobaris
    pars i.
intralobular duct
intralocular
intralocularis
    pars intralaminaris nervi optici i.
intraluminal
intramedullary canal
intramembranous ossification
intrameningeal
intramural
    i. artery
    i. part of male urethra
intramyocardial
intramyometrial
intranasal
intraneural
intranuclear
intraocular
    i. fluid
    i. part of optic nerve
intraocularis
    pars postlaminaris nervi optici i.
    pars prelaminaris nervi optici i.
intraoral
intraorbital groove
intraosseous
intraosteal
intraovarian
intraovular
intraparietal
    i. sulcus
    i. sulcus of Turner
intraparietalis
    sulcus i.
intraparotideus
    plexus i.
intraparotid plexus of facial nerve
intrapatellar bursa
intrapelvic
intrapericardiac
intrapericardial
intraperitoneal viscus
intrapharyngeal space
intrapial
intrapleural
intrapontine
intraprostatic
intraprotoplasmic

intrapulmonales
    nodi lymphoidei i.
intrapulmonalia
    vasa sanguinea i.
intrapulmonary
    i. blood vessel
    i. lymph node
intrarachidian
intrarectal
intrarenal
    i. artery
    i. vein
intrarenales
    arteriae i.
intrarenalium
    rami capsulares arteriorum i.
intraretinal space
intrarrhachidian
intrascrotal
intrasegmental
    i. bronchi
    i. part
    i. part of pulmonary vein
intrasegmentalis, pl. **intrasegmentales**
    bronchi intrasegmentales
    pars i.
intraspinal
intrasplenic
intrastromal
intrasynovial
intratarsal
intratendinous
    i. bursa of elbow
    i. olecranon bursa
intrathecal
intrathecally
intrathoracic
intrathyroid cartilage
intratonsillar
intratubal
intratubular
intratympanic
intrauterina
    rachitis i.
intrauterine amputation
intravagale
    glomus i.
intravaginal space
intravascular
intravesical
intravitelline
intravitreous
intrinsic
    i. fibers
    i. muscle
    i. muscle of foot
    i. muscle of larynx
    i. reflex
    i. sphincter

introflection
introflexion
introgastric
introitus
    i. canalis
    esophageal i.
    i. of facial canal
    vaginal i.
intromission
intromittent organ
intumescence
    tympanic i.
intumescent
intumescentia
    i. cervicalis
    i. ganglioformis
    i. lumbalis
    i. lumbosacralis
    i. tympanica
intussusceptive growth
inulin clearance
invaginate planula
invagination
    basilar i.
inverse symmetry
inversion
    i. of ankle
    visceral i.
inversus
    dextrocardia with situs i.
    situs i.
inverted
    i. pelvis
    i. testis
invertor
investiens
    fascia i.
investing
    i. cartilage
    i. layer
    i. layer of cervical fascia
    i. tissues
investment
    aponeurosis of i.
invisible differentiation
involucra (*pl. of* involucrum)
involucre
involucrin
involucrum, pl. involucra
involuntary muscle

involution
    i. cyst
    senile i.
IO
    inferior oblique
iodinophile
iodinophilous
ion channel
IPJ
    interphalangeal joint
ipsilateral
    i. mentalis muscle
    i. reflex
IPSP
    inhibitory postsynaptic potential
IR
    inferior rectus
irideremia
irides (*pl. of* iris)
iridescent
iridial part of retina
iridica
    stella lentis i.
iridis
    angulus i.
    anulus i.
    atresia i.
    dilator i.
    facies anterior i.
    facies posterior i.
    ligamentum pectinatum i.
    margo ciliaris i.
    margo pupillaris i.
    musculus dilator i.
    plicae i.
    sphincter i.
    stratum pigmenti i.
    stroma i.
iridocorneal
    angle of i.
    i. angle
    i. mesodermal dysgenesis
iridocornealis
    angulus i.
    ligamentum pectinatum anguli i.
    spatia anguli i.
iridodilator
iridopupillary lamina
iris, pl. irides
    angle of i.
    anterior surface of i.
    border of i.

NOTES

iris (*continued*)
    i. ciliary body
    ciliary border of i.
    ciliary margin of i.
    circle of i.
    crypt of i.
    fold of i.
    i. frill
    greater arterial circle of i.
    greater ring of i.
    inner border of i.
    lesser arterial circle of i.
    lesser ring of i.
    major arterial circle of i.
    major circulus arteriosus of i.
    minor arterial circle of i.
    minor circulus arteriosus of i.
    outer border of i.
    pectinate ligament of i.
    pigmented layer of i.
    pillar of i.
    posterior surface of i.
    pupillary border of i.
    pupillary margin of i.
    ring of i.
    stroma of i.
    surface of interior i.
irregular bone
irregulare
    os i.
irritability
    electric i.
    myotatic i.
irritable
irritation cell
irruptive
IRV
    inferior radicular vein
isauxesis
ischemic hypoxia
ischesis
ischia
ischiadic
    i. nerve
    i. plexus
    i. spine
ischiadica
    arteria i.
    spina i.
ischiadicum
    foramen i.
    tuber i.
ischiadicus
    nervus i.
ischial
    i. bone
    i. bursa
    i. ramus

    i. spine
    i. tuberosity
ischiatica
    arteria i.
ischiatic notch
ischii
    corpus ossis i.
    os i.
    ramus ossis i.
ischioanal fossa
ischioanalis
    corpus adiposum fossae i.
    fossa i.
ischiobulbar
ischiocapsulare
    ligamentum i.
ischiocapsular ligament
ischiocavernosus
    musculus i.
ischiocavernous muscle
ischiococcygeal
ischiococcygeus
    musculus i.
ischiofemorale
    ligamentum i.
ischiofemoral ligament
ischiofibular
ischiogluteal
ischiomelus
ischiopagus
ischioperineal
ischiopubic ramus
ischiorectal
    i. fat-pad
    i. fossa
ischiorectalis
    corpus adiposum fossae i.
    fossa i.
ischiosacral
ischiothoracopagus
ischiotibial
ischiovaginal
ischiovertebral
ischium
    body of i.
    ramus of i.
    tuber of i.
island
    blood i.
    i.'s of Calleja
    Langerhans i.
    i. of Reil
islet
    blood i.
    i. cell
    i.'s of Langerhans
    pancreatic i.'s
    principal i.'s
    i. tissue

isobaric
isocapnia
isocellular
isochoric
isochromatic
isochromatophil
isochromatophile
isochronia
isochronous
isocortex
isodactylism
isodynamic
isodynamogenic
isoelectric
isoenergetic
isogenesis
isogenous chondrocyte
isolated dextrocardia
isolecithal ovum
isomeric function
isometric
    i. contraction
    i. exercise
    i. relaxation
isomorphic
isomorphism
isomorphous
isoncotic
iso-osmotic
isoplassonts
isopleth
isopotential
isorrhea
isosbestic
isosmotic
isospore
isothermal
isotonia
isotonic
    i. contraction
    i. exercise
isotonicity
isotropic disk
isotropous
isovolume pressure-flow curve
isovolumetric relaxation
isovolumic relaxation
isthmi (*pl. of* isthmus)
isthmian
isthmic
isthmus, pl. isthmi, isthmuses

    i. of aorta
    i. aortae
    aortic i.
    i. of auditory tube
    i. of cartilage of ear
    i. cartilaginis auricularis
    i. cartilaginis auris
    i. of cartilaginous auricle
    i. of cingular
    i. of eustachian tube
    i. of external acoustic meatus
    i. of fallopian tube
    i. of fauces
    i. faucium
    ganglion isthmi
    i. glandulae thyroideae
    Guyon i.
    i. gyri cinguli
    i. of gyrus fornicatus
    i. of His
    i. of limbic lobe
    i. meatus acustici externi
    oropharyngeal i.
    pharyngeal i.
    i. pharyngis
    i. pharyngonasalis
    i. of pharyngotympanic tube
    i. of pharynx
    pleural i.
    i. prostatae
    i. of prostate
    i. rhombencephali
    rhombencephalic i.
    sphincter of the pharyngeal i.
    i. of thyroid
    thyroid i.
    i. of thyroid gland
    i. tubae auditivae
    i. tubae auditoriae
    i. tubae uterinae
    i. of uterine tube
    i. of uterus
    Vieussens i.
ITB
    iliotibial band
Ito cell
IVC
    inferior vena cava
ivory membrane

NOTES

Jackson
>J. membrane
>J. veil

Jacobson
>J. anastomosis
>J. canal
>J. cartilage
>J. nerve
>J. organ
>J. plexus

Jacquart facial angle
Jacquemet recess
Jacques plexus
James
>J. fiber
>J. tract

James-Lange theory
janiceps
>j. asymmetrus
>j. parasiticus

japonicum
>os j.

Jarjavay ligament
jaundice
>j. of newborn
>nuclear j.
>physiologic j.

jaw
>angle of j.
>j. bone
>j. jerk
>j. joint
>lower j.
>j. reflex
>j. skeleton
>upper j.

jawbone
jejunal
>j. artery
>j. and ileal vein
>j. villus

jejunales
>arteriae j.
>j. et ilei

jejunoileal
jejunum
>arterial arches of j.
>intestinum j.
>proximal j.

jelly
>cardiac j.
>interlaminar j.
>Wharton j.

jerk
>ankle j.

chin j.
>crossed j.
>elbow j.
>jaw j.
>knee j.
>triceps surae j.

jerky respiration
jet ejector pump
j-g complex
Jobert de Lamballe fossa
joint
>abduction of first
> carpometacarpal j.
>abduction of glenohumeral j.
>abduction of metacarpophalangeal j.
>AC j.
>acromioclavicular j.
>adduction of glenohumeral j.
>amphiarthrodial j.
>amphidiarthrodial j.
>ankle j.
>anterior intraoccipital j.
>anterior tibiotalar part of medial
> ligament of ankle j.
>arthrodial j.
>articular disk of
> acromioclavicular j.
>articular disk of distal radioulnar j.
>articular disk of sternoclavicular j.
>articular disk of
> temporomandibular j.
>astragalonavicular j.
>atlantoaxial j.
>atlantooccipital j.
>j. of auditory ossicle
>ball-and-socket j.
>biaxial j.
>bicondylar j.
>bilocular j.
>j. branch
>Budin obstetrical j.
>calcaneocuboid j.
>capitular j.
>j. capsule
>capsule of cricoarytenoid j.
>capsule of cricothyroid j.
>carpal j.
>carpometacarpal j.
>j. cartilage
>cartilaginous j.
>Chopart j.
>Clutton j.'s
>coccygeal j.
>cochlear j.

joint *(continued)*

compartment of thigh for extensors of hip j.
complex j.
composite j.
compound j.
condylar j.
condyloid j.
costochondral j.
costotransverse j.
costovertebral j.
cotyloid j.
cranial synovial j.
cricoarytenoid j.
cricothyroid j.
Cruveilhier j.
cubital j.
cuboideonavicular j.
cuneiform j.
cuneocuboid j.
cuneometatarsal j.
cuneonavicular j.
cylindrical j.
dentoalveolar j.
diarthrodial j.
digital j.
DIP j.
distal interphalangeal j.
distal radioulnar j.
distal tibiofibular j.
j. of ear bone
elbow j.
ellipsoid j.
ellipsoidal j.
enarthrodial j.
extension of carpometacarpal j.
extension of elbow j.
extension of interphalangeal j.
extension of metacarpophalangeal j.
facet j.
j. facet
femoropatellar j.
fibrocartilaginous j.
fibrous j.
fibulotalar j.
j. of foot
j. of free inferior limb
j. of free superior limb
ginglymoid j.
glenohumeral j.
gliding j.
gompholic j.
j. of hand
j. of head of rib
hinge j.
hip j.
humeroradial j.
humeroulnar j.
immovable j.

incudomalleal j.
incudomalleolar j.
incudostapedial j.
j. of inferior limb girdle
inferior radioulnar j.
inferior tibiofibular j.
interarticular j.
intercarpal j.
interchondral j.
intercuneiform j.
intermetacarpal j.
intermetatarsal j.
internal semilunar fibrocartilage of knee j.
interphalangeal j. (IPJ)
intertarsal j.
jaw j.
knee j.
lateral atlantoaxial j.
lateral atlantoepistrophic j.
lateral ligament of temporomandibular j.
ligamentous j.
j. line
Lisfranc j.
lumbosacral j.
Luschka j.
malleoincudal j.
mandibular j.
manubriosternal j.
medial ligament of ankle j.
medial ligament of talocrural j.
medial ligament of temporomandibular j.
median atlantoaxial j.
membrane of j.
meniscus of acromioclavicular j.
metacarpophalangeal j.
metatarsophalangeal j.
midcarpal j.
midcervical apophyseal j.
middle atlantoepistrophic j.
middle carpal j.
middle radioulnar j.
midtarsal j.
mortise j.
movable j.
MP j.
multiaxial j.
neurocentral j.
oblique ligament of elbow j.
j. oil
orbicular zone of hip j.
palmar ligament of metacarpophalangeal j.
j. of pectoral girdle
peg-and-socket j.
j. of pelvic girdle
petrooccipital j.

J

phalangeal j.
PIP j.
pisiform j.
pisotriquetral j.
pivot j.
plane j.
plantar ligament of
 metatarsophalangeal j.
polyaxial j.
posterior intraoccipital j.
posterior tibiotalar part of medial
 ligament of ankle j.
proximal interphalangeal j. (PIPJ)
proximal radioulnar j.
proximal tibiofibular j.
radial collateral ligament of
 elbow j.
radial collateral ligament of
 wrist j.
radiocarpal j.
radiohumeral j.
radioulnar j.
rotary j.
rotatory j.
round ligament of elbow j.
sacciform recess of distal
 radioulnar j.
sacciform recess of elbow j.
sacrococcygeal j.
sacroiliac j.
saddle j.
scaphotrapeziotrapezoid j.
scapuloclavicular j.
schindyletic j.
screw j.
secondary cartilaginous j.
shoulder j.
simple j.
socket j.
sphenooccipital j.
spheroid j.
spheroidal j.
spiral j.
sternal j.
sternoclavicular j.
sternocostal j.
subtalar j.
j. of superior limb girdle
superior radioulnar j.
superior tibiofibular j.
suture j.
synarthrodial j.

synchondrodial j.
syndesmodial j.
syndesmotic j.
synovial j.
talocalcaneal j.
talocalcaneonavicular j.
talocrural j.
talonavicular j.
tarsal j.
tarsometatarsal j.
tectorial membrane of median
 atlantoaxial j.
temporomandibular j.
temporomaxillary j.
thigh j.
tibial collateral ligament of
 ankle j.
tibiocalcaneal part of medial
 ligament of ankle j.
tibiofibular j.
tibionavicular part of medial
 ligament of ankle j.
tibiotalar part of medial ligament
 of ankle j.
transverse tarsal j.
trochoid j.
ulnar collateral ligament of
 elbow j.
ulnar collateral ligament of wrist j.
uncovertebral j.
uniaxial j.
unilocular j.
vein of temporomandibular j.
wedge-and-groove j.
wrist j.
xiphisternal j.
zygapophysial j.
Jolly reaction
Jonnesco fossa
Joule equivalent
juga (*pl. of* jugum)
jugal
 j. bone
 j. ligament
 j. point
jugomaxillary
jugular
 j. bulb
 j. duct
 j. floor
 j. foramen
 j. fossa

NOTES

jugular *(continued)*
    j. ganglion
    j. gland
    j. lymphatic plexus
    j. lymphatic trunk
    j. lymph node
    j. nerve
    j. notch
    j. notch of manubrium
    j. notch of occipital bone
    j. notch of petrous part of temporal bone
    j. notch of sternum
    j. process
    j. process of occipital bone
    j. pulse
    j. sinus
    j. tubercle
    j. tubercle of occipital bone
    j. vein
    j. venous arch
    j. wall
    j. wall of middle ear

jugulare
    foramen j.
    glomus j.
    tuberculum j.

jugularis
    j. anterior
    bulbus venae j.
    fossa j.
    incisura j.
    j. interna
    nervus j.
    plexus j.
    plexus lymphaticus j.
    processus j.
    truncus lymphaticus j.

juguli
    arcus venosus j.

jugulodigastric lymph node
jugulodigastricus
    nodus j.
    nodus lymphoideus j.

juguloomohyoideus
    nodus lymphoideus j.

juguloomohyoid lymph node
jugum, pl. **juga**
    j. alveolare
    juga cerebralia
    j. sphenoidale

juice
    appetite j.
    gastric j.
    intestinal j.
    pancreatic j.

junctio
    j. anorectalis

junction
    amelodental j.
    amelodentinal j.
    amnioembryonic j.
    anorectal j.
    atrioventricular j.
    cardioesophageal j.
    cementodentinal j.
    choledochoduodenal j.
    corneoscleral j.
    costochondral j.
    cystic-choledochal j.
    dentinocemental j.
    dentinoenamel j.
    duodenojejunal j.
    electrotonic j.
    esophagogastric mucosal j.
    fibromuscular j.
    fundic-antral j.
    gap j.
    gastric j.
    gastroesophageal j.
    ileocecal j.
    intercellular j.'s
    intermediate j.
    j. of lip
    manubriogladiolar j.
    manubriosternal j.
    mucocutaneous j.
    muscle-tendon j.
    myoneural j.
    nasolabial j.
    neuroectodermal j.
    neuromuscular j.
    neurosomatic j.
    osteochondral j.
    pancreaticobiliary ductal j.
    pyloroduodenal j.
    rectosigmoid j.
    sacrococcygeal j.
    sclerocorneal j.
    sinotubular j.
    squamocolumnar j.
    sternoclavicular j.
    sternomanubrial j.
    tight j.
    tracheoesophageal j.
    tympanostapedial j.
    ureteropelvic j. (UPJ)
    uterovesical j. (UVJ)
    xiphoid-manubrial j.

junctional
    j. complex
    j. epithelium

junctura
    j. cartilaginea
    j.'s cinguli membri superioris
    j. fibrosa
    j. lumbosacralis

j.'s membri inferioris liberi
j.'s membri superioris liberi
j.'s ossium
j. sacrococcygea
j. synovialis
j.'s tendineae
j.'s tendinum
j.'s zygapophyseales
j.'s zygapophysiales
juncture
    saphenofemoral j.
Jung muscle
justo
    j. major
    j. minor
juvenile pelvis
juxtaarticular nodule
juxta-auricular fossa
juxtacolica
    arteria j.

juxtacolic artery
juxtaepiphysial
juxtaesophageales
    nodi lymphoidei j.
juxtaesophageal pulmonary lymph node
juxtaglomerular
    j. apparatus
    j. body
    j. cells
    j. complex
    j. granule
juxtaintestinales
    lymphonodi j.
    nodi lymphoidei j.
juxtaintestinal mesenteric lymph node
juxtallocortex
juxtaposition
juxtaregional lymph node
juxtarestiform body

J

NOTES

K

K cell

Kaes

line of K.

Kaes-Bechterew

band of K.-B.

kappa granule
karyochrome cell
karyoclasis
karyocyte
karyogamic
karyogamy
karyogenesis
karyogenic
karyokinesis
karyokinetic
karyolymph
karyomicrosome
karyomitosis
karyomitotic
karyomorphism
karyon
karyoplasm
karyoplast
karyopyknosis
karyorrhexis
karyostasis
karyotheca
katathermometer
keeled chest
Keith

K. bundle

K. and Flack node

kelosomia
Kempner diet
Kent bundle
Kent-His bundle
keratin filaments
keratinocyte
keratinosome
keratocricoid
keratocyte
keratogenesis
keratogenetic
keratogenous membrane
keratoglobus
keratoglossus
keratohyal
keratohyalin granule
keratopharyngeus

musculus k.

Kerckring

K. center

K. fold

K. ossicle

valve of K.

K. valve

kernicterus
kern-plasma relation theory
ketogenesis
ketogenic
Kety-Schmidt method
Key-Retzius

foramen of K.-R.

key ridge
KHN

Knoop hardness number

kidney

adipose capsule of k.

anterior inferior segmental artery
of k.

anterior superior segmental artery
of k.

anterior surface of k.

arciform vein of k.

arcuate artery of k.

arcuate vein of k.

arterial segments of k.

arteriole of k.

artery of anterior inferior segment
of k.

artery of anterior superior segment
of k.

artery of inferior segment of k.

artery of posterior segment of k.

artery of superior segment of k.

cake k.

capsule of k.

convoluted tubule of k.

cortex of k.

cortical arches of k.

cortical lobules of k.

cow k.

disk k.

fatty capsule of k.

fibrous capsule of k.

fused k.

hilum of k.

k. hilus

hind k.

horseshoe k.

inferior extremity of k.

inferior pole of k.

inferior segmental artery of k.

infundibula of k.

inguinal ligament of k.

interlobar artery of k.

interlobar vein of k.

interlobular artery of k.

interlobular vein of k.

K

297

kidney *(continued)*
 lateral border of k.
 lateral margin of k.
 left k.
 liver, spleen, and k.'s (LSK)
 k. lobe
 long axis of k.
 medial border of k.
 medulla of k.
 middle k.
 nonrotation of k.
 pancake k.
 papillary foramina of k.
 k. pelvis
 perforating radiate artery of k.
 posterior segmental artery of k.
 posterior surface of k.
 primordial k.
 pyramid of k.
 radiate artery of k.
 right k.
 segmental artery of k.
 stellate vein of k.
 straight venules of k.
 superior extremity of k.
 superior pole of k.
 superior segmental artery of k.
 supernumerary k.
 k. tissue
 ureteric branch of k.
 vein of k.
 venous segments of k.
 venulae rectae of k.
Kiernan space
Kiesselbach
 K. area
 K. plexus
 K. triangle
Kilian line
killer
 k. cells
 natural k. (NK)
Killian bundle
Killian-Jamieson area
kilogram calorie
kilogram-meter
kinematic
 kinematic viscosity
kinemometer
kinesimeter
kinesiology
kinesiometer
kinesis
kinesthesiometer
kinesthetic sense
kinetic
 k. energy
 k. system
kinetochore

kinetogenic
kinetoplasm
kinetoscope
kinetosome
kinocentrum
kinocilium
kinomometer
kinoplasm
kinoplasmic
Kirchner cell
kleeblattschädel
Klein muscle
knee
 anterior region of k.
 articular muscle of k.
 articular surface of k.
 articular vascular network of k.
 k. bone
 capsule of k.
 compartment of thigh for extensors
  of k.
 compartment of thigh for flexors
  of k.
 coronary ligament of k.
 cruciate ligaments of k.
 descending artery of k.
 extension of k.
 flexion of k.
 k. jerk
 k. joint
 lateral ligament of k.
 k. ligament
 ligament of k.
 medial collateral ligament of k.
 medial ligament of k.
 meniscus of k.
 k. phenomenon
 posterior ligament of k.
 posterior region of k.
 posterior surface of k.
 k. reflex
 synovial capsule of k.
 transverse ligament of k.
 vein of k.
kneecap
knee-jerk reflex
knob
 aortic k.
knock-out
knockout mouse
Knoll gland
Knoop
 K. hardness number (KHN)
 K. hardness test
knot
 Hensen k.
 Hubrecht protochordal k.
 primitive k.
 protochordal k.

syncytial k.
vital k.
knuckle
    aortic k.
    k. pad
Kobelt tubules
Koch node
Koeppe nodule
Kohlrausch
    K. fold
    K. muscle
    K. valve
Kohn pore
koilosternia
Kölliker
    K. layer
    K. reticulum
kolytic
koniocortex
Korff fibers
koronion
Krause
    K. bone
    K. end bulbs
    K. gland

K. ligament
K. muscle
K. respiratory bundle
K. valve
Krebs
    K. ornithine cycle
    K. urea cycle
Krebs-Henseleit cycle
Kretschmann space
Krukenberg vein
Kugel anastomotic artery
Kühne
    K. fiber
    K. phenomenon
    K. plate
    K. spindle
Kuhnt space
Kulchitsky cells
Kupffer cells
Kürsteiner canals
kyphos
kyphosis
    sacral k.
    thoracic k.
    k. thoracica

K

NOTES

L1–5
>lumbar segments of spinal cord L.

Labbé
>L. triangle
>vein of L.
>L. vein

labia (*pl. of* labium)
labial
>l. artery
>l. branch of mental nerve
>l. commissure
>l. fold
>l. frenum
>l. gland
>l. part
>l. part of orbicularis oris muscle
>l. sulcus
>l. surface
>l. swelling
>l. tubercle
>l. vein
>l. vestibule

labialis, pl. **labiales**
>labiales anteriores
>facies l.
>glandulae labiales
>l. inferior
>pars l.
>labiales posteriores
>l. superior

labially
labii
>caput angulare quadrati l.
>modiolus l.

labile
>l. current
>l. elements

lability
labiodental sulcus
labiogingival lamina
labioglossolaryngeal
labioglossopharyngeal nerve
labiomental
labionasal
labiopalatine
labiorum
>commissura l.

labioscrotal
>l. folds
>l. swelling

labiotenaculum
labium, pl. **labia**
>l. anterius
>l. anterius ostii uterus
>l. externum cristae iliacae

>l. inferius oris
>l. internum cristae iliacae
>l. laterale lineae asperae
>l. limbi tympanicum
>l. limbi tympanicum laminae
>spiralis ossei
>l. limbi tympanicum limbi spiralis
>ossei
>l. limbi vestibulare
>l. limbi vestibulare laminae spiralis
>ossei
>l. limbi vestibulare limbi spiralis
>ossei
>l. majora
>l. majus
>l. majus pudendi
>l. mediale lineae asperae
>l. minora
>l. minus
>l. minus pudendi
>labia oris
>l. posterius
>l. posterius ostii uterus
>l. superius oris
>l. urethrae
>labia uterus
>l. vocale
>labia vocalia

labrocyte
labrum, pl. **labra**
>acetabular l.
>l. acetabulare
>l. of acetabulum
>articular l.
>l. articulare
>glenoid l.
>l. glenoidale
>l. glenoidale scapulae

labyrinth
>artery of l.
>bony l.
>l. of brain
>cochlear l.
>elliptical recess of bony l.
>ethmoidal l.
>Ludwig l.
>l. membrane
>membranous l.
>osseous l.
>renal l.
>saccular recess of bony l.
>Santorini l.
>semicircular canal of bony l.
>spherical recess of bony l.
>utricle of vestibular l.

L

labyrinth *(continued)*
    utricular recess of bony l.
    utricular recess of membranous l.
    vestibular l.
labyrinthi
    arteria l.
    rami vestibulares arteriae l.
    ramus cochlearis arteriae l.
    venae l.
labyrinthine
    l. artery
    l. placenta
    l. reflex
    l. righting reflex
    l. vein
    l. wall of middle ear
    l. wall of tympanic cavity
labyrinthus
    l. cochlearis
    l. ethmoidalis
    l. membranaceus
    l. osseus
    l. vestibularis
lacerated foramen
lacertus
    l. cordis
    l. fibrosus
    l. of lateral rectus muscle
    l. medius
    l. musculi recti lateralis
lacerum
    foramen l.
lachrymal
lacinate ligament
laciniae tubae
laciniate ligament
laciniatum
    ligamentum l.
lacis cell
lacrimal
    l. apparatus
    l. artery
    l. bay
    l. bone
    l. border of maxilla
    l. canal
    l. canaliculus
    l. caruncle
    l. crest
    l. duct
    l. fascia
    l. fold
    l. fossa
    l. ganglion
    l. gland
    l. groove
    l. hamulus
    l. lake
    l. margin of maxilla

    l. nerve
    l. notch
    l. opening
    l. papilla
    l. part of orbicularis oculi muscle
    l. pathway
    l. process
    l. process of inferior nasal concha
    l. punctum
    l. sac
    l. vein
lacrimale
    os l.
    punctum l.
lacrimali
    ramus anastomoticus arteriae
        meningeae mediae cum arteriae l.
lacrimalis
    ampulla canaliculi l.
    ampulla ductus l.
    apparatus l.
    arteria l.
    canaliculus l.
    caruncula l.
    ductuli excretorii glandulae l.
    ductus l.
    fornix sacci l.
    fossa glandulae l.
    fossa sacci l.
    glandula l.
    hamulus l.
    incisura l.
    lacus l.
    nervus l.
    papilla l.
    pars orbitalis glandulae l.
    pars palpebralis glandulae l.
    plica l.
    processus l.
    rivus l.
    sacculus l.
    saccus l.
    sulcus l.
    vena l.
lacrimoconchalis
    sutura l.
lacrimoconchal suture
lacrimoethmoidal suture
lacrimomaxillaris
    sutura l.
lacrimomaxillary suture
lacrimonasal duct
lacrimoturbinal suture
lactacid oxygen debt
lactate
    excess l.
lactation
lactational

lacteal
> central l.
> l. vessel

lacteus
> dens l.

lactifera
> ampulla l.

lactiferi
> ductus l.
> sinus l.
> tubuli l.

lactiferous
> l. ampulla
> l. duct
> l. gland
> l. sinus

lactifugal
lactifuge
lactigenous
lactigerous
lactis
lactogenesis
lactogenic
> l. factor
> l. hormone

lactotrophic
lacuna, pl. **lacunae**
> cartilage l.
> l. cerebri
> Howship lacunae
> intervillous l.
> lacunae laterales
> lateral venous l.
> l. magna
> lacunae of Morgagni
> Morgagni l.
> muscular l.
> l. musculorum
> l. musculorum retroinguinalis
> osseous l.
> pharyngeal l.
> l. pharyngis
> resorption lacunae
> trophoblastic l.
> urethral l.
> lacunae urethrales
> l. urethralis
> vascular l.
> l. vasorum
> l. vasorum retroinguinalis

lacunare
> ligamentum l.

lacunar ligament
lacunule
lacus
> l. lacrimalis
> l. seminalis

LAD
> left anterior descending artery

lag
> nitrogen l.

lagena, pl. **lagenae**
Laimer
> area of L.
> L. area

Laimer-Haeckerman area
lake
> capillary l.
> lacrimal l.
> lateral l.
> seminal l.
> subchorial l.

Lallouette pyramid
lambda
lambdoid
> l. arc
> l. border of occipital bone
> l. margin of occipital bone
> l. suture
> l. suture line

lambdoidea
> sutura l.

lamella, pl. **lamellae**
> annulate lamellae
> articular l.
> l. of bone
> circumferential l.
> concentric l.
> elastic l.
> glandulopreputial l.
> ground l.
> haversian l.
> intermediate l.
> interstitial l.
> triangular l.
> l. tympanica laminae spiralis ossei
> l. vestibularis laminae spiralis ossei
> vitreous l.

lamellar
> l. bone
> l. granule

lamellate
lamellated corpuscles
lamellipodia

NOTES

lamellipodium
lamellosa
    corpuscula l.
lamina, pl. **laminae**
    l. affixa
    l. alaris
    laminae albae cerebelli
    l. anterior fasciae thoracolumbalis
    anterior limiting l.
    l. anterior vagina musculi recti
      abdominis
    l. arcus vertebrae
    basal l.
    l. basalis
    l. basalis choroideae
    l. basalis corporis ciliaris
    basement l.
    basilar l.
    l. basilaris cochlea
    l. basilaris corporis ciliaris
    l. basilaris ductus cochlearis
    l. cartilaginis cricoideae
    l. cartilaginis lateralis
    l. cartilaginis medialis
    l. cartilaginis thyroideae
    l. choriocapillaris
    l. choroidea
    l. choroidea epithelialis
    l. choroidocapillaris
    l. cinerea
    l. cribrosa
    l. cribrosa ossis ethmoidalis
    l. cribrosa of sclera
    l. cribrosa sclerae
    cribrous l.
    l. of cricoid cartilage
    deep l.
    l. densa
    dental l.
    l. dentata
    dentogingival l.
    l. dextra cartilaginis thyroidea
    l. dorsalis
    l. dura
    l. elastica anterior
    l. elastica posterior
    l. epiphysialis
    episcleral l.
    l. episcleralis
    epithelial l.
    l. externa
    l. externa calvaria
    l. externa cranii
    l. fibrocartilaginea interpubica
    l. fibroreticularis
    fusca l.
    l. fusca sclerae
    hamulus of spiral l.
    hepatic laminae

hook of spiral l.
l. horizontalis ossis palatini
l. interna
l. interna calvaria
l. interna ossium cranii
iridopupillary l.
labiogingival l.
l. lateralis cartilaginis tubae
  auditivae
l. lateralis cartilaginis tubae
  auditoriae
l. lateralis processus pterygoidei
l. of lens
l. of levator palpebrae
limbus of osseous spiral l.
l. limitans anterior
l. limitans anterior corneae
l. limitans posterior corneae
l. lucida
l. medialis cartilaginis tubae
  auditivae
l. medialis cartilaginis tubae
  auditoriae
l. medialis processus pterygoidei
laminae medullares cerebelli
laminae medullares thalami
l. medullaris lateralis corporis
  striati
l. medullaris medialis corporis
  striati
l. membranacea
l. membranacea cartilaginis tubae
  auditivae
l. membranacea cartilaginis tubae
  auditoriae
l. modioli
l. modioli cochlea
l. of modiolus of cochlea
l. muscularis mucosae
l. orbitalis ossis ethmoidalis
osseous spiral l.
l. of palatine bone
l. papyracea
parietal l.
l. parietalis
l. parietalis pericardii
l. parietalis pericardii serosi
l. parietalis tunicae vaginalis testis
periclaustral l.
l. perpendicularis
l. perpendicularis ossis ethmoidalis
l. perpendicularis ossis palatini
posterior l.
l. posterior vaginae musculi recti
  abdominis
l. pretrachealis
l. pretrachealis fasciae cervicalis
l. prevertebralis
l. prevertebralis fasciae cervicalis

primary dental l.
l. profunda
l. profunda fasciae temporalis
l. profunda fasciae thoracolumbalis
l. profunda musculi levatoris
palpebrae superioris
proper l.
l. propria
l. propria mucosae
pterygoid l.
l. quadrigemina
reticular l.
l. retrorectalis fasciae endopelvicae
rostral l.
l. rostralis
secondary spiral l.
l. septi pellucidi
l. of septum pellucidum
l. sinistra cartilaginis thyroidea
l. spiralis ossea
l. spiralis secundaria
superficial l.
l. superficialis
l. superficialis fasciae cervicalis
l. superficialis fasciae temporalis
l. superficialis musculi levatoris
palpebrae superioris
suprachoroid l.
l. suprachoroidea
l. supraneuroporica
l. tecti mesencephali
l. terminalis cerebri
l. of thyroid cartilage
tragal l.
l. tragi
l. of tragus
tympanic labium of limbus of
spiral l.
tympanic lamella of osseous
spiral l.
tympanic lip of limbus of spiral l.
l. vasculosa choroideae
l. ventralis
l. of vertebral arch
vestibular labium of limbus of
spiral l.
vestibular lamella of osseous
spiral l.
vestibular lip of limbus of
spiral l.
visceral l.
l. visceralis

l. visceralis pericardii
l. visceralis tunicae vaginalis testis
l. vitrea
laminar flow
laminated
l. cortex
l. epithelium
lamination
laminin
lamins
lancisi
L. muscle
L. nerve
striae l.
Landsmeer ligament
Landström muscle
Landzert fossa
Lane band
Langenbeck
triangle of L.
L. triangle
Langendorff method
Langer
L. axillary arch
L. line
L. muscle
Langerhans
L. cells
L. granule
L. island
islets of L.
Langhans
Langhans cells
L. layer
L. stria
Langhans-type giant cells
Langley granule
Langmuir trough
language zone
laniary
Lannelongue
L. foramen
L. ligament
Lanterman
L. incisure
L. segments
lanuginosa
hypertrichosis l.
lanuginous
lanugo hair
lanum
Lanz line

NOTES

Lapicque law
Laplace law
large
    l. bowel
    l. calorie
    l. intestine
    l. muscle of helix
    l. pelvis
    l. pudendal lip
    l. saphenous vein
laryngea, pl. **laryngeae**
    bursa subcutanea prominentiae
      laryngeae
    glandulae laryngeae
    l. inferior
    prominentia l.
    protuberantia l.
    l. superior
laryngeal
    l. aditus
    l. aperture
    l. artery
    l. atresia
    l. bursa
    l. cartilage
    l. cavity
    l. commissure
    l. cyst
    l. diaphragm
    l. gland
    l. inlet
    l. lymphoid nodule
    l. mucosa
    l. muscle
    l. nerve
    l. part of pharynx
    l. plexus
    l. pouch
    l. prominence
    l. saccule
    l. sinus
    l. skeleton
    l. tonsil
    l. vein
    l. ventricle
larynges (*pl. of* larynx)
laryngeus, pl. **laryngei**
    folliculi lymphatici laryngei
    l. inferior
    plica nervi laryngei
    l. recurrens
    sinus l.
    l. superior
laryngis
    aditus l.
    appendix ventriculi l.
    cartilagines l.
    cartilago sesamoidea l.
    cavitas l.

    cavum l.
    conus elasticus l.
    corniculum l.
    membrana fibroelastica l.
    musculi l.
    sacculus l.
    tunica mucosa l.
    ventriculus l.
    vestibulum l.
laryngopharyngeal
    l. branch of superior cervical
      ganglion
    l. reflex
laryngopharyngei
    rami l.
laryngopharyngeus
    musculus l.
laryngopharynx
laryngoptosis
laryngotracheal groove
larynx, pl. **larynges**
    aditus of l.
    cartilage of l.
    cavity of l.
    fibroelastic membrane of l.
    infraglottic part of l.
    inlet of l.
    intrinsic muscle of l.
    lateral wall of l.
    lymphatic follicles of l.
    mucosa of l.
    mucous membrane of l.
    muscle of l.
    posterior l.
    saccule of l.
    sesamoid cartilage of l.
    superficial anterior l.
    superior l.
    supraglottic l.
    ventricular band of l.
    vestibule of l.
lash reflex
lata, pl. **latae**
    falciform fold of fascia l.
    fascia l.
    musculus tensor fasciae latae
    tensor muscle of fascia l.
Latarjet
    nerve of L.
    L. nerve
    L. vein
latebra
latent
    l. energy
    l. period
lateral
    l. abdominal/pectoral cutaneous
      branch of intercostal nerve
    l. ampullar nerve

l. ampullary nerve
l. angle of eye
l. angle of scapula
l. angle of uterus
l. antebrachial cutaneous nerve
l. anterior thoracic nerve
l. aperture of fourth ventricle
l. arcuate ligament
l. aspect
l. atlantoaxial joint
l. atlantoepistrophic joint
l. atrial branch of left coronary artery
l. atrial branch of right coronary artery
l. atrial vein
l. axillary lymph node
l. basal branch
l. basal bronchopulmonary segment
l. basal segmental artery
l. bicipital groove
l. border
l. border of foot
l. border of forearm
l. border of humerus
l. border of kidney
l. border of nail
l. border of scapula
l. branch of artery of tuber cinereum
l. branch of pontine artery
l. branch of posterior rami of spinal nerve
l. bronchopulmonary segment S IV
l. calcaneal branch of sural nerve
l. canthus
l. cartilaginous layer
l. cartilaginous plate
l. central palmar space
l. cerebellomedullary cistern
l. cerebral aperture
l. cerebral fissure
l. cerebral fossa
l. cerebral sulcus
l. cervical region
l. check ligament of eyeball
l. circumflex artery of thigh
l. circumflex femoral artery
l. circumflex femoral vein
l. collateral ligament
l. collateral ligament of ankle
l. column

l. column of spinal cord
l. compartment of leg
l. condyle
l. condyle bone
l. condyle of femur
l. condyle of humerus
l. condyle of tibia
l. conus
l. cord of brachial plexus
l. corticospinal tract
l. costal branch of internal thoracic artery
l. costotransverse ligament
l. cranial surface
l. cricoarytenoid muscle
l. crus
l. crus of horizontal part of the facial canal
l. crus of the major alar cartilage of the nose
l. crus of the superficial inguinal ring
l. cuneate nucleus
l. cuneiform bone
l. curvature of spine
l. cutaneous branch
l. cutaneous branch of intercostal nerve
l. cutaneous branch of ventral primary ramus of thoracic spinal nerve
l. cutaneous nerve of calf
l. cutaneous nerve of forearm
l. cutaneous nerve of thigh
l. direct vein
l. division of left liver
l. dorsal cutaneous nerve
l. dorsal cutaneous nerve of foot
l. epicondylar crest
l. epicondylar ridge
l. epicondyle
l. epicondyle of femur
l. epicondyle of humerus
l. femoral cutaneous (LFC)
l. femoral cutaneous nerve
l. femoral tuberosity
l. fillet
l. flexor muscle of forearm
l. foramen
l. fossa of brain
l. frontobasal artery
l. funiculus of spinal cord

L

NOTES

lateral (*continued*)

l. geniculate body
l. ginglymus
l. glossoepiglottic fold
l. great muscle
l. ground bundle
l. hamstring
l. head of malleolus
l. hermaphroditism
l. incision
l. incisor
l. incisor tooth
l. inferior genicular artery
l. inferior hepatic area
l. inguinal fossa
l. intermuscular septum
l. joint of ankle
l. jugular lymph node
l. lacunae of cranial dura mater
l. lacunae of superior sagittal sinus
l. lacunar lymph node
l. lake
l. lamina of cartilage of pharyngotympanic auditory tube
l. ligament of bladder
l. ligament of elbow
l. ligament of knee
l. ligament of malleus
l. ligament of temporomandibular joint
l. ligament of wrist
l. lingual swelling
l. lip of linea aspera
l. lobes of prostate
l. longitudinal arch
l. longitudinal stria
l. lumbar
l. lumbar intertransversarii muscle
l. lumbar intertransverse muscle
l. lumbocostal arch
l. malleolar branch of fibular peroneal artery
l. malleolar facet of talus
l. malleolar ligament
l. malleolar network
l. malleolar subcutaneous bursa
l. malleolar surface of talus
l. malleolus bursa
l. mamillary nucleus of Rose
l. mammary branch
l. mammary branch of lateral cutaneous branch of intercostal nerve
l. mammary branch of lateral cutaneous branch of thoracic spinal nerve
l. mammary branch of lateral thoracic artery
l. mammary vein

l. margin
l. margin of breast
l. margin of kidney
l. margin of orbit
l. mass of atlas
l. mass of ethmoid bone
l. and medial palpebral artery
l. and medial parietal artery
l. and medial posterior choroidal branch of posterior cerebral artery
l. medullary branch of intracranial part of vertebral artery
l. medullary lamina of corpus striatum
l. meniscus
l. midpalmar space
l. nasal branch of anterior ethmoidal nerve
l. nasal branch of facial artery
l. nasal elevation
l. nasal fold
l. nasal process
l. nucleus of medulla oblongata
l. nucleus of thalamus
l. oblique fascia
l. occipital artery
l. occipital sulcus
l. occipitotemporal gyrus
l. orbitofrontal artery
l. palpebral artery
l. palpebral commissure
l. palpebral ligament
l. palpebral raphe
l. part of longitudinal arch of foot
l. part of middle lobe vein of right superior pulmonary vein
l. part of occipital bone
l. part of posterior cervical intertransversarii muscle
l. part of posterior extensor compartment of forearm
l. part of sacrum
l. part of vaginal fornix
l. patellar retinaculum
l. pectoral nerve
l. pelvic wall
l. pelvic wall triangle
l. pericardial lymph node
l. pharyngeal space
l. plantar artery
l. plantar nerve
l. plantar vein
l. plate of cartilaginous auditory tube
l. plate mesoderm
l. plate of pterygoid process
l. pole
l. popliteal nerve

l. posterior cervical intertransversarii muscle
l. posterior nasal artery
l. preoptic nucleus
l. process of calcaneal tuberosity
l. process of malleus
l. process of septal nasal cartilage
l. pterygoid muscle
l. pterygoid nerve
l. pterygoid plate
l. puboprostatic ligament
l. pyramidal tract
l. recess of fourth ventricle
l. rectus (LR)
l. rectus muscle
l. rectus muscle of head
l. rectus tendon
l. region of abdominal region
l. region of neck
l. reticular nucleus
l. retinaculum of patella
l. root of median nerve
l. root of olfactory tract
l. root of optic tract
l. sacral branch of median sacral artery
l. sacral crest
l. sacral vein
l. sacrococcygeal ligament
l. segmental artery
l. semicircular canal
l. spinothalamic tract
l. splanchnic arteries
l. striate artery
l. superior genicular artery
l. superior geniculate artery
l. superior hepatic area
l. supraclavicular nerve
l. supracondylar crest
l. supracondylar ridge
l. supracondylar ridge of humerus
l. supraepicondylar ridge
l. sural cutaneous nerve
l. surface of arm
l. surface of fibula
l. surface of finger
l. surface of leg
l. surface of lower limb
l. surface of ovary
l. surface of pubis
l. surface of testis
l. surface of tibia

l. surface of toe
l. surface of zygomatic bone
l. sympathetic line
l. talocalcaneal ligament
l. tarsal artery
l. temporomandibular ligament
l. thalamic peduncle
l. thoracic artery
l. thoracic vein
l. thorax
l. thyrohyoid ligament
l. thyroid ligament
l. tubercle of posterior process of talus
l. umbilical fold
l. umbilical ligament
l. vastus muscle
l. vein of lateral ventricle
l. venous lacuna
l. ventricle horn
l. vestibular nucleus
l. wall of larynx
l. wall of middle ear
l. wall of orbit
l. wall of tympanic cavity

laterale
caput l.
cornu l.
corpus geniculatum l.
crus l.
ligamentum arcuatum l.
ligamentum collaterale l.
ligamentum costotransversarium l.
ligamentum hyothyroideum l.
ligamentum mallei l.
ligamentum palpebrale l.
ligamentum puboprostaticum l.
ligamentum sacrococcygeum l.
ligamentum talocalcaneare l.
ligamentum talocalcaneum l.
ligamentum thyrohyoideum l.
ligamentum umbilicale l.
os cuneiforme l.
rete malleolare l.
retinaculum patellae l.
segmentum bronchopulmonale basale l.
spatium pharyngeum l.
trigonum colli l.
tuberculum intercondylare mediale et l.

lateralis, pl. **laterales**

NOTES

lateralis *(continued)*
ampullaris l.
angulus oculi l.
arcus lumbocostalis l.
arcus pedis longitudinalis pars l.
arteria circumflexa femoris l.
arteriae malleolares posteriores
  laterales
arteriae nasales posteriores laterales
arteria frontobasalis l.
arteria genus inferior l.
arteria genus superior l.
arteria malleolaris anterior l.
arteria occipitalis l.
arteria plantaris l.
arteria sacralis l.
arteria segmentalis l.
arteria segmentalis basalis l.
arteria tarsea l.
arteria thoracica l.
articulatio atlantoaxialis l.
atrium ventriculi l.
bursa subcutanea malleoli l.
cartilago nasi l.
circumflexa femoris l.
cisterna cerebellomedullaris l.
columna l.
commissura palpebrarum l.
condylus l.
cornu inferius ventriculi l.
cornu posterius ventriculi l.
cricoarytenoideus l.
cristae sacrales laterales
crista sacralis l.
crista supracondylaris l.
crista supraepicondylaris l.
cutaneus antebrachii l.
cutaneus dorsalis l.
cutaneus femoris l.
cutaneus surae l.
digitales plantares communes nervi
  plantaris l.
digitales plantares proprii nervi
  plantaris l.
ductus semicircularis l.
epicondylus l.
facies l.
fasciculus corticospinalis l.
fasciculus pyramidalis l.
fissura cerebri l.
fossa inguinalis l.
fossa malleoli l.
funiculus l.
genus inferior l.
genus superior l.
gyrus occipitotemporalis l.
lacertus musculi recti l.
lacunae laterales
lamina cartilaginis l.

lemniscus l.
ligamentum malleoli l.
ligamentum menisci l.
lymphonodi aortici laterales
lymphonodi cavales laterales
lymphonodi jugulares laterales
lymphonodi pericardiales laterales
lymphonodi vesicales laterales
lymphonodus iliaci communes
  laterales
lymphonodus iliaci externi laterales
malleolus l.
margo l.
meniscus l.
musculus cricoarytenoideus l.
musculus pterygoideus l.
musculus rectus capitis l.
musculus vastus l.
nervus ampullaris l.
nervus cutaneus antebrachii l.
nervus cutaneus brachii l.
nervus cutaneus dorsalis l.
nervus cutaneus femoris l.
nervus cutaneus surae l.
nervus pectoralis l.
nervus plantaris l.
nervus supraclavicularis l.
nodi lymphatici iliaci externi
  laterales
nodi lymphatici vesicales laterales
nodi lymphoidei axillares laterales
nodi lymphoidei jugulares laterales
nodi lymphoidei pericardiales
  laterales
nodus lymphoideus lacunaris l.
norma l.
nucleus lemnisci l.
nucleus preopticus l.
nucleus ventralis l.
palpebrales laterales
pars centralis ventriculi l.
pedunculus thalami l.
plantaris l.
plexus choroideus ventriculi l.
plica glossoepiglottica l.
plica umbilicalis l.
prominentia canalis semicircularis l.
pterygoideus l.
rami calcanei laterales
rami corporis geniculati l.
rami malleolares laterales
rami mammarii laterales
rami mammarii laterales arteriae
  thoracicae l.
rami medullares laterales
rami nasales posteriores superiores
  laterales
rami temporales intermedii arteriae
  occipitalis l.

rami temporales medii arteriae occipitalis l.
ramus anterior l.
ramus basalis l.
ramus choroidei posteriores laterales
ramus choroidei ventriculi l.
ramus costalis l.
ramus cutaneus l.
ramus descendens arteriae circumflexae femoris l.
ramus orbitofrontalis l.
ramus profundus nervi plantaris l.
ramus superficialis nervi plantaris l.
ramus transversus arteriae circumflexae femoris l.
raphe palpebralis l.
rectus capitis l.
regio abdominis l.
regio cervicalis l.
regiones thoracicae anteriores et laterales
sacrales laterales
stria longitudinalis l.
sulcus bicipitalis l.
sulcus occipitalis l.
supraclaviculares laterales
tarsea l.
thoracica l.
tractus corticospinalis l.
tractus pyramidalis l.
tractus spinothalamicus l.
vastus l.
vena atrii l.
venae circumflexae femoris laterales
venae directae laterales
venae sacrales laterales
vena thoracica l.
vena ventriculi lateralis l.
ventriculus l.

laterality
lateris
nevus unius l.
lateroabdominal
lateropharyngeum
spatium l.
latissimus
l. dorsi
l. dorsi muscle
l. dorsi tendon
lattice
latticed layer

latus
metatarsus l.
Laumonier ganglion
Lauth
L. canal
L. ligament
Lavdovsky nucleoid
law
Ambard l.
Arndt l.
Baer l.
Bernoulli l.
biogenetic l.
Boyle l.
Charles l.
Dale-Feldberg l.
Dalton l.
Dalton-Henry l.
Du Bois-Reymond l.
Elliott l.
Fechner-Weber l.
Graham l.
l. of gravitation
Haeckel l.
l. of the heart
Heidenhain l.
Henry l.
Hooke l.
Lapicque l.
Laplace l.
Le Chatelier l.
Marey l.
Mariotte l.
Meltzer l.
l. of the minimum
Müller l.
Newton l.
Nysten l.
l. of partial pressure
Pflüger l.
Poiseuille l.
Ritter l.
Rosenbach l.
Spallanzani l.
l. of specific nerve energy
Starling l.
Thoma l.
Weber l.
Weber-Fechner l.
Weigert l.
Williston l.
Wolff l.

L

NOTES

layer

ameloblastic l.
anterior elastic l.
bacillary l.
basal cell l.
l. of Bechterew
brown l.
cambium l.
l.'s of cerebellar cortex
l.'s of cerebral cortex
Chievitz l.
choriocapillary l.
choroid capillary l.
claustral l.
columnar l.
cremasteric l.
deep l.
enamel l.
ependymal l.
epithelial choroid l.
epitrichial l.
fascial l.
fibrous l.
fillet l.
fusiform l.
germ l.
germinative l.
Henle fiber l.
Henle nervous l.
Huxley l.
infragranular l.
intermediate l.
intimal l.
investing l.
Kölliker l.
Langhans l.
lateral cartilaginous l.
latticed l.
long pitch helicoidal l.
malpighian l.
mantle l.
marginal l.
medial cartilaginous l.
membranous l.
Meynert l.
molecular l.
multiform l.
muscular l.
musculofascial l.
l. of nail
Nitabuch l.
odontoblastic l.
optic l.
osteogenetic l.
outer l.
outermost l.
palisade l.
papillary l.
parietal l.

l. of piriform neuron
plasma l.
plexiform l.
polymorphous l.
posterior elastic l.
pretracheal l.
prevertebral l.
prickle cell l.
Purkinje l.
pyramidal cell l.
Rauber l.
l.'s of retina
l. of rods and cones
rostral l.
Sattler elastic l.
seromuscular l.
short pitch helicoidal l.
l.'s of skin
sluggish l.
somatic l.
spindle-celled l.
spinous l.
splanchnic l.
still l.
subendocardial l.
subendothelial l.
subpapillary l.
subserous l.
superficial l.
suprachoroid l.
Tomes granular l.
vascular l.
visceral l.
Waldeyer zonal l.
Weil basal l.
zonular l.

LCA
left coronary artery
LCX, LCx
left circumflex artery
LE
LE cell
Le
Le Chatelier law
Le Chatelier principle
lead
augmented l.
leaf, pl. **leaves**
leaves of broad ligament
leaves of dura
leaves of mesentery
leaflet
cardiac valve l.
l. tip
leaflike villus
least
l. gluteal muscle
l. splanchnic nerve
leaves (*pl. of* leaf)

Leber plexus
lecithin-cholesterol transferase
lecithoblast
Lee ganglion
Leeuwenhoek canals
left
l. anterior descending artery (LAD)
l. anterior lateral hepatic segment
l. atrioventricular orifice
l. atrioventricular valve
l. atrium
l. atrium of heart
l. auricle
l. auricular appendage
l. brachiocephalic vein
l. branch
l. branch of hepatic artery proper
l. bronchomediastinal trunk
l. bundle of atrioventricular bundle
l. circumflex artery (LCX, LCx)
l. colic artery
l. colic flexure
l. colic lymph node
l. colic vein
l. colon
l. colonic flexure
l. coronary artery (LCA)
l. coronary valve
l. coronary vein
l. crus of atrioventricular bundle
l. crus of atrioventricular trunk
l. crus of diaphragm
l. duct of caudate lobe
l. duct of caudate lobe of liver
l. femoral artery
l. fibrous trigone of heart
l. gastric artery
l. gastric lymph node
l. gastric vein
l. gastroepiploic lymph node
l. gastroepiploic vein
l. gastroomental artery
l. gastroomental lymph node
l. gastroomental vein
l. gonadal vein
l. gutter
l. hepatic artery
l. hepatic duct
l. hepatic vein
l. hilum
l. hypochondriac region
l. hypogastric nerve

l. iliac region
l. inferior abdominal quadrant
l. inferior pulmonary vein
l. intercostal space
l. internal jugular vein
l. kidney
l. lateral division of liver
l. lateral region
l. lower lobe (LLL)
l. lower quadrant (LLQ)
l. lumbar lymph node
l. lumbar trunk
l. main bronchus
l. marginal artery
l. marginal vein
l. margin of heart
l. medial division of liver
l. medial hepatic segment
l. midinguinal line
l. midlingual plane
l. obturator artery
l. ovarian artery
l. ovarian vein
l. ovary
l. part of liver
l. pericardiacophrenic vein
l. perineal artery
l. phrenic vein
l. plate of thyroid cartilage
l. posterior lateral hepatic segment III
l. posterior oblique
l. pterygoid process
l. pulmonary artery
l. rhomboid muscle
l. and right brachiocephalic vein
l. sagittal fissure
l. sigmoid sinus
l. subclavian artery
l. subclavian vein
l. superior abdominal quadrant
l. superior gluteal vein
l. superior intercostal vein
l. superior pulmonary vein
l. suprarenal vein
l. sympathetic trunk
l. testicular artery
l. testicular vein
l. triangular ligament
l. triangular ligament of liver
l. umbilical artery
l. umbilical fold

L

NOTES

left *(continued)*
    l. umbilical vein
    l. upper lobe (LUL)
    l. upper lobe bronchus
    l. upper quadrant (LUQ)
    l. ureter
    l. venous angle
    l. ventricle
    l. ventricular
    l. ventricular inflow tract
    l. ventricular outflow tract
    l. vertebral artery
    l. vomer

left-sidedness
    bilateral l.-s.

leg
    anterior compartment of l.
    anterior region of l.
    anterior surface of l.
    l. of antihelix
    cruciate ligament of l.
    deep fascia of l.
    deep part of posterior flexor
      compartment of l.
    dorsiflexor compartment of l.
    extensor compartment of l.
    fascia of l.
    fibular compartment of l.
    flexor compartment of l.
    interosseous membrane of l.
    interosseous nerve of l.
    lateral compartment of l.
    lateral surface of l.
    medial cutaneous nerve of l.
    musculocutaneous nerve of l.
    peroneal compartment of l.
    plantarflexor compartment of l.
    posterior compartment of l.
    posterior region of l.
    posterior surface of l.
    soleal part of posterior plantar
      flexor compartment of l.
    superficial part of posterior plantar
      flexor compartment of l.
    transverse ligament of l.

Leishman chrome cells
lemmoblast
lemmocyte
lemniscorum
    decussatio l.
lemniscus, pl. lemnisci
    acoustic l.
    auditory l.
    decussation of medial l.
    gustatory l.
    l. lateralis
    medial l.
    l. medialis
    nucleus of lateral l.

    l. spinalis
    stratum interolivare lemnisci
    trigeminal l.
    l. trigeminalis
    trigonum lemnisci

length
    crown-heel l.
    crown-rump l.
    resting l.

length-breadth index
lengthening reaction
length-height index
Lenhossék process
Lenoir facet
lens
    anterior pole of l.
    anterior surface of l.
    axis of l.
    l. capsule
    capsule of l.
    cortex of l.
    crystalline l.
    epithelium of l.
    equator of l.
    lamina of l.
    ligament of l.
    nucleus of l.
    l. pit
    l. placode
    posterior pole of l.
    posterior surface of l.
    radii of l.
    l. star
    suspensory ligament of l.
    l. suture
    l. vesicle

lenticonus
lenticula
lenticular
    l. ansa
    l. apophysis
    l. bone
    l. capsule
    l. fossa
    l. ganglion
    l. loop
    l. nucleus
    l. papilla
    l. process of incus
    l. vesicle

lenticularis
    ansa l.
    fasciculus l.

lenticulooptic
lenticulostriate artery
lenticulothalamic
lentiform
    l. bone
    l. nucleus

lentiformis
discus l.
nucleus l.
lentiglobus
lentis
apparatus suspensorius l.
axis l.
capsula vasculosa l.
cortex l.
ectopia l.
epithelium l.
equator l.
facies anterior l.
facies posterior l.
fibrae l.
nucleus l.
polus anterior l.
polus posterior l.
radii l.
substantia l.
tunica vasculosa l.
vortex l.
lepidic
lepra cells
leprechaunism
leptin
leptocephalous
leptocephaly
leptochromatic
leptodactylous
leptomeningeal space
leptomeninx, pl. **leptomeninges**
parietal layer of leptomeninges
leptomere
lepton
leptoprosopia
leptoprosopic
leptorrhine
leptosomatic
leptosomic
leptotene
LES
lower esophageal sphincter
Lesgaft
L. space
L. triangle
lesser
l. alar cartilage
l. arterial circle of iris
l. circulation
l. cul-de-sac
l. curvature

l. curvature of stomach
l. horn
l. horn of hyoid
l. humerus tubercle
l. internal cutaneous nerve
l. multangular bone
l. occipital nerve
l. omentum
l. ovarian vein
l. palatine artery
l. palatine canal
l. palatine foramen
l. palatine nerve
l. pancreas
l. pelvis
l. peritoneal cavity
l. peritoneal sac
l. petrosal nerve
l. petrosal nerve hiatus
l. rhomboid muscle
l. ring of iris
l. sciatic foramen
l. sciatic notch
l. splanchnic nerve
l. superficial petrosal nerve
l. supraclavicular fossa
L. triangle
l. trochanter
l. trochanter of femur
l. tubercle of humerus
l. tuberosity of humerus
l. tympanic spine
l. vestibular gland
l. wing of sphenoid bone
l. zygomatic muscle
Lesshaft triangle
leuko
leukocyte
acidophilic l.
basophilic l.
eosinophilic l.
globular l.
mast l.
oxyphilic l.
Türk l.
leukocytoclasis
leukocytopoiesis
leukoderma
leukodermatous
leukopathia
leukopoiesis
leukopoietic

L

NOTES

levator
l. angularis oris muscle
l. anguli oris
l. anguli oris muscle
l. ani
l. ani muscle
l. aponeurosis
l. check ligament
l. costarum muscle
l. cushion
l. glandulae thyroideae
l. labii superioris
l. labii superioris alaeque nasi
l. labii superioris alaeque nasi muscle
l. labii superioris muscle
l. muscle of anus
l. muscle of thyroid gland
l. palati muscle
l. palpebrae superioris (LPS)
l. palpebrae superioris muscle
l. palpebrae superioris tendon
l. prostatae muscle
l. prostate
l. scapulae
l. swelling
l. veli palatini
l. veli palatini muscle
l. veli palatini nerve

levatores
l. costarum
l. costarum breves muscle
l. costarum longi muscle
l. prostatae

levatorius
torus l.
l. torus

level
medial joint l.
plasma l.
prevertebral fascia l.
subcuticular l.

levoatriocardinal vein
Lewis
P substance of L.
leydigarche
Leydig cells
LFC
lateral femoral cutaneous
liber
margo l.
libera
tenia l.
liberae
terminationes nervorum l.
liberi
articulationes membri inferioris l.
articulationes membri superioris l.

juncturae membri inferioris l.
juncturae membri superioris l.
liberomotor
lid
lower l.
l. margin
upper l.
Liddell-Sherrington reflex
Lieberkühn
crypt of L.
L. crypts
L. follicle
L. gland
lien
l. accessorius
l. succenturiatus
lienal
l. artery
l. vein
lienalis, pl. lienales
arteria l.
folliculi lymphatici lienales
lymphonodi lienales
nodi lymphoidei lienales
plexus nervosus l.
rami lienales
rami lienales arteriae l.
recessus l.
vena l.
lienculus
lienis
capsula l.
facies renalis l.
facies visceralis l.
hilum l.
margo inferior l.
margo superior l.
porta l.
pulpa l.
segmenta l.
sinus l.
sustentaculum l.
trabeculae l.
tunica fibrosa l.
tunica propria l.
tunica serosa l.
lienopancreatic
lienophrenic ligament
lienorenale
ligamentum l.
lienorenal ligament
lienunculus
Lieutaud
L. body
L. triangle
L. trigone
L. uvula
life, pl. lives
l. cycle

postnatal l.
prenatal l.
vegetative l.
lifespan
ligament
    accessory plantar l.
    accessory volar l.
    acromioclavicular l.
    acromiocoracoid l.
    adipose l.
    alar l.
    alveolodental l.
    annular l.
    anococcygeal l.
    anterior annular l.
    anterior costotransverse l.
    anterior cruciate l. (ACL)
    anterior longitudinal l.
    anterior meniscofemoral l.
    anterior sacrococcygeal l.
    anterior sacroiliac l.
    anterior sacrosciatic l.
    anterior sternoclavicular l.
    anterior talocalcaneal l.
    anterior talofibular l.
    anterior talotibial l.
    anterior tibiofibular l.
    anterior tibiotalar part of deltoid l.
    Arantius l.
    arcuate popliteal l.
    arcuate pubic l.
    arterial l.
    atlantooccipital l.
    l.'s of auditory ossicle
    l. of auricle
    auricular l.
    axis l.
    Bardinet l.
    Barkow l.
    Bellini l.
    Berry l.
    Bertin l.
    Bichat l.
    bifurcate l.
    bifurcated l.
    Bigelow l.
    Botallo l.
    Bourgery l.
    brachioradial l.
    broad uterine l.
    Brodie l.
    Burns l.

    calcaneocuboid l.
    calcaneofibular l.
    calcaneonavicular l.
    calcaneotibial l.
    Caldani l.
    Campbell l.
    Camper l.
    capsular l.
    cardinal l.
    caroticoclinoid l.
    carpometacarpal l.
    caudal l.
    ceratocricoid l.
    check l.
    chondroxiphoid l.
    ciliary l.
    Civinini l.
    Clado l.
    Cleland cutaneous l.
    Cloquet l.
    coccygeal l.
    collateral l.
    Colles l.
    conoid l.
    Cooper l.
    coracoacromial l.
    coracoclavicular l.
    coracohumeral l.
    corniculopharyngeal l.
    coronary l.
    costoclavicular l.
    costocolic l.
    costotransverse l.
    costoxiphoid l.
    cotyloid l.
    Cowper l.
    cricoarytenoid l.
    cricopharyngeal l.
    cricosantorinian l.
    cricothyroid arytenoid l.
    cricotracheal l.
    crucial l.
    cruciate l.
    crural l.
    Cruveilhier l.
    cuboideonavicular l.
    cular l.
    cuneiform l.
    cuneocuboid interosseous l.
    cuneometatarsal interosseous l.
    cuneonavicular l.
    cystoduodenal l.

NOTES

L

ligament *(continued)*

deep dorsal sacrococcygeal l.
deep posterior sacrococcygeal l.
deep transverse metacarpal l.
deep transverse metatarsal l.
deltoid l.
Denonvilliers l.
dentate l.
denticulate l.
Denucé l.
descending part of iliofemoral l.
digital retinacular l.
distal ulnar collateral l.
dorsal calcaneocuboid l.
dorsal carpal arcuate l.
dorsal carpometacarpal l.
dorsal cuboideonavicular l.
dorsal cuneocuboid l.
dorsal cuneonavicular l.
dorsal intercarpal l.
dorsal intercuneiform l.
dorsal metacarpal l.
dorsal metatarsal l.
dorsal radiocarpal l.
dorsal sacroiliac l.
dorsal talonavicular l.
dorsal tarsal l.
dorsal tarsometatarsal l.
duodenocolic l.
duodenorenal l.
l. of epididymis
l. of epididymis inferior and
  superior
epihyal l.
external calcaneoastragaloid l.
external cruciate l.
external lateral l.
extracapsular l.
fabellofibular l.
facet on talus for calcaneonavicular
  part of bifurcate l.
facet on talus for plantar
  calcaneonavicular l.
falciform process of
  sacrotuberous l.
fallopian l.
false l.
femoral l.
Ferrein l.
fibular collateral l.
fissure of round l.
fissure of venous l.
flaval l.
Flood l.
l. of foot
foveal l.
gastrocolic l.
gastrodiaphragmatic l.
gastrohepatic l.

gastrolienal l.
gastropancreatic l.
gastrophrenic l.
gastrosplenic l.
genital l.
genitoinguinal l.
Gerdy l.
Gillette suspensory l.
Gimbernat l.
gingivodental l.
glenohumeral l.
glenoid l.
glossoepiglottic l.
Gruber l.
Günz l.
hammock l.
l. of head of femur
l.'s of head of fibula
Helmholtz axis l.
l. of Henry
Hensing l.
hepatic l.
hepatocolic l.
hepatocystocolic l.
hepatoduodenal l.
hepatoesophageal l.
hepatogastric l.
hepatogastroduodenal l.
hepatophrenic l.
hepatorenal l.
hepatoumbilical l.
Hesselbach l.
Hey l.
Holl l.
Hueck l.
Humphrey l.
Hunter l.
Huschke l.
hyalocapsular l.
hyoepiglottic l.
hypsiloid l.
iliofemoral l.
iliolumbar l.
iliopectineal l.
iliotrochanteric l.
impression for costoclavicular l.
l. of incus
inferior calcaneonavicular l.
inferior glenohumeral l.
inferior pubic l.
inferior transverse l.
inferior transverse scapular l.
infundibuloovarian l.
infundibulopelvic l.
inguinal l.
intercarpal l.
interchondral l.
interclavicular l.
interclinoid l.

intercornual l.
intercostal l.
intercuneiform l.
interdigital l.
interfoveolar l.
intermetacarpal l.
intermetatarsal l.
internal calcaneoastragaloid l.
internal lateral l.
interosseous cuneocuboid l.
interosseous cuneometatarsal l.
interosseous metacarpal l.
interosseous metatarsal l.
interosseous sacroiliac l.
interosseous talocalcaneal l.
interosseous tibiofibular l.
interphalangeal collateral l.
interspinous l.
intertransverse l.
intraarticular sternocostal l.
intracapsular l.
ischiocapsular l.
ischiofemoral l.
Jarjavay l.
jugal l.
knee l.
l. of knee
Krause l.
lacinate l.
laciniate l.
lacunar l.
Landsmeer l.
Lannelongue l.
lateral arcuate l.
lateral collateral l.
lateral costotransverse l.
lateral malleolar l.
lateral palpebral l.
lateral puboprostatic l.
lateral sacrococcygeal l.
lateral talocalcaneal l.
lateral temporomandibular l.
lateral thyrohyoid l.
lateral thyroid l.
lateral umbilical l.
Lauth l.
leaves of broad l.
l. of left superior vena cava
left triangular l.
l. of left vena cava
l. of lens
levator check l.

lienophrenic l.
lienorenal l.
Lisfranc l.
Lockwood l.
long external lateral l.
longitudinal l.
long plantar l.
lumbocostal l.
Luschka l.
Mackenrodt l.
mallear l.
l. of malleus
Mauchart l.
Meckel l.
medial arcuate l.
medial canthal l.
medial collateral l. (MCL)
medial palpebral l.
medial puboprostatic l.
medial talocalcaneal l.
medial thyroid l.
medial umbilical l.
median arcuate l.
median cricothyroid l.
median thyrohyoid l.
median thyroid l.
median umbilical l.
meniscofemoral l.
metacarpal l.
metatarsal interosseous l.
middle costotransverse l.
middle umbilical l.
mucosal suspensory l.
natatory l.
nephrocolic l.
nuchal l.
oblique popliteal l.
occipitoaxial l.
odontoid l.
olecranon l.
orbicular l.
Osborne l.
ovarian l.
l. of ovary
palmar carpal l.
palmar carpometacarpal l.
palmar metacarpal l.
palmar radiocarpal l.
palmar ulnocarpal l.
palpebral l.
pancreaticosplenic l.
patellar l.

L

NOTES

ligament *(continued)*

pectinate l.
pectineal l.
peridental l.
periodontal l.
Petit l.
petroclinoid l.
petrosphenoid l.
phrenicocolic l.
phrenicoesophageal l.
phrenicolienal l.
phrenicosplenic l.
phrenoesophageal l.
phrenogastric l.
phrenosplenic l.
pisohamate l.
pisometacarpal l.
pisounciform l.
pisouncinate l.
plantar calcaneocuboid l.
plantar calcaneonavicular l.
plantar cuboideonavicular l.
plantar cuneocuboid l.
plantar cuneonavicular l.
plantar metatarsal l.
plantar tarsal l.
plantar tarsometatarsal l.
posterior annular l.
posterior costotransverse l.
posterior cricoarytenoid l.
posterior cruciate l.
posterior false l.
posterior leaf of broad l.
posterior longitudinal l.
posterior meniscofemoral l.
posterior occipitoaxial l.
posterior sacroiliac l.
posterior sacrosciatic l.
posterior sternoclavicular l.
posterior talocalcaneal l.
posterior talofibular l.
posterior talotibial l.
posterior tibiofibular l.
posterior tibiotalar part of
  deltoid l.
Poupart inguinal l.
pterygomandibular l.
pterygospinal l.
pterygospinous l.
pubic arcuate l.
pubocapsular l.
pubocervical l.
pubofemoral l.
puboprostatic l.
pubovesical l.
pulmonary l.
quadrate l.
radial collateral l.
radiate carpal l.

radiate sternocostal l.
radiocapitate l.
radiocarpal l.
radiotriquetral l.
radioulnar l.
rectouterine l.
reflected inguinal l.
l. reflecting edge
reflex l.
Retzius l.
rhomboid l.
right prostatic l.
right triangular l.
ring l.
round uterine l.
sacrodural l.
sacrogenital l.
sacrospinous l.
sacrotuberous l.
sacrouterine l.
Scarpa l.
serous l.
sesamoid cartilage of
  cricopharyngeal l.
sheath l.
l. shelving edge
shelving edge of Poupart l.
short plantar l.
Simonart l.
skin l.
Soemmering l.
sphenomandibular l.
spinoglenoid l.
splenocolic l.
splenorenal l.
spring l.
Stanley cervical l.
stellate l.
sternoclavicular l.
sternocostal l.
sternopericardial l.
Struthers l.
l. of Struthers
stylohyoid l.
stylomandibular l.
stylomaxillary l.
superficial dorsal sacrococcygeal l.
superficial posterior
  sacrococcygeal l.
superficial transverse metacarpal l.
superficial transverse metatarsal l.
superior astragalonavicular l.
superior costotransverse l.
superior pubic l.
superior transverse scapular l.
suprascapular l.
supraspinous l.
sutural l.
synovial l.

talocalcaneal interosseous l.
talocalcaneonavicular l.
talofibular l.
talonavicular l.
tarsal interosseous l.
tarsometatarsal l.
temporomandibular l.
tendinotrochanteric l.
Teutleben l.
Thompson l.
thyroepiglottic l.
thyroepiglottidean l.
thyrohyoid l.
tibial collateral l.
tibiocalcaneal part of deltoid l.
tibiofibular l.
tibionavicular part of deltoid l.
l. of Toldt
Toldt l.
transverse acetabular l.
transverse atlantal l.
transverse carpal l.
transverse cervical l.
transverse crural l.
transverse genicular l.
transverse humeral l.
transverse metacarpal l.
transverse metatarsal l.
transverse part of iliofemoral l.
transverse perineal l.
transverse tibiofibular l.
trapezoid l.
Treitz l.
l. of Treitz
triangular l.
tuberosity for coracoclavicular l.
ulnar collateral l.
ulnocarpal l.
umbilical l.
urachal l.
uteropelvic l.
uterosacral l.
uterovesical l.
vaginal l.
Valsalva l.
venous l.
ventral sacrococcygeal l.
ventral sacroiliac l.
ventricular l.
vertebropelvic l.
vesicoumbilical l.
vesicouterine l.

vestibular l.
vocal l.
volar carpal l.
volar radiocarpal l.
web l.
Weitbrecht l.
Whitnall l.
Winslow l.
Wrisberg l.
l. of Wrisberg
yellow l.
Y-shaped l.
Zaglas l.
Zinn l.
ligamenta (*pl. of* ligamentum)
ligamental mucosa
ligamenti teretis uterus
ligamentous
l. falx
l. joint
l. structure
ligamentum, pl. **ligamenta**
l. acromioclaviculare
ligamenta alaria
l. annulare
l. annulare radii
l. annulare stapedis
l. anococcygeum
l. anulare
l. anulare bulbi
l. anulare digitorum
l. anulare stapedis
ligamenta anularia trachealia
l. apicis dentis
l. arcuatum laterale
l. arcuatum mediale
l. arcuatum medianum
l. arcuatum pubis
l. arteriosum
l. auriculare anterius
l. auriculare posterius
l. auriculare superius
ligamenta auricularia
ligamenta basium
l. bifurcatum
l. calcaneocuboideum
l. calcaneocuboideum plantare
l. calcaneofibulare
l. calcaneonaviculare
l. calcaneonaviculare plantare
l. calcaneotibiale
l. capitis costae intraarticulare

L

NOTES

ligamentum *(continued)*

l. capitis costae radiatum
l. capitis femoris
l. capitis fibulae
l. capitis fibulae anterius
l. capitis fibulae posterius
ligamenta capitulorum transversa
l. capsulare
l. cardinale
l. carpi dorsale
l. carpi radiatum
l. carpi transversum
l. carpi volare
ligamenta carpometacarpalia
l. carpometacarpalia dorsalia
ligamenta carpometacarpalia dorsalia
ligamenta carpometacarpalia
 dorsalia/palmaria
ligamenta carpometacarpalia
 palmaria
l. carpometacarpalia palmaria
l. caudale
l. ceratocricoideum
l. collaterale
l. collaterale carpi radiale
l. collaterale carpi radiale
 articulationis radiocarpalis
l. collaterale carpi ulnare
l. collaterale carpi ulnare
 articulationis radiocarpalis
l. collaterale fibulare
l. collaterale laterale
l. collaterale mediale
l. collaterale radiale articulationis
 cubiti
l. collaterale tibiale
l. collaterale ulnare articulationis
 cubiti
ligamenta collateralia
l. colli costae
l. conoideum
l. coracoacromiale
l. coracoclaviculare
l. coracohumerale
l. corniculopharyngeum
l. coronarium hepatis
l. costoclaviculare
l. costotransversarium
l. costotransversarium anterius
l. costotransversarium laterale
l. costotransversarium posterius
l. costotransversarium superius
l. costoxiphoideum
l. cotyloideum
l. cricoarytenoideum posterius
l. cricopharyngeum
l. cricothyroideum
l. cricotracheale
ligamenta cruciata digitorum

ligamenta cruciata genus
l. cruciatum anterius
l. cruciatum atlantis
l. cruciatum cruris
l. cruciatum posterius
l. cruciatum tertium genus
l. cruciforme atlantis
ligamenta cuboideonaviculare
l. cuboideonaviculare
l. cuboideonaviculare dorsale
l. cuboideonaviculare plantare
ligamenta cuboideonavicularia
 plantaria
l. cuneocuboideum
l. cuneocuboideum dorsale
l. cuneocuboideum interosseum
l. cuneocuboideum plantare
ligamenta cuneometatarsalia
 interossea
ligamenta cuneonavicularia
ligamenta cuneonavicularia dorsalia
l. cuneonavicularia dorsalia
ligamenta cuneonavicularia plantaria
l. cuneonavicularia plantaria
l. deltoideum
l. denticulatum
l. ductus venosi
l. duodenorenale
l. epididymidis
l. epididymidis inferius
ligamenta epididymidis inferius et
 superius
l. epididymidis superius
ligamenta extracapsularia
l. falciforme
l. falciforme hepatis
l. flavum
l. fundiforme clitoridis
l. fundiforme penis
l. gastrocolicum
l. gastrolienale
l. gastrophrenicum
l. gastrosplenicum
l. genitoinguinale
ligamenta glenohumeralia
l. glenoidale
l. hepatocolicum
l. hepatoduodenale
l. hepatoesophageum
l. hepatogastricum
l. hepatorenale
l. hyaloideo-capsulare
l. hyaloideo-capsulario
l. hyoepiglotticum
l. hyothyroideum laterale
l. hyothyroideum medium
l. iliofemorale
l. iliolum
l. iliolumbale

l. iliopectineale
l. incudis
l. incudis posterius
l. incudis superius
l. inguinale
ligamenta intercarpalia
l. intercarpalia dorsalia
l. intercarpalia interossea
l. intercarpalia palmaria
l. interclaviculare
ligamenta intercostalia
ligamenta intercuneiformia
ligamenta intercuneiformia dorsalia
l. intercuneiformia dorsalia
ligamenta intercuneiformia interossea
l. intercuneiformia interossea
ligamenta intercuneiformia plantaria
l. intercuneiformia plantaria
l. interfoveolare
l. interspinale
l. intertransversarium
ligamenta intracapsularia
l. ischiocapsulare
l. ischiofemorale
l. laciniatum
l. lacunare
l. laterale articulationis
    temporomandibularis
l. laterale vesicae
l. latum pulmonis
l. latum uterus
l. lienorenale
l. longitudinale
l. longitudinale anterius
l. longitudinale posterius
ligamenta longitudinalia
l. lumbocostale
l. mallei anterius
l. mallei laterale
l. mallei superius
l. malleoli lateralis
l. mediale articulationis talocruralis
l. mediale articulationis
    temporomandibularis
l. mediale puboprostaticum
l. menisci lateralis
l. meniscofemorale
l. meniscofemorale anterius
l. meniscofemorale posterius
ligamenta meniscofemoralia
l. metacarpale transversum
    profundum

l. metacarpale transversum
    superficiale
ligamenta metacarpalia
ligamenta metacarpalia dorsalia
l. metacarpalia dorsalia
l. metacarpalia interossea
ligamenta metacarpalia interossea
l. metacarpalia palmaria
ligamenta metacarpalia palmaria
l. metatarsale transversum
    profundum
l. metatarsale transversum
    superficiale
ligamenta metatarsalia
l. metatarsalia dorsalia
ligamenta metatarsalia dorsalia
l. metatarsalia interossea
ligamenta metatarsalia interossea
ligamenta metatarsalia plantaria
l. metatarsalia plantaria
l. natatorium
ligamenta navicularicuneiformia
l. nuchae
l. orbiculare radii
ligamenta ossiculorum auditorium
ligamenta ossiculorum auditus
l. ovarii proprium
ligamenta palmaria articulationis
    interphalangeae manus
ligamenta palmaria articulationis
    metacarpophalangeae
l. palpebrale externum
l. palpebrale laterale
l. palpebrale mediale
l. patellae
l. pectinatum
l. pectinatum anguli iridocornealis
l. pectinatum iridis
l. pectineale
l. phrenicocolicum
l. phrenicolienale
l. phrenicosplenicum
l. pisohamatum
l. pisometacarpeum
l. plantare longum
ligamenta plantaria articulationis
    interphalangeae pedis
ligamenta plantaria articulationis
    metatarsophalangeae
l. popliteum arcuatum
l. popliteum obliquum
l. pterygospinale

L

NOTES

ligamentum (*continued*)

l. pubicum inferius
l. pubicum superius
l. pubocapsulare
l. pubofemorale
l. puboprostaticum laterale
l. puboprostaticum mediale
l. pubovesicale femininum
l. pubovesicale masculinum
l. pulmonale
l. quadratum
l. radiocarpale dorsale
l. radiocarpale palmare
l. reflexum
l. sacrococcygeum anterius
l. sacrococcygeum dorsale
   superficiale
l. sacrococcygeum laterale
l. sacrococcygeum posterius
   profundum
l. sacrococcygeum posterius
   superficiale
l. sacrodurale
ligamenta sacroiliaca anteriora
ligamenta sacroiliaca interossea
ligamenta sacroiliaca posteriora
l. sacroiliacum posterius
l. sacrospinale
l. sacrospinosum
l. sacrotuberale
l. sacrotuberosum
l. serosum
l. sphenomandibula
l. sphenomandibulare
l. spirale cochlea
l. spirale ductus cochlearis
l. splenorenale
l. sternoclaviculare
l. sternoclaviculare anterius
l. sternoclaviculare posterius
ligamenta sternoclavicularia
l. sternocostale intraarticulare
ligamenta sternocostalia radiata
ligamenta sternopericardiaca
l. stylohyoideum
l. stylomandibulare
l. supraspinale
ligamenta suspensoria mammae
ligamenta suspensoria mammaria
l. suspensorium axillae
l. suspensorium bulbi
l. suspensorium clitoridis
l. suspensorium duodeni
l. suspensorium glandulae
   thyroideae
l. suspensorium ovarii
l. suspensorium penis
l. talocalcaneare
l. talocalcaneare interosseum

l. talocalcaneare laterale
l. talocalcaneare mediale
l. talocalcaneum
l. talocalcaneum laterale
l. talocalcaneum mediale
l. talocalcaneum posterius
l. talofibulare anterius
l. talofibulare posterius
l. talonaviculare
l. talotibiale anterius
l. talotibiale posterius
l. tarsale externum
l. tarsale internum
ligamenta tarsi
l. tarsi dorsalia
ligamenta tarsi dorsalia
ligamenta tarsi interossea
ligamenta tarsi plantaria
l. tarsi plantaria
ligamenta tarsometatarsalia
ligamenta tarsometatarsalia dorsalia
ligamenta tarsometatarsalia plantaria
l. temporomandibulare
l. teres
l. teres femoris
l. teres hepatis
l. teres uterus
l. testis
l. thyroepiglotticum
l. thyrohyoideum laterale
l. thyrohyoideum medianum
l. tibiofibulare anterius
l. tibiofibulare medium
l. tibiofibulare posterius
l. tibionaviculare
ligamenta trachealia
l. transversale cervicis
l. transversalis colli
l. transversum acetabuli
l. transversum atlantis
l. transversum cruris
l. transversum genus
l. transversum humeri
l. transversum pelvis
l. transversum perinei
l. transversum scapulae inferius
l. transversum scapulae superius
l. trapezoideum
l. triangulare
l. triangulare dextrum
l. triangulare dextrum hepatis
l. triangulare sinistrum
l. triangulare sinistrum hepatis
l. tuberculi costae
l. ulnocarpale palmare
l. umbilicale laterale
l. umbilicale mediale
l. umbilicale medianum
l. uteroovaricum

l. venae cavae sinistrae
l. venosum
l. ventriculare
l. vestibulare
l. vocale
ligand-gated channel
ligature
Stannius l.
light cells of thyroid
limb
afferent l.
anterior surface of lower l.
artery of lower l.
artery of upper l.
bone of inferior l.
bone of lower l.
bone of superior l.
bone of upper l.
l. of bony semicircular canal
l. bud
dermatome of lower l.
dermatome of upper l.
efferent l.
flexor retinaculum of lower l.
free part of lower l.
free part of upper l.
l. girdle
l. of helix
inferior l.
joint of free inferior l.
joint of free superior l.
lateral surface of lower l.
lower l.
lymph node of lower l.
lymph node of upper l.
medial l.
pelvic l.
posterior surface of lower l.
region of inferior l.
region of lower l.
region of superior l.
region of upper l.
skeleton of free inferior l.
skeleton of free superior l.
superior l.
synovial joint of free lower l.
synovial joint of free upper l.
thoracic l.
upper l.
vein of lower l.
vein of upper l.
limbal groove

limbi (*pl. of* limbus)
limbic
l. center
l. lobe
l. system
limbicus
gyrus l.
limbus, pl. **limbi**
l. acetabuli
l. alveolaris
l. anterior palpebrae
conjunctival l.
l. of cornea
l. corneae
corneal l.
l. fossae ovalis
l. laminae spiralis osseae
l. membranae tympani
l. of osseous spiral lamina
limbi palpebrales
l. palpebrales anteriores
l. palpebrales posteriores
l. penicillatus
l. posterior palpebrae
l. sphenoidalis
l. of sphenoid bone
l. striatus
tympanic lip of spiral l.
l. of tympanic membrane
vestibular lip of spiral l.
Vieussens l.
limen, pl. **limina**
l. insulae
l. nasi
liminal stimulus
liminometer
limit
elastic l.
Hayflick l.
proportional l.
limitans
membrana l.
sulcus l.
limiting
l. layers of cornea
l. membrane of neural tube
l. membrane of retina
l. sulcus of Reil
l. sulcus of rhomboid fossa
line
accretion l.
alveolonasal l.

L

NOTES

line (*continued*)

Amberg lateral sinus l.
anocutaneous l.
anorectal l.
anterior axillary l. (AAL)
anterior gluteal l.
anterior median l.
arcuate l.
auriculomastoid l.
axillary l.
azygos venous l.
Baillarger l.
base l.
basinasal l.
l. of Bechterew
Brödel bloodless l.
Camper l.
cement l.
Chamberlain l.
Chaussier l.
cleavage l.
costoclavicular l.
Crampton l.
Daubenton l.
dentate l.
Douglas l.
epiphysial l.
Farre l.
Feiss l.
flexure l.
frontal zygomatic suture l.
l. of Gennari
germ l.
gingival l.
glabelloalveolar l.
gluteal l.
gum l.
Haller l.
Hensen l.
highest nuchal l.
Hilton white l.
Holden l.
Hunter l.
Hunter-Schreger l.
iliopectineal l.
incremental l.
inferior nuchal l.
inferior temporal l.
infracostal l.
infraorbitomeatal l.
intercondylar l.
internal oblique l.
interspinal l.
intertrochanteric l.
intertubercular l.
joint l.
l. of Kaes
Kilian l.
lambdoid suture l.

Langer l.
Lanz l.
lateral sympathetic l.
left midinguinal l.
Looser l.
M l.
mamillary l.
mammary l.
McKee l.
medial sympathetic l.
median l.
midaxillary l.
midclavicular l.
midcostal l.
middle axillary l.
midscapular l.
milk l.
Monro l.
Monro-Richter l.
mylohyoid l.
nasobasilar l.
Nélaton l.
nipple l.
nuchal l.
Obersteiner-Redlich l.
oblique l.
Ogston l.
l. of Ogston
omphalospinous l.
orbitomeatal l.
Owen l.
parasternal l.
paravertebral l.
Paris l.
pectate l.
pectinate l.
pectineal l.
l. of pleural reflection
Poirier l.
popliteal l.
postaxillary l.
posterior axillary l.
posterior median l.
Poupart l.
preaxillary l.
pubic hair l.
pubococcygeal l.
Reid base l.
l.'s of Retzius
Richter-Monro l.
right midinguinal l.
Roser-Nélaton l.
rough l.
sagittal suture l.
scapular l.
Schreger l.
Schwalbe l.
semicircular l.
semilunar l.

septal l.
soleal l.
l. for soleus muscle
sphenofrontal suture l.
sphenoparietal suture l.
Spieghel l.
Spigelius l.
spiral l.
squamous suture l.
sternal l.
subcostal l.
superior nuchal l.
superior temporal l.
supracondylar l.
supracrestal l.
sylvian l.
temporal l.
tension l.
tentorial l.
terminal l.
thoracolumbar venous l.
Toldt l.
l. of Toldt
Topinard l.
transverse l.
trapezoid l.
Vesling l.
white l.
Winberger l.
Z l.
Zahn l.
l. of Zahn
zygomaticofrontal suture l.
zygomatic suture l.

linea, pl. **lineae**
l. adminiculum
l. alba
l. alba cervicalis
l. alba hernia
l. anocutanea
l. anorectalis
l. arcuata
l. arcuata ossis ilii
l. arcuata vaginae musculi recti
  abdominis
l. aspera
l. axillaris anterior
l. axillaris media
l. axillaris posterior
lineae distractionis
l. epiphysialis
l. glutea

l. glutea anterior
l. glutea inferior
l. glutea posterior
l. intercondylaris
l. intercondylaris femoris
l. intermedia cristae iliacae
l. interspinalis
l. intertrochanterica
l. intertubercularis
l. mamillaris
l. mammillaris
l. mediana anterior
l. mediana posterior
l. medioaxillaris
l. medioclavicularis
l. musculi solei
l. mylohyoidea
l. nuchae inferior
l. nuchae mediana
l. nuchae superior
l. nuchae suprema
l. obliqua
l. obliqua cartilaginis thyroidea
l. obliqua cartilaginis thyroideae
l. obliqua mandibulae
l. parasternalis
l. paravertebralis
l. pectinata canalis analis
l. pectinea
l. pectinea femoris
l. poplitea
l. postaxillaris
l. preaxillaris
l. scapularis
l. semicircularis
l. semilunaris
l. spiralis
l. splendens
l. sternalis
l. subcostalis
l. supracristalis
l. temporalis inferior
l. temporalis inferior ossis parietalis
l. temporalis ossis frontalis
l. temporalis superior
l. temporalis superior ossis
  parietalis
l. terminalis
l. terminalis pelvis
l. transversa
lineae transversae ossis sacri
l. trapezoidea

NOTES

L

327

linear
  l. acceleration
  l. groove
lingua, pl. **linguae**
  apex linguae
  aponeurosis linguae
  arteria profunda linguae
  l. cerebelli
  copula linguae
  corpus linguae
  dorsales linguae
  dorsum linguae
  facies inferior linguae
  folia linguae
  foramen cecum linguae
  frenulum linguae
  longitudinalis superior linguae
  margo linguae
  musculi linguae
  musculus longitudinalis inferior
    linguae
  musculus longitudinalis superior
    linguae
  musculus transversus linguae
  musculus verticalis linguae
  nucleus fibrosus linguae
  pars anterior linguae
  pars posterior linguae
  pars postsulcalis linguae
  pars presulcalis linguae
  plica fimbriata faciei inferioris
    linguae
  profunda linguae
  radix linguae
  rami dorsales linguae
  raphe linguae
  septum linguae
  sulcus medianus linguae
  sulcus terminalis linguae
  transversus linguae
  tunica mucosa linguae
  vena dorsalis linguae
  venae dorsales linguae
  vena profunda linguae
  verticalis linguae
  vinculum linguae
lingual
  l. aponeurosis
  l. arch
  l. artery
  l. bone
  l. branch
  l. branch of facial nerve
  l. crypt
  l. follicles
  l. frenulum
  l. frenum
  l. ganglion
  l. gingival papilla

  l. gland
  l. gyrus
  l. interdental papilla
  l. lobe
  l. lymph node
  l. mucosa
  l. muscle
  l. plexus
  l. sulcus
  l. surface of tooth
  l. tonsil
  l. vein
linguali
  rami communicantes ganglii
    sublingualis cum nervo l.
  rami communicantes ganglii
    submandibularis cum nervo l.
lingualis, pl. **linguales**
  arteria l.
  ductus l.
  folliculi linguales
  gyrus l.
  nervus l.
  nodi lymphoidei linguales
  papilla l.
  plexus periarterialis arteriae l.
  rami linguales
  rami dorsales linguae arteriae l.
  rami fauciales nervi l.
  rami isthmi faucium nervi l.
  rami linguales nervi l.
  ramus suprahyoideus arteriae l.
  tonsilla l.
  vena l.
linguiform
linguiformis
  lobus l.
lingula, pl. **lingulae**
  l. cerebelli
  l. of left lung
  lung l.
  l. of mandible
  l. mandibulae
  l. pulmonis sinistri
  sphenoidal l.
  l. sphenoidalis
lingular
  l. artery
  l. bronchus
  l. segment
  l. vein
lingularis
  arteria l.
  pars inferior ramus l.
  pars superior ramus l.
  ramus l.
  vena l.
linguofacial
  l. arterial trunk

linguofacialis
> truncus l.

linguogingival groove

linguoplate
> palatal l.

lining
> l. cell
> endothelial l.
> epithelial l.
> membranous l.
> stomach l.

linin network

lip
> acetabular l.
> anterior cervical l.
> articular l.
> l. of cervix
> cleft l.
> commissure of l.
> compressor muscle of l.
> depressor muscle of lower l.
> elevator muscle of upper l.
> frenulum of lower l.
> frenulum of pudendal l.
> glenoidal l.
> Hapsburg jaw and l.
> inferior commissure of l.
> junction of l.
> large pudendal l.
> lower l.
> l.'s of mouth
> pit of l.
> posterior cervical l.
> quadrate muscle of upper l.
> rhombic l.
> small pudendal l.
> l. sulcus
> superior commissure of l.
> transitional zone of l.
> tubercle of upper l.
> tympanic l.
> upper l.
> l. vermilion
> vermilion surface of l.
> vestibular l.

lipid
> anisotropic l.

lipoblast

lipochrome

lipocyte

lipodermoid

lipofuscin

lipogenesis

lipogenic

lipogenous

lipoidosis

lipoid theory of narcosis

lipolysis

lipolytic

lipomeningocele

lipophanerosis

liposome

Lipschütz cell

liquid
> Cotunnius l.

liquor, pl. **liquores**
> l. amnii
> l. cerebrospinalis
> l. cotunnii
> l. entericus
> l. folliculi
> Morgagni l.
> Scarpa l.

liquorrhea

Lisfranc
> L. joint
> L. ligament
> scalene tubercle of L.
> L. tubercle

Lissauer
> L. bundle
> L. fasciculus
> L. marginal zone
> L. tract

lissencephalia

lissencephalic

lissencephaly

lissosphincter

Lister tubercle

lithokelyphopedion

lithokelyphopedium

lithokelyphos

lithopedion

lithopedium

little
> L. area
> l. finger
> l. fossa of cochlear window
> l. fossa of oval vestibular window
> l. fossa of vestibular window
> l. head of humerus
> l. toe

littoral cell

Littré gland

L

NOTES

Littré-Richter hernia
liveborn infant
livedo telangiectatica
liver
l. acinus
anterior part of diaphragmatic surface of l.
anterior portion of left medial segment IV of l.
bare area of l.
l. bed
l. bud
cardiac impression of l.
cardiac impression of diaphragmatic surface of l.
caudate lobe of l.
central vein of l.
colic impression on l.
coronary ligament of l.
duodenal impression on l.
l. edge
esophageal impression on l.
falciform ligament of l.
fibrous appendix of l.
fibrous capsule of l.
fissure for round ligament of l.
fissures of l.
gastric impression on l.
inferior border of l.
inferior margin of l.
interlobular artery of l.
interlobular vein of l.
l., kidneys, and spleen (LKS)
lateral division of left l.
left duct of caudate lobe of l.
left lateral division of l.
left medial division of l.
left part of l.
left triangular ligament of l.
lobe of l.
lobules of l.
notch for round ligament of l.
omental tuberosity of l.
papillary process of caudate lobe of l.
peritoneal ligament of l.
portal lobule of l.
posterior part of the diaphragmatic surface of the l.
quadrate part of l.
renal impression on l.
Riedel lobe of l.
right duct of caudate lobe of l.
right lateral division of l.
right medial division of l.
right part of diaphragmatic surface of l.
right triangle ligament of l.
right triangular ligament of l.

round ligament of l.
segmental artery of l.
segments of l.
serosa of l.
l. shadow
l., spleen, and kidneys (LSK)
stellate cells of l.
superior part of diaphragmatic surface of l.
suprarenal impression on l.
triangular ligament of l.
venous segments of l.
visceral surface of l.
lives (*pl. of* life)
living anatomy
Livingston
triangle of L.
LKS
liver, kidneys, and spleen
L1–L5
segmenta lumbalia L.
segmenta medullae spinalis lumbaria L.
LLL
left lower lobe
LLQ
left lower quadrant
LNPF
lymph node permeability factor
lobar
l. sclerosis
lobares
bronchi l.
lobate
lobe
artery of caudate l.
azygos l.
l. of brain
caudate l.
l.'s of cerebrum
cuneiform l.
ear l.
falciform l.
flocculonodular l.
frontal l.
hepatic l.
Home l.
inferior parietal l.
isthmus of limbic l.
kidney l.
left duct of caudate l.
left lower l. (LLL)
left upper l. (LUL)
limbic l.
lingual l.
l. of liver
l. of lung
l. of mammary gland
middle l.

nervous l.
occipital l.
paracentral l.
parietal l.
l. of pituitary gland
placental l.
posteromedian l.
l. of prostate
prostatic l.
pyramidal l.
quadrate l.
renal l.
Riedel l.
right duct of caudate l.
right lower l. (RLL)
right upper l. (RUL)
Spigelius l.
succenturiate l.
superior parietal l.
temporal l.
thyroid l.
l. of thyroid gland
lobi (*pl. of* lobus)
lobose
lobous
Lobstein ganglion
lobster-claw deformity
lobular
lobulate
lobulated
lobule
anterior lunate l.
l. of auricle
biventral l.
central l.
convoluted part of kidney l.
l.'s of epididymis
falciform l.
hepatic l.
inferior parietal l.
inferior semilunar l.
l.'s of liver
l.'s of mammary gland
medial l.
paracentral l.
posterior lunate l.
primary pulmonary l.
pulmonary l.
quadrangular l.
quadrate l.
renal cortical l.
respiratory l.

simple l.
slender l.
superior parietal l.
superior semilunar l.
l.'s of testis
l.'s of thymus
l.'s of thyroid gland
wing of central l.
lobulet
lobulette
lobulus, pl. **lobuli**
l. auriculae
l. biventer
l. biventralis
l. centralis cerebelli
l. clivi
l. corticalis renalis
l. culminis
l. cuneiformis
lobuli epididymidis
l. folii
l. fusiformis
lobuli glandulae mammariae
lobuli glandulae thyroideae
l. gracilis
l. hepatis
l. paracentralis
l. parietalis inferior
l. parietalis superior
l. quadrangularis
l. quadratus
l. semilunaris inferior
l. semilunaris superior
l. simplex
lobuli testis
lobuli thymi
lobus, pl. **lobi**
l. anterior hypophyseos
l. appendicularis
l. azygos
l. azygos pulmonis dextri
l. caudatus
lobi cerebri
l. clivi
l. falciformis
l. frontalis cerebri
lobi glandulae mammariae
lobi glandulae thyroideae
l. glandularis hypophyseos
l. hepatis dexter
l. hepatis sinister
l. inferior pulmonis

NOTES

L

lobus *(continued)*
    l. inferior pulmonis dextri et sinistri
    l. linguiformis
    l. medius prostatae
    l. medius pulmonis dextri
    l. nervosus
    l. occipitalis cerebri
    l. parietalis cerebri
    l. posterior hypophyseos
    l. pyramidalis glandulae thyroideae
    l. quadratus
    l. renalis
    l. superior pulmonis
    l. superior pulmonis dextri et sinistri
    l. temporalis

localization
    cerebral l.
    germinal l.
    stereotaxic l.

local stimulant

loci *(pl. of* locus)

Lockwood ligament

locomotive

locomotor

locomotorial

locomotorium

locomotory

locular

loculate

loculation

loculus, pl. **loculi**

locus, pl. **loci**
    l. ceruleus
    l. cinereus
    l. ferrugineus
    l. niger
    l. perforatus anticus
    l. perforatus posticus

Loevit cells

Loewenthal
    L. bundle
    L. tract

loin
    quadrate muscle of l.

long
    l. abductor muscle
    l. abductor muscle of thumb
    l. abductor tendon
    l. adductor muscle
    l. axis of body
    l. axis of kidney
    l. bone
    l. buccal nerve
    l. central artery
    l. ciliary artery
    l. ciliary nerve
    l. crus of incus

    l. extensor muscle
    l. extensor muscle of great toe
    l. extensor muscle of thumb
    l. external lateral ligament
    l. fibular muscle
    l. flexor muscle
    l. flexor muscle of great toe
    l. flexor muscle of thumb
    l. gyrus of insula
    l. levatores costarum muscle
    l. limb of incus
    l. muscle of head
    l. muscle of neck
    l. nerve of Bell
    l. palmar muscle
    l. peroneal muscle
    l. pitch helicoidal layer
    l. plantar ligament
    l. posterior ciliary artery
    l. posterior ciliary axis
    l. process of incus
    l. process of malleus
    l. radial extensor muscle
    l. radial extensor muscle of wrist
    l. root of ciliary ganglion
    l. rotator muscle
    l. saphenous nerve
    l. saphenous vein
    l. subscapular nerve
    l. thoracic artery
    l. thoracic nerve
    l. thoracic vein
    l. tract
    l. vinculum

longa, pl. **longae**
    arteria centralis l.
    arteria ciliaris posterior l.
    ciliares posteriores longae

longi
    ciliares l.
    musculi levatores costarum l.
    nervi ciliares l.
    oris pollicis l.
    sulcus tendinis musculi fibularis l.
    sulcus tendinis musculi flexoris hallucis l.
    sulcus tendinis musculi peronei l.
    vagina tendinis musculi extensoris hallucis l.
    vagina tendinis musculi extensoris pollicis l.
    vagina tendinis musculi flexoris hallucis l.
    vagina tendinis musculi flexoris pollicis l.
    vagina tendinum musculi extensoris digitorum pedis l.
    vagina tendinum musculi flexoris digitorum pedis l.

longissimus
l. capitis
l. capitis muscle
l. cervicis
l. cervicis muscle
l. colli
musculus l.
l. thoracis
l. thoracis muscle
longitudinal
l. arch
l. arch of foot
l. arch support
l. arc of skull
l. axis
l. bands of cruciform ligament of atlas
l. canal of modiolus
l. cerebral fissure
l. duct of epoöphoron
l. fissure of cerebrum
l. fold of duodenum
l. layer of muscle coat of small intestine
l. layer of muscular coat
l. layers of muscular tunic
l. ligament
l. medial bundle
l. nerve
l. oval pelvis
l. pharyngeal muscle
l. plane
l. pontine bundle
l. section
l. sulcus
l. sulcus of heart
l. vertebral venous sinus
longitudinale
ligamentum l.
longitudinalia
ligamenta l.
longitudinalis, pl. **longitudinales**
arcus pedis l.
medialis arcus pedis l.
pars lateralis arcus pedis l.
pars medialis arcus pedis l.
sinus vertebrales longitudinales
l. superior linguae
longitype
longum
caput l.
ligamentum plantare l.

os l.
vinculum l.
longus
l. capitis muscle
l. colli muscle
extensor carpi radialis l. (ECRL)
extensor digitorum l. (EDL)
extensor hallucis l. (EHL)
extensor pollicis l. (EPL)
flexor digitorum l. (FDL)
groove for tendon of fibularis l.
groove for tendon of flexor hallucis l.
groove for tendon of peroneus l.
musculus abductor pollicis l.
musculus adductor l.
musculus extensor carpi radialis l.
musculus extensor digitorum l.
musculus extensor hallucis l.
musculus extensor pollicis l.
musculus fibularis l.
musculus flexor digitorum l.
musculus flexor hallucis l.
musculus flexor pollicis l.
musculus palmaris l.
musculus peroneus l.
musculus supinator l.
nervus ciliaris l.
nervus thoracicus l.
palmaris l.
peroneus l.
pollicis l.
thoracicus l.
loop
afferent l.
l. of bowel
bowel l.
bulboventricular l.
capillary l.
cervical l.
efferent l.
gamma l.
Gerdy interatrial l.
Granit l.
Henle l.
Hyrtl l.
ileal l.
inferior root of cervical l.
intestinal l.
lenticular l.
memory l.
nephronic l.

L

NOTES

loop *(continued)*
    peduncular l.
    puborectalis l.
    sigmoid l.
    l.'s of spinal nerve
    subclavian l.
    superior root of cervical l.
    ventricular l.
    Vieussens l.
    l. of Vieussens
Looser
    L. line
    L. zone
lop-ear, lop ear
lordosis
    cervical l.
    l. cervicis
    l. colli
    l. lumbalis
    lumbar l.
lordotic curve
Louis
    angle of L.
    L. angle
Lovén reflex
low
    l. back
    l. flow principle
Löwenberg
    L. canal
    L. scala
lower
    l. airway
    l. alveolar point
    l. dental arcade
    l. esophageal sphincter (LES)
    l. extremity
    l. eyelid
    l. jaw
    l. lateral cutaneous nerve of arm
    l. lid
    l. limb
    l. lip
    l. lobe of lung
    l. motor neuron
    l. pole
    l. pole of testis
    L. ring
    L. tubercle
lowest
    l. lumbar artery
    l. splanchnic nerve
    l. thyroid artery
LPS
    levator palpebrae superioris
LR
    lateral rectus
LSK
    liver, spleen, and kidneys

Lucas groove
lucida
    lamina l.
lucidum
    septum l.
    stratum l.
lucifugal
lucipetal
Lückenschädel
Ludwig
    angle of L.
    L. angle
    depressor nerve of L.
    L. ganglion
    L. labyrinth
    L. nerve
    L. stromuhr
LUL
    left upper lobe
luliberin
lumbale
    tetragon l.
    trigonum l.
lumbalia
    ganglia l.
lumbalis, pl. **lumbales**
    arteria l.
    l. ascendens
    cisterna l.
    costa l.
    l. ima
    intumescentia l.
    lordosis l.
    nervi splanchnici lumbales
    pars l.
    plexus lymphaticus l.
    processus accessorius vertebrae l.
    processus mammillaris vertebrae l.
    ramus dorsalis arteriae l.
    regio l.
    segmentum medullae spinalis l.
    splanchnici lumbales
    trunci lymphatici lumbales
    venae lumbales
    vertebrae lumbales
lumbalium
    plexus nervorum l.
    rami anteriores nervorum l.
    rami ventrales nervorum l.
lumbar
    l. branch of iliolumbar artery
    l. cistern
    l. enlargement of spinal cord
    l. flexure
    l. ganglia
    l. iliocostal muscle
    l. interspinal muscle
    lateral l.
    l. lordosis

l. lordotic curve
l. lymphatic plexus
l. lymph node
l. musculature
l. nervous plexus
l. part
l. part of diaphragm
l. quadrate muscle
l. reflex
l. region
l. rib
l. rotator muscle
l. segments L1–L5 of spinal cord
l. segments of spinal cord L1–5
l. spine
l. splanchnic nerve
l. suture
l. triangle
l. trunks
l. vein
l. vertebrae
lumbarization
lumbi (*pl. of* lumbus)
lumboabdominal
lumbocostal
l. ligament
l. triangle of diaphragm
lumbocostale
ligamentum l.
trigonum l.
lumbocostoabdominal triangle
lumbodorsal fascia
lumboiliac
lumboinguinal nerve
lumbo-ovarian
lumborum
fascia musculi quadrati l.
iliocostalis l.
musculi intertransversarii laterales l.
musculi intertransversarii mediales l.
musculi rotatores l.
musculus iliocostalis l.
musculus interspinalis l.
musculus quadratus l.
pars thoracica muscularis
  iliocostalis l.
pars ventralis musculi
  intertransversarii lateralium l.
partes dorsales musculorum
  intertransversariorum lateralium l.
quadratus l.

lumbosacral
l. angle
l. curve
l. enlargement
l. enlargement of spinal cord
l. joint
l. nerve trunk
l. nervous plexus
l. spine
lumbosacralis
articulatio l.
intumescentia l.
junctura l.
plexus nervosus l.
truncus l.
lumbrical
l. muscle
l. muscle of foot
l. muscle of hand
lumbricales
l. manus
l. pedis
lumbricalis
lumbricals
l. of foot
l. of hand
l. (lumbrical muscles) of foot
l. (lumbrical muscles) of hand
lumbricoid
lumbricus
lumbus, pl. **lumbi**
lumen, pl. **lumina, lumens**
l. of appendix
bile duct l.
bowel l.
l. of bowel
bronchial l.
l. of bronchial artery
cystic duct l.
duct l.
duodenal l.
esophageal l.
gastroduodenal l.
l. of gut
intestinal l.
residual l.
vaginal l.
l. of vein
luminal
luminalis
lunare
lunar periodicity

L

NOTES

lunata
> plica l.

lunate
> l. bone
> l. fissure
> l. sulcus
> l. surface of acetabulum

lunatum
> os l.

lung
> anterior border of l.
> aortic impression of left l.
> apex of l.
> apical segmental artery of superior lobar artery of right l.
> azygos lobe of right l.
> base of l.
> breathing l.
> l. bud
> cardiac impression of l.
> cardiac notch of l.
> costal surface of l.
> descending branch of anterior segmental artery of left and right l.
> descending branch of posterior segmental artery of left and right l.
> diaphragmatic surface of l.
> dynamic compliance of l.
> fissures of l.
> hilum of l.
> l. hilus
> horizontal fissure of right l.
> human l.
> inferior anterior segment of l.
> inferior border of l.
> inferior lateral basal segment of l.
> inferior lobe of left/right l.
> inferior medial basal segment of l.
> infundibulum of l.
> interlobar surfaces of l.
> l. lingula
> lingula of left l.
> lobe of l.
> lower lobe of l.
> main lobe of l.
> medial surface of l.
> mediastinal part of l.
> mediastinal surface of l.
> mesentery of l.
> middle lobar artery of right l.
> middle lobe of right l.
> oblique fissure of l.
> l. parenchyma
> pedicle of l.
> posterior basal segment of l.
> posterior basal segmental artery of left/right l.

> root of l.
> subclavian sulcus of l.
> superior inferior segment of l.
> superior lobe of right/left l.
> superior segment of l.
> l. tissue
> l. unit
> upper lobe of l.
> vertebral part of costal surface of the l.

lunula, pl. **lunulae**
> l. of semilunar cusps of aortic/pulmonary valve
> l. unguis
> l. valvulae semilunaris
> lunulae valvularum semilunarium valvae aortae/trunci pulmonalis

lunule of nail

lupus
> l. erythematosus cell
> l. lymphaticus

LUQ
> left upper quadrant

lura

lural

Luschka
> L. bursa
> crypts of L.
> L. cystic gland
> L. ducts
> foramen of L.
> L. joint
> L. laryngeal cartilage
> L. ligament
> L. and Magendie foramen
> L. nerve
> L. sinus
> L. tonsil

Luse bodies

lusoria
> arteria l.

lusus naturae

lutea
> macula l.

luteae
> fovea centralis maculae l.

luteal
> l. cell
> l. phase

lutein cell

luteinization

luteinize

luteinizing
> l. hormone/follicle-stimulating hormone-releasing
> l. hormone/follicle-stimulating hormone-releasing factor
> l. hormone-releasing hormone

luteogenic

luteoplacental shift
luteotropic hormone
luteotropin
luteum
    corpus l.
    punctum l.
luteus
luxans
    coxa vara l.
Luys
    body of L.
    L. body
    centre médian de L.
    nucleus of L.
luysii
    corpus l.
lymph
    l. capillary
    l. cell
    l. channel
    l. circulation
    l. follicle
    l. gland
    l. node
    l. node of abdominal organ
    l. node of anterior border of
      omental foramen
    l. node of arch of azygos vein
    l. node around cardia of stomach
    l. node of azygos arch
    l. node of elbow
    l. node of head and neck
    l. node of ligamentum arteriosum
    l. node of lower limb
    l. node permeability factor (LNPF)
    l. node system
    l. node of upper limb
    l. nodule
    l. sac
    l. sinus
    l. space
    l. vessel
lymphadenocele
lymphadenoid tissue
lymphangial
lymphangiectodes
lymphangiology
lymphangioma
    l. capillare varicosum
    l. circumscriptum
    l. superficium simplex
lymphangion

lymphatic
    afferent l.
    l. duct
    efferent l.
    l. fistula
    l. follicle
    l. follicles of larynx
    l. follicles of rectum
    l. nodule
    l. plexus
    l. ring of cardia
    l. ring of cardiac part of stomach
    l. sinus
    l. stroma
    superficial l.
    l. system
    l. tissue
    l. valvule
    l. vessel
lymphatica
    valvula l.
    vasa l.
lymphatici
    cortex nodi l.
    hilum nodi l.
    medulla nodi l.
    nodi l.
    l. solitarii
lymphatics
lymphaticum
    systema l.
    vas l.
lymphaticus
    folliculus l.
    lupus l.
    nodulus l.
    nodus l.
    plexus l.
lymphatology
lymphization
lymphoblast
lymphocerastism
lymphocinesia
lymphocinesis
lymphocyte
    B l.
    T l.
lymphocytic series
lymphocytoblast
lymphocytopoiesis
lymphoduct
lymphogenesis

L

NOTES

lymphogenic
lymphogenous
lymphoglandula
lymphoid
    l. cell
    l. ring
    l. series
    l. system
    l. tissue
lymphoidei
    medulla nodi l.
    trabeculae nodi l.
lymphoideum
    systema l.
lymphoideus
    nodus l.
lymphoidocyte
lymphokinesis
lympholeukocyte
lymphology
lymphonodus, pl. **lymphonodi**
    l. abdominis viscerales
    l. anorectales
    l. aortici laterales
    l. appendiculares
    l. arcus vena azygos
    l. axillares
    l. brachiales
    l. bronchopulmonale
    l. cavales laterales
    l. cervicales anteriores
    l. cervicales anteriores profundi
    l. cervicales anteriores superficiales
    l. cervicales laterales superficiales
    l. coeliaci
    l. colici dextri
    l. colici medii
    l. colici sinistri
    l. comitantes nervi accessorii
    l. cubitales
    l. epigastrici inferiores
    l. faciales
    l. gastrici dextri
    l. gastrici sinistri
    l. gastroomentales dextri
    l. gastroomentales sinistri
    l. gluteales
    l. gluteales inferiores
    l. gluteales superiores
    l. hepatici
    hilum lymphonodi
    l. ileocolici
    l. iliaci communes
    l. iliaci communes intermedii
    l. iliaci communes laterales
    l. iliaci communes mediales
    l. iliaci communes promontorii
    l. iliaci communes subaortici
    l. iliaci externi

l. iliaci externi intermedii
l. iliaci externi laterales
l. iliaci externi mediales
l. iliaci interni
l. inguinales inferiores
l. inguinales superficiales
l. inguinales superolaterales
l. inguinales superomediales
l. intercostales
l. interiliaci
l. interpectorales
l. jugulares anteriores
l. jugulares laterales
l. juxta-esophageales pulmonales
l. juxtaintestinales
l. lienales
l. lumbales dextri
l. lumbales intermedii
l. lumbales sinistri
l. mastoidei
l. mediastinales anteriores
l. mediastinales posteriores
l. mesenterici
l. mesocolici
l. obturatorii
l. occipitales
l. pancreatici
l. pancreatici inferiores
l. pancreatici superiores
l. pancreaticoduodenales
l. paracolici
l. pararectales
l. parasternales
l. paratracheales
l. parauterini
l. paravaginales
l. paravesiculares
l. parotidei intraglandulares
l. parotidei profundi
l. parotidei subfasciales
    praeauriculares
l. parotidei superficiales
l. pericardiales laterales
l. phrenici inferiores
l. phrenici superiores
l. popliteales
l. postaortici
l. postcavales
l. postvesiculares
l. preaortici
l. precavales
l. prececales
l. prelaryngeales
l. prepericardiales
l. pretracheales
l. prevertebrales
l. prevesiculares
l. promontorii
l. pulmonales

l. pylorici
l. retrocecales
l. retropharyngeales
l. sacrales
l. sigmoidei
l. splenici
l. subaortici
l. submandibulares
l. submentales
l. superiores centrales
l. supraclaviculares
l. thyroidei
l. tracheobronchiales inferiores
l. tracheobronchiales superiores
l. vesicales laterales

lymphopoiesis
lymphopoietic
lymphostasis
lymphotaxis
lymphotrophy
lyra
    l. davidis
    l. uterina
lyre of David
lysinemia
lysosome
    definitive l.'s
    primary l.'s
    secondary l.'s

NOTES

L

M

    M band
    M line

m

    mass
    meter
    molar
    moles (per liter)

M2 segment of middle cerebral artery
Macewen triangle
Mach number
Mackenrodt ligament
macrencephalia
macrencephaly
macrobiosis
macrobiotic
macrobiotics
macroblepharia
macrobrachia
macrocardia
macrocephalia
macrocephalic
macrocephalous
macrocephaly
macrocheiria, macrochiria
macrocnemia
macrocornea
macrocranium
macrodactylia
macrodactylism
macrodont
macrodontia
macrodontism
macroencephalon
macrogastria
macroglia cell
macroglossia
macrognathia
macrogyria
macromastia
macromazia
macromelia
macromere
macronucleus
macronychia
macropenis
macrophage
    alveolar m.
    m. colony-stimulating factor (M-CSF)
    fixed m.
    free m.
    system of m.'s
macrophagocyte
macrophallus

macrophthalmia
macropodia
macroprosopia
macroprosopous
macrorhinia
macroscelia
macroscopic
    m. anatomy
    m. sphincter
macrosigmoid
macrosis
macrosomia
macrosplanchnic
macrostomia
macrotia
macrotome
macula, pl. maculae
    maculae acusticae
    m. adherens
    m. communicans
    m. communis
    m. corneae
    m. cribrosa
    m. cribrosa inferior
    m. cribrosa media
    m. cribrosa quarta
    m. cribrosa superior
    m. densa
    m. flava
    m. germinativa
    m. lutea
    mongolian m.
    neuroepithelium of m.
    m. pellucida
    m. of retina
    m. retinae
    m. of saccule
    m. sacculi
    m. of utricle
    maculae of utricle and saccule
    m. utriculi
    maculae utriculosaccularis
macular artery
macularis
    fasciculus m.
macule
Magendie
    M. foramen
    M. space
magenstrasse
magma reticulare
magna, pl. magnae
    anastomotica m.
    arteria anastomotica auricularis m.
    arteria pancreatica m.

M

magna *(continued)*
  arteria radicularis m.
  cerebri m.
  chorda m.
  cisterna m.
  cordis m.
  coxa m.
  lacuna m.
  saphena m.
  vena cardiaca m.
  vena cerebri m.
  vena cordis m.
  vena saphena m.
  vertebra m.
magnet
  m. reaction
  m. reflex
magnetic
  m. field
  m. inertia
magni
  ramus posterior nervi auricularis m.
magnification
magnitude
magnocellular
magnum
  foramen m.
  os m.
magnus
  adductor m.
  auricularis m.
  musculus adductor m.
  musculus serratus m.
  nervus auricularis m.
Mahaim
  M. bundle in heart
  M. fiber
maidenhead
Maier sinus
main
  m. fourché
  m. lobe of lung
  m. pancreatic duct (MPD)
  m. pulmonary artery
  m. renal vein
  m. sensory nucleus
  m. stem bronchus
Maissiat band
major
  m. alar cartilage
  anulus iridis m.
  m. arterial circle of iris
  arteria palatina m.
  bursa of teres m.
  m. calices
  camera oculi m.
  canalis palatinus m.
  cartilago alaris m.
  circulus arteriosus iridis m.

  m. circulus arteriosus of iris
  crista tuberculi m.
  curvatura ventriculi m.
  ductus sublingualis m.
  m. duodenal papilla
  forceps m.
  fossa scarpae m.
  fossa supraclavicularis m.
  glandula vestibularis m.
  globus m.
  helicis m.
  hippocampus m.
  incisura ischiadica m.
  justo m.
  musculus helicis m.
  musculus pectoralis m.
  musculus psoas m.
  musculus rectus capitis anticus m.
  musculus rectus capitis posterior m.
  musculus rectus capitis posticus m.
  musculus rhomboideus m.
  musculus teres m.
  musculus zygomaticus m.
  nervus occipitalis m.
  nervus palatinus m.
  nervus petrosus m.
  nervus splanchnicus m.
  occipitalis m.
  palatina m.
  palatinus m.
  papilla duodeni m.
  pectoralis m.
  pelvis justo m.
  petrosus m.
  psoas m.
  rectus capitis posterior m.
  rhomboideus m.
  m. salivary gland
  spina tympanica m.
  splanchnicus m.
  subcutaneous bursa of teres m.
  m. sublingual duct
  sulcus palatinus m.
  teres m.
  trochanter m.
  vena azygos m.
  zygomaticus m.
majora
  labium m.
majoris, pl. **majores**
  bursa subtendinea musculi
    teretis m.
  calices renales majores
  crista tuberculi m.
  crus laterale cartilaginis alaris m.
  crus mediale cartilaginis alaris m.
  foramen bursae omentalis m.
  glandulae salivariae majores
  hiatus canalis nervi petrosi m.

margo zygomaticus alae m.
pars abdominalis musculi
  pectorales m.
pars clavicularis musculi
  pectoralis m.
pars sternocostalis musculi
  pectoralis m.
rami nasales posteriores inferiores
  nervi palatini m.
sulcus nervi petrosi m.

majus

cornu m.
foramen ischiadicum m.
foramen palatinum m.
labium m.
multangulum m.
omentum m.
os multangulum m.

Makeham hypothesis
mala
malabsorption
Malacarne

M. pyramid
M. space

malar

m. arch
m. area
m. bone
m. eminence
m. fold
m. foramen
m. lymph node
m. point
m. process

malare

os m.

malaris

nodus lymphoideus m.

malassimilation
maldigestion
male

m. breast
m. external genitalia
external urethral sphincter of m.
m. gonad
m. hermaphroditism
m. internal genitalia
m. pronucleus
m. pseudohermaphroditism
pubovesical ligament of m.
m. reproductive organ
m. urethra

urethral crest of m.
urethral gland of m.

malformation

Arnold-Chiari m.

malfunction
Malgaigne

M. fossa
M. triangle

malignant down
Mall

M. formula
periportal space of M.
M. ridge

mallear

m. fold
m. ligament
m. process
m. prominence
m. stripe

mallearis

plica m.
prominentia m.
stria m.

mallei (*pl. of* malleus)
malleoincudal joint
malleolar

m. artery
m. articular surface of fibula
m. articular surface of tibia
m. fold
m. groove
m. plexus
m. stria
m. sulcus

malleolaris

m. anterior medialis
sulcus m.

malleolus, pl. **malleoli**

articular facet of lateral m.
articular facet of medial m.
external m.
fossa of lateral m.
inner m.
internal m.
lateral head of m.
m. lateralis
manubrium of m.
medial m.
m. medialis
m. muscle
outer m.

M

NOTES

malleolus (*continued*)
    subcutaneous bursa of lateral m.
    subcutaneous bursa of medial m.
malleus, pl. **mallei**
    anterior ligament of m.
    anterior process of m.
    axis ligament of m.
    m. bone
    caput mallei
    collum mallei
    external ligament of m.
    head of m.
    lateral ligament of m.
    lateral process of m.
    ligament of m.
    long process of m.
    manubrium of m.
    manubrium mallei
    neck of m.
    processus anterior mallei
    processus lateralis mallei
    short process of m.
    slender process of m.
    superior ligament of m.
    tuberculum mallei
malomaxillary suture
malpighian
    m. body
    m. capsule
    m. cell
    m. corpuscles
    m. gland
    m. glomerulus
    m. layer
    m. nodule
    m. pyramid
    m. rete
    m. stigma
    m. stratum
    m. tuft
    m. vesicle
Malpighi vesicle
malposition
malrotation
malum
mamelonated
mamelonation
mamilla, pl. **mamillae**
mamillare
    corpus m.
mamillaris
    linea m.
    nuclei corporis m.
    pedunculus corporis m.
    processus m.
mamillary
    m. body
    m. duct

    m. line
    m. process
    m. tubercle
    m. tubercle of hypothalamus
mamillate
mamillated
mamillation
mamilliform
mamillotegmental fasciculus
mamillotegmentalis
    fasciculus m.
mamillothalamic
    m. fasciculus
    m. tract
mamillothalamicus
    fasciculus m.
mamma, pl. **mammae**
    m. accessoria
    areola mammae
    corpus mammae
    m. erratic
    m. erratica
    ligamenta suspensoria mammae
    m. masculina
    papilla mammae
    retinaculum cutis mammae
    supernumerary m.
    m. virilis
mammaria, pl. **mammariae**
    glandula m.
    ligamenta suspensoria m.
    lobi glandulae mammariae
    lobuli glandulae mammariae
    regio m.
mammarii
    rami m.
mammarius
    plexus m.
mammary
    m. artery
    m. body
    m. branch
    m. duct
    m. fold
    m. gland
    m. line
    perforating arteries of internal m.
    m. plexus
    m. region
    m. ridge
    m. vein
    m. vessel
mammiform
mammilla
mammillare, pl. **mammillares**
    arteriae mammillares
mammillaris
    linea m.

mammillary
    m. artery
    m. process of lumbar vertebra
mammillate
mammillation
mammilliform
mammose
mammosomatotroph
mammotroph
mammotropic, mammotrophic
    m. hormone
manchette
mandible
    alveolar arch of m.
    alveolar part of m.
    angle of m.
    base of m.
    body of m.
    condylar process of m.
    condyle of m.
    coronoid process of m.
    head of m.
    interradicular septa of maxilla
     and m.
    lingula of m.
    mental tubercle of m.
    neck of m.
    oblique line of m.
    pterygoid tuberosity of m.
    ramus of m.
    symphysis of m.
    temporal crest of m.
    yoke of m.
mandibula, pl. **mandibulae**
    angulus mandibulae
    arcus alveolaris mandibulae
    basis mandibulae
    biventer mandibulae
    canalis mandibulae
    caput mandibulae
    collum mandibulae
    corpus mandibulae
    crista temporalis mandibulae
    foramen mandibulae
    incisura mandibulae
    linea obliqua mandibulae
    lingula mandibulae
    musculus biventer mandibulae
    pars alveolaris mandibulae
    processus condylaris mandibulae
    processus coronoideus mandibulae
    ramus marginalis mandibulae

ramus meningeus nervi m.
symphysis mandibulae
tuberculum mentale mandibulae
tuberositas pterygoidea mandibulae
mandibular
    m. arch
    m. artery
    m. articulation
    m. axis
    m. bone
    m. canal
    m. cartilage
    m. condyle
    m. dental arcade
    m. dentition
    m. dentition odontectomy
    m. disk
    m. foramen
    m. fossa
    m. joint
    m. lymph node
    m. nerve
    m. notch
    m. process
    m. ramus
    m. reflex
    m. ridge
    m. symphysis
    m. tongue
    m. tooth
mandibularis
    arcus dentalis m.
    articulatio m.
    fossa m.
    nervus m.
    nodus lymphoideus m.
    ramus communicans ganglii otici
     cum ramo meningeo nervi m.
    ramus meningeus nervi m.
mandibuloacral dysostosis
mandibulofacial
    m. dysostosis
    m. dysplasia
mandibulooculofacial dysmorphia
mandibulo-oculofacialis
    dyscephalia m.-o.
mandibulopharyngeal
mandibulum
maneuver
    Müller m.
    Valsalva m.

M

NOTES

manifesta
    spina bifida m.
maniphalanx
Mann-Bollman fistula
mannose-6-phosphate receptor (MPR)
manometer
    aneroid m.
    dial m.
    differential m.
    mercurial m.
manometric
manometry
mantle
    brain m.
    m. layer
    myoepicardial m.
    m. zone
manubria (*pl. of* manubrium)
manubriogladiolar junction
manubriosternal
    m. joint
    m. junction
    m. symphysis
manubriosternalis
    symphysis m.
    synchondrosis m.
manubrium, pl. **manubria**
    jugular notch of m.
    m. mallei
    m. of malleolus
    m. of malleus
    median section of m.
    m. sterni
    m. of sternum
manus, pl. **manus**
    abductor digiti minimi quinti m.
    articulationes interphalangeae m.
    basis phalangis m.
    digitales dorsales m.
    digiti m.
    digitus m.
    dorsum m.
    facies lateralis digiti m.
    fascia dorsalis m.
    interossei dorsales m.
    ligamenta palmaria articulationis
      interphalangeae m.
    lumbricales m.
    musculi interossei dorsales m.
    musculi interossei dorsalis m.
    musculi lumbricales m.
    musculus abductor digiti minimi m.
    musculus extensor digitorum
      brevis m.
    musculus flexor digiti minimi
      brevis m.
    musculus interosseus dorsalis m.
    musculus lumbricalis m.
    opponens digiti minimi m.

    os naviculare m.
    palma m.
    phalanx distalis m.
    phalanx media pedis et m.
    phalanx proximalis m.
    pulpa digiti m.
    rete venosum dorsale m.
    torus m.
    vagina communis tendinum
      musculorum flexorum m.
    vaginae fibrosae digitorum m.
    vaginae synoviales digitorum m.
    vinculum breve digitorum m.
    vinculum longum digitorum m.
Marcacci muscle
Marchand wandering cell
Marchant zone
Marchi tract
Marcille triangle
Marcus
    M. Gunn phenomenon
    M. Gunn pupil (MGP)
Marey law
margin
    acetabular m.
    m. of acetabulum
    anterior palpebral m.
    articular m.
    ciliary m.
    corneal m.
    costal m.
    costochondral m.
    m. of eyelid
    falciform m.
    m. of fossa ovalis
    free m.
    frontal m.
    gingival m.
    incisal m.
    inferior m.
    inferolateral m.
    infraorbital m.
    intercostal m.
    interosseous m.
    lateral m.
    lid m.
    medial m.
    mesovarian m.
    occipital m.
    m. of orbit
    orbital m.
    palpebral m.
    parietal m.
    m. of piriform aperture
    posterior palpebral m.
    squamosal m.
    squamous m.
    superior m.
    superomedial m.

supraorbital m.
m. of tongue
vermilion m.

marginal

m. arcade
m. artery of colon
m. atrial branch of right coronary artery
m. branch of cingulate sulcus
m. branch of parietooccipital sulcus
m. crest
m. crest of tooth
m. gyrus
m. layer
m. mandibular branch of facial nerve
m. part
m. part of orbicularis oris muscle
m. ridge
m. sinus
m. sinus of placenta
m. sphincter
m. tentorial branch of internal carotid artery
m. tubercle
m. tubercle of zygomatic bone
m. zone

marginalis

crista m.
fasciculus m.
gyrus m.
pars m.
placenta previa m.
ramus tentorii m.
sulcus m.

marginata

placenta m.

margination of placenta

marginis

margo, pl. **margines**

m. acetabularis
m. anterior
m. anterior corporis pancreatis
m. anterior fibulae
m. anterior pulmonis
m. anterior radii
m. anterior testis
m. anterior tibiae
m. anterior ulnae
m. arcuatus hiatus sapheni
m. ciliaris iridis
m. dexter cordis

m. falciformis hiatus sapheni
m. fibularis pedis
m. frontalis ossis parietalis
m. frontalis ossis sphenoidalis
m. incisalis
m. inferior
m. inferior cerebri
m. inferior corporis pancreatis
m. inferior corporis splenis
m. inferior hepatis
m. inferior lienis
m. inferior pancreatis
m. inferior pulmonis
m. inferior splenis
m. infraorbitalis
m. interosseus
m. interosseus fibulae
m. interosseus radii
m. interosseus tibiae
m. interosseus ulnae
m. lacrimalis maxillae
m. lambdoideus ossis occipitalis
m. lambdoideus squamae occipitalis
m. lateralis
m. lateralis antebrachii
m. lateralis humeri
m. lateralis pedis
m. lateralis renis
m. lateralis scapulae
m. lateralis unguis
m. liber
m. liber ovarii
m. liber unguis
m. linguae
m. mastoideus ossis occipitalis
m. mastoideus squamae occipitalis
m. medialis
m. medialis antebrachii
m. medialis cerebri
m. medialis glandulae suprarenalis
m. medialis humeri
m. medialis pedis
m. medialis renis
m. medialis scapulae
m. medialis tibiae
m. mesovaricus
m. mesovaricus ovarii
m. nasalis ossis frontalis
m. occipitalis ossis parietalis
m. occipitalis ossis temporalis
m. occultus unguis
m. orbitalis

M

NOTES

margo (*continued*)
    m. palpebrae
    m. parietalis alaris majoris ossis sphenoidalis
    m. parietalis ossis frontalis
    m. parietalis partis squamosae ossis temporalis
    m. posterior fibulae
    m. posterior partis petrosae ossis temporalis
    m. posterior radii
    m. posterior testis
    m. posterior ulnae
    m. pupillaris iridis
    m. radialis antebrachii
    m. sagittalis ossis parietalis
    m. sphenoidalis ossis temporalis
    m. squamosus
    m. squamosus alaris majoris ossis sphenoidalis
    m. squamosus ossis parietalis
    m. superior
    m. superior corporis pancreatis
    m. superior glandulae suprarenalis
    m. superior hemispherii cerebri
    m. superior lienis
    m. superior pancreatis
    m. superior partis petrosae ossis temporalis
    m. superior scapulae
    m. superior splenis
    m. supraorbitalis
    m. tibialis pedis
    m. ulnaris antebrachii
    m. uterus
    m. zygomaticus
    m. zygomaticus alae majoris
    m. zygomaticus alaris majoris ossis sphenoidalis

Marin Amat syndrome
Mariotte
    M. blind spot
    M. law
mark
marker
    time m.
marmoratus
    status m.
marrow
    bone m.
    m. canal
    m. cell
    gelatinous bone m.
    red bone m.
    spinal m.
    yellow bone m.
marrow-lymph gland

Marshall
    M. oblique vein
    M. vestigial fold
marsupial notch
marsupium
Martegiani
    M. area
    M. funnel
Martinotti cell
maschale
masculina
    mamma m.
    urethra m.
    vagina m.
masculinae
    crista urethralis m.
    glandulae urethrales m.
    musculus sphincter urethrae externus m.
    pars intermedia urethrae m.
    pars intramuralis urethrae m.
    pars membranacea urethrae m.
    pars preprostatica urethrae m.
    pars spongiosa urethrae m.
    tunica muscularis partis intermediae urethrae m.
    tunica muscularis partis prostaticae urethrae m.
    tunica muscularis partis spongiosae urethrae m.
    tunica muscularis urethrae m.
    tunica urethrae m.
masculine
    m. pelvis
    m. uterus
masculinum
    ligamentum pubovesicale m.
    ovarium m.
masculinus
    uterus m.
mass (m)
    injection m.
    inner cell m.
    m. movement
    muscle m.
    m. peristalsis
massa, pl. **massae**
    m. intermedia
    m. lateralis atlantis
masseter
    m. muscle
    musculus m.
    m. reflex
    m. tendon
masseteri
    pars profunda musculi m.
    pars superficialis musculi m.
masseteric
    m. area

m. artery
m. fascia
m. nerve
m. reflex
m. space
m. tuberosity
m. vein

masseterica
arteria m.
fascia m.
tuberositas m.

massetericus
nervus m.

mast
m. cell
m. leukocyte

master
m. gland
M. two-step exercise test

masticating surface

mastication
force of m.
muscle of m.

masticator
m. nerve
m. space

masticatoria
facies m.

masticatorius
nucleus m.

masticatory
m. apparatus
m. force
m. movement
m. muscle
m. nucleus
m. surface
m. system

mastoccipital

mastocyte

mastocytogenesis

mastoid
m. air cell
m. angle of parietal bone
m. antrum
m. border of occipital bone
m. branch of occipital artery
m. branch of posterior auricular
artery
m. branch of posterior tympanic
artery
m. canaliculus

m. cortex
m. emissary vein
m. fontanelle
m. foramen
m. fossa
m. groove
m. lymph node
m. margin of occipital bone
m. notch
m. part
m. part of the temporal bone
m. process
m. process of petrous part of
temporal bone
m. sinuses
m. suture
m. wall of middle ear
m. wall of tympanic cavity

mastoidal

mastoidale

mastoidea, pl. **mastoideae**
cellulae mastoideae
emissaria m.
incisura m.
pars m.
vena emissaria m.

mastoideum
aditus ad antrum m.
antrum m.
emissarium m.
foramen m.
os m.
tegmen m.

mastoideus, pl. **mastoidei**
canaliculus m.
fonticulus m.
lymphonodi mastoidei
nodi lymphoidei mastoidei
processus m.
ramus m.

mastooccipital

mastoparietal

mastosquamous

mastosyrinx

mater
arachnoidea m.
arachnoid mater and pia m.
arterial branch to dura m.
cerebral part of dura m.
cranial arachnoid m.
cranial dura m.
dura m.

M

NOTES

mater *(continued)*
external filum terminale of
dura m.
external filum terminale pia m.
filum of spinal dura m.
internal filum terminale of pia m.
lateral lacunae of cranial dura m.
meningeal layer of dura m.
periosteal layer of dura m.
pia m.
sinuses of dura m.
spinal arachnoid m.
spinal dura m.

material
fibrocartilaginous m.

maternal
m. cotyledon
m. placenta

mating
matrical
matrices (*pl. of* matrix)
matricial
matris
sinus durae m.

matrix, pl. **matrices**
bone m.
cartilage m.
cell m.
crest of nail m.
cytoplasmic m.
m. Gla protein
m. metalloproteinase
mitochondrial m.
m. mitochondrialis
nail m.
territorial m.
m. unguis

matter
gray m.
periventricular white m.
pontine gray m.
white m.

maturation arrest
mature ovarian follicle
maturity
Mauchart ligament
Mauthner
M. cell
M. sheath

maxilla, pl. **maxillae**
alveolar arch of m.
alveolar canal of m.
alveolar foramina of m.
alveolar process of m.
anterior nasal spine of m.
anterior surface of m.
arcus alveolaris maxillae
body of m.
canales alveolares corporis maxillae

conchal crest of body of m.
corpus maxillae
crista conchalis corporis maxillae
crista ethmoidalis maxillae
crista nasalis processus palatini
maxillae
eminentia maxillae
ethmoidal crest of m.
facies anterior corporis maxillae
facies infratemporalis corporis
maxillae
facies nasalis maxillae
foramina alveolaria corporis
maxillae
frontal process of m.
inferior m.
infratemporal surface of body
of m.
lacrimal border of m.
lacrimal margin of m.
margo lacrimalis maxillae
nasal crest of palatine process
of m.
nasal notch of m.
nasal surface of m.
palatine process of m.
processus alveolaris maxillae
processus frontalis maxillae
processus palatinus ossis maxillae
processus zygomaticus maxillae
septa interradicularia mandibulae
et m.
spina nasalis anterior corporis
maxillae
superior m.
tuber maxillae
yoke of m.
zygomatic process of m.

maxillaris, pl. **maxillares**
arcus dentalis m.
arteria m.
hiatus m.
nervus m.
plexus periarterialis arteriae m.
processus m.
rami alveolares superiores
posteriores nervi m.
rami ganglionici nervi m.
rami nasales posteriores superiores
laterales nervi m.
rami nasales posteriores superiores
mediales nervi m.
rami orbitales nervi m.
rami pterygoidei arteriae m.
ramus meningeus medius nervi m.
ramus meningeus nervi m.
sinus m.
vena m.

maxillary
  m. antrum
  m. artery
  m. articulation
  m. crest
  m. dental arcade
  m. dentition
  m. eminence
  m. gland
  m. hiatus
  m. nerve
  m. plexus
  m. process
  m. process of embryo
  m. process of inferior nasal
    concha
  m. ridge
  m. sinus
  m. surface of greater wing of
    sphenoid bone
  m. surface of palatine bone
  m. tooth
  m. torus
  m. tubercle
  m. tuberosity
  m. vein
maxillodental
maxillofacial region
maxillojugal
maxillomandibular
maxillopalatine
maxilloturbinal
maximal stimulus
maximi
  bursa ischiadica musculi glutei m.
  bursa trochanterica musculi
    glutei m.
maximum
  m. breathing capacity (MBC)
  glucose transport m.
  m. occipital point
  transport m.
  tubular m.
  m. urea clearance
  m. velocity
  m. voluntary ventilation (MVV)
maximus
  gluteus m.
  musculus gluteus m.
  sciatic bursa of gluteus m.
  trochanteric bursa of gluteus m.

Mayo
  pyloric vein of M.
  M. vein
Mazzoni corpuscle
MBC
  maximum breathing capacity
MCA
  middle cerebral artery
McKee line
MCL
  medial collateral ligament
MCP
  metacarpophalangeal
McPhail test
M-CSF
  macrophage colony-stimulating factor
M'Dowel
  frenulum of M.
meal
  Boyden m.
mean calorie
meatal
  m. cartilage
  m. spine
meatus, pl. meatus
  acoustic m.
  m. acustici externi
  m. acusticus externus
  m. acusticus externus cartilagineus
  m. acusticus internus
  anterior nasal m.
  atrium of middle nasal m.
  bony part of external acoustic m.
  cartilage of acoustic m.
  cartilaginous part of external
    acoustic m.
  external acoustic m.
  external auditory m.
  external urinary m.
  fundus of internal acoustic m.
  fundus of internal auditory m.
  inferior m.
  internal acoustic m.
  internal auditory m.
  isthmus of external acoustic m.
  middle m.
  nasal m.
  m. nasi
  m. nasi inferior
  m. nasi medius
  m. nasi superior
  nasopharyngeal m.

M

NOTES

351

meatus *(continued)*
    m. nasopharyngeus
    nerve of external acoustic m.
    notch in cartilage of acoustic m.
    opening of external acoustic m.
    opening of internal acoustic m.
    opening of middle nasal m.
    orifice of external acoustic m.
    orifice of internal acoustic m.
    spina m.
    superior m.
    transverse crest of internal
      acoustic m.
    ureteral m.
    m. urinarius
    vertical crest of internal
      acoustic m.
mechanical mixture
mechanicoreceptor
mechanics
    body m.
mechanism
    countercurrent m.
    gating m.
    pressoreceptive m.
    proprioceptive m.
mechanocyte
mechanoreceptor
Meckel
    M. band
    M. cartilage
    M. cave
    M. cavity
    M. diverticulum
    M. ganglion
    M. ligament
    M. plane
    M. space
meconium
    m. ileus
    m. peritonitis
media
    arteria cerebri m.
    arteria colica m.
    arteria collateralis m.
    arteria genus m.
    arteria lobaris m.
    arteria meningea m.
    arteria rectalis m.
    arteria suprarenalis m.
    arteria temporalis m.
    auris m.
    cerebri m.
    colica m.
    collateralis m.
    concha nasalis m.
    cordis m.
    fossa cranii m.
    genus m.

    linea axillaris m.
    macula cribrosa m.
    meningea m.
    nodus rectalis m.
    phalanx m.
    plica umbilicalis m.
    rectalis m.
    scala m.
    suprarenalis m.
    temporalis m.
    tunica m.
    vena colica m.
    vena cordis m.
    vena temporalis m.
    vena thyroidea m.
media *(pl. of* medium)
mediad
mediae
    cellulae ethmoidales m.
    pars corticalis arteriae cerebralis m.
    pars insularis arteriae cerebri m.
    pars sphenoidalis arteriae
      cerebralis m.
    plexus periarterialis arteriae
      cerebri m.
    rami prostatici arteriae rectalis m.
    ramus accessorius arteriae
      meningeae m.
    ramus frontalis arteriae
      meningeae m.
    ramus meningeus accessorius
      arteriae meningeae m.
    ramus orbitalis arteriae
      meningeae m.
    ramus parietalis arteriae
      meningeae m.
    ramus petrosus arteriae
      meningeae m.
    ramus temporalis medius partis
      insularis arteriae cerebrae m.
    ramus temporalis posterior arteriae
      cerebri m.
    sinus ethmoidales m.
    sulcus arteriae temporalis m.
    thyroideae m.
    venae hemorrhoidales m.
    venae hepaticae m.
    venae meningeae m.
    venae rectales m.
medial
    m. accessory olivary nucleus
    m. angle
    m. angle of eye
    m. antebrachial cutaneous nerve
    m. anterior malleolus artery
    m. anterior thoracic nerve
    m. aperture
    m. arcuate ligament
    m. arteriole of retina

m. aspect
m. atrial vein
m. basal branch of pulmonary artery
m. basal bronchopulmonary segment S VII
m. basal segment
m. basal segmental artery
m. bicipital groove
m. border
m. border of foot
m. border of forearm
m. border of humerus
m. border of kidney
m. border of scapula
m. border of suprarenal gland
m. border of tibia
m. brachial cutaneous nerve
m. branch of artery of tuber cinereum
m. branch of pontine artery
m. branch of posterior branch of spinal nerve
m. branch of posterior rami of spinal nerve
m. bronchopulmonary segment S V
m. calcaneal branch of tibial nerve
m. calcaneocuboid
m. canthal ligament
m. canthic fold
m. canthus
m. capsule
m. cartilaginous layer
m. cartilaginous plate
m. central nucleus of thalamus
m. check ligament of eyeball
m. circumflex artery of thigh
m. circumflex femoral artery
m. circumflex femoral vein
m. cluneal nerve
m. collateral artery
m. collateral ligament (MCL)
m. collateral ligament of elbow
m. collateral ligament of knee
m. commisural artery
m. compartment
m. compartment of thigh
m. condyle
m. condyle of femur
m. condyle of humerus
m. condyle of tibia
m. cord of brachial plexus

m. crest of fibula
m. crural cutaneous branch of saphenous nerve
m. crus
m. crus of the horizontal part of the facial canal
m. crus of major alar cartilage of nose
m. crus of the superficial inguinal ring
m. cuneiform bone
m. cutaneous branch of dorsal branch of posterior intercostal artery
m. cutaneous nerve
m. cutaneous nerve of arm
m. cutaneous nerve of forearm
m. cutaneous nerve of leg
m. dorsal cutaneous nerve
m. eminence
m. epicondylar crest
m. epicondylar ridge
m. epicondyle
m. epicondyle of femur
m. epicondyle of humerus
m. femoral circumflex artery
m. femoral tuberosity
m. fillet
m. flexor muscle of forearm
m. foramen
m. forebrain bundle
m. frontobasal artery
m. genicular vein
m. geniculate body
m. great muscle
m. hamstring
m. hamulus
m. head
m. incisor tooth
m. inferior genicular artery
m. inguinal fossa
m. intermuscular septum
m. joint of ankle
m. joint level
m. lacunar lymph node
m. lacunar node
m. lamina of cartilage of pharyngotympanic auditory tube
m. lemniscus
m. ligament of ankle joint
m. ligament of elbow
m. ligament of knee

M

NOTES

medial (*continued*)

m. ligament of talocrural joint
m. ligament of temporomandibular joint
m. ligament of wrist
m. limb
m. lip of linea aspera
m. lobule
m. longitudinal arch
m. longitudinal bundle
m. longitudinal fasciculus
m. longitudinal stria
m. lumbar intertransversarii muscle
m. lumbar intertransverse muscle
m. lumbocostal arch
m. malleolar branch of posterior tibial artery
m. malleolar facet of talus
m. malleolar network
m. malleolar subcutaneous bursa
m. malleolar surface of talus
m. malleolus
m. malleolus of tibia
m. mammary branch
m. margin
m. medullary branch of vertebral artery
m. medullary lamina of corpus striatum
m. meniscus
m. midpalmar space
m. nasal branch of anterior ethmoidal nerve
m. nasal elevation
m. nasal fold
m. nasal process
m. nucleus of thalamus
m. occipital artery
m. occipitotemporal gyrus
m. orbitofrontal artery
m. palpebral artery
m. palpebral commissure
m. palpebral ligament
m. part
m. part of longitudinal arch of foot
m. part of middle lobe vein of right superior pulmonary vein
m. patellar retinaculum
m. pectoral nerve
m. plantar artery
m. plantar nerve
m. plantar vein
m. plateau of tibia
m. plate of cartilaginous auditory tube
m. plate of pterygoid process
m. pole
m. pole of ovary

m. popliteal nerve
m. posterior cervical intertransversarii muscle
m. process of calcaneal tuberosity
m. pterygoid muscle
m. pterygoid nerve
m. pterygoid plate
m. puboprostatic ligament
m. raphe
m. rectus muscle
m. retinaculum of patella
m. root of median nerve
m. root of olfactory tract
m. root of optic tract
m. rotator
m. segmental artery
m. striate artery
m. superior genicular artery
m. supraclavicular nerve
m. supracondylar crest
m. supracondylar ridge
m. supracondylar ridge of humerus
m. supraepicondylar ridge
m. sural cutaneous nerve
m. surface
m. surface of arytenoid cartilage
m. surface of fibula
m. surface of lung
m. surface of ovary
m. surface of testis
m. surface of tibia
m. surface of toe
m. surface of ulna
m. sympathetic line
m. talocalcaneal ligament
m. tarsal artery
m. thyroid ligament
m. tubercle of posterior process of talus
m. umbilical fold
m. umbilical ligament
m. vastus muscle
m. vein of lateral ventricle
m. venule of retina
m. vestibular nucleus
m. wall
m. wall of middle ear
m. wall of orbit
m. wall of tympanic cavity

mediale

caput m.
compartimentum femoris m.
corpus geniculatum m.
crus m.
ligamentum arcuatum m.
ligamentum collaterale m.
ligamentum palpebrale m.
ligamentum puboprostaticum m.
ligamentum talocalcaneare m.

ligamentum talocalcaneum m.
ligamentum umbilicale m.
os cuneiforme m.
rete malleolare m.
retinaculum patellae m.
segmentum basale m.
segmentum bronchopulmonale m.

medialecithal

mediali

ramus communicans ganglii otici cum nervo pterygoideo m.
ramus communicans nervi fibularis communis cum nervo cutaneo surae m.
ramus communicans nervi peronei communis cum nervo cutaneo surae m.

medialis, pl. **mediales**

angulus oculi m.
arcus lumbocostalis m.
m. arcus pedis longitudinalis
arcus pedis longitudinalis pars m.
arteria circumflexa femoris m.
arteriae malleolares posteriores mediales
arteriae palpebrales laterales et mediales
arteriae parietales laterales et mediales
arteria frontobasalis m.
arteria genus inferior m.
arteria genus superior m.
arteria malleolaris anterior m.
arteria occipitalis m.
arteria plantaris m.
arteria segmentalis basalis m.
arteria tarsea m.
bursa subcutanea malleoli m.
circumflexa femoris m.
commissura palpebrarum m.
condylus m.
crista sacralis m.
crista supracondylaris m.
crista supraepicondylaris m.
cutaneous dorsalis m.
cutaneus antebrachii m.
cutaneus brachii m.
cutaneus surae m.
digitales dorsales hallucis lateralis et digiti secunda m.
digitales plantares communes nervi plantaris m.

digitales plantares proprii nervi plantaris m.
eminentia m.
epicondylus m.
facies m.
fasciculus longitudinalis m.
fossa inguinalis m.
genus inferior m.
genus superior m.
gyrus occipitotemporalis m.
lamina cartilaginis m.
lemniscus m.
lymphonodus iliaci communes mediales
lymphonodus iliaci externi mediales
malleolaris anterior m.
malleolus m.
margo m.
meniscus m.
musculus pterygoideus m.
musculus rectus m.
musculus vastus m.
nervus cutaneus antebrachii m.
nervus cutaneus brachii m.
nervus cutaneus dorsalis m.
nervus cutaneus surae m.
nervus pectoralis m.
nervus plantaris m.
nervus supraclavicularis m.
nodi lymphatici iliaci communes mediales
nodi lymphatici iliaci externi mediales
nodus lacunaris m.
nodus lymphoideus lacunaris m.
nucleus corporis geniculati m.
nucleus olivaris accessorius m.
nucleus preopticus m.
palpebrales mediales
pars m.
plantaris m.
plica umbilicalis m.
pterygoideus m.
rami calcanei mediales
rami choroidei posteriores arteriae cerebri posteriores laterales et mediales
rami malleolares mediales
rami mammarii mediales
rami medullares mediales
rami nasales posteriores superiores mediales

M

NOTES

medialis *(continued)*
    rami temporales intermedii mediales
    ramus m.
    ramus basalis m.
    ramus calcarinus arteriae
      occipitalis m.
    ramus choroidei posteriores
      mediales
    ramus communicans cum nervo
      pterygoidei m.
    ramus cutaneus m.
    ramus descendens arteriae
      circumflexae femoris m.
    ramus orbitofrontalis m.
    ramus parietalis arteriae
      occipitalis m.
    ramus parietooccipitalis arteriae
      occipitalis m.
    ramus posterior nervi cutanei
      antebrachii m.
    ramus profundus arteriae
      circumflexae femoris m.
    ramus profundus arteriae
      plantaris m.
    ramus superficialis arteriae
      circumflexae femoris m.
    ramus superficialis arteriae
      plantaris m.
    ramus transversus arteriae
      circumflexae femoris m.
    ramus ulnaris nervi cutanei
      antebrachii m.
    rectus m.
    stria longitudinalis m.
    sulcus bicipitalis m.
    supraclaviculares mediales
    tarsea mediales
    vastus m.
    vena atrii m.
    venae circumflexae femoris
      mediales
    vena ventriculi lateralis m.

medially
median
    m. antebrachial vein
    m. aperture of fourth ventricle
    m. arcuate ligament
    m. atlantoaxial joint
    m. basilic vein
    m. callosal artery
    m. cephalic vein
    m. commissural artery
    m. conjugate
    m. cricothyroid ligament
    m. cubital vein
    m. eminence
    m. fissure
    m. frontal sulcus
    m. furrow of prostate

    m. glossoepiglottic fold
    m. groove of tongue
    m. line
    m. longitudinal raphe of tongue
    m. nerve
    m. palatine suture
    m. sacral artery
    m. sacral crest
    m. sacral vein
    m. sagittal plane
    m. section
    m. section of manubrium
    m. sulcus of fourth ventricle
    m. sulcus of tongue
    m. thyrohyoid ligament
    m. thyroid ligament
    m. tongue bud
    m. umbilical fold
    m. umbilical ligament
    m. vein of forearm
    m. vein of neck

mediana
    m. antebrachii
    arteria callosa m.
    arteria commissuralis m.
    arteria sacralis m.
    articulatio atlantoaxialis m.
    m. basilica
    m. cephalica
    crista sacralis m.
    m. cubiti
    diameter m.
    eminentia m.
    linea nuchae m.
    plica glossoepiglottica m.
    plica umbilicalis m.
    sacralis m.
    sutura palatina m.
    vena sacralis m.

medianae
    membrana tectoria articulationis
      atlantoaxialis m.
    rami sacrales laterales arteriae
      sacralis m.

mediani
    arteria comes nervi m.
    arteria comitans nervi m.
    digitales palmares communes
      nervi m.
    digitales palmares proprii nervi m.
    radix lateralis nervi m.
    radix medialis nervi m.
    rami musculares nervi m.
    ramus palmaris nervi m.

medianum
    centrum m.
    ligamentum arcuatum m.
    ligamentum thyrohyoideum m.

ligamentum umbilicale m.
planum m.
medianus
diameter m.
nervus m.
m. posterior medullae oblongatae
mediastinal
m. branch
m. branch of internal thoracic
artery
m. branch of thoracic aorta
m. part
m. part of lung
m. part of parietal pleura
m. shadow
m. space
m. surface of lung
m. vein
mediastinale
cavum m.
septum m.
mediastinalis, pl. **mediastinales**
pars m.
pleura m.
rami mediastinales
venae mediastinales
mediastinum
anterior m.
m. anterius
inferior m.
m. inferius
m. medium
middle m.
posterior m.
m. posterius
superior m.
m. superius
m. of testis
mediate
medical anatomy
medicephalic
medii
atrium meatus m.
bursae trochantericae musculi
glutei m.
lymphonodi colici m.
nervi clunium m.
nodi lymphoidei colici m.
pars ceratopharyngea musculi
constrictoris pharyngis m.
pars chondropharyngea musculi
constrictoris pharyngei m.

plexus rectales m.
rami terminales arteriae cerebri m.
ramus lobi m.
vena lobi m.
medioaxillaris
linea m.
mediocarpal
m. articulation
mediocarpalis
articulatio m.
medioccipital
medioclavicularis
linea m.
mediocolic sphincter
mediodens
mediodorsal nucleus
mediolateral
mediotarsal
mediotype
medium, pl. **media**
m. artery
foramen lacerum m.
ganglion cervicale m.
ligamentum hyothyroideum m.
ligamentum tibiofibulare m.
mediastinum m.
stratum griseum m.
m. vein
medius
arteriola maculae m.
bronchus lobaris m.
cardiacus cervicalis m.
constrictor pharyngis m.
digitus manus m.
gluteus m.
gyrus temporalis m.
lacertus m.
meatus nasi m.
musculus constrictor pharyngis m.
musculus gluteus m.
musculus scalenus m.
nervus cardiacus cervicalis m.
nodus lymphoideus rectalis m.
plexus lymphaticus sacralis m.
plexus nervosus rectalis m.
processus clinoideus m.
ramus alveolaris superior m.
scalenus m.
sulcus frontalis m.
sulcus temporalis m.

M

NOTES

medius *(continued)*
    trochanteric bursae of gluteus m.
    truncus m.
medulla, pl. **medullae**
    m. of adrenal gland
    m. glandulae suprarenalis
    m. of hair shaft
    inner stripes of renal m.
    inner zone of renal m.
    m. of kidney
    m. of lymph node
    m. nodi lymphatici
    m. nodi lymphoidei
    m. oblongata
    m. ossium
    m. ossium flava
    m. ossium rubra
    outer stripes of renal m.
    outer zone of renal m.
    posterior pyramid of m.
    renal m.
    m. renalis
    m. spinalis
    suprarenal m.
    m. of suprarenal gland
medullar
medullare
    cavum m.
    centrum m.
medullaris
    cavitas m.
    conus m.
    substantia m.
    tubus m.
medullary
    m. arteries of brain
    m. canal
    m. cavity
    m. cell
    m. center
    m. chemoreceptor
    m. cone
    m. cords
    m. fold
    m. groove
    m. layers of thalamus
    m. membrane
    m. plate
    m. pyramid
    m. ray
    m. segment
    m. sheath
    m. space
    m. spinal artery
    m. striae of fourth ventricle
    m. substance
    m. teniae
    m. tube
medullated nerve fiber

medullation
medullocell
medusae
    caput m.
Meeh-Dubois formula
Meeh formula
megacardia
megacaryoblast
megacaryocyte
megacephalia
megacephalic
megacephalous
megacephaly
megacolon
megadactylia
megadactylism
megadactyly
megadolichocolon
megadont
megadontism
megagnathia
megahertz
megakaryoblast
megakaryocyte
megalecithal
megaloblast
megalocardia
megalocephalia
megalocephaly
megalocheiria
megalocornea
megalocyte
megalodactylia
megalodactylism
megalodactyly
megalodontia
megaloencephalic
megaloencephalon
megaloencephaly
megaloenteron
megalogastria
megalokaryocyte
megalomelia
megalonychosis
megalopenis
megalophallus
megalophthalmus
    anterior m.
megalopodia
megalosplanchnic
megalosyndactylia
megalosyndactyly
megaloureter
megalourethra
meganucleus
megaprosopia
megaprosopous
megarectum
megaseme

megasigmoid
megasomia
megaureter
megaurethra
megophthalmus
Meibom gland
meibomian gland
meiosis
meiotic division
Meissner
    M. corpuscle
    M. ganglion
    M. plexus
melanin-pigmented cell
melanoblast
melanocyte
melanocyte-stimulating hormone
melanodendrocyte
melanophage
melanophore
melanosome
melanotroph
melomelia
meloschisis
melotia
Meltzer law
member
    inferior m.
    superior m.
    virile m.
membra (*pl. of* membrum)
membrana, pl. membranae
    m. abdominis
    m. adamantina
    m. adventitia
    m. atlantooccipitalis anterior
    m. atlantooccipitalis posterior
    m. basalis ductus semicircularis
    m. basilaris
    m. capsularis
    m. capsulopupillaris
    m. carnosa
    m. cerebri
    m. choriocapillaris
    m. cordis
    m. cricothyroidea
    m. decidua
    m. eboris
    m. fibroelastica laryngis
    m. fibrosa
    m. fibrosa capsulae articularis

m. flaccida
m. fusca
m. germinativa
m. granulosa
m. hyaloidea
m. hyothyroidea
membranae intercostales
membranae intercostalia
m. intercostalis externa
m. intercostalis interna
m. interossea antebrachii
m. interossea cruris
m. limitans
m. limitans gliae
m. mucosa
m. obturatoria
m. perinei
m. pituitosa
m. preformativa
m. propria ductus semicircularis
m. propria of semicircular duct
m. pupillaris
m. quadrangularis
m. reticularis
m. reticularis organi spiralis
m. serosa
m. serotina
m. stapedis
m. statoconiorum
m. sterni
m. striata
m. succingens
m. suprapleuralis
m. synovialis
m. tectoria
m. tectoria articulationis
  atlantoaxialis medianae
m. tectoria ductus cochlearis
m. tensa
m. thyrohyoidea
m. tympani
m. tympani secundaria
m. versicolor
m. vestibularis
m. vestibularis ductus cochlearis
m. vibrans
m. vitellina
m. vitrea
membranacea, pl. membranaceae
ampulla m.
lamina m.

M

NOTES

membranacea *(continued)*
    musculus sphincter urethrae
      membranaceae
    placenta m.
membranacei
    recessus utricularis labyrinthi m.
membranaceous
membranaceus
    labyrinthus m.
membranae (*pl. of* membrana)
membranate
membrane
    adamantine m.
    adventitious m.
    allantoic m.
    alveolocapillary m.
    alveolodental m.
    anal m.
    anterior atlantooccipital m.
    anterior recess of tympanic m.
    arachnoid m.
    atlantooccipital m.
    basement m.
    basilar m.
    Bichat m.
    Bogros serous m.
    m. bone
    Bowman m.
    m. of brain
    branch of auriculotemporal nerve
      to tympanic m.
    Bruch m.
    Brunn m.
    bucconasal m.
    buccopharyngeal m.
    cell m.
    m. of cervix uterus
    chorioallantoic m.
    chorion m.
    circular layer of tympanic m.
    cloacal m.
    Corti m.
    cricothyroid m.
    cricotracheal m.
    cricovocal m.
    cutaneous layer of tympanic m.
    cuticular m.
    cystic m.
    Debove m.
    deciduous m.
    Descemet m.
    m. of Descemet
    drum m.
    Duddell m.
    egg m.
    elastic m.
    enamel m.
    epipapillary m.
    exocelomic m.

    m. expansion theory
    external intercostal m.
    false m.
    fenestrated m.
    fertilization m.
    fetal m.
    fibrocartilaginous ring of
      tympanic m.
    fibrous m.
    Fielding m.
    flaccid part of tympanic m.
    germ m.
    germinal m.
    glassy m.
    glial m.
    glomerular basement m.
    Henle fenestrated elastic m.
    Heuser m.
    Hunter m.
    Huxley m.
    hyaline m.
    hyaloid m.
    hyoglossal m.
    intercostal m.'s
    intermuscular m.
    internal intercostal m.
    internal limiting m.
    interosseous m.
    ivory m.
    Jackson m.
    m. of joint
    keratogenous m.
    labyrinth m.
    limbus of tympanic m.
    medullary m.
    m. of meninges
    mucous m.
    Nasmyth m.
    nerve of tympanic m.
    Nitabuch m.
    nuclear m.
    obturator m.
    olfactory m.
    oronasal m.
    oropharyngeal m.
    otolithic m.
    outer limiting m.
    ovular m.
    Payr m.
    pericardiopleural m.
    peridental m.
    perineal m.
    periodontal m.
    periorbital m.
    peritoneal m.
    pharyngeal m.
    pial-glial m.
    pituitary m.
    placental m.

plasma m.
pleuropericardial m.
pleuroperitoneal m.
posterior atlantooccipital m.
posterior recess of tympanic m.
postsynaptic m.
m. potential
presynaptic m.
primary egg m.
proligerous m.
proper m.
pupillary m.
quadrangular m.
radiate layer of tympanic m.
Reissner m.
renal brush border m.
reticular m.
Rivinus m.
Ruysch m.
Scarpa m.
schneiderian m.
Schultze m.
Schwalbe m.
secondary egg m.
secondary tympanic m.
semipermeable m.
serous m.
Shrapnell m.
spiral m.
stapedial m.
statoconial m.
statoconic m.
sternal m.
striated m.
superior recess of tympanic m.
suprapleural m.
synovial m.
tectorial m.
tense part of tympanic m.
tensor muscle of tympanic m.
tertiary egg m.
thyrohyoid m.
Toldt m.
Tourtual m.
tympanic m.
m. of tympanum
umbo of tympanic m.
unit m.
urogenital m.
urorectal m.
uteroepichorial m.
vaginal synovial m.

vestibular m.
virginal m.
vitelline m.
vitreous m.
Wachendorf m.
yolk m.
Zinn m.

membrane-coating granule
membraniform
membranocartilaginous
membranoid
membranous
    m. ampulla
    m. ampullae of the semicircular duct
    m. canal
    m. cochlea
    m. labyrinth
    m. lamina of cartilage of pharyngotympanic auditory plate
    m. layer
    m. layer of subcutaneous tissue of abdomen
    m. layer of superficial fascia
    m. layer of superficial fascia of perineum
    m. limb of semicircular duct
    m. lining
    m. neurocranium
    m. ossification
    m. part of interventricular septum
    m. part of male urethra
    m. part of nasal septum
    m. tissue
    m. viscerocranium
    m. wall
    m. wall of middle ear
    m. wall of trachea
    m. wall of tympanic cavity
membri
    articulationes cinguli m.
membrum, pl. **membra**
    m. inferius
    m. muliebre
    m. superius
    m. virile
memory loop
Mendel dorsal reflex of foot
Meniere
    M. disease
    M. syndrome
meningea, pl. **meningeae**

M

NOTES

meningea *(continued)*
    m. anterior
    m. media
    m. posterior
    venae meningeae
meningeal
    m. branch
    m. branch of cavernous part of internal carotid artery
    m. branch of cerebral part of internal carotid artery
    m. branch of intracranial part of vertebral artery
    m. branch of mandibular nerve
    m. branch of maxillary nerve
    m. branch of occipital artery
    m. branch of ophthalmic nerve
    m. branch of spinal nerve
    m. branch of vagus nerve
    m. groove
    m. hernia
    m. layer of dura mater
    m. plexus
    m. vein
    m. vessel
meningei *(pl. of* meningeus)
meningeo
    ramus communicans cum ramo m.
meningeocortical
meninges *(pl. of* meninx)
meningeus, pl. **meningei**
    plexus m.
    rami meningei
    ramus m.
meningioma
    cutaneous m.
meningis
meningocele
meningocortical
meningocyte
meningoencephalocele
meningomyelocele
meningoradicular
meningorrhachidian
meningosis
meningovascular
meninx, pl. **meninges**
    m. fibrosa
    innermost membrane of meninges
    membrane of meninges
    m. primitiva
    m. serosa
    m. tenuis
    m. vasculosa
meniscal bone
menisci *(pl. of* meniscus)
meniscocyte
meniscofemorale
    ligamentum m.

meniscofemoralia
    ligamenta m.
meniscofemoral ligament
meniscus, pl. **menisci**
    m. of acromioclavicular joint
    articular m.
    m. articularis
    m. of knee
    lateral m.
    m. lateralis
    medial m.
    m. medialis
    tactile m.
    m. tactus
    m. of temporomaxillary
menstrual cycle
menstrualis
    decidua m.
mensual
mensuration
mental
    m. branch of inferior alveolar artery
    m. branch of m. nerve
    m. canal
    m. foramen
    m. point
    m. process
    m. protuberance
    m. region
    m. spine
    m. symphysis
    m. tubercle
    m. tubercle of mandible
mentale
    foramen m.
    tuberculum m.
mentalia
    ossicula m.
mentalis, pl. **mentales**
    arteria m.
    m. muscle
    musculus m.
    nervus m.
    protuberantia m.
    rami mentales
    rami labiales inferiores nervi m.
    rami labiales nervi m.
    rami mentales nervi m.
    regio m.
    spina m.
    symphysis m.
menti, pl. **mentum**
    musculus quadratus m.
    musculus transversus m.
    symphysis m.
    transversus m.
mentoanterior

mentolabial
> m. furrow
> m. sulcus

mentolabialis
> sulcus m.

Mercier
> M. bar
> M. valve

mercurial manometer

meridian
> m. of cornea
> m.'s of eyeball

meridiani bulbi oculi

meridianus

meridional
> m. cleavage
> m. fibers
> m. fibers of ciliary muscle

meridionales
> fibrae m.

meristic

Merkel
> M. corpuscle
> M. filtrum ventriculi
> M. fossa
> M. muscle
> M. tactile cell
> M. tactile disk

mermaid deformity

meroacrania

meroanencephaly

meroblastic cleavage

merocrine gland

merogenesis

merogenetic

merogenic

merogony

meromelia

meromicrosomia

Méry gland

mesad

mesal

mesameboid

mesangial cell

mesangium
> extraglomerular m.

mesaraic

mesaticephalic

mesatipellic
> m. pelvis

mesatipelvic

mesaxon

mesectic

mesectoderm

mesencephali
> lamina tecti m.
> tectum m.
> tegmentum m.

mesencephalic
> m. flexure
> m. nucleus of trigeminus
> m. tegmentum
> m. tract of trigeminal nerve
> m. vein

mesencephalicae
> venae m.

mesencephalon
> superior portion of m.

mesenchyma

mesenchymal
> m. cells
> m. epithelium
> m. tissue

mesenchyme
> interzonal m.
> synovial m.

mesenteriale
> intestinum tenue m.

mesenteric
> m. gland
> m. inferior artery
> m. lymph node
> m. portion of small intestine
> m. superior artery
> m. superior vein
> m. triangle

mesenterica
> m. inferior
> m. superior

mesenterici
> lymphonodi m.
> nodi lymphoidei m.

mesentericoparietal
> m. fossa
> m. recess

mesenterii
> radix m.

mesenteriolum processus vermiformis

mesenterium dorsale commune

mesentery
> appendiceal m.
> m. of appendix
> m. of cecum
> leaves of m.

M

NOTES

mesentery *(continued)*
    m. of lung
    root of m.
    m. of sigmoid colon
    m. of transverse colon
    urogenital m.
mesethmoid bone
meshwork
    trabecular m.
mesiad
mesial
    m. occlusion
    m. surface
    m. surface of tooth
mesoappendix
mesoarium
mesoblast
mesoblastema
mesoblastemic
mesoblastic
    m. segment
    m. sensibility
mesocardium, pl. **mesocardia**
    dorsal m.
    ventral m.
mesocarpal
mesocecal
mesocecum
mesocephalic
mesocephalous
mesocolic
    m. lymph node
    m. shelf
    m. tenia
mesocolica
    tenia m.
mesocolici
    lymphonodi m.
    nodi lymphoidei m.
mesocolon
    m. ascendens
    ascending m.
    m. descendens
    descending m.
    sigmoid m.
    m. sigmoideum
    transverse m.
    m. transversum
mesocord
mesocuneiform bone
mesoderm
    branchial m.
    extraembryonic m.
    gastral m.
    intermediate m.
    intraembryonic m.
    lateral plate m.
    paraxial m.
    primary m.

    prostomial m.
    secondary m.
    somatic m.
    somitic m.
    splanchnic m.
    visceral m.
mesodermal segment
mesodermic
mesoduodenal
mesoduodenum
mesoenteriolum
mesoepididymis
mesogaster
mesogastric
mesogastrium
    dorsal m.
    ventral m.
mesogenitale
mesoglia
mesoglial cells
mesogluteal
mesogluteus
mesognathic
mesognathion
mesognathous
mesoileum
mesojejunum
mesolateral fold
mesolobus
mesolymphocyte
mesomelia
mesomere
mesometrium
mesomorph
mesomorphic
mesonephric
    m. duct
    m. fold
    m. rest
    m. ridge
    m. tissue
    m. tubule
mesonephricus
    ductus m.
mesonephroi
mesonephros
    diaphragmatic ligament of m.
    glomerulus of m.
mesoontomorph
mesophragma
mesophryon
mesopneumonium
mesoprocton
mesoprosopic
mesopulmonum
mesorchial
mesorchium
mesorectum
mesorrhine

mesosalpinx
mesoscapulae
    delta m.
mesoseme
mesosigmoid
mesosigmoidopexy
mesosomatous
mesosomia
mesostenium
mesosternum
mesotarsal
mesotendineum
mesotendon
mesothelia
mesothelial cell
mesothelium
mesotropic
mesouranic
mesouterine fold
mesovaria
mesovarian
    m. border of ovary
    m. margin
    m. margin of ovary
mesovaricus
    margo m.
mesovarium
mesuranic
MET
    metabolic equivalent
metabolic
    m. acidosis
    m. alkalosis
    m. equivalent (MET)
    m. pool
metabolimeter
metabolin
metabolism
    basal m.
    carbohydrate m.
    electrolyte m.
    fat m.
    protein m.
    respiratory m.
metabolite
metabolize
metacarpal
    base of m.
    body of m.
    m. bone
    m. head
    head of m.

    m. index
    m. ligament
    shaft of m.
    m. vein
    m. of wrist
metacarpale
    corpus m.
    os m.
metacarpalia
    ligamenta m.
    ossa m. I–V
metacarpalis
    basis ossis m.
    caput ossis m.
    corpus ossis m.
    processus styloideus ossis m. III
metacarpeae dorsales
metacarpea palmares
metacarpi (*pl. of* metacarpus)
metacarpocarpal articulation
metacarpophalangeae
    articulationes m.
    ligamenta palmaria articulationis m.
metacarpophalangeal (MCP, MP)
    m. articulation
    m. joint
    m. of thumb
metacarpus, pl. **metacarpi**
    ossa metacarpi
    spatia interossea metacarpi
metachromasia
metachromatism
metachromophil
metachromophile
metachrosis
metafacial angle
metallophilia
metalloproteinase
    matrix m.
metameric nervous system
metamerism
metamorphosis, pl. **metamorphoses**
    retrograde m.
metamorphotic
metanephric
    m. bud
    m. cap
    m. diverticulum
    m. duct
    m. tubule
metanephrogenic tissue

M

NOTES

metanephrogenous
metanephros, pl. **metanephroi**
metaneutrophil, metaneutrophile
metaphysial, metaphyseal
    m. dysostosis
    m. dysplasia
metaphysis, pl. **metaphyses**
metaplasis
metaplasm
metaplastic
metaplexus
metapophysis
metapore
metarteriole
metasternum
metatarsal
    m. artery
    base of m.
    body of m.
    m. bone
    head of m.
    m. interosseous ligament
    shaft of m.
    tuberosity of fifth m.
    tuberosity of first m.
metatarsale
    corpus m.
    os m.
metatarsalia
    ligamenta m.
    ossa m. I–V
metatarsalis
    arteria m.
    basis ossis m.
    caput ossis m.
metatarsi (*pl. of* metatarsus)
metatarsophalangeae
    articulationes m.
    ligamenta plantaria articulationis m.
metatarsophalangeal (MP, MTP)
    m. articulation
    m. joint
metatarsus, pl. **metatarsi**
    m. abductus
    m. adductovarus
    m. adductus
    m. atavicus
    m. latus
    ossa metatarsi
    spatia interossea metatarsi
    m. varus
metatasophalangeal joint
metathalamus
metatypical
metencephalic
metencephalon
meter (m)
    ventilation m.
    Venturi m.

metergasia
meter-kilogram-second
    m.-k.-s. system
    m.-k.-s. unit
metestrum
metestrus
methacrylate resin
method
    closed circuit m.
    Fick m.
    Gräupner m.
    Hamilton-Stewart m.
    indicator dilution m.
    Kety-Schmidt m.
    Langendorff m.
    micro-Astrup m.
    microsphere m.
    open circuit m.
    Pavlov m.
    Stewart-Hamilton m.
methylation
methylenophil, methylenophile
methylenophilic
methylenophilous
methylmalonic
    m. acidemia
    m. aciduria
metopagus
metopic
    m. point
    m. suture
metopica
    sutura m.
metopion
metopism
metoposcopy
metra
metriocephalic
metrocyte
Mexican hat cell
Meyer
    M. cartilage
    M. sinus
Meyer-Overton theory of narcosis
Meynert
    M. cells
    M. commissures
    M. decussation
    M. fasciculus
    M. layer
    M. retroflex bundle
MGP
    Marcus Gunn pupil
mice (*pl. of* mouse)
Michaelis
    rhomboid of M.
micrencephalia
micrencephalous
micrencephaly

microanatomist
microanatomy
micro-Astrup method
microbiotic
microbody
microbrachia
microcardia
microcentrum
microcephalia
microcephalic
microcephalism
microcephalous
microcephaly
      encephaloclastic m.
      schizencephalic m.
microcheilia, microchilia
microcheiria, microchiria
microcinematography
microcirculation
microdactylia
microdactylous
microdactyly
microdissection
microdont
microdontia
microdontism
microelectrode
microencephaly
microfibril
microfilament
microgastria
microgenia
microgenitalism
microglia cells
microgliacyte
microglial cells
microglossia
micrognathia
micrograph
microgyria
microhepatia
micromelia
micromelica
      rachitis fetalis m.
micromere
micromyelia
micronucleus
micronychia
micropenis
microphage
microphagocyte
microphallus

microphthalmia
      progeria with m.
microphthalmos
microplasia
microplethysmography
micropodia
microprosopia
micropyle
microrespirometer
microscopic
      m. anatomy
      m. sphincter
microscopical
microseme
microsome
microsomia
microspectrophotometry
microspectroscope
microsphere method
microsplanchnic
microsplenia
microstomia
microthelia
microtia
microtonometer
microtubule
microvillus, pl. microvilli
micrurgical
miction
micturate
micturition
      frequency of m.
      m. reflex
midaxillary line
midbody
midbrain
      tectum of m.
      m. tegmentum
      m. vesicle
midcarpal joint
midcervical apophyseal joint
midclavicular line
midcostal line
middle
      m. alveolar artery
      m. atlantoepistrophic joint
      m. axillary line
      m. cardiac cervical nerve
      m. cardiac vein
      m. carpal joint
      m. cerebellar peduncle
      m. cerebral artery (MCA)

M

NOTES

middle *(continued)*
m. cervical cardiac nerve
m. cervical fascia
m. cervical ganglion
m. cervical peduncle
m. clinoid process
m. cluneal nerve
m. colic artery
m. colic lymph node
m. colic vein
m. collateral artery
m. constrictor muscle of pharynx
m. constrictor pharyngeal muscle
m. costotransverse ligament
m. cranial fossa
m. cuneiform bone
m. ear
m. esophageal constriction
m. ethmoidal air cell
m. finger
m. frontal convolution
m. frontal gyrus
m. frontal sulcus
m. genicular artery
m. genicular vein
m. glossoepiglottic fold
m. gluteal muscle
m. greater tubercle of facet of humerus
m. group of mesenteric lymph node
m. hemorrhoidal artery
m. hemorrhoidal plexus
m. hemorrhoidal vein
m. hepatic vein
m. kidney
m. lobar artery
m. lobar artery of right lung
m. lobe
m. lobe branch of right superior pulmonary vein
m. lobe bronchus
m. lobe of heart
m. lobe of prostate
m. lobe of right lung
m. lobe vein
m. macular arteriole
m. meatus
m. mediastinum
m. meningeal artery
m. meningeal branch of maxillary nerve
m. meningeal vein
m. nasal concha
m. palatine suture
m. palmar space
m. phalanges of foot and hand
m. phalanx
m. radioulnar joint

m. rectal artery
m. rectal lymph node
m. rectal nervous plexus
m. rectal vein
m. sacral artery
m. sacral lymphatic plexus
m. scalene muscle
m. sinuses
m. superior alveolar branch of infraorbital nerve
m. supraclavicular nerve
m. suprarenal artery
m. talar articular surface of calcaneus
m. temporal branch of insular part of middle cerebral artery
m. temporal branch of lateral occipital artery
m. temporal convolution
m. temporal gyrus
m. temporal sulcus
m. temporal vein
m. thyroid vein
m. transverse rectal fold
m. trunk
m. trunk of brachial plexus
m. turbinate
m. turbinated bone
m. umbilical fold
m. umbilical ligament
m. vesical artery

midgastric transverse sphincter
midget bipolar cells
midgracile
midgut
midoccipital
midpalmar space
midpelvis
plane of m.
midplane
midriff
midsacral region
midsagittal
m. plane
m. section
midscapular line
midshaft of bone
midsigmoid
m. colon
m. sphincter
midsternum
midtarsal joint
midtemporal area
midthigh
midtransverse colon
migration
m. of ovum
Mikulicz cells

milieu
> m. intérieur
> m. interne

milk
> m. duct
> m. gland
> m. line
> m. ridge
> m. tooth

milk-ejection reflex
millibar
mimetic muscle
minal center of Flemming
miniature stomach
minimae
> cordis m.
> foramina of the venae m.
> venae cardiacae m.
> venae cordis m.

minimal air
minimarum
> foramina venarum m.

minimi
> bursa trochanterica musculi
>   glutei m.
> musculus extensor digiti m.
> musculus opponens digiti m.
> vagina tendinis musculi extensoris
>   digiti m.

minimum
> law of the m.

minimus
> digitus manus m.
> digitus pedis m.
> gluteus m.
> musculus accessorius gluteus m.
> musculus adductor m.
> musculus scalenus m.
> scalenus m.
> trochanteric bursae of gluteus m.

minor
> m. alar cartilage
> anulus iridis m.
> m. arterial circle of iris
> arteria palatina m.
> m. calices
> camera oculi m.
> circulus arteriosus iridis m.
> m. circulus arteriosus of iris
> curvatura ventriculi m.
> m. duodenal papilla
> foramen ischiadicum majus et m.

> forceps m.
> fossa supraclavicularis m.
> globus m.
> helicis m.
> hippocampus m.
> incisura ischiadica m.
> justo m.
> musculus complexus m.
> musculus helicis m.
> musculus iliacus m.
> musculus pectoralis m.
> musculus psoas m.
> musculus rectus capitis anticus m.
> musculus rectus capitis posterior m.
> musculus rectus capitis posticus m.
> musculus rhomboideus m.
> musculus teres m.
> musculus zygomaticus m.
> nervus occipitalis m.
> nervus petrosus m.
> nervus splanchnicus m.
> occipitalis m.
> papilla duodeni m.
> pectoralis m.
> pelvis justo m.
> rectus capitis posterior m.
> rhomboideus m.
> m. salivary gland
> spina tympanica m.
> splanchnicus m.
> m. sublingual duct
> teres m.
> trochanter m.
> zygomaticus m.

minora
> foramina palatina m.
> frenulum of labia m.
> labium m.

minorem
> aditus ad saccum peritonei m.

minores
> calices renales m.
> canales palatini m.
> ductus sublinguales m.
> glandulae salivariae m.
> glandulae vestibulares m.
> nervi palatini m.
> palatinae m.
> palatini m.
> rami tonsillares nervi palatini m.

minoris
> apertura pelvis m.

M

NOTES

369

minoris *(continued)*
    canalis nervi petrosi
      superficialis m.
    crista tuberculi m.
    hiatus canalis nervi petrosi m.
    ramus renalis nervi splanchnici m.
    sulcus nervi petrosi m.

minorum
    frenulum labiorum m.

minus
    cornu m.
    foramen ischiadicum m.
    labium m.
    omentum m.
    os multangulum m.
    pancreas m.

minute
    m. output
    m. volume

miodidymus
miodymus
miolecithal
mionectic
miopragia
miopus
mirabile
    rete m.

mirror
    van Helmont m.

mirror-image cell
MIS
    mitral insufficiency

MIT
    monoiodotyrosine

mitochondria (*pl. of* mitochondrion)
mitochondrial
    m. matrix
    m. sheath

mitochondrialis, pl. **mitochondriales**
    cristae mitochondriales
    matrix m.

mitochondrion, pl. **mitochondria**
    cristae of mitochondria

mitogenesis
mitogenetic
mitosis, pl. **mitoses**
    heterotype m.
    multipolar m.
    somatic m.

mitotic
    m. division
    m. index
    m. period
    m. rate
    m. spindle

mitral
    m. cells
    m. insufficiency (MIS)

    m. orifice
    m. valve

mitralis
    cuspis anterior valvae m.
    cuspis posterior valvae m.
    valva m.

mixed
    m. expired gas
    m. gland
    m. nerve

mixti
    ramus cutaneus nervi m.

mixture
    mechanical m.
    physical m.

mixtus
    nervus m.

mobile
    m. end
    m. part of nasal septum
    punctum m.

mobilization
mobilize
modal alteration
moderator band
modification
modiolus, pl. **modioli**
    m. of angle of mouth
    m. anguli oris
    canales longitudinales modioli
    canalis spiralis modioli
    m. labii
    lamina modioli
    longitudinal canal of m.
    plate of m.
    spiral canal of m.
    spiralis modioli
    spiral vein of m.
    vena spiralis modioli

modulation
modulus
    bulk m.
    m. of volume elasticity
    Young m.

Mohrenheim
    M. fossa
    M. space

moiety
molar (m)
    first permanent m.
    m. gland
    second m.
    sixth-year m.
    third m.
    m. tooth
    m. tubercle
    twelfth-year m.

molare
    tuberculum m.
molaris, pl. molares
    dens m.
    glandulae molares
mole
    blood m.
    cystic m.
    grape m.
    hydatid m.
    hydatidiform m.
    vesicular m.
molecular
    m. layer
    m. layer of cerebellar cortex
    m. layer of cerebral cortex
    m. layer of retina
    m. layers of olfactory bulb
moleculare
    stratum m.
molecule
    cell adhesion m.
moles (per liter) (m)
molimen
molimina
molle
    palatum m.
Moll gland
molt
Monakow
    M. bundle
    M. nucleus
    M. tract
monauchenos
    dicephalus m.
monaxonic
moner
Monge disease
mongolian
    m. fold
    m. macula
    m. spot
monilethrix
moniliform hair
monitor
monoamelia
monoaminergic
monoamniotic twins
monoblast
monobrachius
    acephalus m.
monocardian

monocephalus
monochorial twins
monochorionic
    m. diamniotic placenta
    m. monoamniotic placenta
monochromatophil, monochromatophile
monochromophil, monochromophile
monocranius
monoculus
monocyte
monocytoid cell
monodactylism
monodactyly
monogenesis
monogenetic
monogenous
monogerminal
monoiodotyrosine (MIT)
monolocular
monomelic
monomorphic
monomphalus
mononeme
mononeural
mononeuric
mononuclear phagocyte system
monophasic
monophthalmus
monoplasmatic
monopodia
monops
monoptychial
monopus
    sympus m.
monorchia
monorchid
monorchidic
monorchidism
monorchism
monorhinic
monoscelous
monosomia
monosomous
monospermy
monostratal
monosymmetros
    cephalothoracopagus m.
monosynaptic
monothermia
monovular twins
monoxenous
monozygotic twins

M

NOTES

monozygous
Monro
    bursa of M.
    M. bursa
    M. doctrine
    foramen of M.
    M. foramen
    M. line
    M. sulcus
Monro-Kellie doctrine
Monro-Richter line
mons, pl. montes
    m. pubis
    m. ureteris
    m. veneris
Montgomery
    M. follicles
    M. gland
monticulus, pl. monticuli
    palmar monticuli
montis
Morand
    M. foot
    M. spur
Morgagni
    M. appendix
    M. cartilage
    M. caruncle
    M. columns
    M. concha
    M. crypt
    M. foramen
    foramen of M.
    M. fossa
    M. fovea
    frenulum of M.
    M. frenum
    M. humor
    M. hydatid
    M. lacuna
    lacunae of M.
    M. liquor
    M. nodule
    M. retinaculum
    M. sinus
    M. tubercle
    M. valve
    M. ventricle
morgagnian cyst
Morison
    pouch of M.
    M. pouch
morphogenesis
morphogenetic movement
morphologic element
morphology
morphometric
morphometry
morphon

morphophysiology
morphosis
mortis
    rigor m.
mortise
    ankle m.
    m. joint
    tibiofibular m.
Morton plane
morula
morulation
moruloid
Mosso ergograph
mossy
    m. cell
    m. fibers
mother cell
motile
motility
motion
    extraocular m. (EOM)
motofacient
motoneuron
motor
    m. area
    m. cell
    m. cortex
    m. decussation
    m. endplate
    m. fiber
    m. impersistence
    m. nerve of face
    m. neuron
    m. nuclei
    m. nucleus of facial nerve
    m. nucleus of trigeminus
    m. oculi
    m. plate
    m. root of ciliary ganglion
    m. root of trigeminal nerve
    m. speech center
    m. unit
    m. urgency
    m. zone
motoria
    decussatio m.
    radix m.
motorial
motormeter
moult
mounding
mouse, pl. mice
    knockout m.
mouth
    angle of m.
    m. breathing
    diaphragm of m.
    floor of m.
    gland of m.

lips of m.
modiolus of angle of m.
mucosa of m.
orbicular muscle of m.
roof of m.
vestibule of m.
m. of womb
movable
m. joint
m. testis
movement
active m.
adversive m.
circus m.
extraocular m. (EOM)
fetal m.
mass m.
masticatory m.
morphogenetic m.
muscular m.
neurobiotactic m.
passive m.
pendular m.
protoplasmic m.
reflex m.
streaming m.
translatory m.
vermicular m.
MP
metacarpophalangeal
metatarsophalangeal
MP joint
MPD
main pancreatic duct
mph node
MPR
mannose-6-phosphate receptor
MTP
metatarsophalangeal
mucid
muciferous
muciform
mucigenous
mucilaginous gland
mucin
mucinogen granule
mucinoid
mucinous
muciparous gland
mucoalbuminous cells
mucobuccal fold

mucocutaneous
m. junction
m. muscle
mucoepidermoid
mucoid
mucolysis
mucolytic
mucomembranous
mucoperiosteal
mucoperiosteum
mucosa
alveolar m.
antral m.
m. of bronchi
bronchial m.
buccal m.
bursa m.
cardiac m.
m. of colon
colorectal m.
columnar m.
m. of ductus deferens
duodenal m.
endocervical m.
esophageal m.
m. of esophagus
m. of female urethra
foveolar gastric m.
fundic m.
m. of gallbladder
gastric m.
gastroduodenal m.
gastrointestinal m.
gingival m.
glands of biliary m.
glandula m.
intestinal m.
m. of large intestine
laryngeal m.
m. of larynx
ligamental m.
lingual m.
m. of male urethra
membrana m.
m. of mouth
muscular layer of m.
nasal m.
m. of nose
olfactory region of nasal m.
oral m.
pharyngeal m.

M

NOTES

mucosa *(continued)*
    m. of pharyngotympanic auditory tube
    m. of pharynx
    pyloric m.
    rectal m.
    region of olfactory m.
    region of respiratory m.
    m. of renal pelvis
    respiratory m.
    m. of seminal gland
    m. of seminal vesicle
    m. of small intestine
    m. of stomach
    throat m.
    m. of tongue
    m. of trachea
    tracheal m.
    tunica m.
    m. of tympanic cavity
    m. of ureter
    m. of urinary bladder
    m. of uterine tube
    m. of vagina
    vaginal m.
mucosae
    lamina muscularis m.
    lamina propria m.
    muscularis m.
    pars olfactoria tunicae m.
    pars respiratoria tunicae m.
mucosal
    m. fold
    m. fold of gallbladder
    m. hernia
    m. suspensory ligament
    m. tunic
mucoserous cells
mucostatic
mucous
    m. connective tissue
    m. gland
    m. gland of auditory tube
    m. membrane
    m. membrane of bronchus
    m. membrane of ductus deferens
    m. membrane of esophagus
    m. membrane of female urethra
    m. membrane of gallbladder
    m. membrane of large intestine
    m. membrane of larynx
    m. membrane of male urethra
    m. membrane of nose
    m. membrane of pharyngotympanic auditory tube
    m. membrane of pharynx
    m. membrane of small intestine
    m. membrane of stomach
    m. membrane of tongue

    m. membrane of trachea
    m. membrane of tympanic cavity
    m. membrane of ureter
    m. membrane of urinary bladder
    m. membrane of uterine tube
    m. membrane of vagina
    m. neck cell
    m. sheath of tendon
mucro, pl. **mucrones**
    m. cordis
    m. sterni
mucronate
mucus
    glairy m.
mucus-secreting gland
Mueller muscle
muliebre
    membrum m.
    pudendum m.
muliebria
muliebris
    corpus spongiosum urethrae m.
    dartos m.
    penis m.
    urethra m.
Müller
    M. capsule
    M. duct
    M. fibers
    M. law
    M. maneuver
    M. muscle
    M. radial cells
    M. trigone
    M. tubercle
müllerian
    m. duct
    m. duct inhibitory factor
    m. inhibiting factor
    m. inhibiting substance
    m. regression factor
multangular bone
multangulum majus
multiaxial joint
multicapsular
multicellular
multicolony-stimulating factor
multicuspid tooth
multifetation
multifid
multifidi
multifidus
    m. muscle
    musculus m.
    m. spinae
multiform layer
multiglandular
multilamellar body

multiloba
    placenta m.
multilobar
multilobate
multilobed
multilobular
multilocular
    m. adipose tissue
    m. fat
multimammae
multinodal
multinodular
multinodulate
multinuclear
multinucleate
multipennate muscle
multipennatus
    musculus m.
multiple
    m. epiphysial dysplasia
    m. exostoses
    m. fission
multiplex
    dysplasia epiphysialis m.
multiplicative growth
multiplier
    countercurrent m.
multipolar
    m. cell
    m. mitosis
    m. neuron
multirooted
multisynaptic
multivesicular bodies
mummification
Munro
    pouch of M.
mural cell
muriform
muscarinic receptor
muscle
    m. of abdomen
    abdominal external oblique m.
    abdominal internal oblique m.
    abdominal part of pectoralis
      major m.
    abductor hallucis m.
    abductor longus m.
    abductor magnus m.
    abductor pollicis brevis m.
    abductor pollicis longus m.
    acromial part of deltoid m.

adductor brevis m.
adductor hallucis m.
adductor longus m.
adductor magnus m.
adductor minimus m.
adductor pollicis m.
Aeby m.
agonistic m.
alar part of nasalis m.
Albinus m.
m. of anal triangle
anconeus m.
anorectoperineal m.
antagonistic m.
anterior auricular m.
anterior belly of digastric m.
anterior cervical
  intertransversarii m.
anterior cervical intertransverse m.
anterior fascicle of
  palatopharyngeus m.
anterior scalene m.
anterior serratus m.
anterior splenis m.
anterior tibial m.
antigravity m.
antitragicus m.
antitragus m.
m. of antitragus
aponeurosis of abdominal
  oblique m.
aponeurosis of external oblique m.
aponeurosis of internal oblique m.
aponeurosis of vastus m.
appendicular m.
arrector pili m.
articular m.
articularis cubiti m.
articularis genus m.
aryepiglottic m.
aryepiglottic part of oblique
  arytenoid m.
arytenoid m.
ascending part of trapezius m.
m.'s of auditory ossicle
auricular m.
auricularis anterior m.
auricularis posterior m.
auricularis superior m.
axial m.
axillary arch m.
m. of back

M

NOTES

muscle *(continued)*

m. of back proper
Bell m.
m. belly
belly of digastric m.
belly of omohyoid m.
biceps brachii m.
biceps femoris m.
bipennate m.
Bochdalek m.
Bovero m.
Bowman m.
brachial m.
brachialis m.
brachioradial m.
brachioradialis m.
branch of glossopharyngeal nerve
  to stylopharyngeus m.
branchiomeric m.
Braune m.
brevis m.
bronchoesophageal m.
bronchoesophageus m.
Brücke m.
buccinator m.
bulbocavernosus m.
bulbocavernous m.
bulbospongiosus m.
m. bundle
bursa of extensor carpi radialis
  brevis m.
bursa of semimembranosus m.
bursa of tensor veli palatini m.
c m.
canal for tensor tympani m.
capillary m.
cardiac m.
casserian m.
Casser perforated m.
ceratocricoid m.
ceratoglossus m.
ceratopharyngeal m.
cervical iliocostal m.
cervical interspinal m.
cervical interspinales m.
cervical longissimus m.
cervical rotator m.
Chassaignac axillary m.
check ligament of medial and
  lateral rectus m.
cheek m.
chin m.
chondroglossus m.
ciliary m.
circular pharyngeal m.
clavicular head of pectoralis
  major m.
clavicular part of deltoid m.

clavicular part of pectoralis
  major m.
coccygeal m.
coccygeus m.
m. of coccyx
Coiter m.
common tendinous ring of
  extraocular m.
compressor naris m.
compressor urethra m.
congenerous m.
constrictor m.
coracobrachial m.
coracobrachialis m.
corrugator supercilii m.
cowl m.
Crampton m.
cremaster m.
cremasteric m.
crest of supinator m.
cricoarytenoid m.
cricopharyngeal m.
cricopharyngeus m.
cricothyroid m.
cruciate m.
m. curve
cutaneomucous m.
cutaneous m.
dartos m.
deep flexor m.
deep part of masseter m.
deep part of palpebral part of
  orbicularis oculi m.
deep transverse perineal m.
deltoid m.
depressor anguli oris m.
depressor labii inferioris m.
depressor septi nasi m.
depressor supercilii m.
descending part of trapezius m.
detrusor m.
diaphragmatic m.
digastric m.
dilator pupillae m.
dorsal interosseous m.
dorsal part of intertransversarii
  laterales lumborum m.
dorsal sacrococcygeal m.
dorsal sacrococcygeus m.
Dupré m.
Duverney m.
epicranial m.
epicranius m.
epimeric m.
m. epithelium
epitrochleoanconeus m.
erector spinae m.
eustachian m.
extensor carpi radialis brevis m.

extensor carpi radialis longus m.
extensor carpi ulnaris m.
extensor digiti minimi m.
extensor digitorum brevis m.
extensor digitorum longus m.
extensor hallucis brevis m.
extensor hallucis longus m.
extensor indicis m.
extensor pollicis brevis m.
extensor pollicis longus m.
external abdominal m.
external anal sphincter m.
external intercostal m.
external oblique m.
external obturator m.
external pterygoid m.
external rectus m.
extracostal m.
extraocular m. (EOM)
extrinsic m.
m. of eyeball
facial m.
m.'s of facial expression
fascia of extraocular m.
fascial sheaths of extraocular m.
fascia of omohyoid m.
m. fascicle
femoral m.
fibularis brevis m.
fibularis longus m.
fibularis tertius m.
fixator m.
flat m.
flexor accessorius m.
flexor carpi radialis m.
flexor carpi ulnaris m.
flexor digitorum brevis m.
flexor digitorum longus m.
flexor digitorum profundus m.
flexor digitorum superficialis m.
flexor hallucis brevis m.
flexor hallucis longus m.
flexor pollicis brevis m.
flexor pollicis longus m.
Folius m.
four-headed m.
frontal belly of occipitofrontalis m.
frontalis m.
fusiform m.
Gantzer m.
gastrocnemius m.
Gavard m.

gemellus m.
genioglossal m.
genioglossus m.
geniohyoid m.
glossopalatine m.
glossopharyngeal m.
gluteal m.
gluteus maximus m.
gluteus medius m.
gluteus minimus m.
gracilis m.
great adductor m.
greater pectoral m.
greater psoas m.
greater rhomboid m.
greater trochanter m.
greater zygomatic m.
greatest gluteal m.
groove for tendon of long
  peroneal m.
Guthrie m.
hamstring m.
m. of head
heart m.
helicis major m.
helicis minor m.
Hilton m.
Horner m.
Houston m.
humeroulnar head of flexor
  digitorum superficialis m.
hyoglossal m.
hyoglossus m.
hyoid m.
hypaxial m.
hypothenar m.
iliac m.
iliacus minor m.
iliococcygeal m.
iliococcygeus m.
iliocostal m.
iliocostalis cervicis m.
iliocostalis lumborum m.
iliocostalis thoracis m.
iliohypogastric m.
iliopsoas m.
index extensor m.
inferior belly of omohyoid m.
inferior constrictor pharyngeal m.
inferior gemellus m.
inferior lingual m.
inferior oblique m.

M

NOTES

muscle *(continued)*

inferior part of trapezius m.
inferior posterior serratus m.
inferior rectus m.
inferior retinaculum of extensor m.
inferior tarsal m.
infrahyoid m.
infraspinatus m.
infraspinous m.
innermost intercostal m.
innervated m.
interarytenoid m.
intercostal m.
intercrural fibers of aponeurosis of abdominal oblique m.
intermediate great m.
intermediate vastus m.
internal abdominal m.
internal intercostal m.
internal oblique m.
internal obturator m.
internal pterygoid m.
interosseous m.
interspinal m.
interspinales cervicis m.
interspinales lumborum m.
interspinales thoracis m.
intertransversarii m.
intertransverse m.
intraauricular m.
intrinsic m.
involuntary m.
ipsilateral mentalis m.
ischiocavernous m.
Jung m.
Klein m.
Kohlrausch m.
Krause m.
labial part of orbicularis oris m.
lacertus of lateral rectus m.
lacrimal part of orbicularis oculi m.
Lancisi m.
Landström m.
Langer m.
laryngeal m.
m. of larynx
lateral cricoarytenoid m.
lateral great m.
lateral lumbar intertransversarii m.
lateral lumbar intertransverse m.
lateral part of posterior cervical intertransversarii m.
lateral posterior cervical intertransversarii m.
lateral pterygoid m.
lateral rectus m.
lateral vastus m.
latissimus dorsi m.

m. layer in fatty layer of subcutaneous tissue
least gluteal m.
left rhomboid m.
lesser rhomboid m.
lesser zygomatic m.
levator angularis oris m.
levator anguli oris m.
levator ani m.
levator costarum m.
levatores costarum breves m.
levatores costarum longi m.
levator labii superioris m.
levator labii superioris alaeque nasi m.
levator palati m.
levator palpebrae superioris m.
levator prostatae m.
levator veli palatini m.
line for soleus m.
lingual m.
long abductor m.
long adductor m.
long extensor m.
long fibular m.
long flexor m.
longissimus capitis m.
longissimus cervicis m.
longissimus thoracis m.
longitudinal pharyngeal m.
long levatores costarum m.
long palmar m.
long peroneal m.
long radial extensor m.
long rotator m.
longus capitis m.
longus colli m.
lumbar iliocostal m.
lumbar interspinal m.
lumbar quadrate m.
lumbar rotator m.
lumbrical m.
malleolus m.
Marcacci m.
marginal part of orbicularis oris m.
m. mass
masseter m.
m. of mastication
masticatory m.
medial great m.
medial lumbar intertransversarii m.
medial lumbar intertransverse m.
medial posterior cervical intertransversarii m.
medial pterygoid m.
medial rectus m.
medial vastus m.
mentalis m.

meridional fibers of ciliary m.
Merkel m.
middle constrictor pharyngeal m.
middle gluteal m.
middle scalene m.
mimetic m.
mucocutaneous m.
Mueller m.
Müller m.
multifidus m.
multipennate m.
muscular fascia of extraocular m.
mylohyoid m.
mylopharyngeal m.
nasal m.
nasalis m.
m. of neck
nerve to stapedius m.
nerve to tensor tympani m.
nerve to tensor veli palatini m.
nerve to thyrohyoid m.
m. of notch of helix
oblique abdominal m.
oblique arytenoid m.
oblique auricular m.
oblique capitis m.
oblique part of cricothyroid m.
obliquus capitis inferior m.
obliquus capitis superior m.
obturator externus m.
obturator internus m.
occipital belly of
  occipitofrontalis m.
occipitalis m.
occipitofrontal m.
occipitofrontalis m.
ocular m.
Oehl m.
omohyoid m.
opponens digiti minimi m.
opponens pollicis m.
orbicular m.
orbicularis oculi m.
orbicularis oris m.
orbitalis m.
orbital part of orbicularis oculi m.
palatoglossal m.
palatoglossus m.
palatopharyngeal sphincter m.
palatopharyngeus m.
palatouvularis m.
palmar interossei interosseous m.

palmar interosseous m.
palmaris brevis m.
palmaris longus m.
palpebral part of orbicularis
  oculi m.
panniculus carnosus m.
papillary m.
paraspinal m.
paraspinous m.
pars lacrimalis of orbicularis
  oculi m.
pectinate m.
pectineal m.
pectineus m.
pectoral m.
pectoralis major m.
pectoralis minor m.
pectorodorsal m.
pectorodorsalis m.
pennate m.
penniform m.
perineal m.
peroneal m.
peroneus brevis m.
peroneus longus m.
peroneus tertius m.
pharyngeal constrictor m.
pharyngopalatine m.
pharyngopalatinus m.
m. of pharynx
physical elasticity of m.
physiologic elasticity of m.
piriform m.
piriformis m.
plantar interossei interosseous m.
plantar interosseous m.
plantaris m.
plantar quadrate m.
plantar tendon sheath of fibularis
  longus m.
plantar tendon sheath of peroneus
  longus m.
m. plate
platysma m.
pleuroesophageal m.
pleuroesophageus m.
popliteal m.
popliteus m.
postaxial m.
posterior auricular m.
posterior belly of digastric m.

M

NOTES

muscle (*continued*)

posterior cervical intertransversarii m.
posterior cervical intertransverse m.
posterior cricoarytenoid m.
posterior fascicle of palatopharyngeus m.
posterior scalene m.
posterior tibial m.
Pozzi m.
procerus m.
profundus m.
pronator quadratus m.
pronator teres m.
psoas major m.
psoas minor m.
puboanalis m.
pubococcygeal m.
pubococcygeus m.
puboperinealis m.
puboprostatic m.
puboprostaticus m.
puborectal m.
puborectalis m.
pubovaginal m.
pubovaginalis m.
pubovesical m.
pubovesicalis m.
pupillae sphincter m.
pupillary m.
pyramidal auricular m.
pyramidalis m.
quadrate pronator m.
quadratus femoris m.
quadratus labii m.
quadratus lumborum m.
quadratus plantae m.
quadriceps femoris m.
radial extensor m.
radial flexor m.
rectococcygeal m.
rectococcygeus m.
rectourethral m.
rectouterine m.
rectouterinus m.
rectovesical m.
rectovesicalis m.
rectus abdominis m.
rectus capitis anterior m.
rectus capitis lateralis m.
rectus capitis posterior major m.
rectus capitis posterior minor m.
rectus femoris m.
red m.
Reisseisen m.
retinacula of extensor m.
retinacula of peroneal m.
retinaculum of flexor m.
retractor bulbi m.

rhomboid major m.
rhomboid minor m.
rider's m.
Riolan m.
rotator cuff m.
rotatores cervicis m.
rotatores lumborum m.
rotatores thoracis m.
Rouget m.
round pronator m.
Ruysch m.
sacrococcygeal m.
sacrospinal m.
salpingopharyngeal m.
salpingopharyngeus m.
Santorini m.
sartorius m.
scalene m.
scalenus anterior m.
scalenus medius m.
scalenus minimus m.
scalenus posterior m.
scalp m.
scapular m.
scapulohumeral m.
Sebileau m.
second tibial m.
semicanal for tensor tympani m.
semimembranosus m.
semimembranous m.
semipennate m.
semispinal m.
semispinalis capitis m.
semispinalis cervicis m.
semispinalis thoracis m.
semitendinosus m.
semitendinous m.
serratus anterior m.
serratus posterior inferior m.
serratus posterior superior m.
shawl m.
sheath of rectus m.
sheath of superior oblique m.
short abductor m.
short adductor m.
short anconeus m.
short extensor m.
short fibular m.
short flexor m.
short levator m.
short levatores costarum m.
short palmar m.
short peroneal m.
short radial extensor m.
short rotator m.
shunt m.
Sibson m.
skeletal m.
smaller pectoral m.

smaller psoas m.
smallest scalene m.
smooth m.
Soemmering m.
soleus m.
somatic m.
m. spasm
sphincter m.
spinal m.
spinalis capitis m.
spinalis cervicis m.
spinalis thoracis m.
spinal part of deltoid m.
m. spindle
spindle-shaped m.
splenis cervicis m.
splenius capitis m.
splenius cervicis m.
stapedius m.
sternal m.
sternalis m.
sternochondroscapular m.
sternoclavicular m.
sternocleidomastoid m.
sternocostal head of pectoralis
  major m.
sternocostalis m.
sternocostal part of pectoralis
  major m.
sternohyoid m.
sternomastoid m.
sternothyroid m.
straight part of cricothyroid m.
strap m.
striated m.
styloauricular m.
styloglossus m.
stylohyoid m.
stylopharyngeal m.
stylopharyngeus m.
subanconeus m.
subclavian m.
subclavius m.
subcostal m.
subcrural m.
suboccipital m.
subquadricipital m.
subscapular m.
subscapularis m.
subtendinous bursae of
  gastrocnemius m.
subvertebral m.

superciliary depressor m.
superficial back m.
superficial flexor m.
superficial lingual m.
superficial part of masseter m.
superficial perineal m.
superficial transverse perineal m.
superior auricular m.
superior belly of omohyoid m.
superior constrictor pharyngeal m.
superior gemellus m.
superior oblique m.
superior pharyngeal constrictor m.
superior posterior serratus m.
superior rectus m.
superior retinaculum of extensor m.
superior tarsal m.
supinator m.
supraclavicular m.
suprahyoid m.
supramediastinal m.
supraspinalis m.
supraspinatus m.
supraspinous m.
synergic m.
synergistic m.
tailor's m.
tarsal m.
temporal m.
temporalis m.
temporoparietal m.
temporoparietalis m.
tendinous arch of levator ani m.
tendinous arch of soleus m.
tendinous sheath of extensor carpi
  radialis m.
tendinous sheath of extensor carpi
  ulnaris m.
tendinous sheath of extensor digiti
  minimi m.
tendinous sheath of extensor
  digitorum and extensor indicis m.
tendinous sheath of extensor
  hallucis longus m.
tendinous sheath of extensor
  pollicis longus m.
tendinous sheath of flexor carpi
  radialis m.
tendinous sheath of flexor hallucis
  longus m.
tendinous sheath of flexor pollicis
  longus m.

M

NOTES

muscle *(continued)*

tendinous sheath of superior oblique m.
tendinous sheath of tibialis anterior m.
tendinous sheath of tibialis posterior m.
tensor fasciae latae m.
tensor tarsi m.
tensor trochleae m.
tensor tympani m.
tensor veli palati m.
teres major m.
teres minor m.
m. of terminal notch
Theile m.
thenar m.
third peroneal m.
thoracic interspinal m.
thoracic interspinales m.
thoracic intertransversarii m.
thoracic intertransverse m.
thoracic longissimus m.
thoracic part of iliocostalis lumborum m.
thoracic rotator m.
thoracoappendicular m.
m. of thorax
three-headed m.
m. of thumb
thyroarytenoid m.
thyroepiglottic part of thyroarytenoid m.
thyroepiglottidean m.
thyrohyoid m.
thyroid m.
thyropharyngeal m.
tibial m.
tibialis anterior m.
tibialis posterior m.
m. tissue
Tod m.
m. of tongue
total elasticity of m.
Toynbee m.
tracheal m.
trachealis m.
tracheloclavicular m.
trachelomastoid m.
tragicus m.
m. of tragus
tragus m.
transverse arytenoid m.
transverse auricular m.
transverse part of nasalis m.
transverse part of trapezius m.
transverse perineal m.
transversospinal m.
transversospinales m.

transversus abdominis m.
transversus menti m.
transversus nuchae m.
transversus thoracis m.
trapezius m.
Treitz m.
triangular m.
triceps brachii m.
triceps coxae m.
triceps surae m.
trochlea of superior oblique m.
trunk m.
tubercle of anterior scalene m.
tuberosity for serratus anterior m.
two-bellied m.
two-headed m.
ulnar flexor m.
underlying chest m.
unipennate m.
unstriated m.
unstriped m.
m. of urogenital triangle
uterine m.
m. of uvula
uvular m.
vaginal m.
Valsalva m.
vastus intermedius m.
vastus lateralis m.
vastus medialis m.
ventral part of intertransversarii laterales lumborum m.
ventral sacrococcygeal m.
ventral sacrococcygeus m.
vertical m.
vestigial m.
visceral m.
vocal m.
vocalis m.
voluntary m.
white m.
Wilson m.
zygomatic m.
zygomaticus major m.
zygomaticus minor m.

muscle-tendon

m.-t. attachment
m.-t. junction

muscular

m. artery of ophthalmic artery
m. branch
m. coat
m. coat of bronchi
m. coat of colon
m. coat of ductus deferens
m. coat of esophagus
m. coat of female urethra
m. coat of gallbladder

m. coat of intermediate part of male urethra
m. coat of large intestine
m. coat of pharynx
m. coat of prostatic urethra
m. coat of rectum
m. coat of small intestine
m. coat of spongy part of male urethra
m. coat of stomach
m. coat of trachea
m. coat of ureter
m. coat of urinary bladder
m. coat of uterine tube
m. coat of uterus
m. coat of vagina
m. fascia
m. fascia of extraocular muscle
m. fibril
m. lacuna
m. layer
m. layer of bronchi
m. layer of colon
m. layer of ductus deferens
m. layer of esophagus
m. layer of female urethra
m. layer of gallbladder
m. layer of intermediate part of male urethra
m. layer of large intestine
m. layer of mucosa
m. layer of pharynx
m. layer of prostatic urethra
m. layer of rectum
m. layer of renal pelvis
m. layer of seminal gland
m. layer of small intestine
m. layer of spongy male urethra
m. layer of stomach
m. layer of trachea
m. layer of ureter
m. layer of urinary bladder
m. layer of uterine tube
m. layer of vagina
m. movement
m. part of interventricular septum of heart
m. process of arytenoid cartilage
m. pulley
m. reflex
m. sense

m. space of retroinguinal compartment
m. sphincter supracollicularis
m. substance of prostate
m. system
m. tissue
m. tone
m. triangle
m. triangle of neck
m. trochlea
m. tunic
m. tunic of gallbladder
m. vein
m. wall

musculare
trigonum m.

muscularis, pl. **musculares**
fibrae obliquae tunicae m.
m. mucosae
rami musculares
stratum circulare tunicae m.
stratum longitudinale tunicae m.
trochlea m.
tunica m.

musculature
cervical m.
external m.
lumbar m.
paraspinous m.
paravertebral m.
pelvic m.

musculi (*pl. of* musculus)
musculoaponeurotic
musculocartilaginous stricture
musculocutanei
rami musculares nervi m.

musculocutaneous
m. artery
m. nerve
m. nerve of leg
m. vein

musculocutaneus
nervus m.

musculofascial
m. layer
m. structure
m. wall

musculofascially
musculomembranous
musculophrenic
m. artery
m. vein

M

NOTES

musculophrenica
arteria m.
musculophrenicae
venae m.
musculorum
lacuna m.
musculoskeletal
musculospiral
m. groove
m. nerve
musculotendinous cuff
musculotubal canal
musculotubarii
septum canalis m.
musculotubarius
canalis m.
musculus, pl. musculi
musculi abdominis
m. abductor
m. abductor digiti minimi manus
m. abductor digiti minimi pedis
m. abductor digiti quinti
m. abductor hallucis
m. abductor pollicis brevis
m. abductor pollicis longus
m. accessorius gluteus minimus
m. adductor
m. adductor brevis
m. adductor hallucis
m. adductor longus
m. adductor magnus
m. adductor minimus
m. adductor pollicis
m. anconeus
musculi anorectoperineales
m. antitragicus
musculi arrectores pilorum
m. arrector pili
m. articularis
m. articularis cubiti
m. articularis genus
m. aryepiglotticus
m. arytenoideus obliquus
m. arytenoideus transversus
m. aryvocalis
m. attollens aurem
m. attollens auriculam
m. attrahens aurem
m. attrahens auriculam
musculi auriculares
m. auricularis anterior
m. auricularis posterior
m. auricularis superior
m. azygos uvulae
m. biceps brachii
m. biceps femoris
m. biceps flexor cruris
m. bipennatus
m. biventer

m. biventer mandibulae
m. brachialis
m. brachiocephalicus
m. brachioradialis
m. bronchoesophageus
m. buccinator
m. buccopharyngeus
musculi bulbi
m. bulbocavernosus
m. bulbospongiosus
m. caninus
musculi capitis
m. cephalopharyngeus
m. ceratocricoideus
m. ceratoglossus
m. ceratopharyngeus
m. cervicalis ascendens
musculi cervicis
m. chondroglossus
m. chondropharyngeus
m. ciliaris
m. cleidoepitrochlearis
m. cleidomastoideus
m. cleidooccipitalis
m. coccygeus
musculi colli
m. complexus
m. complexus minor
m. compressor naris
m. compressor urethrae
m. constrictor pharyngis inferior
m. constrictor pharyngis medius
m. constrictor pharyngis superior
m. constrictor urethrae
m. coracobrachialis
m. corrugator cutis ani
m. corrugator supercilii
m. cremaster
m. cricoarytenoideus lateralis
m. cricoarytenoideus posterior
m. cricopharyngeus
m. cricothyroideus
m. cruciatus
m. cutaneomucosus
m. cutaneus
m. deltoideus
m. depressor anguli oris
m. depressor labii inferioris
m. depressor septi
m. depressor supercilii
m. detrusor urinae
m. diaphragma
m. digastricus
m. dilatator
m. dilator
m. dilator iridis
m. dilator naris
m. dilator pupillae
m. dilator pylori gastroduodenalis

m. dilator pylori ilealis
m. dilator tubae
musculi dorsi
musculi dorsi proprii
m. ejaculator seminis
m. epicranius
m. epitrochleoanconeus
m. erector clitoridis
m. erector penis
m. erector spinae
m. extensor
m. extensor brevis digitorum
m. extensor brevis pollicis
m. extensor carpi radialis brevis
m. extensor carpi radialis longus
m. extensor carpi ulnaris
m. extensor coccygis
m. extensor digiti minimi
m. extensor digiti quinti proprius
m. extensor digitorum brevis
m. extensor digitorum brevis
  manus
m. extensor digitorum communis
m. extensor digitorum longus
m. extensor hallucis brevis
m. extensor hallucis longus
m. extensor indicis
m. extensor indicis proprius
m. extensor longus digitorum
m. extensor longus pollicis
m. extensor minimi digiti
m. extensor ossis metacarpi pollicis
m. extensor pollicis brevis
m. extensor pollicis longus
musculi externi bulbi oculi
musculi faciales
musculi faciei
m. fibularis brevis
m. fibularis longus
m. fibularis tertius
m. flexor
m. flexor accessorius
m. flexor brevis digitorum
m. flexor brevis hallucis
m. flexor carpi radialis
m. flexor carpi ulnaris
m. flexor digiti minimi brevis
  manus
m. flexor digiti minimi brevis
  pedis
m. flexor digitorum brevis
m. flexor digitorum longus

m. flexor digitorum profundus
m. flexor digitorum sublimis
m. flexor digitorum superficialis
m. flexor hallucis brevis
m. flexor hallucis longus
m. flexor longus digitorum
m. flexor longus hallucis
m. flexor longus pollicis
m. flexor pollicis brevis
m. flexor pollicis longus
m. flexor profundus
m. frontalis
m. fusiformis
m. gastrocnemius
m. gemellus inferior
m. gemellus superior
m. genioglossus
m. geniohyoglossus
m. geniohyoideus
m. glossopalatinus
m. glossopharyngeus
m. gluteus maximus
m. gluteus medius
m. gluteus quartus
m. helicis major
m. helicis minor
m. hyoglossus
m. hypopharyngeus
m. iliacus
m. iliacus minor
m. iliocapsularis
m. iliococcygeus
m. iliocostalis
m. iliocostalis cervicis
m. iliocostalis dorsi
m. iliocostalis lumborum
m. iliocostalis thoracis
m. iliopsoas
m. incisivus labii inferioris
m. incisivus labii superioris
m. incisurae helicis
musculi infracostales
m. infracostalis
musculi infrahyoidei
m. infraspinatus
musculi intercostales externi
musculi intercostales interni
m. intercostalis externus
m. intercostalis internus
m. intercostalis intimus
musculi interossei
musculi interossei dorsales manus

M

NOTES

musculus (*continued*)

musculi interossei dorsales pedis
musculi interossei dorsalis manus
musculi interossei dorsalis pedis
musculi interossei palmares
musculi interossei plantares
m. interosseus dorsalis manus
m. interosseus dorsalis pedis
m. interosseus palmaris
m. interosseus plantaris
musculi interosseus plantaris
m. interosseus volaris
musculi interspinales
m. interspinalis cervicis
m. interspinalis lumborum
m. interspinalis thoracis
m. intertragicus
musculi intertransversarii
musculi intertransversarii anteriores
  cervicis
musculi intertransversarii laterales
  lumborum
musculi intertransversarii mediales
  lumborum
musculi intertransversarii posteriores
  cervicis
musculi intertransversarii thoracis
m. ischiocavernosus
m. ischiococcygeus
m. keratopharyngeus
musculi laryngis
m. laryngopharyngeus
m. latissimus dorsi
m. levator alae nasi
m. levator anguli oris
m. levator anguli scapulae
m. levator ani
m. levator costae
musculi levatores costarum
musculi levatores costarum brevis
musculi levatores costarum longi
m. levator glandulae thyroideae
musculi levatoris palpebrae
  superioris
m. levator labii inferioris
m. levator labii superioris
m. levator labii superioris alaeque
  nasi
m. levator palati
m. levator palpebrae superioris
m. levator prostatae
m. levator scapulae
m. levator veli palatini
musculi linguae
m. longissimus
m. longissimus capitis
m. longissimus cervicis
m. longissimus dorsi
m. longissimus thoracis

m. longitudinalis inferior
m. longitudinalis inferior linguae
m. longitudinalis superior
m. longitudinalis superior linguae
m. longus capitis
m. longus colli
musculi lumbricales manus
musculi lumbricales pedis
m. lumbricalis manus
m. lumbricalis pedis
m. masseter
m. mentalis
m. multifidus
m. multifidus spinae
m. multipennatus
m. mylohyoideus
m. mylopharyngeus
m. nasalis
m. obliquus auriculae
m. obliquus capitis inferior
m. obliquus capitis superior
m. obliquus externus abdominis
m. obliquus internus abdominis
m. obliquus superior
m. obturator externus
m. obturator internus
m. occipitalis
m. occipitofrontalis
m. omohyoideus
m. opponens
m. opponens digiti minimi
m. opponens digiti quinti
m. opponens minimi digiti
m. opponens pollicis
m. orbicularis
m. orbicularis oculi
m. orbicularis oris
m. orbicularis palpebrarum
m. orbitalis
m. orbitopalpebralis
musculi ossiculorum auditoriorum
musculi ossiculorum auditus
m. palatoglossus
m. palatosalpingeus
m. palatostaphylinus
m. palmaris brevi
m. palmaris brevis
m. palmaris longus
m. papillaris
musculi pectinati
m. pectineus
m. pectoral
m. pectoralis major
m. pectoralis minor
m. pennatus
musculi perinei
m. peroneocalcaneus
m. peroneus brevis
m. peroneus longus

m. peroneus tertius
m. petropharyngeus
m. petrostaphylinus
m. pharyngopalatinus
m. piriformis
m. plana
m. platysma
m. platysma myoides
m. pleuroesophageus
m. popliteus
m. procerus
m. pronator
m. pronator pedis
m. pronator quadratus
m. pronator radii teres
m. pronator teres
m. prostaticus
m. psoas major
m. psoas minor
m. pterygoideus externus
m. pterygoideus internus
m. pterygoideus lateralis
m. pterygoideus medialis
m. pterygopharyngeus
m. pterygospinosus
m. puboanalis
m. pubococcygeus
m. puboperinealis
m. puboprostaticus
m. puborectalis
m. pubovaginalis
m. pubovesicalis
m. pyramidalis
m. pyramidalis auriculae
m. pyramidalis nasi
m. pyriformis
m. quadratus
m. quadratus femoris
m. quadratus labii inferioris
m. quadratus labii superioris
m. quadratus lumborum
m. quadratus menti
m. quadratus plantae
m. quadriceps
m. quadriceps extensor femoris
m. rectococcygeus
musculi rectourethrales
m. rectourethralis
m. rectouterinus
m. rectovesicalis
m. rectus
m. rectus abdominis

m. rectus capitis anterior
m. rectus capitis anticus major
m. rectus capitis anticus minor
m. rectus capitis lateralis
m. rectus capitis posterior major
m. rectus capitis posterior minor
m. rectus capitis posticus major
m. rectus capitis posticus minor
m. rectus externus
m. rectus femoris
m. rectus inferior
m. rectus internus
m. rectus medialis
m. rectus superior
m. rectus thoracis
musculi regionis analis
musculi regionis urogenitalis
m. retrahens aurem
m. retrahens auriculam
m. rhomboatloideus
m. rhomboideus major
m. rhomboideus minor
m. risorius
m. rotator
musculi rotatores
musculi rotatores cervicis
musculi rotatores lumborum
musculi rotatores thoracis
m. sacrococcygeus anterior
m. sacrococcygeus dorsalis
m. sacrococcygeus posterior
m. sacrococcygeus ventralis
m. sacrolumbalis
m. sacrospinalis
m. salpingopharyngeus
m. sartorius
m. scalenus anterior
m. scalenus anticus
m. scalenus medius
m. scalenus minimus
m. scalenus posterior
m. scalenus posticus
m. scansorius
musculi scapulohumerales
m. semimembranosus
m. semipennatus
m. semispinalis
m. semispinalis capitis
m. semispinalis cervicis
m. semispinalis colli
m. semispinalis dorsi
m. semispinalis thoracis

M

NOTES

musculus *(continued)*

m. semitendinosus
m. serratus anterior
m. serratus magnus
m. serratus posterior inferior
m. serratus posterior superior
m. skeleti
m. soleus
m. sphenosalpingostaphylinus
m. sphincter
m. sphincter ampullae
m. sphincter ampullae
   biliaropancreaticae
m. sphincter ampullae
   hepatopancreaticae
m. sphincter ani externus
m. sphincter ani internus
m. sphincter ductus biliaris
m. sphincter ductus choledochi
m. sphincter ductus pancreatici
m. sphincter oris
m. sphincter palatopharyngeus
m. sphincter pupillae
m. sphincter pylori
m. sphincter urethrae
m. sphincter urethrae externus
m. sphincter urethrae externus
   femininae
m. sphincter urethrae externus
   masculinae
m. sphincter urethrae internus
m. sphincter urethrae membranaceae
m. sphincter urethrovaginalis
m. sphincter vaginae
m. sphincter vesicae
m. spinalis
m. spinalis capitis
m. spinalis cervicis
m. spinalis colli
m. spinalis dorsi
m. spinalis thoracis
musculi splenii
m. splenius capitis
m. splenius cervicis
m. splenius colli
m. stapedius
m. sternalis
m. sternochondroscapularis
m. sternoclavicularis
m. sternocleidomastoideus
m. sternofascialis
m. sternohyoideus
m. sternothyroideus
m. styloauricularis
m. styloglossus
m. stylohyoideus
m. stylolaryngeus
m. stylopharyngeus
m. subclavius

musculi subcostales
m. subcostalis
m. subcutaneus colli
musculi suboccipitales
m. subscapularis
m. supinator
m. supinator longus
m. supinator radii brevis
m. supraclavicularis
musculi suprahyoidei
m. supraspinalis
m. supraspinatus
m. suspensorius duodeni
m. tarsalis inferior
m. tarsalis superior
m. temporalis
m. temporoparietalis
m. tensor fasciae femoris
m. tensor fasciae latae
m. tensor palati
m. tensor tarsi
m. tensor tympani
m. tensor veli palatini
m. teres major
m. teres minor
m. tetragonus
musculi thoracis
musculi thoracoappendiculares
m. thyroarytenoideus
m. thyroarytenoideus externus
m. thyroarytenoideus internus
m. thyroepiglotticus
m. thyrohyoideus
m. thyropharyngeus
m. tibialis anterior
m. tibialis anticus
m. tibialis gracilis
m. tibialis posterior
m. tibialis posticus
m. tibialis secundus
m. tibiofascialis anterior
m. tibiofascialis anticus
m. trachealis
m. tracheloclavicularis
m. trachelomastoideus
m. tragicus
m. transversalis abdominis
m. transversalis capitis
m. transversalis cervicis
m. transversalis colli
m. transversalis nasi
musculi transversospinales
m. transversospinalis
m. transversus abdominis
m. transversus auriculae
m. transversus linguae
m. transversus menti
m. transversus nuchae
m. transversus perinei profundus

m. transversus perinei superficialis
m. transversus thoracis
m. trapezius
m. triangularis
m. triangularis labii inferioris
m. triangularis labii superioris
m. triangularis sterni
m. triceps
m. triceps brachii
m. triceps coxa
m. triceps surae
m. triticeoglossus
m. unipennatus
m. vastus externus
m. vastus intermedius
m. vastus internus
m. vastus lateralis
m. vastus medialis
vastus medialis obliquus m. (VMO)
venter occipitalis musculi
m. ventricularis
m. verticalis linguae
m. vocalis
m. zygomaticus
m. zygomaticus major
m. zygomaticus minor
mutual resistance
MVV
maximum voluntary ventilation
myasthenic reaction
myatonia
myatony
myelatelia
myelencephalon
myelic
myelin
m. body
m. figure
m. sheath
myelinated
m. nerve fiber
myelination
myelinic
myelinization
myelinogenesis
myeloblast
myelocele
myelocyst
myelocystic
myelocystocele
myelocystomeningocele

myelocyte
m. A, B, C
myelocytic
myelodysplasia
myelogenesis
myelogenetic
myelogenic
myelogenous
myeloic
myeloid
m. cell
m. series
m. tissue
myelomeningocele
myelomere
myelomono
myelonic
myelopetal
myelopoiesis
myelopoietic
myeloradiculodysplasia
myeloschisis
myelospongium
myenteric
m. nervous plexus
m. reflex
myentericus
plexus m.
plexus nervosus m.
myenteron
myesthesia
mylohyoid
m. branch of inferior alveolar artery
m. bridge
m. fossa
m. groove
m. line
m. muscle
nerve to m.
m. nerve
m. region
m. ridge
mylohyoidea
linea m.
mylohyoideus
musculus m.
nervus m.
ramus m.
sulcus m.
mylopharyngea
pars m.

M

NOTES

mylopharyngeal
    m. muscle
    m. part
    m. part of superior constrictor
      muscle of pharynx
    m. part of superior pharyngeal
      constrictor
mylopharyngeus
    musculus m.
myoarchitectonic
myoblast
myoblastoma
myobradia
myocardia
myocardial
myocardium, pl. **myocardia**
myocele
myochronoscope
myocinesimeter
myocomma
myocommata
myocutaneous
myocyte
    Anitschkow m.
myodermal
myodynamia
myodynamics
myodynamometer
myodystony
myoedema
myoelastic
myoelectric
myoepicardial mantle
myoepithelial cell
myoepithelium
myoesthesia
myoesthesis
myofascial
myofibril
myofibrilla, pl. **myofibrillae**
myofibroblast
myofilaments
myofunctional
myogenesis
myogenetic
myogenic
    m. potential
    m. theory
myogenous
myoglobin
    carbonmonoxy m.
myognathus
myogram
myograph
myographic
myography
myoid cells
myoidema

myoides
    musculus platysma m.
myokinesimeter
myolemma
myologia
myologist
myology
    descriptive m.
myomere
myometer
myometrial
    m. arcuate artery
    m. radial artery
myometrium
myon
myoneme
myoneural junction
myonymy
myopectineal orifice
myopia
myopic
myoplasm
myopolar
myorhythmia
myosalpinx
myoseptum
myosin filament
myosthenometer
myostroma
myotactic
myotasis
myotatic
    m. contraction
    m. irritability
    m. reflex
myothermic
myotome
myotomy
myotone
myotonoid
myotony
myotrophy
myotube
myotubule
myovascular sphincter
myovenous sphincter
myringa
myrinx
myrtiform
    caruncula m.
myrtiformis, pl. **myrtiformes**
    caruncula m.
myxasthenia
myxocyte
myxoid
myxomatous
myxopoiesis

NA
   Nomina Anatomica
nacreous
Nageotte cells
nail
   n. bed
   body of n.
   cuticle of n.
   n. fold
   free border of n.
   germinative layer of n.
   hidden border of n.
   horny layer of n.
   lateral border of n.
   layer of n.
   lunule of n.
   n. matrix
   occult border of n.
   n. plate
   proximal border of n.
   retinacula of n.
   root of n.
   sinus of n.
   n. skin
   wall of n.
   n. wall
NANC
   nonadrenergic, noncholinergic
   NANC neuron
nanism
nanocephalia
nanocephalic
nanocephalous
nanocephaly
nanocormia
nanoid
nanomelia
nanophthalmia
nanophthalmos
nanous
nanus
nape
   transverse muscle of n.
napex
narcosis
   adsorption theory of n.
   colloid theory of n.
   enzyme inhibition theory of n.
   lipoid theory of n.
   Meyer-Overton theory of n.
   oxygen deprivation theory of n.
   permeability theory of n.
   surface tension theory of n.
   thermodynamic theory of n.

narcotic hunger
naris, pl. **nares**
   anterior n.
   compressor muscle of n.
   dilator n.
   external n.
   internal n.
   musculus compressor n.
   musculus dilator n.
   posterior n.
narium
   choana n.
   tuberculum septi n.
nasal
   n. airway
   n. atrium
   n. border of frontal bone
   n. bridge
   n. capsule
   n. cavity
   n. concha
   n. crest
   n. crest of horizontal plate of
     palatine bone
   n. crest of palatine process of
     maxilla
   n. dome cartilage
   n. duct
   n. eminence
   n. foramen
   n. ganglion
   n. gland
   n. height
   n. index
   n. margin of frontal bone
   n. meatus
   n. mucosa
   n. muscle
   n. nerve
   n. notch
   n. notch of maxilla
   n. part of frontal bone
   n. passage
   n. pharynx
   n. pit
   n. placode
   n. point
   n. process
   n. pyramid
   n. region
   n. ridge
   n. sac
   n. septal branch of superior labial
     branch of facial artery
   n. septal cartilage

N

nasal *(continued)*
    n. septum
    n. sill
    n. spine
    n. spine of frontal bone
    n. surface of maxilla
    n. surface of palatine bone
    n. valve
    n. vein
    n. venous arch
    n. venule
    n. venules of retina
    n. vestibule
    n. wall
nasale
    os n.
nasalis, pl. **nasales**
    angulus oculi n.
    atrium meatus medii n.
    concha n.
    crista n.
    nasales externae
    glandulae nasales
    incisura n.
    n. muscle
    musculus n.
    pars alaris musculi n.
    pars transversa musculi n.
    nasales posteriores laterales et septi
    ramus septi posterioris n.
    regio n.
nascent
nasi
    agger n.
    ala n.
    apex n.
    arteria dorsalis n.
    cartilago septi n.
    cavitas n.
    cavum n.
    columella n.
    columna n.
    depressor septi n.
    dorsalis n.
    dorsum n.
    levator labii superioris alaeque n.
    limen n.
    meatus n.
    musculus levator alae n.
    musculus levator labii superioris
     alaeque n.
    musculus pyramidalis n.
    musculus transversalis n.
    pars cartilaginea septi n.
    pars membranacea septi n.
    pars mobilis septi n.
    pars ossea septi n.
    pinna n.
    ponticulus n.

processus posterior cartilaginis
    septi n.
processus sphenoidalis cartilaginis
    septi n.
radix n.
regio olfactoria tunicae mucosae n.
regio respiratoria tunicae
    mucosae n.
septum mobile n.
sulcus olfactorius cavi n.
tunica mucosa n.
vestibulum n.
nasioiniac
nasion
nasion-postcondylar plane
Nasmyth
    N. cuticle
    N. membrane
nasoantral
nasobasilar line
nasobregmatic arc
nasociliaris
    nervus n.
    radix n.
nasociliary
    n. branches of ophthalmic nerve
    n. root
    n. root of ciliary ganglion
nasofrontal
    n. duct
    n. suture
    n. vein
nasofrontalis
    sutura n.
    vena n.
nasogastric
nasojugal fold
nasolabial
    n. crease
    n. fold
    n. groove
    n. junction
    n. lymph node
    n. reflex
    n. sulcus
nasolabialis
    nodus n.
    nodus lymphoideus n.
    sulcus n.
nasolacrimal
    n. canal
    n. duct
    n. sac
nasolacrimalis
    canalis n.
    ductus n.
nasomandibular fixation
nasomaxillaris
    sutura n.

nasomaxillary suture
nasomental reflex
nasooccipital arc
nasooral
nasopalatine
    n. groove
    n. nerve
    n. plexus of Woodruff
    n. recess
nasopalatinus
    nervus n.
nasopharyngeal
    n. area
    n. groove
    n. meatus
    n. passage
nasopharyngeus
    meatus n.
nasopharynx
nasorostral
nasoturbinal
    concha n.
nasus externus
natal cleft
natatorium
    ligamentum n.
natatory ligament
natriferic
naturae
    lusus n.
natural
    n. killer (NK)
    n. killer cells
navel
navicula
navicular
    n. articular surface of talus
    n. bone
    n. bone of hand
    n. fossa
    n. fossa of male urethra
naviculare
    os n.
navicularicuneiformia
    ligamenta n.
navicularis
    tuberositas ossis n.
    valvula fossae n.
naviculocapitate
neate fasciculus
neck
    anterior region of n.

    anterior triangle of n.
    anterolateral n.
    bladder n.
    deep anterior n.
    deep fascia of n.
    dental n.
    fascia of n.
    femoral n.
    n. of femur
    n. of fibula
    fibular n.
    n. of gallbladder
    n. of glans
    n. of glans penis
    n. of hair follicle
    n. of humerus
    innervation of head and n.
    lateral region of n.
    long muscle of n.
    lymph node of head and n.
    n. of malleus
    n. of mandible
    median vein of n.
    muscle of n.
    muscular triangle of n.
    n. of pancreas
    posterior region of n.
    posterior triangle of n.
    radial n.
    n. of radius
    n. reflex
    region of n.
    n. of rib
    n. of scapula
    semispinal muscle of n.
    space four of n.
    spinal muscle of n.
    splenius muscle of n.
    n. of stapes
    superficial n.
    n. of talus
    n. of thigh bone
    n. of tooth
    transverse artery of n.
    transverse nerve of n.
    transverse vein of n.
    triangle of n.
    n. of urinary bladder
    n. of uterus
    n. vein
    n. of womb
necrobiosis

N

NOTES

necrobiotic
necrocytosis
necrometer
necrotomy
neencephalon
NEEP
    negative end-expiratory pressure
negative
    n. base excess
    n. chronotropism
    n. electrotaxis
    n. end-expiratory pressure (NEEP)
    n. feedback
    n. pressure
    n. supporting reaction
    n. taxis
    n. thermotaxis
negatively
    n. bathmotropic
    n. dromotropic
    n. inotropic
Negro phenomenon
Nélaton
    N. fiber
    N. fold
    N. line
    N. sphincter
nematospermia
neobiogenesis
neoblastic
neocerebellum
neocortex
neoencephalon
neofetal
neofetus
neoformation
neogenesis
neogenetic
neokinetic
neomorphism
neonatorum
    ichthyosis congenita n.
    icterus n.
neopallium
neostriatum
neothalamus
nephric
    n. blastema
    n. duct
nephroblastema
nephrocardiac
nephrocele
nephrocelom
nephrocolic ligament
nephrogenetic
nephrogenic
    n. cord
    n. tissue
nephrogenous

nephroid
nephrolumbar ganglion
nephromere
nephron
nephronic loop
nephros
nephrostoma
nephrostome
nephrotome
nephrotomic cavity
nephrotrophic
nephrotropic
Nernst
    N. equation
    N. theory
nerve
    abdominopelvic splanchnic n.
    abducens n.
    abducent n.
    accelerator n.
    accessory phrenic n.
    accessory portion of spinal
      accessory n.
    accommodation of n.
    accompanying artery of sciatic n.
    accompanying vein of
      hypoglossal n.
    acoustic n.
    afferent n.
    alveolar n.
    ampullary n.
    Andersch n.
    anococcygeal n.
    ansa cervicalis n.
    ansa subclavia n.
    anterior abdominal cutaneous
      branch of intercostal n.
    anterior alveolar branch of
      maxillary n.
    anterior ampullar n.
    anterior ampullary n.
    anterior antebrachial n.
    anterior auricular n.
    anterior crural n.
    anterior cutaneous branch of
      femoral n.
    anterior cutaneous branch of
      iliohypogastric n.
    anterior cutaneous branch of
      intercostal n.
    anterior ethmoidal n.
    anterior femoral cutaneous n.
    anterior interosseous n.
    anterior labial n.
    anterior palatine n.
    anterior pectoral cutaneous branch
      of intercostal n.
    anterior rami of cervical n.
    anterior rami of lumbar n.

anterior rami of sacral n.
anterior rami of thoracic n.
anterior ramus of spinal n.
anterior root of spinal n.
anterior scrotal n.
anterior superior alveolar branch of infraorbital n.
anterior supraclavicular n.
aortic n.
area of facial n.
Arnold n.
artery to sciatic n.
articular recurrent n.
auditory tube n.
augmentor n.
auricular branch of vagus n.
auriculotemporal n.
autonomic n.
axillary n.
baroreceptor n.
Bell respiratory n.
Bock n.
brachial plexus n.
buccal branch of facial n.
buccinator n.
canal for lesser palatine n.
cardiac n.
cardiopulmonary splanchnic n.
caroticotympanic n.
carotid branch of glossopharyngeal n.
carotid sinus n.
n. to carotid sinus
celiac branch of vagus n.
n. cell
n. cell body
centrifugal n.
centripetal n.
cerebral n.
cervical branch of facial n.
cervical n. (C1–C8)
cervical splanchnic n.
chorda tympani n.
ciliary n.
circumflex n.
cluneal n.
coccygeal n.
cochlear part of vestibulocochlear n.
cochlear root of vestibulocochlear n.
collateral branch of intercostal n.

comitant artery of median n.
common fibular n.
common palmar digital n.
common peroneal n.
common plantar digital n.
communicating branch of anterior interosseous nerve with ulnar n.
communicating branch of auriculotemporal nerve with facial n.
communicating branch of chorda tympani to lingual n.
communicating branch of facial nerve with glossopharyngeal n.
communicating branch of internal laryngeal nerve with recurrent laryngeal n.
communicating branch of lacrimal nerve with zygomatic n.
communicating branch of median nerve with ulnar n.
communicating branch of otic ganglion with medial pterygoid n.
communicating branch of otic ganglion with meningeal branch of mandibular n.
communicating branch of radial nerve with ulnar n.
communicating branch of spinal n.
communicating branch of superficial radial nerve with ulnar n.
communicating branch of tympanic plexus with auricular branch of vagus n.
companion artery to sciatic n.
companion lymph node of accessory n.
n. conduction
n. conduction velocity
cranial n. (I–XII)
cranial root of accessory n.
crural interosseous n.
cubital n.
cutaneous branch of anterior branch of obturator n.
cutaneous branch of mixed n.
cutaneous cervical n.
Cyon n.
decussation of trochlear n.'s
deep branch of lateral plantar n.
deep branch of radial n.
deep branch of ulnar n.

N

NOTES

nerve *(continued)*

deep fibular n.
deep peroneal n.
deep petrosal n.
deep temporal n.
dental n.
depressor n.
descending branch of
  hypoglossal n.
descending tract of trigeminal n.
diaphragmatic n.
digastric branch of facial n.
n. distribution
dorsal branch of ulnar n.
dorsal digital nerve of deep
  fibular n.
dorsal digital nerve of superficial
  fibular n.
dorsal digital nerve of ulnar n.
dorsal interosseous n.
dorsal lateral cutaneous n.
dorsal medial cutaneous n.
dorsal primary ramus of spinal n.
dorsal rami n.
dorsal root of spinal n.
dorsal scapular n.
dorsal sensory branch of ulnar n.
dural sheath of optic n.
efferent n.
eighth cranial n.
eleventh cranial n.
n. ending
esodic n.
esophageal branch of recurrent
  laryngeal n.
esophageal branch of vagus n.
ethmoidal n.
excitor n.
excitoreflex n.
exodic n.
n. of external acoustic meatus
external branch of superior
  laryngeal n.
external branch of trunk of
  accessory n.
external carotid n.
external nasal branch of
  infraorbital n.
external popliteal n.
external pterygoid n.
external saphenous n.
external sheath of optic n.
external spermatic n.
facial n.
n. fascicle
faucial branch of lingual n.
femoral branch of genitofemoral n.
n. fiber
fibular n.

n. field
fifth cranial n.
first cranial n.
fold of superior laryngeal n.
n. force
fourth cranial n.
fourth lumbar n.
frontal n.
furcal n.
Galen n.
gangliated n.
n. ganglion
ganglion of facial n.
ganglionic branch of lingual n.
ganglionic branch of maxillary n.
ganglionic layer of optic n.
ganglion of intermediate n.
Gaskell n.'s
geniculum of facial n.
genital branch of genitofemoral n.
genital branch of iliohypogastric n.
genitocrural n.
genitofemoral n.
genu of facial n.
glossopharyngeal n.
gluteal n.
n. of Grassi
great auricular n.
greater occipital n.
greater palatine n.
greater petrosal n.
greater splanchnic n.
greater superficial petrosal n.
great sciatic n.
groove for greater petrosal n.
groove for lesser petrosal n.
groove for radial n.
groove for spinal n.
groove for ulnar n.
n. growth factor (NGF)
n. growth factor antiserum
hemorrhoidal n.
hepatic branch of vagus n.
Hering sinus n.
hiatus of canal for greater
  petrosal n.
hiatus of canal for lesser
  petrosal n.
hypogastric n.
hypoglossal n.
iliohypogastric n.
ilioinguinal n.
n. impulse
inferior alveolar n.
inferior anal n.
inferior branch of oculomotor n.
inferior branch of transverse
  cervical n.

inferior cervical cardiac branch of vagus n.
inferior cluneal n.
inferior dental n.
inferior ganglion of glossopharyngeal n.
inferior ganglion of vagus n.
inferior gluteal n.
inferior hemorrhoidal n.
inferior labial branch of mental n.
inferior laryngeal n.
inferior lateral brachial cutaneous n.
inferior maxillary n.
inferior palpebral n.
inferior part of vestibulocochlear n.
inferior rectal n.
inferior root of vestibulocochlear n.
infraoccipital n.
infraorbital branch of maxillary n.
infrapatellar branch of saphenous n.
infratrochlear branch of ophthalmic n.
inguinal n.
inhibitory n.
inner sheath of optic n.
intercarotid n.
intercostal n.
intercostobrachial n.
intercostohumeral n.
intermediary n.
intermediate dorsal cutaneous n.
intermediate supraclavicular n.
internal branch of superior laryngeal n.
internal branch of trunk of accessory n.
internal carotid n.
internal ramus of accessory n.
internal rectal n.
internal saphenous n.
internal sheath of optic n.
interosseous anterior n.
interosseous cruris n.
interosseous dorsalis n.
interosseous posterior n.
intersheath space of optic n.
intervaginal subarachnoid space of optic n.
intracanicular part of optic n.
intracranial part of optic n.

intralaminar part of intralocular part of optic n.
intraocular part of optic n.
intraparotid plexus of facial n.
ischiadic n.
Jacobson n.
jugular n.
labial branch of mental n.
labioglossopharyngeal n.
lacrimal n.
Lancisi n.
laryngeal n.
n. of Latarjet
Latarjet n.
lateral abdominal/pectoral cutaneous branch of intercostal n.
lateral ampullar n.
lateral ampullary n.
lateral antebrachial cutaneous n.
lateral anterior thoracic n.
lateral branch of posterior rami of spinal n.
lateral calcaneal branch of sural n.
lateral cutaneous branch of intercostal n.
lateral cutaneous branch of ventral primary ramus of thoracic spinal n.
lateral dorsal cutaneous n.
lateral femoral cutaneous n.
lateral mammary branch of lateral cutaneous branch of intercostal n.
lateral mammary branch of lateral cutaneous branch of thoracic spinal n.
lateral nasal branch of anterior ethmoidal n.
lateral pectoral n.
lateral plantar n.
lateral popliteal n.
lateral pterygoid n.
lateral root of median n.
lateral supraclavicular n.
lateral sural cutaneous n.
least splanchnic n.
left hypogastric n.
lesser internal cutaneous n.
lesser occipital n.
lesser palatine n.
lesser petrosal n.
lesser splanchnic n.
lesser superficial petrosal n.

N

NOTES

nerve *(continued)*
    levator veli palatini n.
    lingual branch of facial n.
    long buccal n.
    long ciliary n.
    longitudinal n.
    long saphenous n.
    long subscapular n.
    long thoracic n.
    loops of spinal n.
    lowest splanchnic n.
    Ludwig n.
    lumbar splanchnic n.
    lumboinguinal n.
    Luschka n.
    mandibular n.
    marginal mandibular branch of
      facial n.
    masseteric n.
    masticator n.
    maxillary n.
    medial antebrachial cutaneous n.
    medial anterior thoracic n.
    medial brachial cutaneous n.
    medial branch of posterior branch
      of spinal n.
    medial branch of posterior rami of
      spinal n.
    medial calcaneal branch of
      tibial n.
    medial cluneal n.
    medial crural cutaneous branch of
      saphenous n.
    medial cutaneous n.
    medial dorsal cutaneous n.
    medial nasal branch of anterior
      ethmoidal n.
    medial pectoral n.
    medial plantar n.
    medial popliteal n.
    medial pterygoid n.
    medial root of median n.
    medial supraclavicular n.
    medial sural cutaneous n.
    median n.
    meningeal branch of mandibular n.
    meningeal branch of maxillary n.
    meningeal branch of ophthalmic n.
    meningeal branch of spinal n.
    meningeal branch of vagus n.
    mental branch of mental n.
    mesencephalic tract of trigeminal n.
    middle cardiac cervical n.
    middle cervical cardiac n.
    middle cluneal n.
    middle meningeal branch of
      maxillary n.
    middle superior alveolar branch of
      infraorbital n.

    middle supraclavicular n.
    mixed n.
    motor nucleus of facial n.
    motor root of trigeminal n.
    musculocutaneous n.
    musculospiral n.
    n. to mylohyoid
    mylohyoid n.
    nasal n.
    nasociliary branches of
      ophthalmic n.
    nasopalatine n.
    nervus intermedius n.
    ninth cranial n.
    nuclei of cranial n.
    nucleus of abducent n.
    nucleus of oculomotor n.
    nucleus of trochlear n.
    obturator n.
    occipital n.
    ocular n.
    oculomotor n.
    olfactory n.
    olivocochlear n.
    Oort n.
    ophthalmic n.
    optic n.
    orbital branch of maxillary n.
    orbital part of optic n.
    outer sheath of optic n.
    palatine n.
    palmar branch of anterior
      interosseous n.
    palmar branch of median n.
    palmar branch of ulnar n.
    palmar digital n.
    palpebral branch of
      infratrochlear n.
    n. papilla
    parasympathetic n.
    parotid plexus of facial n.
    pathetic n.
    pectoral and abdominal anterior
      cutaneous branch of intercostal n.
    pelvic autonomic n.
    pelvic splanchnic n.
    pericardial branch of phrenic n.
    pericardiophrenic n.
    perineal branch of posterior
      femoral cutaneous n.
    peripheral n.
    peritonsillar n.
    peroneal communicating n.
    petrosal n.
    pharyngeal branch of
      glossopharyngeal n.
    pharyngeal branch of recurrent
      laryngeal n.
    pharyngeal branch of vagus n.

phrenic n.
phrenicoabdominal branch of
  phrenic n.
plantar digital n.
n. plexus
plexus of spinal n.
pneumogastric n.
popliteal communicating n.
posterior ampullar n.
posterior ampullary n.
posterior antebrachial cutaneous n.
posterior auricular n.
posterior brachial cutaneous n.
posterior branch of great
  auricular n.
posterior branch of medial
  antebrachial cutaneous n.
posterior branch of obturator n.
posterior branch of spinal n.
posterior cutaneous femoral n.
posterior ethmoidal n.
posterior inferior nasal branch of
  greater palatine n.
posterior interosseous n.
posterior labial n.
posterior ramus of spinal n.
posterior root of spinal n.
posterior scapular n.
posterior scrotal n.
posterior superior alveolar branch
  of maxillary n.
posterior superior lateral nasal
  branch of maxillary n.
posterior superior medial nasal
  branch of maxillary n.
posterior supraclavicular n.
posterior thoracic n.
posterior tibial n.
postlaminar part of intraocular part
  of optic n.
prelaminar part of intraocular part
  of optic n.
presacral n.
pressor n.
pressoreceptor n.
proper palmar digital n.
proper plantar digital n.
pterygoid n.
n. of pterygoid canal
pterygopalatine n.
pudendal n.
pudic n.

pular n.
radial n.
rectal n.
recurrent branch of spinal n.
recurrent laryngeal n.
recurrent meningeal branch of
  spinal n.
renal branch of lesser
  splanchnic n.
renal branch of vagus n.
n. to rhomboid
right common iliac n.
right hypogastric n.
n. root
root of facial n.
n. rootlet
roots of trigeminal n.
saccular n.
sacral splanchnic n.
saphenous n.
Scarpa n.
sciatic n.
scrotal n.
second cranial n.
secretomotor n.
secretory n.
sensory branch of radial n.
  (SBRN)
sensory root of spinal n.
sensory root of trigeminal n.
seventh cranial n.
short ciliary n.
short saphenous n.
n. signal
sinocarotid n.
sinus n.
sinuvertebral n.
sixth cranial n.
small deep petrosal n.
smallest splanchnic n.
small sciatic n.
n. of smell
somatic n.
spermatic n.
sphenopalatine n.
spinal nucleus of accessory n.
spinal part of accessory n.
spinal root of accessory n.
spinal tract of trigeminal n.
splanchnic n.
stapedial n.
n. to stapedius muscle

NOTES

N

nerve *(continued)*
statoacoustic n.
stylohyoid branch of facial n.
stylopharyngeal branch of
  glossopharyngeal n.
subclavian n.
subcostal n.
subcutaneous temporal n.
sublingual n.
submaxillary n.
suboccipital n.
subscapular n.
sudomotor n.
sulcus for greater palatine n.
sulcus of petrosal n.
superficial branch of lateral
  plantar n.
superficial branch of radial n.
superficial branch of ulnar n.
superficial cervical n.
superficial fibular n.
superficial middle petrosal n.
superficial peroneal n.
superficial radial n.
superficial temporal branch of
  auriculotemporal n.
superior alveolar n.
superior ampullar n.
superior branch of the
  oculomotor n.
superior branch of the transverse
  cervical n.
superior cervical cardiac branch of
  vagus n.
superior cluneal n.
superior dental n.
superior ganglion of
  glossopharyngeal n.
superior ganglion of vagus n.
superior gluteal n.
superior labial branch of
  infraorbital n.
superior laryngeal n.
superior lateral brachial
  cutaneous n.
superior maxillary n.
superior part of
  vestibulocochlear n.
superior pharyngeal constrictor n.
superior root of
  vestibulocochlear n.
supraclavicular n.
supraorbital n.
suprascapular n.
supratrochlear n.
supreme cardiac n.
sural communicating branch of
  common fibular n.

sural communicating branch of
  common peroneal n.
sympathetic n.
temporal branch of facial n.
temporomandibular n.
n. to tensor tympani muscle
n. to tensor veli palatini muscle
tenth cranial n.
tentorial n.
terminal n.
third cranial n.
third occipital n.
thoracic cardiac branch of
  vagus n.
thoracic splanchnic n.
thoracoabdominal n.
thoracodorsal n.
n. to thyrohyoid muscle
tibial communicating n.
Tiedemann n.
n. tissue
tonsillar branch of
  glossopharyngeal n.
tonsillar branch of lesser
  palatine n.
n. tract
transverse cervical n.
transverse colli n.
trifacial n.
trigeminal n.
trigeminus n.
trigone of auditory n.
trigone of hypoglossal n.
trigone of vagus n.
trochlear n.
n. trunk
twelfth cranial n.
tympanic n.
n. of tympanic membrane
ulnar branch of medial antebrachial
  cutaneous n.
ulnar communicating branch of
  superficial radial n.
upper subscapular n.
upper thoracic splanchnic n.
utricular n.
utriculoampullar n.
vagal part of accessory n.
vaginal n.
vagus n.
Valentin n.
vascular circle of optic n.
vasoconstrictor n.
vasomotor n.
vena comitans of hypoglossal n.
venous plexus of canal of
  hypoglossal n.
ventral primary rami of cervical
  spinal n.

ventral primary rami of lumbar
spinal n.
ventral primary rami of sacral
spinal n.
ventral primary rami of thoracic
spinal n.
ventral primary ramus of spinal n.
ventral root of spinal n.
vertebral n.
vestibular part of
vestibulocochlear n.
vestibular root of
vestibulocochlear n.
vestibulocochlear n.
vidian n.
volar interosseous n.
vomeronasal n.
Wrisberg n.
n. of Wrisberg
zygomatic branch of facial n.
zygomaticofacial branch of
zygomatic n.
zygomaticotemporal branch of
zygomatic n.

nervea
tunica n.
nervi (*pl. of* nervus)
nervimotility
nervimotion
nervimotor
nervorum
arteriae n.
nervi n.
rami mammarii mediales ramorum
cutaneorum anteriorum n.
ramus cutaneus anterior (pectoralis
et abdominalis) n.
vasa n.
nervosa
foramina n.
nervosi
pars centralis systematis n.
pars cranialis partis
parasympathetici divisionis
autonomici systematis n.
pars peripherica systematis n.
nervosum
systema n.
Willis centrum n.
nervosus
lobus n.
plexus pudendus n.

nervous
n. force
n. lobe
n. part of retina
n. system
n. tissue
n. tunic of eyeball
nervus, pl. **nervi**
n. abducens
nervi abducentis
n. accessorius
n. acusticus
nervi alveolares superiores
nervi alveolares superiores
anteriores
n. alveolaris inferior
n. ampullaris anterior
n. ampullaris lateralis
n. ampullaris posterior
nervi anales inferiores
nervi anococcygei
n. anococcygeus
n. antebrachii anterior
n. antebrachii posterior
n. articularis
nervi auriculares anteriores
n. auricularis magnus
n. auricularis posterior
n. auriculotemporalis
n. autonomicus
n. axillaris
n. buccalis
n. canalis pterygoidei
nervi cardiaci thoracici
n. cardiacus cervicalis inferior
n. cardiacus cervicalis medius
n. cardiacus cervicalis superior
nervi carotici externi
nervi caroticotympanicus
n. caroticus internus
nervi cavernosi clitoridis
nervi cavernosi penis
nervi cervicales
n. cervicalis superficialis
nervi ciliares breves
nervi ciliares longi
n. ciliaris brevis
n. ciliaris longus
nervi clunium inferiores
nervi clunium medii
nervi clunium superiores
n. coccygeus

N

NOTES

nervus *(continued)*

n. cochlearis
n. communicans fibularis
n. communicans peroneus
nervi craniales
n. cutaneus
n. cutaneus antebrachii lateralis
n. cutaneus antebrachii medialis
n. cutaneus antebrachii posterior
n. cutaneus brachii lateralis
n. cutaneus brachii lateralis inferior
n. cutaneus brachii lateralis superior
n. cutaneus brachii medialis
n. cutaneus brachii posterior
n. cutaneus dorsalis intermedius
n. cutaneus dorsalis lateralis
n. cutaneus dorsalis medialis
n. cutaneus femoris lateralis
n. cutaneus femoris posterior
n. cutaneus surae lateralis
n. cutaneus surae medialis
nervi digitales dorsales
nervi digitales dorsales nervi fibularis profundi
nervi digitales dorsales nervi fibularis superficialis
nervi digitales dorsales nervi ulnaris
nervi digitales dorsales pedis
nervi digitales palmares communes
nervi digitales palmares proprii
nervi digitales plantares communes
nervi digitales plantares proprii
n. dorsalis clitoridis
n. dorsalis penis
n. dorsalis scapulae
nervi erigentes
n. ethmoidalis anterior
n. ethmoidalis posterior
n. facialis
n. femoralis
n. fibularis communis
n. fibularis profundus
n. fibularis superficialis
n. frontalis
n. furcalis
n. genitofemoralis
n. glossopharyngeus
n. gluteus inferior
n. gluteus superior
n. hemorrhoidalis
n. hypogastricus
n. hypoglossus
n. iliohypogastricus
n. ilioinguinalis
n. impar
n. infraorbitalis
n. infratrochlearis

nervi intercostales
nervi intercostobrachiales
n. intermedius nerve
n. interosseus antebrachii anterior
n. interosseus antebrachii posterior
n. interosseus anterior
n. interosseus cruris
n. interosseus dorsalis
n. interosseus posterior
n. ischiadicus
n. jugularis
nervi labiales anteriores
nervi labiales posteriores
n. lacrimalis
n. laryngeus inferior
n. laryngeus recurrens
n. laryngeus superior
n. lingualis
n. mandibularis
n. massetericus
n. maxillaris
n. meatus acustici externi
n. medianus
n. mentalis
n. mixtus
n. musculi tensoris tympani
n. musculi tensoris veli palatini
n. musculocutaneus
n. mylohyoideus
n. nasociliaris
n. nasopalatinus
nervi nervorum
n. obturatorius
n. occipitalis major
n. occipitalis minor
n. occipitalis tertius
n. octavus
n. oculomotorius
n. olfactorii
nervi olfactorii
n. ophthalmicus
n. opticus
nervi palatini minores
n. palatinus major
n. pectoralis lateralis
n. pectoralis medialis
nervi pelvici splanchnici
nervi perineales
n. peroneus communis
n. peroneus profundus
n. peroneus superficialis
n. petrosus major
n. petrosus minor
n. petrosus profundus
n. pharyngeus
nervi phrenici accessorii
n. phrenicus
n. plantaris lateralis
n. plantaris medialis

n. presacralis
n. pterygoideus
nervi pterygopalatini
n. pudendus
n. radialis
rami temporales superficiales nervi
nervi rectales inferiores
n. saccularis
n. saphenus
n. sciaticus
nervi scrotales anteriores
nervi scrotales posteriores
n. spermaticus externus
nervi sphenopalatini
nervi spinales
n. spinosus
nervi splanchnici lumbales
nervi splanchnici pelvini
nervi splanchnici sacrales
n. splanchnicus imus
n. splanchnicus major
n. splanchnicus minor
n. stapedius
n. statoacusticus
n. subclavius
n. subcostalis
n. sublingualis
n. suboccipitalis
nervi subscapulares
n. subscapularis
n. supraclavicularis intermedius
n. supraclavicularis lateralis
n. supraclavicularis medialis
n. supraorbitalis
n. suprascapularis
n. supratrochlearis
n. suralis
nervi temporales profundi
n. tensoris veli palatini
n. tentorii
nervi terminales
n. terminalis
n. thoracicus longus
n. thoracodorsalis
n. tibialis
n. transversus cervicalis
n. transversus colli
n. trigeminus
n. trochlearis
n. tympanicus
n. ulnaris
n. utricularis

n. utriculoampullaris
nervi vaginales
n. vagus
n. vascularis
nervi vascularorum
n. vertebralis
n. vestibularis
n. vestibulocochlearis
n. zygomaticus

nesidioblast
nest

Brunn n.

net

chromidial n.
n. flux

network

acromial arterial n.
arterial n.
arteriolar n.
articular vascular n.
calcaneal arterial n.
chromatin n.
cytokine n.
dorsal carpal n.
n. of heel
lateral malleolar n.
linin n.
medial malleolar n.
patellar n.
peritarsal n.
plantar venous n.
subpapillary n.
trabecular n.

Neubauer artery
Neumann

N. cells
N. sheath

neural

alar plate of n.
n. arch
n. arch of vertebra
n. axis
n. canal
n. crest
n. fold
n. foramen
n. ganglion
n. impulse
n. layer of optic part of retina
n. parenchyma
n. plate
n. segment

NOTES

N

neural *(continued)*
    n. spine
    n. tube
neuramebimeter
neuranagenesis
neurapophysis
neurarchy
neuraxis
neuraxon
neuraxone
neurectomy
    splanchnic n.
neurenteric canal
neurepithelium
neurergic
neurilemma cells
neurility
neurimotility
neurimotor
neurite
neuroanatomy
neurobiology
neurobiotactic movement
neurobiotaxis
neuroblast
neurocardiac
neurocele
neurocentral
    n. joint
    n. suture
    n. synchondrosis
neurochemistry
neurochitin
neurocladism
neurocranium
    cartilaginous n.
    membranous n.
neurocristopathy
neurocyte
neurodendrite
neurodendron
neurodynamic
neuroectoderm
neuroectodermal junction
neuroendocrine
    n. transducer cell
neuroendocrinology
neuroepithelial
    n. body
    n. cells
    n. layer of retina
neuroepithelium
    n. of ampullary crest
    n. cristae ampullaris
    n. of macula
neurofibrae
    n. autonomicae
    n. somaticae
neurofibril

neurofibrillar
neurofilament
neuroganglion
neurogastric
neurogenesis
neurogenetic
neurogenic theory
neurogenous
neuroglia cells
neurogliacyte
neuroglial, neurogliar
neurogram
neurohemal organ
neurohistology
neurohumor
neurohumoral
    n. secretion
    n. transmission
neurohypophysial
neurohypophysis
neuroid
neurokeratin
neurolemma cells
neurologic
neurological
neurolymph
neuromelanin
neuromere
neuromimetic
neuromuscular
    n. cell
    n. junction
    n. spindle
    n. system
neuron
    autonomic motor n.
    bipolar n.
    efferent n.
    gamma motor n.
    ganglionic motor n.
    Golgi type I, II n.
    intercalary n.
    internuncial n.
    layer of piriform n.
    lower motor n.
    motor n.
    multipolar n.
    NANC n.
    nonadrenergic, noncholinergic n.
    postganglionic motor n.
    preganglionic motor n.
    premotor n.
    pseudounipolar n.
    unipolar n.
    upper motor n.
    visceral motor n.
neuronal
neurone
neuronephric

neuronophage
neuropeptide Y
neurophilic
neurophysiology
neuropile
neuroplasm
neuropodia
neuropore
    caudal n.
    rostral n.
neurosecretion
neurosecretory
    n. cells
    n. substance
neurosomatic junction
neurosplanchnic
neurospongium
neurotaxis
neurotendinous
    n. organ
    n. spindle
neurothele
neurotization
neurotize
neurotonic reaction
neurotransmission
neurotransmitter
neurotrophic
neurotrophy
neurotropic attraction
neurotropism
neurotropy
neurotubule
neurovascular
    n. bundle
    n. sheath
neurovegetative
neurovisceral
neurula, pl. neurulae
neurulation
neutrophil
    band n.
    n. granule
    stab n.
neutrophile
neutrophilic
neutrophilous
nevi (pl. of nevus)
nevocyte
nevoid hypertrichosis
nevose

nevous
nevus, pl. nevi
    n. anemicus
    n. cell
    n. unius lateris
newborn
    jaundice of n.
    postnatal pit of n.
newtonian
    n. constant of gravitation
    n. flow
    n. fluid
    n. viscosity
Newton law
newton-meter
NGF
    nerve growth factor
niche
    enamel n.
nicotinic receptor
nicotinomimetic
nidal
nidation
nidus, pl. nidi
    n. avis
    n. hirundinis
Niemann-Pick cell
niger
    locus n.
    nucleus n.
nigra, pl. nigrae
    rami substantiae nigrae
    substantia n.
nigrostriatal
nigrum
    pigmentum n.
    tapetum n.
ninth cranial nerve
nipple
    areola of n.
    n. line
Nissl
    N. bodies
    N. degeneration
    N. granule
    N. substance
Nitabuch
    N. layer
    N. membrane
    N. stria
nitrate respiration

NOTES

nitric
    n. oxide (NO)
    n. oxide synthase
nitrogen
    n. cycle
    n. equivalent
    n. lag
nitrogenous equilibrium
NK
    natural killer
    NK cell
N-methyltransferase
    phenylethanolamine N.-m. (PNMT)
NO
    nitric oxide
    NO synthase
noble cells
Noble-Collip procedure
nociceptive reflex
nociceptor
nocifensor
nocturnal emission
nodal tissue
node
    abdominal lymph n.
    accessory nerve lymph n.
    anorectal lymph n.
    anterior axillary lymph n.
    anterior deep cervical lymph n.
    anterior jugular lymph n.
    anterior mediastinal lymph n.
    anterior superficial cervical
     lymph n.
    anterior tibial lymph n.
    aortic lymph n.
    apical axillary lymph n.
    appendicular lymph n.
    artery to atrioventricular n.
    artery to the sinoatrial n.
    n. of Aschoff and Tawara
    atrioventricular n.
    axillary lymph n.
    n. of azygos arch
    bifurcation lymph n.
    Bouchard n.
    brachial lymph n.
    brachiocephalic lymph n.
    branch to atrioventricular n.
    branch to sinuatrial n.
    bronchopulmonary lymph n.
    buccal lymph n.
    buccinator n.
    Calot n.
    carinal lymph n.
    caval n.
    cecal mesocolic lymph n.
    celiac lymph n.
    central axillary lymph n.
    central mesenteric lymph n.

central superior mesenteric
  lymph n.
cervical paratracheal lymph n.
Cloquet n.
n. of Cloquet
colic lymph n.
common iliac lymph n.
coronary n.
cortex of lymph n.
cubital lymph n.
cystic lymph n.
deep anterior cervical lymph n.
deep cervical jugulodigastric n.
deep inguinal lymph n.
deep lateral cervical lymph n.
deep parotid lymph n.'s
deep popliteal lymph n.
diaphragmatic n.
epitrochlear n.
external iliac lymph n.
facial lymph n.
femoral n.
fibular lymph n.
Flack n.
foraminal lymph n.
gastroduodenal lymph n.
gastroepiploic lymph n.
gluteal lymph n.'s
hemal n.
hemolymph n.
Hensen n.
hepatic lymph n.
hilar lymph n.
hilum of lymph n.
His-Tawara n.
humeral axillary lymph n.
ileocolic lymph n.'s
iliac lymph n.
inferior epigastric lymph n.
inferior mesenteric lymph n.
inferior phrenic lymph n.
inferior tracheobronchial lymph n.
infraauricular deep parotid
  lymph n.
infraauricular subfascial parotid
  lymph n.
infrahyoid lymph n.
infrarenal n.
inguinal lymph n.
intercostal lymph n.
interiliac lymph n.'s
intermediate lacunar lymph n.
intermediate lumbar lymph n.
internal iliac lymph n.
interpectoral lymph n.
intraglandular deep parotid
  lymph n.
intrapulmonary lymph n.
jugular lymph n.

jugulodigastric lymph n.
juguloomohyoid lymph n.
juxtaesophageal pulmonary
  lymph n.'s
juxtaintestinal mesenteric lymph n.
juxtaregional lymph n.
Keith and Flack n.
Koch n.
lateral axillary lymph n.
lateral jugular lymph n.
lateral lacunar lymph n.
lateral pericardial lymph n.
left colic lymph n.
left gastric lymph n.'s
left gastroepiploic lymph n.
left gastroomental lymph n.
left lumbar lymph n.'s
n. of ligamentum arteriosum
lingual lymph n.
lumbar lymph n.'s
lymph n.
malar lymph n.
mandibular lymph n.
mastoid lymph n.
medial lacunar n.
medial lacunar lymph n.
medulla of lymph n.
mesenteric lymph n.
mesocolic lymph n.
middle colic lymph n.
middle group of mesenteric
  lymph n.
middle rectal lymph n.
mph n.'s
nasolabial lymph n.
obturator lymph n.
occipital lymph n.
oraminal n.
pancreatic lymph n.
pancreaticoduodenal lymph n.
pancreaticosplenic lymph n.
paramammary lymph n.
parapharyngeal lymph n.
pararectal lymph n.
parasternal lymph n.'s
paratracheal lymph n.
parauterine lymph n.
paravaginal lymph n.
paravesical lymph n.
parietal lymph n.
parotid lymph n.
pectoral axillary lymph n.

pelvic lymph n.
periaortic n.
perigastric lymph n.
peroneal lymph n.
pharyngeal lymph n.
phrenic lymph n.
popliteal lymph n.
posterior axillary lymph n.
posterior mediastinal lymph n.
posterior tibial lymph n.
preaortic lymph n.
preauricular deep parotid lymph n.
preauricular subfascial parotid
  lymph n.
prececal lymph n.
prelaryngeal lymph n.
prepericardial lymph n.
presymphyseal lymph n.
pretracheal lymph n.
prevertebral lymph n.
primitive n.
promontorial common iliac n.
promontory lymph n.
proximal deep inguinal lymph n.
pulmonary lymph n.
pyloric lymph n.
Ranvier n.
regional n.
retroauricular lymph n.
retrocecal lymph n.
retroperitoneal lymph n.
retropharyngeal lymph n.
retropyloric lymph n.
retrorectal lymph n.
right colic lymph n.
right gastric lymph n.
right gastroepiploic lymph n.
right gastroomental lymph n.
right lumbar lymph n.
Rosenmüller n.
n. of Rouviere
S-A n.
sacral lymph n.
scalene lymph n.
Schmorl n.
sentinel n.
sigmoid lymph n.
sinoatrial n. (S-A)
sinus n.
Sister Mary Joseph lymph n.
splenic lymph n.
subaortic lymph n.

N

NOTES

node (*continued*)
 subclavian lymph n.
 subdigastric n.
 sublingual lymph n.
 submandibular lymph n.
 submental lymph n.
 subpyloric lymph n.
 subscapular axillary lymph n.
 succulent n.
 superficial inguinal lymph n.
 superficial lateral cervical lymph n.
 superficial parotid lymph n.
 superficial popliteal lymph n.
 superior gastric lymph n.
 superior mesenteric lymph n.
 superior phrenic lymph n.
 superior rectal lymph n.
 superior tracheobronchial lymph n.
 supraclavicular lymph n.
 suprapyloric lymph n.
 Tawara n.
 thoracic lymph n.
 thyroid lymph n.
 tibial lymph n.
 trabeculae of lymph n.
 tracheal lymph n.
 tracheobronchial n.
 Troisier n.
 visceral n.
 visceral lymph n.
 vital n.
nodi (*pl. of* nodus)
nodose ganglion
nodosity
nodule
 aggregated lymphatic n.
 aggregated lymphoid n.
 Albini n.
 Arantius n.
 Bianchi n.
 Bohn n.
 fibrous n.
 gastric lymphoid n.
 Hoboken n.
 juxtaarticular n.
 Koeppe n.
 laryngeal lymphoid n.
 lymph n.
 lymphatic n.
 malpighian n.
 Morgagni n.
 primary n.
 rantius n.
 secondary n.
 n. of semilunar valve
 solitary lymphatic n.
 splenic lymph n.
nodulus, pl. **noduli**
 n. caroticus

 n. lymphaticus
 noduli lymphoidei aggregati appendicis vermiformis
 noduli lymphoidei solitarii
 n. valvulae semilunaris
 noduli valvularum semilunarium
nodus, pl. **nodi**
 n. atrioventricularis
 n. foraminis
 n. jugulodigastricus
 n. lacunaris medialis
 nodi lymphatici
 nodi lymphatici centrales
 nodi lymphatici colici
 nodi lymphatici comitantes nervi accessorii
 nodi lymphatici iliaci communes mediales
 nodi lymphatici iliaci externi laterales
 nodi lymphatici iliaci externi mediales
 nodi lymphatici mesenterici inferiores
 nodi lymphatici mesenterici superiores
 nodi lymphatici pancreatici superiores
 nodi lymphatici paravesiculares
 nodi lymphatici postcavales
 nodi lymphatici postvesiculares
 nodi lymphatici prevesiculares
 nodi lymphatici vesicales laterales
 n. lymphaticus
 nodi lymphoidei abdominis
 nodi lymphoidei abdominis viscerales
 nodi lymphoidei accessorii
 nodi lymphoidei anorectales
 nodi lymphoidei appendiculares
 nodi lymphoidei axillares
 nodi lymphoidei axillares anteriores
 nodi lymphoidei axillares apicales
 nodi lymphoidei axillares centrales
 nodi lymphoidei axillares humerales
 nodi lymphoidei axillares laterales
 nodi lymphoidei axillares pectorales
 nodi lymphoidei axillares posteriores
 nodi lymphoidei axillares subscapulares
 nodi lymphoidei brachiales
 nodi lymphoidei brachiocephalici
 nodi lymphoidei bronchopulmonale
 nodi lymphoidei capitis et colli
 nodi lymphoidei centrales
 nodi lymphoidei cervicales anteriores

nodi lymphoidei cervicales
anteriores profundi
nodi lymphoidei cervicales
anteriores superficiales
nodi lymphoidei cervicales laterales
profundi
nodi lymphoidei cervicales laterales
superficiales
nodi lymphoidei coeliaci
nodi lymphoidei colici dextri
nodi lymphoidei colici medii
nodi lymphoidei colici sinistri
nodi lymphoidei cubitales
nodi lymphoidei epigastrici
inferiores
nodi lymphoidei faciales
nodi lymphoidei gastrici dextri
nodi lymphoidei gastrici sinistri
nodi lymphoidei gastroomentales
dextri
nodi lymphoidei gastroomentales
sinistri
nodi lymphoidei gluteales
nodi lymphoidei hepatici
nodi lymphoidei ileocolici
nodi lymphoidei iliaci communes
nodi lymphoidei iliaci communes
promontorii
nodi lymphoidei iliaci externi
nodi lymphoidei iliaci interni
nodi lymphoidei inguinales profundi
nodi lymphoidei inguinales
superficiales
nodi lymphoidei intercostales
nodi lymphoidei interiliaci
nodi lymphoidei interpectorales
nodi lymphoidei intrapulmonales
nodi lymphoidei jugulares anteriores
nodi lymphoidei jugulares laterales
nodi lymphoidei juxtaesophageales
nodi lymphoidei juxtaesophageales
pulmonales
nodi lymphoidei juxtaintestinales
nodi lymphoidei lienales
nodi lymphoidei linguales
nodi lymphoidei lumbales dextri
nodi lymphoidei lumbales
intermedii
nodi lymphoidei lumbales sinistri
nodi lymphoidei mastoidei
nodi lymphoidei mediastinales
anteriores

nodi lymphoidei mediastinales
posteriores
nodi lymphoidei membri inferioris
nodi lymphoidei membri superioris
nodi lymphoidei mesenterici
nodi lymphoidei mesenterici
inferiores
nodi lymphoidei mesenterici
superiores
nodi lymphoidei mesocolici
nodi lymphoidei obturatorii
nodi lymphoidei occipitales
nodi lymphoidei pancreatici
nodi lymphoidei
pancreaticoduodenales
nodi lymphoidei pancreaticolienales
nodi lymphoidei pancreaticosplenales
nodi lymphoidei paracolici
nodi lymphoidei paramammarii
nodi lymphoidei pararectales
nodi lymphoidei parasternales
nodi lymphoidei paratracheales
nodi lymphoidei parauterini
nodi lymphoidei paravaginales
nodi lymphoidei parietales
nodi lymphoidei parotidei
intraglandulares
nodi lymphoidei parotidei profundi
nodi lymphoidei parotidei profundi
infraauriculares
nodi lymphoidei parotidei profundi
preauriculares
nodi lymphoidei parotidei
superficiales
nodi lymphoidei pelvis
nodi lymphoidei pericardiales
laterales
nodi lymphoidei phrenici inferiores
nodi lymphoidei phrenici superiores
nodi lymphoidei popliteales
nodi lymphoidei precaecales
nodi lymphoidei prelaryngeales
nodi lymphoidei prepericardiaci
nodi lymphoidei pretracheales
nodi lymphoidei prevertebrales
nodi lymphoidei promontorii
nodi lymphoidei pylorici
nodi lymphoidei rectales superiores
nodi lymphoidei retrocecales
nodi lymphoidei retropharyngeales
nodi lymphoidei retropylorici
nodi lymphoidei sacrales

N

NOTES

nodus (*continued*)
    nodi lymphoidei sigmoidei
    nodi lymphoidei splenici
    nodi lymphoidei subaortici
    nodi lymphoidei submandibulares
    nodi lymphoidei submentales
    nodi lymphoidei subpylorici
    nodi lymphoidei superiores centrales
    nodi lymphoidei supraclaviculares
    nodi lymphoidei thoracis
    nodi lymphoidei thyroidei
    nodi lymphoidei tracheobronchiales inferiores
    nodi lymphoidei tracheobronchiales superiores
    n. lymphoideus
    n. lymphoideus arcus venae azygos
    n. lymphoideus buccinatorius
    n. lymphoideus cysticus
    n. lymphoideus fibularis
    n. lymphoideus foraminalis
    n. lymphoideus jugulodigastricus
    n. lymphoideus juguloomohyoideus
    n. lymphoideus lacunaris intermedius
    n. lymphoideus lacunaris lateralis
    n. lymphoideus lacunaris medialis
    n. lymphoideus ligamenti arteriosi
    n. lymphoideus malaris
    n. lymphoideus mandibularis
    n. lymphoideus nasolabialis
    n. lymphoideus proximalis profundus
    n. lymphoideus rectalis medius
    n. lymphoideus suprapyloricus
    n. lymphoideus tibialis posterior
    n. nasolabialis
    nodi paramammarii
    n. rectalis media
    nodi retropylorici
    n. sinuatrialis
    nodi subpylorici
    n. tibialis anterior
    n. tibialis posterior
noeud vital
Nomina Anatomica (NA)
nomogram
    Radford n.
    Siggaard-Andersen n.
nomotopic
nonadrenergic,
    n. noncholinergic (NANC)
    n. noncholinergic neuron
noncellular
noncholinergic
    nonadrenergic, n. (NANC)
noncommunicating hydrocephalus
nondeciduous placenta

nondisjunction
    primary n.
    secondary n.
nonlamellar bone
nonmedullated fibers
nonmyelinated
non-newtonian fluid
nonnucleated
nonoptic part of retina
nonosteogenic fibroma
nonpedunculated hydatid
nonrotation
    n. of intestine
    n. of kidney
nonsexual generation
nonshivering thermogenesis
nonspecific system
nonsuppressible insulinlike activity (NSILA)
nonvascular
nonviable
norma, pl. **normae**
    n. anterior
    n. basilaris
    n. facialis
    n. frontalis
    n. inferior
    n. lateralis
    n. occipitalis
    n. sagittalis
    n. superior
    n. temporalis
    n. ventralis
    n. verticalis
normobaric
normocapnia
normocephalic
normosthenuria
normotensive
normothermia
normotonic
normotopia
normotopic
normoxia
nose
    ala of n.
    alar artery of n.
    apex of n.
    artery of n.
    bridge of n.
    cartilage of n.
    cleft n.
    dilator muscle of n.
    dorsal artery of n.
    dorsum of n.
    elevator muscle of upper lip and wing of n.
    external artery of n.
    hairs of vestibule of n.

lateral crus of the major alar
  cartilage of the n.
medial crus of major alar cartilage
  of n.
mucosa of n.
mucous membrane of n.
olfactory region of mucosa of n.
olfactory region of tunica mucosa
  of n.
olfactory sulcus of n.
posterior septal artery of n.
posterior septal branch of n.
respiratory region of tunica mucosa
  of n.
root of n.
sesamoid cartilage of n.
threshold of n.
tip of n.
vestibule of n.
wing of n.

nostril
notal
notancephalia
notanencephalia
notch

acetabular n.
n. of acetabulum
angular n.
antegonial n.
aortic n.
n. of apex of heart
auricular n.
cardiac n.
n. of cardiac apex
cardial n.
n. in cartilage of acoustic meatus
cerebellar n.
clavicular n.
coracoid n.
costal n.
cotyloid n.
craniofacial n.
digastric n.
ethmoidal n.
fibular n.
frontal n.
gastric n.
greater ischiatic n.
greater sacrosciatic n.
greater sciatic n.
hamular n.
iliosciatic n.

inferior thyroid n.
inferior vertebral n.
infraorbital n.
interarytenoid n.
interclavicular n.
intercondylar n.
intercondyloid n.
interlobar n.
intertragic n.
interventricular n.
intervertebral n.
ischiatic n.
jugular n.
lacrimal n.
lesser sciatic n.
n. for ligamentum teres
mandibular n.
marsupial n.
mastoid n.
muscle of terminal n.
nasal n.
pancreatic n.
parietal n.
parotid n.
popliteal n.
preoccipital n.
presternal n.
pterygoid n.
pterygomaxillary n.
radial n.
rib n.
rivinian n.
Rivinus n.
n. for round ligament of liver
sacroiliac n.
sacrosciatic n.
scapular n.
sciatic n.
semilunar n.
Sibson n.
sigmoid n.
sphenopalatine n.
sternal n.
sternoclavicular n.
superior thyroid n.
superior vertebral n.
supraorbital n.
suprascapular n.
suprasternal n.
tentorial n.
n. of tentorium
thyroid n.

N

NOTES

411

notch *(continued)*
    trochlear n.
    tympanic n.
    ulnar n.
    umbilical n.
    vertebral n.
notched
notencephalocele
notochord
notochordal
    n. canal
    n. plate
    n. sheath
NSILA
    nonsuppressible insulinlike activity
nucha
nuchae
    fascia n.
    ligamentum n.
    musculus transversus n.
    transversus n.
nuchal
    n. fascia
    n. ligament
    n. line
    n. plane
    n. region
    n. tubercle
nuchalis
    regio n.
Nuck
    canal of N.
    N. diverticulum
    patent canal of N.
nuclear
    n. bag
    n. bag fiber
    n. chain fiber
    n. envelope
    n. hyaloplasm
    n. jaundice
    n. layers of retina
    n. membrane
    n. pore
    n. sap
    n. spindle
nuclear-cytoplasm
nucleated
nuclei *(pl. of* nucleus)
nucleiform
nucleochylema
nucleochyme
nucleofugal
nucleoid
    Lavdovsky n.
nucleolar
    n. organizer
    n. satellite
    n. zone

nucleoli *(pl. of* nucleolus)
nucleoloid
nucleolonema
nucleolus, pl. nucleoli
nucleomicrosome
nucleopetal
nucleoplasm
nucleoreticulum
nucleorrhexis
nucleosome
nucleospindle
nucleus, pl. nuclei
    abducens n.
    n. abducentis
    n. of abducent nerve
    accessory cuneate n.
    accessory olivary nuclei
    n. accumbens septi
    n. acusticus
    n. alae cinereae
    almond n.
    ambiguous n.
    n. ambiguus
    n. amygdalae
    amygdaloid n.
    nuclei anteriores thalami
    n. anterodorsalis
    anterodorsal thalamic n.
    n. anteromedialis
    anteromedial thalamic n.
    n. anteroventralis
    anteroventral thalamic n.
    arcuate n.
    nuclei arcuati
    n. arcuatus
    n. arcuatus thalami
    auditory n.
    n. basalis of Ganser
    Bechterew n.
    benular n.
    Blumenau n.
    branchiomotor nuclei
    Burdach n.
    caudate n.
    n. caudatus
    n. centralis lateralis thalami
    n. centralis tegmenti superior
    centromedian n.
    n. centromedianus
    Clarke n.
    cochlear n.
    n. colliculi inferioris
    n. corporis geniculati medialis
    nuclei corporis mamillaris
    nuclei of cranial nerve
    cuneate n.
    n. cuneatus
    n. cuneatus accessorius
    n. of Darkschewitsch

deep cerebellar nuclei
Deiters n.
n. dentatus cerebelli
diploid n.
dorsal n.
dorsal accessory olivary n.
n. dorsalis
n. dorsalis corporis trapezoidei
n. dorsalis nervi vagi
dorsomedial hypothalamic n.
n. dorsomedialis hypothalami
Edinger-Westphal n.
emboliform n.
n. emboliformis
external cuneate n.
facial motor n.
n. fasciculi gracilis
n. fastigii
n. fibrosus linguae
n. filiformis
n. funiculi cuneati
n. funiculi gracilis
n. gelatinosus
n. gigantocellularis medullae
  oblongatae
n. globosus
n. of Goll
gustatory n.
n. habenulae
hilum of dentate n.
hilum of olivary n.
hypoglossal n.
inferior olivary n.
inferior salivary n.
inferior vestibular n.
intercalated n.
n. intercalatus
intermediolateral n.
n. intermediolaterali
intermediomedial n.
n. intermediomedialis
interpeduncular n.
n. interpeduncularis
n. interpositus
interstitial n.
n. interstitialis
nuclei intralaminares thalami
lateral cuneate n.
n. lateralis medullae oblongatae
n. of lateral lemniscus
lateral preoptic n.
lateral reticular n.

lateral vestibular n.
n. lemnisci lateralis
n. of lens
lenticular n.
lentiform n.
n. lentiformis
n. lentis
n. of Luys
main sensory n.
n. of mamillary body
n. masticatorius
masticatory n.
medial accessory olivary n.
n. of medial geniculate body
n. medialis centralis thalami
medial vestibular n.
mediodorsal n.
Monakow n.
motor nuclei
n. motorius nervi trigemini
nuclei nervi cochlearis
n. nervi facialis
n. nervi hypoglossi
n. nervi oculomotorii
n. nervi trochlearis
nuclei nervi vestibulocochlearis
nuclei nervorum cranialium
n. niger
oculomotor n.
n. of oculomotor nerve
n. olivaris
n. olivaris acces
n. olivaris accessorius medialis
Onuf n.
nuclei of origin
parabrachial nuclei
nuclei parabrachiale
n. paracentralis thalami
paraventricular n.
n. paraventricularis
Perlia n.
pontine nuclei
nuclei pontis
n. posterior hypothalami
posterior hypothalamic n.
posterior periventricular n.
n. preopticus lateralis
n. preopticus medialis
n. pulposus
n. pyramidalis
nuclei raphe
raphe nuclei

N

NOTES

nucleus *(continued)*
>red n.
>reduction n.
>reoptic n.
>n. reticularis thalami
>rhombencephalic gustatory n.
>Roller n.
>roof n.
>n. ruber
>n. salivatorius inferior
>n. salivatorius superior
>Schwalbe n.
>secondary sensory nuclei
>segmentation n.
>n. sensorius principalis nervi trigemini
>n. sensorius superior nervi trigemini
>shadow n.
>sole nuclei
>n. of solitary tract
>somatic motor nuclei
>special visceral efferent nuclei
>special visceral motor nuclei
>sperm n.
>spherical n.
>n. spinalis nervi accessorii
>Spitzka n.
>Staderini n.
>Stilling n.
>n. sub
>subthalamic n.
>superior olivary n.
>superior salivary n.
>superior vestibular n.
>supraoptic n.
>tail of caudate n.
>nuclei tegmenti
>n. tegmenti pontis caudalis
>n. tegmenti pontis oralis
>terminal nuclei
>nuclei terminales
>nuclei terminationis
>thalamic gustatory n.
>thoracic n.
>n. thoracicus
>n. tractus mesencephali nervi trigemini
>n. tractus solitarii
>n. tractus spinalis nervi trigemini
>trochlear n.
>n. of trochlear nerve
>tuberal nuclei
>nuclei tuberales
>tubercle of cuneate n.
>vein of caudate n.
>n. ventralis anterior thalami
>n. ventralis corporis trapezoidei
>n. ventralis intermedius thalami
>n. ventralis lateralis
>n. ventralis posterior intermedius thalami
>n. ventralis posterolateralis thalami
>n. ventralis posteromedialis thalami
>ventral tier thalamic nuclei
>ventrobasal n.
>ventromedial n.
>vestibular nuclei
>nuclei vestibulares

Nuel space
Nuhn gland
null cells
number
>Brinell hardness n.
>Knoop hardness n. (KHN)
>Mach n.
>Reynolds n.
>transport n.
>wave n.

nurse cells
nutricia, pl. **nutriciae**
>arteria radii n.
>nutriciae humeri

nutricium
>foramen n.

nutricius
>canalis n.

nutrient
>n. artery
>n. artery of femur
>n. artery of fibula
>n. artery of humerus
>n. artery of radius
>n. artery of tibia
>n. artery of ulna
>n. canal
>n. foramen
>n. vessel

nutrition
nutritional energy
nutritive equilibrium
nutriture
nycterine
nycterohemeral
nyctohemeral
nymphocaruncularis
>sulcus n.

nymphocaruncular sulcus
nymphohymenal sulcus
nystagmus
>congenital n.

Nysten law

oaponeurotic system
oarium
oat cell
O'Beirne
    O. sphincter
    O. valve
obeliac
obeliad
obelion
Obersteiner-Redlich
    O.-R. line
    O.-R. zone
obese
obex
obliqua
    chorda o.
    diameter o.
    fissura o.
    linea o.
oblique
    o. abdominal muscle
    o. arytenoid muscle
    o. auricular muscle
    o. bundle of pons
    o. capitis muscle
    o. cord
    o. cord of interosseous membrane
      of forearm
    o. diameter
    o. facial cleft
    o. fibers of muscular layer of
      stomach
    o. fissure
    o. fissure of lung
    o. head
    o. hernia
    inferior o. (IO)
    left posterior o.
    o. ligament of elbow joint
    o. line
    o. line of mandible
    o. line of thyroid cartilage
    o. muscle of auricle
    o. part
    o. part of cricothyroid muscle
    o. pericardial sinus
    o. popliteal ligament
    o. ridge of trapezium
    o. section
    o. sinus of pericardium
    o. tendon
    o. vein of left atrium
obliqui
    pars aryepiglottica musculi
      arytenoidei o.

obliquum
    caput o.
    ligamentum popliteum o.
obliquus
    adductor pollicis o.
    arytenoideus o.
    o. auriculae
    o. capitis inferior
    o. capitis inferior muscle
    o. capitis superior
    o. capitis superior muscle
    o. externus abdominis
    o. inferior bulbi
    o. internus abdominis
    musculus arytenoideus o.
    o. superior bulbi
oblong
    o. fovea of arytenoid cartilage
    o. pit of arytenoid cartilage
oblongata
    anterior column of medulla o.
    anterior median fissure of
      medulla o.
    foramen cecum medullae o.'s
    lateral nucleus of medulla o.
    medulla o.
    pons o.
    posterior median fissure of
      medulla o.
    posterior median sulcus of
      medulla o.
    pyramid of medulla o.
    sensory decussation of medulla o.
    vein of medulla o.
oblongatae
    fissura mediana anterior
      medullae o.
    medianus posterior medullae o.
    nucleus gigantocellularis
      medullae o.
    nucleus lateralis medullae o.
    pyramis medullae o.
    raphe medullae o.
    venae medullae o.
obsolescence
obstetric
    o. conjugate
    o. conjugate diameter
    o. conjugate of pelvic outlet
obstructive hydrocephaly
obturator
    o. artery
    o. branch of pubic branch of
      inferior epigastric vein
    o. canal

O

obturator *(continued)*
- o. crest
- o. externus muscle
- o. fascia
- o. foramen
- o. groove
- o. internus muscle
- o. internus tendon
- o. lymph node
- o. membrane
- o. nerve
- o. tubercle

obturatoria
- arteria o.
- crista o.
- fascia o.
- membrana o.
- vena o.

obturatoriae
- ramus posterior arteriae o.
- ramus pubicus arteriae o.
- venae o.

obturatorii
- lymphonodi o.
- nodi lymphoidei o.
- rami musculares rami anterioris nervi o.
- rami musculares rami posterioris nervi o.
- ramus cutaneus rami anterioris nervi o.
- ramus posterior nervi o.

obturatorium
- tuberculum o.

obturatorius
- canalis o.
- nervus o.
- ramus o.
- sulcus o.

obturatum
- foramen o.

obtuse margin of heart

occipital
- o. angle of parietal bone
- o. artery
- o. aspect
- o. belly
- o. belly of occipitofrontalis muscle
- o. border
- o. border of parietal bone
- o. border of temporal bone
- o. branch
- o. cerebral vein
- o. condyle
- o. diploic vein
- o. emissary vein
- o. fontanelle
- o. foramen
- o. groove
- o. gyri
- o. horn
- o. lobe
- o. lymph node
- o. margin
- o. margin of temporal bone
- o. nerve
- o. operculum
- o. plane
- o. plexus
- o. point
- o. pole
- o. protuberance
- o. region
- o. region of head
- o. segment
- o. sinus
- o. somite
- o. squama
- o. suture
- o. triangle

occipitale
- emissarium o.
- os o.
- planum o.

occipitales
- lymphonodi o.
- nodi lymphoidei o.
- rami o.
- venae o.
- venae encephali o.

occipitalis
- arteria o.
- condylus o.
- craniopagus o.
- diploica o.
- emissaria o.
- incisura jugularis ossis o.
- o. major
- margo lambdoideus ossis o.
- margo lambdoideus squamae o.
- margo mastoideus ossis o.
- margo mastoideus squamae o.
- o. minor
- o. muscle
- musculus o.
- norma o.
- pars basilaris ossis o.
- pars lateralis ossis o.
- plexus periarterialis arteriae o.
- processus jugularis ossis o.
- rami sternocleidomastoidei arteriae o.
- ramus auricularis arteriae o.
- ramus descendens arteriae o.
- ramus mastoideus arteriae o.
- ramus meningeus arteriae o.
- sinus o.
- squama o.

sulcus arteriae o.
o. tertius
torus o.
o. torus
tuberculum jugulare ossis o.
tuberculum pharyngeum partis
  basilaris ossis o.
vena emissaria o.
venter o.
occipitalization
occipitis
rami occipitales arteriae o.
occipitoatlantal articulation
occipitoatloid
occipitoaxial ligament
occipitoaxoid
occipitobregmatic
occipitocollicular tract
occipitofacial
occipitofrontal
o. diameter
o. fasciculus
o. muscle
occipitofrontalis
fasciculus o.
o. muscle
musculus o.
venter frontalis musculi o.
occipitomastoidea
sutura o.
occipitomastoid suture
occipitomental diameter
occipitoparietal suture
occipitopontine tract
occipitopontinus
tractus o.
occipitosphenoidal suture
occipitotectal tract
occipitotemporalis
ramus o.
sulcus o.
occipitotemporal sulcus
occipitothalamic radiation
occiput
occludens
zonula o.
occlusal
o. force
o. surface
o. surface of tooth
occlusion
mesial o.

occulta
spina bifida o.
occult border of nail
octavus
nervus o.
ocular
o. adnexa
o. cup
o. fundus
o. globe
o. humor
o. hypertelorism
o. muscle
o. nerve
o. region
o. tendon
o. vesicle
ocularis
clivus o.
foveola o.
oculi, sing. **oculus**
abducens o.
adnexa o.
attollens o.
axis externus bulbi o.
axis internus bulbi o.
bulbus o.
equator bulbi o.
fundus o.
meridiani bulbi o.
motor o.
musculi externi bulbi o.
musculus orbicularis o.
pars lacrimalis musculi
  orbicularis o.
pars orbitalis musculi orbicularis o.
pars palpebralis musculi
  orbicularis o.
pars profunda partis palpebralis
  musculi orbicularis o.
polus anterior bulbi o.
polus posterior bulbi o.
spatium intervaginale bulbi o.
sphincter o.
tapetum o.
tendo o.
tunica albuginea o.
tunica externa o.
tunica vasculosa o.
tutamina o.
vagina o.
venae choroideae o.

O

NOTES

oculoauriculovertebral dysplasia
oculocardiac reflex
oculocephalic reflex
oculocephalogyric reflex
oculocerebrorenal
oculocutaneous
oculodentodigital dysplasia
oculodermal
oculofacial
oculomandibulodyscephaly
oculomotor
    o. foramen
    o. nerve
    o. nucleus
    o. response
    o. root of ciliary ganglion
    o. sulcus
    o. system
oculomotorii
    nucleus nervi o.
    ramus inferior nervi o.
    ramus superior nervi o.
    sulcus nervi o.
oculomotorius
    nervus o.
oculonasal
oculopupillary
oculovertebral dysplasia
oculozygomatic
oculus (*sing. of* oculi)
Oddi
    sphincter of O.
    O. sphincter
odermal cloaca
Odland body
odontectomy
    mandibular dentition o.
odontoblast
odontoblastic layer
odontoclast
odontogenesis
odontogeny
odontoid
    check ligament of o.
    o. ligament
    o. process
    o. process of epistropheus
    o. vertebra
odontoideum
    os o.
odontoplast
odontosis
odorant binding protein
odoriferous gland
Oehl muscle
Ogston
    line of O.
    O. line
Ogura fossa

oil
    o. gland
    joint o.
olecrani
    bursa intratendinea o.
    bursa subcutanea o.
    fossa o.
olecranon
    bursa of o.
    o. bursa
    o. fossa
    o. ligament
    o. process
    o. of ulna
olfaction
olfactoria
    fila o.
olfactoriae
    glandulae o.
    striae o.
olfactorii
    nervi o.
    nervus o.
    vena gyri o.
olfactorium
    trigonum o.
    tuberculum o.
olfactorius
    bulbus o.
    sulcus o.
    tractus o.
olfactory
    o. angle
    o. area
    o. bulb
    o. bundle
    o. cortex
    o. epithelium
    o. foramen
    o. gland
    o. glomerulus
    o. groove
    o. groove of nasal cavity
    o. membrane
    o. nerve
    o. organ
    o. peduncle
    o. pit
    o. placode
    o. pyramid
    o. receptor cells
    o. region of mucosa of nose
    o. region of nasal mucosa
    o. region of tunica mucosa of
      nose
    o. root
    o. striae
    o. sulcus
    o. sulcus of nasal cavity

o. sulcus of nose
o. tract
o. trigone
o. tubercle
olfactus
organum o.
oligoamnios
oligodactylia
oligodactyly
oligodendria
oligodendroblast
oligodendrocyte
oligodendroglia cells
oligodipsia
oligodontia
oligohydramnios
oligolecithal
oligonephronic
oligoplastic
oligopnea
oligosynaptic
oligotrophia
oligotrophy
olisthetic vertebra
oliva, pl. **olivae**
o. inferior
siliqua olivae
o. superior
olivare
corpus o.
olivaris
hilum nuclei o.
nucleus o.
olivary
o. body
o. eminence
olive
inferior o.
superior o.
olivifugal
olivipetal
olivocerebellaris
tractus o.
olivocerebellar tract
olivocochlear
o. bundle
o. nerve
olivopontocerebellar
olivospinal tract
ollicular stigma
Ollier theory
omandibularis

omenta (*pl. of* omentum)
omental
o. appendage
o. appendix
o. branch
o. bursa
o. eminence of pancreas
o. foramen
o. hernia
o. sac
o. tenia
o. tuber
o. tuberosity of liver
omentale
foramen o.
tuber o.
omentales
appendices o.
rami o.
omentali
rami o.
omentalis
bursa o.
foramen recessus superioris
bursae o.
recessus inferior o.
recessus superior bursae o.
tenia o.
vestibulum bursae o.
omentulum
omentum, pl. **omenta**
colic o.
gastric o.
gastrocolic o.
gastrohepatic o.
gastrosplenic o.
greater o.
lesser o.
o. majus
o. minus
pancreaticosplenic o.
splenogastric o.
omnivorous
omoclavicular
o. fossa
o. triangle
omoclaviculare
trigonum o.
omohyoidei
venter inferior musculi o.
venter superior musculi o.

O

NOTES

omohyoideus
    musculus o.
omohyoid muscle
omothyroid
omotracheale
    trigonum o.
omotracheal triangle
omphalic
omphaloangiopagous twins
omphaloangiopagus
omphalocele
omphaloenteric
omphalomesenteric
    o. artery
    o. duct
    o. vein
    o. vessel
omphalomesentericus
    ductus o.
omphalopagus
omphalos
omphalosite
omphalospinous line
omphalovesical
omphalus
oncograph
oncography
oncometer
oncometric
oncometry
oncotic pressure
one-horned uterus
ontogenesis
ontogenetic
ontogenic homeostasis
ontogeny
Onuf nucleus
onychia
onychoid
onychostroma
oocyte
    primary o.
    secondary o.
oogenesis
oogenetic
oogenic
oogenous
oogonium, pl. **oogonia**
ookinesia
ookinesis
oolemma
oophoron
oophorus
    cumulus o.
ooplasm
Oort nerve
ootheca
ootid
Opalski cell

open circuit method
opening
    aortic o.
    cardiac o.
    o. of carotid canal
    o. contraction
    cornu of falciform margin of
      saphenous o.
    o. of coronary sinus
    crest of cochlear o.
    o. for dorsal artery of penis
    o. for dorsal nerve of penis
    duodenal o.
    esophageal o.
    o. of external acoustic meatus
    external urethral o.
    falciform margin of saphenous o.
    femoral o.
    o. of frontal sinus
    horns of saphenous o.
    ileocecal o.
    inferior horn of falciform margin
      of saphenous o.
    inferior horn of saphenous o.
    o. of inferior vena cava
    intercartilaginous part of glottic o.
    intermembranous part of glottic o.
    internal acoustic o.
    o. of internal acoustic meatus
    internal urethral o.
    lacrimal o.
    o. of left parotid duct
    o. of middle nasal meatus
    oral o.
    orbital o.
    o. of papillary duct
    piriform o.
    pulmonary o.
    o. of pulmonary trunk
    o. of pulmonary vein
    o. for right maxillary sinus
    saphenous o.
    o. of smallest cardiac vein
    o. of sphenoidal sinus
    superior horn of falciform margin
      of saphenous o.
    superior horn of saphenous o.
    o. of superior vena cava
    tendinous o.
    ureteral o.
    ureteric o.
    urethral o.
    urinary o.
    o. of uterus
    vaginal o.
    o. of vestibular canaliculus
opercula (*pl. of* **operculum**)
opercular fold

opercularis
    pars o.
operculi
operculum, pl. **opercula**
    o. ilei
    occipital o.
    orbital o.
    parietal o.
    trophoblastic o.
ophryon
ophryospinal angle
ophthalmic
    o. artery
    o. nerve
    o. plexus
    o. vein
ophthalmica
    arteria o.
    vesicula o.
ophthalmicae
    arteriae musculares arteriae o.
    plexus periarterialis arteriae o.
ophthalmici
    ramus meningeus recurrens nervi o.
ophthalmicus
    caliculus o.
    nervus o.
ophthalmovascular
opiate receptor
opisthenar
opisthiobasial
opisthion
opisthionasial
opisthotic
opodidymus
oppilative
opponens
    o. digiti minimi manus
    o. digiti minimi muscle
    musculus o.
    o. pollicis
    o. pollicis muscle
opposer
    o. muscle of little finger
    o. muscle of thumb
opsiuria
optic
    o. axis
    o. canal
    o. capsule
    o. chiasm
    o. commissure

    o. cup
    o. decussation
    o. disk
    o. fissure
    o. foramen
    o. ganglion
    o. groove
    o. layer
    o. nerve
    o. nerve hypoplasia
    o. papilla
    o. part of retina
    o. placode
    o. radiation
    o. recess
    o. sheath
    o. stalk
    stratum ganglionare nervi o.
    o. tract
    o. vesicle
optica
    radiatio o.
optical
    o. cavity
    o. righting reflex
optici
    circulus vasculosus nervi o.
    discus nervi o.
    papilla nervi o.
    pars canalis nervi o.
    pars intracaniculus nervi o.
    pars intracranialis nervi o.
    pars intralaminaris nervi o.
    pars intraocularis nervi o.
    pars orbitalis nervi o.
    pars postlaminaris nervi o.
    pars prelaminaris nervi o.
    radix lateralis tractus o.
    radix medialis tractus o.
    rami tractus o.
    spatia intervaginalia nervi o.
    spatium intervaginale
      subarachnoidale nervi o.
    vaginae nervi o.
    vagina externa nervi o.
    vagina interna nervi o.
opticociliary vessel
opticopupillary
opticostriate region
opticum
    chiasm o.
    chiasma o.

NOTES

O

opticum *(continued)*
    foramen o.
    stratum o.
opticus
    axis o.
    canalis o.
    nervus o.
    porus o.
    recessus o.
    tractus o.
optomeninx
ora, pl. **orae**
    o. serrata
    o. serrata retinae
oral
    o. cavity
    o. cavity proper
    o. fissure
    o. mucosa
    o. opening
    o. pharynx
    o. plate
    o. region
    o. tooth
    o. vestibule
oralis
    nucleus tegmenti pontis o.
    regio o.
oraminal node
Orbeli effect
orbicular
    o. bone
    o. ligament
    o. ligament of radius
    o. muscle
    o. muscle of eye
    o. muscle of mouth
    o. process
    o. zone
    o. zone of hip joint
orbiculare
    os o.
orbicularis
    o. ciliaris
    musculus o.
    o. oculi muscle
    o. oris muscle
    zona o.
orbiculus ciliaris
orbit
    aperture of o.
    apex of o.
    deep o.
    fat body of o.
    floor of o.
    inferior wall of o.
    lateral margin of o.
    lateral wall of o.
    margin of o.

    medial wall of o.
    of o.
    recti muscles of o.
    roof of o.
    superficial o.
    superior margin of o.
    superior wall of o.
orbita
orbitae
    aditus o.
    corpus adiposum o.
    paries inferior o.
    paries lateralis o.
    paries medialis o.
    paries superior o.
orbital
    o. apex
    o. axis
    o. branch of maxillary nerve
    o. branch of middle meningeal artery
    o. branch of pterygopalatine ganglion
    o. canal
    o. cavity
    o. eminence
    o. eminence of zygomatic bone
    o. fasciae
    o. fat body
    o. fat-pad
    o. floor
    o. gyri
    o. height
    o. index
    o. lamina of ethmoid bone
    o. layer of ethmoid bone
    o. margin
    o. margin of eyelid
    o. opening
    o. operculum
    o. part
    o. part of frontal bone
    o. part of lacrimal gland
    o. part of optic nerve
    o. part of orbicularis oculi muscle
    o. plane
    o. plate
    o. plate of ethmoid bone
    o. process
    o. process of palatine bone
    o. pyramid
    o. region
    o. ridge
    o. rim
    o. septum
    o. sulci
    o. surface
    o. tubercle
    o. tubercle of zygomatic bone

o. vein
o. wall
o. width

orbitale
planum o.
septum o.

orbitales
fasciae o.
gyri o.
rami o.
sulci o.

orbitalis
ala o.
eminentia o.
facies o.
margo o.
o. muscle
musculus o.
pars o.
processus o.
ramus o.
regio o.

orbitofrontal
o. artery
o. cortex

orbitomeatal
o. line
o. plane

orbitonasal index
orbitopagus
orbitopalpebralis
musculus o.

orbitosphenoid cartilage
orchidic
orchis, pl. **orchises**
ordinate
orexigenic
organ
accessory o.
annulospiral o.
Chievitz o.
circumventricular o.'s
o. of Corti
Corti o.
enamel o.
end o.
external female genital o.
external male genital o.
floating o.
genital o.'s
Golgi tendon o.
gustatory o.

o. of hearing
internal female genital o.
internal male genital o.
intromittent o.
Jacobson o.
lymph node of abdominal o.
male reproductive o.
neurohemal o.
neurotendinous o.
olfactory o.
pelvic o.
ptotic o.
reproductive o.
reticular membrane of spinal o.
o. of Rosenmüller
sense o.
o. of smell
spiral o.
subcommissural o.
subfornical o. (SFO)
supernumerary o.
target o.
o. of taste
o. of touch
urinary o.
vestibular o.
vestibulocochlear o.
vestigial o.
o. of vision
vomeronasal o.
wandering o.
Weber o.
o. of Zuckerkandl

organa
o. genitalia
o. genitalia feminina externa
o. genitalia feminina interna
o. genitalia masculina externa
o. genitalia masculina interna
o. oculi accessoria
o. sensuum
o. urinaria

organelle
cell o.

organic
organism
organization
organizer
nucleolar o.
primary o.
procentriole o.

organogenesis

O

NOTES

organogenetic
organogenic
organogeny
organography
organoid
organoleptic
organology
organomegaly
organon
organonomy
organonymy
organophilic
organophilicity
organotrophic
organotropic
organotropism
organotropy
organum
    o. auditus
    o. genitalia feminina externa
    o. genitalia feminina interna
    o. genitalia masculina externa
    o. genitalia masculina interna
    o. gustatorium
    o. gustus
    o. olfactus
    o. spirale
    o. tactus
    o. vestibulocochleare
    o. visus
    o. vomeronasale
orgasm
orgasmic
orgastic
orifice
    anal o.
    aortic o.
    atrioventricular o.
    auriculoventricular o.
    cardiac o.
    cardial o.
    esophagogastric o.
    eustachian o.
    o. of external acoustic meatus
    external urethral o.
    filling internal urethral o.
    floor of o.
    frenulum of ileal o.
    gastroduodenal o.
    heart o.
    hymenal o.
    ileal o.
    o. of ileal papilla
    ileocecal o.
    o. of inferior vena cava
    internal acoustic o.
    o. of internal acoustic meatus
    internal urethral o.
    left atrioventricular o.

    mitral o.
    myopectineal o.
    pharyngeal o.
    pulmonary o.
    pyloric o.
    right atrioventricular o.
    segmental o.
    o. of superior vena cava
    tricuspid o.
    o. tube
    tympanic o.
    ureteral o.
    ureteric o.
    uterine o.
    o. of uterus
    vaginal o.
    o. of vermiform appendix
    vesical o.
    vesicourethral o.
    voiding internal urethral o.
orificial
orificium, pl. orificia
    o. externum uterus
    o. internum uterus
    o. ureteris
    o. urethrae externum
    o. vaginae
origin
    aponeurosis of o.
    deep o.
    ectal o.
    ental o.
    nuclei of o.
    superficial o.
oris
    angulus o.
    cavitas o.
    cavum o.
    claustrum o.
    depressor anguli o.
    diaphragma o.
    glandulae o.
    labia o.
    labium inferius o.
    labium superius o.
    levator anguli o.
    modiolus anguli o.
    musculus depressor anguli o.
    musculus levator anguli o.
    musculus orbicularis o.
    musculus sphincter o.
    pars labialis musculi orbicularis o.
    pars marginalis musculi orbicularis o.
    o. pollicis longi
    rima o.
    sphincter o.
    tunica mucosa o.
    vestibulum o.

ormis
Ornish
> O. prevention diet
> O. reversal diet

ornithine cycle
orodigitofacial dysostosis
orofacial
orolingual
oronasal membrane
oropharyngeal
> o. isthmus
> o. membrane
> o. passage

oropharynx
> posterior wall of o.

orpus geniculatum externum
orthoarteriotony
orthocephalic
orthocephalous
orthochromatic
orthochromophil, orthochromophile
orthodromic
orthogenesis
orthogenic
orthognathic
orthognathous
orthograde
orthomolecular
orthoscope
orthothanasia
orthotopic
orthotropic
orum
os, pl. **ossa**
> o. acromiale
> anterior lip of uterine o.
> o. basilare
> o. breve
> o. calcis
> o. capitatum
> ossa carpi
> o. centrale
> o. centrale tarsi
> o. clitoridis
> o. clitoris
> o. coccygis
> o. costale
> o. coxa
> ossa cranii
> o. cuboideum
> o. cuneiforme intermedium
> o. cuneiforme laterale

o. cuneiforme mediale
ossa digitorum
endocervical o.
o. epitympanicum
o. ethmoidale
external o.
ossa faciei
o. femoris
o. frontale
o. hamatum
o. hyoideum
o. iliacum
o. ilium
o. incisivum
o. intermetatarseum
internal o.
o. interparieta
o. interparietale
o. irregulare
o. ischii
o. japonicum
o. lacrimale
o. longum
o. lunatum
o. magnum
o. malare
o. mastoideum
ossa membri inferioris
ossa membri superioris
o. metacarpale
ossa metacarpalia I–V
ossa metacarpi
o. metatarsale
ossa metatarsalia I–V
ossa metatarsi
o. multangulum majus
o. multangulum minus
o. nasale
o. naviculare
o. naviculare manus
o. occipitale
o. odontoideum
o. orbiculare
o. palatinum
o. parietale
ossa pedis
o. penis
o. pisiforme
o. planum
o. pneumaticum
o. premaxillare
o. pterygoideum

NOTES

O

os *(continued)*
    o. pubis
    o. pubis spline
    o. pyramidale
    o. sacrum
    o. scaphoideum
    ossa sesamoidea
    o. sesamoideum
    o. sphenoidale
    o. subtibiale
    o. suprasternale
    ossa suprasternalia
    o. suturarum
    ossa suturarum
    o. sylvii
    ossa tarsalia
    ossa tarsi
    o. temporale
    o. tibiale posterius
    o. tibiale posticum
    o. trapezium
    o. trapezoideum
    o. triangulare
    o. tribasilare
    o. trigonum
    o. triquetrum
    o. unguis
    o. uteri externum
    o. uteri internum
    o. uterus
    o. vesalianum
    o. vesalianum pedis
    o. zygomaticum
Osborne ligament
oscheal
oscillation
oscillograph
oscillography
oscillometer
oscillometric
oscillometry
oscitate
oscitation
osculum, pl. **oscula**
Osler triad
osmesis
osmication
osmification
osmiophilic
osmiophobic
osmogram
osmolal clearance
osmoregulatory
osmose
osmosis
    reverse o.
osmotic
    o. diuresis
    o. pressure

osphresis
ospinosus
ossa (*pl. of* os)
ossea
    lamina spiralis o.
osseae
    limbus laminae spiralis o.
ossei
    canales semicircularis o.
    labium limbi tympanicum laminae spiralis o.
    labium limbi tympanicum limbi spiralis o.
    labium limbi vestibulare laminae spiralis o.
    labium limbi vestibulare limbi spiralis o.
    lamella tympanica laminae spiralis o.
    lamella vestibularis laminae spiralis o.
    recessus ellipticus labyrinthi o.
    recessus saccularis labyrinthi o.
    recessus sphericus labyrinthi o.
    recessus utricularis labyrinthi o.
ossein, osseine
osseocartilaginous
osseomucin
osseotendinous
osseous
    o. ampulla
    o. cell
    o. labyrinth
    o. lacuna
    o. part of skeletal system
    o. portion
    o. spiral lamina
    o. tissue
osseum
    palatum o.
    septum nasi o.
osseus
    labyrinthus o.
ossicle
    Andernach o.
    auditory o.
    Bertin o.
    epactal o.
    joint of auditory o.
    Kerckring o.
    ligaments of auditory o.
    muscles of auditory o.
    tin o.
ossicular chain
ossiculum, pl. **ossicula**
    ossicula auditus
    ossicula mentalia
ossiferous

ossific
    o. center
ossification
    o. center
    center of o.
    endochondral o.
    intramembranous o.
    membranous o.
    point of o.
    primary center of o.
    primary point of o.
    secondary center of o.
    secondary point of o.
ossificationis
    centrum o.
    punctum o.
ossified edge
ossiform
ossify
ossis
    processus articularis superior o.
ossium
    juncturae o.
    medulla o.
    substantia compacta o.
osteal
osteanagenesis
osteanaphysis
ostein, osteine
ostembryon
osteoanagenesis
osteoblast
osteoblastic
osteocalcin
osteocartilaginous body
osteochondral junction
osteochondrogenic cell
osteochondromatosis
osteochondrous
osteoclast
osteoclastic
osteocollagenous fibers
osteocranium
osteocyte
osteodentin
osteodesmosis
osteoepiphysis
osteogen
osteogenesis
osteogenetic
    o. fiber
    o. layer

osteogenic
    o. cell
    o. tissue
osteogenous
osteogeny
osteography
osteohypertrophy
osteoid tissue
osteologia
osteology
osteomere
osteometry
osteone
osteonectin
osteopedion
osteopetrosis acro-osteolytica
osteophage
osteoplaque
osteoplast
osteoplastic
osteoponin
osteoprogenitor cell
osteosclerosis
osteosis
osteotrophy
osteotympanic
ostial sphincter
ostium, pl. **ostia**
    o. abdominale tubae uterina
    o. abdominale tubae uterinae
    o. anatomicum
    o. aortae
    aortic o.
    o. appendicis vermiformis
    o. arteriosum
    o. atrioventriculare dextrum
    o. atrioventriculare sinistrum
    o. cardiacum
    o. histologicum
    o. ileale
    o. ileocecale
    pharyngeal o.
    o. pharyngeum tubae auditivae
    o. pharyngeum tubae auditoriae
    o. primum
    o. pyloricum
    o. secundum
    o. sinus coronarii
    o. trunci pulmonalis
    o. tympanicum tubae auditivae
    o. of ureter
    o. ureteris

NOTES

O

ostium (*continued*)
   o. urethrae externum
   o. urethrae internum
   o. urethrae internum evacuans
   o. uteri externum
   o. uteri internum
   o. uterinum tubae uterinae
   o. uterus
   o. vaginae
   o. venae cavae inferioris
   o. venae cavae superioris
   ostia venarum pulmonalium
   o. venosum
   o. venosum cordis
   o. of vermiform appendix
ostosis
Ostwald solubility coefficient
otic
   o. barotrauma
   o. capsule
   o. depression
   o. ganglion
   o. placode
   o. vesicle
otici
   radix parasympathica ganglii o.
   radix sympathica ganglii o.
oticum
   ganglion o.
otitic hydrocephalus
otocephaly
otoconia
otocranial
otocranium
otocyst
otoganglion
otogenous
otolites
otolith
otolithic membrane
otomandibular dysostosis
otopalatodigital
otopharyngeal tube
otosalpinx
otosclerosis
otosteal
Ottoson potential
otubarius
outer
   o. aspect
   o. border
   o. border of iris
   o. canthus
   o. cone fiber
   o. convexity
   o. layer
   o. limiting membrane
   o. lip of iliac crest
   o. malleolus

   o. rim
   o. sheath of optic nerve
   o. spiral sulcus
   o. stripes of renal medulla
   o. surface
   o. table bone of skull
   o. table of frontal bone
   o. zone of renal medulla
Outerbridge ridge
outermost
   o. covering
   o. layer
outflow tract
outlet
   conjugate diameter of pelvic o.
   gastric o.
   obstetric conjugate of pelvic o.
   pelvic plane of o.
   plane of o.
   thoracic o.
   vaginal o.
out of phase
output
   cardiac o.
   minute o.
   stroke o.
ova (*pl. of* ovum)
oval
   o. area of Flechsig
   o. corpuscle
   o. fasciculus
   o. foramen
   o. foramen of heart
   o. fossa
   o. window
ovale
   centrum o.
   foramen o.
   valve of foramen o.
   venous plexus of foramen o.
ovalis
   annulus o.
   fenestra o.
   fossa o.
   limbus fossae o.
   margin of fossa o.
   plexus venosus foraminis o.
   rete foraminis o.
   valvula foraminis o.
ovalocyte
ovaria (*pl. of* ovarium)
ovarian
   o. branch of uterine artery
   o. bursa
   o. cortex
   o. cycle
   o. duct
   o. fimbria
   o. follicle

o. fossa
o. hernia
o. ligament
o. nervous plexus
o. stroma
o. tube
o. vein

ovarica
arteria o.
bursa o.
o. dextra
fimbria o.
fossa o.
o. sinistra

ovaricae
rami ureterici arteriae o.
ramus tubarius arteriae o.

ovaricus
cumulus o.
plexus nervosus o.
ramus o.

ovarii
cortex o.
extremitas tubaria o.
extremitas uterina o.
facies lateralis o.
facies medialis o.
hilum o.
ligamentum suspensorium o.
margo liber o.
margo mesovaricus o.
rete o.
stratum granulosum o.
stroma o.
tunica albuginea o.

ovarium, pl. **ovaria**
o. masculinum

ovary
accessory o.
cortex of o.
free border of o.
hilum of o.
hilus of o.
lateral surface of o.
left o.
ligament of o.
medial pole of o.
medial surface of o.
mesovarian border of o.
mesovarian margin of o.
proper ligament of o.
right o.

stroma of o.
suspensory ligament of o.
third o.
tubal extremity of o.
tunica albuginea of o.
uterine extremity of o.

overproduction theory
ovi
ovicidal
oviducal
oviduct
fimbriated o.

oviductal
oviferous
oviform
ovigenesis
ovigenetic
ovigenic
ovigenous
ovigerous
ovocenter
ovocyte
ovogenesis
ovogonium
ovoid
fetal o.

ovoidalis
articulatio o.

ovoplasm
ovotestis
ovula (*pl. of* ovulum)
ovular
o. membrane
o. transmigration

ovulation
ovulatory
ovule
graafian o.

ovulum, pl. **ovula**
ovum, pl. **ova**
alecithal o.
blighted o.
centrolecithal o.
fertilized o.
isolecithal o.
migration of o.
Peters o.
serrata ova
telolecithal o.

Owen
contour lines of O.

NOTES

429

Owen *(continued)*
    interglobular space of O.
    O. line
oxidation-reduction system
oxide
    nitric o. (NO)
oxycephalia
oxycephalic
oxycephalous
oxycephaly
oxychromatic
oxychromatin
oxygen
    o. affinity anoxia
    o. affinity hypoxia
    o. capacity
    o. consumption
    o. debt
    o. deficit
    o. deprivation theory of narcosis
    hyperbaric o.
    o. poisoning
    o. toxicity
    o. utilization coefficient
oxygenate
oxygenation
    apneic o.
    hyperbaric o.
oxyntic
    o. cell
    o. gland
oxyphil
    o. cells
    o. chromatin
    o. granule
oxyphile
oxyphilic leukocyte
oxyrhine
oxytalan

P

P cell
P substance of Lewis

p
PA

pulmonary artery
pacchionian

p. bodies
p. corpuscles
p. depressions
p. foramen
p. gland
p. granulation
pachychromatic
pachygnathous
pachygyria
pachymeninx
pacinian corpuscles
pad

abdominal fat p.
antimesenteric fat p.
fat p.
knuckle p.
periarterial p.
pharyngoesophageal p.
retromolar p.
retropatellar fat p.
retrosternal fat p.
sucking p.
suctorial p.
thumb p.
volar p.
Paget cells
pagetoid cells
paired allosome
palata
palatal

p. index
p. linguoplate
p. process
p. reflex
p. triangle
p. vein
palate

arch of p.
bony p.
Byzantine arch p.
cleft p.
elevator muscle of soft p.
hard p.
pendulous p.
primary p.
primitive p.
secondary p.
soft p.

tensor muscle of soft p.
uvula of soft p.
palati

canal for lesser p.
elevation of levator p.
musculus levator p.
musculus tensor p.
raphe p.
trigonum p.
velum pendulum p.
palatiform
palatin
palatina

aponeurosis p.
p. ascendens
crista p.
p. descendens
p. externa
p. major
ruga p.
tonsilla p.
uvula p.
vena p.
palatinae

glandulae p.
p. minores
spinae p.
palatine

p. aponeurosis
p. arch
p. artery
p. block
bursa of tensor veli p.
p. crest
p. crest of horizontal process of
    palatine bone
p. durum
p. fold
p. foramen
p. gland
p. groove
p. index
p. nerve
p. papilla
p. process
p. process of maxilla
p. protuberance
p. raphe
p. reflex
p. ridge
p. spine
p. surface of horizontal plate of
    palatine bone
p. tonsil
p. uvula

palatine *(continued)*
    p. vein
    p. velum
palatini
    arcus p.
    bursa musculi tensoris veli p.
    crista conchalis ossis p.
    crista ethmoidalis ossis p.
    crista nasalis laminae horizontalis
      ossis p.
    crista palatina laminae horizontalis
      ossis p.
    facies maxillaris ossis p.
    facies nasalis ossis p.
    facies palatina laminae horizontalis
      ossis p.
    lamina horizontalis ossis p.
    lamina perpendicularis ossis p.
    levator veli p.
    p. minores
    musculus levator veli p.
    musculus tensor veli p.
    nervus musculi tensoris veli p.
    nervus tensoris veli p.
    processus orbitalis ossis p.
    processus pyramidalis ossis p.
    processus sphenoidalis ossis p.
    spina nasalis posterior laminae
      horizontalis ossis p.
    sulci p.
    tensoris veli p.
    tensor veli p.
    veli p.
palatinum
    os p.
    velum p.
palatinus
    p. major
    processus p.
    sulcus p.
    p. torus
palatoethmo
palatoethmoidalis
    sutura p.
palatoethmoidal suture
palatoglossal
    p. arch
    p. fold
    p. muscle
palatoglossus
    arcus p.
    p. muscle
    musculus p.
palatomaxillaris
    sutura p.
palatomaxillary
    p. index
    p. suture
palatonasal

palatopharyngeal
    p. arch
    p. sphincter
    p. sphincter muscle
palatopharyngei
    fasciculus anterior musculi p.
    fasciculus posterior musculi p.
palatopharyngeus
    arcus p.
    p. muscle
    musculus sphincter p.
palatosalpingeus
    musculus p.
palatoschisis
palatostaphylinus
    musculus p.
palatouvularis muscle
palatovaginal
    p. canal
    p. groove
palatovaginalis
    canalis p.
    sulcus p.
palatum
    p. durum
    p. fissum
    p. molle
    p. osseum
paleencephalon
pale globe
paleocerebellum
paleocortex
paleokinetic
paleostriatal
paleostriatum
paleothalamus
Palfyn sinus
palinal
palingenesis
palisade layer
pallial
pallidal
pallidi
    rami globi p.
pallidum
pallidus
    globus p.
pallii
    impressio petrosa p.
pallium
    petrosal impression of p.
palm
    cup of p.
    p. of hand
palma, palmae, pl. **palmae**
    p. manus
palmar
    p. aponeurosis

p. branch of anterior interosseous nerve
p. branch of median nerve
p. branch of ulnar nerve
p. carpal branch of radial artery
p. carpal branch of ulnar artery
p. carpal ligament
p. carpal tendinous sheath
p. carpometacarpal ligament
carpometacarpal ligament dorsal and p.
p. crease
p. digital artery
p. digital nerve
p. digital vein
p. fascia
p. interossei interosseous muscle
p. interosseous artery
p. interosseous muscle
p. ligament of interphalangeal joint of hand
p. ligament of metacarpophalangeal joint
p. metacarpal artery
p. metacarpal ligament
p. metacarpal vein
p. monticuli
p. plate
p. radiocarpal ligament
p. space
p. surfaces of finger
p. ulnocarpal ligament

palmare
ligamentum radiocarpale p.
ligamentum ulnocarpale p.

palmares
digitales p.
interossei p.
metacarpea p.
musculi interossei p.
ramus perforans arteriae metacarpalium p.
vaginae tendinum carpales p.
venae digitales p.
venae metacarpeae p.

palmaria
ligamenta carpometacarpalia p.
ligamenta metacarpalia p.
ligamentum carpometacarpalia p.
ligamentum intercarpalia p.
ligamentum metacarpalia p.

palmaris
aponeurosis p.
arteria metacarpalis p.
arteria metacarpea p.
p. brevis
p. brevis muscle
facies digitalis p.
p. longus
p. longus muscle
p. longus tendon
musculus interosseus p.

palmarium
rami perforantes arteriarum metacarpalium p.

palmatae
plicae p.

palmate
p. fold
p. fold of cervical canal

palmatus
penis p.

palpebra, pl. **palpebrae**
aponeurosis of superior levator p.
p. inferior
lamina of levator palpebrae
limbus anterior palpebrae
limbus posterior palpebrae
margo palpebrae
p. superior
tarsus inferior palpebrae
tarsus superior palpebrae

palpebral
p. artery
p. branch of infratrochlear nerve
p. commissure
p. conjunctiva
p. fascia
p. fissure
p. fold
p. furrow
p. ligament
p. margin
p. part
p. part of lacrimal gland
p. part of orbicularis oculi muscle
p. raphe
p. region
p. rim
p. vein

palpebrales
arteriae p.
p. inferiores

NOTES

P

palpebrales *(continued)*
p. laterales
limbi p.
p.   mediales
rami p.
p. superiores
venae p.

palpebralis
pars p.

palpebrarum
facies anterior p.
facies posterior p.
musculus orbicularis p.
p. rima
rima p.
tendo p.
tunica conjunctiva p.

palpebronasal fold

palpebronasalis
plica p.

palpebrum
commissura lateralis p.
commissura medialis p.

palsy
Bell p.

pampiniform
p. body
p. vein
p. venous plexus

pampiniforme
corpus p.

pampiniformis
plexus venosus p.

panblastic

pancake kidney

pancreas, pl. **pancreata**
p. accessorium
accessory p.
alpha cells of p.
annular p.
anterior border of body of p.
anteroinferior surface of p.
anterosuperior surface of body
of p.
artery to tail of p.
Aselli p.
beta cell of p.
body of p.
delta cell of p.
p. divisum
dorsal p.
endocrine part of p.
exocrine part of p.
gamma cell of p.
head of p.
inferior border of body of p.
lesser p.
p. minus
neck of p.

omental eminence of p.
posterior surface of p.
retroperitoneal p.
small p.
superior border of body of p.
tail of p.
unciform p.
uncinate process of p.
ventral p.
Willis p.
Winslow p.

pancreatic
p. branch
p. digestion
p. diverticula
p. duct
p. islets
p. juice
p. lymph node
p. nervous plexus
p. notch
p. ranula
p. sphincter
p. tissue
p. vein

pancreaticae
venae p.

pancreatici
lymphonodi p.
musculus sphincter ductus p.
nodi lymphoidei p.
rami p.

pancreaticobiliary
p. ductal junction
p. sphincter
p. tract

pancreaticoduodenal
p. arterial arcades
p. lymph node
p. vein

pancreaticoduodenales
lymphonodi p.
nodi lymphoidei p.
venae p.

pancreaticoduodenalis
arteria p.

pancreaticoenteric recess

pancreaticogastric fold

pancreaticolienales
nodi lymphoidei p.

pancreaticosplenales
nodi lymphoidei p.

pancreaticosplenic
p. ligament
p. lymph node
p. omentum

pancreaticus
collum p.

ductus p.
plexus nervosus p.
pancreatis
    arteria caudae p.
    caput p.
    cauda p.
    corpus p.
    facies anteroinferior corporis p.
    facies anterosuperioris corporis p.
    facies posterior p.
    incisura p.
    margo anterior corporis p.
    margo inferior p.
    margo inferior corporis p.
    margo superior p.
    margo superior corporis p.
    pars endocrina p.
    pars exocrina p.
    pars tecta p.
    processus uncinatus p.
    tuber omentale p.
pancreatomegaly
pancreatropic
pancreolithotomy
panduraformis
    placenta p.
Paneth granular cells
panniculus, pl. **panniculi**
    p. adiposus
    p. adiposus telae subcutaneae
      abdominis
    p. carnosus
    p. carnosus muscle
pannus, pl. **panni**
Pansch fissure
pantamorphia
pantamorphic
pantanencephalia
pantanencephaly
Papez circuit
papilla, pl. **papillae**
    acoustic p.
    anal p.
    Bergmeister p.
    bile p.
    p. of breast
    circumvallate p.
    clavate p.
    conic p.
    papillae conicae
    conical p.
    papillae corii

papillae of corium
cribriform area of the renal p.
dental p.
dentinal p.
p. dentis
dermal p.
p. of dermis
papillae dermis
p. ductus parotidei
p. duodeni major
p. duodeni minor
excavatio p.
filiform p.
papillae filiformes
papillae foliatae
foliate p.
fungiform p.
papillae fungiformes
gingival p.
p. gingivalis
hair p.
ileal p.
p. ilealis
p. incisiva
incisive p.
inferior lacrimal p.
interdental p.
p. interdentalis
interproximal p.
lacrimal p.
p. lacrimalis
lenticular p.
lingual gingival p.
lingual interdental p.
p. lingualis
major duodenal p.
p. mammae
minor duodenal p.
nerve p.
p. nervi optici
optic p.
orifice of ileal p.
palatine p.
parotid p.
p. parotidea
p. of parotid gland
p. pili
renal p.
papillae renales
p. renalis
retrocuspid p.
retromolar p.

NOTES

P

papilla *(continued)*
    sphincter p.
    sublingual p.
    superior lacrimal p.
    tactile p.
    papillae of tongue
    urethral p.
    p. urethralis
    p. vallata
    papillae vallatae
    vallate p.
    vascular p.
    p. of Vater
    Vater p.
papillare
    corpus p.
papillaris
    areola p.
    musculus p.
    processus p.
papillary
    p. duct
    p. foramina of kidney
    p. layer
    p. muscle
    p. process
    p. process of caudate lobe of liver
papillate
papillation
papilliferous
papilliform
papillula, pl. **papillulae**
papyracea
    lamina p.
papyraceous plate
papyraceus
    fetus p.
para
    cordis p.
para-aminohippurate clearance
para-aortica
    corpora p.-a.
paraaortic body
parabiosis
parabiotic
parabrachiale
    nuclei p.
parabrachial nuclei
paracentesis
    p. abdominis
    p. tunicae vaginalis
    p. tympani
    p. vesicae
paracentral
    p. branch of callosomarginal artery
    p. branch of pericallosal artery
    p. fissure
    p. lobe

    p. lobule
    p. nucleus of thalamus
paracentrales
    ramus p.
paracentralis
    arteria p.
    gyrus p.
    lobulus p.
paracephalus
    acephalus p.
paracervical
paracervix
parachordal
    p. cartilage
    p. plate
parachute
    p. deformity
    p. mitral valve
paracolic
    p. gutter
    p. recess
paracolici
    lymphonodi p.
    nodi lymphoidei p.
    sulci p.
paracolpium
paracortex
paracrine
paracystic pouch
paracystium
paracytic
paradental
paradentium
paradidymal
paradidymides, pl. **paradidymis**
paradox
    Weber p.
paradoxica
paradoxical contraction
paraduodenal
    p. fold
    p. fossa
    p. recess
paraduodenalis
    plica p.
    recessus p.
paraesophageal
parafollicular cells
paraganglion, pl. **paraganglia**
paraganglionic cells
paragene
paragenital tubules
paraglenoid
    p. groove
    p. sulcus
paraglenoidalis
    sulcus p.
paraglottic space
paragnathus

parahepatic
parahippocampal gyrus
parahippocampalis
    gyrus p.
    uncus gyri p.
parahypophysis
parajejunal fossa
parajejunalis
    fossa p.
parallelae
    striae p.
parallelism
paraluteal cell
paralysis, pl. **paralyses**
    diver's p.
paramammarii
    nodi p.
    nodi lymphoidei p.
paramammary lymph node
paramastoideus
    processus p.
paramastoid process
paramedial sulcus
paramedian pontine branch of pontine
  artery
paramesial
paramesonephric duct
paramesonephricus
    ductus p.
parametria (*pl. of* parametrium)
parametrial
parametric
parametrium, pl. **parametria**
paramolar
paramorphia
paramorphic
paranasal
    p. cell
    p. sinuses
paranasales
    sinus p.
paranephric
    p. body
    p. fat
paraneural
paraneurone
paranuclear body
paranucleate
paranucleolus
paranucleus
paraomphalic
paraoral

paraovarian
parapancreatic
paraperitoneal
parapharyngeal
    p. lymph node
    p. space
parapharyngeum
    spatium p.
paraphysis, pl. **paraphyses**
parapineal
paraplasm
paraplastic
pararectal
    p. fossa
    p. lymph node
    p. pouch
pararectales
    lymphonodi p.
    nodi lymphoidei p.
pararectalis
    fossa p.
pararenal
parasaccular
parasacral
parasagittal
    p. plane
    p. section
parasecretion
paraseptal cartilage
parasinoidal sinuses
parasite
parasiticus
    craniopagus p.
    janiceps p.
parasol insertion
parasomnia
paraspinal muscle
paraspinous
    p. fascia
    p. muscle
    p. musculature
parasternal
    p. hernia
    p. line
    p. lymph nodes
    p. tissue
parasternales
    lymphonodi p.
    nodi lymphoidei p.
parasternalis
    linea p.
parastriate cortex

NOTES

P

parasympathetic
   p. ganglia
   p. nerve
   p. part
   p. part of autonomic division of peripheral nervous system
   p. root of ciliary ganglion
   p. root of otic ganglion
   p. root of pelvic ganglion
   p. root of pterygopalatine ganglion
   p. root of submandibular ganglion
parasympathetica
   ganglia p.
parasympathica
   pars p.
parasympatholytic
parasympathomimetic
parasympathotonia
paratenon
paraterminal
   p. body
   p. gyrus
paraterminale
   corpus p.
paraterminalis
   gyrus p.
parathyroid
   p. artery
   p. gland
   p. hormone
   p. hormonelike protein
   water-clear cell of p.
parathyroidea
   glandula p.
parathyrotrophic
parathyrotropic
paratonsillar vein
paratracheales
   lymphonodi p.
   nodi lymphoidei p.
paratracheal lymph node
paratripsis
paraumbilical
   p. hernia
   p. vein
paraumbilicales
   venae p.
paraurethral
   p. duct
   p. gland
paraurethrales
   ductus p.
parauterine lymph node
parauterini
   lymphonodi p.
   nodi lymphoidei p.
paravaginales
   lymphonodi p.
   nodi lymphoidei p.

paravaginal lymph node
paravalvular
paravenous
paraventricular
   p. nucleus
   p. vein
paraventricularis
   nucleus p.
paravertebral
   p. ganglia
   p. gutter
   p. line
   p. musculature
paravertebralis
   linea p.
paravesical
   p. fossa
   p. lymph node
   p. pouch
   p. space
paravesicalis
   fossa p.
paravesiculares
   lymphonodi p.
   nodi lymphatici p.
paraxial mesoderm
paraxon
parchment heart
parencepha
parencephalocele
parencephalous
parenchyma
   p. glandulae thyroideae
   lung p.
   neural p.
   p. prostatae
   p. of prostate
   p. testis
   p. of thyroid gland
parenchymal
   p. cell
   p. tissue
parenchymatous
   p. cell of corpus pineale
   p. tissue
parent
   p. artery
   p. cell
parenteral absorption
parepicele
parepididymis
pareunia
paries, pl. **parietes**
   p. anterior gastris
   p. anterior vaginae
   p. anterior ventriculi
   p. caroticus cavi tympani
   p. externus ductus cochlearis
   p. inferior orbitae

p. jugularis cavi tympani
p. labyrinthicus cavi tympani
p. lateralis orbitae
p. mastoideus cavi tympani
p. medialis orbitae
p. membranaceus cavi tympani
p. membranaceus tracheae
p. posterior gastris
p. posterior vaginae
p. posterior ventriculi
p. superior orbitae
p. tegmentalis cavi tympani
p. tympanicus ductus cochlearis
p. vestibularis ductus cochlearis

parietal
p. abdominal fascia
p. angle
p. area
p. border
p. border of frontal bone
p. border of sphenoid bone
p. border of squamous part of temporal bone
p. branch
p. branch of medial occipital artery
p. branch of middle meningeal artery
p. branch of superficial temporal artery
p. cell
p. eminence
p. emissary vein
p. eye
p. foramen
p. lamina
p. layer
p. layer of leptomeninges
p. layer of serous pericardium
p. layer of tunica vaginalis of testis
p. lobe
p. lymph node
p. margin
p. margin of frontal bone
p. margin of greater wing of sphenoid
p. notch
p. operculum
p. pelvic fascia
p. peritoneum

p. plate
p. pleura
p. region
p. tuber
p. vein
p. wall

parietale
emissarium p.
foramen p.
os p.
peritoneum p.
tuber p.

parietales
arteriae p.
nodi lymphoidei p.
rami p.
venae p.

parietalis
angulus frontalis ossis p.
angulus mastoideus ossis p.
angulus occipitalis ossis p.
angulus sphenoidalis ossis p.
decidua p.
eminentia p.
emissaria p.
facies externa ossis p.
facies interna ossis p.
fascia abdominalis p.
fascia pelvis p.
incisura p.
lamina p.
linea temporalis inferior ossis p.
linea temporalis superior ossis p.
margo frontalis ossis p.
margo occipitalis ossis p.
margo sagittalis ossis p.
margo squamosus ossis p.
pars costalis pleurae p.
pars diaphragmatica pleurae p.
pars mediastinalis pleurae p.
pleura p.
ramus p.
vena emissaria p.

parietes (*pl. of* paries)
parietis
parieto
parietocolic fold
parietofrontal
parietomastoidea
sutura p.
parietomastoid suture

NOTES

P

parietooccipital
>p. branch of anterior cerebral artery
>p. branch of posterior cerebral artery
>p. fissure
>p. projection
>p. region
>p. sulcus
>p. suture

parietooccipitales

parietooccipitalis
>arteria p.
>fissura p.
>ramus marginalis sulci p.
>sulcus p.

parietoperitoneal fold

parietopontine tract

parietopontinus
>tractus p.

parietosphenoid

parietosplanchnic

parietosquamosal

parietotemporal region

parietovisceral

Paris line

paroccipital process

parodontium

parolfactoria
>area p.

parolfactorius posterior

parolfactory area

parolivary

Parona space

paroophori
>ductuli p.
>tubuli p.

paroöphoron

parorchidium

parorchis

parosteal

parotic

parotid
>p. bed
>p. branch
>p. capsule
>p. duct
>p. fascia
>p. gland
>p. lymph node
>p. notch
>p. papilla
>p. plexus
>p. plexus of facial nerve
>p. recess
>p. sheath
>p. space
>p. vein

parotidea
>fascia p.
>glandula p.
>papilla p.

parotideae
>pars profunda glandulae p.
>pars superficialis glandulae p.
>venae p.

parotidei
>papilla ductus p.
>rami p.

parotideomasseterica
>fascia p.

parotideomasseteric fascia

parotideus
>ductus p.
>recessus p.

parotidis
>pars profunda glandulae p.
>processus retromandibularis glandulae p.
>socia p.

parotidoauricularis

parotis
>glandula p.

parovarian

parovarium

pars, pl. partes
>p. abdominalis aortae
>p. abdominalis ductus thoracici
>p. abdominalis ductus thoracicus
>p. abdominalis esophagi
>p. abdominalis musculi pectorales majoris
>p. abdominalis plexus visceralis et ganglii visceralis
>p. abdominalis ureteris
>p. acromialis musculi deltoidei
>p. alaris musculi nasalis
>p. alveolaris mandibulae
>p. amorpha
>p. annularis vaginae fibrosae
>p. anterior commissurae anterioris
>p. anterior commissurae rostralis
>p. anterior faciei diaphragmatis hepatis
>p. anterior fornicis vaginae
>p. anterior linguae
>p. anularis vaginae fibrosae digitorum manus et pedis
>p. aryepiglottica musculi arytenoidei obliqui
>p. ascendens aortae
>p. ascendens duodeni
>p. ascendens musculi trapezii
>p. atlantica arteriae vertebralis
>p. autonomica
>p. autonomica systematis nervosi peripherici

p. basalis
p. basalis arteriae pulmonalis
p. basalis arteriarum lobarium inferiorum pulmonis sinistri et dextri
p. basilaris
p. basilaris ossis occipitalis
p. basilaris pontis
p. buccopharyngea musculi constrictoris pharyngei superioris
p. canalis nervi optici
p. cardiaca gastricae
p. cardiaca ventriculi
p. cartilaginea septi nasi
p. cartilaginea systematis skeletalis
p. cartilaginea tubae auditivae
p. cartilaginea tubae auditoriae
p. caudalis
p. cavernosa
p. cavernosa arteriae carotidis internae
p. cavernosa arteriae carotis internae
p. ceca retinae
p. centralis systematis nervosi
p. centralis ventriculi lateralis
p. ceratopharyngea
p. ceratopharyngea musculi constrictoris pharyngis medii
p. cerebralis arteriae carotidis internae
p. cerebralis arteriae carotis internae
p. cervical
p. cervicalis arteriae carotidis internae
p. cervicalis arteriae carotis internae
p. cervicalis arteriae vertebralis
p. cervicalis ductus thoracici
p. cervicalis esophagi
p. cervicalis medullae spinalis
p. chondropharyngea
p. chondropharyngea musculi constrictoris pharyngei medii
p. ciliaris retinae
p. clavicularis
p. clavicularis musculi deltoidei
p. clavicularis musculi pectoralis majoris
p. coccygea medullae spinalis
p. cochlearis

p. cochlearis nervi vestibulocochlearis
p. coeliacoduodenalis musculi ligamenti suspensorii duodeni
p. convoluta lobuli corticalis renis
p. corneoscleralis
p. corneoscleralis reticuli trabecularis sclerae
partes corporis humani
p. corticalis
p. corticalis arteriae cerebralis mediae
p. costalis diaphragmatis
p. costalis pleurae parietalis
p. cranialis partis parasympathetici divisionis autonomici systematis nervosi
p. craniocervicalis plexuum et gangliorum visceralium
p. cricopharyngea
p. cricopharyngea musculi constrictoris pharyngis inferioris
p. cruciformis vaginae fibrosae
p. cuneiformis vomeris
p. cupularis
p. cupularis recessus epitympanici
p. cystica
p. descendens aortae
p. descendens duodeni
p. descendens ligamenti iliofemoralis
p. descendens musculi trapezii
p. dextra faciei diaphragmaticae hepatis
p. diaphragmatica pleurae parietalis
p. distalis
p. distalis prostatae
p. distalis urethrae prostaticae
partes dorsales musculorum intertransversariorum lateralium lumborum
p. dorsalis pontis
p. endocrina pancreatis
p. exocrina pancreatis
p. extraocularis arteriae et venae centralis retinae
p. flaccida membranae tympanae
p. flaccida membranae tympani
p. frontalis corporis callosi
p. funicularis ductus deferentis
partes genitales femininae externae
partes genitales masculinae externae

NOTES

pars *(continued)*
p. glossopharyngea
p. glossopharyngea musculi constrictoris pharyngis superioris
p. granulosa
p. hepatica
p. hepatis dextra
p. hepatis sinistra
p. horizontalis duodeni
p. iliaca fasciae iliopsoaticae
p. inferior
p. inferior duodeni
p. inferior ganglii vestibularis
p. inferior ramus lingularis
p. inferior venae lingularis venae pulmonalis superioris sinistrae
p. infraclavicularis plexus brachialis
p. infralobaris
p. infralobaris venae posterioris venae pulmonalis superioris dextrae
p. infrasegmentalis
p. infundibularis
p. inguinalis ductus deferentis
p. insularis
p. insularis arteriae cerebri mediae
p. intercartilaginea rimae glottidis
p. intermedia
p. intermedia adenohypophyseos
p. intermedia urethrae masculinae
p. intermembranacea rimae glottidis
partes intersegmentales venarum pulmonum
p. intersegmentalis
p. intracaniculus nervi optici
p. intracranialis
p. intracranialis arteriae vertebralis
p. intracranialis nervi optici
p. intralaminaris nervi optici
p. intralaminaris nervi optici intralocularis
p. intralobaris
p. intralobaris intersegmentalis venae posterioris lobi superioris pulmonis dextri
p. intramuralis urethrae masculinae
p. intraocularis nervi optici
p. intrasegmentalis
p. intrasegmentalis venae pulmonum
p. iridica retinae
p. labialis
p. labialis musculi orbicularis oris
p. lacrimalis musculi orbicularis oculi
p. lacrimalis of orbicularis oculi muscle
p. laryngea pharyngis
p. lateralis arcus pedis longitudinalis

p. lateralis compartimenti antebrachii posterioris extensorum
p. lateralis fornicis vaginae
p. lateralis fornix vaginae
p. lateralis musculi intertransversarii posteriores cervicis
p. lateralis musculorum intertransversariorum posteriorum cervicis
p. lateralis ossis occipitalis
p. lateralis ossis sacri
p. lateralis venae lobi medii venae pulmonalis dextri superioris
p. lateralis venae pulmonalis
p. libera membri inferioris
p. libera membri superioris
p. lumbalis
p. lumbalis diaphragmatis
p. lumbalis medullae spinalis
p. marginalis
p. marginalis musculi orbicularis oris
p. mastoidea
p. mastoidea ossis temporalis
p. medialis
p. medialis arcus pedis longitudinalis
p. medialis musculi intertransversarii posteriores cervicis
p. medialis musculorum intertransversariorum posteriorum cervicis
p. medialis venae lobi medii venae pulmonis dextri superioris
p. medialis venae pulmonis
p. mediastinalis
p. mediastinalis pleurae parietalis
p. mediastinalis pulmonis
p. membranacea septi atriorum
p. membranacea septi interventricularis
p. membranacea septi nasi
p. membranacea urethrae masculinae
p. mobilis septi nasi
p. muscularis septi interventricularis
p. muscularis septi interventricularis cordis
p. mylopharyngea
p. mylopharyngeus musculi constrictoris pharyngis superioris
p. nasalis ossis frontalis
p. nasalis pharyngis
p. nervosa hypophyseos
p. nervosa retinae
p. obliqua musculi cricothyroidei
p. occipitalis corporis callosi
p. olfactoria tunicae mucosae

p. opercularis
p. optica retinae
p. oralis pharyngis
p. orbitalis
p. orbitalis glandulae lacrimalis
p. orbitalis musculi orbicularis oculi
p. orbitalis nervi optici
p. orbitalis ossis frontalis
p. ossea septi nasi
p. ossea systematis skeletalis
p. ossea tubae auditivae
p. ossea tubae auditoriae
p. palpebralis
p. palpebralis glandulae lacrimalis
p. palpebralis musculi orbicularis oculi
p. parasympathica
p. patens arteriae umbilicalis
p. pelvica
p. pelvica ductus deferentes
p. pelvica plexus visceralis et ganglii visceralis
p. pelvica ureteris
p. peripherica
p. peripherica systematis nervosi
p. perpendicularis
p. petrosa
p. petrosa arteriae carotidis internae
p. petrosa arteriae carotis internae
p. petrosa ossis temporalis
p. phallica
p. pharyngea hypophyseos
p. phrenicocoeliaca musculi ligamenti suspensorii duodeni
p. pialis fili terminalis
p. pigmentosa
p. plana
p. postcommunicalis
p. postcommunicalis arteriae cerebri anterioris
p. postcommunicalis arteriae cerebri posterioris
p. posterior commissurae anterioris
p. posterior faciei diaphragmatis hepatis
p. posterior facies diaphragmatis hepatis
p. posterior fornicis vaginae
p. posterior fornix vaginae
p. posterior linguae
p. postlaminaris nervi optici

p. postlaminaris nervi optici intraocularis
p. postsulcalis
p. postsulcalis linguae
p. precommunicalis
p. precommunicalis arteriae cerebri anterioris
p. precommunicalis arteriae cerebri posterioris
p. prelaminaris nervi optici
p. prelaminaris nervi optici intraocularis
p. preprostatica urethrae masculinae
p. presulcalis
p. presulcalis linguae
p. prevertebralis arteriae prevertebralis
p. prima duodeni
p. profunda
p. profunda compartimenti antebrachii anterioris
p. profunda compartimenti cruris posterioris
p. profunda glandulae parotideae
p. profunda glandulae parotidis
p. profunda musculi masseteri
p. profunda musculi sphincteri ani externi
p. profunda partis palpebralis musculi orbicularis oculi
p. prostatica urethrae
p. proximalis prostatae
p. proximalis urethrae prostaticae
p. psoatica fasciae iliopsoaticae
p. pterygopharyngea
p. pterygopharyngea musculi constrictoris pharyngis superioris
p. pylorica
p. pylorica gastris
p. pylorica ventriculi
p. quadrata
p. quadrata hepatis
p. radiata lobuli corticalis renis
p. recta
p. recta musculi cricothyroidei
p. respiratoria tunicae mucosae
p. retrolentiformis capsulae internae
p. sacralis medullae spinalis
p. scrotalis ductus deferentis
p. secundum duodeni
p. sellaris

NOTES

P

443

pars *(continued)*

p. solealis compartimenti cruris posterioris
p. sphenoidalis
p. sphenoidalis arteriae cerebralis mediae
p. spinalis fili terminalis
p. spinalis musculi deltoidei
p. spinalis nervi accessorii
p. spongiosa urethrae masculinae
p. squamosa ossis temporalis
p. sternalis diaphragmatis
p. sternocostalis
p. sternocostalis musculi pectoralis majoris
p. subcutanea
p. subcutanea musculi sphincteri ani externi
p. sublentiformis capsulae internae
p. superficialis
p. superficialis compartimenti antebrachii anterioris
p. superficialis compartimenti cruris posterioris
p. superficialis glandulae parotideae
p. superficialis musculi masseteri
p. superficialis musculi sphincteri ani externi
p. superior duodeni
p. superior faciei diaphragmaticae hepatis
p. superior facies diaphragmatis hepatis
p. superior ganglii vestibularis
p. superior ramus lingularis
p. superior venae lingularis venae pulmonis superioris sinistri
p. supraclavicularis plexus brachialis
p. sympathica
p. sympathica divisionis autonomicae systematis nervosei peripherici
p. tecta
p. tecta duodeni
p. tecta pancreatis
p. tecta renalis
p. tecta ureteralis
p. tensa membranae tympani
p. terminalis
p. terminalis ilei
p. thoracica
p. thoracica aortae
p. thoracica ductus thoracici
p. thoracica esophagi
p. thoracica medullae spinalis
p. thoracica muscularis iliocostalis lumborum
p. thoracica plexuum et ganglionorum visceralium
p. thoracica tracheae
p. thyroepiglottica musculi thyroarytenoidei
p. thyropharyngea
p. thyropharyngea musculi constrictoris pharyngis inferioris
p. tibiocalcanea
p. tibiocalcanea ligamenti collateralis medialis articulationis talocruralis
p. tibiocalcanea ligamenti deltoidei
p. tibionavicularis
p. tibionavicularis ligamenti collateralis medialis articulationis talocrucalis
p. tibiotalaris anterior
p. tibiotalaris anterior ligamenti collateralis medialis articulationis talocruralis
p. tibiotalaris posterior
p. tibiotalaris posterior ligamenti collateralis medialis articulationis talocruralis
p. transversa
p. transversa ligamenti iliofemoralis
p. transversa musculi nasalis
p. transversa musculi trapezii
p. transversa rami sinistri venae portae hepatis
p. transversaria
p. transversaria arteriae vertebralis
p. triangularis
p. tricipitalis compartimenti cruris posterioris
p. tuberalis
p. tympanica ossis temporalis
p. umbilicalis rami sinistri venae portae hepatis
p. uterina
p. uterina placentae
p. uterina tubae uterinae
p. uvealis
p. uvealis reticuli trabecularis sclerae
p. vagalis
p. vagalis nervi accessorii
p. ventralis musculi intertransversarii lateralium lumborum
p. ventralis pontis
p. vertebralis faciei costalis pulmonis
p. vestibularis nervi vestibulocochlearis

pars phrenicocoeliaca musculi ligamenti suspensorii duodeni
part

abdominal p.
alar p.

anterior tibiotalar p.
autonomic p.
basal p.
basilar p.
buccopharyngeal p.
ceratopharyngeal p.
cervical p.
chondropharyngeal p.
clavicular p.
corneoscleral p.
cortical p.
cricopharyngeal p.
cupular p.
cupulate p.
deep p.
hidden p.
p.'s of human body
inferior p.
infralobar p.
infrasegmental p.
insular p.
intermediate p.
intersegmental p.
intracranial p.
intralobar p.
intrasegmental p.
labial p.
lumbar p.
marginal p.
mastoid p.
medial p.
mediastinal p.
mylopharyngeal p.
oblique p.
p. of optic nerve in canal
orbital p.
palpebral p.
parasympathetic p.
pelvic p.
petrous p.
p. petrous of temporal bone
postcommunical p.
posterior tibiotalar p.
postsulcal p.
precommunical p.
presulcal p.
pterygopharyngeal p.
quadrate p.
soft p.
sphenoidal p.
sternocostal p.
subcutaneous p.

terminal p.
thoracic p.
thyropharyngeal p.
tibiocalcaneal p.
tibionavicular p.
transverse p.
umbilical p.
uterine p.
uveal p.
vertebral p.
partes (*pl. of* pars)
parthenogenesis
partial
    p. anencephaly
    p. anodontia
    p. pressure
partialis
    previa p.
particle
    elementary p.
particulate
partner
    contralateral p.
parumbilical
parva
    saphena p.
    vena cordis p.
    vena saphena p.
parvae
    sinus alae p.
parvicellular
parvicollis
    uterus p.
parvus
passage
    boutons en p.
    nasal p.
    nasopharyngeal p.
    oropharyngeal p.
Passavant
    P. bar
    P. cushion
    P. fold
    P. ridge
passive
    p. congestion
    p. hyperemia
    p. length-tension curve
    p. movement
    p. vasoconstriction
    p. vasodilation
passivism

NOTES

P

Pasteur effect
patagium, pl. **patagia**
 cervical p.
patch
 Peyer p.
patella, pl. **patellae**
 p. alta
 anterior surface of p.
 apex of p.
 articular surface of p.
 p. baja
 base of p.
 basis patellae
 p. bone
 facies anterior patellae
 facies articularis patellae
 lateral retinaculum of p.
 ligamentum patellae
 medial retinaculum of p.
 rete patellae
patellar
 p. anastomosis
 p. dome
 p. facet cartilage
 p. fossa of vitreous
 p. ligament
 p. network
 p. retinaculum
 p. surface of femur
 p. synovial fold
 p. tendon
 p. tendon reflex
 p. tracking
patellare
 rete p.
patellaris
 plica synovialis p.
patelliform
patelloadductor reflex
patellofemoral articulation
patency
 probe p.
patent
 p. canal of Nuck
 p. ductus arteriosus
 p. part of umbilical artery
pathetic nerve
pathologic
 p. absorption
 p. physiology
 p. sphincter
pathophysiology
pathopoiesis
pathway
 atrio-His p.
 Embden-Meyerhof p.
 Embden-Meyerhof-Parnas p.
 lacrimal p.
 pentose phosphate p.

 phosphogluconate p.
 taste p.
pattern
 rugal p.
paucisynaptic
Pauling theory
pavement epithelium
Pavlov
 P. method
 P. pouch
 P. reflex
 P. stomach
Payr membrane
PDA
 posterior descending artery
PDGF
 platelet-derived growth factor
PEA
 pulseless electrical activity
peak
 p. expiratory flow
 widow p.
pear-shaped area
Pecquet
 P. cistern
 P. duct
 P. reservoir
pecqueti
 receptaculum p.
pectate line
pecten
 anal p.
 p. analis
 p. ossis pubis
pectinata
 zona p.
pectinate
 p. fiber
 p. ligament
 p. ligament of iridocorneal angle
 p. ligament of iris
 p. line
 p. muscle
 p. zone
pectinati
 musculi p.
pectinatum
 ligamentum p.
pectinea
 linea p.
pectineal
 p. fascia
 p. hernia
 p. ligament
 p. line
 p. line of femur
 p. line of pubis
 p. muscle

pectineale
    ligamentum p.
pectineus
    p. muscle
    musculus p.
pectiniforme
    septum p.
pectiniform septum
pectora (*pl. of* pectus)
pectoral
    p. and abdominal anterior
      cutaneous branch of intercostal
      nerve
    p. axillary lymph node
    p. branch of thoracoacromial artery
    p. fascia
    p. girdle
    p. gland
    p. groove
    p. muscle
    musculus p.
    p. reflex
    p. region
    p. ridge
    p. vein
pectorale
    cingulum p.
pectorales
    nodi lymphoidei axillares p.
    rami p.
    venae p.
pectoralis
    articulationes cinguli p.
    fascia p.
    p. major
    p. major muscle
    p. minor
    p. minor muscle
    regio p.
pectoris
pectorodorsalis muscle
pectorodorsal muscle
pectus, pl. **pectora**
    p. carinatum
    p. excavatum
    p. recurvatum
pedal system
pedes (*pl. of* pedis)
pedicel
pedicellate
pedicellation

pedicle
    p. of arch of vertebra
    p. of lung
    renal p.
    p. of spleen
    vascular p.
pedicterus
pediculate
pediculus, pl. **pediculi**
    p. arcus vertebrae
pediphalanx
pedis, pl. **pedes**
    abductor digiti minimi quinti p.
    arcus venosus dorsalis p.
    arteria arcuata p.
    arteria dorsalis p.
    arteria plantaris profunda arteriae
      dorsalis p.
    articulationes p.
    articulationes interphalangeae p.
    basis phalangis p.
    calcar p.
    caput phalangis manus et p.
    digitales dorsales p.
    digiti p.
    dorsalis p.
    dorsum p.
    facies digitalis dorsalis manus
      et p.
    facies lateralis digiti p.
    facies medialis digiti p.
    fascia dorsalis p.
    interossei dorsales p.
    ligamenta plantaria articulationis
      interphalangeae p.
    lumbricales p.
    margo fibularis p.
    margo lateralis p.
    margo medialis p.
    margo tibialis p.
    musculi interossei dorsales p.
    musculi interossei dorsalis p.
    musculi lumbricales p.
    musculus abductor digiti minimi p.
    musculus flexor digiti minimi
      brevis p.
    musculus interosseus dorsalis p.
    musculus lumbricalis p.
    musculus pronator p.
    nervi digitales dorsales p.
    ossa p.
    os vesalianum p.

NOTES

P

pedis *(continued)*
pars anularis vaginae fibrosae
digitorum manus et p.
phalanx distalis p.
phalanx proximalis p.
planta p.
pollex p.
ramus plantaris profundus arteriae
dorsalis p.
rete venosum dorsale p.
trochlea phalangis manus et p.
tuberositas phalangis distalis manus
et p.
vaginae fibrosae digitorum p.
vaginae synoviales digitorum p.
vaginae tendinum digitorum p.
venae digitales dorsales p.
vincula tendinum digitorum manus
et p.

peduncle
cerebellar p.
cerebral p.
p. of corpus callosum
decussation of superior
cerebellar p.
p. of flocculus
inferior cerebellar p.
inferior thalamic p.
lateral thalamic p.
p. of mamillary body
middle cerebellar p.
middle cervical p.
olfactory p.
superior cerebellar p.
ventral thalamic p.

peduncular
p. ansa
p. loop

pedunculares
rami p.
venae p.

peduncularis
ansa p.

pedunculate
pedunculated hydatid
pedunculomamillaris
fasciculus p.

pedunculus, pl. **pedunculi**
p. cerebellaris inferior
p. cerebellaris superior
p. cerebri
p. corporis callosi
p. corporis mamillaris
p. flocculi
pes pedunculi
p. of pineal body
p. thalami inferior
p. thalami lateralis

p. thalami ventralis
p. vitellinus

peg
rete p.

peg-and-socket
p.-a.-s. articulation
p.-a.-s. joint

pellicular
pelliculous
pellucida
macula p.
zona p.

pellucidi
cavum septi p.
lamina septi p.

pellucidum
anterior vein of septum p.
cavity of septum p.
lamina of septum p.
posterior vein of septum p.
septum p.
vein of septum p.

pellucid zone
pelma
pelmatic
pelves (*pl. of* pelvis)
pelvic
p. appendix
p. autonomic nerve
p. axis
p. bone
p. brim
p. canal
p. colon
p. colon of Waldeyer
p. diaphragm
p. direction
p. fascia
p. floor
p. fossa
p. ganglia
p. girdle
p. inclination
p. index
p. limb
p. lymph node
p. musculature
p. nervous plexus
p. organ
p. part
p. part of ductus deferens
p. part of peripheral autonomic
plexuses and ganglion
p. part of ureter
p. part of the urogenital sinus
p. peritoneal cavity
p. peritoneum
p. plane of greatest dimension
p. plane of inlet

p. plane of least dimension
p. plane of outlet
p. promontory
p. skeleton
p. splanchnic nerve
p. strait
p. surface of sacrum
p. viscera
p. wall
pelvica
fascia p.
ganglia p.
pars p.
pelvic-femoral angle
pelvici
articulationes cinguli p.
cingulum p.
pelvicorum
radices parasympathicae
gangliorum p.
pelvicus
plexus nervosus p.
pelvifemoral
pelvina
cavitas p.
foramina sacralia p.
ganglia p.
pelvini
nervi splanchnici p.
splanchnici p.
pelvinum
colon p.
pelvinus
plexus p.
pelvioprostatic capsule
pelvirectal sphincter
pelvis, pl. **pelves**
aditus p.
ampullary type of renal p.
android p.
anthropoid p.
arcus tendineus fasciae p.
axis of p.
brachypellic p.
branching type of renal p.
cavitas abdominis et p.
cavum p.
contracted p.
cordate p.
cordiform p.
Deventer p.
diaphragm of p.

diaphragma p.
dolichopellic p.
dwarf p.
false p.
fascia superior diaphragmatis p.
flat p.
funnel-shaped p.
p. of gallbladder
greater p.
gynecoid p.
heart-shaped p.
inclinatio p.
inclination of p.
inverted p.
p. justo major
p. justo minor
juvenile p.
kidney p.
large p.
lesser p.
ligamentum transversum p.
linea terminalis p.
longitudinal oval p.
masculine p.
mesatipellic p.
mucosa of renal p.
muscular layer of renal p.
nodi lymphoidei p.
p. plana
platypellic p.
platypelloid p.
renal p.
p. renalis
reniform p.
Robert p.
round p.
small p.
spider p.
p. spuria
transverse ligament of p.
transverse oval p.
trigonum parietale laterale p.
true p.
ureteric p.
p. vera
pelvisacral
pelvivertebral angle
pendelluft
pendular movement
pendulous palate
penes (*pl. of* penis)
penial

NOTES

penicillary
penicillate
penicillatus
    limbus p.
penicillus, pl. **penicilli**
penile
    p. raphe
    p. reflex
    p. shaft
    p. sheath
    p. skin
    p. urethra
penis, pl. **penes**
    arteria bulbi p.
    arteria dorsalis p.
    arteriae helicinae p.
    arteriae perforantes p.
    arteria profunda p.
    artery of p.
    artery of bulb of p.
    bifid p.
    body of p.
    bulb of p.
    bulbi p.
    bulbus p.
    cavernosae p.
    cavernosi p.
    cavernous body of p.
    cavernous nerve of p.
    cavernous plexus of p.
    cavernous vein of p.
    collum glandis p.
    compressor venae dorsalis p.
    corona glandis p.
    corona of glans p.
    corpus cavernosum p.
    corpus spongiosum p.
    crus of p.
    crus corporis cavernosi p.
    deep artery of p.
    deep dorsal vein of p.
    deep fascia of p.
    dorsal artery of p.
    dorsal nerve of p.
    dorsal vein of p.
    dorsum p.
    double p.
    erectile elements of the p.
    facies urethralis p.
    fascia of deep p.
    fascia of superficial p.
    p. femineus
    foreskin of p.
    frenulum of p.
    fundiform ligament of p.
    glans p.
    helicine artery of p.
    hemisphere of bulb of p.
    ligamentum fundiforme p.
    ligamentum suspensorium p.
    p. muliebris
    musculus erector p.
    neck of glans p.
    nervi cavernosi p.
    nervus dorsalis p.
    opening for dorsal artery of p.
    opening for dorsal nerve of p.
    os p.
    p. palmatus
    perforating artery of p.
    prepuce of p.
    preputium p.
    profunda p.
    profundae p.
    radix p.
    raphe of p.
    root of p.
    scapus p.
    septum of glans p.
    shaft of p.
    spongy body of p.
    subcutaneous tissue of p.
    superficial dorsal vein of p.
    superficial fascia of p.
    suspensory ligament of p.
    tela subcutanea p.
    trabeculae corporis spongiosi p.
    urethral surface of p.
    vein of bulb of p.
    vena bulbi p.
    venae cavernosae p.
    venae profundae p.
    vena profunda p.
penischisis
pennate muscle
pennatus
    musculus p.
penniform muscle
penoscrotal hypospadias
pentadactyle
pentose phosphate pathway
peptic
    p. cell
    p. digestion
    p. gland
peptide
    atrial natriuretic p. (ANP)
    calcitonin gene-related p.
peptidergic
peptogenic
peptogenous
peracephalus
perarticulation
peraxillary
perceptorium
percreta
    placenta p.

perforans
  ramus p.
perforantes
  arteriae p.
  rami p.
  venae p.
perforata
  habenulae p.
  zona p.
perforated
  p. layer of sclera
  p. space
perforating
  p. arteries of internal mammary
  p. arteries of peroneal
  p. artery of deep femoral artery
  p. artery of foot
  p. artery of hand
  p. artery of internal thoracic artery
  p. artery of penis
  p. branch
  p. branch of anterior interosseous
    artery
  p. branch of deep palmar arch
  p. branch of fibular artery
  p. branch of internal thoracic
    artery
  p. branch of palmar metacarpal
    artery
  p. branch of peroneal artery
  p. branch of plantar metatarsal
    artery
  p. fibers
  p. radiate artery of kidney
  p. vein
perforation
perforator
  hunterian p.
perforatorium
perfusate
perfuse
perfusion
periacinal
periacinous
perianal
  p. reflex
  p. skin
periaortic node
periapex
periapical tissue
periappendicular
periareolar

periarterial
  p. pad
  p. plexus
  p. plexus of anterior cerebral
    artery
  p. plexus of ascending pharyngeal
    artery
  p. plexus of choroid artery
  p. plexuses of coronary artery
  p. plexus of facial artery
  p. plexus of inferior phrenic artery
  p. plexus of inferior thyroid artery
  p. plexus of internal thoracic
    artery
  p. plexus of lingual artery
  p. plexus of maxillary artery
  p. plexus of middle cerebral artery
  p. plexus of occipital artery
  p. plexus of ophthalmic artery
  p. plexus of popliteal artery
  p. plexus of posterior auricular
    artery
  p. plexus of subclavian artery
  p. plexus of superficial temporal
    artery
  p. plexus of superior thyroid artery
  p. plexus of testicular artery
  p. plexus of thyroid artery
  p. plexus of vertebral artery
periarterialis
  plexus p.
periarthric
periarticular
periatrial
periauricular
periaxial
periaxillary
periaxonal
periblast
peribronchial
peribronchiolar
peribuccal
peribulbar
peribursal
pericallosa
  arteria p.
pericallosal
  p. artery
  p. vein
pericanalicular
pericapillary cell
pericardia (*pl. of* pericardium)

NOTES

P

pericardiaca
    cavitas p.
    pleura p.
pericardiacae
    venae p.
pericardiaci
    villi p.
pericardiacophrenic
    p. artery
    p. vein
pericardiacophrenica
    arteria p.
pericardiacophrenicae
    venae p.
pericardiac vein
pericardial
    p. baffle
    p. branch of phrenic nerve
    p. branch of thoracic aorta
    p. cavity
    p. pleura
    p. reflection
    p. sac
    p. space
    p. vein
    p. villi
pericardialis
    cavitas p.
pericardii
    cavum p.
    lamina parietalis p.
    lamina visceralis p.
    sinus obliquus p.
    sinus transversus p.
pericardioperitoneal canal
pericardiophrenic nerve
pericardiopleural
    p. fold
    p. membrane
pericardium, pl. **pericardia**
    p. fibrosa
    p. fibrosum
    fibrous p.
    oblique sinus of p.
    parietal layer of serous p.
    serosa of serous p.
    p. serosum
    serous p.
    transverse sinus of p.
    visceral layer of serous p.
pericecal
pericemental attachment
pericentral
perichondral bone
perichondrial
perichondrium
perichord
perichordal
perichoroidal space

perichoroideale
    spatium p.
perichoroideum
    spatium p.
perichoroid space
perichrome
periclaustral lamina
pericolic
periconchal sulcus
pericorneal
pericorpuscular synapse
pericostal suture
pericranial
pericranium
pericystic
pericystium
pericyte
    capillary p.
pericytial
pericytic venule
peridental
    p. ligament
    p. membrane
peridentium
periderm
periderma
peridermal
peridermic
peridesmic
peridesmium
perididymis
peridural
peridurale
    spatium p.
perienteric
periependymal
periesophageal
perietalis
    tunica serosa pleurae p.
perifuse
perifusion
periganglionic
perigastric lymph node
perigemmal
periglottic
periglottis
perihepatic
perihilar area
perikaryon, pl. **perikarya**
perikeratic
perikyma, pl. **perikymata**
perilaryngeal
perilenticular space
periligamentous
perilimbal
perilymph
perilymphangial
perilymphatic
    p. cavity

p. duct
p. fluid
p. space
perilymphatica
cisterna p.
perilymphatici
ductus p.
perilymphaticum
spatium p.
perilymphaticus
ductus p.
perimetric
perimetrium, pl. **perimetria**
perimyelis
perimysial
perimysium, pl. **perimysia**
p. externum
p. internum
perinea (*pl. of* perineum)
perineal
p. area
p. artery
p. body
p. branch of posterior cutaneous
nerve of thigh
p. branch of posterior femoral
cutaneous nerve
p. fascia
p. flexure of anal canal
p. flexure of rectum
p. hypospadias
p. membrane
p. muscle
p. raphe
p. region
p. sinus
p. space
p. tissue
perineales
nervi p.
rami p.
perinealis
arteria p.
rami labiales posteriores arteriae p.
rami scrotales posteriores
arteriae p.
regio p.
perinealium
rami musculares nervorum p.
perinei
centrum tendineum p.
fascia p.

ligamentum transversum p.
membrana p.
musculi p.
raphe p.
tela subcutanea p.
perineoscrotal
perineovaginal
perinephria (*pl. of* perinephrium)
perinephrial
perinephric
p. capsule
p. fascia
perinephrium, pl. **perinephria**
perineum, pl. **perinea**
central tendon of p.
deep transverse muscle of p.
dermatome of female pelvis and p.
dermatome of male pelvis and p.
membranous layer of superficial
fascia of p.
subcutaneous tissue of p.
superficial investing fascia of p.
superficial transverse muscle of p.
transverse ligament of p.
perineural
perineurial
perineurium, pl. **perineuria**
perineuronal satellite
perinuclear space
periocular
period
effective refractory p.
functional refractory p.
latent p.
mitotic p.
refractory p.
relative refractory p.
silent p.
synthesis p.
total refractory p.
periodicity
lunar p.
periodontal
p. ligament
p. ligament fiber
p. membrane
periodontium, pl. **periodontia**
periomphalic
perionychium, pl. **perionychia**
perionyx
periophthalmic
perioral

NOTES

P

periorbit
periorbita
periorbital
    p. membrane
    p. soft tissue
periost
periostea (*pl. of* periosteum)
periosteal
    p. bone
    p. bud
    p. layer of dura mater
periosteous
periosteum, pl. **periostea**
    alveolar p.
    p. alveolare
    p. cranii
periotic
    p. bone
    p. cartilage
    p. duct
    p. space
periovular
peripalpebral
peripapillary
peripenial
peripharyngeal space
peripharyngeum
    spatium p.
peripherad
peripheral
    p. chemoreceptor
    p. dysostosis
    p. nerve
    p. nervous system
    p. pulse
    p. reflex
    p. resistance
    p. vessel
peripheralis
peripherica
    pars p.
peripherici
    divisio autonomica systematis
      nervosi p.
    pars autonomica systematis
      nervosi p.
    pars sympathica divisionis
      autonomicae systematis
      nervosei p.
periphericum
    systema nervosum p.
peripherocentral
periphery
peripolar cell
peripolesis
periportal space of Mall
periproctic
periprostatic
peripylic

peripyloric
perirectal
perirenal
    p. fascia
    p. fat capsule
    p. space
    p. tissue
perirenalis
    capsula adiposa p.
perirhinal
perisalpinx
periscleral space
perisinuous
perisinusoidal space
perispinal area
perisplanchnic
perisplenic
perispondylic
perissodactyl
perissodactylous
peristalsis
    mass p.
peristaltic valve
peristole
peristolic
peristomal
peristomatous
peristriate area
perisylvian
peritarsal network
peritendineum, pl. **peritendinea**
peritenon
perithelial cell
perithelium, pl. **perithelia**
    Eberth p.
perithoracic
peritomy
peritonea (*pl. of* peritoneum)
peritoneal
    p. cavity
    p. fossa
    p. ligament of liver
    p. membrane
    p. reflection
    p. sac
    p. space
    p. surface
    p. villi
peritoneales
    villi p.
peritonealis
    cavitas p.
peritonei
    cavum p.
    processus vaginalis p.
    tunica serosa p.
peritoneocutaneous reflex
peritoneopericardial

peritoneovenous
peritoneum, pl. peritonea
    intestinal p.
    parietal p.
    p. parietale
    pelvic p.
    processus vaginalis of p.
    serosa of p.
    serous coat of p.
    serous layer of p.
    urogenital p.
    p. urogenitale
    vaginal process of p.
    visceral p.
    p. viscerale
peritonitis
    meconium p.
peritonsillar nerve
peritracheal
peritrichal
peritrichate
peritrichic
peritrichous
peritrochanteric
peritubular contractile cells
perityphlic
periumbilical
periungual
periureteral
periurethral tissue
periuterine
periuvular
perivascular
    p. cell
    p. fibrous capsule
    p. lymph space
perivascularis
    capsula fibrosa p.
perivenous
periventricular
    p. fibers
    p. white matter
periventriculares
    fibrae p.
perivertebral
perivesical fascia
perivisceral cavity
perivitelline space
Perlia
    convergence nucleus of P.
    P. nucleus

permanens
    dens p.
permanent
    p. cartilage
    p. tooth
permanetes
permeability theory of narcosis
permeable
permeant
permeate
perobrachius
perocephalus
perochirus
perodactylia
perodactyly
peromelia
peromely
perone
peronea
    arteria p.
peroneae
    ramus communicans arteriae p.
    venae p.
peroneal
    p. anastomotic ramus
    p. artery
    p. bone
    p. border of foot
    p. communicating branch
    p. communicating nerve
    p. compartment of leg
    p. lymph node
    p. muscle
    perforating arteries of p.
    p. pulley
    p. retinaculum
    p. sulcus
    p. tendon
    p. tendon sheath
    p. trochlea of calcaneus
    p. tubercle
    p. vein
    p. vessel
peronealis
    spina p.
    trochlea p.
peronei
    rami malleolares laterales arteriae
        fibularis p.
peroneocalcaneus
    musculus p.

NOTES

P

peroneorum
    compartimentum cruris laterale p.
    retinaculum musculorum p.
peroneotibial
peroneus
    p. brevis
    p. brevis muscle
    p. communis
    p. longus
    p. longus muscle
    nervus communicans p.
    p. profundus
    ramus communicans p.
    p. superficialis
    p. tertius
    p. tertius muscle
peropus
perosplanchnia
peroxisome
perpendicular
    p. of ethmoid plate
    p. fasciculus
    p. plate of ethmoid
    p. plate of ethmoid bone
    p. plate of palatine bone
perpendicularis
    agger p.
    lamina p.
    pars p.
per saltum
persistens
    sutura frontalis p.
persistent
    p. anterior hyperplastic primary vitreous
    p. atrioventricular canal
    p. frontal suture
    p. posterior hyperplastic primary vitreous
    p. truncus arteriosus
perspiration
    insensible p.
perspiratory gland
perversus
    situs p.
pes
    p. abductus
    p. adductus
    p. anserinus
    p. cavus
    p. equinovalgus
    p. equinovarus
    p. hippocampi
    p. pedunculi
    p. planus
    p. pronatus
    p. valgus
    p. varus
pessary cell

Peters ovum
petiolate
petiolated
petiole
petioled
petiolus epiglottidis
Petit
    P. aponeurosis
    P. canal
    P. ligament
    P. lumbar triangle
    P. sinus
petrobasilar suture
petroccipital
petroclinoid ligament
petromastoid
petrooccipital
    p. fissure
    p. joint
    p. synchondrosis
petrooccipitalis
    fissura p.
    synchondrosis p.
petropharyngeus
    musculus p.
petrosa, pl. petrosae
    apex partis petrosae
    fossula p.
    pars p.
    vena p.
petrosal
    p. bone
    p. branch of middle meningeal artery
    p. foramen
    p. fossa
    p. fossula
    p. ganglion
    p. impression of pallium
    p. nerve
    p. vein
    p. venous sinus
petrosalpingostaphylinus
petrosi
    receptaculum ganglii p.
petrosomastoid
petrosphenobasilar suture
petrosphenoidal fissure
petrosphenoid ligament
petrospheno-occipital suture of Gruber
petrosquamosa
    fissura p.
petrosquamosal
petrosquamous
    p. fissure
    p. suture
    p. venous sinus
petrostaphylinus
    musculus p.

petrosum
    foramen p.
petrosus
    p. major
    p. profundus
    ramus p.
petrotympanic
    p. fissure
    p. tissue
petrotympanica
    fissura p.
petrous
    p. apex
    p. apex cell
    p. ganglion
    p. part
    p. part of internal carotid artery
    p. part of temporal bone
    p. pyramid
    p. pyramid air cell
    p. ridge
    p. tip
Peyer
    P. gland
    P. patch
peyerianum
    agmen p.
Peyrot thorax
Pflüger law
PGR
    psychogalvanic response
phacocyst
phagocyte
phagocytic pneumonocyte
phagocytoblast
phagocytose
phagocytosis
phagolysosome
phagosome
phalangeal
    p. articulation
    p. bone
    p. cells
    p. joint
    p. tuft
phalangis
    basis p.
    corpus p.
    trochlea p.
phalanx, pl. **phalanges**
    base of p.
    body of p.

    cutaneous ligament of p.
    distal p.
    p. distalis
    p. distalis manus
    p. distalis pedis
    head of p.
    p. media
    p. media pedis et manus
    middle p.
    proximal head of p.
    p. proximalis
    p. proximalis manus
    p. proximalis pedis
    shaft of p.
    terminal p.
    tuberosity of p.
    ungual p.
phalli (*pl. of* phallus)
phallica
    crista p.
    pars p.
phallic tubercle
phalliform
phalloid
phallus, pl. **phalli**
phanerosis
    fatty p.
phantom
pharmacodynamic
pharmacoendocrinology
pharmacokinetic
pharmacologic
pharmacological
pharyngea
    aponeurosis p.
    p. ascendens
    bursa p.
    tonsilla p.
pharyngeae
    glandulae p.
    venae p.
pharyngeal
    p. aperture
    p. arches
    p. branch
    p. branch of artery of pterygoid canal
    p. branch of ascending pharyngeal artery
    p. branch of descending palatine artery

NOTES

P

pharyngeal *(continued)*
p. branch of glossopharyngeal nerve
p. branch of inferior thyroid artery
p. branch of internal mammary artery
p. branch of pterygopalatine ganglion
p. branch of recurrent laryngeal nerve
p. branch of vagus nerve
p. bursa
caudal p.
p. cell
p. constrictor muscle
p. fistula
p. fornix
p. gland
p. grooves
p. hypophysis
p. isthmus
p. lacuna
p. lymphatic ring
p. lymph node
p. membrane
p. mucosa
p. nervous plexus
p. opening of auditor
p. opening of eustachian tube
p. opening of pharyngotympanic auditory tube
p. orifice
p. ostium
p. pituitary
p. pouch
p. raphe
p. recess
p. reflex
p. region
p. ridge
p. space
p. tonsil
p. tubercle
p. tubercle of basilar part of occipital bone
p. vein
pharyngeales
glandulae p.
rami p.
pharyngealis
fossulae tonsillarum palatini et p.
tonsilla p.
pharyngectomy
pharyngei
rami p.
pharynges (*pl. of* pharynx)
pharyngeum
tuberculum p.

pharyngeus
nervus p.
plexus p.
plexus nervosus p.
ramus p.
recessus p.
pharyngis
anulus lymphoideus p.
cavitas p.
cavum p.
fornix p.
isthmus p.
lacuna p.
pars laryngea p.
pars nasalis p.
pars oralis p.
raphe p.
tela submucosa p.
tunica mucosa p.
tunica muscularis p.
pharyngobasilar
p. fascia
p. fold
pharyngobasilaris
fascia p.
pharyngobranchial ducts
pharyngobranchialis
ductus p. III, IV
pharyngoepiglottic fold
pharyngoepiglottidean
pharyngoesophageal
p. constriction
p. cushions
p. pad
p. sphincter
pharyngoesophagealis
constrictio p.
pharyngoglossal
pharyngoglossus
pharyngolaryngeal
pharyngomaxillary space
pharyngonasal cavity
pharyngonasalis
isthmus p.
pharyngooral
pharyngopalatine
p. arch
p. muscle
pharyngopalatinus
p. muscle
musculus p.
pharyngostaphylinus
pharyngotympanic
p. auditory tube
p. groove
pharynx, pl. **pharynges**
cavity of p.

ceratopharyngeal part of middle pharyngeal constrictor muscle of p.

chondropharyngeal part of middle pharyngeal constrictor muscle of p.

constrictor muscle of p.

cricopharyngeal part of inferior constrictor muscle of p.

inferior constrictor muscle of p.

isthmus of p.

laryngeal part of p.

middle constrictor muscle of p.

mucosa of p.

mucous membrane of p.

muscle of p.

muscular coat of p.

muscular layer of p.

mylopharyngeal part of superior constrictor muscle of p.

nasal p.

oral p.

posterior p.

pterygopharyngeal part of superior constrictor muscle of p.

region of p.

superior constrictor muscle of p.

thyropharyngeal part of inferior pharyngeal constrictor muscle of p.

vault of p.

wall of p.

phase

luteal p.

out of p.

p. rule

short luteal p.

supernormal recovery p.

synaptic p.

phasic reflex

phenology

phenomenology

phenomenon, pl. **phenomena**

Aschner p.

Ashley p.

cogwheel p.

Cushing p.

dip p.

hunting p.

knee p.

Kühne p.

Marcus Gunn p.

Negro p.

rebound p.

reclotting p.

Ritter-Rollet p.

Schiff-Sherrington p.

Sherrington p.

steal p.

Wever-Bray p.

phenozygous

phenylethanolamine N-methyltransferase (PNMT)

pheochrome cell

pheochromoblast

pheomelanin

pheomelanogenesis

pheomelanosome

Philippe triangle

philoprogenitive

philtrum, pl. **philtra**

phlebarteriectasia

phlebectasia

phlebodynamics

phlebography

phleboid

phlebostasis

phlegm

phocomelia

phocomely

phonomyoclonus

phonomyography

phonoreceptor

phosphastat

phosphate tetany

phosphogluconate pathway

photobiology

photohemotachometer

photokinesis

photokinetic

photokymograph

photoperceptive

photoperiodism

photopsin

photoreaction

photoreceptive

photoreceptor cells

photosynthesis

phototaxis

photothermal

phototonus

phototropism

phrenic

p. artery

NOTES

phrenic *(continued)*
    p. ganglia
    p. lymph node
    p. nerve
    p. pleura
    p. plexus
    p. surface of spleen
    p. vein
phrenica
    ganglia p.
    pleura p.
phrenicae
    p. inferiores
    p. superiores
phrenicoabdominal branch of phrenic nerve
phrenicoabdominales
    rami p.
phrenicoceliac part of suspensory muscle ligament of duodenum
phrenicocolic ligament
phrenicocolicum
    ligamentum p.
phrenicocostal sinus
phrenicoesophageal ligament
phrenicogastric
phrenicoglottic
phrenicohepatic
phrenicolienale
    ligamentum p.
phrenicolienal ligament
phrenicomediastinalis
    recessus p.
phrenicomediastinal recess
phrenicopleural fascia
phrenicopleuralis
    fascia p.
phrenicosplenic ligament
phrenicosplenicum
    ligamentum p.
phrenicus
    nervus p.
    plexus p.
phrenocolic
phrenoesophageal ligament
phrenogastric ligament
phrenoglottic
phrenograph
phrenohepatic
phrenopericardial angle
phrenosplenic ligament
phrygian cap
phthinoid chest
phylogenesis
phylogenetic
phylogenic
phylogeny
physaliform

physaliformis
    ecchondrosis p.
physaliphora
    ecchondrosis p.
physaliphorous cell
physeal plate
physical
    p. anthropology
    p. elasticity of muscle
    p. fitness
    p. mixture
physiogenic
physiologic
    p. age
    p. anatomy
    p. congestion
    p. dead space
    p. elasticity of muscle
    p. equilibrium
    p. excavation
    p. homeostasis
    p. hypertrophy
    p. icter
    p. incompatibility
    p. jaundice
    p. sphincter
    p. unit
physiological
    p. anatomy
    p. homeostasis
    p. sphincter
physiologicoanatomical
physiology
    comparative p.
    developmental p.
    general p.
    hominal p.
    pathologic p.
physiopathologic
physiopathology
physique
physis
phytophagous
Pi
pia
    p. mater
    p. mater spinalis
pial
    p. filament
    p. funnel
    p. part of filum terminale
pial-glial membrane
piarachnoid
PICA
    posterior inferior communicating artery
Pick cell
piece
    end p.
    principal p.

pieds terminaux
piesimeter
　　Hales p.
piesis
pigment
　　p. cell
　　p. epithelium
　　p. epithelium of optic retina
pigmentary
pigmented
　　p. layer of ciliary body
　　p. layer of iris
　　p. layer of retina
　　p. part of retina
pigmentosa
　　pars p.
pigmentum nigrum
pigmy
Pignet formula
pileous gland
pileus
pili, sing. **pilus**
　　bulbus p.
　　collum folliculi p.
　　cuticula vaginae folliculi p.
　　folliculus p.
　　musculus arrector p.
　　papilla p.
　　radix p.
　　scapus p.
pillar
　　anterior p.
　　bladder p.
　　p. cells
　　Corti p.
　　p. of diaphragm
　　p.'s of fauces
　　p. of fornix
　　p. of iris
　　posterior p.
　　tonsillar p.
piloerection
piloid
pilomotor
　　p. fibers
　　p. reflex
pilonidal sinus
pilorum
　　arrectores p.
　　cruces p.
　　flumina p.

　　musculi arrectores p.
　　vortices p.
pilosebaceous
pilus (*sing. of* pili)
pineal
　　p. body
　　p. cells
　　p. eye
　　p. gland
　　p. habenula
　　p. recess
　　p. stalk
pineale
　　chief cell of corpus p.
　　corpus p.
　　parenchymatous cell of corpus p.
pinealis
　　recessus p.
pinealocyte
piniform
pinna, pl. **pinnae**
　　ear p.
　　p. nasi
pinnal
pinocyte
pinocytosis
pinocytotic vesicle
pinosome
pinus
PIP
　　proximal interphalangeal
　　　　PIP joint
pipe bone
PIPJ
　　proximal interphalangeal joint
Pirie bone
piriform
　　p. aperture
　　p. area
　　p. cortex
　　p. fossa
　　p. muscle
　　p. opening
　　p. process
　　p. recess
　　p. sill
　　p. sinus
piriformis
　　apertura p.
　　bursa of p.
　　bursa musculi p.
　　p. muscle

NOTES

P

461

piriformis *(continued)*
    musculus p.
    recessus p.
piriformium
    stratum neuronorum p.
Pirogoff
    P. angle
    P. triangle
pisiform
    p. bone
    p. joint
pisiforme
    os p.
pisiformis
    articulatio ossis p.
pisohamate
    p. bone
    p. ligament
pisohamatum
    ligamentum p.
pisometacarpal ligament
pisometacarpeum
    ligamentum p.
pisotriquetral joint
pisounciform ligament
pisouncinate ligament
pit
    anal p.
    p. of atlas for dens
    auditory p.
    central p.
    commissural lip p.
    p. for dens of atlas
    gastric p.
    granular p.
    p. of head of femur
    inferior costal p.
    lens p.
    p. of lip
    nasal p.
    olfactory p.
    primitive p.
    pterygoid p.
    p. of stomach
    sublingual p.
    superior costal p.
    suprameatal p.
    trochlear p.
pith
pithecoid
Pitot tube
Pitres area
pituicyte
pituita
pituitaria
    glandula p.
pituitarium
pituitary
    p. body

    p. diverticulum
    p. fossa
    p. gland
    p. gonadotropic hormone
    p. growth hormone
    hyaline bodies of p.
    p. membrane
    pharyngeal p.
    p. stalk
pituitosa
    membrana p.
pituitous
pivot joint
placenta, pl. **placentae**
    accessory p.
    p. accreta
    p. accreta vera
    adherent p.
    annular p.
    battledore p.
    bidiscoidal p.
    p. biloba
    p. bipartita
    chorioallantoic p.
    choriovitelline p.
    p. circumvallata
    cotyledonary p.
    deciduous p.
    dichorionic diamniotic p.
    p. diffusa
    p. dimidiata
    disperse p.
    p. duplex
    endotheliochorial p.
    endothelio-endothelial p.
    epitheliochorial p.
    p. extrachorales
    p. fenestrata
    fetal p.
    p. fetalis
    hemochorial p.
    hemoendothelial p.
    horseshoe p.
    incarcerated p.
    p. increta
    labyrinthine p.
    marginal sinus of p.
    p. marginata
    margination of p.
    maternal p.
    p. membranacea
    monochorionic diamniotic p.
    monochorionic monoamniotic p.
    p. multiloba
    nondeciduous p.
    p. panduraformis
    pars uterina placentae
    p. percreta
    p. previa

p. previa centralis
p. previa marginalis
p. reflexa
p. reniformis
retained p.
Schultze p.
p. spuria
succenturiate p.
supernumerary p.
syndesmochorial p.
p. triloba
p. tripartita
p. triplex
twin p.
p. uterina
p. velamentosa
villous p.
zonary p.
placental
p. barrier
p. circulation
p. growth hormone
p. lobe
p. membrane
p. plasmodium
p. presentation
p. septum
p. tissue
p. villus
placentation
placode
auditory p.
lens p.
nasal p.
olfactory p.
optic p.
otic p.
plafond
tibial p.
plagiocephalic
plagiocephalism
plagiocephalous
plagiocephaly
plana (*pl. of* planum)
plane
Addison clinical p.
Aeby p.
auriculoinfraorbital p.
canthomeatal p.
coronal p.
datum p.
Daubenton p.

eye-ear p.
first parallel pelvic p.
fourth parallel pelvic p.
Frankfort p.
Frankfort horizontal p.
frontal p.
horizontal p.
infraorbitomeatal p.
p. of inlet
interspinal p.
interspinous p.
intertubercular p.
p. joint
p. of least pelvic dimension
left midlingual p.
longitudinal p.
Meckel p.
median sagittal p.
p. of midpelvis
midsagittal p.
Morton p.
nasion-postcondylar p.
nuchal p.
occipital p.
orbital p.
orbitomeatal p.
p. of outlet
parasagittal p.
p. of pelvic canal
sagittal p.
second parallel pelvic p.
sternal p.
subcostal p.
subcutaneous p.
supracrestal p.
supracristal p.
supraorbitomeatal p.
suprasternal p.
p. suture
temporal p.
third parallel pelvic p.
transpyloric p.
transtubercular p.
transverse p.
umbilical p.
vertical p.
wide p.
planimeter
planithorax
planocellular
planovalgus
planta, pl. **plantae**

NOTES

P

planta *(continued)*
  musculus quadratus plantae
  p. pedis
  quadratus plantae
plantar
  p. aponeurosis
  p. arterial arch
  p. aspect of foot
  p. calcaneocuboid ligament
  p. calcaneonavicular ligament
  p. cuboideonavicular ligament
  p. cuneocuboid ligament
  p. cuneonavicular ligament
  p. digital artery
  p. digital nerve
  p. digital vein
  p. fascia
  p. flexion of foot
  p. flexor
  p. interossei interosseous muscle
  p. interosseous muscle
  p. ligament of interphalangeal joint of foot
  p. ligament of metatarsophalangeal joint
  p. metatarsal artery
  p. metatarsal ligament
  p. metatarsal vein
  p. quadrate muscle
  p. reflex
  p. region
  p. response
  p. space
  p. surface
  p. surface of foot
  p. surface of toe
  p. tarsal ligament
  p. tarsometatarsal ligament
  p. tendon sheath of fibularis longus muscle
  p. tendon sheath of peroneus longus muscle
  p. venous arch
  p. venous network
plantare
  ligamentum calcaneocuboideum p.
  ligamentum calcaneonaviculare p.
  ligamentum cuboideonaviculare p.
  ligamentum cuneocuboideum p.
  rete venosum p.
plantares
  digitales p.
  interossei p.
  musculi interossei p.
  ramus perforans arteriae metatarsearum p.
  venae digitales p.
  venae metatarseae p.
plantarflexion of ankle

plantarflexor compartment of leg
plantaria
  ligamenta cuboideonavicularia p.
  ligamenta cuneonavicularia p.
  ligamenta intercuneiformia p.
  ligamenta metatarsalia p.
  ligamenta tarsi p.
  ligamenta tarsometatarsalia p.
  ligamentum cuneonavicularia p.
  ligamentum intercuneiformia p.
  ligamentum metatarsalia p.
  ligamentum tarsi p.
plantaris
  aponeurosis p.
  arcus venosus p.
  arteria metatarsalis p.
  arteria metatarsea p.
  facies digitalis p.
  p. lateralis
  p. medialis
  p. muscle
  musculi interosseus p.
  musculus interosseus p.
  regio p.
  p. tendon
  vagina tendinis musculi fibularis longi p.
  vagina tendinis musculi peronei longi p.
plantarium
  rami perforantes arteriarum metatarsearum p.
planula, pl. **planulae**
  invaginate p.
planum, pl. **plana**
  articulatio plana
  plana coronalia
  coxa plana
  plana frontalia
  plana horizontalia
  p. interspinale
  p. intertuberculare
  p. medianum
  musculus plana
  p. occipitale
  p. orbitale
  os p.
  pars plana
  pelvis plana
  p. popliteum
  plana sagittalia
  p. semilunatum
  p. sphenoidale
  p. sternale
  p. subcostale
  p. supracristale
  sutura plana
  p. temporale

p. transpyloricum
plana transversalia
planus
pes p.
plasma
p. cell
p. concentration
p. layer
p. level
p. membrane
target p.
plasmablast
plasmacyte
plasmacytoblast
plasmagene
plasmalemma
plasmapheretic
plasmatogamy
plasmid
plasmodia
plasmodial trophoblast
plasmodiotrophoblast
plasmodium
placental p.
plasmogamy
plasmogen
plasmolemma
plasmolysis
plasmolytic
plasmosin
plasmosome
plastic
p. anatomy
Bingham p.
p. corpuscle
plasticity
plastid
plastogamy
plastron
plate
anal p.
axial p.
blood p.
cardiogenic p.
chorionic p.
cloacal p.
cribriform p.
cutis p.
end p.
epiphysial p.
ethmovomerine p.
floor p.

foot p.
foramen in cribriform p.
frontal p.
growth p.
Kühne p.
lateral cartilaginous p.
lateral pterygoid p.
medial cartilaginous p.
medial pterygoid p.
medullary p.
membranous lamina of cartilage of pharyngotympanic auditory p.
p. of modiolus
motor p.
muscle p.
nail p.
neural p.
notochordal p.
oral p.
orbital p.
palmar p.
papyraceous p.
parachordal p.
parietal p.
perpendicular of ethmoid p.
physeal p.
polar p.
prechordal p.
pterygoid p.
quadrigeminal p.
secondary spiral p.
segmental p.
sieve p.
sole p.
spiral p.
tarsal p.
terminal p.
tympanic p.
urethral p.
ventral p.
vertical p.
visceral p.
volar p.
wing p.
plateau
tibial p.
ventricular p.
platelet
platelet-derived growth factor (PDGF)
platybasia
platycephaly
platycnemia

NOTES

P

platycnemic
platycnemism
platycrania
platyhieric
platymeric
platymorphia
platyopia
platyopic
platypellic pelvis
platypelloid pelvis
platyrrhine
platyrrhiny
platysma, pl. platysmas, platysmata
    p. muscle
    musculus p.
platyspondylia
platyspondylisis
platystencephaly
pleomastia
pleomazia
pleomorphic
pleomorphism
pleomorphous
pleonasm
pleonectic
plesiomorphic
plesiomorphism
plesiomorphous
plethora
plethoric
pleura, pl. pleurae
    adipose folds of p.
    cavum pleurae
    cervical p.
    p. costalis
    costal part of parietal p.
    cupula pleurae
    cupula of p.
    p. diaphragmatica
    diaphragmatic part of parietal p.
    dome of p.
    fatty fold of p.
    p. mediastinalis
    mediastinal part of parietal p.
    parietal p.
    p. parietalis
    p. pericardiaca
    pericardial p.
    phrenic p.
    p. phrenica
    plicae adiposae pleurae
    p. pulmonalis
    pulmonary p.
    serosa of parietal p.
    serosa of visceral p.
    visceral p.
    p. visceralis
pleural
    p. cavity

    p. cupula
    p. fluid
    p. isthmus
    p. pressure
    p. recess
    p. reflection
    p. sac
    p. sinuses
    p. space
    p. surface
    p. symphysis
    p. villi
pleurales
    recessus p.
    villi p.
pleuralis
    cavitas p.
pleurapophysis
pleurocentrum
pleuroesophageal muscle
pleuroesophageus
    p. muscle
    musculus p.
pleuropericardial
    p. canals
    p. hiatus
    p. membrane
pleuroperitoneal
    p. canal
    p. cavity
    p. fold
    p. hiatus
    p. membrane
    p. space
pleuropulmonary
pleurovisceral
plexal
plexiform
    p. layer
    p. layer of cerebral cortex
    p. layers of retina
plexogenic
plexus, pl. plexus, plexuses
    abdominal aortic p.
    acromial p.
    annular p.
    p. annularis
    p. of anterior cerebral artery
    anterior coronary periarterial p.
    anterior divisions of brachial p.
    aortic lymphatic p.
    p. aorticus
    areolar venous p.
    p. arteriae choroideae
    arterial p.
    articular vascular p.
    ascending pharyngeal p.
    Auerbach mesenteric p.
    autonomic p.

p. autonomicus brachialis
axillary lymphatic p.
basilar venous p.
Batson p.
biliary p.
brachial autonomic p.
p. card
cardiac p.
p. cardiacus profundus
p. caroticus communis
p. caroticus externus
p. caroticus internus
carotid venous p.
p. cavernosi concharum
cavernous nervous p.
celiac p.
cervical p.
p. cervicalis
choroid p.
p. of choroid artery
p. choroideus
p. choroideus ventriculi lateralis
p. choroideus ventriculi quarti
p. choroideus ventriculi tertii
ciliary ganglionic p.
coccygeal p.
p. coccygeus
colonic myenteric p.
common carotid nervous p.
communicating branch of facial
   nerve with tympanic p.
communicating branch of
   intermediate nerve with
   tympanic p.
p. coronarii cordis
p. coronarius cordis
coronary p.
Cruveilhier p.
cystic p.
deep cardiac p.
deferential p.
p. dental
p. dentalis inferior
p. of ductus deferens
enteric nervous p.
p. entericus
esophageal nervous p.
Exner p.
external carotid nervous p.
external iliac lymphatic p.
external mastoid p.
external maxillary p.

facial p.
femoral nervous p.
ganglia of autonomic p.
p. gangliosus ciliaris
p. gastrici systematis autonomici
gastric nervous p.
p. gulae
Haller p.
Heller p.
hemorrhoidal p.
hepatic nervous p.
p. hypogastricus inferior
p. hypogastricus superior
hypoglossal canal venous p.
ileocolic p.
p. iliaci
iliac nervous p.
p. iliacus externus
inferior dental branch of inferior
   dental p.
inferior dental nervous p.
inferior gingival branch of inferior
   dental p.
inferior hemorrhoidal plexuses
inferior hypogastric nervous p.
inferior mesenteric nervous p.
inferior rectal nervous p.
inferior thyroid p.
inferior trunk of brachial p.
inferior vesical venous p.
infraclavicular part of brachial p.
inguinal lymphatic p.
intermesenteric nervous p.
internal carotid nervous p.
internal carotid venous p.
internal mammary p.
internal maxillary p.
internal thoracic lymphatic p.
intracavernous p.
p. intraparotideus
p. intraparotideus nervi facialis
ischiadic p.
Jacobson p.
Jacques p.
p. jugularis
jugular lymphatic p.
Kiesselbach p.
laryngeal p.
lateral cord of brachial p.
Leber p.
lingual p.
lumbar lymphatic p.

NOTES

plexus *(continued)*

lumbar nervous p.
lumbosacral nervous p.
lymphatic p.
p. lymphaticus
p. lymphaticus axillaris
p. lymphaticus iliacus externus
p. lymphaticus inguinalis
p. lymphaticus jugularis
p. lymphaticus lumbalis
p. lymphaticus sacralis medius
malleolar p.
p. mammarius
p. mammarius internus
mammary p.
p. maxillaris externus
p. maxillaris internus
maxillary p.
medial cord of brachial p.
Meissner p.
meningeal p.
p. meningeus
p. mesentericus inferior
p. mesentericus superior
p. of middle cerebral artery
middle hemorrhoidal p.
middle rectal nervous p.
middle sacral lymphatic p.
middle trunk of brachial p.
myenteric nervous p.
p. myentericus
nerve p.
p. nervorum gastricorum
p. nervorum lumbalium
p. nervorum spinalium
p. nervosus aorticus abdominalis
p. nervosus aorticus thoracicus
p. nervosus arteriae carotidis internae
p. nervosus cardiacus
p. nervosus cardiacus superficialis
p. nervosus caroticus communis
p. nervosus caroticus externus
p. nervosus cavernosus
p. nervosus celiacus
p. nervosus cervicalis posterior
p. nervosus deferentialis
p. nervosus dentalis inferior
p. nervosus dentalis superior
p. nervosus entericus
p. nervosus esophageus
p. nervosus femoralis
p. nervosus hepaticus
p. nervosus hypogastricus inferior
p. nervosus hypogastricus superior
p. nervosus iliacus
p. nervosus intermesentericus
p. nervosus lienalis
p. nervosus lumbosacralis

p. nervosus mesentericus inferior
p. nervosus mesentericus superior
p. nervosus myentericus
p. nervosus ovaricus
p. nervosus pancreaticus
p. nervosus pelvicus
p. nervosus pharyngeus
p. nervosus prostaticus
p. nervosus pulmonalis
p. nervosus rectalis inferiores
p. nervosus rectalis medius
p. nervosus rectalis superior
p. nervosus renalis
p. nervosus splenicus
p. nervosus submucosus
p. nervosus subserosus
p. nervosus suprarenalis
p. nervosus tympanicus
p. nervosus uretericus
p. nervosus uterovaginalis
occipital p.
ophthalmic p.
ovarian nervous p.
pampiniform venous p.
pancreatic nervous p.
parotid p.
pelvic nervous p.
p. pelvinus
periarterial p.
p. periarterialis
p. periarterialis arteriae auricularis posterioris
p. periarterialis arteriae cerebri anterioris
p. periarterialis arteriae cerebri mediae
p. periarterialis arteriae choroideae
p. periarterialis arteriae facialis
p. periarterialis arteriae lingualis
p. periarterialis arteriae maxillaris
p. periarterialis arteriae occipitalis
p. periarterialis arteriae ophthalmicae
p. periarterialis arteriae pharyngeae ascendentis
p. periarterialis arteriae phrenicae inferioris
p. periarterialis arteriae popliteae
p. periarterialis arteriae subclaviae
p. periarterialis arteriae temporalis superficialis
p. periarterialis arteriae testicularis
p. periarterialis arteriae thoracicae internae
p. periarterialis arteriae thyroideae superioris
p. periarterialis arteriae vertebralis
pharyngeal nervous p.
p. pharyngeus

p. pharyngeus ascendens
phrenic p.
p. phrenicus
popliteal p.
p. popliteus
porta hepatis p.
posterior auricular p.
posterior cervical nervous p.
posterior cord of brachial p.
posterior coronary p.
posterior divisions of trunks of
  brachial p.
prostatic nervous p.
prostaticovesical p.
p. prostaticovesicalis
prostaticovesical venous p.
prostatic venous p.
pterygoid venous p.
p. pudendalis
p. pudendus nervosus
pulmonary branch of pulmonary
  nerve p.
pulmonary nervous p.
Quénu hemorrhoidal p.
Ranvier p.
p. rectales inferiores
p. rectales medii
p. rectalis superior
rectal venous p.
Remak p.
renal p.
p. renalis
renal nervous p.
rete p.
p. sacralis
sacral venous p.
p. of Santorini
Santorini p.
Sappey p.
sciatic p.
solar p.
spermatic p.
p. of spinal nerve
spinal nerve p.
splenic nervous p.
Stensen p.
stroma p.
subclavian p.
p. subclavius
subepithelial p.
submucosal nervous p.
submucosal venous p.

suboccipital venous p.
subpleural mediastinal p.
subserous nervous p.
superficial cardiac nervous p.
superficial temporal p.
superior dental branch of superior
  dental p.
superior dental nervous p.
superior gingival branch of
  superior dental p.
superior hemorrhoidal p.
superior hypogastric nervous p.
superior mesenteric nervous p.
superior rectal nervous p.
superior thyroid p.
superior trunk of brachial p.
supraclavicular part of brachial p.
suprarenal nervous p.
sympathetic carotid p.
testicular p.
thoracic aortic nervous p.
p. thyroideus inferior
p. thyroideus superior
trunk of inferior brachial p.
trunk of middle brachial p.
trunk of superior brachial p.
trunk of upper brachial p.
tubal branch of tympanic p.
tympanic nervous p.
p. tympanicus
unpaired thyroid venous p.
ureteric nervous p.
p. uretericus
uterine venous p.
uterovaginal nervous p.
vaginal venous p.
vascular p.
p. vascularis cavernosus conchae
p. vasculosus
p. venosus
p. venosus areolaris
p. venosus basilaris
p. venosus canalis hypoglossi
p. venosus caroticus internus
p. venosus foraminis ovalis
p. venosus pampiniformis
p. venosus prostaticovesicalis
p. venosus prostaticus
p. venosus pterygoideus
p. venosus rectalis
p. venosus sacralis
p. venosus suboccipitalis

NOTES

P

plexus *(continued)*
p. venosus thyroideus impar
p. venosus uterinus
p. venosus vaginalis
p. venosus vertebralis
p. venosus vertebralis externus anterior
p. venosus vertebralis externus posterior
p. venosus vertebralis internus ante
p. venosus vertebralis internus posterior
p. venosus vesicalis
p. venosus vesicalis inferior
venous p.
vertebral venous p.
vesical nervous p.
vesicoprostatic p.
vesicular venous p.
p. viscerales
Walther p.

plica, pl. **plicae**
plicae adiposae
plicae adiposae pleurae
plicae alares
plicae alares plicae synovialis infrapatellaris
plicae ampullares tubae uterinae
p. ampullaris
p. anterior faucium
p. aryepiglottica
p. axillaris
plicae cecales
p. cecalis vascularis
p. chordae tympani
plicae ciliares
plicae circulares
plicae circulares intestini tenuis
p. duodenalis inferior
p. duodenalis superior
p. duodenojejunalis
p. duodenomesocolica
p. epigastrica
p. epiglottica
p. fimbriata
p. fimbriata faciei inferioris linguae
plicae gastricae
plicae gastropancreaticae
p. glossoepiglottica lateralis
p. glossoepiglottica mediana
p. gubernatrix
p. hypogastrica
p. ileocecalis
p. incudis
p. inguinalis
p. interdigitalis
p. interureterica
plicae iridis
p. lacrimalis

p. longitudinalis duodeni
p. lunata
plicae malleares anterior et posterior
p. mallearis
p. membranae tympani
plicae mucosae vesicae biliaris
p. nervi laryngei
p. nervi laryngei superioris
p. palatina transversa
plicae palmatae
plicae palmatae canalis cervicis uterus
p. palpebronasalis
p. paraduodenalis
p. posterior faucium
plicae recti
p. rectouterina
p. rectovaginalis
p. salpingopalatina
p. salpingopharyngea
plicae semilunares of colon
p. semilunaris
p. semilunaris coli
p. semilunaris of conjunctiva
p. semilunaris of eye
p. sigmoidea
p. spiralis ductus cystici
p. stapedialis
p. stapedis
p. sublingualis
p. synovialis
p. synovialis infrapatellaris
p. synovialis patellaris
plicae transversales recti
p. triangularis
plicae tubariae
plicae tubariae tubae uterinae
p. tubopalatina
plicae tunicae mucosae vesicae felleae
p. umbilicalis lateralis
p. umbilicalis media
p. umbilicalis medialis
p. umbilicalis mediana
p. urachi
p. ureterica
p. uterovesicalis
p. venae cavae sinistrae
p. ventricularis
p. vesicalis transversa
p. vesicouterina
p. vestibularis
p. vestibuli
p. villosa
p. vocalis

plicate
plug
epithelial p.

plumose
pluriglandular
plurilocular
plurinuclear
pluripotent cells
pluripotential
pneocardiac reflex
pneodynamics
pneopneic reflex
pneumatic
    p. bone
    p. space
pneumaticum
    os p.
pneumatized bone
pneumatoenteric recess
pneumatometer
pneumatoscope
pneumobulbar
pneumocardial
pneumocyte
pneumodynamics
pneumoenteric recess
pneumogastric nerve
pneumogram
pneumograph
pneumometer
pneumometry
pneumonic
pneumonocyte
    granular p.
    phagocytic p.
pneumoscope
pneumotachogram
pneumotachograph
    Fleisch p.
    Silverman-Lilly p.
pneumotachometer
pneusis
PNMT
    phenylethanolamine N-methyltransferase
pocket
    Rathke p.
    Seessel p.
    Tröltsch p.
pocularis
    sinus p.
poculum diogenis
podalic
podocyte
pogonion
poikilotherm

poikilothermal
poikilothermic
poikilothermism
poikilothermous
poikilothermy
point
    alveolar p.
    apophysary p.
    apophysial p.
    auricular p.
    boiling p.
    Capuron p.
    cold-rigor p.
    craniometric p.
    p. of elbow
    heat-rigor p.
    jugal p.
    lower alveolar p.
    malar p.
    maximum occipital p.
    mental p.
    metopic p.
    nasal p.
    occipital p.
    p. of ossification
    preauricular p.
    spinal p.
    subnasal p.
    Sudeck critical p.
    supraauricular p.
    supranasal p.
    supraorbital p.
    sylvian p.
    Weber p.
    zygomaxillary p.
Poirier
    P. gland
    P. line
Poiseuille
    P. law
    P. space
    P. viscosity coefficient
poisoning
    oxygen p.
polar
    p. body
    p. cell
    p. frontal artery
    p. globule
    p. hypogenesis
    p. plate

NOTES

P

polar *(continued)*
    p. temporal artery
    p. zone
polarity
polarization
pole
    abapical p.
    animal p.
    caudal p.
    cephalic p.
    frontal p.
    germinal p.
    inferior p.
    lateral p.
    lower p.
    medial p.
    occipital p.
    rostral p.
    superior p.
    temporal p.
    upper p.
    vegetal p.
    vegetative p.
    vitelline p.
poli (*pl. of* polus)
polkissen of Zimmermann
poll
pollex pedis
pollicis, pl. **pollices**
    adductor p.
    arteria princeps p.
    articulatio carpometacarpalis p.
    articulatio carpometacarpea p.
    p. brevis
    caput obliquum musculi
      adductoris p.
    caput transversum musculi
      adductoris p.
    p. longus
    musculus adductor p.
    musculus extensor brevis p.
    musculus extensor longus p.
    musculus extensor ossis
      metacarpi p.
    musculus flexor longus p.
    musculus opponens p.
    opponens p.
    princeps p.
    tendinous sheath of abductor
      pollicis longus and extensor p.
    vagina tendinum musculorum
      abductoris longi et extensoris
      brevis p.
polocyte
polus, pl. **poli**
    p. anterior bulbi oculi
    p. anterior lentis
    p. frontalis cerebri
    p. inferior

    p. inferior renis
    p. inferior testis
    poli lienalis inferior et superior
    p. occipitalis cerebri
    p. posterior bulbi oculi
    p. posterior lentis
    poli renales inferior et superior
    poli renalis inferior et superior
    p. superior
    p. superior renis
    p. superior testis
    p. temporalis cerebri
polyadenous
polyaxial joint
polyblast
polyblennia
polycentric
polycheiria
polychiria
polychromasia
polychromatic cell
polychromatophil, polychromatophile
    p. cell
polychromatophilia
polychromatophilic
polychromatosis
polychromophil
polychromophilia
polychylia
polydactylia
polydactylism
polydactylous
polydactyly
polydentia
polyembryony
polyergic
polyganglionic
polyglandular
polygnathus
polygyria
polyhedral
polykaryocyte
polymastia
polymazia
polymelia
polymeria
polymeric
polymetacarpalia
polymetacarpalism
polymetatarsalia
polymetatarsalism
polymorphic
polymorphism
polymorphocellular
polymorphous layer
polyneural
polynuclear
polynucleate
polyodontia

polyonychia
polyorchidism
polyorchism
polyotia
polyovular ovarian follicle
polyovulatory
polyp
polyphalangism
polypi (*pl. of* polypus)
polyplast
polyplastic
polypodia
polyporous
polyposa
     decidua p.
polyposia
polyptychial
polypus, pl. polypi
polyribosomes
polyscelia
polysomes
polysomia
polyspermia
polyspermism
polyspermy
polystichia
polysymbrachydactyly
polysynaptic
polysyndactyly
polythelia
polyunguia
polyzygotic twins
POMC
   proopiomelanocortin
ponderal index
pons, pl. pontes
    anterior part of p.
    arteries of p.
    p. basilaris pontis
    basilar part of p.
    p. cerebelli
    dorsal part of p.
    p. hepatis
    oblique bundle of p.
    p. oblongata
    p. tarini
    tegmentum of p.
    transverse fibers of p.
    p. varolii
    vein of p.
    ventral part of p.

pontem
    rami ad p.
ponticulus
    p. hepatis
    p. nasi
    p. promontorii
pontile
pontine
    p. angle
    p. artery
    p. cistern
    p. flexure
    p. gray matter
    p. nuclei
    p. vein
pontis
    arteriae p.
    basis p.
    brachium p.
    cisterna p.
    fasciculi longitudinales p.
    fasciculus obliquus p.
    nuclei p.
    pars basilaris p.
    pars dorsalis p.
    pars ventralis p.
    pons basilaris p.
    rami laterales arteriae p.
    rami mediales arteriae p.
    raphe p.
    sulcus basilaris p.
    venae p.
pontobulbare
pontocerebellar
    p. cistern
    p. recess
pontocerebellaris
    cisterna p.
ponto-geniculo-occipital spike
pontomedullary groove
pool
    metabolic p.
poples
poplitea
    arteria p.
    fossa p.
    linea p.
    vena p.
popliteae
    plexus periarterialis arteriae p.
popliteal
    p. arch

NOTES

P

popliteal *(continued)*
 p. artery
 p. communicating nerve
 p. fascia
 p. fossa
 p. groove
 p. line
 p. lymph node
 p. muscle
 p. notch
 p. plane of femur
 p. plexus
 p. pulse
 p. region
 p. space
 p. surface of femur
 p. vein
popliteales
 lymphonodi p.
 nodi lymphoidei p.
popliteum
 planum p.
popliteus
 bursa of p.
 groove for p.
 p. muscle
 musculus p.
 plexus p.
 sulcus p.
 p. tendon
poralis posterior
pore
 auditory p.
 external acoustic p.
 external auditory p.
 gustatory p.
 interalveolar p.
 internal acoustic p.
 Kohn p.
 nuclear p.
 skin p.
 slit p.
 sweat p.
 taste p.
pori (*pl. of* porus)
porion, pl. **poria**
porosis, pl. **poroses**
porosity
porotic
porous
porrigo decalvans
porta, pl. **portae**
 p. hepatis
 p. hepatis plexus
 p. lienis
 p. pulmonis
 p. renis
portacaval
 p. anastomosis

portal
 anterior intestinal p.
 p. canals
 p. fissure
 p. hypophysial circulation
 p. lobule of liver
 posterior intestinal p.
 p. sinus
 p. system
 p. triad
 p. vascular bed
 p. vein
 velopharyngeal p.
portalis
 vena p.
portal-systemic anastomosis
Porter fascia
portio, pl. **portiones**
 p. intermedia
 p. major nervi trigemini
 p. minor nervi trigemini
 p. supravaginalis
 p. supravaginalis cervicis
 p. vaginalis
 p. vaginalis cervicis
portion
 osseous p.
portiplexus
portobilioarterial
portosystemic
porus, pl. **pori**
 p. acusticus externus
 p. acusticus internus
 p. crotaphyticobuccinatorius
 p. gustatorius
 p. opticus
 p. sudoriferus
position
 anatomic p.
 anatomical p.
 energy of p.
positive
 p. chronotropism
 p. electrotaxis
 p. feedback
 p. supporting reaction
 p. taxis
 p. thermotaxis
positively
 p. bathmotropic
 p. dromotropic
 p. inotropic
positive-negative pressure breathing
postacetabular
postanal
 p. dimple
 p. gut
postaortici
 lymphonodi p.

postauricular
    p. area
    p. sulcus
postaxial muscle
postaxillaris
    linea p.
postaxillary line
postbrachial
postcapillary venule
postcardinal vein
postcardiotomy
postcava
postcaval
postcavales
    lymphonodi p.
    nodi lymphatici p.
postcentral
    p. area
    p. fissure
    p. gyrus
    p. sulcal artery
    p. sulcus
    p. vein
postcentralis
    arteria sulci p.
    gyrus p.
    sulcus p.
postclavicular
postcloacal
postcoital
postcoitus
postcommunicalis
    pars p.
postcommunical part
postcommunicating
    p. part of anterior cerebral artery
    p. part of posterior cerebral artery
postcordial
postcostal anastomosis
postcricoid
    p. region
    p. space
postcubital
postductal
posterior
    p. abdominal wall
    p. alveolar artery
    alveolaris superior p.
    ampullaris p.
    p. ampullar nerve
    p. ampullary nerve
    p. angle of rib

p. annular ligament
p. antebrachial cutaneous nerve
p. antebrachial region
anterior and p. (A&P)
p. anterior jugular vein
p. arch of atlas
area intercondylaris p.
arteria alveolaris superior p.
arteria auricularis p.
arteria cecalis p.
arteria cerebelli inferior p.
arteria cerebri p.
arteria choroidea p.
arteria circumflexa humeri p.
arteria communicans p.
arteria conjunctivalis p.
arteriae radiculares anterior et p.
arteria ethmoidalis p.
arteria gastrica p.
arteria intercostalis p.
arteria interossea p.
arteria meningea p.
arteria parietales p.
arteria parietalis p.
arteria recurrens tibialis p.
arteria segmentalis p.
arteria spinalis p.
arteria temporalis p.
arteria tibialis p.
arteria tympanica p.
p. articular aorta
p. articular facet of dens
p. articular surface of dens
p. atlantooccipital membrane
p. auricular artery
p. auricular branch of external
  carotid artery
p. auricular groove
auricularis p.
p. auricular muscle
p. auricular nerve
p. auricular plexus
p. auricular vein
p. axillary fold
p. axillary line
p. axillary lymph node
p. basal branch
p. basal bronchopulmonary segment
S X
p. basal segmental artery of
  left/right lung
p. basal segment of lung

NOTES

P

posterior *(continued)*
p. belly of digastric muscle
p. border of eyelid
p. border of fibula
p. border of petrous part of temporal bone
p. border of radius
p. border of testis
p. border of ulna
p. brachial cutaneous nerve
p. brachial region
p. branch of great auricular nerve
p. branch of inferior pancreaticoduodenal artery
p. branch of lateral cerebral sulcus
p. branch of medial antebrachial cutaneous nerve
p. branch of medial cutaneous nerve of forearm
p. branch of obturator artery
p. branch of obturator nerve
p. branch of recurrent ulnar artery
p. branch of renal artery
p. branch of right branch of portal vein
p. branch of right hepatic duct
p. branch of right superior pulmonary vein
p. branch of spinal nerve
p. branch of superior thyroid artery
p. branch of ulnar recurrent artery
p. bronchopulmonary segment S II
p. calcaneal articular surface
camera oculi p.
canales semicircularis p.
p. canaliculus of chorda tympani
p. cardinal vein
p. carpal region
p. cecal artery
p. central convolution
p. central gyrus
cerebelli inferior p.
p. cerebellomedullary cistern
p. cerebral artery
p. cerebral commissure
cerebri p.
p. cervical intertransversarii muscle
p. cervical intertransverse muscle
p. cervical lip
p. cervical nervous plexus
p. cervical region
p. chamber
p. chamber of eye
p. chamber of eyeball
p. choroidal artery
circumflexa humeri p.
p. circumflex humeral artery
p. circumflex humeral vein

cisterna cerebellomedullaris p.
p. clinoid process
columna p.
p. column of spinal cord
commissura labiorum p.
p. communicating artery
p. compartment of arm
p. compartment of forearm
p. compartment of leg
p. compartment of thigh
p. condyloid foramen
p. conjunctival artery
p. cord of brachial plexus
p. coronary plexus
p. costotransverse ligament
p. cranial fossa
p. crest of stapes
cricoarytenoideus p.
p. cricoarytenoid ligament
p. cricoarytenoid muscle
crista lacrimalis p.
p. cruciate ligament
p. crural region
p. crus of stapes
p. cubital region
p. cusp
cuspis p.
p. cusp of left atrioventricular valve
p. cusp of mitral valve
p. cusp of right atrioventricular valve
p. cusp of tricuspid valve
p. cutaneous femoral nerve
p. cutaneous nerve of arm
p. cutaneous nerve of forearm
p. cutaneous nerve of thigh
cutaneus antebrachii p.
cutaneus brachii p.
cutaneus femoris p.
p. dental artery
p. descending artery (PDA)
p. descending coronary artery
diploica temporalis p.
p. divisions of trunks of brachial plexus
ductus semicircularis p.
p. elastic layer
p. ethmoidal air cell
p. ethmoidal branch of ophthalmic artery
ethmoidalis p.
p. ethmoidal nerve
p. extremity
p. extremity of spleen
p. facial vein
facies antebrachialis p.
facies brachialis p.
facies cruralis p.

facies cubitalis p.
facies femoralis p.
p. false ligament
p. fascicle of palatopharyngeus
  muscle
p. fontanelle
fonticulus p.
foramen ethmoidale anterior et p.
foramen ischiadicum anterior et p.
foramina sacralia anterior et p.
forceps p.
p. fornix of vagina
fossa cranii p.
p. funiculus
funiculus p.
p. gastric artery
p. gastric branch of posterior vagal
  trunk
p. glandular branch of superior
  thyroid artery
p. great vessel
p. hepatic segment I
p. horn
p. humeral circumflex artery
p. hypothalamic nucleus
incisura cerebelli p.
p. inferior cerebellar artery
p. inferior communicating artery
  (PICA)
p. inferior iliac spine
p. inferior nasal branch of greater
  palatine nerve
p. intercondylar area
p. intercondylar area of tibia
p. intercostal artery
p. intercostal vein
p. intermediate groove
p. intermediate sulcus
p. intermuscular septum
p. internal orbital canal
interossea p.
interosseous p.
p. interosseous artery
p. interosseous nerve
p. interosseous vein
p. interventricular branch of right
  coronary artery
p. interventricular groove
p. interventricular sulcus
p. intestinal portal
p. intraoccipital joint
p. intraoccipital synchondrosis

p. knee region
p. labial branch of internal
  perineal artery
p. labial commissure
p. labial nerve
p. labial vein
p. lacrimal crest
p. lamina
lamina elastica p.
p. larynx
p. lateral nasal artery
p. layer of rectus abdominis sheath
p. leaf of broad ligament
p. ligament of fibular head
p. ligament of head of fibula
p. ligament of incus
p. ligament of knee
p. limb of internal capsule
p. limb of stapes
p. limiting lamina of cornea
p. limiting layer of cornea
linea axillaris p.
linea glutea p.
linea mediana p.
p. lip of external os of uterus
p. lobe of hypophysis
p. longitudinal bundle
p. longitudinal ligament
p. lunate lobule
p. marginal vein
p. margin of heart
p. medial nucleus of thalamus
p. median fissure
p. median fissure of medulla
  oblongata
p. median fissure of spinal cord
p. median line
p. median sulcus
p. median sulcus of medulla
  oblongata
p. median sulcus of spinal cord
p. mediastinal artery
p. mediastinal lymph node
p. mediastinum
p. medullary velum
membrana atlantooccipitalis p.
meningea p.
p. meningeal artery
p. meniscofemoral ligament
musculus auricularis p.
musculus cricoarytenoideus p.
musculus sacrococcygeus p.

NOTES

P

posterior *(continued)*

musculus scalenus p.
musculus tibialis p.
p. naris
p. nasal aperture
p. nasal spine
p. nasal spine of horizontal plate
of palatine bone
p. neck region
nervus ampullaris p.
nervus antebrachii p.
nervus auricularis p.
nervus cutaneus antebrachii p.
nervus cutaneus brachii p.
nervus cutaneus femoris p.
nervus ethmoidalis p.
nervus interosseus p.
nervus interosseus antebrachii p.
nodus lymphoideus tibialis p.
nodus tibialis p.
p. notch of cerebellum
p. occipitoaxial ligament
p. oropharyngeal wall
p. palatine arch
p. palatine foramen
p. palatine spine
p. palpebral margin
p. pancreaticoduodenal artery
p. parietal artery
parolfactorius p.
p. parolfactory sulcus
p. parotid vein
pars tibiotalaris p.
p. part of anterior commissure of
brain
p. part of the diaphragmatic
surface of the liver
p. part of tongue
p. part of vaginal fornix
p. perforated substance
p. pericallosal vein
p. periventricular nucleus
p. peroneal artery
p. pharynx
p. pillar
p. pillar of fauces
p. pillar of fornix
plexus nervosus cervicalis p.
plexus venosus vertebralis
externus p.
plexus venosus vertebralis
internus p.
plicae malleares anterior et p.
p. pole of eyeball
p. pole of lens
poralis p.
p. primary division
p. process of septal cartilage
p. process of talus

processus clinoideus p.
p. pyramid of medulla
p. rachischisis
rachischisis p.
p. radicular artery
radix p.
ramus basalis p.
ramus interventricularis p.
p. ramus of lateral cerebral sulcus
p. ramus of lateral sulcus of
cerebrum
ramus meningeus p.
p. ramus of spinal nerve
p. recess of tympanic membrane
recessus membranae tympani p.
p. rectus sheath
p. rectus sheath wall
recurrens tibialis p.
regio antebrachialis p.
regio antebrachii p.
regio brachialis p.
regio carpalis p.
regio cervicalis p.
regio colli p.
regio cruralis p.
regio cubitalis p.
regio femoralis p.
regio femoris p.
regio genus p.
p. region of arm
p. region of elbow
p. region of forearm
p. region of knee
p. region of leg
p. region of neck
p. region of thigh
p. region of wrist
p. renal segment
p. root
p. root of spinal nerve
p. sacroiliac ligament
p. sacrosciatic ligament
p. scalene muscle
scalenus p.
p. scapular nerve
p. scrotal branch of internal
pudendal artery
p. scrotal branch of perineal artery
p. scrotal nerve
p. scrotal vein
p. segmental artery
p. segmental artery of kidney
p. segment of eyeball
p. semicircular canal
p. septal artery of nose
p. septal branch of nose
p. septal branch of sphenopalatine
artery
p. septal space

sinus intercavernosi anterior et p.
p. sinus of tympanic cavity
p. spinal artery
spinalis p.
spina nasalis p.
p. spinocerebellar tract
p. sternoclavicular ligament
substantia perforata p.
sulcus intermedius p.
sulcus interventricularis p.
sulcus lateralis p.
p. superior alveolar artery
p. superior alveolar branch of
  maxillary nerve
superior anterior et p.
p. superior iliac spine (PSIS)
p. superior lateral nasal branch of
  maxillary nerve
p. superior lateral nasal branch of
  pterygopalatine ganglion
p. superior medial nasal branch of
  maxillary nerve
p. superior medial nasal branch of
  pterygopalatine ganglion
p. supraclavicular nerve
p. surface of arm
p. surface of arytenoid cartilage
p. surface of cornea
p. surface of elbow
p. surface of eyelid
p. surface of fibula
p. surface of forearm
p. surface of iris
p. surface of kidney
p. surface of knee
p. surface of leg
p. surface of lens
p. surface of lower limb
p. surface of pancreas
p. surface of petrous part of
  temporal bone
p. surface of prostate
p. surface of radius
p. surface of scapula
p. surface of shaft of humerus
p. surface of suprarenal gland
p. surface of thigh
p. surface of tibia
p. surface of ulna
synchondrosis intraoccipitalis p.
p. talar articular surface of
  calcaneus

p. talocalcaneal ligament
p. talofibular ligament
p. talotibial ligament
p. temporal branch of middle
  cerebral artery
p. temporal diploic vein
p. thoracic nerve
tibialis p.
p. tibialis tendon
p. tibial lymph node
p. tibial muscle
p. tibial nerve
p. tibial recurrent artery
p. tibial tendon
p. tibial vein
p. tibiofibular ligament
p. tibiotalar
p. tibiotalar part
p. tibiotalar part of deltoid
  ligament
p. tibiotalar part of medial
  ligament of ankle joint
p. tooth
tractus spinocerebellaris p.
p. triangle of neck
p. tubercle of atlas
p. tubercle of cervical vertebrae
tympanica p.
p. tympanic artery
p. ulnar recurrent artery
p. urethral valve
p. vaginal fornix
p. vaginal trunk
p. vein of left ventricle
p. vein of septum pellucidum
vena auricularis p.
vena circumflexa humeri p.
vena facialis p.
vena septi pellucidi p.
p. ventriculi sinistri cordis
p. vestibular branch of
  vestibulocochlear artery
p. wall of middle ear
p. wall of oropharynx
p. wall of stomach
p. wall of tympanic cavity
p. wall of vagina
posteriora
  ligamenta sacroiliaca p.
posteriores
  arteriae intercostales p. I et II
  arteriae intercostales p. III–XI

NOTES

P

posteriores *(continued)*
  cellulae ethmoidales p.
  conjunctivales p.
  intercostales p.
  labiales p.
  limbus palpebrales p.
  lymphonodi mediastinales p.
  nervi labiales p.
  nervi scrotales p.
  nodi lymphoidei axillares p.
  nodi lymphoidei mediastinales p.
  rami alveolares superiores p.
  rami labiales p.
  rami scrotales p.
  rami temporales p.
  scrotales p.
  sinus ethmoidales p.
  tibiales p.
  venae intercostales p.
  venae labiales p.
  venae scrotales p.
  venae tibiales p.

posterioris
  apex cornus p.
  bulbus cornus p.
  cervix columnae p.
  pars postcommunicalis arteriae
    cerebri p.
  pars precommunicalis arteriae
    cerebri p.
  pars profunda compartimenti
    cruris p.
  pars solealis compartimenti
    cruris p.
  pars superficialis compartimenti
    cruris p.
  pars tricipitalis compartimenti
    cruris p.
  plexus periarterialis arteriae
    auricularis p.
  rami celiaci trunci vagi p.
  rami dentales arteriae alveolaris
    superioris p.
  rami gastrici posteriores trunci
    vagalis p.
  rami malleolares mediales arteriae
    tibialis p.
  rami mastoidei arteriae
    auricularis p.
  rami mastoidei arteriae
    tympanicae p.
  rami occipitales arteriae
    auricularis p.
  rami occipitales nervi auricularis p.
  rami perineales nervi cutanei
    femoris p.
  ramus auricularis arteriae
    auricularis p.

  ramus circumflexus fibularis arteriae
    tibialis p.
  ramus circumflexus peronealis
    arteriae tibialis p.
  ramus nervi oculomotorii arteriae
    communicantis p.
  ramus prelaminaris rami spinalis
    rami dorsalis arteriae intercostalis
    p.
  ramus pterygoideus arteriae
    temporalis profundae p.
  ramus stapedius arteriae
    tympanicae p.
  segmentum P1, P3, P4, arteriae
    cerebri p.
  vagina tendinis musculi tibialis p.
  vena cornus p.

posteriorum
  ramus cutaneus medialis rami
    dorsalis arteriarum
    intercostalium p. III–XI
  ramus dorsalis arteriarum
    intercostalium p. III–XI
  ramus dorsalis venarum
    intercostalium p. IV–XI

posterius
  compartimentum antebrachii p.
  compartimentum brachii p.
  compartimentum cruris p.
  compartimentum femoris p.
  cornu p.
  corpus quadrigeminum p.
  foramen cecum p.
  foramen ethmoidale p.
  foramen lacerum p.
  labium p.
  ligamentum auriculare p.
  ligamentum capitis fibulae p.
  ligamentum costotransversarium p.
  ligamentum cricoarytenoideum p.
  ligamentum cruciatum p.
  ligamentum incudis p.
  ligamentum longitudinale p.
  ligamentum meniscofemorale p.
  ligamentum sacroiliacum p.
  ligamentum sternoclaviculare p.
  ligamentum talocalcaneum p.
  ligamentum talofibulare p.
  ligamentum talotibiale p.
  ligamentum tibiofibulare p.
  mediastinum p.
  os tibiale p.
  rete carpi p.
  segmentum basale p.
  segmentum bronchopulmonale p.
  segmentum hepatis p.
  segmentum renale p.
  trigonum cervicale p.
  venae vestibulares anterius et p.

posteroanterior
posteroexternal
posterointernal
posterolateral
    p. central artery
    p. fissure
    p. fontanelle
    p. groove
    p. sulcus
posterolaterales
    arteriae centrales p.
posterolateralis
    fissura p.
    fonticulus p.
posteromedial
    p. central artery
    p. frontal branch of callosomarginal artery
posteromediales
    arteriae centrales p.
posteromedialis
    ramus frontalis p.
posteromedian lobe
posteroparietal
posterosuperior
posterotemporal region
postesophageal
postestrum
postestrus
postganglionic
    p. motor neuron
    p. parasympathetic fiber
    p. sympathetic fiber
postglenoid foramen
postglomerular arteriole
posthepatic
posthippocampal fissure
posthyoid
posticum
    os tibiale p.
    tibiale p.
posticus
    locus perforatus p.
    musculus scalenus p.
    musculus tibialis p.
    tibialis p.
postischial
postlaminar part of intraocular part of optic nerve
postlingual fissure
postlunate fissure
postmastoid

postmedian
postmediastinal
postmediastinum
postminimus
postmortem rigidity
postnarial
postnaris
postnasal
postnatal
    p. life
    p. pit of newborn
postocular
postoral arches
postorbital
postpalatine
postpharyngeal space
postpyloric sphincter
postpyramidal fissure
postrema
    area p.
    camera p.
postremal chamber of eyeball
postrhinal fissure
postrolandic
postsacral
postscapular
postsphenoid bone
postsplenic
post-styloid space
postsulcal
    p. part
    p. part of tongue
postsulcalis
    pars p.
postsynaptic membrane
posttarsal
posttecta
post-term infant
posttibial
posttransverse
postural
    p. contraction
    p. reflex
postuterine
postvalvar
postvalvular
postvesiculares
    lymphonodi p.
    nodi lymphatici p.
potential
    action p.
    bioelectric p.

NOTES

P

potential *(continued)*
- biotic p.
- demarcation p.
- p. energy
- evoked p.
- excitatory junction p.
- excitatory postsynaptic p.
- generator p.
- inhibitory junction p.
- inhibitory postsynaptic p. (IPSP)
- injury p.
- membrane p.
- myogenic p.
- Ottoson p.
- S p.
- spike p.
- transmembrane p.

potentiometer
pouch
- antral p.
- branchial p.
- Broca p.
- deep perineal p.
- Douglas p.
- p. of Douglas
- endodermal p.
- gastric p.
- Hartmann p.
- p. of Hartmann
- Heidenhain p.
- hepatorenal p.
- hypophyseal p.
- laryngeal p.
- Morison p.
- p. of Morison
- p. of Munro
- paracystic p.
- pararectal p.
- paravesical p.
- Pavlov p.
- pharyngeal p.
- Prussak p.
- Rathke p.
- rectouterine p.
- rectovaginouterine p.
- rectovesical p.
- renal p.
- scleral p.
- Seessel p.
- superficial perineal p.
- suprapatellar p.
- ultimobranchial p.
- uterovesical p.
- vesicouterine p.
- Willis p.

poundal
Poupart
- P. inguinal ligament
- P. line

Pozzi muscle
P1–P4 segment of posterior cerebral artery
practical
- p. anatomy
- p. unit

praeauriculares
- lymphonodi parotidei subfasciales p.

praecox
- pubertas p.

pravesicalis
praxis
preanal
preaortici
- lymphonodi p.

preaortic lymph node
preauricular
- p. cyst
- p. deep parotid lymph node
- p. fossa
- p. groove
- p. point
- p. subfascial parotid lymph node
- p. sulcus

preauriculares
- nodi lymphoidei parotidei profundi p.

preauricularis
- vena p.

preaxial
preaxillaris
- linea p.

preaxillary line
precaecales
- nodi lymphoidei p.

precapillary anastomosis
precardiac
precardinal
precartilage
precavales
- lymphonodi p.

prececales
- lymphonodi p.

prececal lymph node
prececocolica
- fascia p.

prececocolic fascia
precentral
- p. area
- p. cerebellar vein
- p. gyrus
- p. sulcal artery
- p. sulcus

precentralis
- arteria sulci p.
- gyrus p.
- sulcus p.

precervical sinus
prechiasmatic sulcus

prechiasmaticus
  sulcus p.
prechordal plate
precocious puberty
precocity
precollagenous fibers
precommissural
  p. bundle
  p. septal area
  p. septum
precommunical
  p. part
  p. segment of anterior cerebral artery
  p. segment of posterior cerebral artery
precommunicalis
  pars p.
precommunicating
  p. part of anterior cerebral artery
  p. part of posterior cerebral artery
precordial
precordium, pl. **precordia**
precostal anastomosis
precuneal branch of anterior cerebral artery
precunealis
  arteria p.
precuneate
precuneus
precursory cartilage
predecidual
predigestion
predorsal bundle
preductal
preepiglottic
preexcitation
preformativa
  membrana p.
prefrontal
  p. area
  p. cortex
  p. vein
prefrontales
  venae p.
preganglionic
  p. motor neuron
  p. parasympathetic fiber
  p. sympathetic fiber
pregnancy
  bigeminal p.

  p. cells
  compound p.
  twin p.
pregranulosa cells
prehallux
prehelicine
prehensile
prehyoid gland
preinduction
preinterparietal bone
prelacrimal
prelaminar
  p. branch of spinal branch of dorsal branch of posterior intercostal artery
  p. part of intraocular part of optic nerve
prelaryngeales
  lymphonodi p.
  nodi lymphoidei p.
prelaryngeal lymph node
prelimbic
preload
  ventricular p.
premature
prematurity
premaxilla
premaxillare
  os p.
premaxillary
  p. bone
  p. suture
premelanosome
premenstrual
premenstruum
premolar
  p. tooth
premolares
  dentes p.
premolaris
  dens p.
premotor
  p. area
  p. cortex
  p. neuron
premyeloblast
premyelocyte
prenaris, pl. **prenares**
prenatal life
prenodular fissure
preoccipital notch

NOTES

P

preoptic
>p. area
>p. region

preoral gut

preosteoblast

prepalatal

prepancreatic
>p. arch
>p. artery

prepancreatica
>arteria p.

prepapillary
>p. bile duct
>p. sphincter

preparation
>heart-lung p.

prepatellar bursa

prepatellaris
>bursa subcutanea p.
>bursa subfascialis p.
>bursa subtendinea p.

prepericardiaci
>nodi lymphoidei p.

prepericardiales
>lymphonodi p.

prepericardial lymph node

preperitoneal
>p. fat
>p. space

prepiriform gyrus

preplacental

prepontine cistern

prepotential

preprostate urethral sphincter

preprostatic
>p. part of male urethra
>p. sphincter

prepubic fascia

prepuce
>p. of clitoris
>frenulum of p.
>p. of penis

preputia (*pl. of* preputium)

preputial
>p. gland
>p. sac
>p. space

preputiale
>sebum p.

preputiales
>glandulae p.

preputii
>frenulum p.
>smegma p.
>vinculum p.

preputium, pl. **preputia**
>p. clitoridis
>p. penis

prepyloric
>p. sphincter
>p. vein

prepylorica
>vena p.

prepyramidal tract

prerectal

prerenal

preretinal

preRolandic artery

prerubral field

presacral
>p. fascia
>p. nerve
>p. space

presacralis
>fascia p.
>nervus p.

presbyopia

presentation
>placental p.

presomite embryo

presphenoid bone

presphygmic

prespinal

presplenic fold

press
>wedge p.

pressor
>p. fiber
>p. nerve

pressoreceptive mechanism

pressoreceptor
>p. nerve
>p. system

pressosensitive

pressosensitivity
>reflexogenic p.

pressure
>abdominal p.
>atmospheric p.
>back p.
>barometric p.
>blood p.
>capillary wedge p.
>central venous p.
>cerebrospinal p.
>detrusor p.
>diastolic blood p. (DBP)
>differential blood p.
>Donders p.
>effective osmotic p.
>p. epiphysis
>gauge p.
>hydrostatic p.
>internal jugular p. (IJP)
>law of partial p.
>negative p.
>negative end-expiratory p. (NEEP)

oncotic p.
osmotic p.
partial p.
pleural p.
pulmonary p.
pulse p.
standard p.
systolic p.
transmural p.
transpulmonary p.
transthoracic p.
ventricular filling p.
wedge p.
zero end-expiratory p.
pressure-volume index
presternal
p. notch
p. region
presternalis
regio p.
presternum
prestriate area
prestyloid space
presulcal
p. part
p. part of tongue
presulcalis
pars p.
presumptive region
presymphyseal lymph node
presynaptic membrane
pretarsal space
pretecta
pretectal
p. area
p. region
pretectum
preterm infant
prethyroid
prethyroideal
prethyroidean
pretibial
pretracheal
p. layer
p. layer of cervical fascia
p. lymph node
p. space
pretracheales
lymphonodi p.
nodi lymphoidei p.
pretrachealis
lamina p.

pretrematic
pretympanic
preventricular artery
prevertebral
p. fascia level
p. ganglia
p. layer
p. layer of cervical fascia
p. lymph node
p. part of vertebral artery
prevertebrales
lymphonodi p.
nodi lymphoidei p.
prevertebralis
lamina p.
pars prevertebralis arteriae p.
prevesical space
prevesiculares
lymphonodi p.
nodi lymphatici p.
previa
central placenta p.
p. partialis
placenta p.
spontaneous correction of
placenta p.
total placenta p.
vasa p.
previllous
p. chorion
p. embryo
priapus
prickle
p. cell
p. cell layer
prima
costa p.
primae
processus uncinatus vertebrae
thoracicae p.
sulcus arteriae subclaviae costae p.
primal
primarium
centrum ossificationis p.
punctum ossificationis p.
primarius
folliculus ovaricus p.
primary
p. anophthalmia
p. bronchus
p. cementum
p. center of ossification

NOTES

P

primary (*continued*)
- p. choana
- p. constriction
- p. curvature of vertebral column
- p. dental lamina
- p. dentition
- p. digestion
- p. digit of foot
- p. egg membrane
- p. embryonic cell
- p. fissure of cerebellum
- p. hair
- p. hydrocephalus
- p. hypogonadism
- p. interatrial foramen
- p. labial groove
- p. lysosome
- p. mesoderm
- p. nodule
- p. nondisjunction
- p. oocyte
- p. organizer
- p. ossification center
- p. ovarian follicle
- p. palate
- p. point of ossification
- p. pulmonary lobule
- p. spermatocyte
- p. tooth
- p. villus
- p. visual area
- p. visual cortex
- p. vitreous

primi
- tuberositas ossis metatarsalis p.

primitiva
- meninx p.

primitive
- p. aorta
- p. choana
- p. chorion
- p. costal arches
- p. furrow
- p. groove
- p. knot
- p. node
- p. palate
- p. perivisceral cavity
- p. pit
- p. reticular cell
- p. ridge
- p. streak

primordia

primordial
- p. cartilage
- p. germ cell
- p. kidney
- p. ovarian follicle

primordium

primum
- interatrial foramen p.
- ostium p.
- septum p.

primus
- digitus manus p.
- digitus pedis p.

princeps
- p. cervicis
- p. cervicis artery
- p. pollicis
- p. pollicis artery

Princeteau tubercle

principal
- p. artery of thumb
- p. islets
- p. piece

principes

principle
- azygos vein p.
- Bernoulli p.
- Fick p.
- Le Chatelier p.
- low flow p.

prism
- enamel p.'s

prisma, pl. **prismata**
- prismata adamantina

prismatic

proacrosomal granule

proal

proamnion

proatlas

probe patency

proboscides

proboscis

proboscises

procaryote

procaryotic

procedure
- Eden-Lange p.
- Noble-Collip p.

procelia

procelous

procentriole organizer

procephalic

procerus
- p. muscle
- musculus p.

process
- accessory p.
- acromial p.
- acromion p.
- alar p.
- alveolar p.
- anterior clinoid p.
- apical p.
- articular p.
- arytenoid p.

ascending p.
auditory p.
basilar p.
Burns falciform p.
capitular p.
caudate p.
chordal p.
ciliary p.
Civinini p.
clinoid p.
cochleariform p.
condylar p.
condyloid p.
conoid p.
coracoid p.
corniculate p.
coronoid p.
costal pit of transverse p.
cribriform p.
Deiters p.
dendritic p.
dental p.
ensiform p.
ethmoidal p.
falciform p.
Folli p.
follian p.
foot p.
foramen of transverse p.
frontal p.
frontonasal p.
frontosphenoidal p.
funicular p.
glenoid p.
globular p.
hamular p.
head p.
inferior articular p.
Ingrassia p.
intercondylar p.
intrajugular p.
jugular p.
lacrimal p.
lateral nasal p.
lateral plate of pterygoid p.
left pterygoid p.
Lenhossék p.
malar p.
mallear p.
mamillary p.
mandibular p.
mastoid p.

maxillary p.
medial nasal p.
medial plate of pterygoid p.
mental p.
middle clinoid p.
nasal p.
odontoid p.
olecranon p.
orbicular p.
orbital p.
palatal p.
palatine p.
papillary p.
paramastoid p.
paroccipital p.
piriform p.
posterior clinoid p.
progressive p.
pterygoid p.
pterygospinous p.
pyramidal p.
radial styloid p.
Rau p.
Ravius p.
right pterygoid p.
sheath of styloid p.
sphenoid p.
sphenoidal p.
spinous p.
Stieda p.
styloid p.
superior articular p.
superior articulating p.
supracondylar p.
supraepicondylar p.
Tomes p.
transverse p.
trochlear p.
uncinate p.
ungual p.
vaginal p.
vermiform p.
vestige of vaginal p.
vocal p.
xiphoid p.
zygomatic p.

processus, pl. **processus**
p. accessorius
p. accessorius vertebrae lumbalis
p. alveolaris
p. alveolaris maxillae
p. anterior mallei

NOTES

487

processus *(continued)*
- p. articularis
- p. articularis inferior
- p. articularis superior
- p. articularis superior ossis
- p. articularis superior ossis sacri
- p. ascendens
- p. brevis
- p. calcaneus ossis cuboidei
- p. caudatus
- p. ciliaris
- p. clinoideus
- p. clinoideus anterior
- p. clinoideus medius
- p. clinoideus posterior
- p. cochleariformis
- p. condylaris
- p. condylaris mandibulae
- p. coracoideus
- p. coronoideus
- p. coronoideus mandibulae
- p. coronoideus ulnae
- p. costalis
- p. ethmoidalis conchae nasalis inferioris
- p. falciformis
- p. falciformis ligamenti sacrotuberalis
- p. ferreini
- p. frontalis maxillae
- p. frontalis ossis zygomatici
- p. gracilis
- p. intrajugularis
- p. jugularis
- p. jugularis ossis occipitalis
- p. lacrimalis
- p. lacrimalis conchae nasalis inferioris
- p. lateralis mallei
- p. lateralis tali
- p. lateralis tuberis calcanei
- p. lenticularis incudis
- p. mamillaris
- p. mammillaris vertebrae lumbalis
- p. mastoideus
- p. mastoideus partis petrosae ossis temporalis
- p. maxillaris
- p. maxillaris conchae nasalis inferioris
- p. medialis tuberis calcanei
- p. muscularis cartilaginis arytenoideae
- p. orbitalis
- p. orbitalis ossis palatini
- p. palatinus
- p. palatinus ossis maxillae
- p. papillaris

- p. papillaris lobi caudati hepatis
- p. paramastoideus
- p. posterior cartilaginis septi nasi
- p. posterior tali
- p. pterygoideus
- p. pterygoideus ossis sphenoidalis
- p. pterygospinosus
- p. pyramidalis
- p. pyramidalis ossis palatini
- p. ravii
- p. retromandibularis
- p. retromandibularis glandulae parotidis
- p. sphenoidalis cartilaginis septi nasi
- p. sphenoidalis ossis palatini
- p. spinosus
- p. spinosus vertebrae
- p. styloideus ossis metacarpalis III
- p. styloideus ossis temporalis
- p. styloideus radii
- p. styloideus ulnae
- p. supraepicondylaris humeri
- p. temporalis ossis zygomatici
- p. transversus
- p. transversus vertebrae
- p. trochleariformis
- p. trochlearis
- p. uncinatus ossis ethmoidalis
- p. uncinatus pancreatis
- p. uncinatus vertebrae cervicalis
- p. uncinatus vertebrae thoracicae primae
- p. vaginalis ossis sphenoidalis
- p. vaginalis peritonei
- p. vaginalis of peritoneum
- p. vermiformis
- p. vocalis cartilaginis arytenoideae
- p. xiphoideus
- p. zygomaticus
- p. zygomaticus maxillae
- p. zygomaticus ossis frontalis
- p. zygomaticus ossis temporalis

procheilon
prochondral
prochordal
procollagen
procreate
procreation
procreative
proctatresia
proctodeal
proctodeum, pl. **proctodea**
procurvation
proencephalon
proerythroblast
proestrum
proestrus

profile
    facial p.
    urethral pressure p.
profilometer
profunda
    arteria auricularis p.
    arteria cervicalis p.
    arteria circumflexa iliaca p.
    arteria circumflexa ilium p.
    arteria temporalis p.
    p. brachii
    p. brachii artery
    bursa infrapatellaris p.
    cerebri media p.
    cervicalis p.
    p. cervicalis artery
    circumflexa ilium p.
    p. clitoridis
    dorsalis clitoridis p.
    dorsalis penis p.
    faciei p.
    fascia cervicalis p.
    fascia pelvis p.
    fascia penis p.
    p. femoris
    p. femoris artery
    p. femoris vein
    lamina p.
    p. linguae
    p. linguae artery
    pars p.
    p. penis
    vena cerebri media p.
    vena cervicalis p.
    vena circumflexa iliaca p.
    vena circumflexa ilium p.
    vena colli p.
    vena dorsalis clitoridis p.
    vena dorsalis penis p.
    vena faciei p.
profundae
    p. penis
    rami labiales anteriores arteriae
        pudendae externae p.
    rami scrotales anteriores arteriae
        pudendae externae p.
    temporales p.
    venae cerebri p.
    venae temporales p.
profundarum
    rami inguinales arteriarum
        pudendarum externarum p.

profundi
    lymphonodi parotidei p.
    lymphonodus cervicales
        anteriores p.
    nervi digitales dorsales nervi
        fibularis p.
    nervi temporales p.
    nodi lymphoidei cervicales
        anteriores p.
    nodi lymphoidei cervicales
        laterales p.
    nodi lymphoidei inguinales p.
    nodi lymphoidei parotidei p.
    rami musculares nervi fibularis p.
    rami perforantes arcus palmaris p.
    temporales p.
profundum
    ligamentum metacarpale
        transversum p.
    ligamentum metatarsale
        transversum p.
    ligamentum sacrococcygeum
        posterius p.
    spatium perinei p.
    stratum album p.
    stratum griseum p.
    vas lymphaticum p.
profundus
    annulus inguinalis p.
    arcus palmaris p.
    arcus plantaris p.
    arcus venosus palmaris p.
    arcus volaris p.
    arteria plantaris p.
    p. muscle
    musculus flexor p.
    musculus flexor digitorum p.
    musculus transversus perinei p.
    nervus fibularis p.
    nervus peroneus p.
    nervus petrosus p.
    nodus lymphoideus proximalis p.
    peroneus p.
    petrosus p.
    plexus cardiacus p.
    ramus plantaris p.
    p. tendon
    transversus perinei p.
progenitalis
progenitor
progeny

NOTES

P

progeria
>p. with cataract
>p. with microphthalmia

progeroid
progestational hormone
proglossis
progonoma
progressive process
projection
>basilar p.
>p. fibers
>parietooccipital p.
>p. system
>verticosubmental p.

prokaryote
prokaryotic
prolabial
prolabium
prolactin
>p. cell

prolactin-inhibiting
>p.-i. factor
>p.-i. hormone

prolactin-releasing
>p.-r. factor
>p.-r. hormone

prolactoliberin
prolactostatin
prolapsus
>p. ani
>p. recti

proliferate
proliferation
proliferative
proliferous
prolific
proligerous
>p. disk
>p. membrane

proligerus
>discus p.

prominence
>Ammon p.
>canine p.
>cardiac p.
>p. of facial canal
>forebrain p.
>hepatic p.
>hypothenar p.
>laryngeal p.
>p. of lateral semicircular canal
>mallear p.
>sacral p.
>spiral p.
>styloid p.
>subcutaneous bursa of the
>laryngeal p.
>thenar p.

>p. of venous valvular sinus
>vertebral p.

prominens
>vas p.
>vertebra p.

prominentia, pl. prominentiae
>p. canalis facialis
>p. canalis semicircularis lateralis
>p. laryngea
>p. mallearis
>p. spiralis
>p. spiralis ductus cochlearis
>p. styloidea

promontoria (*pl. of* promontorium)
promontorial common iliac node
promontorii
>lymphonodus iliaci communes p.
>nodi lymphoidei p.
>nodi lymphoidei iliaci
>communes p.
>ponticulus p.
>subiculum p.
>sulcus p.

promontorium, pl. promontoria
>p. cavi tympani
>p. ossis sacri

promontory
>p. lymph node
>pelvic p.
>sacral p.
>p. of sacrum
>tympanic p.
>p. of tympanic cavity
>p. of tympanum

pronasion
pronate
pronation
>p. of foot
>p. of forearm

pronator
>musculus p.
>p. quadratus
>p. quadratus muscle
>p. ridge
>p. teres
>p. teres muscle
>p. tuberosity

pronatoria
>tuberositas p.

pronatus
>pes p.

prone
pronephric
>p. duct
>p. tubule

pronephroi
pronephros
>glomerulus of p.

pronograde

pronormoblast
pronucleus, pl. **pronuclei**
    female p.
    male p.
proopiomelanocortin (POMC)
prootic
propagate
propagation
propagative
propalinal
proper
    p. cochlear artery
    p. fasciculi
    p. hepatic artery
    hepatic artery p.
    intermediate branch of hepatic
      artery p.
    p. lamina
    left branch of hepatic artery p.
    p. ligament of ovary
    p. membrane
    p. membrane of semicircular duct
    muscle of back p.
    oral cavity p.
    p. palmar digital artery
    p. palmar digital nerve
    p. plantar digital artery
    p. plantar digital nerve
    right branch of hepatic artery p.
    p. substance
properitoneal fat
proplasia
proplasmacyte
proplexus
proportional limit
propria
    arteria cochlearis p.
    arteria digitalis palmaris p.
    arteria digitalis plantaris p.
    arteria hepatica p.
    cavitas oris p.
    hepatica p.
    lamina p.
    tunica p.
propriae
    digitales palmares p.
    digitales plantares p.
    glandulae p.
    ramus dexter arteriae hepaticae p.
    ramus intermedius arteriae
      hepaticae p.
    ramus sinister arteriae hepaticae p.

proprii
    fasciculi p.
    musculi dorsi p.
    nervi digitales palmares p.
    nervi digitales plantares p.
proprioception
proprioceptive
    p. mechanism
    p. reflex
proprioceptor
propriospinal
proprium
    ligamentum ovarii p.
proprius
    extensor indicis p. (EIP)
    fasciculus anterior p.
    fasciculus lateralis p.
    indicis p.
    musculus extensor digiti quinti p.
    musculus extensor indicis p.
    sacculus p.
proptosis
prorsad
prosecretion granule
prosector
prosectorium
prosencephalon
prosopagus
prosopalgia
prosoplasia
prosopoanoschisis
prosopopagus
prosoposchisis
prosopothoracopagus
prostata
prostatae
    apex p.
    basis p.
    facies anterior p.
    facies inferolateralis p.
    facies posterior p.
    fascia p.
    isthmus p.
    levatores p.
    lobus medius p.
    musculus levator p.
    parenchyma p.
    pars distalis p.
    pars proximalis p.
    substantia glandularis p.
    substantia muscularis p.

NOTES

P

prostate
  amyloid bodies of p.
  anterior surface of p.
  apex of p.
  base of p.
  distal part of p.
  elevator muscle of p.
  fascia of p.
  female p.
  p. gland
  glandular substance of p.
  inferolateral surface of p.
  isthmus of p.
  lateral lobes of p.
  levator p.
  lobe of p.
  median furrow of p.
  middle lobe of p.
  muscular substance of p.
  parenchyma of p.
  posterior surface of p.
  proximal part of p.
prostatic
  p. branch of inferior vesical artery
  p. branch of middle rectal artery
  p. capsule
  p. duct
  p. ductule
  p. fluid
  p. fossa
  p. ganglion
  p. lobe
  p. nervous plexus
  p. sheath
  p. sinus
  p. tissue
  p. urethra
  p. utricle
  p. venous plexus
  p. vesicle
prostatica
  glandula p.
  vesica p.
prostaticae
  pars distalis urethrae p.
  pars proximalis urethrae p.
prostatici
  ductuli p.
  ductus p.
prostaticovesical
  p. plexus
  p. venous plexus
prostaticovesicalis
  plexus p.
  plexus venosus p.
prostaticus
  musculus p.
  plexus nervosus p.
  plexus venosus p.

  sinus p.
  utriculus p.
prosthion
prostomial mesoderm
protean
protective laryngeal reflex
protein
  androgen binding p.
  bone Gla p.
  matrix Gla p.
  p. metabolism
  odorant binding p.
  parathyroid hormonelike p.
  vitamin D–binding p.
proteometabolic
proteometabolism
proteopectic
proteopepsis
proteopexic
proteopexis
protochordal knot
protoderm
protoduodenum
protoneuron
protopathic
protoplasm
  totipotential p.
protoplasmic
  p. astrocyte
  p. movement
protovertebra
protovertebral
protractor
protuberance
  Bichat p.
  external occipital p.
  internal occipital p.
  mental p.
  occipital p.
  palatine p.
  transverse occipital p.
protuberantia
  p. laryngea
  p. mentalis
  p. occipitalis externa
  p. occipitalis interna
Proust space
provertebra
provesicalis
  fossa p.
provisional cortex
proximad
proximal
  p. border of nail
  p. centriole
  p. colon
  p. deep inguinal lymph node
  p. end of rib
  p. end of ulna

p. head of phalanx
p. interphalangeal (PIP)
p. interphalangeal joint (PIPJ)
p. jejunum
p. medial striate artery
p. part of prostate
p. part of prostatic urethra
p. phalanx of foot
p. phalanx of hand
p. phalanx of thumb
p. radioulnar articulation
p. radioulnar joint
p. spiral septum
p. tibia
p. tibiofibular joint
p. ureter
p. urethral sphincter
proximalis
articulatio radioulnaris p.
phalanx p.
proximate
prozygosis
Prussak
P. fiber
P. pouch
P. space
psalteria (*pl. of* psalterium)
psalterial cord
psalterii
cavum p.
psalterium, pl. **psalteria**
psammoma body
pseudoacephalus
pseudoachondroplasia
pseudoachondroplastic spondyloepiphysial
dysplasia
pseudocartilage
pseudocele
pseudodipsia
pseudofluctuation
pseudoganglion
Cloquet p.
pseudohermaphrodite
pseudohermaphroditism
male p.
pseudohydrocephaly
pseudolobster-claw deformity
pseudometaplasia
pseudoplastic fluid
pseudostratified epithelium
pseudotruncus arteriosus

pseudounipolar
p. cell
p. neuron
pseudoventricle
pseudovomiting
pseudoxanthoma cell
PSIS
posterior superior iliac spine
psoas
p. fascia
p. major
p. major muscle
p. minor muscle
p. minor tendon
psoatic part of iliopsoas fascia
psychocardiac reflex
psychogalvanic
p. reaction
p. reflex
p. response (PGR)
psychogalvanometer
psychokym
psychomotor
psychophysiologic
psychophysiology
psychosomatic
psychrometer
sling p.
psychrometry
pterion
pterygium colli
pterygoid
p. branch of maxillary artery
p. branch of posterior deep
temporal artery
p. canal
p. chest
p. depression
p. fissure
p. fossa
p. fovea
p. hamulus
p. lamina
p. nerve
p. notch
p. pit
p. plate
p. process
p. process of sphenoid bone
p. ridge of sphenoid bone
p. tubercle
p. tuberosity

NOTES

P

pterygoid *(continued)*
    p. tuberosity of mandible
    p. venous plexus
pterygoidea
    fissura p.
    fossa p.
    fovea p.
    incisura p.
    tuberositas p.
pterygoidei
    arteria canalis p.
    canalis p.
    lamina lateralis processus p.
    lamina medialis processus p.
    nervus canalis p.
    rami p.
    ramus pharyngeus arteriae
      canalis p.
    sulcus hamuli p.
    vena canalis p.
pterygoideum
    os p.
pterygoideus
    canalis p.
    hamulus p.
    p. lateralis
    p. medialis
    nervus p.
    plexus venosus p.
    processus p.
pterygoma
    fissura p.
pterygomandibular
    p. ligament
    p. raphe
    p. space
pterygomandibularis
    raphe p.
pterygomaxillare
pterygomaxillaris
    fissura p.
pterygomaxillary
    p. fissure
    p. fossa
    p. notch
    p. region
pterygomeningeal artery
pterygomeningealis
    arteria p.
pterygopalatina
    fissura p.
    fossa p.
pterygopalatine
    p. canal
    p. fossa
    p. ganglion
    p. groove
    p. nerve

pterygopalatini
    nervi p.
    radix sensoria ganglii p.
    radix sympathica ganglii p.
    rami nasales posteriores superiores
      laterales ganglii p.
    rami nasales posteriores superiores
      mediales ganglii p.
    ramus orbitalis ganglii p.
    ramus pharyngeus ganglii p.
pterygopalatinum
    ganglion p.
pterygopalatinus
    sulcus p.
pterygopharyngea
    pars p.
pterygopharyngeal
    p. part
    p. part of superior constrictor
      muscle of pharynx
    p. space
pterygopharyngeus
    musculus p.
pterygoquadrate
pterygospinale
    ligamentum p.
pterygospinal ligament
pterygospinosus
    musculus p.
    processus p.
pterygospinous
    p. ligament
    p. process
ptotic organ
ptyocrinous
pubarche
puberal
pubertal
pubertas praecox
puberty
    precocious p.
pubes (*pl. of* pubis)
pubic
    p. angle
    p. arch
    p. arcuate ligament
    p. body
    p. bone
    p. branch of inferior epigastric
      artery
    p. branch of inferior epigastric
      vein
    p. branch of obturator artery
    p. crest
    p. hair
    p. hair line
    p. region
    p. spine

p. symphysis
p. tubercle

pubica
    crista p.
    regio p.
    symphysis p.

pubicum
    tuberculum p.

pubioplasty
pubiotomy
pubis, pl. **pubes**
    arcus p.
    body of p.
    corpus ossis p.
    inferior ramus of p.
    lateral surface of p.
    ligamentum arcuatum p.
    mons p.
    os p.
    pecten ossis p.
    pectineal line of p.
    ramus inferior ossis p.
    ramus superior ossis p.
    spina p.
    symphysial surface of p.
    symphysis p.

puboanalis
    p. muscle
    musculus p.

pubocapsulare
    ligamentum p.

pubocapsular ligament
pubocervical
    p. fascia
    p. ligament

pubococcygeal
    p. line
    p. muscle

pubococcygeus
    p. muscle
    musculus p.

pubofemorale
    ligamentum p.

pubofemoral ligament
puboperinealis
    p. muscle
    musculus p.

puboprostatic
    p. ligament
    p. muscle

puboprostaticum
    ligamentum mediale p.

puboprostaticus
    p. muscle
    musculus p.

puborectalis
    p. loop
    p. muscle
    musculus p.

puborectal muscle
pubourethral triangle
pubovaginalis
    p. muscle
    musculus p.

pubovaginal muscle
pubovesical
    p. ligament
    p. ligament of female
    p. ligament of male
    p. muscle

pubovesicale
pubovesicalis
    p. muscle
    musculus p.

pubovesicocervical fascia
pudenda (*pl. of* pudendum)
pudendae externae
pudendal
    p. artery
    p. canal
    p. cleavage
    p. cleft
    p. nerve
    p. sac
    p. slit
    p. vein
    p. vessel

pudendalis
    canalis p.
    plexus p.

pudendi
    fissura p.
    frenulum labiorum p.
    labium majus p.
    labium minus p.
    rima p.
    vestibulum p.

pudendum, pl. **pudenda**
    p. femininum
    pudenda interna
    p. muliebre

pudendus
    nervus p.

NOTES

P

pudic
>p. nerve
>p. vessel

puerile respiration
pular nerve
pulley
>annular p.
>cruciate p.
>cruciform p.
>p. of humerus
>muscular p.
>peroneal p.
>p. of talus
>p. tendon

pulmo
>p. dexter
>p. sinister

pulmoaortic
pulmometer
pulmometry
pulmonale
>atrium p.
>glomus p.
>ligamentum p.

pulmonales
>lymphonodi juxtaesophageales p.
>nodi lymphoidei
> juxtaesophageales p.
>rami p.
>venae p.

pulmonalis
>arteria p.
>bifurcatio trunci p.
>p. inferior dextra
>p. inferior sinistra
>lunulae valvularum semilunarium
> valvae aortae/trunci p.
>ostium trunci p.
>pars basalis arteriae p.
>pars lateralis venae p.
>pleura p.
>plexus nervosus p.
>rami pulmonales plexi nervosi p.
>sinus trunci p.
>sulcus p.
>p. superior dextra
>p. superior sinistra
>truncus p.
>valva trunci p.
>valvula semilunaris anterior valvae
> trunci p.
>valvula semilunaris dextra valvae
> trunci p.
>valvula semilunaris sinistra valvae
> trunci p.

pulmonalium
>ostia venarum p.

pulmonary
>p. alveolus

>p. apex
>p. artery (PA)
>p. atresia
>p. branch of autonomic nervous
> system
>p. branch of pulmonary nerve
> plexus
>p. cavity
>p. circulation
>p. cone
>p. conus
>p. epithelial cell
>p. groove
>inferior lingular branch of lingular
> branch of left p.
>p. ligament
>p. lobule
>p. lymph node
>p. nervous plexus
>p. opening
>p. orifice
>p. outflow tract
>p. pleura
>p. pressure
>p. ridge
>p. segment
>p. sinus
>p. sulcus
>p. surface of heart
>p. transpiration
>p. trunk
>p. valve
>p. vein
>p. ventilation
>p. vesicle
>p. vessel

pulmones
pulmonic valve
pulmonis
>alveoli p.
>apex p.
>basis p.
>facies costalis p.
>facies interlobares p.
>facies medialis p.
>facies mediastinalis p.
>fissura obliqua p.
>hilum p.
>impressio cardiaca p.
>ligamentum latum p.
>lobus inferior p.
>lobus superior p.
>margo anterior p.
>margo inferior p.
>pars medialis venae p.
>pars mediastinalis p.
>pars vertebralis faciei costalis p.
>porta p.
>radix p.

pulmonocoronary reflex
pulmonum
    pars intrasegmentalis venae p.
    partes intersegmentales venarum p.
pulp
    artery of p.
    p. canal
    p. cavity
    p. cavity of crown
    coronal p.
    crown p.
    dental p.
    dentinal p.
    digital p.
    p. of finger
    radicular p.
    red p.
    root p.
    splenic p.
    p. of toe
    tooth p.
    vertebral p.
    white p.
pulpa
    p. alba splenica
    p. coronale
    p. coronalis
    p. den
    p. dentis
    p. digiti manus
    p. lienis
    p. radicularis
    p. rubra splenica
pulpal wall
pulpar cell
pulparis
    cavitas p.
pulposus
    nucleus p.
pulsate
pulsatile
pulsation
    suprasternal p.
pulsator
pulse
    dorsalis pedis p.
    femoral p.
    jugular p.
    peripheral p.
    popliteal p.
    p. pressure
    radial p.

    saphenous p.
    tibial p.
    trigeminal p.
    venous p.
pulseless electrical activity (PEA)
pulsion hernia
pulsus
pulvinar
pump
    Carrel-Lindbergh p.
    jet ejector p.
    sodium p.
    sodium-potassium p.
puna
puncta (*pl. of* punctum)
punctata
    chondrodysplasia p.
    chondrodystrophia congenita p.
    dysplasia epiphysialis p.
punctate basophilia
puncti
punctiform
punctum, pl. **puncta**
    p. coxale
    p. fixa
    inferior lacrimal puncta
    lacrimal p.
    p. lacrimale
    p. luteum
    p. mobile
    p. ossificationis
    p. ossificationis primarium
    p. ossificationis secundarium
    superior lacrimal puncta
pupil
    ciliary muscle of p.
    dilator of p.
    dilator muscle of p.
    Marcus Gunn p. (MGP)
    sphincter muscle of p.
pupilla, pl. **pupillae**
    dilator pupillae
    musculus dilator pupillae
    musculus sphincter pupillae
    sphincter pupillae
    pupillae sphincter muscle
pupillaris
    membrana p.
    zona p.
pupillary
    p. axis
    p. border of iris

NOTES

P

497

pupillary *(continued)*
    p. margin of iris
    p. membrane
    p. muscle
    p. reflex
    p. zone
pupillodilator fiber
Purkinje
    P. cell
    P. corpuscles
    P. fiber
    P. layer
    P. system
pus
    p. cell
    p. corpuscle
putamen
putrefaction
putrefactive
putrefy
putrescence
putrescent
putrid
Puusepp reflex
pyelocaliceal, pyelocalyceal
    p. system
pyeloileocutaneous
pyelolymphatic
pygoamorphus
pygodidymus
pygomelus
pygopagus
pyknic
pyknodysostosis
pyknomorphous
pylon
pylori *(pl. of* pylorus)
pyloric
    p. antrum
    p. artery
    p. branch of anterior vagal trunk
    p. canal
    p. cap
    p. constriction
    p. lymph gland
    p. lymph node
    p. mucosa
    p. orifice
    p. part of stomach
    p. region
    p. ring
    p. sphincter
    p. valve
    p. vein
    p. vein of Mayo
    p. vestibule
pylorica
    pars p.

pyloricae
    glandulae p.
pylorici
    lymphonodi p.
    nodi lymphoidei p.
pyloricum
    antrum p.
    ostium p.
pyloricus
    canalis p.
pyloroduodenal junction
pylorus, pl. **pylori**
    dilator muscle of p.
    gastric p.
    musculus sphincter pylori
    sphincter pylori
    sphincter muscle of p.
    p. of stomach
    valvula pylori
pyocyte
pyramid
    anterior p.
    base of renal p.
    cartilaginous p.
    cerebellar p.
    Ferrein p.
    p. of kidney
    Lallouette p.
    Malacarne p.
    malpighian p.
    p. of medulla oblongata
    medullary p.
    nasal p.
    olfactory p.
    orbital p.
    petrous p.
    renal p.
    p. of thyroid
    p. of tympanum
    p. of vestibule
pyramidal
    p. auricular muscle
    p. cell layer
    p. cells
    p. decussation
    p. eminence
    p. fibers
    p. lobe
    p. lobe of thyroid
    p. lobe of thyroid gland
    p. muscle of auricle
    p. process
    p. process of palatine bone
    p. process of sphenoid
    p. process of thyroid
    p. radiation
    p. tract
pyramidale
    os p.

pyramidales
    fibrae p.
pyramidalis
    p. auriculae
    eminentia p.
    p. muscle
    musculus p.
    nucleus p.
    processus p.
    radiatio p.
    tractus p.
pyramides renales
pyramidis
    fasciculus circumolivaris p.

pyramidum
    decussatio p.
pyramis
    p. medullae oblongatae
    p. renalis
    p. tympani
    p. vestibuli
pyriform
pyriformis
    musculus p.
pyroninophilia
pyrrhol, pyrrol
    p. cell

NOTES

P

Q
- Q bands
- Q disks

Q̇
- blood flow

quader

quadrangular
- q. cartilage
- q. lobule
- q. lobule of cerebellum
- q. membrane
- q. space

quadrangularis
- lobulus q.
- membrana q.

quadrant
- left inferior abdominal q.
- left lower q. (LLQ)
- left superior abdominal q.
- left upper q. (LUQ)
- right inferior abdominal q.
- right lower q. (RLQ)
- right superior abdominal q.
- right upper q. (RUQ)

quadrata
- pars q.

quadrate
- q. femoral tubercle
- q. gyrus
- q. ligament
- q. lobe
- q. lobule
- q. muscle of loin
- q. muscle of sole
- q. muscle of thigh
- q. muscle of upper lip
- q. part
- q. part of liver
- q. pronator muscle
- q. tubercle of femur

quadratum
- foramen q.
- ligamentum q.

quadratus
- q. femoris
- q. femoris bursa
- q. femoris muscle
- q. labii muscle
- lobulus q.
- lobus q.
- q. lumborum
- q. lumborum fascia
- q. lumborum muscle
- musculus q.
- musculus pronator q.

- q. plantae
- q. plantae muscle
- pronator q.

quadriceps
- q. artery of femur
- q. femoris
- q. femoris muscle
- q. femoris tendon
- q. muscle of thigh
- musculus q.
- q. reflex

quadricuspid pulmonary valve

quadridigitate

quadrigemina
- corpora q.
- lamina q.

quadrigeminal
- q. artery
- q. bodies
- q. cistern
- q. plate

quadrigeminalis
- arteria q.
- cisterna q.

quadrigeminum

quadrigeminus

quadrilateral
- q. cartilage
- q. septum
- q. space

quadriparesis

quadripolar

quadrisect

quadrisection

quadritubercular

quadruplet

qualitative alteration

quantitative alteration

quarta
- crista q.
- macula cribrosa q.

quarti
- apertura lateralis ventriculi q.
- foramen lateralis ventriculi q.
- plexus choroideus ventriculi q.
- recessus lateralis ventriculi q.
- striae medullares ventriculi q.
- sulcus medianus ventriculi q.
- tegmen ventriculi q.
- tela choroidea ventriculi q.
- tenia ventriculi q.

quartisect

quartus
- digitus manus q.
- digitus pedis q.

quartus *(continued)*
    musculus gluteus q.
    ventriculus q.
Quatrefages angle
Quénu hemorrhoidal plexus
quickening
quiescent
quin-2
quinquedigitate
quinquetubercular
quinti
    extensor digiti q. (EDQ)
    flexor digiti q. (FDM)
    musculus abductor digiti q.

    musculus opponens digiti q.
    tuberositas ossis metatarsalis q.
quintuplet
quintus
    digitus manus q.
    digitus pedis q.
    ventriculus q.
quotient
    Ayala q.
    cognitive laterality q.
    growth q.
    respiratory q.
    spinal q.

rachial
rachides (*pl. of* rachis)
rachidial
rachidian
rachiopagus, rachipagus
rachis, pl. **rachides, rachises**
rachischisis
    posterior r.
    r. posterior
rachitis
    r. fetalis
    r. fetalis annularis
    r. fetalis micromelica
    r. intrauterina
    r. uterina
Radford nomogram
radiad
radial
    r. acceleration
    r. aspect
    r. bone
    r. border of forearm
    r. bursa
    r. collateral artery
    r. collateral ligament
    r. collateral ligament of elbow
      joint
    r. collateral ligament of wrist
    r. collateral ligament of wrist joint
    r. eminence of wrist
    r. extensor muscle
    r. flexor muscle
    r. flexor muscle of wrist
    r. fossa
    r. fossa of humerus
    r. foveola
    r. groove
    r. head
    r. index artery
    r. neck
    r. nerve
    r. notch
    r. notch of ulna
    r. part of posterior compartment of
      forearm
    r. pulse
    r. recurrent artery
    r. reflex
    r. shaft
    r. styloid
    r. styloid process
    r. tuberosity
    r. tunnel
    r. vein

radiale
    ligamentum collaterale carpi r.
radiales
    venae r.
radiali
    arteria collateralis r.
    vagina tendinum musculorum
      extensorum carpi r.
radialis
    arteria collateralis r.
    arteria recurrens r.
    arteria volaris indicis r.
    collateralis r.
    digitales dorsales nervi r.
    eminentia carpi r.
    flexor carpi r. (FCR)
    fossa r.
    incisura r.
    r. indicis
    r. indicis artery
    musculus flexor carpi r.
    nervus r.
    rami musculares nervi r.
    ramus carpalis dorsalis arteriae r.
    ramus carpalis palmaris arteriae r.
    ramus carpeus dorsalis arteriae r.
    ramus carpeus palmaris arteriae r.
    ramus communicans ulnaris nervi r.
    ramus palmaris superficialis
      arteriae r.
    ramus profundus nervi r.
    ramus superficialis nervi r.
    recurrens r.
    sulcus bicipitalis r.
    sulcus nervi r.
    vagina tendinis musculi flexoris
      carpi r.
radialium
    vagina tendinum musculorum
      extensorum carpi r.
radian
radiata
    corona r.
    ligamenta sternocostalia r.
    zona r.
radiatae
    arteriae corticales r.
radiate
    r. artery of kidney
    r. carpal ligament
    r. crown
    r. layer of tympanic membrane
    r. ligament of head of rib
    r. ligament of rib

R

radiate *(continued)*
    r. ligament of wrist
    r. sternocostal ligament
radiatio, pl. **radiationes**
    r. acustica
    r. corporis callosi
    r. optica
    r. pyramidalis
radiation
    acoustic r.
    r. of corpus callosum
    geniculocalcarine r.
    Gratiolet r.
    occipitothalamic r.
    optic r.
    pyramidal r.
    Wernicke r.
radiatum
    ligamentum capitis costae r.
    ligamentum carpi r.
radices (*pl. of* radix)
radicis
radicle
    tertiary r.
radicula
radicular
    r. artery
    r. pulp
radiculares
    rami r.
radicularia
    fila r.
radicularis
    pulpa r.
radii (*pl. of* radius)
radiobicipital
    r. reflex
radiocapitate ligament
radiocarpal
    r. articulation
    r. joint
    r. ligament
radiocarpalis
    articulatio r.
    ligamentum collaterale carpi radiale
      articulationis r.
    ligamentum collaterale carpi ulnare
      articulationis r.
radiocarpea
    articulatio r.
radiodigital
radiohumeral joint
radiologic
    r. anatomy
    r. sphincter
radiological
    r. anatomy
    r. sphincter
radiomuscular

radiopalmar
radioreceptor
radiotelemetry
radiotriquetral ligament
radioulnar
    r. articular disk
    r. articulation
    r. joint
    r. ligament
    r. syndesmosis
radioulnaris
    discus articularis r.
    syndesmosis r.
radius, pl. **radii**
    annular ligament of r.
    anterior border of r.
    anterior oblique line of r.
    anterior surface of r.
    articular circumference of head
      of r.
    articular pit of head of r.
    body of r.
    carpal articular surface of r.
    distal end of r.
    dorsal tubercle of r.
    r. fixus
    head of r.
    interosseous border of r.
    interosseous crest of r.
    neck of r.
    nutrient artery of r.
    orbicular ligament of r.
    posterior border of r.
    posterior surface of r.
    scaphoid r.
    shaft of r.
    styloid process of r.
    suprastyloid crest of r.
    tubercle of r.
    tuberosity of r.
radix, pl. **radices**
    r. accessoria
    r. anterior
    r. anterior nervi spinalis
    r. arcus vertebrae
    arteria nutriens radii
    r. brevis ganglii ciliaris
    r. buccalis
    caput radii
    circumferentia articularis capitis
      radii
    r. clinica
    r. clinica dentis
    r. cochlearis
    collum radii
    corpus radii
    radices craniales
    r. cranialis nervi accessorii
    crista suprastyloidea radii

r. dentis
r. dorsalis
r. dorsalis nervi spinalis
facies anterior radii
facies articularis carpi radii
facies posterior radii
fovea articularis capitis radii
r. inferior ansae cervicalis
r. inferior nervi vestibulocochlearis
r. lateralis nervi mediani
r. lateralis tractus optici
radii of lens
radii lentis
ligamentum annulare radii
ligamentum orbiculare radii
r. linguae
r. longa ganglii ciliaris
margo anterior radii
margo interosseus radii
margo posterior radii
r. medialis nervi mediani
r. medialis tractus optici
r. mesenterii
r. motoria
r. motoria nervi spinalis
r. motoria nervi trigemini
r. nasi
r. nasociliaris
r. nasociliaris ganglii ciliaris
r. nervi facialis
r. nervi oculomotorii ad ganglion
  ciliare
radices nervi trigemini
r. oculomotoria ganglii ciliaris
radices parasympathicae gangliorum
  pelvicorum
r. parasympathica ganglii ciliaris
r. parasympathica ganglii otici
r. parasympathica ganglii
  submandibularis
r. penis
r. pili
r. posterior
r. posterior nervi spinalis
processus styloideus radii
r. pulmonis
r. sensoria
r. sensoria ganglii ciliaris
r. sensoria ganglii pterygopalatini
r. sensoria ganglii sublingualis
r. sensoria ganglii submandibularis
r. sensoria nervi spinalis

r. sensoria nervi trigemini
r. spinalis nervi accessorii
r. superior
r. superior ansae cervicalis
r. superior nervi vestibulocochlearis
r. sympathica ganglii ciliaris
r. sympathica ganglii otici
r. sympathica ganglii pterygopalatini
r. sympathica ganglii sublingualis
r. sympathica ganglii
  submandibularis
tuber radii
tuberculum dorsale radii
tuberositas radii
r. unguis
r. ventralis
r. ventralis nervi spinalis
r. vestibularis
Rahn-Otis sample
RAIR
    rectoanal inhibitory reflex
Raji cell
ramal
rami (*pl. of* ramus)
ramification
ramify
ramituberis cinerei
ramose
ramous
ramp
ramulus, pl. **ramuli**
ramus, pl. **rami**
    r. accessorius arteriae meningeae
      mediae
    r. acetabularis
    r. acromialis arteriae suprascapularis
    r. acromialis arteriae
      thoracoacromialis
    rami ad pontem
    rami alveolar
    rami alveolares superiores
    rami alveolares superiores
      posteriores
    rami alveolares superiores
      posteriores nervi maxillaris
    r. alveolaris superior medius
    r. alveolaris superior medius nervi
      infraorbitalis
    r. anastomoticus
    r. anastomoticus arteriae meningeae
      mediae cum arteriae lacrimali
    r. anterior arteriae renalis

R

NOTES

ramus *(continued)*

r. anterior ascendens
r. anterior descendens
rami anteriores nervorum cervicalium
rami anteriores nervorum lumbalium
rami anteriores nervorum sacralium
rami anteriores nervorum thoracis
r. anterior lateralis
r. anterior nervi spinalis
r. anterior sulci lateralis cerebri
r. apicalis
r. apicalis lobi inferioris
r. apicalis lobi inferioris arteriae pulmonalis dextrae
r. apicalis venae pulmonalis dextrae superioris
r. apicoposterior
r. apicoposterior venae pulmonalis sinistrae superioris
rami articulares
rami articulares arteriae descendentis genicularis
r. ascendens arteriae superficialis cervicalis
r. ascendens sulci lateralis cerebri
ascending r.
rami atriales
r. atrialis anastomoticus ramus circumflexus arteriae coronariae sinistrae
r. atrialis intermedius arteriae coronariae dextrae
r. atrialis intermedius arteriae coronariae sinistrae
rami auriculares anteriores
rami auriculares anteriores arteriae temporalis superficialis
r. auricularis arteriae auricularis posterioris
r. auricularis arteriae occipitalis
r. auricularis nervi vagi
r. basalis anterior
r. basalis anterior venae basalis superioris
r. basalis lateralis
r. basalis medialis
r. basalis posterior
r. basalis tentorii arteriae carotidis internae
r. bone
border of r.
rami bronchiales
rami bronchiales segmentorum
rami buccales
rami buccales nervi facialis
rami calcanei
rami calcanei laterales
rami calcanei laterales nervi suralis

rami calcanei mediales
rami calcanei mediales nervi tibialis
r. calcarinus
r. calcarinus arteriae occipitalis medialis
rami capsulae internae
rami capsulares
rami capsulares arteriae renalis
rami capsulares arteriorum intrarenalium
rami cardiaci cervicales inferiores
rami cardiaci cervicales inferiores nervi vagi
rami cardiaci cervicales superiores
rami cardiaci cervicales superiores nervi vagi
rami cardiaci thoracici
rami cardiaci thoracici gangliorum thoracicorum
rami cardiaci thoracici nervi vagi
r. cardiacus
rami caroticotympanici
r. carpalis dorsalis arteriae radialis
r. carpalis dorsalis arteriae ulnaris
r. carpalis palmaris arteriae radialis
r. carpalis palmaris arteriae ulnaris
r. carpeus dorsalis arteriae radialis
r. carpeus dorsalis arteriae ulnaris
r. carpeus palmaris arteriae radialis
r. carpeus palmaris arteriae ulnaris
rami caudae nuclei caudati
rami caudati
rami celiaci
rami celiaci nervi vagi
rami celiaci trunci vagi posterioris
rami centrales anteromediales
cephalic arterial rami
r. cervicalis nervi facialis
r. chiasmaticus
rami choroidei
rami choroidei posteriores arteriae cerebri posteriores laterales et mediales
r. choroidei posteriores laterales
r. choroidei posteriores mediales
r. choroidei ventriculi lateralis
r. choroidei ventriculi tertii
r. cingularis
r. cingularis arteriae callosomarginalis
r. circumflexus
r. circumflexus arteriae coronariae sinistrae
r. circumflexus fibulae
r. circumflexus fibularis arteriae tibialis posterioris
r. circumflexus peronealis arteriae tibialis posterioris

r. clavicularis
r. clavicularis arteriae
thoracoacromialis
rami clivales partis cerebralis
arteriae carotidis internae
r. clivi
r. cochlearis
r. cochlearis arteriae labyrinthi
r. cochlearis arteriae
vestibulocochlearis
r. colicus arteriae ileocolicae
r. collateralis
r. collateralis arteriarum
intercostalium posteriorum III–XI
r. collateralis nervorum
intercostalium
r. colli
r. colli nervi facialis
r. communicans
r. communicans arteriae fibularis
r. communicans arteriae peroneae
r. communicans cum chorda
tympani
r. communicans cum nervo
laryngeo inferiore
r. communicans cum nervo
pterygoidei medialis
r. communicans cum nervo ulnari
r. communicans cum ramo
auriculari nervi vagalis
r. communicans cum ramo
meningeo
r. communicans fibularis nervi
fibularis communis
r. communicans ganglii otici cum
chorda tympani
r. communicans ganglii otici cum
nervo auriculotemporali
r. communicans ganglii otici cum
nervo pterygoideo mediali
r. communicans ganglii otici cum
ramo meningeo nervi mandibularis
r. communicans nervi facialis cum
nervo glossopharyngeo
r. communicans nervi facialis cum
plexu tympanico
r. communicans nervi fibularis
communis cum nervo cutaneo
surae mediali
r. communicans nervi
glossopharyngei cum ramo
auriculari nervi vagi

r. communicans nervi intermedii
cum plexu tympanico
r. communicans nervi interossei
antebrachii anterioris cum nervi
ulnari
r. communicans nervi lacrimalis
cum nervo zygomatico
r. communicans nervi laryngei
interni cum nervo laryngeo
recurrente
r. communicans nervi laryngei
recurrentis cum ramo laryngeo
interno
r. communicans nervi laryngei
superioris cum nervo laryngeo
recurrenti
r. communicans nervi lingualis cum
chorda tympani
r. communicans nervi mediani cum
nervo ulnari
r. communicans nervi nasociliaris
cum ganglio ciliari
r. communicans nervi peronei
communis cum nervo cutaneo
surae mediali
r. communicans nervi radialis cum
nervi ulnari
r. communicans peroneus
r. communicans peroneus nervi
peronei communis
r. communicans plexus tympanici
cum ramo auriculari nervi vagi
r. communicans ulnaris
r. communicans ulnaris nervi
radialis
rami communicantes albi
rami communicantes ganglii
sublingualis cum nervo linguali
rami communicantes ganglii
submandibularis cum nervo
linguali
rami communicantes grisei
rami communicantes nervi
auriculotemporalis cum nervo
faciali
rami communicantes nervi lingualis
cum nervo hypoglosso
rami communicantes nervorum
spinalium
rami communicantes of sympathetic
part of autonomic division of
nervous system

NOTES

ramus *(continued)*
rami corporis amygdaloidei
r. corporis callosi dorsalis
rami corporis geniculati lateralis
r. costalis lateralis
r. costalis lateralis arteriae
thoracicae internae
r. cricothyroideus
r. cricothyroideus arteriae thyroideae
superioris
rami cutanei anteriores nervi
femoralis
rami cutanei anteriores pectoralis et
abdominalis nervorum
intercostalium
rami cutanei cruris mediales nervi
sapheni
r. cutaneus anterior abdominalis
nervi intercostalis
r. cutaneus anterior nervi
iliohypogastrici
r. cutaneus anterior (pectoralis et
abdominalis) nervorum
r. cutaneus anterior pectoralis et
abdominalis nervorum
thoracicorum
r. cutaneus anterior pectoralis nervi
intercostalis
r. cutaneus lateralis
r. cutaneus lateralis
abdominalis/pectoralis nervorum
intercostalium
r. cutaneus lateralis nervi
iliohypogastrici
r. cutaneus lateralis ramorum
posteriorum arteriae intercostalium
r. cutaneus medialis
r. cutaneus medialis rami dorsalis
arteriarum intercostalium
posteriorum III–XI
r. cutaneus medialis ramorum
dorsalium nervorum thoracicorum
r. cutaneus nervi mixti
r. cutaneus rami anterioris nervi
obturatorii
r. deltoideus
r. deltoideus arteriae profundae
brachii
r. deltoideus arteriae
thoracoacromialis
dental rami
rami dentales
rami dentales arteriae alveolaris
inferioris
rami dentales arteriae alveolaris
superioris posterioris
rami dentales inferiores
rami dentales inferiores plexus
dentalis inferioris

rami dentales superiores
rami dentales superiores plexus
dentalis superioris
r. descendens arteriae circumflexae
femoris lateralis
r. descendens arteriae circumflexae
femoris medialis
r. descendens arteriae occipitalis
r. descendens arteriae segmentalis
anterioris pulmonis dextri et
sinistri
r. descendens arteriae segmentalis
posterioris pulmonis dextri et
sinistri
r. descendens rami superficialis
arteriae transversae cervicis
r. dexter
r. dexter arteriae hepaticae propriae
r. dexter venae portae hepatis
r. digastricus nervi facialis
rami dorsales arteriae intercostalis
supremae
r. dorsales arteriae subcostalis
rami dorsales arteriae subcostalis
rami dorsales arteriarum
intercostalium posteriorum primae
et secundae
rami dorsales linguae
rami dorsales linguae arteriae
lingualis
rami dorsales nervi ulnaris
r. dorsalis arteriae lumbalis
r. dorsalis arteriarum intercostalium
posteriorum III–XI
r. dorsalis nervi spinalis
r. dorsalis nervorum spinalium
r. dorsalis venarum intercostalium
posteriorum IV–XI
dorsal lateral cutaneous branch
of r.
dorsal lateral muscular branch
of r.
dorsal medial cutaneous branch
of r.
rami duodenales
rami duodenales arteriae
pancreaticoduodenalis superioris
anterioris
rami epiploicae
rami epiploici
rami esophageales
rami esophageales aortae thoracicae
rami esophageales arteriae gastricae
sinistrae
rami esophageales arteriae
thyroideae inferioris
rami esophageales gangliorum
thoracicorum

rami esophageales partis thoracicae
  aortae
rami esophagei
rami esophagei nervi laryngei
  recurrentis
rami esophagei nervi vagi
r. externus
r. externus nervi laryngei superioris
r. externus trunci nervi accessorii
rami fauciales
rami fauciales nervi lingualis
r. femoralis
r. femoralis nervi genitofemoralis
r. frontalis anteromedialis
r. frontalis anteromedialis arteriae
  callosomarginalis
r. frontalis arteriae meningeae
  mediae
r. frontalis arteriae temporalis
  superficialis
r. frontalis intermediomedialis
  arteriae callosomarginalis
r. frontalis interomedialis
r. frontalis posteromedialis
r. frontalis posteromedialis arteriae
  callosomarginalis
r. ganglii trigeminalis
rami ganglionares
r. ganglionares trigeminales arteriae
  carotidis internae
rami ganglionici nervi maxillaris
r. ganglionis trigemini
rami gastrici anteriores nervi vagi
rami gastrici anteriores trunci
  vagalis anterioris
rami gastrici posteriores nervi vagi
rami gastrici posteriores trunci
  vagalis posterioris
r. genitalis
r. genitalis nervi genitofemoralis
rami gingivales inferiores
rami gingivales inferiores plexus
  dentalis inferioris
rami gingivales superiores
rami gingivales superiores plexus
  dentalis superioris
rami glandulares
r. glandulares
  anterior/lateralis/posterior arteriae
  thyroideae superioris
rami glandulares arteriae facialis

rami glandulares arteriae thyroideae
  inferioris
rami glandulares ganglii
  submandibularis
r. glandularis anterior arteriae
  thyroideae superioris
r. glandularis posterior arteriae
  thyroideae superioris
rami globi pallidi
rami hepatici
rami hepatici nervi vagi
rami hepatici trunci vagi anterior
r. hypothalamicus
r. iliacus arteriae iliolumbalis
r. inferior arteriae gluteae
  superioris
inferior dental rami
rami inferiores nervi transversi
  cervicalis colli
rami inferiores nervi transversi
  colli
r. inferior nervi oculomotorii
r. inferior ossis pubis
inferior pubic r.
r. infrahyoideus
r. infrahyoideus arteriae thyroideae
  superioris
r. infrapatellaris
r. infrapatellaris nervi sapheni
rami inguinales
rami inguinales arteriarum
  pudendarum externarum
  profundarum
rami intercostales anteriores
rami intercostales anteriores arteriae
  thoracicae internae
rami interganglionares
rami interganglionares trunci
  sympathici
r. intermedius arteriae hepaticae
  propriae
r. internus
r. internus nervi laryngei superioris
r. internus trunci nervi accessorii
rami interventriculares septales
rami interventriculares septales
  arteriae coronariae sinistrae/dextrae
r. interventricularis anterior
r. interventricularis anterior arteriae
  coronariae sinistrae
r. interventricularis posterior

NOTES

ramus *(continued)*

r. interventricularis posterior arteriae coronariae dextrae

ischial r.

ischiopubic r.

r. of ischium

rami isthmi faucium

rami isthmi faucium nervi lingualis

rami labiales anteriores

rami labiales anteriores arteriae pudendae externae profundae

rami labiales inferiores

rami labiales inferiores nervi mentalis

rami labiales nervi mentalis

rami labiales posteriores

rami labiales posteriores arteriae perinealis

rami labiales posteriores arteriae pudendae internae

rami labiales superiores

rami labiales superiores nervi infraorbitalis

r. labialis inferior arteriae facialis

r. labialis superior arteriae facialis

rami laryngopharyngei

rami laryngopharyngei ganglii cervicalis superioris

rami laterales arteriae pontis

rami laterales arteriarum centralium anterolateralium

rami laterales arteriarum tuberis cinerei

rami laterales rami sinistri venae portae hepatis

rami laterales ramorum dorsalium nervorum spinalis

r. lateralis ductus hepatici sinistri

r. lateralis interventricularis anterioris arteriae coronariae sinistrae

r. lateralis nasi arteriae facialis

r. lateralis nervi supraorbitalis

r. lateralis rami lobaris medii arteriae pulmonalis dextrae

r. lateralis ramorum dorsalium nervorum thoracicorum

rami lienales

rami lienales arteriae lienalis

rami linguales

rami linguales nervi glossopharyngei

rami linguales nervi hypoglossi

rami linguales nervi lingualis

r. lingualis nervi facialis

r. lingularis

r. lingularis inferior

r. lingularis superior

r. lingularis venae pulmonis sinistrae superioris

rami lobi caudati rami sinistri venae portae hepatis

r. lobi medii

r. lobi medii arteriae pulmonalis dextrae

r. lobi medii venae pulmonalis dextrae superioris

r. lumbalis arteriae iliolumbalis

rami malleolares laterales

rami malleolares laterales arteriae fibularis peronei

rami malleolares mediales

rami malleolares mediales arteriae tibialis posterioris

rami mammarii

rami mammarii laterales

rami mammarii laterales arteriae thoracicae lateralis

rami mammarii laterales ramorum cutaneorum lateralis nervorum thoracicorum

rami mammarii laterales ramorum cutaneorum lateralium nervorum intercostalium

rami mammarii mediales

rami mammarii mediales ramorum cutaneorum anteriorum nervorum

rami mammarii mediales ramorum perforantium arteriae thoracicae internae

r. of mandible

mandibular r.

r. marginalis dexter arteriae coronariae dextrae

r. marginalis mandibulae

r. marginalis mandibulae nervi facialis

r. marginalis sinister arteriae coronariae sinistrae

r. marginalis sulci cinguli

r. marginalis sulci parietooccipitalis

r. marginalis tentorii arteriae carotidis internae

r. marginalis tentorii partis cavernosae arteriae carotidis internae

rami mastoidei arteriae auricularis posterioris

rami mastoidei arteriae tympanicae posterioris

r. mastoideus

r. mastoideus arteriae occipitalis

r. meatus acustic

r. meatus acustici interni

rami mediales arteriae pontis

rami mediales arteriarum centralium anterolateralium

rami mediales arteriarum tuberis cinerei
rami mediales rami sinistri venae portae hepatis
r. medialis
r. medialis ductus hepatici sinistri
r. medialis nervi supraorbitalis
r. medialis rami lobaris medii arteriae pulmonalis dextrae
r. medialis ramorum dorsalium nervorum spinalis
rami mediastinales
rami mediastinales aortae thoracicae
rami mediastinales arteriae thoracicae internae
rami medullares laterales
rami medullares laterales partis intracranialis arteriae vertebralis
rami medullares mediales
rami medullares mediales arteriae vertebralis
r. membranae tympani
rami membranae tympani nervi auriculotemporalis
rami meningei
r. meningeus
r. meningeus accessorius
r. meningeus accessorius arteriae meningeae mediae
r. meningeus anterior arteriae ethmoidalis anterioris
r. meningeus anterior arteriae vertebralis
r. meningeus arteriae carotidis internae
r. meningeus arteriae carotis internae
r. meningeus arteriae occipitalis
r. meningeus medius nervi maxillaris
r. meningeus nervi mandibula
r. meningeus nervi mandibularis
r. meningeus nervi maxillaris
r. meningeus nervi vagi
r. meningeus nervorum spinalium
r. meningeus partis cavernosae arteriae carotidis internae
r. meningeus partis cerebralis arteriae carotidis internae
r. meningeus partis intracranialis arteriae vertebralis
r. meningeus posterior

r. meningeus recurrens nervi ophthalmici
rami mentales
rami mentales nervi mentalis
r. mentalis arteriae alveolaris inferioris
rami musculares
rami musculares arteriae vertebralis
r. musculares nervi accessorii
rami musculares nervi axillaris
rami musculares nervi fibularis profundi
rami musculares nervi fibularis superficialis
rami musculares nervi interossei antebrachii anterior
rami musculares nervi mediani
rami musculares nervi musculocutanei
rami musculares nervi radialis
rami musculares nervi tibialis
rami musculares nervi ulnaris
rami musculares nervorum intercostalium
rami musculares nervorum perinealium
rami musculares nervorum spinalium
rami musculares partis supraclavicularis plexus brachialis
rami musculares rami anterioris nervi obturatorii
rami musculares rami posterioris nervi obturatorii
r. musculi stylopharyngei
r. musculi stylopharyngei nervi glossopharyngei
r. mylohyoideus
r. mylohyoideus arteriae alveolaris inferioris
rami nasales anteriores laterales arteriae ethmoidalis anterioris
rami nasales externi
rami nasales externi nervi ethmoidalis anterioris
rami nasales externi nervi infraorbitalis
rami nasales interni
rami nasales interni nervi ethmoidalis anterioris
rami nasales interni nervi infraorbitalis

NOTES

ramus *(continued)*

rami nasales laterales nervi ethmoidalis anterioris
rami nasales mediales nervi ethmoidalis anterioris
rami nasales posteriores inferiores
rami nasales posteriores inferiores nervi palatini majoris
rami nasales posteriores superiores laterales
rami nasales posteriores superiores laterales ganglii pterygopalatini
rami nasales posteriores superiores laterales nervi maxillaris
rami nasales posteriores superiores mediales
rami nasales posteriores superiores mediales ganglii pterygopalatini
rami nasales posteriores superiores mediales nervi maxillaris
r. nervi oculomotorii arteriae communicantis posterioris
r. nodi atrioventricularis
r. nodi sinuatrialis
r. nodi sinuatrialis arteriae coronariae dextrae
rami nucleorum hypothalamicorum
r. obturatorius
r. obturatorius arteriae epigastricae inferioris
r. obturatorius rami pubici arteriae epigastricae inferioris
rami occipitales
rami occipitales arteriae auricularis posterioris
rami occipitales arteriae occipitis
rami occipitales nervi auricularis posterioris
r. occipitotemporalis
rami omentales
rami omentali
rami orbitales
rami orbitales nervi maxillaris
r. orbitalis
r. orbitalis arteriae meningeae mediae
r. orbitalis ganglii pterygopalatini
r. orbitofrontalis lateralis
r. orbitofrontalis medialis
r. ossis ischii
rami ovarici arteriae uterinae
r. ovaricus
r. palmaris nervi interossei antebrachii anterioris
r. palmaris nervi mediani
r. palmaris nervi ulnaris
r. palmaris profundus arteriae ulnaris

r. palmaris superficialis arteriae radialis
rami palpebrales
rami palpebrales nervi infratrochlearis
rami pancreatici
rami pancreatici arteriae pancreaticoduodenalis superioris
rami pancreatici arteriae splenicae
r. paracentrales
rami paracentrales arteriae callosomarginalis
rami parietales
r. parietalis
r. parietalis arteriae meningeae mediae
r. parietalis arteriae occipitalis medialis
r. parietalis arteriae temporalis superficialis
rami parietooccipitales arteriae cerebri anteriores
r. parietooccipitalis arteriae occipitalis medialis
rami parotidei
r. parotidei arteriae temporalis superficialis
rami parotidei nervi auriculotemporalis
rami parotidei venae facialis
rami pectorales
rami pectorales arteriae thoracoacromialis
rami pedunculares
r. perforans
r. perforans arteriae fibularis
r. perforans arteriae interossei anterioris
r. perforans arteriae metacarpalium palmares
r. perforans arteriae metatarsearum plantares
rami perforantes
rami perforantes arcus palmaris profundi
rami perforantes arteriae thoracicae internae
rami perforantes arteriarum metacarpalium palmarium
rami perforantes arteriarum metatarsearum plantarium
rami pericardiaci aortae thoracicae
rami perineales
rami perineales nervi cutanei femoris posterioris
peroneal anastomotic r.
r. petrosus
r. petrosus arteriae meningeae mediae

rami pharyngeales
rami pharyngeales arteriae
pharyngeae ascendentis
rami pharyngeales arteriae
thyroideae inferioris
rami pharyngei
r. pharyngei nervi glossopharyngei
rami pharyngei nervi laryngei
recurrentis
rami pharyngei nervi vagi
r. pharyngeus
r. pharyngeus arteriae canalis
pterygoidei
r. pharyngeus arteriae palatinae
descendentis
r. pharyngeus ganglii pterygopalatini
rami phrenicoabdominales
r. plantaris profundus
r. plantaris profundus arteriae
dorsalis pedis
r. posterior arteriae obturatoriae
r. posterior arteriae
pancreaticoduodenalis inferioris
r. posterior arteriae recurrentis
ulnaris
r. posterior arteriae renalis
r. posterior arteriae thyroideae
superioris
r. posterior ascendens
r. posterior descendens
r. posterior ductus hepatici dextri
r. posterior nervi auricularis magni
r. posterior nervi cutanei
antebrachii medialis
r. posterior nervi obturatorii
r. posterior nervi spinalis
r. posterior rami dextri venae
portae hepatis
r. posterior sulci lateralis cerebri
r. posterior sulcus lateralis cerebri
r. posterior venae pulmonalis
dextrae superioris
rami precuneales arteriae cerebri
anterioris
r. prelaminaris rami spinalis rami
dorsalis arteriae intercostalis
posterioris
rami profundi arteriae transversae
cervicis
r. profundus arteriae circumflexae
femoris medialis

r. profundus arteriae gluteae
superioris
r. profundus arteriae plantaris
medialis
r. profundus arteriae scapularis
descendentis
r. profundus arteriae transversae
colli
r. profundus arteria scapularis
descendens
r. profundus nervi plantaris lateralis
r. profundus nervi radialis
r. profundus nervi ulnaris
rami prostatici arteriae rectalis
mediae
rami prostatici arteriae vesicalis
inferioris
rami pterygoidei
rami pterygoidei arteriae maxillaris
r. pterygoideus arteriae temporalis
profundae posterioris
r. pubicus arteriae epigastricae
inferioris
r. pubicus arteriae obturatoriae
r. pubicus venae epigastricae
inferioris
rami pulmonales
rami pulmonales plexi nervosi
pulmonalis
rami pulmonales systematis
autonomici
rami pulmonales thoracici
gangliorum thoracicorum
r. pyloricus trunci vagalis anterioris
rami radiculares
rami mammarii mediales rami
rami renales nervi vagi
r. renalis nervi splanchnici minoris
rami sacrales laterales arteriae
sacralis medianae
r. saphenus
r. saphenus arteriae descendentis
genicularis
rami scrotales anteriores
rami scrotales anteriores arteriae
pudendae externae profundae
rami scrotales posteriores
rami scrotales posteriores arteriae
perinealis
rami scrotales posteriores arteriae
pudendae internae
rami septales

NOTES

513

ramus *(continued)*

rami septales anteriores arteriae
ethmoidalis anterioris

r. septi nasi arteriae labialis
superioris

r. septi posterioris nasalis

r. sinister

r. sinister arteriae hepaticae
propriae

r. sinister venae portae hepatis

r. sinus carotici

r. sinus carotici nervi
glossopharyngei

r. sinus cavernosi

r. sinus cavernosi arteriae carotidis
arteriae

r. sinus cavernosi arteriae carotidis
internae

r. sinus cavernosi partis cavernosae
arteriae carotidis internae

rami splenici arteriae splenicae

r. stapedius

r. stapedius arteriae stylomastoideae

r. stapedius arteriae tympanicae
posterioris

rami sternales

rami sternales arteriae thoracicae
internae

rami sternocleidomastoidei arteriae
occipitalis

r. sternocleidomastoideus

r. sternocleidomastoideus arteriae
thyroideae superioris

r. stylohyoideus

r. stylohyoideus nervi facialis

r. subapicalis

rami subendocardiales fasciculi
atrioventricularis

rami subscapulares

r. subscapulares arteriae axillaris

rami substantiae nigrae

r. subsuperior

r. superficialis arteriae circumflexae
femoris medialis

r. superficialis arteriae gluteae
superioris

r. superficialis arteriae plantaris
medialis

r. superficialis arteriae transversae
cervicis

r. superficialis arteriae transversae
colli

r. superficialis nervi plantaris
lateralis

r. superficialis nervi radialis

r. superficialis nervi ulnaris

superior r.

r. superior

r. superior arteriae gluteae
superioris

superior dental rami

rami superiores nervi transversi
colli

r. superior lobi inferioris

r. superior nervi oculomotorii

r. superior nervi transversalis
cervicalis colli

r. superior ossis pubis

superior pubic r.

r. superior venae pulmonalis
dextrae/sinistrae inferioris

r. suprahyoideus

r. suprahyoideus arteriae lingualis

suprapubic rami

r. sympathicus ad ganglion
submandibulare

r. sympathicus sympatheticus ad
ganglion submandibulare

rami temporales

rami temporales anteriores

rami temporales intermedii arteriae
occipitalis lateralis

rami temporales intermedii mediales

rami temporales medii arteriae
occipitalis lateralis

rami temporales nervi facialis

rami temporales posteriores

rami temporales superficiales nervi

rami temporales superficiales nervi
auriculotemporalis

r. temporalis anterior

r. temporalis medius partis insularis
arteriae cerebrae mediae

r. temporalis posterior arteriae
cerebri mediae

r. tentorii

r. tentorii basalis

r. tentorii marginalis

rami terminales arteriae cerebri
medii

rami thalamici

r. thalamicus

rami thymici

rami thymici arteriae thoracicae
internae

r. thyrohyoideus

r. thyrohyoideus ansae cervicalis

r. tonsillae cerebellae

rami tonsillares nervi
glossopharyngei

rami tonsillares nervi palatini
minores

r. tonsillaris

r. tonsillaris arteriae facialis

rami tracheales

rami tracheales arteriae thyroideae
inferioris

R

rami tracheales nervi laryngei
recurrentis
rami tractus optici
r. transversus
r. transversus arteriae circumflexae
femoris lateralis
r. transversus arteriae circumflexae
femoris medialis
r. tubarius
r. tubarius arteriae ovaricae
r. tubarius arteriae uterinae
r. tubarius plexus tympanici
r. ulnaris nervi cutanei antebrachii
medialis
rami ureterici
rami ureterici arteriae ovaricae
rami ureterici arteriae renalis
rami ureterici arteriae suprarenalis
inferioris
rami ureterici arteriae testicularis
rami ureterici partis patentis
arteriae umbilicalis
rami ventrales nervorum
cervicalium
rami ventrales nervorum lumbalium
rami ventrales nervorum sacralium
rami ventrales nervorum thoracis
r. ventralis
r. ventralis nervi spinalis
r. vermis superior
rami vestibulares
rami vestibulares arteriae labyrinthi
r. vestibularis posterior arteriae
vestibulocochlearis
rami zygomatici
rami zygomatici nervi facialis
r. zygomaticofacialis
r. zygomaticofacialis nervi
zygomatici
r. zygomaticotemporalis
r. zygomaticotemporalis nervi
zygomatici

ranina
arteria r.

ranine
r. anastomosis
r. artery
r. vein

raninus
arcus r.

Ranke
R. angle
R. complex

rantius nodule

ranula
pancreatic r.

ranular

Ranvier
R. crosses
R. disks
R. node
R. plexus
R. segment

raphe
amniotic r.
r. anococcygea
anogenital r.
r. corporis callosi
iliococcygeal r.
lateral palpebral r.
r. linguae
medial r.
r. medullae oblongatae
r. musculi iliococcygeus
nuclei r.
r. nuclei
r. palati
palatine r.
palpebral r.
r. palpebralis lateralis
penile r.
r. of penis
perineal r.
r. perinei
pharyngeal r.
r. pharyngis
r. pontis
pterygomandibular r.
r. pterygomandibularis
r. retinae
scrotal r.
r. scroti
r. of scrotum
septal r.
Stilling r.

rapid decompression

Rapoport-Luebering shunt

rarefaction

RAS
reticular activating system

rate
basal metabolic r. (BMR)

NOTES

rate *(continued)*
    glomerular filtration r. (GFR)
    growth r.
    mitotic r.
    respiration r.
    respiratory r.
    shear r.
    slew r.
    steroid metabolic clearance r.
    steroid production r.
    steroid secretory r.
    voiding flow r.

Rathke
    R. bundle
    R. diverticulum
    R. pocket
    R. pouch

ratio
    absolute terminal innervation r.
    body-weight r.
    extraction r.
    flux r.
    functional terminal innervation r.
    hand r.
    respiratory exchange r.
    r. scale
    ventilation/perfusion r. ($\dot{V}a/\dot{Q}$)

Rauber layer
Rau process
ravii
    processus r.
Ravius process
ray
    medullary r.
Raynaud syndrome
reaction
    alarm r.
    Brunn r.
    r. center
    decidual r.
    r. of degeneration
    eosinopenic r.
    fright r.
    galvanic skin r.
    hunting r.
    Jolly r.
    lengthening r.
    magnet r.
    myasthenic r.
    negative supporting r.
    neurotonic r.
    positive supporting r.
    psychogalvanic r.
    shortening r.
    supporting r.
    r. time
reactive
    r. astrocyte
    r. cell

rebound phenomenon
recalcification
recapitulation theory
receptaculum, pl. **receptacula**
    r. chyli
    r. ganglii petrosi
    r. pecqueti
receptor
    adrenergic r.
    alpha-adrenergic r.'s
    ANP clearance r.
    beta-adrenergic r.'s
    cholinergic r.
    dopamine r.
    mannose-6-phosphate r. (MPR)
    muscarinic r.
    nicotinic r.
    opiate r.
    ryanodine r.
    scavenger r.
    sensory r.
    stretch r.
recess
    r. of anterior ischioanal fossa
    cecal r.
    cerebellopontine r.
    cochlear r.
    costodiaphragmatic r.
    costomediastinal r.
    cupular part of epitympanic r.
    duodenojejunal r.
    elliptical r.
    epitympanic r.
    frontal r.
    hepatoenteric r.
    hepatorenal r.
    Hyrtl epitympanic r.
    inferior duodenal r.
    inferior ileocecal r.
    inferior omental r.
    infundibular r.
    intersigmoid r.
    Jacquemet r.
    mesentericoparietal r.
    nasopalatine r.
    r. of omental bursa
    optic r.
    pancreaticoenteric r.
    paracolic r.
    paraduodenal r.
    parotid r.
    pharyngeal r.
    phrenicomediastinal r.
    pineal r.
    piriform r.
    pleural r.
    pneumatoenteric r.
    pneumoenteric r.
    pontocerebellar r.

Reichert cochlear r.
retrocecal r.
retroduodenal r.
Rosenmüller r.
sacciform r.
sphenoethmoidal r.
spherical r.
splenic r.
subhepatic r.
subphrenic r.
subpopliteal r.
superior duodenal r.
superior ileocecal r.
superior omental r.
suprabullar r.
suprapineal r.
supratonsillar r.
triangular r.
Tröltsch r.
tubotympanic r.
r.'s of tympanic cavity
vertebromediastinal r.

recessus, pl. **recessus**
    r. anterior membranae tympanicae
    r. cochlearis
    r. costodiaphragmaticus
    r. costomediastinalis
    r. duodenalis inferior
    r. duodenalis superior
    r. ellipticus
    r. ellipticus labyrinthi ossei
    r. epitympanicus
    r. hepatorenalis
    r. hepatorenalis recessus subhepatici
    r. ileocecalis inferior
    r. ileocecalis superior
    r. inferior omentalis
    r. infundibuliformis
    r. intersigmoideus
    r. lateralis ventriculi quarti
    r. lienalis
    r. membranae tympani anterior
    r. membranae tympanicae
    r. membranae tympani posterior
    r. membranae tympani superior
    r. opticus
    r. paraduodenalis
    r. parotideus
    r. pharyngeus
    r. phrenicomediastinalis
    r. pinealis
    r. piriformis

    r. pleurales
    r. posterior membranae tympanicae
    r. retrocecalis
    r. retroduodenalis
    r. sacciformis articulationis
    r. sacciformis articulationis
       radioulnaris distalis
    r. saccularis labyrinthi ossei
    r. sphenoethmoidalis
    r. sphericus labyrinthi ossei
    r. splenicus
    r. subhepaticus
    r. subphrenici
    r. subphrenicus
    r. subpopliteus
    r. superior bursae omentalis
    r. superior membranae tympanicae
    r. suprapinealis
    r. triangularis
    r. utricularis labyrinthi membranacei
    r. utricularis labyrinthi ossei
    r. vertebromediastinalis

recipiomotor
reciprocal
    r. inhibition
    r. innervation
reclotting phenomenon
recollection
reconstitution
recovery
    creep r.
recruiting response
recruitment
recta (*pl. of* rectum)
rectae
    arteriolae r.
    venae r.
rectal
    r. ampulla
    r. artery
    r. canal
    r. column
    r. fascia
    r. fold
    r. hernia
    r. mucosa
    r. nerve
    r. reflex
    r. shelf
    r. sinuses
    r. site
    r. valve

NOTES

rectal (*continued*)
- r. vein
- r. venous plexus
- r. verge

rectalis, pl. **rectales**
- rectales inferior
- rectales inferiores
- r. media
- plexus venosus r.
- r. superior

rectally

recti
- ampulla r.
- corpora cavernosa r.
- flexura perinealis r.
- flexura sacralis r.
- folliculi lymphatici r.
- r. muscles of orbit
- plicae r.
- plicae transversales r.
- prolapsus r.
- stratum circulare tunicae muscularis r.
- stratum longitudinale tunicae muscularis r.
- tunica muscularis r.

recto
- hernia in r.

rectoanal inhibitory reflex (RAIR)

rectococcygeal muscle

rectococcygeus
- r. muscle
- musculus r.

rectoperineal

rectosacral fascia

rectosacralis
- fascia r.

rectosigmoid
- r. junction
- r. sphincter
- r. vein

rectourethrales
- musculi r.

rectourethralis
- musculus r.

rectourethral muscle

rectouterina
- excavatio r.
- plica r.

rectouterine
- r. fold
- r. ligament
- r. muscle
- r. pouch

rectouterinus
- r. muscle
- musculus r.

rectovaginal
- r. fascia

- r. fold
- r. septum

rectovaginale
- septum r.

rectovaginalis
- plica r.

rectovaginouterine pouch

rectovesical
- r. fascia
- r. fold
- r. muscle
- r. pouch
- r. septum
- r. space

rectovesicale
- septum r.

rectovesicalis
- excavatio r.
- r. muscle
- musculus r.

rectovestibular

rectum, pl. **recta, rectums**
- ampulla of r.
- conjugata recta
- flexure of r.
- intestinum r.
- lymphatic follicles of r.
- muscular coat of r.
- muscular layer of r.
- pars recta
- perineal flexure of r.
- sacral flexure of r.
- transverse fold of r.
- vas r.
- vasa recta

rectus
- r. abdominis
- r. abdominis muscle
- r. abdominis sheath
- r. capitis anterior
- r. capitis anterior muscle
- r. capitis lateralis
- r. capitis lateralis muscle
- r. capitis posterior major
- r. capitis posterior major muscle
- r. capitis posterior minor
- r. capitis posterior minor muscle
- r. fascia
- r. femoris
- r. femoris muscle
- r. femoris tendon
- gyrus r.
- r. inferior
- inferior r. (IR)
- r. inferior bulbi
- lateral r. (LR)
- r. lateralis bulbi
- r. medialis
- r. muscle of abdomen

r. muscle of thigh
musculus r.
sinus r.
superior r.
r. superior
r. superior bulbi
tubulus r.
tubulus renalis r.
tubulus seminiferus r.
recurrens
arteria interossea r.
interossea r.
laryngeus r.
nervus laryngeus r.
r. radialis
r. tibialis anterior
r. tibialis posterior
r. ulnaris
recurrent
r. branch of spinal nerve
r. interosseous artery
r. laryngeal nerve
r. meningeal branch of spinal
nerve
r. radial artery
r. ulnar artery
recurrente
ramus communicans nervi laryngei
interni cum nervo laryngeo r.
recurrenti
ramus communicans nervi laryngei
superioris cum nervo laryngeo r.
recurrentis
rami esophagei nervi laryngei r.
rami pharyngei nervi laryngei r.
rami tracheales nervi laryngei r.
recurvatum
genu r.
pectus r.
red
r. blood cell
r. bone marrow
r. fibers
r. muscle
r. nucleus
r. pulp
r. pulp cords
r. pulp of spleen
redifferentiation
reduced enamel epithelium
reduction
r. deformity

r. division
r. nucleus
redux
testis r.
Reed cells
Reed-Sternberg cells
reel foot
reentry theory
reflect
reflected inguinal ligament
reflection
r. coefficient
costal line of pleural r.
line of pleural r.
pericardial r.
peritoneal r.
pleural r.
sternal line of pleural r.
vertebral line of pleural r.
reflex
abdominal r.
abdominocardiac r.
Achilles tendon r.
acromial r.
adductor r.
allied r.
anal r.
ankle r.
antagonistic r.
aortic r.
Aschner r.
Aschner-Dagnini r.
atriopressor r.
attitudinal r.
auriculopressor r.
Bainbridge r.
biceps femoris r.
bladder r.
body righting r.
bulbocavernosus r.
bulbomimic r.
cardiac depressor r.
carotid sinus r.
chain r.
chin r.
ciliary r.
clasping r.
cochleo-orbicular r.
cochleopalpebral r.
cochleopapillary r.
cochleostapedial r.
r. control

NOTES

reflex *(continued)*
  coordinated r.
  corneomandibular r.
  corneomental r.
  corneopterygoid r.
  coronary r.
  cortex r.
  cranial r.
  cremasteric r.
  crossed extension r.
  deep r.
  deep tendon r. (DTR)
  defense r.
  deglutition r.
  delayed r.
  depressor r.
  diffused r.
  diving r.
  dorsal r.
  dorsocuboidal r.
  elbow r.
  enterogastric r.
  epigastric r.
  faucial r.
  femoral r.
  finger-thumb r.
  flexor r.
  foveolar r.
  gag r.
  Galant abdominal r.
  galvanic skin r.
  gastrocnemius r.
  gastrocolic r.
  gastroileac r.
  gastroileal r.
  gastropancreatic r.
  Gault cochleopalpebral r.
  gluteal r.
  Gonda r.
  Hering-Breuer r.
  Hirschberg r.
  ileogastric r.
  infraspinatus r.
  inguinal r.
  r. inhibition
  innate r.
  intestinointestinal r.
  intrinsic r.
  ipsilateral r.
  jaw r.
  knee r.
  knee-jerk r.
  labyrinthine r.
  labyrinthine righting r.
  laryngopharyngeal r.
  lash r.
  Liddell-Sherrington r.
  r. ligament
  Lovén r.

  lumbar r.
  magnet r.
  mandibular r.
  masseter r.
  masseteric r.
  micturition r.
  milk-ejection r.
  r. movement
  muscular r.
  myenteric r.
  myotatic r.
  nasolabial r.
  nasomental r.
  neck r.
  nociceptive r.
  oculocardiac r.
  oculocephalic r.
  oculocephalogyric r.
  optical righting r.
  palatal r.
  palatine r.
  patellar tendon r.
  patelloadductor r.
  Pavlov r.
  pectoral r.
  penile r.
  perianal r.
  peripheral r.
  peritoneocutaneous r.
  pharyngeal r.
  phasic r.
  pilomotor r.
  plantar r.
  pneocardiac r.
  pneopneic r.
  postural r.
  proprioceptive r.
  protective laryngeal r.
  psychocardiac r.
  psychogalvanic r.
  pulmonocoronary r.
  pupillary r.
  Puusepp r.
  quadriceps r.
  radial r.
  radiobicipital r.
  rectal r.
  rectoanal inhibitory r. (RAIR)
  renointestinal r.
  renorenal r.
  retrobulbar pupillary r.
  righting r.
  rooting r.
  sacral r.
  scapular r.
  scapulohumeral r.
  skin r.
  skin-muscle r.
  sole r.

R

Somagyi r.
somatointestinal r.
spinal r.
static r.
statokinetic r.
statotonic r.
stretch r.
suckling r.
superficial r.
supinator longus r.
supporting r.
suprapatellar r.
suprapubic r.
supraumbilical r.
swallowing r.
synchronous r.
tarsophalangeal r.
tendo Achilles r.
tendon r.
tibioadductor r.
trace conditioned r.
tracheal r.
triceps surae r.
trigeminus r.
ulnar r.
unconditioned r.
urinary r.
utricular r.
vascular r.
vasopressor r.
venorespiratory r.
vesical r.
vesicointestinal r.
vestibulospinal r.
viscerocardiac r.
viscerosensory r.
viscerotrophic r.
vomiting r.
withdrawal r.
zygomatic r.
reflexa
decidua r.
placenta r.
tunica r.
reflexogenic pressosensitivity
reflexogenous
reflexograph
reflexometer
reflexophil
reflexophile
reflexum
ligamentum r.

reflux
refractory period
regenerate
regeneration
regio, pl. **regiones**
regiones abdominis
r. abdominis lateralis
r. analis
r. antebrachialis anterior
r. antebrachialis posterior
r. antebrachii anterior
r. antebrachii posterior
r. axillaris
r. brachialis anterior
r. brachialis posterior
r. brachii anterior
r. buccalis
r. calcanea
regiones capitis
r. carpalis anterior
r. carpalis posterior
regiones cervicales
r. cervicalis anterior
r. cervicalis lateralis
r. cervicalis posterior
r. colli posterior
regiones corporis
r. cruralis posterior
r. cruris anterior
r. cubitalis anterior
r. cubitalis posterior
r. deltoidea
regiones dorsales
regiones dorsi
r. epigastrica
r. facialis
r. femoralis posterior
r. femoris
r. femoris anterior
r. femoris posterior
r. frontalis capitis
r. genus anterior
r. genus posterior
r. glutealis
r. hypochondriaca
r. infraclavicularis
r. inframammaria
r. infraorbitalis
r. infrascapularis
r. inguinalis
r. lateralis abdominis
r. lumbalis

NOTES

regio *(continued)*
    r. mammaria
    regiones membri inferioris
    regiones membri superioris
    r. mentalis
    r. nasalis
    r. nuchalis
    r. occipitalis capitis
    r. olfactoria tunicae mucosae nasi
    r. oralis
    r. orbitalis
    r. parietalis capitis
    r. pectoralis
    r. perinealis
    r. plantaris
    r. presternalis
    r. pubica
    r. respiratoria tunicae mucosae nasi
    r. sacralis
    r. scapularis
    r. sternocleidomastoidea
    r. suralis
    r. talocruralis
    r. tarsalis
    r. temporalis capitis
    regiones thoracicae anteriores et
      laterales
    r. umbilicalis
    r. urogenitalis
    r. vertebralis
    r. zygomatica

region
    abdominal r.
    anal r.
    ankle r.
    antebrachial r.
    anterior antebrachial r.
    anterior brachial r.
    anterior carpal r.
    anterior cervical r.
    anterior crural r.
    anterior cubital r.
    anterior knee r.
    anterior and lateral thoracic r.
    artery of chiasmal r.
    axillary r.
    r. of back
    r. of body
    brain r.
    Broca r.
    buccal r.
    calcaneal r.
    carpal r.
    centroparietal r.
    r. of chest
    ciliary r.
    clavicular r.
    crural r.
    cubital r.

deltoid r.
dorsal r.
epigastric r.
extrapolar r.
face r.
r. of face
falcine r.
femoral r.
frontal r.
genitourinary r.
gluteal r.
r. of head
heel r.
hilar r.
hyoid r.
hypochondriac r.
hypogastric r.
iliac r.
r. of inferior limb
infraclavicular r.
inframammary r.
infraorbital r.
infrascapular r.
infratemporal r.
inguinal r.
interscapular r.
lateral cervical r.
lateral region of abdominal r.
left hypochondriac r.
left iliac r.
left lateral r.
r. of lower limb
lumbar r.
mammary r.
maxillofacial r.
mental r.
midsacral r.
mylohyoid r.
nasal r.
r. of neck
nuchal r.
occipital r.
ocular r.
r. of olfactory mucosa
opticostriate r.
oral r.
orbital r.
palpebral r.
parietal r.
parietooccipital r.
parietotemporal r.
pectoral r.
perineal r.
pharyngeal r.
r. of pharynx
plantar r.
popliteal r.
postcricoid r.
posterior antebrachial r.

posterior brachial r.
posterior carpal r.
posterior cervical r.
posterior crural r.
posterior cubital r.
posterior knee r.
posterior neck r.
posterotemporal r.
preoptic r.
presternal r.
presumptive r.
pretectal r.
pterygomaxillary r.
pubic r.
pyloric r.
r. of respiratory mucosa
retrocardiac r.
right hypochondriac r.
right iliac r.
right lateral r.
rolandic r.
sacral r.
sacrococcygeal r.
scapular r.
sternocleidomastoid r.
subareolar r.
subauricular r.
subhyoid r.
submaxillary r.
submental r.
suboccipital r.
superior labial r.
r. of superior limb
superior palpebral r.
supraclavicular r.
supraorbital r.
suprapubic r.
sural r.
temporal r.
thoracic r.
thyroid r.
trabecular r.
umbilical r.
r. of upper limb
urogenital r.
vertebral r.
vestibular r.
volar r.
Wernicke r.
zygomatic r.

regional
    r. anatomy
    r. node
regiones (*pl. of* regio)
regulation
regurgitant
regurgitate
regurgitation
rehydration
Reichert
    R. cartilage
    R. cochlear recess
Reid base line
Reil
    R. ansa
    R. band
    circular sulcus of R.
    island of R.
    limiting sulcus of R.
    R. ribbon
    R. triangle
Reinke
    R. crystalloids
    R. space
Reisseisen muscle
Reissner
    R. canal
    R. fiber
    R. membrane
relational threshold
relative
    r. dehydration
    r. humidity
    r. refractory period
    r. viscosity
relax
relaxation
    r. factor
    isometric r.
    isovolumetric r.
    isovolumic r.
    r. response
released substance
releasing
    r. factor
    r. hormone
Remak
    R. fibers
    R. ganglia
    R. plexus
remineralization
remodeling

NOTES

renal
- r. agenesis
- r. artery
- r. branch of lesser splanchnic nerve
- r. branch of vagus nerve
- r. brush border membrane
- r. collar
- r. columns
- r. corpuscle
- r. cortex
- r. cortical lobule
- r. fascia
- r. fossa
- r. ganglia
- r. impression
- r. impression on liver
- r. impression of spleen
- r. labyrinth
- r. lobe
- r. medulla
- r. nervous plexus
- r. papilla
- r. pedicle
- r. pelvis
- r. plexus
- r. portal system
- r. pouch
- r. pyramid
- r. segments
- r. sinus
- r. surface of spleen
- r. surface of suprarenal gland
- r. vein

renalia
- ganglia r.
- segmenta r.

renalis, pl. **renales**
- area cribrosa papillae r.
- arteria r.
- columnae renales
- cortex r.
- fascia r.
- hilum r.
- impressio r.
- lobulus corticalis r.
- lobus r.
- medulla r.
- papilla r.
- papillae renales
- pars tecta r.
- pelvis r.
- plexus r.
- plexus nervosus r.
- pyramides renales
- pyramis r.
- rami capsulares arteriae r.
- rami ureterici arteriae r.
- ramus anterior arteriae r.

- ramus posterior arteriae r.
- sinus r.
- stria externa medullae r.
- stria interna medullae r.
- tunica mucosa pelvis r.
- tunica muscularis pelvis r.
- typus ampullaris pelvis r.
- typus dendriticus pelvis r.
- venae renales
- zona externa medullae r.
- zona interna medullae r.

renculus
renes
renicapsule
renicardiac
reniculus, pl. **reniculi**
reniformis
- placenta r.

reniform pelvis
renin
renin-angiotensin system
reniportal
renis
- adeps r.
- arteriae arcuatae r.
- arteriae perforantes radiatae r.
- arteria interlobulares r.
- arteria segmenti anterioris inferioris r.
- arteria segmenti anterioris superioris r.
- arteria segmenti inferioris r.
- arteria segmenti posterioris r.
- arteria segmenti superioris r.
- basis pyramidis r.
- capsula adiposa r.
- capsula fibrosa r.
- corpusculum r.
- cortex r.
- ectopia r.
- extremitas inferior r.
- extremitas superior r.
- facies anterior r.
- facies posterior r.
- foramina papillaria r.
- interlobares r.
- interlobulares r.
- margo lateralis r.
- margo medialis r.
- pars convoluta lobuli corticalis r.
- pars radiata lobuli corticalis r.
- polus inferior r.
- polus superior r.
- porta r.
- tunica fibrosa r.
- vasa recta r.
- venae arcuatae r.
- venae interlobares r.

venae interlobulares r.
venulae rectae r.

renocutaneous
renogastric
renointestinal reflex
renopulmonary
renorenal reflex
renotrophic
renotrophin
renotropic
renotropin
renovascular
Renshaw cells
renunculus
reoptic nucleus
replacement bone
replicate
replication
replisome
repolarization
repositio
reposition
reproduction
asexual r.
cytogenic r.
sexual r.
somatic r.
reproductive
r. cycle
r. organ
r. system
repulsion
reserve
r. air
breathing r.
r. force
reservoir
fecal r.
ileal r.
Pecquet r.
r. of spermatozoa
urinary r.
residual
r. air
r. capacity
r. cleft
r. lumen
r. volume
resilience
resin
methacrylate r.

resistance
airway r.
expiratory r.
mutual r.
peripheral r.
synaptic r.
total peripheral r.
resistor
resorb
resorption
bone r.
r. lacunae
respirable
respiration
abdominal r.
aerobic r.
anaerobic r.
artificial r.
assisted r.
Cheyne-Stokes r.
cogwheel r.
controlled r.
costal r.
external r.
forced r.
internal r.
interrupted r.
jerky r.
nitrate r.
puerile r.
r. rate
sulfate r.
thoracic r.
tissue r.
respiratoria
glottis r.
rima r.
respiratorii
bronchioli r.
respiratorium
systema r.
respiratorius
apparatus r.
respiratory
r. acidosis
r. airway
r. alkalosis
r. apparatus
r. bronchioles
r. burst
r. capacity
r. center

R

NOTES

respiratory *(continued)*
- r. coefficient
- r. dead space
- r. epithelium
- r. exchange ratio
- r. frequency
- r. insufficiency
- r. lobule
- r. metabolism
- r. minute volume
- r. mucosa
- r. quotient
- r. rate
- r. region of mucosa of nasal cavity
- r. region of tunica mucosa of nose
- r. system
- r. tract

respire

respirometer
- Dräger r.
- Wright r.

response
- Cushing r.
- depletion r.
- evoked r.
- flight or fight r.
- galvanic skin r. (GSR)
- Henry-Gauer r.
- immune r.
- oculomotor r.
- plantar r.
- psychogalvanic r. (PGR)
- recruiting r.
- relaxation r.
- sonomotor r.
- unconditioned r.

rest
- mesonephric r.
- wolffian r.

restiform
- r. body
- r. eminence

restiforme
- corpus r.

restiformis
- eminentia r.

resting
- r. length
- r. saliva
- r. tidal volume
- r. wandering cell

restored cycle

restructured cell

resuscitate

resuscitation

retained
- r. placenta
- r. testis

retardation

retch

retching

rete, pl. **retia**
- r. acromiale
- r. acromiale arteriae thoracoacromialis
- r. arteriosum
- r. articulare cubiti
- r. articulare genus
- r. calcaneum
- r. canalis hypoglossi
- r. carpale dorsale
- r. carpi dorsale
- r. carpi posterius
- r. cords
- r. cutaneum corii
- r. foraminis ovalis
- Haller r.
- r. halleri
- r. malleolare laterale
- r. malleolare mediale
- malpighian r.
- r. mirabile
- r. ovarii
- r. patellae
- r. patellare
- r. peg
- r. plexus
- r. ridge
- r. subpapillare
- r. testis
- r. vasculosum articulare
- r. venosum
- r. venosum dorsale manus
- r. venosum dorsale pedis
- r. venosum plantare

retention

retial

reticula (*pl. of* reticulum)

reticular
- r. activating system (RAS)
- r. cartilage
- r. cell
- r. fibers
- r. formation
- r. lamina
- r. layer of corium
- r. membrane
- r. membrane of spinal organ
- r. nuclei of brainstem
- r. nucleus of thalamus
- r. substance
- r. tissue

reticulare
- magma r.

R

reticularis
  r. cell
  fetal r.
  formatio r.
  membrana r.
  substantia r.
  zona r.
reticulated
  r. bone
  r. corpuscle
reticulin
reticulocyte
reticuloendothelial
  r. cell
  r. system
reticuloendothelium
reticulofilamentosa
  substantia r.
reticulospinalis
  tractus r.
reticulospinal tract
reticulum, pl. **reticula**
  agranular endoplasmic r.
  cistern of cytoplasmic r.
  Ebner r.
  endoplasmic r.
  Golgi internal r.
  granular endoplasmic r.
  Kölliker r.
  rough-surfaced endoplasmic r.
  sarcoplasmic r.
  smooth-surfaced endoplasmic r.
  stellate r.
  trabecular r.
  r. trabeculare
  r. trabeculare sclerae
  uveal part of trabecular r.
retiform
  r. cartilage
  r. tissue
retina
  central artery of r.
  central vein of r.
  cerebral layer of r.
  ciliary part of r.
  cone cell of r.
  foveola of r.
  ganglionic layer of r.
  granular layers of r.
  horizontal cells of r.
  inferior nasal arteriole of r.
  inferior nasal venule of r.

inferior temporal arteriole of r.
inferior temporal venule of r.
iridial part of r.
layers of r.
limiting membrane of r.
macula of r.
medial arteriole of r.
medial venule of r.
molecular layer of r.
nasal venules of r.
nervous part of r.
neural layer of optic part of r.
neuroepithelial layer of r.
nonoptic part of r.
nuclear layers of r.
optic part of r.
pigmented layer of r.
pigmented part of r.
pigment epithelium of optic r.
plexiform layers of r.
rod cell of r.
superior nasal arteriole of r.
superior nasal venule of r.
superior temporal arteriole of r.
superior temporal venule of r.
temporal arteriole of r.
temporal venules of r.
retinaculi
retinaculum, pl. **retinacula**
  antebrachial flexor r.
  r. of articular capsule of hip
  r. capsulae articularis coxae
  caudal r.
  r. caudale
  r. cutis
  r. cutis mammae
  deep part of flexor r.
  dorsal r.
  extensor r.
  retinacula of extensor muscle
  flexor r.
  r. of flexor muscle
  r. flexorum
  inferior extensor r.
  inferior fibular r.
  inferior peroneal r.
  lateral patellar r.
  medial patellar r.
  Morgagni r.
  r. musculorum extensorum
  r. musculorum extensorum inferius
  r. musculorum extensorum superius

NOTES

527

retinaculum *(continued)*
    r. musculorum fibularium
    r. musculorum fibularium inferius
    r. musculorum fibularium superius
    r. musculorum flexorum
    r. musculorum flexorum membri
      inferioris
    r. musculorum peroneorum
    r. musculorum peroneorum inferius
    r. musculorum peroneorum superius
    retinacula of nail
    r. patellae laterale
    r. patellae mediale
    patellar r.
    peroneal r.
    retinacula of peroneal muscle
    r. of skin
    superior extensor r.
    superior fibular r.
    superior peroneal r.
    r. tendinum
    retinacula unguis
    vesical r.
retinae
    arteria centralis r.
    arteriola medialis r.
    centralis r.
    fovea centralis r.
    foveola r.
    macula r.
    ora serrata r.
    pars ceca r.
    pars ciliaris r.
    pars extraocularis arteriae et venae
      centralis r.
    pars iridica r.
    pars nervosa r.
    pars optica r.
    raphe r.
    stratum cerebrale r.
    stratum ganglionare r.
    stratum moleculare r.
    stratum neuroepitheliale r.
    stratum nucleare externum r.
    stratum nucleare externum et
      internum r.
    stratum nucleare internum r.
    stratum pigmenti r.
    superior venula nasalis r.
    vasa sanguinea r.
    vena centralis r.
    venula medialis r.
retinal
    r. artery
    r. blood vessel
    r. cone
    r. fold
    r. rod
    r. vein

retinochoroid
retinoid
retoperithelium
retract
retractile
retraction
retractor bulbi muscle
retrad
retrahens
    r. aurem
    r. auriculam
retroadductor space
retroauricular
    r. lymph node
    r. sulcus
retrobuccal
retrobulbar
    r. fat
    r. pupillary reflex
retrocalcaneal bursa
retrocardiac
    r. region
    r. space
retrocaval ureter
retrocecal
    r. appendix
    r. lymph node
    r. recess
retrocecalis, pl. **retrocecales**
    lymphonodi retrocecales
    nodi lymphoidei retrocecales
    recessus r.
retrocentral sulcus
retrocervical
retrocochlear
retrocolic
retrocursive
retrocuspid papilla
retroduodenal
    r. artery
    r. fossa
    r. recess
retroduodenalis
    arteria r.
    recessus r.
retroesophageal
    r. aorta
    r. artery
    r. space
retroflex fasciculus
retroflexus
    fasciculus r. (FRF)
retroglandular sulcus
retrograde metamorphosis
retrogression
retrohyoid bursa
retrohyoidea
    bursa r.

retroinguinale
  spatium r.
retroinguinalis
  lacuna musculorum r.
  lacuna vasorum r.
retroinguinal space
retroiridian
retrolental space
retrolenticular limb of internal capsule
retrolingual
retromammary space
retromandibular
  r. fossa
  r. process of parotid gland
  r. vein
retromandibularis
  fossa r.
  processus r.
  vena r.
retromanubrial
retromastoid
retromolar
  r. fossa
  r. pad
  r. papilla
  r. triangle
  r. trigone
retromolare
  trigonum r.
retromolaris
  fossa r.
retromorphosis
retromylohyoid space
retronasal
retroocular
retroparotid space
retropatellar fat pad
retroperitoneal
  r. cavity
  r. lymph node
  r. pancreas
  r. part of duodenum
  r. space
  r. vein
retroperitoneale
  spatium r.
retroperitoneales
  venae r.
retroperitoneum
retropharyngeal
  r. lymph node
  r. space

retropharyngeales
  lymphonodi r.
  nodi lymphoidei r.
retropharyngeum
  spatium r.
retropharynx
retroplacental
retroplasia
retropubic space
retropubicum
  spatium r.
retropylorici
  nodi r.
  nodi lymphoidei r.
retropyloric lymph node
retrorectal
  r. lamina of endopelvic fascia
  r. lamina of hypogastric sheath
  r. lymph node
retrosternal
  r. fat pad
  r. gland
  r. hernia
  r. thyroid
retrotarsal fold
retrouterine
retrovaginal septum
retrovisceral
  r. fascia
  r. space
retrozonulare
  spatium r.
retrozonular space
return
  venous r.
returning cycle
retzii
  cavum r.
Retzius
  calcification lines of R.
  R. cavity
  R. foramen
  R. gyrus
  gyrus olfactorius lateralis of R.
  gyrus olfactorius medialis of R.
  R. ligament
  lines of R.
  sheath of Key and R.
  space of R.
  R. space
  R. striae
  R. vein

R

NOTES

reuniens
    canaliculus r.
    canalis r.
    r. duct
    ductus r.
    saccus r.
    sinus r.
reunient
revehens, pl. **revehentes**
    venae r.
reverberating circuit
reversal
    adrenaline r.
    epinephrine r.
reverse
    r. Eck fistula
    r. osmosis
revivescence
revivification
Reynolds number
rhabdocyte
rhabdoid
rhabdosphincter
rhagiocrine cell
rhaphe
rheobase
rheobasic
rheocardiography
rheoencephalogram
rheoencephalography
rheogram
rheology
rheometer
rheometry
rheopexy
rheostat
rheotaxis
rheotropism
rhinal
    r. fissure
    r. sulcus
rhinalis
    sulcus r.
rhinencephalic
rhinencephalon
rhinion
rhinoanemometer
rhinocele
rhinocephalia
rhinocephaly
rhinodymia
rhinomanometer
rhinomanometry
rhinopharyngeal
rhinopharynx
rhizoid
rhizomelia
rhodophylactic
rhodophylaxis

rhombencephali
    isthmus r.
    tegmentum r.
rhombencephalic
    r. gustatory nucleus
    r. isthmus
rhombencephalon
    tegmentum of r.
    ventricle of r.
rhombic
    r. groove
    r. lip
rhomboatloideus
    musculus r.
rhomboid
    r. fossa
    r. impression
    r. ligament
    r. major muscle
    r. of Michaelis
    r. minor muscle
    nerve to r.
rhomboidal
rhomboidea, pl. **rhomboideae**
    fossa r.
    sulcus limitans fossae rhomboideae
rhomboideus
    r. major
    r. minor
rhombomere
rhonchus, pl. **rhonchi**
rhopheocytosis
rhythm
    basic electrical r. (BER)
    circadian r.
    circus r.
    diurnal r.
    ultradian r.
rib
    angle of anterior r.
    angle of posterior r.
    anterior angle of r.
    articular facet of head of r.
    articular facet of tubercle of r.
    articular surface of head of r.
    articular surface of tubercle of r.
    bicipital r.
    bifid r.
    body of r.
    r. cage
    cervical r.
    costal groove of r.
    costochondral articulation of r.
    crest of body of r.
    crest of head of r.
    crest of neck of r.
    elevator muscle of r.
    false r.
    first r.

floating r.
groove for r.
head of r.
inferior articular facet of r.
intraarticular ligament of head
  of r.
joint of head of r.
lumbar r.
neck of r.
r. notch
posterior angle of r.
proximal end of r.
radiate ligament of r.
radiate ligament of head of r.
rudimentary r.
superior articular facet of r.
superior margin of r.
true r.
tubercle of articular facet of r.
vertebral r.
vertebrochondral r.
vertebrosternal r.
Zahn r.

ribbon
Reil r.
Ribes ganglion
rib-vertebral angle
rice diet
Richard fringe
Richter-Monro line
rider's
r. bone
r. muscle
r. tendon
ridge
alveolar r.
basal r.
bicipital r.
bladder neck r.
bulbar r.
bulboventricular r.
crest of alveolar r.
dermal r.
epicondylar r.
epidermal r.
epipericardial r.
external oblique r.
r. formed by esophagus
ganglion r.
genital r.
gluteal r.
gonadal r.

intertrochanteric r.
interureteric r.
key r.
lateral epicondylar r.
lateral supracondylar r.
lateral supraepicondylar r.
Mall r.
mammary r.
mandibular r.
marginal r.
maxillary r.
medial epicondylar r.
medial supracondylar r.
medial supraepicondylar r.
mesonephric r.
milk r.
mylohyoid r.
nasal r.
orbital r.
Outerbridge r.
palatine r.
Passavant r.
pectoral r.
petrous r.
pharyngeal r.
primitive r.
pronator r.
pulmonary r.
rete r.
skin r.
sphenoidal r.
superciliary r.
supraorbital r.
taste r.
temporal r.
transverse palatine r.
trapezoid r.
triangular r.
trochlear r.
ureteric r.
urogenital r.
vomerine r.
wolffian r.
Ridley
R. circle
R. sinus
ridleyi
circulus venosus r.
Riedel
R. lobe
R. lobe of liver

R

NOTES

Riedel *(continued)*
R. lobe right and left fibrous
rings of heart
Rieder cells
rietale
right
r. anterior lateral hepatic segment
r. anterior medial hepatic segment
r. aortic arch
r. atrial branch of right coronary
artery
r. atrioventricular orifice
r. atrioventricular valve
r. atrium
r. atrium of heart
r. auricle
r. auricular appendage
r. border of heart
r. brachiocephalic vein
r. branch
r. branch of hepatic artery proper
r. branch of portal vein
r. bronchomediastinal trunk
r. bundle of atrioventricular bundle
r. colic artery
r. colic flexure
r. colic lymph node
r. colic vein
r. colon
r. colonic flexure
r. common iliac nerve
r. coronary valve
r. crus of atrioventricular bundle
r. crus of atrioventricular trunk
r. crus of diaphragm
r. dome of diaphragm
r. duct of caudate lobe
r. duct of caudate lobe of liver
r. ear (AD)
r. femoral artery
r. fibrous trigone
r. fibrous trigone of heart
r. flexural artery
r. gastric artery
r. gastric lymph node
r. gastric vein
r. gastroepiploic artery
R. gastroepiploic lymph node
r. gastroepiploic vein
r. gastroomental artery
r. gastroomental lymph node
r. gastroomental vein
r. gonadal vein
r. gutter
r. hepatic artery
r. hepatic duct
r. hepatic vein
r. hilum
r. hypochondriac region

r. hypogastric nerve
r. iliac region
r. inferior abdominal quadrant
r. inferior gluteal vein
r. inferior pulmonary vein
r. internal jugular vein
r. kidney
r. lateral division of liver
r. lateral region
r. lower lobe (RLL)
r. lower quadrant (RLQ)
r. lumbar lymph node
r. lumbar trunk
r. lumbosacral trunk
r. lymphatic duct
r. lymphatic trunk
r. main bronchus
r. marginal branch of right
coronary artery
r. margin of heart
r. medial division of liver
r. middle suprarenal artery
r. midinguinal line
r. obturator artery
r. ovarian artery
r. ovarian vein
r. ovary
r. part of diaphragmatic surface of
liver
r. pericardiacophrenic vein
r. phrenic vein
r. plate of thyroid cartilage
r. posterior lateral hepatic segment
r. posterior medial hepatic segment
r. prostatic ligament
r. pterygoid process
r. pulmonary artery
r. sagittal fissure
r. septal valve
r. sigmoid sinus
r. subclavian artery
r. subclavian vein
r. superior abdominal quadrant
r. superior intercostal vein
r. superior pulmonary vein
r. suprarenal vein
r. sympathetic trunk
r. testicular artery
r. testicular vein
r. triangle ligament of liver
r. triangular ligament
r. triangular ligament of liver
r. umbilical fold
r. upper lobe (RUL)
r. upper quadrant (RUQ)
r. ureter
r. venous angle
r. ventricle
r. ventricular hypoplasia

R

r. ventricular inflow tract
r. ventricular outflow tract
righting reflex
right/left
r./l. pulmonary surfaces of heart
r./l. ventricles of heart
rigidity
cadaveric r.
clasp-knife r.
postmortem r.
rigor mortis
rim
acetabular r.
helical r.
intercartilaginous r.
orbital r.
outer r.
palpebral r.
scleral r.
rima, pl. **rimae**
glottidis r.
r. glottidis
r. oris
r. palpebrarum
palpebrarum r.
r. pudendi
r. respiratoria
r. vestibuli
r. vocalis
r. vulvae
rimose
rimula
Rindfleisch
R. cells
R. fold
ring
abdominal r.
amnion r.
anterior limiting r.
aortic r.
apex of external r.
atrioventricular r.
Bickel r.
Cannon r.
cardiac lymphatic r.
cartilaginous r.
ciliary r.
common annular r.
common tendinous r.
conjunctival r.
constriction r.
crural r.

deep inguinal r.
external inguinal r.
femoral r.
fibrocartilaginous r.
fibrous r.
r. finger
hymenal r.
Imlach r.
inguinal r.
intercrural fibers of superficial r.
internal abdominal r.
internal inguinal r.
r. of iris
lateral crus of the superficial inguinal r.
r. ligament
Lower r.
lymphoid r.
medial crus of the superficial inguinal r.
pharyngeal lymphatic r.
pyloric r.
Schatzki r.
Schwalbe r.
subcutaneous r.
superficial inguinal r.
tendinous r.
tonsillar r.
tracheal r.
tympanic r.
umbilical r.
vascular r.
Vieussens r.
Waldeyer throat r.
Zinn r.
Riolan
R. anastomosis
R. arc
R. arcade
R. arch
R. bone
R. bouquet
R. muscle
riosus cerebri
riparian
risorius
musculus r.
Ritter
R. law
R. opening tetanus
Ritter-Rollet phenomenon

NOTES

rivini
    incisura r.
rivinian
    r. foramen
    r. notch
    r. segment
Rivinus
    R. canal
    R. duct
    R. gland
    R. incisure
    R. membrane
    R. notch
    R. segment
rivus lacrimalis
RLL
    right lower lobe
RLQ
    right lower quadrant
Robert pelvis
rod
    r. cell of retina
    Corti r.
    r. disks
    enamel r.
    r. fiber
    r. granule
    r. nuclear cell
    retinal r.
Röhrer index
Rohr stria
rolandic
    r. cortex
    r. region
    r. sulcal artery
    r. vein
Rolando
    R. area
    R. cells
    R. column
    fasciculus of R.
    fissure of R.
    R. fissure
    R. gelatinous substance
    R. tubercle
    R. vein
roll
    scleral r.
Roller nucleus
Rolleston rule
roof
    r. of fourth ventricle
    r. of mouth
    r. nucleus
    r. of orbit
    r. of skull
    r. of tympanic cavity
    r. of tympanum
roofplate

root
    accessory nerve r.
    ansa cervicalis r.
    anterior nerve r.
    aortic r.
    r. apex
    r. canal of tooth
    ciliary ganglion r.
    r. of ciliary ganglion
    clinical r.
    conjoined nerve r.
    cranial r.'s
    dorsal nerve r.
    facial r.
    facial nerve r.
    r. of facial nerve
    r. filaments
    r. of foot
    r. foramen
    glossopharyngeal nerve r.
    hair r.
    inferior r.
    r. of inferior nasal concha
    r. of lung
    r. of mesentery
    r. of nail
    nasociliary r.
    nerve r.
    r. of nose
    olfactory r.
    r. of penis
    posterior r.
    r. pulp
    r. sheath
    spinal r.
    superior r.
    r. tip
    tip of tooth r.
    r. of tongue
    r. of tooth
    r.'s of trigeminal nerve
    trigeminal nerve r.
    vagus nerve r.
    ventral nerve r.
rooting reflex
rootlet
    nerve r.
Rose
    lateral mamillary nucleus of R.
Rosenbach law
Rosenmüller
    R. body
    fossa of R.
    R. fossa
    R. gland
    R. node
    organ of R.
    R. recess
    R. valve

R

Rosenthal
R. ascending vein
basal vein of R.
R. canal
R. fiber
Roser-Nélaton line
rosette-forming cells
rostra (*pl. of* rostrum)
rostrad
rostral
r. aspect
r. lamina
r. layer
r. neuropore
r. pole
rostralis
lamina r.
pars anterior commissurae r.
rostrally
rostrate
rostriform
rostrum, pl. **rostra**
r. corporis callosi
r. of sinus
r. of sphenoid
sphenoidal r.
r. sphenoidale
r. of the sphenoid bone
rotameter
rotary joint
rotation
r. center
center of r.
external r.
internal r.
intestinal r.
rotator
r. cuff muscle
r. cuff of shoulder
medial r.
musculus r.
rotatores
r. cervicis muscle
r. lumborum muscle
musculi r.
r. spinae
r. thoracis muscle
rotatory joint
Roth
vas aberrans of R.
R. vas aberrans

rotoplasmatic
rotunda
fenestra r.
fossula r.
rotundum
dorsum r.
foramen r.
rotundus
fasciculus r.
Rouget
R. bulb
R. cell
R. muscle
Rouget-Neumann sheath
rough line
rough-surfaced endoplasmic reticulum
Roughton-Scholander apparatus
round
r. eminence
r. foramen
r. ligament of elbow joint
r. ligament of femur
r. ligament of liver
r. ligament of uterus
r. pelvis
r. pronator muscle
r. uterine ligament
r. window
Rouviere
node of R.
ruber
nucleus r.
Rubner laws of growth
rubra
medulla ossium r.
rubriblast
rubrobulbar tract
rubroreticulares
fasciculi r.
rubroreticular tract
rubrospinal
r. decussation
r. tract
rubrospinalis
tractus r.
rudiment
rudimentary
r. bone
r. disk space
r. eye
r. rib

NOTES

rudimentary *(continued)*
    r. uterus
    r. vagina
rudimentum, pl. **rudimenta**
    r. hippocampi
Ruffini
    R. corpuscles
    flower-spray organ of R.
rufflec canal
ruga, pl. **rugae**
    rugae fold
    rugae of gallbladder
    gastric r.
    r. gastrica
    r. gastricae
    r. palatina
    r. of stomach
    stomach r.
    transverse mucosal r.
    rugae of vagina
    vaginal r.
    r. vaginales
    rugae vaginales
    rugae vesicae biliaris
rugal
    r. columns of vagina

    r. fold
    r. pattern
    r. pattern of stomach
rugarum
    columnae r.
rugose
rugosity
rugous
RUL
    right upper lobe
rule
    phase r.
    Rolleston r.
RUQ
    right upper quadrant
Russell
    hooked bundle of R.
    uncinate bundle of R.
    uncinate fasciculus of R.
Ruysch
    R. membrane
    R. muscle
    R. tube
    R. vein
ryanodine receptor

S

S potential

S-A

sinoatrial node
S-A node

sac

abdominal s.
air s.
allantoic s.
alveolar s.
amniotic s.
aortic s.
caudal s.
chorionic s.
conjunctival s.
cupular blind s.
dental s.
dural s.
endolymphatic s.
fibrous s.
fornix of lacrimal s.
fossa of lacrimal s.
greater peritoneal s.
heart s.
hernial s.
Hilton s.
lacrimal s.
s. of lesser omental bursa
lesser peritoneal s.
lymph s.
nasal s.
nasolacrimal s.
omental s.
pericardial s.
peritoneal s.
pleural s.
preputial s.
pudendal s.
scrotal s.
serous s.
subarachnoid s.
superior recess of lesser
    peritoneal s.
synovial s.
tear s.
tooth s.
vestibular blind s.
vitelline s.
yolk s.

saccate
saccharometabolic
saccharometabolism
sacci (*pl. of* saccus)
sacciform

s. recess

s. recess of distal radioulnar joint
s. recess of elbow joint

saccular

s. duct
s. gland
s. nerve
s. recess of bony labyrinth
s. spot

saccularis

ductus s.
nervus s.

sacculated
sacculation of colon
saccule

air s.
s. of ear
laryngeal s.
s. of larynx
macula of s.
maculae of utricle and s.
vestibular s.

sacculocochlear
sacculus, pl. **sacculi**

s. alveolaris
s. communis
s. endolymphaticus
s. lacrimalis
s. laryngis
macula sacculi
s. proprius
s. vestibuli

saccus, pl. **sacci**

s. conjunctiva
s. conjunctivalis
s. endolymphaticus
s. lacrimalis
s. reuniens
s. vaginalis

sacra (*pl. of* sacrum)
sacrad
sacral

s. ala
s. bone
s. canal
s. cornu
s. crest
s. cul-de-sac
s. flexure
s. flexure of rectum
s. foramen
s. hiatus
s. horn
s. index
s. kyphosis
s. lymph node

S

sacral *(continued)*
    s. part of spinal cord
    s. prominence
    s. promontory
    s. reflex
    s. region
    s. spine
    s. splanchnic nerve
    s. triangle
    s. tuberosity
    s. vein
    s. venous plexus
    s. vertebra
sacrale
    cornu s.
    foramen s.
sacrales
    s. laterales
    lymphonodi s.
    nervi splanchnici s.
    nodi lymphoidei s.
    splanchnici s.
    vertebrae s.
sacralia
    ganglia s.
sacralis
    ala s.
    ansa s.
    canalis s.
    crista s.
    hiatus totalis s.
    s. mediana
    plexus s.
    plexus venosus s.
    regio s.
    segmentum medullae spinalis s.
    tuberositas s.
sacralium
    rami anteriores nervorum s.
    rami ventrales nervorum s.
sacred bone
sacri
    apex ossis s.
    basis ossis s.
    facies auricularis ossis s.
    facies dorsalis ossis s.
    facies pelvica ossis s.
    lineae transversae ossis s.
    pars lateralis ossis s.
    processus articularis superior
      ossis s.
    promontorium ossis s.
sacrococcygea
    articulatio s.
    junctura s.
    symphysis s.
sacrococcygeal
    s. articulation
    s. cyst

    s. disk
    s. joint
    s. junction
    s. muscle
    s. region
    s. sinus
sacrococcygeus
    s. dorsalis
    s. ventralis
sacrodurale
    ligamentum s.
sacrodural ligament
sacrogenital
    s. fold
    s. ligament
sacrohorizontal angle
sacroiliac
    s. articulation
    s. joint
    s. notch
sacroiliaca
    articulatio s.
sacrolumbalis
    musculus s.
sacrolumbar
sacropelvic surface of ilium
sacroperineal
sacrosciatic notch
sacrospinale
    ligamentum s.
sacrospinalis
    musculus s.
sacrospinal muscle
sacrospinosum
    ligamentum s.
sacrospinous ligament
sacrotuberale
    ligamentum s.
sacrotuberalis
    processus falciformis ligamenti s.
sacrotuberosum
    ligamentum s.
sacrotuberous ligament
sacrouterine
    s. fold
    s. ligament
sacrovaginal fold
sacrovertebral
sacrovesical fold
sacrum, pl. **sacra**
    ala of s.
    apex of s.
    assimilation s.
    auricular surface of s.
    base of s.
    dorsal surface of s.
    lateral part of s.
    os s.
    pelvic surface of s.

promontory of s.
superior articular facet of s.
superior articular process of s.
superior facet of s.
transverse line of s.
transverse ridge of s.
wing of s.

saddle
s. area
s. head
s. joint
tubercle of s.
Turkish s.

safety
triangle of s.

safranophil, safranophile

sagitta

sagittal
s. area
s. axis
s. border of parietal bone
s. fontanelle
s. groove
s. plane
s. section
s. sulcus
s. suture
s. suture line
s. synostosis
s. venous sinus

sagittalia
plana s.

sagittalis
s. inferior sinus
norma s.
s. superior sinus
sutura s.

salient

saliva
chorda s.
ganglionic s.
resting s.
sublingual s.
sympathetic s.

salivant

salivaria
glandula s.

salivaris
caruncula s.

salivary
s. corpuscle
s. duct

s. gland
s. gland hormone

salivate

salivator

salpinges (*pl. of* salpinx)

salpingian

salpingopalatina
plica s.

salpingopalatine fold

salpingopharyngea
plica s.

salpingopharyngeal
s. fascia
s. fold
s. muscle

salpingopharyngeus
s. muscle
musculus s.

salpinx, pl. **salpinges**
s. uterina

salt
s. action
s. depletion
s. wasting

saltation

saltatory conduction

saltum
per s.

Salus arch

salvatella vein

sample
end-tidal s.
Haldane-Priestley s.
Rahn-Otis s.

sand
brain s.

Sandström body

sanguifacient

sanguiferous

sanguification

sanguine

sanguineous

sanguineum
vas s.

sanguis

Santorini
S. canal
cartilage of S.
S. cartilage
concha of S.
S. concha
S. duct

S

NOTES

Santorini (*continued*)
S. fissure
incisura S.
S. incisure
S. labyrinth
S. major caruncle
S. minor caruncle
S. muscle
S. parietal vein
plexus of S.
S. plexus
S. tubercle
S. vein

sap
nuclear s.

saphena
s. accessoria
s. magna
s. parva

sapheni
cornu inferius marginis falciformis
hiatus s.
crus inferius marginis falciformis
hiatus s.
crus superius marginis falciformis
hiatus s.
margo arcuatus hiatus s.
margo falciformis hiatus s.
rami cutanei cruris mediales
nervi s.
ramus infrapatellaris nervi s.

saphenofemoral
s. juncture

saphenous
s. branch of descending genicular
artery
s. hiatus
s. nerve
s. opening
s. pulse
s. system
s. vein

saphenus
cornu inferius hiatus s.
cornu superius hiatus s.
hiatus s.
nervus s.
ramus s.

sapiens
Homo s.

sapientiae
dens s.

saponification
saponify
Sappey
S. fiber
S. plexus
S. vein

sarcoblast

sarcode
sarcogenic cell
sarcoglia
sarcolemma
sarcolemmal
sarcolemmic
sarcolemmous
sarcology
sarcomere
sarcoplasm
sarcoplasmic reticulum
sarcoplast
sarcopoietic
sarcosome
sarcotubules
sarcous
sartorii
bursae subtendineae musculi s.

sartorius
s. bursa
s. muscle
musculus s.
subtendinous bursa of s.
s. tendon

SASMAS
skin-adipose superficial
musculoaponeurotic system

satellite
nucleolar s.
perineuronal s.

satellite-rich heterochromatin
Sattler elastic layer
saturation
satyri
apex s.

sauriasis
sauriderma
sauriosis
sauroderma
Savage perineal body
SBRN
sensory branch of radial nerve

SC
supracondylar

S&C
sclera and conjunctiva

scala, pl. **scalae**
cochlear scalae
Löwenberg s.
s. media
s. tympani
s. vestibuli

scale
homigrade s.
ratio s.

scalene
s. fascia
s. hiatus
s. lymph node

s. muscle
s. tubercle
s. tubercle of Lisfranc
scalene-vertebral triangle
scalenus
s. anterior
s. anterior muscle
s. anticus
s. anticus syndrome
s. medius
s. medius muscle
s. minimus
s. minimus muscle
s. posterior
s. posterior muscle
scalp
epicranial aponeurosis of s.
s. hair
s. muscle
subcutaneous s.
scalpriform
scansorius
musculus s.
scapha, pl. **scaphae**
eminence of s.
eminentia scaphae
scaphocepha
scaphocephalic
scaphocephalism
scaphocephalous
scaphocephaly
scaphoid
s. fossa
s. fossa of sphenoid bone
s. radius
s. scapula
s. tuberosity
scaphoidea
facies s.
fossa s.
scaphoidei
tuberculum ossis s.
scaphoideum
os s.
scaphotrapeziotrapezoid (STT)
s. joint
scapula, pl. **scapulae**
acromion scapulae
alar s.
angle of inferior s.
angle of superior s.
angulus inferior scapulae

angulus lateralis scapulae
angulus superior scapulae
arteria circumflexa scapulae
arteria dorsalis scapulae
cavitas glenoidalis scapulae
circumflexa scapulae
circumflex artery of s.
collum scapulae
coracoid process of s.
costal surface of s.
deltoid tubercle of spine of s.
dorsalis scapulae
dorsal nerve of s.
dorsal surface of s.
dorsum scapulae
s. elevata
elevator muscle of s.
facies costalis scapulae
facies dorsalis scapulae
facies posterior scapulae
glenoid cavity of s.
glenoid labrum of s.
incisura scapulae
inferior angle of s.
infraglenoid tubercle of s.
labrum glenoidale scapulae
lateral angle of s.
lateral border of s.
levator scapulae
margo lateralis scapulae
margo medialis scapulae
margo superior scapulae
medial border of s.
musculus levator scapulae
musculus levator anguli scapulae
neck of s.
nervus dorsalis scapulae
posterior surface of s.
scaphoid s.
spina scapulae
spine of s.
superior angle of s.
superior border of s.
supraglenoid tubercle of s.
transverse artery of s.
transverse vein of s.
tuberculum deltoideum spinae
scapulae
tuberculum infraglenoidale scapulae
tuberculum supraglenoidale scapulae
vena transversa scapulae
vertebral border of s.

S

NOTES

scapular
    s. area
    s. artery
    s. line
    s. muscle
    s. notch
    s. reflex
    s. region
scapularis
    linea s.
    regio s.
scapulary
scapuloclavicular
    s. articulation
    s. joint
scapulohumeral
    s. muscle
    s. reflex
scapulohumerales
    musculi s.
scapulothoracic bursa
scapus
    s. penis
    s. pili
Scarpa
    S. fascia
    S. fluid
    S. foramen
    foramina of S.
    S. ganglion
    S. habenula
    S. hiatus
    S. ligament
    S. liquor
    S. membrane
    S. nerve
    S. sheath
    S. triangle
scavenger
    s. cell
    s. receptor
Schacher ganglion
Schatzki ring
schematic
schematograph
Schiff-Sherrington phenomenon
Schilling band cell
schindylesis
schindyletic joint
schisto
schistocelia
schistocormia
schistorrhachis
schistosomia
schistosternia
schistothorax
schizamnion
schizaxon
schizencephalic microcephaly

schizencephaly
schizogenesis
schizogony
schizogyria
schizotonia
Schlemm canal
Schmidt-Lanterman
    S.-L. clefts
    S.-L. incisure
Schmorl
    S. body
    S. furrow
    S. node
schneiderian membrane
Scholander apparatus
Schreger line
Schüller duct
Schultze
    S. cells
    comma bundle of S.
    comma tract of S.
    S. fold
    S. membrane
    S. placenta
Schütz
    S. bundle
    tract of S.
Schwalbe
    S. corpuscle
    S. foramen
    S. line
    S. membrane
    S. nucleus
    S. ring
    S. sheath
    S. space
Schwann
    S. cells
    sheath of S.
    S. white substance
Schweigger-Seidel
    sheath of S.-S.
sciatic
    s. artery
    s. bursa of gluteus maximus
    s. foramen
    s. nerve
    s. notch
    s. plexus
    s. spine
sciaticus
    nervus s.
scission
scissiparity
scissura, pl. **scissurae**
scissure
sclera, pl. **sclerae**
    calcar sclerae
    s. and conjunctiva (S&C)

corneoscleral part of trabecular tissue of s.

foramen of s.

lamina cribrosa of s.

lamina cribrosa sclerae

lamina fusca sclerea

pars corneoscleralis reticuli trabecularis sclerae

pars uvealis reticuli trabecularis sclerae

perforated layer of s.

reticulum trabeculare sclerae

sinus venosus sclerae

substantia propria of s.

sulcus of s.

superficial s.

suprachoroid lamina of s.

trabecular tissue of s.

uveal part of trabecular tissue of s.

venous sinus of s.

scleral

s. bed

s. canal

s. pouch

s. rim

s. roll

s. spur

s. sulcus

s. vein

s. venous sinus

sclerales

venae s.

scleroblastema

sclerochoroidal

scleroconjunctival

sclerocornea

sclerocorneal junction

scleromere

sclerophthalmia

sclerosis, pl. **scleroses**

endocardial s.

lobar s.

sclerotica

tunica s.

sclerotic coat

scoliosis

scotopsin

screw

s. arteries

s. joint

scriptorius

calamus s.

scrobiculus cordis

scroll bone

scrota (*pl. of* scrotum)

scrotal

s. area

s. artery

s. compartment

s. nerve

s. part of ductus deferens

s. raphe

s. sac

s. septum

s. swelling

s. vein

scrotales

s. anteriores

s. posteriores

scroti

raphe s.

septum s.

scrotiform

scrotum, pl. **scrota, scrotums**

raphe of s.

superficial fascia of s.

testes in s.

scultetus

scuta (*pl. of* scutum)

scutate

scute

tympanic s.

scutiform

scutum, pl. **scuta**

scybalum, pl. **scybala**

scyphiform

scyphoid

SDA

specific dynamic action

sea-blue histiocyte

sebaceae

glandulae s.

sebaceous

s. follicles

s. gland

sebaceus

sebiagogic

sebiferous

Sebileau

S. hollow

S. muscle

sebiparous

S

NOTES

sebum preputiale
second
- s. cervical vertebra
- s. cranial nerve
- s. cuneiform bone
- s. finger
- s. gas effect
- s. incisor
- s. lumbar artery
- s. molar
- s. parallel pelvic plane
- s. part of duodenum
- s. temporal convolution
- s. tibial muscle
- s. toe
- s. tooth

secondarium
secondary
- s. anophthalmia
- s. cartilage
- s. cartilaginous joint
- s. cementum
- s. center of ossification
- s. choana
- s. constriction
- s. curvatures of vertebral column
- s. dentition
- s. digestion
- s. egg membrane
- s. fissure of cerebellum
- s. hypogonadism
- s. interatrial foramen
- s. lysosome
- s. mesoderm
- s. nodule
- s. nondisjunction
- s. oocyte
- s. ossification center
- s. palate
- s. point of ossification
- s. sensory cortex
- s. sensory nuclei
- s. spermatocyte
- s. spiral lamina
- s. spiral plate
- s. tympanic membrane
- s. villus
- s. visual area
- s. visual cortex
- s. vitreous
- s. X zone

secreta
secrete
secretion
- cytocrine s.
- gastric s.
- gland of internal s.
- neurohumoral s.

secretomotor nerve

secretomotory
secretory
- s. canaliculus
- s. duct
- s. granule
- s. nerve

sectio, pl. **sectiones**
section
- axial s.
- cross s.
- diagonal s.
- frontal s.
- longitudinal s.
- median s.
- midsagittal s.
- oblique s.
- parasagittal s.
- sagittal s.
- transverse s.

sectorial
secunda
- arteriae intercostales posteriores prima et s.

secundae
- rami dorsales arteriarum intercostalium posteriorum primae et s.

secundaria
- lamina spiralis s.
- membrana tympani s.

secundarium
- centrum ossificationis s.
- punctum ossificationis s.

secundina, pl. **secundinae**
secundines
secundum
- interatrial foramen s.
- ostium s.
- septum s.

secundus
- digitus manus s.
- digitus pedis s.
- musculus tibialis s.

sedigitate
Seessel
- S. pocket
- S. pouch

segment
- anterior basal bronchopulmonary s.
- anterior bronchopulmonary s.
- anterior inferior renal s.
- anterior ocular s.
- anterior superior renal s.
- apical bronchopulmonary s.
- apicoposterior bronchopulmonary s.
- bronchopulmonary s.
- cardiac s.
- cranial s.
- frontal s.

hepatic venous s.'s
s. I
inferior lingular
  bronchopulmonary s.
inferior renal s.
interannular s.
intermaxillary s.
internodal s.
Lanterman s.'s
lateral basal bronchopulmonary s.
lateral bronchopulmonary s. S IV
left anterior lateral hepatic s.
left medial hepatic s.
left posterior lateral hepatic s. III
lingular s.
s.'s of liver
medial basal s.
medial basal bronchopulmonary s.
  S VII
medial bronchopulmonary s. S V
medullary s.
mesoblastic s.
mesodermal s.
neural s.
occipital s.
posterior basal bronchopulmonary s.
  S X
posterior bronchopulmonary s. S II
posterior hepatic s. I
posterior renal s.
pulmonary s.
Ranvier s.
renal s.'s
right anterior lateral hepatic s.
right anterior medial hepatic s.
right posterior lateral hepatic s.
right posterior medial hepatic s.
rivinian s.
Rivinus s.
s.'s of spinal cord
s.'s of spleen
subapical s.
subsuperior s.
superior s.
superior lingular
  bronchopulmonary s. S IV
superior renal s.
sympathetic s.
uterine s.
segmenta (*pl. of* segmentum)
segmental
  s. artery of kidney

s. artery of liver
s. bronchus
s. medullary artery
s. orifice
s. plate
s. sphincter
s. zone
segmentalis, pl. **segmentales**
  arteriae medullares segmentales
  bronchus s.
segmentation
  s. cavity
  s. nucleus
segmented cell
segmenti
  arteria s.
segmentorum
  rami bronchiales s.
segmentum, pl. **segmenta**
  s. I, II, III, IV
  s. A1, A2 arteriae cerebri
    anterioris
  s. anterius inferius
  s. apicale
  s. basale mediale
  s. basale posterius
  s. bronchopulmonale
  s. bronchopulmonale anterius S III
  s. bronchopulmonale apicale S I
  s. bronchopulmonale apicoposterius
  s. bronchopulmonale basale anterius
  s. bronchopulmonale basale laterale
  s. bronchopulmonale basale mediale
    S VII
  s. bronchopulmonale basale
    posterius S X
  s. bronchopulmonale lingulare
    superius
  s. bronchopulmonale mediale
  s. bronchopulmonale posterius
  s. bronchopulmonalis
  s. cardiacum
  segmenta cervicalia C1–C5
  segmenta cervicalia medullae
    spinalis
  segmenta coccygea medullae
    spinalis
  segmenta hepatis
  s. hepatis anterius laterale dextrum
  s. hepatis anterius laterale sinistrum
  s. hepatis anterius mediale dextrum
  s. hepatis mediale sinistrum

S

NOTES

545

segmentum *(continued)*
s. hepatis posterius
s. hepatis posterius laterale dextrum
s. hepatis posterius laterale
  sinistrum
s. hepatis posterius mediale
  dextrum
s. internodale
segmenta lienis
s. lingulare bronchopulmonale
  inferius S V
segmenta lumbalia L1–L5
segmenta medullae spinalis
segmenta medullae spinalis C1–Co
s. medullae spinalis cervicalia
segmenta medullae spinalis
  cervicalia C1–C8
s. medullae spinalis coccygeum
s. medullae spinalis lumbalis
segmenta medullae spinalis
  lumbaria L1–L5
s. medullae spinalis sacralis
s. medullae spinalis thoracica
s. oculare anterius
s. P1, P3, P4 arteriae cerebri
  posterioris
s. renale anterius inferius
s. renale anterius superius
s. renale posterius
segmenta renalia
segmenta sacralia medullae spinalis
s. subapicale
s. superius
segmenta thoracica medullae
  spinalis
Seiler cartilage
selene unguium
sella
diaphragm of s.
dorsum s.
s. turcica
sellae
diaphragma s.
dorsum s.
foramen diaphragmatis s.
tuberculum s.
sellar diaphragm
sellaris
articulatio s.
pars s.
Selye
adaptation syndrome of S.
semen
semi
glandula s.
semicanal
s. of auditory tube
s. for tensor tympani muscle
semicanalis, pl. **semicanales**

s. musculi tensoris tympani
s. tubae auditivae
s. tubae auditoriae
semicartilaginous
semicircular
s. canal
s. canal of bony labyrinth
s. duct
s. line
semicircularis, pl. **semicirculares**
crura ossea canales semicirculares
ductus semicirculares
epithelium ductus s.
linea s.
membrana basalis ductus s.
membrana propria ductus s.
tenia s.
semicircularium
ampullae osseae canalium s.
crista ampullaris ductuum s.
ductuum s.
venae ductuum s.
semicrista incisiva
semidecussation
semilunar
s. bone
s. cartilage
s. conjunctival fold
s. cusp
cusp of anterior s.
cusp of left s.
cusp of right s.
s. fascia
s. fasciculus
s. fibrocartilage
s. fold of colon
s. ganglion
s. hiatus
s. line
s. notch
s. nucleus of Flechsig
s. valve
semilunare
velum s.
semilunaris
fasciculus s.
hiatus s.
linea s.
lunula valvulae s.
nodulus valvulae s.
plica s.
valvula s.
semilunarium
noduli valvularum s.
semilunatum
planum s.
semimembranosus, pl. **semimembranosi**
bursa of s.
bursa musculi semimembranosi

s. muscle
musculus s.
s. tendon
semimembranous
s. bursa
s. muscle
semina
seminal
s. capsule
s. colliculus
s. duct
s. fluid
s. gland
s. granule
s. hillock
s. lake
s. vesicle
seminalis
colliculus s.
ductus excretorius vesiculae s.
glandula s.
lacus s.
tunica mucosa vesiculae s.
vesicula s.
semination
seminiferous
s. epithelium
s. tubule
seminis
musculus ejaculator s.
semiorbicular
semioval center
semiovale
centrum s.
Vicq d'Azyr centrum s.
semipennate
s. muscle
semipennatus
musculus s.
semipenniform
semipermeable membrane
semispinal
s. muscle
s. muscle of head
s. muscle of neck
s. muscle of thorax
semispinalis
s. capitis
s. capitis muscle
s. cervicis
s. cervicis muscle
musculus s.

s. thoracis
s. thoracis muscle
semisulcus
semitendinosus
s. muscle
musculus s.
s. tendon
semitendinous muscle
senile involution
sensate
sense
kinesthetic s.
muscular s.
s. organ
sensibility
deep s.
electromuscular s.
mesoblastic s.
sensible temperature
sensiferous
sensigenous
sensitivity
sensitized cell
sensomobile
sensomobility
sensomotor
sensor
sensoria (*pl. of* sensorium)
sensorial areas
sensoriglandular
sensorimotor area
sensorimuscular
sensorineural
sensorium, pl. **sensoria**
decussatio sensoria
epicritic s.
radix sensoria
sensorivascular
sensorivasomotor
sensory
s. areas
s. branch of radial nerve (SBRN)
s. cortex
s. crossway
s. decussation of medulla oblongata
s. function
s. inattention
s. receptor
s. root of ciliary ganglion
s. root of pterygopalatine ganglion
s. root of spinal nerve
s. root of sublingual ganglion

S

NOTES

sensory *(continued)*
s. root of submandibular ganglion
s. root of trigeminal nerve
s. speech center
s. tract
s. urgency
sensuum
organa s.
sentinel
s. fold
s. gland
s. node
separans
funiculus s.
separation of amino acid
septa *(pl. of* septum)
septal
s. area
s. artery
s. bone
s. branch
s. cell
s. cusp
cusp of s.
s. cusp of right atrioventricular
valve
s. cusp of tricuspid valve
s. line
s. nasal cartilage
s. raphe
septales
rami s.
rami interventriculares s.
septalis
cuspis s.
septate
s. uterus
vagina s.
s. vagina
septi
arteria nasalis posterior s.
falx s.
musculus depressor s.
nasales posteriores laterales et s.
nucleus accumbens s.
septomarginal
s. fasciculus
s. trabecula
s. tract
septomarginalis
fasciculus s.
trabecula s.
septonasal
septulum, pl. **septula**
s. testi
septula of testis
s. testis
septum, pl. **septa**
s. accessorium

alveolar s.
anterior intermuscular s.
anteromedial intermuscular s.
aortopulmonary s.
atrioventricular s.
s. atrioventriculare
s. of auditory tube
Bigelow s.
bony nasal s.
bulbar s.
s. bulbi urethrae
s. canalis musculotubarii
cartilage of nasal s.
cartilaginous s.
caudal s.
s. cervicale intermedium
Cloquet s.
comblike s.
s. of corpora cavernosa of clitoris
s. corporum cavernosorum
s. corporum cavernosorum clitoridis
crural s.
depressor muscle of s.
distal spiral s.
s. endovenosum
endovenous s.
femoral s.
s. femorale
s. of frontal sinuses
gingival s.
s. glandis
s. of glans
s. of glans penis
hanging s.
interalveolar s.
s. interalveolare
septa interalveolaria
interatrial s.
s. interatriale
intercavernosus s.
intermuscular s.
s. intermusculare
s. intermusculare vastoadductorium
interpulmonary s.
interradicular septa
septa interradicularia
s. interradicularia mandibulae et
maxilla
interventricular s.
s. interventriculare
lateral intermuscular s.
s. linguae
s. lucidum
medial intermuscular s.
s. mediastinale
s. membranaceum ventriculorum
membranous part of
interventricular s.
membranous part of nasal s.

s. mobile nasi
mobile part of nasal s.
s. musculare ventriculorum
s. of musculotubal canal
nasal s.
s. nasi osseum
orbital s.
s. orbitale
pectiniform s.
s. pectiniforme
s. pellucidum
s. of pharyngotympanic auditory
 tube
placental s.
posterior intermuscular s.
precommissural s.
s. primum
proximal spiral s.
quadrilateral s.
rectovaginal s.
s. rectovaginale
rectovesical s.
s. rectovesicale
retrovaginal s.
scrotal s.
s. scroti
s. secundum
sinus s.
s. sinuum frontalium
s. sinuum sphenoidalium
s. of sphenoidal sinuses
spiral bulbar s.
s. spurium
subarachnoid s.
supravaginal s.
tarsus orbital s.
s. of testis
s. of tongue
transparent s.
transverse s.
s. tubae
urogenital s.
urorectal s.
ventricular s.
septus
 uterus s.
sequestration
 bronchopulmonary s.
serial blood gas
series
 cytic s.
 erythrocytic s.

lymphocytic s.
lymphoid s.
myeloid s.
thrombocytic s.
serofibrinous
serofibrous
seromembranous
seromucosa
 glandula s.
seromucous gland
seromuscular
 s. coat
 s. layer
serosa, pl. **serosae**
 s. of colon
 s. of esophagus
 s. of gallbladder
 glandula s.
 s. of large intestine
 s. of liver
 membrana s.
 meninx s.
 s. of parietal pleura
 s. of peritoneum
 s. of serous pericardium
 s. of small intestine
 s. of the spleen
 s. of stomach
 tunica s.
 s. of urinary bladder
 s. of uterine tube
 s. of uterus
 s. of visceral pleura
serosal
 s. fold
 s. surface
seroserosal
seroserous
serosi
 lamina parietalis pericardii s.
 tunica serosa pericardii s.
serosum
 ligamentum s.
 pericardium s.
serotina
 decidua s.
 membrana s.
serotinus
 dens s.
serous
 s. cell
 s. coat

S

NOTES

serous *(continued)*
s. coat of peritoneum
s. demilunes
s. gland
s. layer of peritoneum
s. ligament
s. membrane
s. pericardium
s. sac
s. tunic
serrata
ora s.
s. ova
sutura s.
serrated
serrate suture
serration
serratus
s. anterior
s. anterior muscle
aponeurosis of posterior superior s.
s. posterior inferior
s. posterior inferior muscle
s. posterior superior
s. posterior superior muscle
Serres
S. angle
S. gland
serrulate
serrulated
Sertoli
S. cell
S. columns
servation
Servetus circulation
servomechanism
sesamoid
s. bone
s. cartilage
s. cartilage of cricopharyngeal
ligament
s. cartilage of larynx
s. cartilage of nose
sesamoidea
ossa s.
sesamoideum
os s.
sessile hydatid
seta, pl. **setae**
setaceous
setiferous
setigerous
seventh cranial nerve
sex
s. cell
s. chromosome imbalance
s. cords
s. hormone
s. steroid-binding globulin

sexdigitate
sexual
s. gland
s. reproduction
SFO
subfornical organ
shadow
s. cells
Gumprecht s.
hilar s.
liver s.
mediastinal s.
s. nucleus
shadow-casting
shaft
s. of bone
s. of clavicle
femoral s.
s. of femur
s. of fibula
fibular s.
hair s.
humeral s.
s. of humerus
medulla of hair s.
s. of metacarpal
s. of metatarsal
penile s.
s. of penis
s. of phalanx
radial s.
s. of radius
s. of tibia
tibial s.
s. of ulna
ulnar s.
shaggy chorion
shallow breathing
sham feeding
shank bone
Sharpey fiber
shawl muscle
shear
s. flow
s. rate
s. stress
shearing edge
sheath
adventitial s.
anterior layer of rectus
abdominis s.
anterior tarsal tendinous s.
anulus of fibrous s.
arcuate line of rectus s.
axillary s.
carotid s.
carpal tendinous s.
caudal s.
common flexor s.

common peroneal tendon s.
connective tissue s.
cruciform part of fibrous digital s.
crural s.
cuticle of root s.
dentinal s.
dorsal carpal tendinous s.
dural s.
enamel rod s.
external rectus s.
external root s.
s. of eyeball
fascial s.
femoral s.
fibrous tendon s.
fibular tarsal tendinous s.
flexor s.
glial s.
Huxley s.
infundibuliform s.
internal root s.
intertubercular tendon s.
s. of Key and Retzius
s. ligament
Mauthner s.
medullary s.
mitochondrial s.
myelin s.
Neumann s.
neurovascular s.
notochordal s.
optic s.
palmar carpal tendinous s.
parotid s.
penile s.
peroneal tendon s.
posterior layer of rectus
  abdominis s.
posterior rectus s.
s. process of sphenoid bone
prostatic s.
rectus abdominis s.
s. of rectus muscle
retrorectal lamina of hypogastric s.
root s.
Rouget-Neumann s.
Scarpa s.
Schwalbe s.
s. of Schwann
s. of Schweigger-Seidel
s. of styloid process
s. of superior oblique muscle

synovial s.
synovial tendon s.
tail s.
tendinous s.
tendon s.
s. of thyroid gland
tibial tarsal tendinous s.
transseptal s.
vascular s.
venous s.
s. of vessel
Waldeyer s.
sheathed artery
shelf, gen. and pl. **shelves**
dental s.
mesocolic s.
rectal s.
vocal s.
shell
cytotrophoblastic s.
shelving edge of Poupart ligament
Sherrington phenomenon
shield
embryonic s.
shift
luteoplacental s.
shin bone
ship
Fabricius s.
shivering thermogenesis
shock
break s.
short
s. abductor muscle
s. abductor muscle of thumb
s. adductor muscle
s. anconeus muscle
s. bone
s. central artery
s. ciliary artery
s. ciliary nerve
s. circumferential artery
s. extensor muscle
s. extensor muscle of great toe
s. extensor muscle of thumb
s. fibular muscle
s. flexor muscle
s. flexor muscle of great toe
s. flexor muscle of little finger
s. flexor muscle of little toe
s. flexor muscle of thumb
s. gastric branch of lienal artery

S

NOTES

short *(continued)*
s. gastric vein
s. gyri of insula
s. head
s. head of biceps brachii
s. head of biceps femoris
s. levatores costarum muscle
s. levator muscle
s. limb of incus
s. luteal phase
s. palmar muscle
s. peroneal muscle
s. pitch helicoidal layer
s. plantar ligament
s. posterior ciliary artery
s. posterior ciliary axis
s. process of malleus
s. radial extensor muscle
s. radial extensor muscle of wrist
s. root of ciliary ganglion
s. rotator muscle
s. saphenous nerve
s. saphenous vein
s. vinculum
shortening reaction
shoulder
back, arm, neck, s. (BANS)
s. blade
s. girdle
s. joint
rotator cuff of s.
Shrapnell membrane
shunt
Dickens s.
hexose monophosphate s.
s. muscle
Rapoport-Luebering s.
Warburg-Lipmann-Dickens-
Horecker s.
sialaden
sialic
sialine
Siamese twins
Sibson
S. aortic vestibule
S. aponeurosis
S. fascia
S. groove
S. muscle
S. notch
sickle cell
sieve
s. bone
s. plate
Siggaard-Andersen nomogram
sigma effect
sigmoid
s. artery
s. bladder

s. colon
s. flexure
s. fold
s. fossa
s. groove
s. loop
s. lymph node
s. mesocolon
s. notch
s. sulcus
s. vein
s. venous sinus
sigmoidal branch of inferior mesenteric artery
sigmoidea, pl. **sigmoideae**
arteriae sigmoideae
flexura s.
plica s.
venae sigmoideae
sigmoidei
lymphonodi s.
nodi lymphoidei s.
sulcus sinus s.
sigmoideum
colon s.
mesocolon s.
sigmoideus
sinus s.
sigmoidovesical
signal
nerve s.
signet ring cells
silent period
silhouette
cardiothoracic s.
diaphragmatic s.
heart s.
siliqua olivae
sill
nasal s.
piriform s.
silver cell
Silverman-Lilly pneumotachograph
simian
s. crease
s. fissure
Simonart
S. bands
S. ligament
S. thread
simple
s. crus of semicircular duct
s. fission
s. joint
s. lobule
s. squamous epithelium
simplex
articulatio s.

lobulus s.
lymphangioma superficium s.

sincipita
sincipital
sinciput
sinew
singulare

foramen s.

singular foramen
sinister

bronchus principalis s.
ductus hepaticus s.
ductus lobi caudati s.
lobus hepatis s.
pulmo s.
ramus s.
ventriculus s.

sinistra

arteria colica s.
arteria coronaria s.
arteria gastrica s.
arteria gastroepiploica s.
arteria gastroomentalis s.
arteria pulmonalis s.
auricula atrii s.
colica s.
divisio lateralis s.
flexura coli s.
gastrica s.
gastroepiploica s.
hemicardia s.
intercostalis superior s.
ovarica s.
pars hepatis s.
pulmonalis inferior s.
pulmonalis superior s.
suprarenalis s.
testicularis s.
umbilicalis s.
valva atrioventricularis s.
vena colica s.
vena gastrica s.
vena gastroomentalis s.
vena intercostalis superior s.
vena ovarica s.
vena pulmonalis inferior s.
vena pulmonalis superior s.
vena suprarenalis s.
vena testicularis s.
vena umbilicalis s.

sinistrad

sinistrae

cuspis anterior valvae
atrioventricularis s.
cuspis posterior valvae
atrioventricularis s.
ligamentum venae cavae s.
pars inferior venae lingularis venae
pulmonalis superioris s.
plica venae cavae s.
rami esophageales arteriae
gastricae s.
ramus atrialis anastomoticus ramus
circumflexus arteriae coronariae s.
ramus atrialis intermedius arteriae
coronariae s.
ramus circumflexus arteriae
coronariae s.
ramus interventricularis anterior
arteriae coronariae s.
ramus lateralis interventricularis
anterioris arteriae coronariae s.
ramus marginalis sinister arteriae
coronariae s.
venae brachiocephalicae dextrae
et s.
venae hepaticae s.

sinistrae/dextrae

rami interventriculares septales
arteriae coronariae s.

sinistral
sinistri

impressio aortica pulmonis s.
incisura cardiaca pulmonis s.
lingula pulmonis s.
lobus inferior pulmonis dextri et s.
lobus superior pulmonis dextri
et s.
lymphonodi colici s.
lymphonodi gastrici s.
lymphonodi gastroomentales s.
lymphonodi lumbales s.
nodi lymphoidei colici s.
nodi lymphoidei gastrici s.
nodi lymphoidei gastroomentales s.
nodi lymphoidei lumbales s.
pars superior venae lingularis
venae pulmonis superioris s.
ramus descendens arteriae
segmentalis anterioris pulmonis
dextri et s.

NOTES

S

sinistri (*continued*)
    ramus descendens arteriae
      segmentalis posterioris pulmonis
      dextri et s.
    ramus lateralis ductus hepatici s.
    ramus medialis ductus hepatici s.
    trabeculae carneae ventriculorum
      dextri et s.
    venae posteriores ventriculi s.
    vena obliqua atrii s.
    vena posterior ventriculi s.
sinistrocerebral
sinistrous
sinistrum
    atrium cordis s.
    ligamentum triangulare s.
    ostium atrioventriculare s.
    segmentum hepatis anterius
      laterale s.
    segmentum hepatis mediale s.
    segmentum hepatis posterius
      laterale s.
    trigonum fibrosum s.
sinoatrial node (S-A)
sinoauricular
sinocarotid nerve
sinopulmonary
sinospiral muscle bundle
sinotubular junction
sinovagin
sinuatrial
    s. chamber
    s. nodal branch of right coronary
      artery
    s. node artery
sinuatrialis
    nodus s.
    ramus nodi s.
sinus, pl. **sinus, sinuses**
    air s.
    s. alae parvae
    anal sinuses
    s. anales
    anterior sinuses
    s. anteriores
    s. aortae
    aortic s.
    Arlt s.
    s. barotrauma
    basilar venous s.
    branchial s.
    Breschet s.
    s. caroticus
    carotid cavernous s.
    s. cavernosus
    cavernous venous s.
    cerebral sinuses
    cervical s.
    cilia of paranasal s.

s. circularis
circular venous s.
coccygeal s.
confluence of sinuses
s. coronarius
coronary s.
costomediastinal s.
costophrenic s.
cranial venous s.
cribriform s.
dermal s.
ductal s.
s. durae matris
dural sinuses
sinuses of dura mater
endodermal s.
Englisch s.
s. epididymidis
s. of epididymis
ethmoidal s.
s. ethmoidales
s. ethmoidales anteriores
s. ethmoidales mediae
s. ethmoidales posteriores
s. frontalis
frontal paranasal s.
groove for inferior petrosal s.
groove for sigmoid s.
groove for superior petrosal s.
groove for superior sagittal s.
groove for transverse s.
Guérin s.
Huguier s.
inferior longitudinal s.
inferior petrosal s.
inferior sagittal s.
s. intercavernosi
s. intercavernosi anterior et
  posterior
intercavernous sinuses
intercavernous venous s.
jugular s.
s. lactiferi
lactiferous s.
laryngeal s.
s. laryngeus
lateral lacunae of superior
  sagittal s.
left sigmoid s.
s. lienis
longitudinal vertebral venous s.
Luschka s.
lymph s.
lymphatic s.
Maier s.
marginal s.
mastoid sinuses
s. maxillaris
maxillary s.

Meyer s.
middle sinuses
Morgagni s.
s. of nail
s. nerve
nerve to carotid s.
s. nerve of Hering
s. node
oblique pericardial s.
s. obliquus pericardii
occipital s.
s. occipitalis
opening of coronary s.
opening of frontal s.
opening for right maxillary s.
opening of sphenoidal s.
Palfyn s.
paranasal sinuses
s. paranasales
parasinoidal sinuses
pelvic part of the urogenital s.
perineal s.
Petit s.
petrosal venous s.
petrosquamous venous s.
s. petrosus inferior
s. petrosus superior
phrenicocostal s.
pilonidal s.
piriform s.
pleural sinuses
s. pocularis
portal s.
s. posterior cavi tympani
precervical s.
prominence of venous valvular s.
prostatic s.
s. prostaticus
pulmonary s.
s. of pulmonary trunk
rectal sinuses
s. rectus
renal s.
s. renalis
s. reuniens
Ridley s.
right sigmoid s.
rostrum of s.
sacrococcygeal s.
sagittalis inferior s.
s. sagittalis inferior
sagittalis superior s.

s. sagittalis superior
sagittal venous s.
scleral venous s.
s. septum
septum of frontal sinuses
septum of sphenoidal sinuses
s. sigmoideus
sigmoid venous s.
sphenoidal s.
s. sphenoidalis
sphenoidal paranasal s.
s. sphenoparietalis
sphenoparietal venous s.
splenic s.
straight sphenoidal s.
straight venous s.
subarachnoidal s.
sulcus for transverse s.
superior longitudinal s.
superior petrosal s.
superior sagittal s.
suprapubic s.
tarsal s.
s. tarsi
tentorial s.
terminal s.
s. terminalis
thyroglossal s.
s. tonsillaris
Tourtual s.
s. tract
transverse pericardial s.
transverse venous s.
s. transversus
s. transversus pericardii
s. trunci pulmonalis
s. tubercle
tympani s.
tympanic s.
s. unguis
urogenital s.
s. urogenitalis
uterine s.
uteroplacental s.
Valsalva s.
valve of coronary s.
s. of the vena cava
s. venarum cavarum
s. venosus
venosus sclerae s.
s. venosus sclerae

S

NOTES

555

sinus *(continued)*
    venous sinuses
    s. vertebrales longitudinales
sinusoid
    uterine s.
sinusoidal
    s. capillary
    s. space
sinuum
    confluens s.
sinuvertebral nerve
siphon
    carotid s.
sireniform
sirenomelia
Sister Mary Joseph lymph node
site
    rectal s.
sitotaxis
sitotropism
situs
    s. inversus
    s. inversus viscerum
    s. perversus
    s. solitus
    s. transversus
sixth
    s. cranial nerve
    s. ventricle
sixth-year molar
skein
    s. cell
    choroid s.
skeletal
    s. muscle
    s. muscle fibers
    s. muscle tissue
    s. of spine
    s. structure
    s. system
skeletale
    systema s.
skeletalis
    pars cartilaginea systematis s.
    pars ossea systematis s.
skeleti
    musculus s.
skeletology
skeleton
    appendicular s.
    s. appendiculare
    articulated s.
    axial s.
    s. axiale
    cardiac s.
    cardiac fibrous s.
    s. of eyelid
    facial s.
    s. of foot

    s. of free inferior limb
    s. of free superior limb
    gill arch s.
    s. of heart
    jaw s.
    laryngeal s.
    pelvic s.
    spine s.
    thoracic s.
    s. thoracicus
    s. thoracis
    visceral s.
Skene
    Bartholin, urethral, and S. (BUS)
    S. duct
    S. gland
    S. tubules
skin
    appendages of s.
    s. folding
    s. furrow
    s. grooves
    s. heart
    hidden nail s.
    layers of s.
    s. ligament
    nail s.
    penile s.
    perianal s.
    s. pore
    s. reflex
    retinaculum of s.
    s. ridge
    s. sulci
    s. tag
    s. of tooth
    underlying s.
skin-adipose superficial
    musculoaponeurotic system (SASMAS)
skin-muscle reflex
skull
    base of s.
    bone of s.
    cloverleaf s.
    external base of s.
    inner table bone of s.
    internal base of s.
    longitudinal arc of s.
    outer table bone of s.
    roof of s.
    steeple s.
    towers s.
skullcap
slender
    s. lobule
    s. process of malleus
slew rate
sliding filament hypothesis
sling psychrometer

slit
  filtration s.
  s. pore
  pudendal s.
  vulvar s.
sluggish layer
sluice
SMA
  superior mesenteric artery
small
  s. arteries
  s. bowel
  s. calorie
  s. canal of chorda tympani
  s. cardiac vein
  s. cleaved cell
  s. deep petrosal nerve
  s. intestine
  s. pancreas
  s. pelvis
  s. pudendal lip
  s. saphenous vein
  s. sciatic nerve
  s. trochanter
smaller
  s. muscle of helix
  s. pectoral muscle
  s. posterior rectus muscle of head
  s. psoas muscle
smallest
  s. cardiac vein
  s. scalene muscle
  s. splanchnic nerve
smegma
  s. clitoridis
  s. preputii
smell
  nerve of s.
  organ of s.
smell-brain
smooth
  s. chorion
  s. muscle
  s. muscle tissue
  s. muscular sphincter
smooth-surfaced endoplasmic reticulum
smudge cells
sneeze
snuffbox
  anatomic s.
  anatomical s.
socia parotidis

socket
  eye s.
  s. joint
  tooth s.
sodium-potassium pump
sodium pump
Soemmering
  S. area
  S. arterial vein
  S. external radial vein
  S. foramen
  S. ganglion
  S. ligament
  S. muscle
  S. spot
soft
  s. palate
  s. part
solar
  s. ganglia
  s. plexus
sole
  s. of foot
  s. nuclei
  s. plate
  quadrate muscle of s.
  s. reflex
soleal
  s. line
  s. part of posterior plantar flexor
    compartment of leg
solei
  arcus tendineus musculi s.
  linea musculi s.
sole-plate ending
soleus
  s. muscle
  musculus s.
solitarii
  folliculi lymphatici s.
  lymphatici s.
  noduli lymphoidei s.
  nucleus tractus s.
solitarius
  fasciculus s.
  funiculus s.
  tractus s.
solitary
  s. bundle
  s. follicles
  s. foramen
  s. gland

S

NOTES

solitary (*continued*)
    s. lymphatic nodule
    s. nodules of intestine
    s. tract
solitus
    situs s.
solum
solvent drag
soma
Somagyi reflex
somatic
    s. arteries
    s. cells
    s. layer
    s. mesoderm
    s. mitosis
    s. motor nuclei
    s. muscle
    s. nerve
    s. nerve fiber
    s. reproduction
    s. sensory cortex
somaticae
    neurofibrae s.
somaticosplanchnic
somaticovisceral
somatochrome
somatogenic
somatointestinal reflex
somatology
somatomegaly
somatopagus
somatoplasm
somatopleural
somatopleure
somatosensory
    s. area
    s. cortex
somatotopic
somatotopy
somatotroph
somatotrophic
somatotropic hormone
somatotropin release-inhibiting factor
somatotropin-releasing factor
somatotype
somatotypology
somesthetic
    s. area
    s. system
somite
    s. cavity
    occipital s.
somitic mesoderm
Sondergaard
    S. cleft
    S. groove
Sondermann canal
sonomotor response

sorius dorsalis
soroche
    chronic s.
space
    air s.
    alveolar dead s.
    anatomical dead s.
    anatomic dead s.
    antecubital s.
    axillary s.
    Bogros s.
    Böttcher s.
    Bowman s.
    Broca s.
    Burns s.
    capsular s.
    cartilage s.
    cavernous s.
    central palmar s.
    Chassaignac s.
    Cloquet s.
    Colles s.
    corneal s.
    Cotunnius s.
    cranial extradural s.
    dead s.
    deep perineal s.
    disk s.
    Disse s.
    s. of Donders
    endolymphatic s.
    epicardial s.
    epidural s.
    episcleral s.
    epitympanic s.
    extraperitoneal s.
    filtration s.
    s. of Fontana
    Fontana s.
    s. of Forel
    s. four of neck
    gingival s.
    haversian s.
    Henke s.
    hepatorenal recess of subhepatic s.
    His perivascular s.
    iliocostal s.
    infraglottic s.
    intercellular s.
    intercostal s.
    interdigital s.
    interfascial s.
    interglobular s.
    interosseous metacarpal s.
    interosseous metatarsal s.
    interpeduncular joint s.
    interpleural s.
    interseptovalvular s.
    intersphincteric s.

interstitial s.
interventricular s.
intervillous s.
intrafascial s.
intrapharyngeal s.
intraretinal s.
intravaginal s.
s.'s of iridocorneal angle
Kiernan s.
Kretschmann s.
Kuhnt s.
lateral central palmar s.
lateral midpalmar s.
lateral pharyngeal s.
left intercostal s.
leptomeningeal s.
Lesgaft s.
lymph s.
Magendie s.
Malacarne s.
masseteric s.
masticator s.
Meckel s.
medial midpalmar s.
mediastinal s.
medullary s.
middle palmar s.
midpalmar s.
Mohrenheim s.
Nuel s.
palmar s.
paraglottic s.
parapharyngeal s.
paravesical s.
Parona s.
parotid s.
perforated s.
pericardial s.
perichoroid s.
perichoroidal s.
perilenticular s.
perilymphatic s.
perineal s.
perinuclear s.
periotic s.
peripharyngeal s.
perirenal s.
periscleral s.
perisinusoidal s.
peritoneal s.
perivascular lymph s.
perivitelline s.

pharyngeal s.
pharyngomaxillary s.
physiologic dead s.
plantar s.
pleural s.
pleuroperitoneal s.
pneumatic s.
Poiseuille s.
popliteal s.
postcricoid s.
posterior septal s.
postpharyngeal s.
post-styloid s.
preperitoneal s.
preputial s.
presacral s.
prestyloid s.
pretarsal s.
pretracheal s.
prevesical s.
Proust s.
Prussak s.
pterygomandibular s.
pterygopharyngeal s.
quadrangular s.
quadrilateral s.
rectovesical s.
Reinke s.
respiratory dead s.
retroadductor s.
retrocardiac s.
retroesophageal s.
retroinguinal s.
retrolental s.
retromammary s.
retromylohyoid s.
retroparotid s.
retroperitoneal s.
retropharyngeal s.
retropubic s.
retrovisceral s.
retrozonular s.
s. of Retzius
Retzius s.
rudimentary disk s.
Schwalbe s.
sinusoidal s.
spinal epidural s.
subaponeurotic s.
subarachnoid s.
subchorial s.
subdural s.

S

NOTES

space *(continued)*
  subgaleal s.
  subgingival s.
  subhepatic s.
  submucosal s.
  subperitoneal s.
  subphrenic s.
  subumbilical s.
  superficial perineal s.
  suprahepatic s.
  supralevator s.
  suprasternal s.
  Tarin s.
  temporal s.
  s. of Tenon
  Tenon s.
  thenar s.
  tissue s.
  s. of Traube
  Traube semilunar s.
  Trautmann triangular s.
  triangular s.
  vertebral epidural s.
  Virchow-Robin s.
  Waldeyer s.
  web s.
  Westberg s.
  zonular s.
Spallanzani law
span
sparing action
spasm
  muscle s.
spasmodica
  tabes s.
spasticity
  clasp-knife s.
spatial
spatium, pl. **spatia**
  spatia anguli iridocornealis
  s. endolymphaticum
  s. episclerale
  s. extradurale
  s. extraperitoneale
  s. intercostale
  s. interfasciale
  s. interglobula
  spatia interglobularia
  spatia interossea metacarpi
  spatia interossea metatarsi
  s. intervaginale bulbi oculi
  s. intervaginale subarachnoidale
    nervi optici
  spatia intervaginalia nervi optici
  s. lateropharyngeum
  s. parapharyngeum
  s. perichoroideale
  s. perichoroideum
  s. peridurale

  s. perilymphaticum
  s. perinei profundum
  s. perinei superficiale
  s. peripharyngeum
  s. pharyngeum laterale
  s. retroinguinale
  s. retroperitoneale
  s. retropharyngeum
  s. retropubicum
  s. retrozonulare
  s. subarachnoideum
  s. subdurale
  s. supraspinale
  spatia zonularia
special
  s. anatomy
  s. somatic afferent column
  s. visceral column
  s. visceral efferent nuclei
  s. visceral motor nuclei
specific
  s. compliance
  s. dynamic action (SDA)
spectrin
speech center
Spence
  tail of S.
sperm
  s. aster
  s. cell
  s. nucleus
spermagglutination
sperm-aster
spermatic
  s. artery
  s. cord
  s. duct
  s. fascia
  s. filament
  s. nerve
  s. plexus
  s. vein
  s. vesicle
  s. vessel
spermatica
  chorda s.
  s. externa
spermaticus, pl. **spermatici**
  funiculus s.
  tunicae funiculi spermatici
spermatid
spermatoblast
spermatocele
spermatocyst
spermatocytal
spermatocyte
  primary s.
  secondary s.
spermatocytogenesis

spermatogenesis
spermatogenetic
spermatogenic
spermatogenous
spermatogeny
spermatogone
spermatogonium
spermatoid
spermatology
spermatolysis
spermatolytic
spermatophore
spermatopoietic
spermatozoa (*pl. of* spermatozoon)
spermatozoal
spermatozoan
spermatozoon, pl. **spermatozoa**
    reservoir of spermatozoa
spermia (*pl. of* spermium)
spermiduct
spermiogenesis
spermium, pl. **spermia**
spermolysis
sphenethmoid
sphenion
sphenobasilar
sphenoccipital
sphenocephaly
sphenoethmoid
sphenoethmoidal
    s. recess
    s. suture
    s. synchondrosis
sphenoethmoidalis
    recessus s.
    sutura s.
    synchondrosis s.
sphenofrontal
    s. suture
    s. suture line
sphenofrontalis
    sutura s.
sphenoid
    s. angle
    body of s.
    s. bone
    s. crest
    s. emissary foramen
    s. emissary vein
    frontal margin of s.
    infratemporal crest of greater wing
     of s.

infratemporal surface of greater
  wing of s.
parietal margin of greater wing
  of s.
s. part of middle cerebral artery
s. process
s. process of septal nasal cartilage
pyramidal process of s.
rostrum of s.
spinous process of s.
squamosal margin of greater wing
  of s.
sphenoida
sphenoidal
    s. angle of parietal bone
    s. border of temporal bone
    s. concha
    s. crest
    s. emissary foramen
    s. fissure
    s. fontanelle
    s. lingula
    s. margin of temporal bone
    s. paranasal sinus
    s. part
    s. process
    s. process of palatine bone
    s. ridge
    s. rostrum
    s. sinus
    s. sinus aperture
    s. spine
    s. turbinate
    s. turbinated bone
    s. wing
    s. yoke
sphenoidale
    jugum s.
    os s.
    planum s.
    rostrum s.
sphenoidales
    conchae s.
sphenoidalis
    ala major ossis s.
    ala minor ossis s.
    apertura sinus s.
    corpus ossis s.
    crista infratemporalis alaris majoris
     ossis s.
    facies infratemporalis alaris majoris
     ossis s.

S

NOTES

sphenoidalis (*continued*)
    facies maxillaris alaris majoris ossis s.
    fonticulus s.
    foramen ovale ossis s.
    fossa scaphoidea ossis s.
    limbus s.
    lingula s.
    margo frontalis ossis s.
    margo parietalis alaris majoris ossis s.
    margo squamosus alaris majoris ossis s.
    margo zygomaticus alaris majoris ossis s.
    pars s.
    processus pterygoideus ossis s.
    processus vaginalis ossis s.
    sinus s.
    spina ossis s.

sphenoidalium
    septum sinuum s.

sphenomalar suture

sphenomandibula
    ligamentum s.

sphenomandibulare
    ligamentum s.

sphenomandibular ligament

sphenomaxillaris
    sutura s.

sphenomaxillary
    s. fissure
    s. fossa
    s. ganglion
    s. suture

sphenooccipital
    s. joint
    s. suture
    s. synchondrosis

sphenooccipitalis
    synchondrosis s.

sphenoorbitalis
    sutura s.

sphenoorbital suture

sphenopalatina
    arteria s.
    incisura s.

sphenopalatine
    s. branch of internal maxillary artery
    s. foramen
    s. ganglion
    s. ganglionectomy
    s. nerve
    s. notch
    s. notch of palatine bone

sphenopalatini
    nervi s.

sphenopalatinum
    foramen s.

sphenoparietal
    sutura s.
    s. suture
    s. suture line
    s. venous sinus

sphenoparietalis
    sinus s.
    sutura s.

sphenopetrosa
    fissura s.
    synchondrosis s.

sphenopetrosal
    s. fissure
    s. suture
    s. synchondrosis

sphenopetrous synchondrosis
sphenorbital
sphenosalpingostaphylinus
    musculus s.

sphenosquamosa
    sutura s.

sphenosquamosal
sphenosquamous suture
sphenotemporal suture
sphenotic
    s. center
    s. foramen

sphenoturbinal
sphenovomeriana
    sutura s.

sphenovomerine suture
sphenozygomatica
    sutura s.

sphenozygomatic suture
sphere
    attraction s.

spherica
    fovea s.

spherical
    s. nucleus
    s. recess
    s. recess of bony labyrinth

sphericus
spheroid
    s. articulation
    s. joint

spheroidal joint
spheroidea
    articulatio s.

spherospermia
spherule
sphincter
    s. ampullae hepatopancreaticae
    anal s.
    anatomic s.
    anatomical s.
    angular s.

s. angularis
s. ani
s. ani externus
s. ani internus
s. ani tertius
annular s.
antral s.
s. antri
basal s.
bicanalicular s.
s. of biliaropancreatic ampulla
Boyden s.
canalicular s.
cardioesophageal s.
choledochal s.
colic s.
s. constrictor cardiae
cricopharyngeal s.
deep part of external anal s.
dilator muscle of ileocecal s.
s. ductus choledochi
duodenal s.
duodenojejunal s.
esophageal s.
external anal s.
external rectal s.
external urethral s.
extrinsic s.
first duodenal s.
functional s.
s. of gastric antrum
gastroesophageal s.
Glisson s.
Henle s.
s. of hepatic flexure of colon
hepatopancreatic s.
s. of hepatopancreatic ampulla
Hyrtl s.
ileal s.
ileocecocolic s.
iliopelvic s.
inferior esophageal s.
inguinal s.
s. intermedius
internal anal s.
internal rectal s.
internal urethral s.
intrinsic s.
s. iridis
lower esophageal s. (LES)
macroscopic s.
marginal s.

mediocolic s.
microscopic s.
midgastric transverse s.
midsigmoid s.
s. muscle
s. muscle of common bile duct
s. muscle of pancreatic duct
s. muscle of pupil
s. muscle of pylorus
s. muscle of urethra
s. muscle of urinary bladder
musculus s.
myovascular s.
myovenous s.
Nélaton s.
O'Beirne s.
s. oculi
Oddi s.
s. of Oddi
s. oris
ostial s.
palatopharyngeal s.
pancreatic s.
pancreaticobiliary s.
s. papilla
pathologic s.
pelvirectal s.
s. of the pharyngeal isthmus
pharyngoesophageal s.
physiologic s.
physiological s.
postpyloric s.
prepapillary s.
preprostate urethral s.
preprostatic s.
prepyloric s.
proximal urethral s.
s. pupillae
s. pylori
pyloric s.
radiologic s.
radiological s.
rectosigmoid s.
segmental s.
smooth muscular s.
striated muscular s.
subcutaneous part of external
  anal s.
subcutaneous portion of external
  anal s.
superficial part of external anal s.
superior esophageal s.

NOTES

sphincter *(continued)*
    supracollicular s.
    s. of third portion of duodenum
    unicanalicular s.
    upper esophageal s. (UES)
    s. urethrae
    s. urethrae externus
    urethrovaginal s.
    urinary s.
    s. vaginae
    Varolius s.
    velopharyngeal s.
    s. vesicae
    s. vesicae biliaris
    s. vesicae felleae
    s. vesicae urinariae
sphincteral
sphincterial
sphincteric
sphincteroid tract of ileum
spicula (*pl. of* spiculum)
spicular
spicule
spiculum, pl. **spicula**
spider
    s. cell
    s. finger
    s. pelvis
Spieghel line
Spigelius
    S. line
    S. lobe
spike
    ponto-geniculo-occipital s.
    s. potential
spiloma
spilus
spina, pl. **spinae**
    s. angularis
    s. bifida
    s. bifida aperta
    s. bifida cystica
    s. bifida manifesta
    s. bifida occulta
    s. dorsalis
    erector spinae
    spinae geniorum inferior et
      superior
    s. helicis
    s. iliaca anterior inferior
    s. iliaca anterior superior
    s. iliaca posterior inferior
    s. iliaca posterior superior
    s. ischiadica
    s. meatus
    s. mentalis
    s. mentalis inferior et superior
    multifidus spinae
    musculus erector spinae

    musculus multifidus spinae
    s. nasalis anterior
    s. nasalis anterior corporis maxillae
    s. nasalis ossis frontalis
    s. nasalis posterior
    s. nasalis posterior laminae
      horizontalis ossis palatini
    s. ossis sphenoidalis
    spinae palatinae
    s. peronealis
    s. pubis
    rotatores spinae
    s. scapulae
    s. suprameatalis
    s. suprameatica
    s. suprameatum
    s. trochlearis
    s. tympanica major
    s. tympanica minor
spinal
    s. arachnoid mater
    s. artery
    s. branch
    s. canal
    s. column
    s. cord
    s. dura mater
    s. epidural space
    s. fluid
    s. ganglion
    s. induction
    s. marrow
    s. muscle
    s. muscle of head
    s. muscle of neck
    s. muscle of thorax
    s. nerve plexus
    s. nucleus of accessory nerve
    s. nucleus of trigeminus
    s. part of accessory nerve
    s. part of arachnoid
    s. part of deltoid muscle
    s. part of filum terminale
    s. point
    s. quotient
    s. reflex
    s. root
    s. root of accessory nerve
    s. tract of trigeminal nerve
    s. vein
spinale
    ganglion s.
spinalis, pl. **spinales**
    s. anterior
    arachnoidea mater s.
    canalis centralis medullae s.
    s. capitis
    s. capitis muscle
    s. cervicis muscle

dura mater s.
filum durae matris s.
fissura mediana anterior
   medullae s.
funiculi medullae s.
ganglion sensorium nervi s.
hydrocele s.
lemniscus s.
medulla s.
musculus s.
nervi spinales
pars cervicalis medullae s.
pars coccygea medullae s.
pars lumbalis medullae s.
pars sacralis medullae s.
pars thoracica medullae s.
pia mater s.
s. posterior
radix anterior nervi s.
radix dorsalis nervi s.
radix motoria nervi s.
radix posterior nervi s.
radix sensoria nervi s.
radix ventralis nervi s.
rami laterales ramorum dorsalium
   nervorum s.
ramus anterior nervi s.
ramus dorsalis nervi s.
ramus medialis ramorum dorsalium
   nervorum s.
ramus posterior nervi s.
ramus ventralis nervi s.
segmenta cervicalia medullae s.
segmenta coccygea medullae s.
segmenta medullae s.
segmenta sacralia medullae s.
segmenta thoracica medullae s.
sulcus medianus posterior
   medullae s.
sulcus nervi s.
s. thoracis
s. thoracis muscle
venae medullae s.

spinalium
ansae nervorum s.
plexus nervorum s.
rami communicantes nervorum s.
rami musculares nervorum s.
ramus dorsalis nervorum s.
ramus meningeus nervorum s.

spinate

spindle
aortic s.
s. cell
central s.
cleavage s.
s. fiber
His s.
Kühne s.
mitotic s.
muscle s.
neuromuscular s.
neurotendinous s.
nuclear s.
spindle-celled layer
spindle-shaped muscle
spine
alar s.
angular s.
anterior inferior iliac s.
anterior nasal s. (ANS)
anterior-superior iliac s. (ASIS)
s. cell
cervical s.
cleft s.
crest of scapular s.
dendritic s.
dorsal s.
erector muscle of s.
greater tympanic s.
s. of helix
hemal s.
Henle s.
s. of Henle
iliac s.
ischiadic s.
ischial s.
lateral curvature of s.
lesser tympanic s.
lumbar s.
lumbosacral s.
meatal s.
mental s.
nasal s.
neural s.
palatine s.
posterior inferior iliac s.
posterior nasal s.
posterior palatine s.
posterior superior iliac s. (PSIS)
pubic s.
sacral s.
s. of scapula

S

NOTES

spine *(continued)*
    sciatic s.
    skeletal of s.
    s. skeleton
    sphenoidal s.
    s. of sphenoid bone
    Spix s.
    superior s.
    superior iliac s.
    suprameatal s.
    thoracic s.
    tibial s.
    trochlear s.
spinifugal
spinipetal
spinobulbar
spinocerebellar tracts
spinocerebellum
spinocollicular
spinocostalis
spinoglenoid ligament
spin-olivary tract
spinomuscular
spinoneural artery
spinosa
    stria s.
spinose
spinosum
    foramen s.
spinosus
    nervus s.
    processus s.
    sulcus s.
spinotectalis
    tractus s.
spinothalamic tract
spinothalamicus
    tractus s.
spinotransversarius
spinous
    s. layer
    s. process
    s. process of sphenoid
    s. process of tibia
    s. process of vertebra
spiracle
spiral
    s. arteriole
    s. bulbar septum
    s. canal of cochlea
    s. canal of modiolus
    s. crest
    s. crest of cochlear duct
    s. fold of cystic duct
    s. foraminous tract
    s. ganglion
    s. ganglion of cochlea
    s. groove
    s. joint

    s. ligament of cochlea
    s. ligament of cochlear duct
    s. line
    s. membrane
    s. modiolar artery
    s. organ
    s. plate
    s. prominence
    s. prominence of cochlear duct
    s. tubule
    s. valve
    s. valve of cystic duct
    s. vein
    s. vein of modiolus
spirale
    organum s.
    vas s.
spiralis
    crista s.
    hamulus laminae s.
    linea s.
    membrana reticularis organi s.
    s. modioli
    prominentia s.
    valvula s.
spiro-index
spitting
spittle
Spitzer theory
Spitzka
    S. marginal tract
    S. marginal zone
    S. nucleus
Spix spine
splanchnapophysial
splanchnapophysis
splanchnic
    s. afferent column
    s. cavity
    s. circulation
    s. efferent column
    s. ganglion
    s. layer
    s. mesoderm
    s. nerve
    s. neurectomy
    s. wall
splanchnicectomy
splanchnici
    s. lumbales
    nervi pelvici s.
    s. pelvini
    s. sacrales
splanchnicum
    ganglion s.
    ganglion thoracicum s.
splanchnicus
    s. imus

s. major
s. minor
splanchnocele
splanchnocranium
splanchnocystica
dysencephalia s.
splanchnography
splanchnologia
splanchnology
splanchnomegaly
splanchnomicria
splanchnopleural
splanchnopleure
splanchnopleuric
splanchnoskeletal
splanchnoskeleton
splanchnosomatic
splay
splayfoot
spleen
s. accessorius
accessory s.
anterior extremity of s.
capsule of s.
colic impression of s.
colic surface of s.
fibrous capsule of s.
gastric impression on s.
gastric surface of s.
hilum of s.
inferior border of s.
liver, kidneys, and s. (LKS)
pedicle of s.
phrenic surface of s.
posterior extremity of s.
red pulp of s.
renal impression of s.
renal surface of s.
segments of s.
serosa of the s.
superior border of s.
trabeculae of s.
venae cavernosae of s.
visceral surface of s.
white pulp of s.
splendens
linea s.
spleneolus
splenetic
splenia (*pl. of* splenium)
splenial gyrus

splenic
s. branch of splenic artery
s. capsule
s. cells
s. cords
s. corpuscles
s. flexure
s. flexure of colon
s. hilum
s. lymph follicles
s. lymph node
s. lymph nodule
s. nervous plexus
s. pulp
s. recess
s. recess of omental bursa
s. sinus
s. tissue
s. trabeculae
s. vein
splenica
arteria s.
extremitas anterior s.
extremitas posterior s.
facies pancreatica s.
flexura colica s.
pulpa alba s.
pulpa rubra s.
vena s.
splenicae
rami pancreatici arteriae s.
rami splenici arteriae s.
trabeculae s.
splenici
lymphonodi s.
nodi lymphoidei s.
spleniculus
splenicum
hilum s.
splenicus
plexus nervosus s.
recessus s.
spleniform
splenii
musculi s.
splenis
s. cervicis muscle
facies colica s.
facies gastrica s.
facies renalis s.
facies visceralis s.
margo inferior s.

S

NOTES

splenis *(continued)*
    margo inferior corporis s.
    margo superior s.
    tunica fibrosa s.
    tunica serosa s.
spleniserrate
splenium, pl. **splenia**
    s. corporis callosi
splenius
    s. capitis
    s. capitis muscle
    s. cervicis
    s. cervicis muscle
    s. muscle of head
    s. muscle of neck
splenocolic ligament
splenogastric omentum
splenoid
splenolymphatic
splenolymph gland
splenonephric
splenopancreatic
splenophrenic
splenorenal
    s. angle
    s. ligament
splenorenale
    ligamentum s.
splenotomy
splenule
splenulus, pl. **splenuli**
splenunculus, pl. **splenunculi**
spline
    os pubis s.
    stapes s.
splint bone
split
    s. brain
    s. hand
spodophorous
spondyloschisis
spondylothoracic
spondylous
spongioblast
spongiocyte
spongiosa
    decidua s.
    substantia s.
spongiose part of male urethra
spongiosi
    cavernae corporis s.
    tunica albuginea corporis s.
spongiosum
    bulb of corpus s.
    cavernous space of corporus s.
    caverns of corpus s.
    cavity of corpus s.
    fibrous tunic of corpus s.
    stratum s.

    trabeculae of corpus s.
    tunica albuginea of corpus s.
spongy
    s. body of penis
    s. bone
    s. layer of female urethra
    s. layer of vagina
    s. part of male urethra
    s. spot
    s. substance
spontaneous
    s. amputation
    s. correction of placenta previa
    s. generation
sporoplasm
sporulation
spot
    acoustic s.
    blind s.
    blue s.
    corneal s.
    germinal s.
    Mariotte blind s.
    mongolian s.
    saccular s.
    Soemmering s.
    spongy s.
    utricular s.
    yellow s.
spreading depression
Sprengel deformity
spring ligament
sprout
    syncytial s.
spur
    calcarine s.
    Morand s.
    scleral s.
    vascular s.
spuria, pl. **spuriae**
    chordae tendineae spuriae
    costae spuriae
    glottis s.
    pelvis s.
    placenta s.
    vertebrae spuriae
spurious
spurium
    corpus luteum s.
    septum s.
squama, pl. **squamae**
    s. alveolaris
    frontal s.
    s. frontalis
    occipital s.
    s. occipitalis
    temporal s.
    s. temporalis
squamate

squamatization
squamocellular
squamocolumnar junction
squamofrontal
squamomast
squamomastoid suture
squamooccipital
squamoparietalis
    sutura s.
squamoparietal suture
squamopetrosal
squamosa, pl. **squamosae**
    sutura s.
squamosal
    s. border
    s. border of parietal bone
    s. margin
    s. margin of greater wing of
      sphenoid
    s. suture
squamosomastoidea
    sutura s.
squamosomastoid suture
squamosoparietal suture
squamosphenoid suture
squamosus
    margo s.
squamotemporal
squamotympanic fissure
squamous
    s. alveolar cells
    s. border
    s. border of parietal bone
    s. border of sphenoid bone
    s. cell
    s. epithelium
    s. margin
    s. part of frontal bone
    s. part of occipital bone
    s. part of temporal bone
    s. suture
    s. suture of cranium
    s. suture line
squamozygomatic
square wave stimuli
squarrose
squarrous
stab
    s. cell
    s. neutrophil
Staderini nucleus
staff cell

stagnant
    s. anoxia
    s. hypoxia
stagnation
stalk
    allantoic s.
    body s.
    connecting s.
    connective tissue s.
    s. of epiglottis
    infundibular s.
    optic s.
    pineal s.
    pituitary s.
    yolk s.
stalked hydatid
STA-MCA
    superficial temporary artery-middle
      cerebral artery
standard
    s. atmosphere
    s. bicarbonate
    s. pressure
    s. urea clearance
Stanley cervical ligament
Stannius ligature
stapedes (*pl. of* stapes)
stapedial
    s. branch of posterior tympanic
      artery
    s. branch of stylomastoid artery
    s. crus
    s. fold
    s. membrane
    s. nerve
    s. tendon
stapedialis
    plica s.
stapedii (*pl. of* stapedius)
stapediovestibular
stapedis
    basis s.
    caput s.
    crus anterius s.
    crus posterius s.
    ligamentum annulare s.
    ligamentum anulare s.
    membrana s.
    plica s.
stapedius, pl. **stapedii**
    s. muscle
    musculus s.

S

NOTES

stapedius *(continued)*
    nervus s.
    ramus s.
    s. tendon
stapes, pl. **stapedes, stapes**
    annular ligament of s.
    anterior crest of s.
    anterior crus of s.
    anterior limb of s.
    base of s.
    s. bone
    capitulum of s.
    crus of s.
    fixator muscle of base of s.
    fold of s.
    s. footplate
    head of s.
    neck of s.
    posterior crest of s.
    posterior crus of s.
    posterior limb of s.
    s. spline
    s. superstructure
staphyline
staphylion
star
    lens s.
    Verheyen s.'s
    Winslow s.'s
starch equivalent
Starling
    S. curve
    S. hypothesis
    S. law
starting friction
starvation
stathmokinesis
static
    s. compliance
    s. friction
    s. hysteresis
    s. reflex
    s. system
statoacoustic nerve
statoacusticus
    nervus s.
statoconia
statoconial membrane
statoconic membrane
statoconiorum
    membrana s.
statokinetic reflex
statoliths
statosphere
statotonic reflex
stature
status
    s. dysraphicus
    s. marmoratus

statvolt
staurion
steal phenomenon
steatogenesis
steeple skull
Steidele complex
stella, pl. **stellae**
    s. lentis hyaloidea
    s. lentis iridica
stellatae
    venae s.
    venulae s.
stellate
    s. cells of cerebral cortex
    s. cells of liver
    s. ganglion
    s. ligament
    s. reticulum
    s. vein
    s. vein of kidney
    s. venule
stellatum
    ganglion s.
stellula, pl. **stellulae**
    stellulae vasculosae
    stellulae verheyenii
    stellulae winslowii
stem
    s. bronchus
    s. cell
    descending brain s.
    infundibular s.
    tectum of brain s.
    transposition of arterial s.
stenion
stenobregmatic
stenocephalia
stenocephalic
stenocephalous
stenocephaly
stenocrotaphia
stenocrotaphy
Steno duct
stenosis, pl. **stenoses**
    double aortic s.
    hypertrophic pyloric s.
    subaortic s.
    subvalvar s.
stenothorax
Stensen
    S. canal
    S. duct
    S. foramen
    S. plexus
    S. vein
stephanial
stephanion
stercoraceous
stereocilia

stereocilium
stereotactic
stereotaxic localization
stereotaxis
stereotropic
stereotropism
sterna (*pl. of* sternum)
sternad
sternal
    s. angle
    s. articular surface of clavicle
    s. bar
    s. border
    s. branch of internal thoracic artery
    s. cartilage
    s. depression
    s. end of clavicle
    s. extremity of clavicle
    s. facet of clavicle
    s. fissure
    s. joint
    s. line
    s. line of pleural reflection
    s. membrane
    s. muscle
    s. notch
    s. part of diaphragm
    s. plane
    s. synchondroses
sternale
    planum s.
sternales
    rami s.
    synchondroses s.
sternalis
    incisura jugularis s.
    linea s.
    s. muscle
    musculus s.
Sternberg cells
Sternberg-Reed cells
sternebra, pl. **sternebrae**
sternen
sterni
    angulus s.
    corpus s.
    manubrium s.
    membrana s.
    mucro s.
    musculus triangularis s.

sternochondroscapularis
    musculus s.
sternochondroscapular muscle
sternoclavicular
    s. angle
    s. articular disk
    s. articulation
    s. joint
    s. junction
    s. ligament
    s. muscle
    s. notch
sternoclaviculare
    ligamentum s.
sternoclavicularia
    ligamenta s.
sternoclavicularis
    articulatio s.
    discus articularis s.
    musculus s.
sternocleidal
sternocleidomastoid
    s. branch of occipital artery
    s. branch of superior thyroid artery
    s. muscle
    s. region
    s. vein
sternocleidomastoidea
    regio s.
    vena s.
sternocleidomastoideus,
  pl. **sternocleidomastoidei**
    musculus s.
    ramus s.
sternocostal
    s. articulation
    s. head of pectoralis major muscle
    s. joint
    s. ligament
    s. part
    s. part of pectoralis major muscle
    s. surface of heart
    s. triangle of diaphragm
sternocostalis, pl. **sternocostales**
    articulationes sternocostales
    s. muscle
    pars s.
sternofascialis
    musculus s.
sternoglossal
sternohyoideus
    musculus s.

S

NOTES

sternohyoid muscle
sternoid
sternomanubrial junction
sternomastoid
    s. artery
    s. muscle
sternopagia
sternopericardiaca
    ligamenta s.
sternopericardial ligament
sternoschisis
sternothyroideus
    musculus s.
sternothyroid muscle
sternotracheal
sternovertebral
sternum, pl. **sterna**
    body of s.
    clavicular notch of s.
    costal notch of s.
    jugular notch of s.
    manubrium of s.
sternutation
steroid
    s. metabolic clearance rate
    s. production rate
    s. secretory rate
steroidogenesis
Stewart-Hamilton method
stichochrome cell
Stieda process
stigma, pl. **stigmata**
    malpighian s.
    ollicular s.
stillborn infant
Stilling
    S. canal
    S. column
    S. gelatinous substance
    S. nucleus
    S. raphe
still layer
stilus (*var. of* stylus)
stimulant
    diffusible s.
    general s.
    local s.
stimulation
stimulator
stimulus, pl. **stimuli**
    adequate s.
    conditioned s.
    heterologous s.
    homologous s.
    inadequate s.
    liminal s.
    maximal s.
    square wave stimuli
    subliminal s.

    subthreshold s.
    summation of stimuli
    supramaximal s.
    threshold s.
    s. threshold
    unconditioned s.
stippled epiphysis
stippling
stirrup
stoke
stoma, pl. **stomas, stomata**
    Fuchs stomas
stomach
    anterior wall of s.
    bare area of s.
    bed of s.
    body of s.
    cardia of s.
    cardiac gland of s.
    cardiac part of s.
    cardial part of s.
    chief cell of s.
    dumping s.
    fornix of s.
    foveolar cells of s.
    fundus of s.
    greater curvature of s.
    hourglass contraction of s. (HCS)
    lesser curvature of s.
    s. lining
    lymphatic ring of cardiac part
      of s.
    lymph node around cardia of s.
    miniature s.
    mucosa of s.
    mucous membrane of s.
    muscular coat of s.
    muscular layer of s.
    oblique fibers of muscular layer
      of s.
    Pavlov s.
    pit of s.
    posterior wall of s.
    pyloric part of s.
    pylorus of s.
    s. ruga
    ruga of s.
    rugal pattern of s.
    serosa of s.
    surface mucous cells of s.
    theca cells of s.
    s. tooth
    s. wall
stomachal
stomachic
stomal
stomas (*pl. of* stoma)
stomata (*pl. of* stoma)
stomatal

stomatic
stomatocyte
stomatodeum
stomatognathic
    s. system
stomocephalus
stomodeal
stomodeum
stool
storiform
strabismic
straight
    s. artery
    s. conjugate
    s. gyrus
    s. part of cricothyroid muscle
    s. seminiferous tubule
    s. sphenoidal sinus
    s. tubule of testis
    s. venous sinus
    s. venules of kidney
strain gauge
strait
    inferior s.
    pelvic s.
    superior pelvic s.
strap
    s. cell
    s. muscle
strata (*pl. of* stratum)
stratification
stratified
    s. ciliated columnar epithelium
    s. squamous epithelium
stratiform fibrocartilage
stratum, pl. **strata**
    s. aculeatum
    s. album profundum
    s. basale
    s. cerebrale retinae
    s. cinereum colliculi superioris
    s. circulare membranae tympani
    s. circulare musculi detrusoris vesicae
    s. circulare tunicae muscularis
    s. circulare tunicae muscularis coli
    s. circulare tunicae muscularis intestini tenuis
    s. circulare tunicae muscularis recti
    s. circulare tunicae muscularis ventriculi
    s. compactum

s. corneum epidermidis
s. corneum unguis
s. cutaneum membranae tympani
s. cylindricum
s. disjunctum
s. fibrosum
s. fibrosum capsulae articularis
s. fibrosum panniculi adiposi telae subcutaneae
s. fibrosum vaginae tendinis
s. functionale
s. ganglionare nervi optic
s. ganglionare retinae
s. gangliosum cerebelli
s. germinativum
s. germinativum unguis
s. granulosum cerebelli
s. granulosum epidermidis
s. granulosum folliculi ovarici vesiculosi
s. granulosum ovarii
s. griseum
s. griseum colliculi superioris
s. griseum medium
s. griseum profundum
s. helicoidale brevis gradus
s. helicoidale longi gradus
s. interolivare lemnisci
s. longitudinale tunicae muscularis
s. longitudinale tunicae muscularis coli
s. longitudinale tunicae muscularis intestini tenuis
s. longitudinale tunicae muscularis recti
s. longitudinale tunicae muscularis ventriculi
s. lucidum
malpighian s.
s. moleculare
s. moleculare cerebelli
s. moleculare retinae
s. musculosum panniculi adiposi telae subcutaneae
s. neuroepitheliale retinae
s. neuronorum piriformium
s. nucleare externum et internum retinae
s. nucleare externum retinae
s. nucleare internum retinae
s. opticum
s. papillare corii

S

NOTES

stratum (*continued*)
    s. pigmenti bulbi
    s. pigmenti corporis ciliaris
    s. pigmenti iridis
    s. pigmenti retinae
    s. plexiforme externum
    s. radiatum membranae tympani
    s. reticulare corii
    s. reticulare cutis
    s. spinosum epidermidis
    s. spongiosum
    s. subcutaneum
    s. synoviale
    s. zonale
streak
    primitive s.
stream
    hair s.
streaming movement
streblodactyly
Streeter bands
strength
    biting s.
strength-duration curve
stress
    shear s.
    tensile s.
    yield s.
stress-strain curve
stretch
    s. receptor
    s. reflex
stria, pl. **striae**
    acoustic striae
    auditory striae
    brown striae
    striae ciliares
    s. externa medullae renalis
    s. fornicis
    Gennari s.
    s. interna medullae renalis
    striae lancisi
    Langhans s.
    lateral longitudinal s.
    s. longitudinalis lateralis
    s. longitudinalis medialis
    s. mallearis
    malleolar s.
    medial longitudinal s.
    striae medullares ventriculi quarti
    s. medullaris thalami
    Nitabuch s.
    striae olfactoriae
    olfactory striae
    striae parallelae
    Retzius striae
    Rohr s.
    s. spinosa
    s. tecta

    terminal s.
    s. terminalis
    s. of thalamus
    s. vascularis of cochlea
    s. vascularis of cochlear duct
    s. vascularis ductus cochlearis
    s. ventriculi tertii
striata, pl. **striatae**
    membrana s.
    venae striatae
    zona s.
striatal
striate
    s. area
    s. artery
    s. body
    s. cortex
striated
    s. border
    s. duct
    s. membrane
    s. muscle
    s. muscular sphincter
striati
    cauda s.
    lamina medullaris lateralis corporis s.
    lamina medullaris medialis corporis s.
striation
    basal s.
striatonigral
striatum
    corpus s.
    lateral medullary lamina of corpus s.
    medial medullary lamina of corpus s.
    vein of corpus s.
striatus
    limbus s.
stricture
    musculocartilaginous s.
string
    auditory s.'s
stripe
    s. of Gennari
    Hensen s.
    mallear s.
    vascular s.
stroke
    s. output
    s. volume
    s. work index
stroma, pl. **stromata**
    s. glandulae thyroideae
    s. iridis
    s. of iris
    lymphatic s.

ovarian s.
s. ovarii
s. of ovary
s. plexus
s. of thyroid gland
s. of vitreous
s. vitreum
stromal tissue
stromic
stromuhr
Ludwig s.
strophocephaly
strophosomia
Stroud pectinated area
structura
structurae oculi accessoriae
structural
structure
accessory visual s.
collecting s.
elbow s.
fine s.
hilar s.
ligamentous s.
musculofascial s.
skeletal s.
subcortical s.
subtentorial s.
superficial s.
supraglottic s.
underlying s.
Struthers
ligament of S.
S. ligament
STT
scaphotrapeziotrapezoid
stye
styliform
styloauricularis
musculus s.
styloauricular muscle
styloglossus
s. muscle
musculus s.
stylohyal
stylohyoid
s. branch of facial nerve
s. ligament
s. muscle
stylohyoideum
ligamentum s.

stylohyoideus
musculus s.
ramus s.
styloid
s. cornu
s. process
s. process of fibula
s. process of radius
s. process of temporal bone
s. process of third metacarpal bone
s. process of ulna
s. prominence
radial s.
styloidea
prominentia s.
styloidei
vagina processus s.
stylolaryngeus
musculus s.
stylomandibulare
ligamentum s.
stylomandibular ligament
stylomastoid
s. artery
s. foramen
s. vein
stylomastoidea
arteria s.
vena s.
stylomastoideae
ramus stapedius arteriae s.
stylomastoideum
foramen s.
stylomaxillary ligament
stylopharyngeal
s. branch of glossopharyngeal
nerve
s. muscle
stylopharyngei
ramus musculi s.
stylopharyngeus
s. muscle
musculus s.
stylopodium
stylostaphyline
stylosteophyte
stylus, stilus
sub
nucleus s.
subabdominal
subabdominoperitoneal
subacromial bursa

S

NOTES

subacromialis
    bursa s.
subanal
subanconeus muscle
subaortic
    s. lymph node
    s. stenosis
subaortici
    lymphonodi s.
    lymphonodus iliaci communes s.
    nodi lymphoidei s.
subapicale
    segmentum s.
subapicalis
    ramus s.
subapical segment
subaponeurotic space
subarachnoid
    s. cavity
    s. sac
    s. septum
    s. space
subarachnoidal
    s. cistern
    s. sinus
subarachnoidale
    cavum s.
subarachnoidea
    cavum s.
subarachnoideae
    cisternae s.
subarachnoideale
    cavum s.
subarachnoideales
subarachnoideum
    spatium s.
subarcuata
    fossa s.
subarcuate fossa
subarcuatus
    hiatus s.
subareolar region
subastragalar
subaural
subauricular region
subaxial
subaxillary
subbasal
subbrachycephalic
subcalcarine
subcallosa
    area s.
subcallosal
    s. area
    s. gyrus
subcallosus
    fasciculus s.
    gyrus s.
subcapsular cortex

subcardinal vein
subcarinal angle
subcartilaginous
subcecal fossa
subcellular
subception
subchondral bone
subchorial
    s. lake
    s. space
subchorionic
subchoroidal
subclavia
    ansa s.
    arteria s.
    vena s.
subclaviae
    plexus periarterialis arteriae s.
    sulcus arteriae s.
    sulcus costae arteriae s.
    sulcus venae s.
subclavian
    s. artery
    s. duct
    s. groove
    s. loop
    s. lymphatic trunk
    s. lymph node
    s. muscle
    s. nerve
    s. plexus
    s. sulcus
    s. sulcus of lung
    s. triangle
    s. vein
subclavianus
    sulcus s.
subclavicular
subclavii
    sulcus musculi s.
subclavius
    groove for s.
    s. muscle
    musculus s.
    nervus s.
    plexus s.
    sulcus s.
    truncus s.
subcollateral gyrus
subcommissural organ
subconjunctival
subcoracoid bursa
subcortex
subcortical structure
subcostal
    s. angle
    s. arch
    s. area
    s. artery

s. groove
s. line
s. muscle
s. nerve
s. plane
s. vein

subcostale
planum s.

subcostales
musculi s.

subcostalis
arteria s.
linea s.
musculus s.
nervus s.
rami dorsales arteriae s.
ramus dorsales arteriae s.

subcostosternal
subcranial
subcruralis
subcrural muscle
subcrureus
subcutanea
s. abdominis
pars s.
tela s.

subcutaneae
stratum fibrosum panniculi adiposi
telae s.
stratum musculosum panniculi
adiposi telae s.

subcutaneous
s. acromial bursa
s. adipose tissue
s. bursa of the laryngeal
prominence
s. bursa of lateral malleolus
s. bursa of medial malleolus
s. bursa of teres major
s. bursa of tibial tuberosity
s. bursa of tuberosity of tibia
s. calcaneal bursa
s. connective tissue
s. fascia
s. fat
s. infrapatellar bursa
s. olecranon bursa
s. part
s. part of external anal sphincter
s. plane
s. portion of external anal
sphincter

s. prepatellar bursa
s. ring
s. scalp
s. temporal nerve
s. tissue of penis
s. tissue of perineum
s. vein of abdomen
s. venous arch at root of finger

subcutaneously
subcutaneum
stratum s.

subcutaneus
subcuticular
s. fat
s. level

subcutis
subdeltoid bursa
subdeltoidea
bursa s.

subdermic
subdiaphragmatic
subdigastric
s. lymph field
s. node

subdorsal
subdural
s. cavity
s. cleavage
s. cleft
s. fluid
s. space

subdurale
cavum s.
spatium s.

subendocardial
s. branch of atrioventricular
bundles
s. conducting system of heart
s. layer

subendothelial layer
subendothelium
subendymal
subependymal
subepicardial fat
subepidermal
subepidermic
s. bulla

subepithelial
subepithelial plexus
subepithelium
subfascial
s. prepatellar bursa

NOTES

577

subfertility
subfissure
subfolium
subfornical organ (SFO)
subgaleal space
subgemmal
subgerminal cavity
subgingival space
subglenoid
subglossal
subglottic
   s. area
   s. cavity
subgranular
subhepatic
   s. recess
   s. space
subhepatici
   recessus hepatorenalis recessus s.
subhepaticus
   recessus s.
subhyaloid
subhyoid
   s. bursa
   s. region
subhyoidean
subicular
subiculum, pl. subicula
   s. promontorii
subiliac
subilium
subinguinal
   s. fossa
   s. triangle
subintegumental
subintimal
subjacent
subject
   human s.
subjugal
sublabial
sublenticular limb of internal capsule
subliminal
   s. stimulus
   s. thirst
sublimis
   chiasm of flexor s.
   flexor digitorum s. (FDS)
   musculus flexor digitorum s.
   s. tendon
sublingual
   s. artery
   s. bursa
   s. caruncula
   s. crescent
   s. cyst
   s. duct
   s. fold
   s. fossa

   s. ganglion
   s. gland
   s. lymph node
   s. nerve
   s. papilla
   s. pit
   s. saliva
   s. vein
sublinguale
   ganglion s.
sublingualis
   arteria s.
   bursa s.
   caruncula s.
   fovea s.
   glandula s.
   nervus s.
   plica s.
   radix sensoria ganglii s.
   radix sympathica ganglii s.
   vena s.
sublobular
   s. vein
sublumbar
subluminal
submammary
submandibular
   s. duct
   s. fossa
   s. ganglion
   s. lymph node
   s. salivary gland
   s. triangle
submandibulare
   ganglion s.
   ramus sympathicus ad ganglion s.
   ramus sympathicus sympatheticus
     ad ganglion s.
   trigonum s.
submandibulares
   lymphonodi s.
   nodi lymphoidei s.
submandibularis
   ductus s.
   fossa s.
   fovea s.
   glandula s.
   radix parasympathica ganglii s.
   radix sensoria ganglii s.
   radix sympathica ganglii s.
   rami glandulares ganglii s.
submarginal
submaxilla
submaxillaris
   ductus s.
   fovea s.
submaxillary
   s. duct
   s. fossa

s. ganglion
s. gland
s. nerve
s. region
s. triangle
submedial
submedian
submembranous
submental
s. artery
s. gland
s. lymph node
s. region
s. triangle
s. vein
submentale
trigonum s.
submentales
lymphonodi s.
nodi lymphoidei s.
submentalis
arteria s.
vena s.
submicronic
submicroscopic
submitral area
submucosa
tela s.
tunica s.
submucosal
s. connective tissue
s. nervous plexus
s. space
s. venous plexus
s. vessel
submucosus
plexus nervosus s.
submucous
subnasal point
subneural apparatus
subnucleus
suboccipital
s. muscle
s. nerve
s. part of vertebral artery
s. region
s. triangle
s. venous plexus
suboccipitales
musculi s.

suboccipitalis
nervus s.
plexus venosus s.
suboccipitobregmatic diameter
suborbital
subpapillare
rete s.
subpapillary
s. layer
s. network
s. zone
subparietalis
sulcus s.
subparietal sulcus
subpatellar
subpectoral
subpelviperitoneal
subpericardial
subperiosteal bone
subperiosteally
subperitoneal
s. fascia
s. space
subperitonealis
fascia s.
subperitoneoabdominal
subperitoneopelvic
subpetrosal
subpharyngeal
subphrenic
s. recess
s. space
subphrenici
recessus s.
subphrenicus
recessus s.
subpial
subplacental
subplasmalemmal dense zone
subpleural mediastinal plexus
subplexal
subpopliteal recess
subpopliteus
recessus s.
subpreputial
subpubic
s. angle
s. hernia
subpubicus
angulus s.
subpulmonary

S

NOTES

subpylorici
    nodi s.
    nodi lymphoidei s.
subpyloric lymph node
subpyramidal fossa
subquadricipital muscle
subretinal fluid
subsartorial
    s. canal
    s. fascia
subscapular
    s. angle
    s. axillary lymph node
    s. branch of axillary artery
    s. bursa
    s. fossa
    s. muscle
    s. nerve
    s. vein
subscapulares
    nervi s.
    nodi lymphoidei axillares s.
    rami s.
subscapularis
    arteria s.
    bursa subtendinea musculi s.
    fascia s.
    fossa s.
    s. muscle
    musculus s.
    nervus s.
    subtendinous bursa of s.
    s. tendon
subscleral
subsclerotic
subsegmental area
subseptale
    foramen s.
subseptate uterus
subseptus
    uterus s.
subserosa
    tela s.
subserosal
subserosus
    plexus nervosus s.
subserous
    s. fascia
    s. layer
    s. nervous plexus
subsigmoid fossa
subspinous
substage
substance
    alpha s.
    anterior perforated s.
    basophil s.
    central gray s.
    chromidial s.

    chromophil s.
    compact s.
    cortical s.
    filar s.
    gelatinous s.
    gray s.
    ground s.
    H s.
    innominate s.
    interspongioplastic s.
    s. of lens of eye
    medullary s.
    müllerian inhibiting s.
    neurosecretory s.
    Nissl s.
    posterior perforated s.
    proper s.
    released s.
    reticular s.
    Rolando gelatinous s.
    Schwann white s.
    spongy s.
    Stilling gelatinous s.
    threshold s.
    tigroid s.
    white s.
substantia, pl. **substantiae**
    s. adamantina
    s. alba
    s. basophilia
    s. cinerea
    s. compacta
    s. compacta ossium
    s. corticalis
    s. eburnea
    s. ferruginea
    s. fundamentalis
    s. gelatinosa
    s. gelatinosa centralis
    s. glandularis prostatae
    s. grisea
    s. grisea centralis
    s. innominata
    s. lentis
    s. medullaris
    s. muscularis prostatae
    s. nigra
    s. ossea dentis
    s. perforata anterior
    s. perforata posterior
    s. propria of cornea
    s. propria corneae
    s. propria membranae tympani
    s. propria of sclera
    s. reticularis
    s. reticulofilamentosa
    s. spongiosa
    s. trabecularis
    s. vitrea

S

substernal
    s. angle
    s. gland
substernomastoid
substratum
substructure
subsuperior
    ramus s.
    s. segment
subsuperius
subsurface cisterna
subtalar
    s. articulation
    s. joint
subtalaris
    articulatio s.
subtarsal
subtegumental
subtendinous
    s. bursae of gastrocnemius muscle
    s. bursa of gastrocnemius
    s. bursa of iliacus
    s. bursa of infraspinatus
    s. bursa of latissimus dorsi
    s. bursa of sartorius
    s. bursa of subscapularis
    s. bursa of tibialis anterior
    s. bursa of trapezius
    s. bursa of triceps brachii
    s. iliac bursa
    s. prepatellar bursa
subtentorial structure
subterminal
subthalamic nucleus
subthalamus
subthreshold stimulus
subthyroideus
subtibiale
    os s.
subtrapezial
subtrochanteric
subtrochlear
subtuberal
subtympanic
subumbilical space
subungual
subunguial
suburethral
subvaginal
subvalvar stenosis
subvalvular

subvertebral muscle
subvesical
    s. duct
    s. fascia
subvitrinal
subxiphoid
subzonal
subzygomatic
succagogue
succedaneous
    s. dentition
    s. tooth
succedaneus
    dens s.
succenturiate
    s. lobe
    s. placenta
succenturiatus
    lien s.
successional tooth
succi (*pl. of* succus)
succingens
    membrana s.
succinic acid cycle
succulent node
succus, pl. **succi**
sucking
    s. cushion
    s. pad
suckling reflex
Sucquet
    S. anastomosis
    S. canal
Sucquet-Hoyer
    S.-H. anastomosis
    S.-H. canal
suction
suctorial pad
sudanophilia
sudanophobic zone
sudation
Sudeck critical point
sudomotor
    s. fibers
    s. nerve
sudoriferae
    corpus glandulae s.
    glandulae s.
sudoriferous
    s. duct
    s. gland

NOTES

sudoriferus
    ductus s.
    porus s.
sudorific
sudoriparous gland
sudorometer
suffocate
suffocation
sulcal artery
sulcate
sulcated
sulci (*pl. of* sulcus)
sulciform
sulcomarginal tract
sulcular
    s. epithelium
    s. fluid
sulculus, pl. **sulculi**
sulcus, pl. **sulci**
    alveolobuccal s.
    alveololabial s.
    alveololingual s.
    s. ampullaris
    ampullary s.
    s. angle
    s. angularis
    anterior interventricular s.
    anterior parolfactory s.
    anterolateral s.
    s. anthelicis transversus
    aortic s.
    s. aorticus
    s. arteriae occipitalis
    s. arteriae subclaviae
    s. arteriae subclaviae costae primae
    s. arteriae temporalis mediae
    s. arteriae vertebralis
    sulci arteriosi
    artery of central s.
    artery of postcentral s.
    artery of precentral s.
    atrioventricular s.
    s. for auditory tube
    s. auriculae anterior
    basilar s.
    s. basilaris pontis
    s. bicipitalis lateralis
    s. bicipitalis medialis
    s. bicipitalis radialis
    s. bicipitalis ulnaris
    sulci in brain
    buccal s.
    calcaneal s.
    s. calcanei
    calcarine s.
    s. calcarinus
    callosal s.
    s. callosomarginalis
    cardiohepatic s.

    s. caroticus
    carotid s.
    s. carpi
    central s.
    s. centralis
    cerebellar sulci
    cerebral s.
    sulci cerebri
    chiasmatic s.
    cingulate s.
    s. cinguli
    s. of cingulum
    s. circularis insulae
    collateral s.
    s. collateralis
    coronal s.
    s. coronarius
    coronary s.
    s. of corpus collasum
    s. costae
    s. costae arteriae subclaviae
    costophrenic s.
    s. cruris helicis
    sulci cutis
    deltopectoral s.
    s. ethmoidalis
    external spiral s.
    fimbriodentate s.
    s. fimbriodentatus
    s. frontalis inferior
    s. frontalis medius
    s. frontalis superior
    s. frontomarginalis
    gingival s.
    s. gingivalis
    gingivobuccal s.
    gingivolabial s.
    gingivolingual s.
    s. gluteus
    s. for greater palatine nerve
    s. hamuli pterygoidei
    s. of heart
    s. hippocampi
    hypothalamic s.
    s. hypothalamicus
    inferior frontal s.
    inferior petrosal s.
    inferior temporal s.
    s. infraorbitalis
    infrapalpebral s.
    s. infrapalpebralis
    inner spiral s.
    s. intermedius anterior
    s. intermedius posterior
    internal spiral s.
    interparietal s.
    intertubercular s.
    s. intertubercularis
    s. interventricularis anterior

s. interventricularis cordis
s. interventricularis posterior
s. intragracilis
intraparietal s.
s. intraparietalis
labial s.
labiodental s.
s. lacrimalis
lateral cerebral s.
s. lateralis anterior
s. lateralis cerebri
s. lateralis posterior
lateral occipital s.
s. limitans
s. limitans fossae rhomboideae
lingual s.
lip s.
longitudinal s.
lunate s.
s. lunatus cerebri
malleolar s.
s. malleolaris
marginal branch of cingulate s.
marginal branch of
  parietooccipital s.
s. marginalis
s. medialis cruris cerebri
median frontal s.
s. medianus linguae
s. medianus posterior medullae
  spinalis
s. medianus ventriculi quarti
s. of meningeal artery
mentolabial s.
s. mentolabialis
middle frontal s.
middle temporal s.
s. for middle temporal artery
Monro s.
s. musculi subclavii
s. mylohyoideus
nasolabial s.
s. nasolabialis
s. nervi oculomotorii
s. nervi petrosi majoris
s. nervi petrosi minoris
s. nervi radialis
s. nervi spinalis
s. nervi ulnaris
nymphocaruncular s.
s. nymphocaruncularis
nymphohymenal s.

s. obturatorius
s. of occipital artery
s. occipitalis lateralis
s. occipitalis superior
s. occipitalis transversus
occipitotemporal s.
s. occipitotemporalis
oculomotor s.
s. olfactorius
s. olfactorius cavi nasi
olfactory s.
orbital sulci
sulci orbitales
outer spiral s.
sulci palatini
s. palatinus
s. palatinus major
s. palatovaginalis
s. palpebralis superior
sulci paracolici
paraglenoid s.
s. paraglenoidalis
paramedial s.
parietooccipital s.
s. parietooccipitalis
s. parolfactorius anterior
periconchal s.
peroneal s.
s. of petrosal nerve
s. for pharyngotympanic tube
s. popliteus
postauricular s.
postcentral s.
s. postcentralis
s. posterior auriculae
posterior branch of lateral
  cerebral s.
posterior intermediate s.
posterior interventricular s.
posterior median s.
posterior parolfactory s.
posterior ramus of lateral
  cerebral s.
posterolateral s.
preauricular s.
precentral s.
s. precentralis
prechiasmatic s.
s. prechiasmaticus
s. promontorii
s. promontorii cavitatis tympanicae

NOTES

sulcus *(continued)*
    s. of promontory of tympanic cavity
    s. of pterygoid hamulus
    s. pterygopalatinus
    s. pulmonalis
    pulmonary s.
    retroauricular s.
    retrocentral s.
    retroglandular s.
    rhinal s.
    s. rhinalis
    sagittal s.
    s. of sclera
    scleral s.
    sigmoid s.
    s. sinus petrosi inferioris
    s. sinus petrosi superioris
    s. sinus sagittalis superioris
    s. sinus sigmoidei
    s. sinus transversi
    skin sulci
    s. spinosus
    s. spiralis externus
    s. spiralis internus
    subclavian s.
    s. of subclavian artery
    s. subclavianus
    s. subclavius
    subparietal s.
    s. subparietalis
    superior frontal s.
    superior longitudinal s.
    superior occipital s.
    superior petrosal s.
    superior temporal s.
    supraacetabular s.
    s. supraacetabularis
    talar s.
    s. tali
    sulci temporales trans
    s. temporalis inferior
    s. temporalis medius
    s. temporalis superior
    s. tendinis musculi fibularis longi
    s. tendinis musculi flexoris hallucis longi
    s. tendinis musculi peronei longi
    terminal s.
    s. terminalis
    terminalis s.
    s. terminalis atrii dextri
    s. terminalis cordis
    s. terminalis linguae
    tonsillolingual s.
    transverse occipital s.
    s. for transverse sinus
    transverse temporal sulci
    s. tubae auditoriae

    Turner s.
    tympanic s.
    s. tympanicus
    s. of umbilical vein
    s. for vena cava
    s. venae cavae
    s. venae cavae cranialis
    s. venae subclaviae
    s. venae umbilicalis
    sulci venosi
    s. ventralis
    s. for vertebral artery
    s. verticalis
    vomeral s.
    s. vomeralis
    s. vomeris
    s. vomerovaginalis
    Waldeyer s.
sulfate respiration
sulfation factor
summation of stimuli
superacromial
superactivity
superanal
supercilia (*pl. of* supercilium)
superciliaris
    arcus s.
superciliary
    s. arch
    s. depressor muscle
    s. ridge
supercilii
    depressor s.
    musculus corrugator s.
    musculus depressor s.
supercilium, pl. **supercilia**
    corrugator supercilia
superexcitation
superfetation
superficial
    s. abdominal fascia
    s. anterior cervic
    s. anterior larynx
    s. anterior wall
    s. back muscle
    s. blood vessel
    s. brachial artery
    s. branch
    s. branch of lateral plantar nerve
    s. branch of medial circumflex femoral artery
    s. branch of medial plantar artery
    s. branch of radial nerve
    s. branch of superior gluteal artery
    s. branch of transverse cervical artery
    s. branch of ulnar nerve
    s. cardiac nervous plexus
    s. cerebral vein

s. cervical artery
s. cervical nerve
s. circumflex iliac artery
s. circumflex iliac vein
s. cleavage
s. and deep external pudendal
artery
s. dorsal sacrococcygeal ligament
s. dorsal vein
s. dorsal vein of clitoris
s. dorsal vein of penis
s. dorsum of hand
s. ectoderm
s. epigastric artery
s. epigastric vein
s. fascia of penis
s. fascia of scrotum
s. femoral arch
s. fibular nerve
s. flexor muscle
s. flexor muscle of finger
s. head
s. head of flexor pollicis brevis
s. inguinal fascia
s. inguinal lymph node
s. inguinal ring
s. investing fascia of perineum
s. lamina
s. lateral cervical lymph node
s. layer
s. layer of deep cervical fascia
s. layer of the levator palpebrae
superioris
s. layer of temporal fascia
s. lingual muscle
s. lymphatic
s. lymph vessel
s. middle cerebral vein
s. middle petrosal nerve
s. neck
s. orbit
s. origin
s. palmar arterial arch
s. palmar branch of radial artery
s. palmar vein
s. palmar venous arch
s. palm of hand
s. parotid lymph node
s. part of anterior flexor
compartment of forearm
s. part of external anal sphincter
s. part of masseter muscle

s. part of parotid gland
s. part of posterior plantar flexor
compartment of leg
s. perineal artery
s. perineal fascia
s. perineal muscle
s. perineal pouch
s. perineal space
s. perineal vein
s. peroneal nerve
s. popliteal lymph node
s. posterior sacrococcygeal ligament
s. radial nerve
s. reflex
s. sclera
s. structure
s. temporal branch of
auriculotemporal nerve
s. temporal branch of external
carotid artery
s. temporalis artery
s. temporal plexus
s. temporal vein
s. temporary artery-middle cerebral
artery (STA-MCA)
s. transverse fiber
s. transverse metacarpal ligament
s. transverse metatarsal ligament
s. transverse muscle of perineum
s. transverse perineal muscle
s. vertebral vein
s. volar artery

superficiale
ligamentum metacarpale
transversum s.
ligamentum metatarsale
transversum s.
ligamentum sacrococcygeum
dorsale s.
ligamentum sacrococcygeum
posterius s.
spatium perinei s.
vas lymphaticum s.

superficialis, pl. **superficiales**
annulus inguinalis s.
arcus venosus palmaris s.
arcus volaris s.
arteria brachialis s.
arteria cervicalis s.
arteria circumflexa iliaca s.
arteria circumflexa ilium s.
arteria epigastrica s.

NOTES

superficialis (*continued*)
- arteria temporalis s.
- brachialis s.
- cerebri media s.
- circumflexa ilium s.
- dorsales clitoridis superficiales
- dorsalis penis superficiales
- epigastrica s.
- fascia investiens perinei s.
- fascia pelvis s.
- fascia penis s.
- fibrae intercrurales anuli inguinalis s.
- flexor digitorum s. (FDS)
- iacus s.
- lamina s.
- lymphonodi cervicales laterales superficiales
- lymphonodi inguinales superficiales
- lymphonodi parotidei superficiales
- lymphonodus cervicales anteriores superficiales
- musculus flexor digitorum s.
- musculus transversus perinei s.
- nervi digitales dorsales nervi fibularis s.
- nervus cervicalis s.
- nervus fibularis s.
- nervus peroneus s.
- nodi lymphoidei cervicales anteriores superficiales
- nodi lymphoidei cervicales laterales superficiales
- nodi lymphoidei inguinales superficiales
- nodi lymphoidei parotidei superficiales
- pars s.
- peroneus s.
- plexus nervosus cardiacus s.
- plexus periarterialis arteriae temporalis s.
- rami auriculares anteriores arteriae temporalis s.
- rami musculares nervi fibularis s.
- ramus frontalis arteriae temporalis s.
- ramus parietalis arteriae temporalis s.
- ramus parotidei arteriae temporalis s.
- temporales superficiales
- temporalis s.
- transversus perinei s.
- vena cerebri media s.
- vena circumflexa iliaca s.
- vena circumflexa ilium s.
- venae cerebri superficiales
- venae dorsales clitoridis superficiales
- venae dorsales penis superficiales
- vena epigastrica s.
- venae temporales superficiales
- s. volae

superficies
superfuse
superfusion
supergenual
superificialis
- caput humeroulnare musculi flexoris digitorum s.

superimpregnation
superior
- s. aberrant ductule
- aditus glottidis s.
- s. alveolar artery
- s. alveolar nerve
- s. ampullar nerve
- anastomotica s.
- s. anastomotic vein
- s. angle of scapula
- s. annulus
- s. anterior et posterior
- apertura pelvis s.
- apertura thoracis s.
- arcus dentalis s.
- arcus palpebralis s.
- area vestibularis s.
- arteria collateralis ulnaris s.
- arteriae lobares inferior et s.
- arteria epigastrica s.
- arteria glutea s.
- arteria hypophysialis s.
- arteria labialis s.
- arteria laryngea s.
- arteria lingularis s.
- arteria mesenterica s.
- arteria pancreaticoduodenalis s.
- arteria phrenica s.
- arteria rectalis s.
- arteria segmentalis s.
- arteria suprarenalis s.
- arteria thoracica s.
- arteria thyroidea s.
- arteria tympanica s.
- arteria vesicalis s.
- arteriola macularis s.
- arteriola nasalis retinae s.
- arteriola temporalis retinae s.
- s. articular facet
- s. articular facet of atlas
- s. articular facet of rib
- s. articular facet of sacrum
- s. articular pit of atlas
- s. articular process
- s. articular process of sacrum
- s. articular process of vertebra

s. articular surface of atlas
s. articular surface of tibia
s. articulating process
s. aspect
s. astragalonavicular ligament
auricularis s.
s. auricular muscle
s. azygos vein
s. basal vein
s. belly of omohyoid muscle
s. border
s. border of body of pancreas
s. border of petrous part of
temporal bone
s. border of scapula
s. border of spleen
s. border of suprarenal gland
s. branch
s. branch of the oculomotor nerve
s. branch of the pubic bone
s. branch of the right and left
inferior pulmonary vein
s. branch of the superior gluteal
artery
s. branch of the transverse cervical
nerve
bronchus lobaris s.
s. bulb of internal jugular vein
s. bursa of biceps femoris
bursa musculi bicipitis femoris s.
cardiacus cervicalis s.
s. carotid ganglion
s. carotid triangle
s. cerebellar artery
s. cerebellar peduncle
s. cerebellar vein
cerebelli s.
s. cerebral vein
s. cervical cardiac branch of vagus
nerve
s. cervical chain
s. cervical ganglion
s. choroid vein
cisterna s.
s. cluneal nerve
collateralis ulnaris s.
s. colliculus
colliculus s.
s. commissure
s. commissure of lip
concha nasalis s.
s. constrictor muscle of pharynx

s. constrictor pharyngeal muscle
s. coronary artery
s. costal facet
s. costal facet of vertebra
s. costal pit
s. costotransverse ligament
s. crus
s. cul-de-sac
cutaneus brachii lateralis s.
s. dental arch
s. dental branch of superior dental
plexus
s. dental nerve
s. dental nervous plexus
s. dental rami
s. dorsal surface of tongue
ductulus aberrans s.
s. duodenal flexure
s. duodenal fold
s. duodenal fossa
s. duodenal recess
epigastrica s.
s. epigastric artery
s. epigastric vein
s. esophageal sphincter
s. extensor retinaculum
s. extensor retinaculum of ankle
extremitas s.
s. extremity
s. extremity of kidney
s. eyelid
s. facet of sacrum
s. facet of trochlear of talus
s. facial index
fascia diaphragmatis urogenitalis s.
s. fascia of pelvic diaphragm
s. fascia of urogenital diaphragm
fasciculus longitudinalis s.
s. fibular retinaculum
s. fibular retinaculum of ankle
fissura orbitalis s.
flexura duodeni s.
s. flexure of duodenum
s. fornix
fornix conjunctivae s.
s. fovea
fovea costalis s.
s. frontal convolution
s. frontal gyrus
s. frontal sulcus
s. ganglion of glossopharyngeal
nerve

S

NOTES

superior *(continued)*
    s. ganglion of vagus nerve
    s. gastric lymph node
    gemellus s.
    s. gemellus muscle
    s. gingival branch of superior dental plexus
    glutea s.
    s. gluteal artery
    s. gluteal nerve
    s. gluteal vein
    gluteus s.
    gyrus frontalis s.
    gyrus temporalis s.
    s. hemorrhoidal artery
    s. hemorrhoidal plexus
    s. hemorrhoidal vein
    s. horn
    s. horn of falciform margin of saphenous opening
    s. horn of saphenous opening
    s. horn of thyroid cartilage
    s. hypogastric nervous plexus
    s. hypophysial artery
    s. ileocecal recess
    s. iliac spine
    incisura thyroidea s.
    s. inferior segment of lung
    s. intercostal artery
    s. intercostal vein
    s. internal parietal artery
    s. labial branch of facial artery
    s. labial branch of infraorbital nerve
    labialis s.
    s. labial region
    s. labial vein
    s. lacrimal duct
    s. lacrimal gland
    s. lacrimal papilla
    s. lacrimal puncta
    laryngea s.
    s. laryngeal artery
    s. laryngeal cavity
    s. laryngeal nerve
    s. laryngeal vein
    laryngeus s.
    s. larynx
    s. lateral brachial cutaneous nerve
    s. lateral cutaneous nerve of arm
    s. lateral genicular artery
    s. lateral genicular vein
    s. ligament of epididymis
    ligament of epididymis inferior and s.
    s. ligament of incus
    s. ligament of malleus
    s. limb
    s. limb of ansa cervicalis

    linea nuchae s.
    linea temporalis s.
    s. lingular branch of lingular branch of superior lobar left pulmonary artery
    s. lingular bronchopulmonary segment S IV
    s. lobe of heart
    s. lobe of right/left lung
    lobulus parietalis s.
    lobulus semilunaris s.
    s. longitudinal fasciculus
    s. longitudinal muscle of tongue
    s. longitudinal sinus
    s. longitudinal sulcus
    s. lumbar facet of vertebra
    macula cribrosa s.
    s. macular arteriole
    s. macular venule
    s. margin
    s. margin of aorta
    s. margin of cerebral hemisphere
    s. margin of orbit
    s. margin of rib
    margo s.
    s. maxilla
    s. maxillary nerve
    s. meatus
    meatus nasi s.
    s. medial genicular artery
    s. medial genicular vein
    s. mediastinum
    s. medullary velum
    s. member
    mesenterica s.
    s. mesenteric artery (SMA)
    s. mesenteric ganglion
    s. mesenteric lymph node
    s. mesenteric nervous plexus
    s. mesenteric vein
    musculus auricularis s.
    musculus constrictor pharyngis s.
    musculus gemellus s.
    musculus longitudinalis s.
    musculus obliquus s.
    musculus obliquus capitis s.
    musculus rectus s.
    musculus serratus posterior s.
    musculus tarsalis s.
    s. nasal arteriole of retina
    s. nasal concha
    s. nasal retinal arteriole
    s. nasal retinal venule
    s. nasal venule of retina
    nervus cardiacus cervicalis s.
    nervus cutaneus brachii lateralis s.
    nervus gluteus s.
    nervus laryngeus s.
    norma s.

s. nuchal line
nucleus centralis tegmenti s.
nucleus salivatorius s.
s. oblique muscle
s. oblique muscle of head
s. oblique tendon
obliquus capitis s.
s. occipital gyrus
s. occipital sulcus
oliva s.
s. olivary nucleus
s. olive
s. omental recess
s. ophthalmic vein
s. orbital fissure
s. ossicular chain
palpebra s.
s. palpebral arterial arch
s. palpebral region
s. palpebral vein
s. pancreaticoduodenal artery
s. parathyroid gland
s. parietal gyrus
s. parietal lobe
s. parietal lobule
s. part of diaphragmatic surface of
  liver
s. part of duodenum
s. part of lingular vein of left
  superior pulmonary vein
s. part of vestibular ganglion
s. part of vestibulocochlear nerve
pedunculus cerebellaris s.
s. pelvic aperture
s. pelvic strait
s. peroneal retinaculum
s. petrosal sinus
s. petrosal sulcus
s. pharyngeal constrictor muscle
s. pharyngeal constrictor nerve
s. phrenic artery
s. phrenic lymph node
s. phrenic vein
plexus hypogastricus s.
plexus mesentericus s.
plexus nervosus dentalis s.
plexus nervosus hypogastricus s.
plexus nervosus mesentericus s.
plexus nervosus rectalis s.
plexus rectalis s.
plexus thyroideus s.
plica duodenalis s.

s. pole
s. pole of calyx
s. pole of kidney
s. pole of testis
s. pole of thyroid
poli lienalis inferior et s.
poli renales inferior et s.
poli renalis inferior et s.
polus s.
s. portion of mesencephalon
s. posterior serratus muscle
processus articularis s.
s. pubic ligament
s. pubic ramus
s. pulmonary vein
s. quadrigeminal brachium
s. radioulnar joint
radix s.
ramus s.
s. ramus
ramus lingularis s.
ramus vermis s.
s. recess of lesser peritoneal sac
s. recess of omental bursa
s. recess of tympanic membrane
recessus duodenalis s.
recessus ileocecalis s.
recessus membranae tympani s.
s. rectal artery
s. rectal fold
rectalis s.
s. rectal lymph node
s. rectal nervous plexus
s. rectal vein
rectus s.
s. rectus
s. rectus muscle
s. renal segment
s. retinaculum of extensor muscle
s. root
s. root of ansa cervicalis
s. root of cervical loop
s. root of vestibulocochlear nerve
s. sagittal sinus
s. salivary nucleus
s. segment
s. segmental artery
s. segmental artery of kidney
s. segment of lung
s. semilunar lobule
serratus posterior s.
sinus petrosus s.

S

NOTES

superior *(continued)*
sinus sagittalis s.
spinae geniorum inferior et s.
spina iliaca anterior s.
spina iliaca posterior s.
spina mentalis inferior et s.
s. spine
sulcus frontalis s.
sulcus occipitalis s.
sulcus palpebralis s.
sulcus temporalis s.
s. suprarenal artery
s. surface of talus
tarsalis s.
s. tarsal muscle
tarsus s.
s. tarsus
tela choroidea s.
s. temporal arteriole of retina
s. temporal branch of retinal artery
s. temporal convolution
s. temporal fissure
s. temporal gyrus
s. temporal line
s. temporal line of parietal bone
s. temporal retinal arteriole
s. temporal retinal venule
s. temporal sulcus
s. temporal venule of retina
s. thalamostriate vein
s. thoracic aperture
s. thoracic artery
s. thoracic facet of vertebra
s. thyroid artery
thyroidea s.
s. thyroid gland
s. thyroid notch
s. thyroid plexus
s. thyroid tubercle
s. thyroid vein
s. tibial articulation
s. tibiofibular joint
s. tracheobronchial lymph node
s. transverse scapular ligament
truncus s.
s. trunk
s. trunk of brachial plexus
s. turbinate
s. turbinated bone
tympanic s.
s. tympanic artery
s. ulnar collateral artery
s. vagal ganglion
s. veins of cerebellar hemisphere
s. vein of vermis
vena anastomotica s.
vena azygos minor s.
vena basalis s.
vena cava s.

s. vena cava
vena choroidea s.
vena hemorrhoidalis s.
vena labialis s.
vena laryngea s.
vena mesenterica s.
vena ophthalmica s.
vena rectalis s.
vena thalamostriata s.
vena thyroidea s.
vena vermis s.
venula macularis s.
s. venula nasalis retinae
venula nasalis retinae s.
venula temporalis retinae s.
s. vermian branch of superior
cerebellar artery
s. vertebral notch
s. vesical artery
s. vestibular area
s. vestibular nucleus
s. wall of orbit
zygapophysis s.

superiores
alveolares s.
arteriae lobares s.
cerebelli s.
cerebri s.
clunium s.
epigastrica s.
lymphonodi phrenici s.
lymphonodi tracheobronchiales s.
lymphonodus gluteales s.
lymphonodus pancreatici s.
nervi alveolares s.
nervi clunium s.
nodi lymphatici mesenterici s.
nodi lymphatici pancreatici s.
nodi lymphoidei mesenterici s.
nodi lymphoidei phrenici s.
nodi lymphoidei rectales s.
nodi lymphoidei
tracheobronchiales s.
palpebrales s.
phrenicae s.
rami alveolares s.
rami cardiaci cervicales s.
rami dentales s.
rami gingivales s.
rami labiales s.
venae cerebelli s.
venae epigastricae s.
venae gluteae s.
venae hemispherii cerebelli s.
venae palpebrales s.
venae phrenicae s.
vesicales s.

superioris
arteriae membri s.

articulationes cinguli membri s.
brachium colliculi s.
caput infraorbitale quadrati labii s.
caput zygomaticum quadrati labii s.
cingulum membri s.
cisterna s.
deep layer of levator palpebrae s.
frenulum labii s.
incisivi labii s.
juncturae cinguli membri s.
lamina profunda musculi levatoris
  palpebrae s.
lamina superficialis musculi
  levatoris palpebrae s.
levator labii s.
levator palpebrae s. (LPS)
musculi levatoris palpebrae s.
musculus incisivus labii s.
musculus levator labii s.
musculus levator palpebrae s.
musculus quadratus labii s.
musculus triangularis labii s.
nodi lymphoidei membri s.
ossa membri s.
ostium venae cavae s.
pars buccopharyngea musculi
  constrictoris pharyngei s.
pars glossopharyngea musculi
  constrictoris pharyngis s.
pars lateralis venae lobi medii
  venae pulmonalis dextri s.
pars libera membri s.
pars medialis venae lobi medii
  venae pulmonis dextri s.
pars mylopharyngeus musculi
  constrictoris pharyngis s.
pars pterygopharyngea musculi
  constrictoris pharyngis s.
plexus periarterialis arteriae
  thyroideae s.
plica nervi laryngei s.
rami dentales superiores plexus
  dentalis s.
rami gingivales superiores plexus
  dentalis s.
rami laryngopharyngei ganglii
  cervicalis s.
rami pancreatici arteriae
  pancreaticoduodenalis s.
ramus apicalis venae pulmonalis
  dextrae s.

ramus apicoposterior venae
  pulmonalis sinistrae s.
ramus basalis anterior venae
  basalis s.
ramus cricothyroideus arteriae
  thyroideae s.
ramus externus nervi laryngei s.
ramus glandulares
  anterior/lateralis/posterior arteriae
  thyroideae s.
ramus glandularis anterior arteriae
  thyroideae s.
ramus glandularis posterior arteriae
  thyroideae s.
ramus inferior arteriae gluteae s.
ramus infrahyoideus arteriae
  thyroideae s.
ramus internus nervi laryngei s.
ramus lingularis venae pulmonis
  sinistrae s.
ramus lobi medii venae pulmonalis
  dextrae s.
ramus posterior arteriae
  thyroideae s.
ramus posterior venae pulmonalis
  dextrae s.
ramus profundus arteriae gluteae s.
ramus septi nasi arteriae labialis s.
ramus sternocleidomastoideus
  arteriae thyroideae s.
ramus superficialis arteriae
  gluteae s.
ramus superior arteriae gluteae s.
regiones membri s.
stratum cinereum colliculi s.
stratum griseum colliculi s.
sulcus sinus petrosi s.
sulcus sinus sagittalis s.
superficial layer of the levator
  palpebrae s.
tuberculum labii s.
vagina musculorum obliqui s.
vagina tendinis musculi obliqui s.
venae membri s.

superiorum
decussatio pedunculorum
  cerebellarium s.
superius
brachium quadrigeminum s.
cornu s.
ductulus aberrans s.
frenulum veli medullaris s.

S

NOTES

superius *(continued)*
    ganglion cervicale s.
    ganglion mesentericum s.
    ligamenta epididymidis inferius
      et s.
    ligamentum auriculare s.
    ligamentum costotransversarium s.
    ligamentum epididymidis s.
    ligamentum incudis s.
    ligamentum mallei s.
    ligamentum pubicum s.
    ligamentum transversum scapulae s.
    mediastinum s.
    membrum s.
    retinaculum musculorum
      extensorum s.
    retinaculum musculorum
      fibularium s.
    retinaculum musculorum
      peroneorum s.
    segmentum s.
    segmentum bronchopulmonale
      lingulare s.
    segmentum renale anterius s.
    tuberculum thyroideum s.
    velum medullare s.
supermedial
supermotility
supernormal recovery phase
supernumerary
    s. breast
    s. kidney
    s. mamma
    s. organ
    s. placenta
superolateral aspect
superolaterales
    lymphonodus inguinales s.
superomediales
    lymphonodus inguinales s.
superomedial margin
superpetrosal
superstructure
    stapes s.
supertension
supinate
supination
    s. of foot
    s. of forearm
supinator
    s. crest
    s. crest of ulna
    s. longus reflex
    s. muscle
    musculus s.
supinatorius
    crista musculi s.
supine
supplemental air

supplementary motor cortex
support
    longitudinal arch s.
supporting
    s. cell
    s. reaction
    s. reflex
supraacetabular
    s. groove
    s. sulcus
supraacetabularis
    sulcus s.
supraacromial
supraanal
supraarytenoid cartilage
supraauricular point
supraaxillary
suprabuccal
suprabullar recess
supracallosal gyrus
supracallosus
    gyrus s.
supracardinal vein
supracerebellar
supracerebral
suprachiasmatica
    arteria s.
suprachiasmatic artery
suprachoroid
    s. lamina
    s. lamina of sclera
    s. layer
suprachoroidea
    lamina s.
supraciliary canal
supraclavicular
    s. area
    s. fossa
    s. lymph node
    s. muscle
    s. nerve
    s. part of brachial plexus
    s. region
    s. triangle
supraclaviculares
    s. intermedii
    s. laterales
    lymphonodi s.
    s. mediales
    nodi lymphoidei s.
supraclavicularis
    musculus s.
supracollicularis
    muscular sphincter s.
supracollicular sphincter
supracondylar (SC)
    s. line
    s. process
    s. process of humerus

supracondyloid
supracostal
supracotyloid
supracrestal
  s. line
  s. plane
supracristale
  planum s.
supracristalis
  linea s.
supracristal plane
supradiaphragmatic
supraduodenal artery
supraduodenalis
  arteria s.
supraepicondylar process
supraglenoid
  s. tubercle
  s. tubercle of scapula
supraglenoidale
  tuberculum s.
supraglottic
  s. larynx
  s. structure
suprahepatic
  s. caval cuff
  s. space
suprahyoid
  s. branch of lingual artery
  s. gland
  s. muscle
  s. triangle
suprahyoideus, pl. suprahyoidei
  musculi suprahyoidei
  ramus s.
suprainguinal
suprainterparietal bone
supraintestinal
supralevator space
supralumbar
supramalleolar
supramammary
supramandibular
supramarginal
  s. convolution
  s. gyrus
supramarginalis
  gyrus s.
supramastoid
  s. crest
  s. fossa

supramastoidea
  crista s.
supramaxilla
supramaxillary
supramaximal stimulus
suprameatal
  s. pit
  s. spine
  s. triangle
suprameatalis
  foveola s.
  spina s.
suprameatica
  foveola s.
  spina s.
suprameatum
  spina s.
supramediastinal muscle
supramental
supranasal point
supraneural
supraneuroporica
  lamina s.
supranuclear
supraoptic
  s. artery
  s. canal
  s. commissure
  s. nucleus
supraoptica, pl. supraopticae
  arteria s.
  commissurae supraopticae
supraopticohypophyseal axis
supraopticohypophysialis
  tractus s.
supraopticohypophysial tract
supraorbital
  s. arch
  s. area
  s. artery
  s. canal
  s. foramen
  s. margin
  s. nerve
  s. notch
  s. point
  s. region
  s. ridge
  s. vein
supraorbitale
  foramen s.

S

NOTES

supraorbitalis
    arteria s.
    incisura s.
    margo s.
    nervus s.
    ramus lateralis nervi s.
    ramus medialis nervi s.
    vena s.
supraorbitomeatal plane
suprapatellar
    s. bursa
    s. pouch
    s. reflex
suprapatellaris
    bursa s.
suprapelvic
supraphysiologic
supraphysiological
suprapinealis
    recessus s.
suprapineal recess
suprapiriform foramen
suprapleuralis
    membrana s.
suprapleural membrane
suprapubic
    s. rami
    s. reflex
    s. region
    s. sinus
suprapyloric
    s. lymph node
suprapyloricus
    nodus lymphoideus s.
suprarenal
    s. area
    s. artery
    s. body
    s. capsule
    s. cortex
    s. epithelium
    s. ganglion
    s. gland
    s. impression
    s. impression on liver
    s. medulla
    s. nervous plexus
    s. vein
suprarenalis
    centralis glandulae s.
    cortex glandulae s.
    facies anterior glandulae s.
    facies posterior glandulae s.
    facies renalis glandulae s.
    glandula s.
    impressio s.
    s. inferior
    margo medialis glandulae s.
    margo superior glandulae s.

    s. media
    medulla glandulae s.
    plexus nervosus s.
    s. sinistra
    vena centralis glandulae s.
suprascapular
    s. artery
    s. ligament
    s. nerve
    s. notch
    s. vein
suprascapularis
    arteria s.
    nervus s.
    ramus acromialis arteriae s.
    vena s.
suprascleral
suprasellar cisterna
supraspinal
supraspinale
    ligamentum s.
    spatium s.
supraspinalis
    s. muscle
    musculus s.
supraspinata
    fossa s.
supraspinatus
    s. fossa
    s. muscle
    musculus s.
    s. tendon
supraspinous
    s. fossa
    s. ligament
    s. muscle
suprastapedial
suprasternal
    s. bone
    s. fossa
    s. notch
    s. plane
    s. pulsation
    s. space
suprasternale
    os s.
suprasternalia
    ossa s.
suprastyloid crest of radius
suprasylvian
suprasymphysary
supratempora
supratemporal
supratentorial part of central nervous system
suprathoracic
supratonsillar
    s. fossa
    s. recess

supratonsillaris
  fossa s.
supratragic tubercle
supratragicum
  tuberculum s.
supratrochlear
  s. artery
  s. depression
  s. nerve
  s. vein
supratrochleares
  venae s.
supratrochlearis
  arteria s.
  nervus s.
supraturbinal
supratympanic
supraumbilical reflex
supravaginal
  s. part of cervix
  s. septum
supravaginalis
  portio s.
supravalvar
supravalvular
supraventricular
  s. crest
supraventricularis
  crista s.
supravesical fossa
supravesicalis
  fossa s.
  fovea s.
suprema
  arteria intercostalis s.
  concha nasalis s.
  intercostalis s.
  linea nuchae s.
  vena intercostalis s.
supremae
  rami dorsales arteriae
    intercostalis s.
supreme
  s. cardiac nerve
  s. intercostal artery
  s. intercostal vein
  s. nasal concha
  s. turbinated bone
sura
surae
  musculus triceps s.
  triceps s.

sural
  s. artery
  s. communicating branch of
    common fibular nerve
  s. communicating branch of
    common peroneal nerve
  s. region
suralis, pl. surales
  arteria s.
  nervus s.
  rami calcanei laterales nervi s.
  regio s.
surface
  s. anatomy
  anterior s.
  articular s.
  axial s.
  buccal s.
  cerebral s.
  costal s.
  diaphragmatic s.
  dorsal s.
  epicardial s.
  s. epithelium
  external s.
  glenoid s.
  grinding s.
  incisal s.
  s. of interior iris
  internal s.
  labial s.
  lateral cranial s.
  masticating s.
  masticatory s.
  medial s.
  mesial s.
  s. mucous cells of stomach
  occlusal s.
  orbital s.
  s. of orbital zygomatic bone
  outer s.
  peritoneal s.
  plantar s.
  pleural s.
  posterior calcaneal articular s.
  serosal s.
  temporal s.
  s. tension theory of narcosis
  thenar s.
  ventral s.
  vertebral s.
  s. vessel

S

NOTES

surface (*continued*)
    visceral s.
    volar s.
surface-active
surfactant
surgical
    s. anatomy
    s. neck of humerus
surrenal
suspension
suspensory
    s. ligament of axilla
    s. ligament of breast
    s. ligament of clitoris
    s. ligament of Cooper
    s. ligament of duodenum
    s. ligament of esophagus
    s. ligament of eye
    s. ligament of eyeball
    s. ligament of gonad
    s. ligament of lens
    s. ligament of ovary
    s. ligament of penis
    s. ligament of testis
    s. ligament of thyroid gland
    s. muscle of duodenum
    s. retinaculum of breast
sustentacular cell
sustentaculum, pl. **sustentacula**
    s. lienis
    s. tali
sutura, pl. **suturae**
    s. coronalis
    suturae cranii
    s. ethmoidolacrimalis
    s. ethmoidomaxillaris
    s. frontalis
    s. frontalis persistens
    s. frontoethmoidalis
    s. frontolacrimalis
    s. frontomaxillaris
    s. frontonasalis
    s. frontozygomatica
    s. incisiva
    s. infraorbitalis
    s. intermaxillaris
    s. internasalis
    s. interparietalis
    s. lacrimoconchalis
    s. lacrimomaxillaris
    s. lambdoidea
    s. metopica
    s. nasofrontalis
    s. nasomaxillaris
    s. occipitomastoidea
    s. palatina mediana
    s. palatina transversa
    s. palatoethmoidalis
    s. palatomaxillaris

    s. parietomastoidea
    s. plana
    s. sagittalis
    s. serrata
    s. sphenoethmoidalis
    s. sphenofrontalis
    s. sphenomaxillaris
    s. spheno-orbitalis
    s. sphenoparietal
    s. sphenoparietalis
    s. sphenosquamosa
    s. sphenovomeriana
    s. sphenozygomatica
    s. squamoparietalis
    s. squamosa
    s. squamosomastoidea
    s. temporozygomatica
    s. zygomaticofrontalis
    s. zygomaticomaxillaris
    s. zygomaticotemporalis
sutural
    s. bone
    s. ligament
suturarum
    os s.
    ossa s.
suture
    anterior palatine s.
    biparietal s.
    coronal s.
    cranial s.
    dentate s.
    ethmoidolacrimal s.
    ethmoidomaxillary s.
    false s.
    frontal s.
    frontoethmoidal s.
    frontolacrimal s.
    frontomalar s.
    frontomaxillary s.
    frontonasal s.
    frontoparietal s.
    frontosphenoid s.
    frontozygomatic s.
    harmonic s.
    incisive s.
    infraorbital s.
    intermaxillary s.
    internasal s.
    interpalpebral s.
    interparietal s.
    s. joint
    lacrimoconchal s.
    lacrimoethmoidal s.
    lacrimomaxillary s.
    lacrimoturbinal s.
    lambdoid s.
    lens s.
    lumbar s.

malomaxillary s.
mastoid s.
median palatine s.
metopic s.
middle palatine s.
nasofrontal s.
nasomaxillary s.
neurocentral s.
occipital s.
occipitomastoid s.
occipitoparietal s.
occipitosphenoidal s.
palatoethmoidal s.
palatomaxillary s.
parietomastoid s.
parietooccipital s.
pericostal s.
persistent frontal s.
petrobasilar s.
petrosphenobasilar s.
petrosquamous s.
plane s.
premaxillary s.
sagittal s.
serrate s.
sphenoethmoidal s.
sphenofrontal s.
sphenomalar s.
sphenomaxillary s.
sphenooccipital s.
sphenoorbital s.
sphenoparietal s.
sphenopetrosal s.
sphenosquamous s.
sphenotemporal s.
sphenovomerine s.
sphenozygomatic s.
squamomastoid s.
squamoparietal s.
squamosal s.
squamosomastoid s.
squamosoparietal s.
squamosphenoid s.
squamous s.
temporozygomatic s.
transverse palatine s.
tympanomastoid s.
uteroparietal s.
wedge-and-groove s.
zygomaticofrontal s.
zygomaticomaxillary s.
zygomaticotemporal s.

Suzanne gland
swallow
swallowing
    s. reflex
    s. threshold
sweat
    s. duct
    s. gland
    s. pore
sweating
sweep
swelling
    arytenoid s.
    genital s.
    labial s.
    labioscrotal s.
    lateral lingual s.
    levator s.
    scrotal s.
sylvian
    s. angle
    s. aqueduct
    s. area
    s. artery
    s. fissure
    s. fossa
    s. line
    s. point
    s. valve
    s. vein
    s. ventricle
sylvii
    aqueductus s.
    caro quadrata s.
    os s.
    vallecula s.
Sylvius
    angle of S.
    fissure of S.
    fossa of S.
symblepharon
symbrachydactyly
Symington anococcygeal body
symmelia
symmetry
    inverse s.
sympathetic
    s. branch to submandibular
      ganglion
    s. carotid plexus
    s. chain
    s. formative cell

S

NOTES

sympathetic *(continued)*
    s. ganglia
    s. imbalance
    s. nerve
    s. nerve fiber
    s. nervous system
    s. part of autonomic division of peripheral nervous system
    s. root of ciliary ganglion
    s. root of otic ganglion
    s. root of pterygopalatine ganglion
    s. root of sublingual ganglion
    s. root of submandibular ganglion
    s. saliva
    s. segment
    s. trunk
sympathetoblast
sympathic
sympathica
    pars s.
sympathici
    ganglia trunci s.
    rami interganglionares trunci s.
sympathicoblast
sympathicolytic
sympathicomimetic
sympathicotropic cells
sympathicus
    truncus s.
sympathin
sympathoadrenal
sympathoblast
sympathochromaffin cell
sympathogonia
sympatholytic
sympathomimetic
sympathoparalytic
sympexis
symphalangism
symphalangy
symphyseal
symphyses (*pl. of* symphysis)
symphysialis
    facies s.
symphysial surface of pubis
symphysic teratosis
symphysion
symphysis, pl. **symphyses**
    amphiarthrotic pubic s.
    eminentia s.
    intervertebral s.
    s. intervertebralis
    s. of mandible
    s. mandibulae
    mandibular s.
    manubriosternal s.
    s. manubriosternalis
    mental s.
    s. mentalis

    s. menti
    pleural s.
    pubic s.
    s. pubica
    s. pubis
    s. sacrococcygea
    s. xiphosternalis
symplasmatic
symplast
symplastic tissue
sympodia
symport
symporter
sympus
    acephalus s.
    s. apus
    s. dipus
    s. monopus
synadelphus
synaphoceptors
synapse
    axoaxonic s.
    axodendritic s.
    axosomatic s.
    electrotonic s.
    pericorpuscular s.
synapsis
synaptic
    s. boutons
    s. cleft
    s. conduction
    s. endings
    s. phase
    s. resistance
    s. terminal
    s. trough
    s. vesicle
synaptinemal complex
synaptology
synaptosome
synarthrodia
synarthrodial joint
synarthrosis, pl. **synarthroses**
syncephalus asymmetros
syncephaly
synchondrodial joint
synchondrosis, pl. **synchondroses**
    anterior intraoccipital s.
    s. arycorniculata
    arycorniculate s.
    cranial synchondroses
    synchondroses cranii
    s. epiphyseos
    s. intraoccipitalis anterior
    s. intraoccipitalis posterior
    s. manubriosternalis
    neurocentral s.
    petrooccipital s.
    s. petrooccipitalis

posterior intraoccipital s.
sphenoethmoidal s.
s. sphenoethmoidalis
sphenooccipital s.
s. sphenooccipitalis
s. sphenopetrosa
sphenopetrosal s.
sphenopetrous s.
sternal synchondroses
synchondroses sternales
s. xiphosternalis
synchorial
synchronia
synchronism
synchronous reflex
synchrony
syncinesis
synclinal
syncytia (*pl. of* syncytium)
syncytial
s. bud
s. knot
s. sprout
s. trophoblast
syncytiotrophoblast
syncytium, pl. **syncytia**
syndactyle
syndactylia
syndactylism
syndactylous
syndactyly
syndesmochorial placenta
syndesmodial joint
syndesmologia
syndesmosis, pl. **syndesmoses**
radioulnar s.
s. radioulnaris
tibiofibular s.
s. tibiofibularis
s. tympanostapedia
tympanostapedial s.
s. tympanostapedialis
syndesmotic joint
syndrome
Boerhaave s.
Costen s.
Frey s.
Horner s.
Marin Amat s.
Meniere s.
Raynaud s.
scalenus anticus s.

synencephalocele
synergetic
synergia
synergic
s. control
s. muscle
synergism
synergist
synergistic muscle
synergy
syngamy
syngenesis
syngenetic
syngnathia
synkinesis
synkinetic
synonychia
synophthalmia
synophthalmus
synorchidism
synorchism
synoscheos
synosteology
synostosis
cranial s.
sagittal s.
tribasilar s.
synotia
synovia (*pl. of* synovium)
synovial
s. bursa
s. capsule
s. capsule of knee
s. cavity
s. cell
s. crypt
s. flap
s. fluid
s. fold
s. frena
s. frenula
s. fringe
s. ganglion
s. gland
s. hernia
s. joint
s. joint of free lower limb
s. joint of free upper limb
s. joint of thorax
s. ligament
s. membrane
s. mesenchyme

S

NOTES

synovial *(continued)*
    s. sac
    s. sheath
    s. sheaths of digits of foot
    s. sheaths of digits of hand
    s. sheaths of toe
    s. tendon sheath
    s. trochlear bursa
    s. tufts
    s. villi
synoviale
    stratum s.
synovialis, pl. **synoviales**
    articulatio s.
    bursa s.
    junctura s.
    membrana s.
    plica s.
    vagina s.
    villi synoviales
synoviparous
synovium, pl. **synovia**
synpolydactyly
synthase
    nitric oxide s.
    NO s.
synthermal
synthesis, pl. **syntheses**
synthesis period
synthorax
syntrophism
syntrophoblast
syntropic
syntropy
syringadenosus
syringeal
syringes (*pl. of* syrinx)
syringocele
syringoencephalomyelia
syringoid
syringomeningocele
syringomyelocele
syrinx, pl. **syringes**
syssarcosic
syssarcosis
syssarcotic
system
    absorbent s.
    alimentary s.
    arch-loop-whorl s.
    atrioventricular conduction s.
     (AVCS)
    autonomic division of nervous s.
    autonomic nervous s. (ANS)
    autonomic part of peripheral
     nervous s.
    blood-vascular s.
    bony part of skeletal s.
    bulbosacral s.

caliceal s.
cardiovascular s.
cartilaginous part of skeletal s.
centimeter-gram-second s.
central nervous s.
cerebrospinal s.
chromaffin s.
circulatory s.
collecting s.
conduction s.
cranial part of parasympathetic part
  of autonomic division of
  nervous s.
craniosacral division of autonomic
  nervous s.
dermal s.
dermoid s.
digestive s.
ecological s.
endocrine s.
esthesiodic s.
excretory s.
extrapyramidal motor s.
feedback s.
foot-pound-second s.
gamma motor s.
gastric plexuses of autonomic s.
genital s.
genitourinary s.
glandular s.
haversian s.
heart conduction s.
hematopoietic s.
hepatic portal s.
His-Purkinje s.
His-Tawara s.
hypophysial portal s.
hypophysioportal s.
hypothalamohypophysial portal s.
hypoxia warning s.
integumentary s.
intermediary s.
kinetic s.
limbic s.
lymphatic s.
lymph node s.
lymphoid s.
s. of macrophages
masticatory s.
metameric nervous s.
meter-kilogram-second s.
mononuclear phagocyte s.
muscular s.
nervous s.
neuromuscular s.
nonspecific s.
oaponeurotic s.
oculomotor s.
osseous part of skeletal s.

oxidation-reduction s.
parasympathetic part of autonomic division of peripheral nervous s.
pedal s.
peripheral nervous s.
portal s.
pressoreceptor s.
projection s.
pulmonary branch of autonomic nervous s.
Purkinje s.
pyelocaliceal s.
rami communicantes of sympathetic part of autonomic division of nervous s.
renal portal s.
renin-angiotensin s.
reproductive s.
respiratory s.
reticular activating s. (RAS)
reticuloendothelial s.
saphenous s.
skeletal s.
skin-adipose superficial musculoaponeurotic s. (SASMAS)
somesthetic s.
static s.
stomatognathic s.
supratentorial part of central nervous s.
sympathetic nervous s.
sympathetic part of autonomic division of peripheral nervous s.
thoracolumbar nervous s.
urinary s.
urogenital s.

uropoietic s.
vascular s.
vegetative nervous s.
vertebral-basilar s.
vertebral venous s.
visceral motor s.
visceral nervous s.
systema
  s. alimentarium
  s. cardiovasculare
  s. conducens cordis
  s. digestorium
  s. genitalia
  s. lymphaticum
  s. lymphoideum
  s. nervosum
  s. nervosum autonomicum
  s. nervosum centrale
  s. nervosum periphericum
  s. respiratorium
  s. skeletale
  s. urinarium
  s. urogenitale
systematic
  s. anatomy
systematize
systemic
  s. anatomy
  s. circulation
systole
systoletail fold
systolic
  ejection fraction s.
  s. pressure
systolometer

NOTES

T
- T cell
- T lymphocyte
- T tubule

TA
- Terminologia Anatomica

tabatière anatomique
tabes spasmodica
tablature
table
- Aub-DuBois t.
- vitreous t.

tabular
tachogram
tachograph
tachography
tachometer
tachyauxesis
tachycardia
- ventricular t. (VT)

tactile
- t. cell
- t. corpuscle
- t. disk
- t. elevations
- t. meniscus
- t. papilla

tactiles
- toruli t.

tactometer
tactor
tactus
- corpusculum t.
- meniscus t.
- organum t.

taenia
tag
- epiploic t.
- skin t.

tail
- t. of axillary breast
- t. bone
- t. bud
- t. of caudate nucleus
- t. of dentate gyrus
- t. of epididymis
- t. fold
- t. of helix
- t. of pancreas
- t. sheath
- t. of Spence
- t. vertebrae

tailbone area
tailgut
tailor's muscle

talar
- t. articular surface of calcaneus
- t. component
- t. sulcus
- t. tendon

tali (*pl. of* talus)
talipedic
talipes
talipomanus
talocalcanea
- articulatio t.

talocalcaneal
- t. interosseous ligament
- t. joint

talocalcanean
talocalcaneare
- ligamentum t.

talocalcaneonavicular
- t. articulation
- t. joint
- t. ligament

talocalcaneonavicularis
- articulatio t.

talocalcaneum
- ligamentum t.

talocrucalis
- pars tibionavicularis ligamenti collateralis medialis articulationis t.

talocrural
- t. articulation
- t. joint
- t. joint of ankle

talocruralis
- articulatio t.
- ligamentum mediale articulationis t.
- pars tibiocalcanea ligamenti collateralis medialis articulationis t.
- pars tibiotalaris anterior ligamenti collateralis medialis articulationis t.
- pars tibiotalaris posterior ligamenti collateralis medialis articulationis t.
- regio t.

talofibular ligament
talonavicular
- t. articulation
- t. joint
- t. ligament

talonaviculare
- ligamentum t.

taloscaphoid

talotibial
talus, pl. **tali**
    articular surface of t.
    body of t.
    calcaneal articular surface of t.
    caput tali
    collum tali
    corpus tali
    facies articularis calcanea tali
    facies articularis navicularis tali
    facies articularis partis
      calcaneonavicularis ligamenti
      bifurcati tali
    facies ligamenti calcaneonavicularis
      plantaris tali
    facies malleolaris lateralis tali
    facies malleolaris medialis tali
    facies superior tali
    head of t.
    t. head
    interosseous groove of t.
    lateral malleolar facet of t.
    lateral malleolar surface of t.
    lateral tubercle of posterior process
      of t.
    medial malleolar facet of t.
    medial malleolar surface of t.
    medial tubercle of posterior process
      of t.
    navicular articular surface of t.
    neck of t.
    posterior process of t.
    processus lateralis tali
    processus posterior tali
    pulley of t.
    sulcus tali
    superior facet of trochlear of t.
    superior surface of t.
    sustentaculum tali
    trochlea tali
    trochlea of t.
    tuberculum laterale processus
      posterioris tali
    tuberculum mediale processus
      posterioris tali
tanned red cells
tanycyte
tapeta (*pl. of* tapetum)
tapetochoroidal
tapetoretinal
tapetum, pl. **tapeta**
    t. alveoli
    t. nigrum
    t. oculi
tapinocephalic
tapinocephaly
tar
tarda
    cutanea t.

target
    t. cell
    t. gland
    t. organ
    t. plasma
Tarin
    T. space
    T. tenia
    T. valve
tarini
    pons t.
    valvula semilunaris t.
    velum t.
tarsal
    t. arch
    t. bone
    t. canal
    t. cartilage
    t. fold
    t. gland
    t. interosseous ligament
    t. joint
    t. muscle
    t. plate
    t. sinus
    t. tunnel
tarsale
tarsalia
    ossa t.
tarsalis, pl. **tarsales**
    glandulae tarsales
    t. inferior
    regio t.
    t. superior
tarsea
    t. lateralis
    t. mediales
tarsen
tarseus
    arcus t.
tarsi (*pl. of* tarsus)
tarsomegaly
tarsometata
tarsometatarsal
    t. articulation
    t. joint
    t. ligament
tarsometatarsales
    articulationes t.
tarsometatarsalia
    ligamenta t.
tarsometatarseae
    articulationes t.
tarsoorbital
tarsophalangeal reflex
tarsotarsal
tarsotibial
tarsus, pl. **tarsi**
    t. inferior

inferior t.
t. inferior palpebrae
ligamenta tarsi
musculus tensor tarsi
t. orbital septum
os centrale tarsi
ossa tarsi
sinus tarsi
superior t.
t. superior
t. superior palpebrae
tart cell
taste
t. bud
t. bulb
t. cell
t. corpuscle
t. hairs
organ of t.
t. pathway
t. pore
t. ridge
tautomeric fibers
Tawara
T. node
node of Aschoff and T.
taxis
negative t.
positive t.
TBW
total body water
T-cell growth factor
TE
tracheoesophageal
tear sac
teat
tecta (*pl. of* tectum)
tectal tract
tecti
tectiform
tectobulbaris
tractus t.
tectobulbar tract
tectocephalic
tectocephaly
tectology
tectopontine tract
tectopontinus
tractus t.
tectoria
membrana t.

tectorial
t. membrane
t. membrane of cochlear duct
t. membrane of median atlantoaxial
joint
tectorium
tectospinal
t. decussation
t. tract
tectospinalis
tractus t.
tectum, pl. **tecta**
t. of brain stem
t. mesencephali
t. of midbrain
pars tecta
stria tecta
tenia tecta
zona tecta
teeth (*pl. of* tooth)
tegmen, pl. **tegmina**
t. cruris
t. mastoideum
t. tympani
t. ventriculi quarti
tegmenta (*pl. of* tegmentum)
tegmental
t. decussation
t. fields of Forel
t. root of tympanic cavity
t. wall of middle ear
t. wall of tympanic cavity
tegmenti
decussationes t.
nuclei t.
tractus centralis t.
tegmentum, pl. **tegmenta**
t. mesencephali
mesencephalic t.
midbrain t.
t. of pons
t. rhombencephali
t. of rhombencephalon
tegmina (*pl. of* tegmen)
tegminis
tela, pl. **telae**
t. choroidea inferior
t. choroidea superior
t. choroidea ventriculi quarti
t. choroidea ventriculi tertii
t. conjunctiva
t. elastica

T

NOTES

tela *(continued)*
t. subcutanea
t. subcutanea penis
t. subcutanea perinei
t. submucosa
t. submucosa pharyngis
t. subserosa
tenia telae
t. vasculosa

telangiectatica
livedo t.

telangion
telecanthus
telemeter
telemetry
telencephalic
t. flexure
t. vesicle

telencephalization
telencephalon
teleorganic
telereceptor
telergy
telodendron
telogen
teloglia
telokinesia
telolecithal ovum
telophase
telotism
temperature
sensible t.

temple
tempo
venae articulares t.

tempora *(pl. of* tempus)
temporal
t. aponeurosis
t. apophysis
t. arcade
t. area
t. arteriole of retina
t. artery
t. branch of facial nerve
t. canal
t. canthus
t. cortex
t. crest of mandible
t. diploic vein
t. dispersion
t. fascia
t. field
t. fossa
t. gyrus
t. horn
t. horn of lateral ventricle
t. line
t. line of frontal bone
t. lobe

t. muscle
t. plane
t. pole
t. process of zygomatic bone
t. region
t. region of head
t. ridge
t. space
t. squama
t. surface
t. vein
t. venules of retina

temporale
os t.
planum t.

temporales
t. profundae
t. profundi
rami t.
t. superficiales

temporalis
ala t.
angulus oculi t.
apex partis petrosae ossis t.
arteria polaris t.
eminentia articularis ossis t.
facies anterior partis petrosae
ossis t.
facies articularis fossae
mandibularis ossis t.
facies inferior partis petrosae
ossis t.
facies posterior partis petrosae
ossis t.
fascia t.
fossa t.
incisura jugularis ossis t.
lamina profunda fasciae t.
lamina superficialis fasciae t.
lobus t.
margo occipitalis ossis t.
margo parietalis partis squamosae
ossis t.
margo posterior partis petrosae
ossis t.
margo sphenoidalis ossis t.
margo superior partis petrosae
ossis t.
t. media
t. muscle
musculus t.
norma t.
pars mastoidea ossis t.
pars petrosa ossis t.
pars squamosa ossis t.
pars tympanica ossis t.
processus mastoideus partis petrosae
ossis t.
processus styloideus ossis t.

processus zygomaticus ossis t.
squama t.
t. superficialis
t. tendon
tuberculum articulare ossis t.
temporally
temporary
　t. cartilage
　t. tooth
temporis
temporoauricular
temporofrontal tract
temporohyoid
temporomalar
temporomandibular
　t. articular disk
　t. articulation
　t. joint
　t. ligament
　t. nerve
temporomandibulare
　ligamentum t.
temporomandibulares
　venae articulares t.
temporomandibularis
　articulatio t.
　discus articularis t.
　ligamentum laterale articulationis t.
　ligamentum mediale articulationis t.
temporomaxillary
　t. articulation
　t. joint
　meniscus of t.
　t. vein
temporooccipital
temporoparietalis
　t. muscle
　musculus t.
temporoparietal muscle
temporopontine tract
temporopontinus
　tractus t.
temporosphenoid
temporozygomatica
　sutura t.
temporozygomatic suture
temps utile
tempus, pl. **tempora**
tenacious

tenacity
　cellular t.
tenaculum, pl. **tenacula**
　tenacula tendinum
tendinea
　inscriptio t.
　intersectio t.
tendineae
　chordae t.
　false chordae t.
　juncturae t.
tendines (*pl. of* tendo)
tendineus
　arcus t.
　hiatus t.
tendinis
　stratum fibrosum vaginae t.
　theca t.
　vagina fibrosa t.
　vagina mucosa t.
　vagina synovialis t.
tendinotrochanteric ligament
tendinous
　t. arch
　t. arch of levator ani
　t. arch of levator ani muscle
　t. arch of pelvic fascia
　t. arch of soleus muscle
　t. center
　t. chiasm
　t. chiasm of the digital tendon
　t. galea
　t. inscription
　t. intersect
　t. intersection
　t. intersections of rectus abdominis
　t. opening
　t. ring
　t. sheath
　t. sheath of abductor pollicis
　longus and extensor pollicis
　t. sheath of extensor carpi radialis
　muscle
　t. sheath of extensor carpi ulnaris
　muscle
　t. sheath of extensor digiti minimi
　muscle
　t. sheath of extensor digitorum and
　extensor indicis muscle
　t. sheath of extensor digitorum
　longus muscle of foot

T

NOTES

tendinous (*continued*)
t. sheath of extensor hallucis longus muscle
t. sheath of extensor pollicis longus muscle
t. sheath of flexor carpi radialis muscle
t. sheath of flexor digitorum longus muscle of foot
t. sheath of flexor hallucis longus muscle
t. sheath of flexor pollicis longus muscle
t. sheath of superior oblique muscle
t. sheath of tibialis anterior muscle
t. sheath of tibialis posterior muscle

tendinum
chiasma t.
juncturae t.
retinaculum t.
tenacula t.
vincula t.

tendo, pl. **tendines**
t. Achilles
t. Achillis reflex
t. calcaneus
t. conjunctivus
t. cricoesophageus
t. oculi
t. palpebrarum

tendocalcaneus
tendomucin
tendomucoid
tendon
abductor pollicis longus t.
Achilles t.
adductor magnus t.
anterior tibial t.
anterior tibialis t.
biceps t.
t. bundle
bursa of calcaneal t.
calcaneal t.
calcanean t.
t. cartilage
t. cells
central perineum t.
common annular t.
common extensor t.
communis t.
conjoined t.
conjoint t.
coronary t.
cricoesophageal t.
digital extensor t.
erector spinae t.
extensor carpi radialis brevis t.

extensor carpi radialis longus t.
extensor carpi ulnaris t.
extensor digiti minimi t.
extensor digitorum longus t.
extensor hallucis longus t.
extensor indicis t.
extensor pollicis brevis t.
extensor pollicis longus t.
fibularis longus t.
fibularis tertius t.
flexor carpi radialis t.
flexor carpi ulnaris t.
flexor digitorum longus t.
flexor digitorum profundus t.
flexor digitorum superficialis t.
flexor hallucis brevis t.
flexor hallucis longus t.
flexor pollicis longus t.
Gerlach annular t.
gracilis t.
groove for tibialis posterior t.
hamstring t.
heel t.
iliopsoas t.
infraspinatus t.
t. insertion
intermediate digastric t.
intermediate omohyoid t.
lateral rectus t.
latissimus dorsi t.
levator palpebrae superioris t.
long abductor t.
masseter t.
mucous sheath of t.
oblique t.
obturator internus t.
ocular t.
palmaris longus t.
patellar t.
peroneal t.
plantaris t.
popliteus t.
posterior tibial t.
posterior tibialis t.
profundus t.
psoas minor t.
pulley t.
quadriceps femoris t.
rectus femoris t.
t. reflex
rider's t.
sartorius t.
semimembranosus t.
semitendinosus t.
t. sheath
stapedial t.
stapedius t.
sublimis t.
subscapularis t.

superior oblique t.
supraspinatus t.
talar t.
temporalis t.
tendinous chiasm of the digital t.
Todaro t.
trefoil t.
triceps t.
vastus intermedius t.
vastus lateralis t.
vastus medialis t.
vincula of t.
Zinn t.
t. of Zinn
tendovaginal
tenia, pl. **teniae**
teniae acusticae
t. choroidea
colic teniae
t. fimbriae
t. fornicis
t. of fornix
t. of fourth ventricle
free t.
t. hippocampi
t. libera
t. libera coli
medullary teniae
mesocolic t.
t. mesocolica
omental t.
t. omentalis
t. semicircularis
Tarin t.
t. tecta
t. telae
t. terminalis
t. thalami
thalamic t.
teniae of Valsalva
t. ventriculi quarti
t. ventriculi tertii
tenial
teniform
tenioid
teniola corporis callosi
tenofibril
Tenon
capsule of T.
T. capsule
space of T.
T. space

tenontography
tenontology
tenoreceptor
tensa
membrana t.
tense
t. part of tympanic membrane
tensile stress
tensiometer
tension
arterial t.
t. line
tissue t.
tensor
distal femoral t.
t. fasciae latae muscle
t. muscle of fascia lata
t. muscle of soft palate
t. muscle of tympanic membrane
t. tarsi muscle
t. trochleae muscle
t. tympani
t. tympani canal
t. tympani muscle
t. veli palati muscle
t. veli palatini
tensoris, pl. **tensores**
t. tympani
t. veli palatini
tenth cranial nerve
tentoria (*pl. of* tentorium)
tentorial
t. basal branch of internal carotid
artery
t. line
t. marginal branch of cavernous
part of internal carotid artery
t. nerve
t. notch
t. sinus
tentorii
incisura t.
nervus t.
ramus t.
tentorium, pl. **tentoria**
t. cerebelli
t. of cerebellum
t. of hypophysis
incisura of t.
notch of t.
tenue
intestinum t.

NOTES

tenuis
>folliculi lymphatici solitarii
>>intestini t.
>
>glandulae intestini t.
>meninx t.
>plicae circulares intestini t.
>stratum circulare tunicae muscularis
>>intestini t.
>
>stratum longitudinale tunicae
>>muscularis intestini t.
>
>tunica mucosa intestini t.
>tunica muscularis intestini t.
>tunica serosa intestini t.

teras, pl. **terata**
teratic
teratism
teratogen
teratogenesis
teratogenetic
teratogenic
teratogenicity
teratogeny
teratoid
teratologic
teratology
teratomatous cyst
teratosis
>atresic t.
>ceasmic t.
>ectogenic t.
>ectopic t.
>hypergenic t.
>symphysic t.

teres, pl. **teretes**
>eminentia t.
>fissure for ligamentum t.
>funiculus t.
>ligamentum t.
>t. major
>t. major muscle
>t. minor
>t. minor muscle
>musculus pronator t.
>musculus pronator radii t.
>notch for ligamentum t.
>pronator t.

teretis
>caput humerale musculi
>>pronatoris t.
>
>caput ulnare musculi pronatoris t.
>fissura ligamenti t.

tergal
tergo
>vis a t.

terminad
terminal
>t. aorta
>axon t.
>t. bar

t. bile duct
t. boutons
t. branch of middle cerebral artery
t. bronchiole
t. cisternae
t. crest
t. filum
t. ganglion
t. ileum
t. line
t. nerve
t. nerve corpuscles
t. notch of auricle
t. nuclei
t. part
t. phalanx
t. plate
t. sinus
t. stria
t. sulcus
t. sulcus of tongue
synaptic t.
t. thread
t. vein
t. ventricle
t. web

terminale
>dural part of filum t.
>filum t.
>ganglion t.
>pial part of filum t.
>spinal part of filum t.
>velum t.

terminales
>nervi t.
>nuclei t.

terminalia
>corpuscula nervosa t.

terminalis
>bronchiolus t.
>crista t.
>linea t.
>nervus t.
>pars t.
>pars pialis fili t.
>pars spinalis fili t.
>sinus t.
>stria t.
>sulcus t.
>t. sulcus
>tenia t.
>vena t.
>ventriculus t.

terminatio, pl. **terminationes**
>terminationes nervorum liberae

termination
terminationis
>nuclei t.

terminaux
    bouton t.
    pieds t.
term infant
Terminologia Anatomica (TA)
terminus, pl. **termini**
    termini generales
Terrien valve
territorial matrix
Terson gland
tertiary
    t. carina
    t. cortex
    t. egg membrane
    t. peristaltic activity
    t. radicle
    t. villus
    t. vitreous
tertii
    plexus choroideus ventriculi t.
    ramus choroidei ventriculi t.
    stria ventriculi t.
    tela choroidea ventriculi t.
    tenia ventriculi t.
tertius
    dens molaris t.
    digitus manus t.
    digitus pedis t.
    musculus fibularis t.
    musculus peroneus t.
    nervus occipitalis t.
    occipitalis t.
    peroneus t.
    sphincter ani t.
    trochanter t.
    ventriculus t.
Tesla current
tessellated
test
    breath-holding t.
    comb-growth t.
    Corner-Allen t.
    CO_2-withdrawal seizure t.
    dehydrocholate t.
    ether t.
    exercise t.
    Knoop hardness t.
    Master two-step exercise t.
    McPhail t.
    two-step exercise t.
    vaginal cornification t.
    vaginal mucification t.

testes (*pl. of* testis)
testi
    septulum t.
testicle
testicular
    t. appendage
    t. artery
    t. cord
    t. duct
    t. plexus
    t. vein
testicularis
    arteria t.
    t. dextra
    plexus periarterialis arteriae t.
    rami ureterici arteriae t.
    t. sinistra
testiculus
testis, pl. **testes**
    aberrans t.
    anterior border of t.
    appendix t.
    t. cords
    cryptorchid t.
    descensus paradoxus t.
    ductuli efferentes t.
    ductulus efferens t.
    dystopia transversa externa t.
    dystopia transversa interna t.
    ectopia t.
    ectopic t.
    efferent ductule of t.
    extremitas inferior t.
    extremitas superior t.
    facies lateralis t.
    facies medialis t.
    gubernaculum t.
    inferior pole of t.
    inverted t.
    lamina parietalis tunicae
      vaginalis t.
    lamina visceralis tunicae
      vaginalis t.
    lateral surface of t.
    ligamentum t.
    lobules of t.
    lobuli t.
    lower pole of t.
    margo anterior t.
    margo posterior t.
    medial surface of t.
    mediastinum of t.

NOTES

T

testis *(continued)*
    movable t.
    parenchyma t.
    parietal layer of tunica vaginalis
      of t.
    polus inferior t.
    polus superior t.
    posterior border of t.
    t. redux
    retained t.
    rete t.
    testes in scrotum
    septula of t.
    septulum t.
    septum of t.
    straight tubule of t.
    superior pole of t.
    suspensory ligament of t.
    trabecula t.
    tunica albuginea t.
    tunica vaginalis t.
    tunica vasculosa t.
    undescended t.
    upper pole of t.
    vaginal process of t.
    vascular layer of t.
    visceral layer of tunica vaginalis
      of t.
testium
    bursula t.
tetanic
    t. contraction
    t. convulsion
tetanization
tetanize
tetanode
tetanometer
tetanomotor
tetanus
    acoustic t.
    cathodal duration t. (CaDTe)
    complete t.
    incomplete t.
    Ritter opening t.
tetany
    phosphate t.
tetra amelia
tetrabrachius
    dicephalus dipus t.
tetrachirus
tetradactyl
tetragon, tetragonum
    t. lumbale
tetragonus
    musculus t.
tetramastia
tetramastous
tetramelus
tetrameric

tetramerous
tetraparesis
tetraperomelia
tetraphocomelia
tetrapus
tetrascelus
tetrastichiasis
tetrotus
Teutleben ligament
textiform
textural
texture
textus
thalamencephalic
thalamencephalon
thalami (*pl. of* thalamus)
thalamic
    t. gustatory nucleus
    t. tenia
thalamici
    rami t.
thalamicus
    fasciculus t.
    ramus t.
thalamocortical
thalamomamillaris
    fasciculus t.
thalamostriate vein
thalamus, pl. **thalami**
    anterior nuclei of t.
    anterior tubercle of t.
    anterodorsal nucleus of t.
    anteromedial nucleus of t.
    anteroventral nucleus of t.
    intralaminar nuclei of t.
    laminae medullares thalami
    lateral nucleus of t.
    medial central nucleus of t.
    medial nucleus of t.
    medullary layers of t.
    nuclei anteriores thalami
    nuclei intralaminares thalami
    nucleus arcuatus thalami
    nucleus centralis lateralis thalami
    nucleus medialis centralis thalami
    nucleus paracentralis thalami
    nucleus reticularis thalami
    nucleus ventralis anterior thalami
    nucleus ventralis intermedius
      thalami
    nucleus ventralis posterior
      intermedius thalami
    nucleus ventralis posterolateralis
      thalami
    nucleus ventralis posteromedialis
      thalami
    paracentral nucleus of t.
    posterior medial nucleus of t.
    reticular nucleus of t.

stria of t.
stria medullaris thalami
tenia thalami
tuberculum anterius thalami
ventral anterior nucleus of t.
ventral posterior intermediate
　nucleus of t.
ventral posterior lateral nucleus
　of t.
ventral posterolateral nucleus of t.
ventral posteromedial nucleus of t.
thanatobiologic
thaumatropy
thebesian
t. foramen
t. valve
t. vein
Thebesius
vein of T.
theca, pl. **thecae**
t. cells of stomach
t. cordis
t. externa
t. folliculi
t. interna
t. interna cone
t. lutein cell
t. tendinis
t. vertebralis
thecal cell
Theile
T. canal
T. gland
T. muscle
thele
thelia
thelium
thenad
thenal
thenar
t. cleft
t. eminence
t. fascia
t. flap
t. muscle
t. prominence
t. space
t. surface
t. web
thenaris
eminentia t.

thenen
theorem
Bernoulli t.
theory
Altmann t.
Bowman t.
Burn and Rand t.
Cannon t.
de Bordeau t.
emergency t.
gate-control t.
germ layer t.
Helmholtz-Gibbs t.
James-Lange t.
kern-plasma relation t.
membrane expansion t.
myogenic t.
Nernst t.
neurogenic t.
Ollier t.
overproduction t.
Pauling t.
recapitulation t.
reentry t.
Spitzer t.
therapeutic incompatibility
therapy
hyperbaric oxygen t.
therencephalous
thermacogenesis
thermal
thermoanesthesia
thermodynamic theory of narcosis
thermoexcitory
thermogenesis
nonshivering t.
shivering t.
thermogenetic
thermogenic action
thermogenous
thermoinhibitory
thermointegrator
thermolysis
thermolytic
thermoreceptor
thermosteresis
thermostromuhr
thermosystaltic
thermosystaltism
thermotactic
thermotaxic

T

NOTES

thermotaxis
    negative t.
    positive t.
thermotonometer
thermotropism
thickness
thigh
    adductor compartment of t.
    anterior compartment of t.
    anterior region of t.
    anterior surface of t.
    biceps muscle of t.
    t. bone
    deep artery of t.
    deep fascia of t.
    deep vein of t.
    extensor compartment of t.
    flexor compartment of t.
    t. joint
    lateral circumflex artery of t.
    lateral cutaneous nerve of t.
    medial circumflex artery of t.
    medial compartment of t.
    perineal branch of posterior cutaneous nerve of t.
    posterior compartment of t.
    posterior cutaneous nerve of t.
    posterior region of t.
    posterior surface of t.
    quadrate muscle of t.
    quadriceps muscle of t.
    rectus muscle of t.
thigmotaxis
thigmotropism
third
    t. corpuscle
    t. cranial nerve
    t. cuneiform bone
    t. finger
    t. molar
    t. occipital nerve
    t. ovary
    t. parallel pelvic plane
    t. part of duodenum
    t. peroneal muscle
    t. temporal convolution
    t. toe
    t. tonsil
    t. trochanter
    t. ventricle
third-year molar tooth
thirst
    false t.
    insensible t.
    subliminal t.
Thiry fistula
Thiry-Vella fistula
thixolabile

thixotropic
    t. fluid
thixotropy
Thoma law
Thompson ligament
thoracal
thoraces (*pl. of* thorax)
thoracic
    t. aorta
    t. aortic nervous plexus
    t. axis
    t. cage
    t. cardiac branch of thoracic ganglion
    t. cardiac branch of vagus nerve
    t. cavity
    t. compliance
    t. constriction of esophagus
    t. fascia
    t. ganglia
    t. girdle
    t. index
    t. inferior vena cava
    t. inlet
    t. interspinales muscle
    t. interspinal muscle
    t. intertransversarii muscle
    t. intertransverse muscle
    t. kyphosis
    t. limb
    t. longissimus muscle
    t. lymph node
    t. nucleus
    t. outlet
    t. part
    t. part of iliocostalis lumborum muscle
    t. part of peripheral autonomic plexuses and ganglion
    t. part of thoracic duct
    t. part of trachea
    t. pulmonary branch of thoracic ganglion
    t. region
    t. respiration
    t. rotator muscle
    t. skeleton
    t. spinal cord
    t. spine
    t. splanchnic ganglion
    t. splanchnic nerve
    t. vein
    t. wall
thoracica
    aorta t.
    ganglia t.
    t. interna
    kyphosis t.
    t. lateralis

pars t.
segmentum medullae spinalis t.
t. suprema ai
thoracicae
t. internae
rami esophageales aortae t.
rami mediastinales aortae t.
rami pericardiaci aortae t.
vertebrae t.
thoracici
arcus ductus t.
cardiaci t.
nervi cardiaci t.
pars abdominalis ductus t.
pars cervicalis ductus t.
pars thoracica ductus t.
rami cardiaci t.
thoracicoabdominal
thoracicoacromial
thoracicohumeral
thoracicorum
cutanei anterioris ramorum
ventralium nervorum t.
rami cardiaci thoracici
gangliorum t.
rami esophageales gangliorum t.
rami mammarii laterales ramorum
cutaneorum lateralis nervorum t.
rami pulmonales thoracici
gangliorum t.
ramus cutaneus anterior pectoralis
et abdominalis nervorum t.
ramus cutaneus medialis ramorum
dorsalium nervorum t.
ramus lateralis ramorum dorsalium
nervorum t.
thoracicus
ductus t.
t. longus
nucleus t.
pars abdominalis ductus t.
plexus nervosus aorticus t.
skeleton t.
thoracis
articulationes t.
cavea t.
cavitas t.
cavum t.
compages t.
iliocostalis t.
longissimus t.
musculi t.

musculi intertransversarii t.
musculi rotatores t.
musculus iliocostalis t.
musculus interspinalis t.
musculus longissimus t.
musculus rectus t.
musculus semispinalis t.
musculus spinalis t.
musculus transversus t.
nodi lymphoidei t.
rami anteriores nervorum t.
rami ventrales nervorum t.
semispinalis t.
skeleton t.
spinalis t.
transversus t.
thoracoabdominal
t. nerve
thoracoacromial
t. artery
t. trunk
t. vein
thoracoacromialis
arteria t.
rami pectorales arteriae t.
ramus acromialis arteriae t.
ramus clavicularis arteriae t.
ramus deltoideus arteriae t.
rete acromiale arteriae t.
vena t.
thoracoappendiculares
musculi t.
thoracoappendicular muscle
thoracoceloschisis
thoracodelphus
thoracodorsal
t. artery
t. nerve
thoracodorsalis
arteria t.
nervus t.
thoracoepigastrica
vena t.
thoracoepigastricae
venae t.
thoracoepigastric vein
thoracogastroschisis
thoracolumbalis
fascia t.
lamina anterior fasciae t.
lamina profunda fasciae t.

T

NOTES

thoracolumbar
    t. aponeurosis
    t. fascia
    t. nervous system
    t. venous line
thoracomelus
thoracometer
thoracopagus
thoracoparacephalus
thoracoschisis
thoradelphus
thorax, pl. **thoraces**
    infrasternal angle of t.
    interspinal muscle of t.
    lateral t.
    muscle of t.
    Peyrot t.
    semispinal muscle of t.
    spinal muscle of t.
    synovial joint of t.
    transverse muscle of t.
thorn
    dendritic t.
thread
    Simonart t.
    terminal t.
three-cornered bone
three-headed muscle
threshold
    absolute t.
    t. body
    differential t.
    t. differential
    galvanic t.
    t. of nose
    t. pads of anal canal
    relational t.
    stimulus t.
    t. stimulus
    t. substance
    swallowing t.
thrix
throat
    ears, nose, and t. (ENT)
    t. mucosa
thromboblast
thrombocyte
thrombocytic series
thromboplastid
throwback
thumb
    t. abduction
    adductor muscle of t.
    carpometacarpal joint of t.
    chief artery of t.
    distal phalanx of t.
    t. finger
    hitchhiker t.'s
    long abductor muscle of t.

    long extensor muscle of t.
    long flexor muscle of t.
    metacarpophalangeal of t.
    muscle of t.
    opposer muscle of t.
    t. pad
    principal artery of t.
    proximal phalanx of t.
    short abductor muscle of t.
    short extensor muscle of t.
    short flexor muscle of t.
thymi (*pl. of* thymus)
thymic
    t. agenesis
    t. branch of internal thoracic artery
    t. corpuscle
    t. duct
    t. hypoplasia
    t. vein
thymicae
    arteriae t.
    venae t.
thymici
    rami t.
thymicolymphatic
thymocyte
thymokinetic
thymus, pl. **thymi, thymuses**
    cortex of t.
    cortex thymi
    t. gland
    lobules of t.
    lobuli thymi
thymus-dependent zone
thyroaplasia
thyroarytenoidei
    pars thyroepiglottica musculi t.
thyroarytenoideus
    musculus t.
thyroarytenoid muscle
thyrocervical
    t. arterial trunk
    t. artery
thyrocervicalis
    truncus t.
thyroepiglottic
    t. ligament
    t. part of thyroarytenoid muscle
thyroepiglotticum
    ligamentum t.
thyroepiglotticus
    musculus t.
thyroepiglottidean
    t. ligament
    t. muscle
thyroesophageus
thyroglossal
    t. cyst
    t. diverticulum

t. duct
t. sinus
thyroglossus
ductus t.
thyrohyal
thyrohyoid
t. branch of ansa cervicalis
t. ligament
t. membrane
t. muscle
thyrohyoidea
membrana t.
thyrohyoideus
musculus t.
ramus t.
thyroid
accessory t.
t. articular surface of cricoid
  cartilage
t. axis
t. body
t. capsule
t. colloid
t. diverticulum
t. eminence
t. foramen
t. gland
t. ima artery
inferior pole of t.
isthmus of t.
t. isthmus
light cells of t.
t. lobe
t. lymph node
t. muscle
t. notch
pyramid of t.
pyramidal lobe of t.
pyramidal process of t.
t. region
retrosternal t.
superior pole of t.
t. tissue
t. vein
thyroidal
t. hernia
thyroidea
t. accessoria
cartilago t.
glandula t.
t. ima
t. ima ai

t. inferior
lamina dextra cartilaginis t.
lamina sinistra cartilaginis t.
linea obliqua cartilaginis t.
t. superior
thyroideae
capsula fibrosa glandulae t.
cornu inferius cartilaginis t.
cornu superius cartilaginis t.
folliculi glandulae t.
isthmus glandulae t.
lamina cartilaginis t.
levator glandulae t.
ligamentum suspensorium
  glandulae t.
linea obliqua cartilaginis t.
lobi glandulae t.
lobuli glandulae t.
lobus pyramidalis glandulae t.
t. mediae
musculus levator glandulae t.
parenchyma glandulae t.
stroma glandulae t.
thyroidei
lymphonodi t.
nodi lymphoidei t.
thyroideum
foramen t.
thyroid-stimulating
t.-s. hormone (TSH)
t.-s. hormone-releasing factor
t.-s. immunoglobulin (TSI)
thyrolaryngeal fascia
thyrolingual duct
thyropalatine
thyropharyngea
pars t.
thyropharyngeal
t. muscle
t. part
t. part of inferior pharyngeal
  constrictor muscle of pharynx
thyropharyngeus
musculus t.
thyrotrophic
thyrotropic hormone
thyrotropin
thyrotropin-releasing
t.-r. factor
t.-r. hormone
thyroxine
tibia, pl. **tibiae**

T

NOTES

tibia *(continued)*
anterior border of t.
anterior intercondylar area of t.
area intercondylaris anterior tibiae
area intercondylaris posterior tibiae
arteria nutricia tibiae
arteria nutriens tibiae
body of t.
bursa subcutanea tuberositas tibiae
bursa subcutanea tuberositatis tibiae
condylus lateralis tibiae
condylus medialis tibiae
corpus tibiae
facies articularis fibularis tibiae
facies articularis inferior tibiae
facies articularis malleoli tibiae
facies articularis malleoli medialis
   tibiae
facies articularis superior tibiae
facies lateralis tibiae
facies medialis tibiae
facies posterior tibiae
fibular articular facet of t.
fibular articular surface of t.
inferior articular surface of t.
interosseous border of t.
lateral condyle of t.
lateral surface of t.
malleolar articular surface of t.
margo anterior tibiae
margo interosseus tibiae
margo medialis tibiae
medial border of t.
medial condyle of t.
medial malleolus of t.
medial plateau of t.
medial surface of t.
nutrient artery of t.
posterior intercondylar area of t.
posterior surface of t.
proximal t.
shaft of t.
spinous process of t.
subcutaneous bursa of tuberosity
   of t.
superior articular surface of t.
tuberositas tibiae
t. valga
t. vara
tibiad
tibial
t. border of foot
t. collateral ligament
t. collateral ligament of ankle joint
t. communicating nerve
t. compartment
t. condyle
t. crest
t. intertendinous bursa

t. lymph node
t. muscle
t. nutrient artery
t. plafond
t. plateau
t. pulse
t. shaft
t. spine
t. tarsal tendinous sheath
t. tubercle
t. tuberosity
t. vein
tibiale
ligamentum collaterale t.
t. posticum
tibiales
t. anteriores
t. posteriores
tibialis
t. anterior
t. anterior muscle
arteria nutriens t.
nervus t.
t. posterior
t. posterior muscle
t. posticus
rami calcanei mediales nervi t.
rami musculares nervi t.
vaginae tendinum tarsales t.
tibioadductor reflex
tibiocalcanea
pars t.
tibiocalcaneal
t. part
t. part of deltoid ligament
t. part of medial ligament of
   ankle joint
tibiocalcanean
tibiofascialis
tibiofemoral
t. fossa
t. index
tibiofibular
t. articulation
t. joint
t. ligament
t. mortise
t. syndesmosis
tibiofibularis
articulatio t.
syndesmosis t.
tibionavicular
t. part
t. part of deltoid ligament
t. part of medial ligament of
   ankle joint
tibionaviculare
ligamentum t.

tibionavicularis
    pars t.
tibioperoneal
tibioscaphoid
tibiotalar
    t. part of medial ligament of
      ankle joint
    posterior t.
tibiotarsal articulation
tic douloureux
ticorenalia
tidal
    t. air
    t. volume
Tiedemann
    T. gland
    T. nerve
tight junction
tigroid
    t. bodies
    t. substance
time
    circulation t.
    forced expiratory t.
    inertia t.
    t. marker
    reaction t.
    transit t.
    utilization t.
tin ossicle
tip
    t. of auricle
    t. of ear
    t. of elbow
    leaflet t.
    t. of nose
    petrous t.
    root t.
    t. of tongue
    t. of tooth root
    Woolner t.
tissue
    acinar t.
    adenoid t.
    adipose t.
    adjacent t.
    adventitial t.
    appendiceal t.
    areolar t.
    basement t.
    t. basophil
    t. bed

bone t.
cancellous t.
cardiac muscle t.
cartilaginous t.
cavernous t.
chondroid t.
chordal t.
chorionic t.
chromaffin t.
compact t.
compression of t.
connective t.
cribriform t.
dartoic t.
dartoid t.
decidual t.
deciduous t.
dermal t.
t. displaceability
t. displacement
elastic t.
endometrial t.
endothelial t.
episcleral t.
epithelial t.
epivaginal connective t.
erectile t.
extraperitoneal t.
fatty layer of subcutaneous t.
fibrinous t.
fibrohyaline t.
fibrous connective t.
fibrous layer in or on deep aspect
    of fatty layer of subcutaneous t.
t. fluid
gelatinous t.
gingival t.
glandular t.
Haller vascular t.
hard t.
heart t.
hematopoietic t.
hemopoietic t.
indifferent t.
interstitial t.
intertubular t.
investing t.'s
islet t.
kidney t.
lung t.
lymphadenoid t.
lymphatic t.

NOTES

T

tissue *(continued)*
  lymphoid t.
  membranous t.
  mesenchymal t.
  mesonephric t.
  metanephrogenic t.
  mucous connective t.
  multilocular adipose t.
  muscle t.
  muscle layer in fatty layer of
    subcutaneous t.
  muscular t.
  myeloid t.
  nephrogenic t.
  nerve t.
  nervous t.
  nodal t.
  osseous t.
  osteogenic t.
  osteoid t.
  pancreatic t.
  parasternal t.
  parenchymal t.
  parenchymatous t.
  periapical t.
  perineal t.
  periorbital soft t.
  perirenal t.
  periurethral t.
  petrotympanic t.
  placental t.
  prostatic t.
  t. respiration
  reticular t.
  retiform t.
  skeletal muscle t.
  smooth muscle t.
  t. space
  splenic t.
  stromal t.
  subcutaneous adipose t.
  subcutaneous connective t.
  submucosal connective t.
  symplastic t.
  t. tension
  thyroid t.
  tonsillar t.
  underlying t.
  uterine t.
  vaginal t.
tissular
Todaro tendon
Tod muscle
toe
  abductor muscle of great t.
  abductor muscle of little t.
  adductor muscle of great t.
  t. bone
  bursa of great t.

  dorsal digital vein of t.
  dorsal nerve of t.
  fibrous digital sheaths of t.
  fourth t.
  great t.
  lateral surface of t.
  little t.
  long extensor muscle of great t.
  long flexor muscle of great t.
  medial surface of t.
  plantar surface of t.
  pulp of t.
  second t.
  short extensor muscle of great t.
  short flexor muscle of great t.
  short flexor muscle of little t.
  synovial sheaths of t.
  third t.
  webbed t.'s
toenail
Toldt
  T. fascia
  T. ligament
  ligament of T.
  T. line
  line of T.
  T. membrane
  white line of T.
tomentum cerebri
Tomes
  T. fibers
  T. granular layer
  T. process
tone
  muscular t.
tongue
  anterior part of t.
  apex of t.
  base of t.
  bifid t.
  blind foramen of t.
  body of t.
  t. bone
  t. of cerebellum
  cleft t.
  deep artery of t.
  dorsal vein of t.
  dorsum of t.
  fimbriated fold of inferior surface
    of t.
  foramen cecum of t.
  frenulum of t.
  inferior longitudinal muscle of t.
  inferior surface of t.
  mandibular t.
  margin of t.
  median groove of t.
  median longitudinal raphe of t.
  median sulcus of t.

mucosa of t.
mucous membrane of t.
muscle of t.
papillae of t.
posterior part of t.
postsulcal part of t.
presulcal part of t.
root of t.
septum of t.
superior dorsal surface of t.
superior longitudinal muscle of t.
terminal sulcus of t.
tip of t.
transverse muscle of t.
vertical muscle of t.

tonic
t. contraction
t. control
t. convulsion

tonicity
tonofibril
tonofilament
tonometer
tonoplast
tonotropic
tonsil
t. air cells
cerebellar t.
eustachian t.
faucial t.
Gerlach t.
laryngeal t.
lingual t.
Luschka t.
palatine t.
pharyngeal t.
third t.
tubal t.

tonsilla, pl. **tonsillae**
t. adenoidea
t. cerebelli
t. intestinalis
t. lingualis
t. palatina
t. pharyngea
t. pharyngealis
t. tubaria

tonsillar
t. branch of facial artery
t. branch of glossopharyngeal nerve
t. branch of lesser palatine nerve
t. crypt

t. fold
t. fossa
t. fossulae
t. hernia
t. pillar
t. ring
t. tissue

tonsillaris, pl. **tonsillares**
crypta t.
fossa t.
ramus t.
sinus t.

tonsillary
tonsillolingual sulcus
tonus
tooth, pl. **teeth**
accessory root of t.
acoustic t.
anterior t.
apex of cusp of t.
apical foramen of t.
approximal surface of t.
auditory t.
baby t.
back t.
bicuspid t.
buccal root of t.
t. bud
canine t.
cavity of t.
t. cement
central incisor t.
cervical margin of t.
cervical zone of t.
cervix of t.
cheek t.
cingulum of t.
clinical root of t.
contact surface of t.
Corti auditory t.
crown of t.
cusp of t.
cuspidate t.
cutting t.
deciduous t.
distal surface of t.
eye t.
facial surface of t.
t. germ
Huschke auditory t.
incisal surface of t.
incisor t.

T

NOTES

tooth *(continued)*
    interproximal surface of t.
    lateral incisor t.
    lingual surface of t.
    mandibular t.
    marginal crest of t.
    maxillary t.
    medial incisor t.
    mesial surface of t.
    milk t.
    molar t.
    multicuspid t.
    neck of t.
    occlusal surface of t.
    oral t.
    permanent t.
    posterior t.
    premolar t.
    primary t.
    t. pulp
    root of t.
    root canal of t.
    t. sac
    second t.
    skin of t.
    t. socket
    stomach t.
    succedaneous t.
    successional t.
    temporary t.
    third-year molar t.
    tubercle of t.
    vestibular surface of t.
    wisdom t.
toothed vertebra
topical
Topinard
    T. facial angle
    T. line
topistic
topographic anatomy
topography
topology
toponym
toponymy
torcular herophili
tori (*pl. of* torus)
toric
Tornwaldt cyst
torr
torsion
    angle of femoral t.
torso
torticollis
    congenital t.
tortuous
torulus, pl. **toruli**
    toruli tactiles
torus, pl. **tori**

    t. frontalis
    levatorius t.
    t. levatorius
    t. manus
    maxillary t.
    t. occipitalis
    occipitalis t.
    palatinus t.
    tubarius t.
    t. tubarius
    t. uretericus
    t. uterinus
total
    t. body water (TBW)
    t. elasticity of muscle
    t. energy
    t. facial index
    t. lung capacity
    t. peripheral resistance
    t. placenta previa
    t. refractory period
totipotence
totipotency
totipotent cell
totipotential protoplasm
touch
    t. cell
    t. corpuscle
    organ of t.
Tourtual
    T. membrane
    T. sinus
Touton giant cell
towers skull
toxicity
    oxygen t.
Toynbee
    T. corpuscles
    T. muscle
trabecula, pl. **trabeculae**
    trabeculae carneae
    trabeculae carneae of right and left ventricle
    trabeculae carneae ventriculorum dextri et sinistri
    trabeculae of corpora cavernosa
    trabeculae corporis spongiosi penis
    trabeculae corporum cavernosorum
    trabeculae of corpus spongiosum
    trabeculae cranii
    trabeculae lienis
    trabeculae of lymph node
    trabeculae nodi lymphoidei
    septomarginal t.
    t. septomarginalis
    trabeculae of spleen
    splenic trabeculae
    trabeculae splenicae
    t. testis

trabecular
- t. bone
- t. meshwork
- t. network
- t. region
- t. reticulum
- t. tissue of sclera
- t. vein
- t. zone

trabeculare
- reticulum t.

trabecularis
- substantia t.

trabeculate

trabeculation

trace conditioned reflex

trach

trachea
- annular ligaments of t.
- bifurcation of t.
- carina of t.
- internal surface of t.
- membranous wall of t.
- mucosa of t.
- mucous membrane of t.
- muscular coat of t.
- muscular layer of t.
- thoracic part of t.

tracheae
- bifurcatio t.
- carina t.
- paries membranaceus t.
- pars thoracica t.
- tunica mucosa t.
- tunica muscularis t.

tracheal
- t. bifurcation
- t. branch
- t. cannula
- t. carina
- t. cartilage
- t. fistula
- t. gland
- t. lymph node
- t. mucosa
- t. muscle
- t. reflex
- t. ring
- t. tree
- t. triangle
- t. vein
- t. wall

tracheales
- cartilagines t.
- glandulae t.
- rami t.
- venae t.

trachealia
- ligamenta t.
- ligamenta anularia t.

trachealis
- t. muscle
- musculus t.

trachelalis

trachelian

trachelobregmatic diameter

tracheloclavicularis
- musculus t.

tracheloclavicular muscle

trachelomastoideus
- musculus t.

trachelomastoid muscle

trachelooccipitalis

trachelos

tracheloschisis

tracheobiliary fistula

tracheobronchial
- t. groove
- t. node
- t. tree

tracheobronchomegaly

tracheoesophageal (TE)
- t. fistula
- t. junction

tracheolaryngeal

tracheopharyngeal

trachoma gland

trachychromatic

tracing

tracking
- patellar t.

tract
- alimentary t.
- anterior corticospinal t.
- anterior pyramidal t.
- anterior spinocerebellar t.
- anterior spinothalamic t.
- Arnold t.
- association t.
- atrio-His t.
- auditory t.
- biliary t.
- Burdach t.
- cerebellorubral t.

T

NOTES

623

tract *(continued)*
cerebellothalamic t.
Collier t.
corticobulbar t.
corticopontine t.
corticospinal t.
crossed pyramidal t.
cuneocerebellar t.
deiterospinal t.
dentatothalamic t.
digestive t.
direct pyramidal t.
dorsal spinocerebellar t.
dorsolateral t.
ental t.
fastigiobulbar t.
Flechsig t.
frontopontine t.
frontotemporal t.
gastrointestinal t.
geniculocalcarine t.
genital t.
genitourinary t.
t. of Goll
Gowers t.
habenulointerpeduncular t.
hepatic outflow t.
Hoche t.
hypothalamohypophysial t.
iliopubic t.
iliotibial t.
intestinal t.
James t.
lateral corticospinal t.
lateral pyramidal t.
lateral root of olfactory t.
lateral root of optic t.
lateral spinothalamic t.
left ventricular inflow t.
left ventricular outflow t.
Lissauer t.
Loewenthal t.
long t.
mamillothalamic t.
Marchi t.
medial root of olfactory t.
medial root of optic t.
Monakow t.
t. of Münzer and Wiener
nerve t.
nucleus of solitary t.
occipitocollicular t.
occipitopontine t.
occipitotectal t.
olfactory t.
olivocerebellar t.
olivospinal t.
optic t.
outflow t.

pancreaticobiliary t.
parietopontine t.
posterior spinocerebellar t.
prepyramidal t.
pulmonary outflow t.
pyramidal t.
respiratory t.
reticulospinal t.
right ventricular inflow t.
right ventricular outflow t.
rubrobulbar t.
rubroreticular t.
rubrospinal t.
t. of Schütz
sensory t.
septomarginal t.
sinus t.
solitary t.
spinocerebellar t.'s
spin-olivary t.
spinothalamic t.
spiral foraminous t.
Spitzka marginal t.
sulcomarginal t.
supraopticohypophysial t.
tectal t.
tectobulbar t.
tectopontine t.
tectospinal t.
temporofrontal t.
temporopontine t.
tuberoinfundibular t.
Türck t.
upper gastrointestinal t. (UGI)
upper respiratory t.
urinary t.
urogenital t.
uveal t.
ventral spinocerebellar t.
ventral spinothalamic t.
vestibulospinal t.
Waldeyer t.
traction epiphysis
tractus
t. centralis tegmenti
t. cerebellorubralis
t. cerebellothalamicus
t. corticobulbaris
t. corticopontini
t. corticospinalis
t. corticospinalis anterior
t. corticospinalis lateralis
t. descendens nervi trigemini
t. dorsolateralis
t. fastigiobulbaris
t. frontopontinus
t. habenulopeduncularis
t. iliopubicus
t. iliotibi

t. iliotibialis
t. mesencephalicus nervi trigemini
t. occipitopontinus
t. olfactorius
t. olivocerebellaris
t. opticus
t. parietopontinus
t. pyramidalis
t. pyramidalis anterior
t. pyramidalis lateralis
t. reticulospinalis
t. rubrospinalis
t. solitarius
t. spinalis nervi trigemini
t. spinocerebellaris anterior
t. spinocerebellaris posterior
t. spinotectalis
t. spinothalamicus
t. spinothalamicus anterior
t. spinothalamicus lateralis
t. spiralis foraminosus
t. spiralis foraminulosus
t. supraopticohypophysialis
t. tectobulbaris
t. tectopontinus
t. tectospinalis
t. tegmentalis centralis
t. temporopontinus
t. tuberoinfundibularis
t. vestibulospinalis

tragal lamina
tragi (*pl. of* tragus)
tragica
incisura t.
tragicus
t. muscle
musculus t.
tragion
tragus, pl. **tragi**
hairs of t.
lamina tragi
lamina of t.
muscle of t.
t. muscle
trans
sulci temporales t.
transbasal
transcalent
transcarpal
transcellular
t. fluid
t. water

transcendental anatomy
transcortical
transcytosis
transducer cell
transduction
transethmoidal
transfer
embryo t.
transferase
lecithin-cholesterol t.
transformation
Haldane t.
transfusion
twin-twin t.
transiliac
transilient
transinsular
transischiac
transisthmian
transition
cervicothoracic t.
transitional
t. cell
t. convolution
t. epithelium
t. gyrus
t. vertebra
t. zone
t. zone of lip
transit time
translatory movement
transmembrane potential
transmigration
direct ovular t.
external ovular t.
indirect ovular t.
internal ovular t.
ovular t.
transmission
duplex t.
neurohumoral t.
transmural pressure
transnexus channel
transocular
transonance
transparent septum
transparietal
transpirable
transpiration
pulmonary t.
transpire
transplacental

T

NOTES

625

transplantar
transpleural
transport
    active t.
    axoplasmic t.
    hydrogen t.
    t. maximum
    t. number
    vesicular t.
transpose
transposition of arterial stem
transpulmonary pressure
transpyloric plane
transpyloricum
    planum t.
transsacral
transscrotal
transsegmental
transseptal
    t. fibers
    t. sheath
transsphenoidal
transsynaptic
transtentorial
transthalamic
transthermia
transthoracic pressure
transtubercular plane
transudate
transudation
transude
transurethral
transvaginal
transvenous
transversa
    articulatio tarsi t.
    t. colli
    crista t.
    diameter t.
    t. faciei
    foramen intermesocolica t.
    fossa intermesocolica t.
    ligamenta capitulorum t.
    linea t.
    pars t.
    plica palatina t.
    plica vesicalis t.
    sutura palatina t.
transversae
    fibrae pontis t.
transversalia
    plana t.
transversalis
    arcus pedis t.
    crista t.
    fascia t.
    t. fascia
transversaria
    pars t.

transversarial part of vertebral artery
transversarium
    foramen t.
transverse
    t. acetabular ligament
    t. anthelicine groove
    t. arch
    t. arch of foot
    t. artery of face
    t. artery of neck
    t. artery of scapula
    t. arytenoid muscle
    t. atlantal ligament
    t. auricular muscle
    t. branch of lateral femoral
    circumflex artery
    t. carpal ligament
    t. cervical artery
    t. cervical ligament
    t. cervical nerve
    t. cervical vein
    t. circumflex vessel
    t. colli nerve
    t. colon
    t. commissure
    t. costal facet
    t. costal facet of vertebra
    t. crest
    t. crest of internal acoustic meatus
    t. crural ligament
    deep t.
    t. diameter
    t. disk
    t. ductule of epoöphoron
    t. facial
    t. facial artery
    t. facial vein
    t. fascia
    t. fasciculi
    t. fiber
    t. fibers of pons
    t. fissure of cerebellum
    t. fissure of cerebrum
    t. fold of rectum
    t. foramen
    t. fornix
    t. genicular ligament
    t. head
    t. horizontal axis
    t. humeral ligament
    t. intermesocolic fossa
    t. ligament of acetabulum
    t. ligament of atlas
    t. ligament of elbow
    t. ligament of knee
    t. ligament of leg
    t. ligament of pelvis
    t. ligament of perineum
    t. line

t. line of sacrum
t. mesocolon
t. metacarpal ligament
t. metatarsal ligament
t. mucosal ruga
t. muscle of abdomen
t. muscle of auricle
t. muscle of chin
t. muscle of nape
t. muscle of thorax
t. muscle of tongue
t. nerve of neck
t. occipital protuberance
t. occipital sulcus
t. oval pelvis
t. palatine fold
t. palatine ridge
t. palatine suture
t. pancreatic artery
t. part
t. part of iliofemoral ligament
t. part of left branch of portal vein
t. part of nasalis muscle
t. part of trapezius muscle
t. pelvic inlet
t. pericardial sinus
t. perineal ligament
t. perineal muscle
t. plane
t. process
t. process of vertebra
t. rectal fold
t. rhombencephalic flexure
t. ridge of sacrum
t. scapular artery
t. section
t. septum
t. sinus of pericardium
t. sulcus of heart
t. tarsal articulation
t. tarsal joint
t. temporal convolutions
t. temporal gyri
t. temporal sulci
t. tibiofibular ligament
t. vein
t. vein of face
t. vein of neck
t. vein of scapula
t. venous sinus
t. vesical fold

transversi
 fasciculi t.
 foramen processus t.
 fovea costalis processus t.
 gyri temporales t.
 sulcus sinus t.
transversocostal
transversospinales
 t. muscle
 musculi t.
transversospinalis
 musculus t.
transversospinal muscle
transversourethralis
transversovertical index
transversum
 caput t.
 colon t.
 ligamentum carpi t.
 mesocolon t.
 velum t.
transversus
 t. abdominis
 t. abdominis muscle
 arytenoideus t.
 t. auriculae
 t. colli
 t. linguae
 t. menti
 t. menti muscle
 musculus arytenoideus t.
 t. nuchae
 t. nuchae muscle
 t. perinei profundus
 t. perinei superficialis
 processus t.
 ramus t.
 sinus t.
 situs t.
 sulcus anthelicis t.
 sulcus occipitalis t.
 t. thoracis
 t. thoracis muscle
trapezia (*pl. of* trapezium)
trapezial
trapeziform
trapezii
 bursa subtendinea musculi t.
 pars ascendens musculi t.
 pars descendens musculi t.
 pars transversa musculi t.
 tuberculum ossis t.

T

NOTES

trapeziometacarpal
trapezium, pl. **trapezia**
    t. bone
    oblique ridge of t.
    os t.
    tubercle of t.
trapezius
    bursa of t.
    t. muscle
    musculus t.
    subtendinous bursa of t.
trapezoid
    t. body
    t. bone
    t. ligament
    t. line
    t. ridge
trapezoidea
    linea t.
trapezoidei
    nucleus dorsalis corporis t.
    nucleus ventralis corporis t.
trapezoideum
    corpus t.
    ligamentum t.
    os t.
Traube
    T. semilunar space
    space of T.
Traube-Hering
    T.-H. curve
    T.-H. wave
Trautmann
    triangle of T.
    T. triangle
    T. triangular space
tree
    biliary t.
    bronchial t.
    tracheal t.
    tracheobronchial t.
trefoil tendon
Treitz
    T. arch
    T. fascia
    T. fossa
    ligament of T.
    T. ligament
    T. muscle
Treves fold
triad
    hepatic t.
    Osler t.
    portal t.
tri amelia
triangle
    anal t.
    anterior t.
    Assézat t.

auricular t.
ausculatory t.
t. of auscultation
axillary t.
Béclard t.
Bonwill t.
Calot t.
t. of Calot
cardiohepatic t.
carotid t.
cephalic t.
cervical t.
clavipectoral t.
cystohepatic t.
deltoideopectoral t.
deltopectoral t.
digastric t.
Elaut t.
t. of elbow
facial t.
Farabeuf t.
t. of Farabeuf
femoral t.
t. of fillet
frontal t.
Gerhardt t.
Gombault t.
Grynfeltt t.
Henle t.
Hesselbach t.
inferior carotid t.
inferior lumbar t.
inferior occipital t.
infraclavicular t.
inguinal t.
interscalene t.
Kiesselbach t.
Labbé t.
Langenbeck t.
t. of Langenbeck
lateral pelvic wall t.
Lesgaft t.
Lesser t.
Lesshaft t.
Lieutaud t.
t. of lingual artery
t. of Livingston
lumbar t.
lumbocostoabdominal t.
Macewen t.
Malgaigne t.
Marcille t.
mesenteric t.
muscle of anal t.
muscle of urogenital t.
muscular t.
t. of neck
occipital t.
omoclavicular t.

omotracheal t.
palatal t.
Petit lumbar t.
Philippe t.
Pirogoff t.
pubourethral t.
Reil t.
retromolar t.
sacral t.
t. of safety
scalene-vertebral t.
Scarpa t.
subclavian t.
subinguinal t.
submandibular t.
submaxillary t.
submental t.
suboccipital t.
superior carotid t.
supraclavicular t.
suprahyoid t.
suprameatal t.
tracheal t.
Trautmann t.
t. of Trautmann
umbilicomammillary t.
urogenital t.
t. of vertebral artery
vesical t.
t. of Ward
Weber t.
triangular
t. aponeurosis
t. bone
t. crest
t. disk of wrist
t. fascia
t. fold
t. fossa
t. fossa of auricle
t. fovea of arytenoid cartilage
t. lamella
t. ligament
t. ligament of liver
t. muscle
t. pit of arytenoid cartilage
t. recess
t. ridge
t. space
t. uterus
triangulare
ligamentum t.

os t.
velum t.
triangularis
crista t.
eminentia fossae t.
musculus t.
pars t.
plica t.
recessus t.
uterus t.
triangulum
triatriatum
cor t.
tribasilare
os t.
tribasilar synostosis
tribology
tribrachia
tribrachius
dicephalus dipus t.
tributary
tricarboxylic acid cycle
tricephalus
triceps
t. brachii
t. brachii muscle
t. bursa
t. coxae muscle
t. muscle of arm
t. muscle of calf
t. muscle of hip
musculus t.
t. surae
t. surae jerk
t. surae muscle
t. surae reflex
t. tendon
trichangion
trichion
trichohyalin
trichomegaly
trichotomy
trichterbrust
tricipital
tricorn
tricornute
tricuspid
t. atresia
t. orifice
t. valve
tricuspidal
valva t.

NOTES

tricuspidalis
    cuspis anterior valvae t.
    cuspis posterior valvae t.
    cuspis septalis valvae t.
    valva t.
    valvula t.
tricuspidate
tridactylous
trident hand
tridermic
tridigitate
tridymus
trifacial nerve
trifid
trifurcation
trigastric
trigeminal
    t. cave
    t. cavity
    t. crest
    t. ganglion
    t. gland
    t. impression
    t. lemniscus
    t. nerve
    t. nerve root
    t. pulse
trigeminale
    cavum t.
    ganglion t.
trigeminales
trigeminalis
    impressio t.
    lemniscus t.
    ramus ganglii t.
trigemini
    nucleus motorius nervi t.
    nucleus sensorius principalis
     nervi t.
    nucleus sensorius superior nervi t.
    nucleus tractus mesencephali
     nervi t.
    nucleus tractus spinalis nervi t.
    portio major nervi t.
    portio minor nervi t.
    radices nervi t.
    radix motoria nervi t.
    radix sensoria nervi t.
    ramus ganglionis t.
    tractus descendens nervi t.
    tractus mesencephalicus nervi t.
    tractus spinalis nervi t.
trigeminus
    descending nucleus of t.
    mesencephalic nucleus of t.
    motor nucleus of t.
    t. nerve
    nervus t.

    t. reflex
    spinal nucleus of t.
trigger
trigona (*pl. of* trigonum)
trigonal
trigone
    t. of auditory nerve
    t. of bladder
    collateral t.
    deltoideopectoral t.
    t. of fillet
    t. of habenula
    habenular t.
    t. of hypoglossal nerve
    inguinal t.
    t. of lateral ventricle
    Lieutaud t.
    Müller t.
    olfactory t.
    retromolar t.
    right fibrous t.
    urogenital t.
    t. of vagus nerve
    vertebrocostal t.
trigonid
trigonocephalic
trigonocephaly
trigonum, pl. **trigona**
    t. acustici
    t. auscultationis
    t. caroticum
    t. cerebrale
    t. cervicale
    t. cervicale anterius
    t. cervicale posterius
    t. clavipectorale
    t. collaterale
    t. colli
    t. colli anterius
    t. colli laterale
    t. cystohepaticum
    t. deltoideopectorale
    t. deltopectorale
    t. femorale
    t. femoris
    trigona fibrosa cordis
    t. fibrosum dextrum
    t. fibrosum sinistrum
    t. habenulae
    t. inguinale
    t. lemnisci
    t. lumbale
    t. lumbale inferius
    t. lumbocostale
    t. lumbocostale diaphragmatis
    t. musculare
    t. musculare regionis cervicalis
     anterioris
    t. nervi hypoglossi

t. nervi vagi
t. olfactorium
t. omoclaviculare
t. omotracheale
os t.
t. palati
t. parietale laterale pelvis
t. retromolare
t. sternocostale diaphragmatis
t. submandibulare
t. submentale
t. ventriculi
t. vesicae

triiniodymus
trilaminar blastoderm
trilateral
triloba
placenta t.
trilobate
trilobed
trilocular
triloculare
cor t.
trilogy
trinity
hepatic t.
triophthalmos
triorchism
triotus
tripartita
placenta t.
triphalangia
triplet
triplex
placenta t.
triploblastic
tripod
Haller t.
vital t.
tripodia
triprosopus
triquetral bone
triquetrous cartilage
triquetrum
t. bone
os t.
triradial
triradiate
triradius
trisplanchnic
tristichia
trisulcate

triticea
cartilago t.
triticeal cartilage
triticeoglossus
musculus t.
triticeous
triticeum
corpus t.
trizonal
trochanter
greater t.
lesser t.
t. major
t. minor
small t.
t. tertius
third t.
trochanterian
trochanteric
t. bursa
t. bursae of gluteus medius
t. bursae of gluteus minimus
t. bursa of gluteus maximus
t. crest
t. fossa
trochanterica
bursa t.
bursa subcutanea t.
fossa t.
trochantin
trochantinian
trochlea, pl. **trochleae**
t. femoris
t. fibularis calcanei
t. humeri
t. of humerus
muscular t.
t. muscularis
t. musculi obliqui superioris bulbi
t. peronealis
trochleae of phalanges of hand and foot
t. phalangis
t. phalangis manus et pedis
t. of superior oblique muscle
t. tali
t. of talus
vagina synovialis trochleae
trochlear
t. fossa
t. fovea
t. nerve

NOTES

trochlear *(continued)*
    t. notch
    t. nucleus
    t. pit
    t. process
    t. ridge
    t. spine
    t. synovial bursa
trochleariform
trochleariformis
    processus t.
trochlearis
    fossa t.
    fovea t.
    incisura t.
    nervus t.
    nucleus nervi t.
    processus t.
    spina t.
trochlearium
    decussatio nervorum t.
trochleiform
trochoid
    t. articulation
    t. joint
trochoidal articulation
trochoidea
    articulatio t.
Troisier
    T. ganglion
    T. node
Trolard vein
Tröltsch
    T. corpuscles
    T. fold
    T. pocket
    T. recess
trophectoderm
trophic
trophicity
trophism
trophoblast
    plasmodial t.
    syncytial t.
trophoblastic
    t. lacuna
    t. operculum
trophocyte
trophoderm
trophodynamics
trophoplasm
trophoplast
trophospongia
trophotaxis
trophotropic
trophotropism
tropism
trough
    gingival t.

    Langmuir t.
    synaptic t.
    vestibular t.
true
    t. conjugate
    t. glottis
    t. hermaphroditism
    t. knot of umbilical cord
    t. muscle of back
    t. pelvis
    t. rib
    t. vertebra
    t. vocal cord
truncal
truncate
truncus, pl. **trunci**
    t. arteriosus
    t. arteriosus communis
    t. brachiocephalicus
    t. celiacus
    t. corporis callosi
    t. costocervicalis
    t. fascicularis atrioventricularis
    t. inferior
    t. inferior plexus brachialis
    t. linguofacialis
    t. lumbosacralis
    trunci lymphatici intestinales
    trunci lymphatici lumbales
    t. lymphaticus bronchiomediastinalis
    t. lymphaticus jugularis
    t. medius
    t. medius plexus brachialis
    t. nervi accessorii
    trunci plexus brachialis
    t. pulmonalis
    t. subclavius
    t. superior
    t. superior plexus brachialis
    t. sympathicus
    t. thyrocervicalis
    t. vagalis
trunk
    accessory nerve t.
    anterior gastric branch of anterior
      vagal t.
    anterior vaginal t.
    atrioventricular t.
    t. of atrioventricular bundle
    bifurcation of pulmonary t.
    brachiocephalic arterial t.
    bronchomediastinal lymphatic t.
    celiac arterial t.
    celiac branch of posterior vagal t.
    communicating branch of
      sympathetic t.
    communicating rami of
      sympathetic t.
    t. of corpus callosum

costocervical arterial t.
ganglion of sympathetic t.
gastric branch of anterior vagal t.
gastric branch of posterior vagal t.
hepatic branch of anterior vagal t.
inferior t.
t. of inferior brachial plexus
interganglionic branch of
  sympathetic t.
intestinal t.'s
jugular lymphatic t.
left bronchomediastinal t.
left crus of atrioventricular t.
left lumbar t.
left sympathetic t.
linguofacial arterial t.
lumbar t.'s
lumbosacral nerve t.
middle t.
t. of middle brachial plexus
t. muscle
nerve t.
opening of pulmonary t.
posterior gastric branch of posterior
  vagal t.
posterior vaginal t.
pulmonary t.
pyloric branch of anterior vagal t.
right bronchomediastinal t.
right crus of atrioventricular t.
right lumbar t.
right lumbosacral t.
right lymphatic t.
right sympathetic t.
sinus of pulmonary t.
subclavian lymphatic t.
superior t.
t. of superior brachial plexus
sympathetic t.
thoracoacromial t.
thyrocervical arterial t.
upper t.
t. of upper brachial plexus
vagal nerve t.
t. valve
valve of pulmonary t.

TSH
thyroid-stimulating hormone
TSI
thyroid-stimulating immunoglobulin
tub
tuba, pl. **tubae**

t. acustica
t. auditiva
t. auditoria
dilator tubae
t. eustachiana
t. eustachii
t. fallopii
laciniae tubae
musculus dilator tubae
septum tubae
t. uterina
tubal
t. air cell of pharyngotympanic
  tube
t. air cells
t. branch
t. branch of ovarian artery
t. branch of tympanic plexus
t. branch of uterine artery
t. cartilage
t. extremity
t. extremity of ovary
t. gland of pharyngotympanic tube
t. tonsil
tubaria
tonsilla t.
tubariae
glandulae t.
plicae t.
tubarius
ramus t.
torus t.
t. torus
tube
abdominal ostium of uterine t.
air cell of auditory t.
alar lamina of neural t.
ampullary fold of uterine t.
ampulla of uterine t.
auditory t.
basal lamina of neural t.
basal plate of neural t.
bony part of auditory t.
bony part of pharyngotympanic t.
bronchial t.
canal for pharyngotympanic t.
cardiac t.
cartilage of auditory t.
cartilage of pharyngotympanic t.
cartilaginous part of auditory t.
cartilaginous part of
  pharyngotympanic t.

T

NOTES

tube (*continued*)
    cartilaginous plate of auditory t.
    cathode ray t.
    collecting t.
    digestive t.
    dorsal plate of neural t.
    eustachian t.
    fallopian t.
    Ferrein t.
    fimbriae of uterine t.
    gland of auditory t.
    gland of eustachian t.
    groove for auditory t.
    Haldane t.
    infundibulum of fallopian t.
    infundibulum of uterine t.
    isthmus of auditory t.
    isthmus of eustachian t.
    isthmus of fallopian t.
    isthmus of pharyngotympanic t.
    isthmus of uterine t.
    lateral lamina of cartilage of
      pharyngotympanic auditory t.
    lateral plate of cartilaginous
      auditory t.
    limiting membrane of neural t.
    medial lamina of cartilage of
      pharyngotympanic auditory t.
    medial plate of cartilaginous
      auditory t.
    medullary t.
    mucosa of pharyngotympanic
      auditory t.
    mucosa of uterine t.
    mucous gland of auditory t.
    mucous membrane of
      pharyngotympanic auditory t.
    mucous membrane of uterine t.
    muscular coat of uterine t.
    muscular layer of uterine t.
    neural t.
    orifice t.
    otopharyngeal t.
    ovarian t.
    pharyngeal opening of eustachian t.
    pharyngeal opening of
      pharyngotympanic auditory t.
    pharyngotympanic auditory t.
    Pitot t.
    Ruysch t.
    semicanal of auditory t.
    septum of auditory t.
    septum of pharyngotympanic
      auditory t.
    serosa of uterine t.
    sulcus for auditory t.
    sulcus for pharyngotympanic t.
    tubal air cell of
      pharyngotympanic t.

    tubal gland of pharyngotympanic t.
    tympanic opening of eustachian t.
    tympanic opening of
      pharyngotympanic auditory t.
    uterine opening of uterine t.
    uterine ostium of uterine t.
    uterine part of uterine t.
    ventral plate of neural t.
    Venturi t.
    vesicular appendices of uterine t.

tuber, pl. **tubera**
    t. anterius
    ashen t.
    t. calcanei
    t. calcis
    t. cinereum
    t. cochlea
    t. corporis callosi
    t. dorsale
    eustachian t.
    frontal t.
    t. frontale
    gray t.
    t. ischiadicum
    t. of ischium
    t. maxillae
    omental t.
    t. omentale
    t. omentale hepatis
    t. omentale pancreatis
    parietal t.
    t. parietale
    t. radii
    t. valvulae
    t. vermis
    t. zygomaticum
tuberales
    nuclei t.
tuberalis
    pars t.
tuberal nuclei
tubercle
    accessory t.
    acoustic t.
    adductor t.
    amygdaloid t.
    t. of anterior scalene muscle
    areolar t.
    articular t.
    t. of articular facet of rib
    ashen t.
    auricular t.
    calcaneal t.
    carotid t.
    Chassaignac t.
    conoid t.
    corniculate t.
    crest of greater t.
    crest of lesser t.

crown t.
t. of cuneate nucleus
cuneiform t.
Darwin t.
darwinian t.
dental t.
dorsal radius t.
epiglottic t.
t. of epiglottis
genial t.
genital t.
Gerdy knee t.
Ghon t.
gracile t.
gray t.
greater t.
greater humerus t.
iliac t.
t. of iliac crest
inferior thyroid t.
intercolumnar t.
intercondylar t.
intervenous t.
jugular t.
labial t.
lesser humerus t.
Lisfranc t.
Lister t.
Lower t.
mamillary t.
marginal t.
maxillary t.
mental t.
molar t.
Morgagni t.
Müller t.
nuchal t.
t. of nucleus gracilis
obturator t.
olfactory t.
orbital t.
peroneal t.
phallic t.
pharyngeal t.
Princeteau t.
pterygoid t.
pubic t.
quadrate femoral t.
t. of radius
Rolando t.
t. of saddle
Santorini t.

scalene t.
t. of scaphoid bone
sinus t.
superior thyroid t.
supraglenoid t.
supratragic t.
tibial t.
t. of tooth
t. of trapezium
t. of trapezium bone
t. of upper lip
wedge-shaped t.
Whitnall orbital t.
Wrisberg t.
tuberculation
tuberculum, pl. tubercula
t. adductorium
t. adductorium femoris
t. anterius atlantis
t. anterius thalami
t. anterius vertebrarum cervicalium
tubercula areolae
t. articulare
t. articulare ossis temporalis
t. auriculae
t. calcanei
t. caroticum
t. cinereum
t. conoideum
t. conoideum claviculare
t. corniculatum
t. coronae
t. costae
t. cuneatus
t. cuneiforme
t. deltoideum spinae scapulae
t. dentis
t. dorsale
t. dorsale radii
t. epiglotticum
t. hypoglossi
t. iliacum
t. impar
t. infraglenoidale
t. infraglenoidale scapulae
t. intercondylare
t. intercondylare mediale et laterale
t. intervenosum
t. intervenosum atrii dextri
t. jugulare
t. jugulare ossis occipitalis
t. labii superioris

T

NOTES

tuberculum *(continued)*
t. laterale processus posterioris tali
t. majus humeri
t. mallei
t. marginale ossis zygomatici
t. mediale processus posterioris tali
t. mentale
t. mentale mandibulae
t. minus humeri
t. molare
t. musculi scaleni anterioris
t. nuclei cuneati
t. nuclei gracilis
t. obturatorium
t. olfactorium
t. orbitale ossis zygomatici
t. ossis scaphoidei
t. ossis trapezii
t. pharyngeum
t. pharyngeum partis basilaris ossis occipitalis
t. posterius atlantis
t. posterius vertebrarum cervicalium
t. pubicum
t. sellae
t. septi narium
t. supraglenoidale
t. supraglenoidale scapulae
t. supratragicum
t. thyroideum inferius
t. thyroideum superius
t. of trapezium bone
tuberiferous
tuberoinfundibularis
tractus t.
tuberoinfundibular tract
tuberose
tuberositas
t. coracoidea
t. costalis
t. deltoidea
t. deltoidea humeri
t. glutea
t. iliaca
t. ligamenti coracoclavicularis
t. masseterica
t. musculi serrati anterioris
t. ossis cuboidei
t. ossis metatarsalis primi
t. ossis metatarsalis quinti
t. ossis navicularis
t. phalangis distalis
t. phalangis distalis manus et pedis
t. pronatoria
t. pterygoidea
t. pterygoidea mandibulae
t. radii
t. sacralis
t. tibiae

t. ulnae
t. unguicularis
tuberosity
t. of anterior calcaneus
bicipital t.
calcaneal t.
t. for coracoclavicular ligament
coracoid t.
costal t.
t. of cuboid bone
deltoid t.
t. of distal phalanx of hand and foot
t. of fifth metatarsal
t. of fifth metatarsal bone
t. of first metatarsal
t. of first metatarsal bone
gluteal t.
iliac t.
infraglenoid t.
ischial t.
lateral femoral t.
lateral process of calcaneal t.
masseteric t.
maxillary t.
medial femoral t.
medial process of calcaneal t.
t. of navicular bone
t. of phalanx
pronator t.
pterygoid t.
radial t.
t. of radius
sacral t.
scaphoid t.
t. of scaphoid bone
t. for serratus anterior muscle
t. of sesamoid bone
subcutaneous bursa of tibial t.
tibial t.
t. of ulna
ungual t.
tuberous
tubi (*pl. of* tubus)
tuboabdominal
tuboligamentous
tuboovarian, tubo-ovarian
tubopalatina
plica t.
tuboperitoneal
tubotympanal
tubotympanic
t. canal
t. recess
tubouterine
tubovaginal
tubular
t. gland
t. maximum

tubule
>Albarran y Dominguez t.
>collecting t.
>connecting t.
>convoluted seminiferous t.
>dental t.
>dentinal t.
>discharging t.
>Henle t.'s
>Kobelt t.'s
>mesonephric t.
>metanephric t.
>paragenital t.'s
>pronephric t.
>seminiferous t.
>Skene t.'s
>spiral t.
>straight seminiferous t.
>T t.
>uriniferous t.
>wolffian t.'s

tubuli (*pl. of* tubulus)
tubuliform
tubulin
tubuloacinar gland
tubuloalveolar gland
tubuloglomerular feedback
tubuloracemose
tubulose
tubulous
tubulus, pl. **tubuli**
>tubuli biliferi
>tubuli dentales
>tubuli epoöphori
>tubuli galactophori
>tubuli lactiferi
>tubuli paroophori
>t. rectus
>t. renalis contortus
>t. renalis rectus
>t. seminiferus contortus
>t. seminiferus rectus

tubus, pl. **tubi**
>t. digestorius
>t. medullaris
>t. vertebralis

tuft
>malpighian t.
>phalangeal t.
>synovial t.'s

tufted cell
Tulpius valve

tunic
>t. adventitia
>Bichat t.
>Brücke t.
>circular layers of muscular t.
>longitudinal layers of muscular t.
>mucosal t.
>muscular t.
>serous t.
>vascular t.

tunica, pl. **tunicae**
>t. abdominalis
>t. adventitia
>t. albuginea
>t. albuginea of corpora cavernosa
>t. albuginea corporis spongiosi
>t. albuginea corporum cavernosorum
>t. albuginea of corpus spongiosum
>t. albuginea oculi
>t. albuginea ovarii
>t. albuginea of ovary
>t. albuginea testis
>t. carnea
>t. conjunctiva
>t. conjunctiva bulbi
>t. conjunctiva palpebrarum
>t. dartos
>t. elastica
>t. externa
>t. externa oculi
>t. externa thecae folliculi
>t. extima
>t. fibromusculocartilaginea bronchi
>t. fibrosa
>t. fibrosa bulbi
>t. fibrosa hepatis
>t. fibrosa lienis
>t. fibrosa renis
>t. fibrosa splenis
>tunicae funiculi spermatici
>t. interna bulbi
>t. interna thecae folliculi
>t. intima
>t. media
>t. mucosa
>t. mucosa bronchi
>t. mucosa cavitatis tympani
>t. mucosa coli
>t. mucosa ductus deferentis
>t. mucosa esophagi
>t. mucosa gastrica
>t. mucosa intestini crassi

T

NOTES

tunica (*continued*)

t. mucosa intestini tenuis
t. mucosa laryngis
t. mucosa linguae
t. mucosa nasi
t. mucosa oris
t. mucosa pelvis renalis
t. mucosa pharyngis
t. mucosa tracheae
t. mucosa tubae auditivae
t. mucosa tubae auditoriae
t. mucosa tubae uterinae
t. mucosa ureteris
t. mucosa urethrae femininae
t. mucosa uterus
t. mucosa vaginae
t. mucosa ventriculi
t. mucosa vesicae biliaris
t. mucosa vesicae felleae
t. mucosa vesicae urinariae
t. mucosa vesiculae seminalis
t. muscularis
t. muscularis bronchiorum
t. muscularis coli
t. muscularis ductus deferentis
t. muscularis esophagi
t. muscularis gastrica
t. muscularis glandulae vesiculosae
t. muscularis intestini crassi
t. muscularis intestini tenuis
t. muscularis partis intermediae urethrae masculinae
t. muscularis partis prostaticae urethrae masculinae
t. muscularis partis spongiosae urethrae masculinae
t. muscularis pelvis renalis
t. muscularis pharyngis
t. muscularis recti
t. muscularis tracheae
t. muscularis tubae uterinae
t. muscularis ureteris
t. muscularis urethrae femininae
t. muscularis urethrae masculinae
t. muscularis uterus
t. muscularis vaginae
t. muscularis ventriculi
t. muscularis vesicae biliaris
t. muscularis vesicae felleae
t. muscularis vesicae urinariae
t. nervea
t. propria
t. propria corii
t. propria lienis
t. reflexa
t. sclerotica
t. serosa
t. serosa coli
t. serosa esophagi

t. serosa gastricae
t. serosa hepatis
t. serosa intestini crassi
t. serosa intestini tenuis
t. serosa lienis
t. serosa pericardii serosi
t. serosa peritonei
t. serosa pleurae perietalis
t. serosa pleurae visceralis
t. serosa splenis
t. serosa tubae uterinae
t. serosa uterus
t. serosa ventriculi
t. serosa vesicae biliaris
t. serosa vesicae felleae
t. serosa vesicae urinariae
t. spongiosa urethrae femininae
t. spongiosa vaginae
t. submucosa
t. urethrae masculinae
t. vaginalis
t. vaginalis communis
t. vaginalis testis
t. vasculosa
t. vasculosa bulbi
t. vasculosa lentis
t. vasculosa oculi
t. vasculosa testis
t. vitrea

tunnel

carpal t.
t. cells
Corti t.
radial t.
tarsal t.
ulnar t.

tunnius canal
turbinal
turbinate

inferior t.
middle t.
sphenoidal t.
superior t.

turbinated

t. body
t. bone
t. crest

turcica

diaphragm of sella t.
hypoglossal fossa of sella t.
hypophyseal fossa of sella t.
sella t.

Türck

T. bundle
T. column
T. tract

turgometer
turgor vitalis

Türk
> T. cell
> T. leukocyte

Turkish saddle

Turner
> intraparietal sulcus of T.
> T. sulcus

turricephaly

tutamen, pl. **tutamina**
> tutamina cerebri
> tutamina oculi

twelfth cranial nerve

twelfth-year molar

twin
> t. cone
> t. placenta
> t. pregnancy

twinning

twins
> conjoined asymmetrical t.
> conjoined equal t.
> conjoined unequal t.
> dichorial t.
> diovular t.
> dizygotic t.
> enzygotic t.
> fraternal t.
> heterologous t.
> identical t.
> incomplete conjoined t.
> monoamniotic t.
> monochorial t.
> monovular t.
> monozygotic t.
> omphaloangiopagous t.
> polyzygotic t.
> Siamese t.
> uniovular t.

twin-twin transfusion

two-bellied muscle

two-headed muscle

two-step exercise test

tylion, pl. **tylia**

tympana (*pl. of* tympanum)

tympanae
> pars flaccida membranae t.

tympanal

tympani
> anterior canaliculus of chorda t.
> anulus fibrocartilagineus
>   membranae t.

apertura tympanica canaliculi
  chordae t.
canaliculus of chorda t.
cavum t.
chorda t.
communicating branch of otic
  ganglion to chorda t.
fold of chorda t.
fundus t.
limbus membranae t.
membrana t.
musculus tensor t.
nervus musculi tensoris t.
paracentesis t.
paries caroticus cavi t.
paries jugularis cavi t.
paries labyrinthicus cavi t.
paries mastoideus cavi t.
paries membranaceus cavi t.
paries tegmentalis cavi t.
pars flaccida membranae t.
pars tensa membranae t.
plica chordae t.
plica membranae t.
posterior canaliculus of chorda t.
promontorium cavi t.
pyramis t.
ramus communicans cum chorda t.
ramus communicans ganglii otici
  cum chorda t.
ramus communicans nervi lingualis
  cum chorda t.
ramus membranae t.
scala t.
semicanalis musculi tensoris t.
t. sinus
sinus posterior cavi t.
small canal of chorda t.
stratum circulare membranae t.
stratum cutaneum membranae t.
stratum radiatum membranae t.
substantia propria membranae t.
tegmen t.
tensor t.
tensoris t.
tunica mucosa cavitatis t.
tympanic aperture of canaliculus
  for chorda t.
tympanic opening of canal for
  chorda t.
tympanic opening of canaliculus
  for chorda t.

T

NOTES

tympani *(continued)*
umbo membranae t.
vein of scala t.
vena scalae t.

tympanic
t. air cell
t. antrum
t. aperture of canaliculus for chorda tympani
t. artery
t. attic
t. body
t. canal
t. canaliculus
t. cavity
t. ganglion
t. gland
t. groove
t. incisure
t. intumescence
t. labium of limbus of spiral lamina
t. lamella of osseous spiral lamina
t. lip
t. lip of limbus of spiral lamina
t. lip of spiral limbus
t. membrane
t. nerve
t. nervous plexus
t. notch
t. opening of canal for chorda tympani
t. opening of canaliculus for chorda tympani
t. opening of eustachian tube
t. opening of pharyngotympanic auditory tube
t. orifice
t. part of temporal bone
t. plate
t. plate of temporal bone
t. promontory
t. ring
t. scute
t. sinus
t. sulcus
t. superior
t. surface of cochlear duct
t. vein
t. wall of cochlear duct

tympanica
t. anterior
cavitas t.
crista t.
incisura t.
t. inferior
intumescentia t.
t. posterior

tympanicae
cellulae t.
recessus anterior membranae t.
recessus membranae t.
recessus posterior membranae t.
recessus superior membranae t.
sulcus promontorii cavitatis t.
venae t.

tympanichord
tympanichordal
tympanici
ramus tubarius plexus t.

tympanico
ramus communicans nervi facialis cum plexu t.
ramus communicans nervi intermedii cum plexu t.

tympanicum
ganglion t.
labium limbi t.

tympanicus
anulus t.
canaliculus t.
nervus t.
plexus t.
plexus nervosus t.
sulcus t.

tympanoeustachian
tympanohyal bone
tympanomalleal
tympanomandibular
tympanomastoid
t. cavity
t. fissure
t. suture

tympanomastoidea
fissura t.

tympanosquamosa
fissura t.

tympanosquamosal
tympanosquamous fissure
tympanostapedia
syndesmosis t.

tympanostapedial
t. junction
t. syndesmosis

tympanostapedialis
syndesmosis t.

tympanotemporal
tympanum, pl. **tympana, tympanums**
cord of t.
membrane of t.
promontory of t.
pyramid of t.
roof of t.

type
t. 1, 2 dextrocardia
t. II cells

typhlon

typus
    t. ampullaris pelvis renalis
    t. dendriticus pelvis renalis

Tyrrell fascia
Tyson gland
Tzanck cells

NOTES

U
    unit
UES
    upper esophageal sphincter
UFA
    unesterified free fatty acid
UGI
    upper gastrointestinal tract
ulna, pl. **ulnae**
    anterior border of u.
    anterior surface of u.
    articular circumference of head
        of u.
    body of u.
    coronoid process of u.
    distal u.
    head of u.
    interosseous border of u.
    interosseous crest of u.
    medial surface of u.
    nutrient artery of u.
    olecranon of u.
    posterior border of u.
    posterior surface of u.
    proximal end of u.
    radial notch of u.
    shaft of u.
    styloid process of u.
    supinator crest of u.
    tuberosity of u.
ulnad
ulnae
    arteria nutricia u.
    arteria nutriens u.
    caput u.
    circumferentia articularis capitis u.
    corpus u.
    crista musculi supinatoris u.
    facies anterior u.
    facies medialis u.
    facies posterior u.
    incisura semilunaris u.
    margo anterior u.
    margo interosseus u.
    margo posterior u.
    processus coronoideus u.
    processus styloideus u.
    tuberositas u.
ulnar
    u. aspect
    u. border of forearm
    u. branch of medial antebrachial
        cutaneous nerve
    u. bursa
    u. collateral ligament

u. collateral ligament of elbow
    joint
u. collateral ligament of wrist
u. collateral ligament of wrist joint
u. communicating branch of
    superficial radial nerve
u. eminence of wrist
u. extensor muscle of wrist
u. flexor muscle
u. flexor muscle of wrist
u. groove
u. head
u. margin of forearm
u. notch
u. recurrent artery
u. reflex
u. shaft
u. styloid bone
u. tunnel
u. vein
ulnare
    caput u.
    ligamentum collaterale carpi u.
ulnares
    digitales palmares communes
        nervi u.
    venae u.
ulnari
    ramus communicans cum nervo u.
    ramus communicans nervi interossei
        antebrachii anterioris cum nervi
        u.
    ramus communicans nervi mediani
        cum nervo u.
    ramus communicans nervi radialis
        cum nervi u.
ulnaris
    arteria recurrens u.
    caput humerale musculi flexoris
        carpi u.
    caput ulnare musculi flexoris
        carpi u.
    digitales dorsales nervi u.
    digitales palmares proprii nervi u.
    eminentia carpi u.
    extensor carpi u. (ECU)
    flexor carpi u. (FCU)
    incisura u.
    musculus extensor carpi u.
    musculus flexor carpi u.
    nervi digitales dorsales nervi u.
    nervus u.
    rami dorsales nervi u.
    rami musculares nervi u.
    ramus carpalis dorsalis arteriae u.

U

ulnaris *(continued)*
    ramus carpalis palmaris arteriae u.
    ramus carpeus dorsalis arteriae u.
    ramus carpeus palmaris arteriae u.
    ramus communicans u.
    ramus palmaris nervi u.
    ramus palmaris profundus
      arteriae u.
    ramus posterior arteriae
      recurrentis u.
    ramus profundus nervi u.
    ramus superficialis nervi u.
    recurrens u.
    sulcus bicipitalis u.
    sulcus nervi u.
    vagina tendinis musculi extensoris
      carpi u.
ulnocarpal ligament
ulnoradial
ultimobranchial
    u. artery
    u. body
    u. pouch
ultrabrachycephalic
ultradian rhythm
ultradolichocephalic
ultrafiltration coefficient
ultrastructural anatomy
ultrastructure
ultromotivity
umbilical
    u. artery
    u. circulation
    u. cord
    u. fascia
    u. fissure
    u. fold
    u. fossa
    u. ligament
    u. notch
    u. part
    u. part of left branch of portal
      vein
    u. plane
    u. prevesical fascia
    u. region
    u. ring
    u. vesicle
umbilicalis
    anulus u.
    arteria u.
    canalis u.
    chorda arteriae u.
    fascia u.
    fossa venae u.
    funiculus u.
    incisura u.
    pars patens arteriae u.

    rami ureterici partis patentis
      arteriae u.
    regio u.
    u. sinistra
    sulcus venae u.
    vena u.
    vesicula u.
umbilicate
umbilicated
umbilication
umbilici (*pl. of* umbilicus)
umbilicomammillary triangle
umbilicovesical fascia
umbilicus, pl. **umbilici**
    canalis u.
umbo, pl. **umbones**
    u. membranae tympani
    u. of tympanic membrane
umbonis
uncal artery
uncalis
    arteria u.
unci (*pl. of* uncus)
unciform
    u. bone
    u. fasciculus
    u. pancreas
unciforme
uncinate
    u. bundle of Russell
    u. fasciculus of Russell
    u. groove
    u. gyrus
    u. process
    u. process of cervical vertebra
    u. process of ethmoid bone
    u. process of first thoracic vertebra
    u. process of pancreas
uncinatum
uncinatus
    fasciculus u.
    gyrus u.
uncompensated
    u. acidosis
    u. alkalosis
unconditioned
    u. reflex
    u. response
    u. stimulus
uncoossified
uncovertebral joint
uncus, pl. **unci**
    u. band of Giacomini
    u. gyri parahippocampalis
    vein of u.
underhorn
underlying
    u. chest muscle
    u. fascia

u. skin
u. structure
u. tissue
undershoot
underventilation
undescended testis
undifferentiated cell
unequal cleavage
unesterified free fatty acid (UFA)
ungual
u. phalanx
u. process
u. tuberosity
ungues (*pl. of* unguis)
unguicularis
tuberositas u.
unguinal
unguis, pl. **ungues**
u. avis
corpus u.
cristae matricis u.
Haller u.
u. incarnatus
lunula u.
margo lateralis u.
margo liber u.
margo occultus u.
matrix u.
os u.
radix u.
retinacula u.
sinus u.
stratum corneum u.
stratum germinativum u.
vallum u.
unguium
arcus u.
selene u.
uniarticular
uniaxial joint
unicameral
unicamerate
unicanalicular sphincter
unicellular gland
unicollis
uterus bicornate u.
unicommissural aortic valve
unicornis
uterus u.
unicornous
unicorn uterus
unidirectional flux

uniforate
uniform
unigerminal
uniglandular
unilaminar
unilaminate
unilateral hermaphroditism
unilobar
unilocular
u. fat
u. joint
uninhibited neurogenic bladder
uninuclear
uninucleate
unioval
uniovular twins
unipennate muscle
unipennatus
musculus u.
unipolar
u. cell
u. neuron
uniport
uniporter
uniseptate
unit (U)
alpha u.
anatomic mass u. (amu)
base u.
centimeter-gram-second u. (CGS)
electromagnetic u.
electrostatic u. (esu)
u. of energy
u. fibrils
foot-pound-second u.
u. of force
gravitational u.
u. of heat
lung u.
u. of magnetic field intensity
u. membrane
meter-kilogram-second u.
motor u.
physiologic u.
practical u.
volume u. (VU)
u. of work
uniting
u. canal
u. cartilage
u. duct
unmedullated

U

NOTES

unmyelinated fibers
unpaired thyroid venous plexus
unphysiologic
unstriated muscle
unstriped muscle
UPJ
    ureteropelvic junction
upper
    u. abdomen
    u. airway
    u. dental arcade
    u. esophageal constriction
    u. esophageal sphincter (UES)
    u. esophagus
    u. extremity
    u. extremity of fibula
    u. eyelid
    u. gastrointestinal tract (UGI)
    u. jaw
    u. jaw bone
    u. lateral cutaneous nerve of arm
    u. lid
    u. limb
    u. lip
    u. lobe of lung
    u. motor neuron
    u. pole
    u. pole of testis
    u. respiratory tract
    u. subscapular nerve
    u. thoracic splanchnic nerve
    u. trunk
uppermost
upraopticus hypothalami
upsiloid
uptake
urachal
    u. fold
    u. fossa
    u. ligament
    u. sinus of bladder
urachus, pl. urachi
    plica urachi
uragogue
uraniscochasm
uraniscus
uranoschisis
urceiform
urceolate
urea
    u. clearance
    u. cycle
ureagenesis
ureapoiesis
ureotelic
uresiesthesia
uresis
ureter
    abdominal part of u.

    circumcaval u.
    constrictions of u.
    left u.
    mucosa of u.
    mucous membrane of u.
    muscular coat of u.
    muscular layer of u.
    ostium of u.
    pelvic part of u.
    proximal u.
    retrocaval u.
    right u.
ureteral
    u. branch
    u. meatus
    u. opening
    u. orifice
ureteralis
    pars tecta u.
ureteric
    u. branch
    u. branch of inferior suprarenal
      artery
    u. branch of kidney
    u. branch of ovarian artery
    u. branch of patent part of
      umbilical artery
    u. branch of renal artery
    u. branch of testicular artery
    u. bud
    u. fold
    u. nervous plexus
    u. opening
    u. orifice
    u. pelvis
    u. ridge
ureterica
    plica u.
ureterici
    rami u.
uretericus
    plexus u.
    plexus nervosus u.
    torus u.
ureteris
    mons u.
    orificium u.
    ostium u.
    pars abdominalis u.
    pars pelvica u.
    tunica mucosa u.
    tunica muscularis u.
ureteropelvic junction (UPJ)
ureterovesical
urethra, pl. urethrae
    arteria bulbi urethrae
    bulb of u.
    bulbus urethrae
    cavernous portion of u.

compressor urethrae
corpus cavernosum urethrae
distal part of prostatic u.
external opening of u.
female u.
u. feminina
fossa navicularis urethrae
fossa terminalis urethrae
gland of female u.
gland of male u.
hemispherium bulbi urethrae
intermediate part of male u.
intramural part of male u.
labium urethrae
male u.
u. masculina
membranous part of male u.
mucosa of female u.
mucosa of male u.
mucous membrane of female u.
mucous membrane of male u.
u. muliebris
muscular coat of female u.
muscular coat of intermediate part
of male u.
muscular coat of prostatic u.
muscular coat of spongy part of
male u.
muscular layer of female u.
muscular layer of intermediate part
of male u.
muscular layer of prostatic u.
muscular layer of spongy male u.
musculus compressor urethrae
musculus constrictor urethrae
musculus sphincter urethrae
navicular fossa of male u.
pars prostatica urethrae
penile u.
preprostatic part of male u.
prostatic u.
proximal part of prostatic u.
septum bulbi urethrae
sphincter urethrae
sphincter muscle of u.
spongiose part of male u.
spongy layer of female u.
spongy part of male u.
u. virilis

urethral
  u. artery
  u. carina of vagina
  u. crest
  u. crest of female
  u. crest of male
  u. diverticulum
  u. gland
  u. gland of female
  u. gland of male
  u. groove
  u. lacuna
  u. opening
  u. papilla
  u. plate
  u. pressure profile
  u. surface of penis
  u. valve
  u. wall

urethrales
  lacunae u.

urethralis
  anulus u.
  arteria u.
  colliculus u.
  crista u.
  glandulae u.'s
  habenula u.
  lacuna u.
  papilla u.

urethrobulbar
urethropenile
urethroperineal
urethroperineoscrotal
urethroprostatic
urethrorectal
urethrovaginal
  u. sphincter
urethrovaginalis
  musculus sphincter u.
urethrovesical
URF
  uterine relaxing factor
urgency
  motor u.
  sensory u.
uricolytic index
uricotelic
uriesthesia
urinae
  detrusor u.
  musculus detrusor u.
urinaria
  cystis u.

U

NOTES

urinaria *(continued)*
organa u.
vesica u.
urinariae
cervix vesicae u.
fundus vesicae u.
sphincter vesicae u.
tunica mucosa vesicae u.
tunica muscularis vesicae u.
tunica serosa vesicae u.
urinarium
systema u.
urinarius
meatus u.
urinary
u. apparatus
u. bladder
u. opening
u. organ
u. reflex
u. reservoir
u. sphincter
u. system
u. tract
urinate
urination
urine
uriniferous
u. tubule
urinific
uriniparous
urinogenital
urinogenous
urinosexual
urocyst
urocystic
urocystis
urodynamics
urogenital
u. apparatus
u. canal
u. cleft
u. diaphragm
u. fold
u. membrane
u. mesentery
u. peritoneum
u. region
u. ridge
u. septum
u. sinus
u. system
u. tract
u. triangle
u. trigone
u. vestibule
urogenitale
diaphragma u.

peritoneum u.
systema u.
urogenitalis
alveus u.
apparatus u.
musculi regionis u.
regio u.
sinus u.
urogenous
urogonadotropin
uropoiesis
uropoietic system
urorectal
u. fold
u. membrane
u. septum
uteri *(pl. of* uterus)
uterina
arteria u.
lyra u.
ostium abdominale tubae u.
pars u.
placenta u.
rachitis u.
salpinx u.
tuba u.
uterinae
ampulla tubae u.
fimbriae tubae u.
glandulae u.
infundibulum tubae u.
isthmus tubae u.
ostium abdominale tubae u.
ostium uterinum tubae u.
pars uterina tubae u.
plicae ampullares tubae u.
plicae tubariae tubae u.
rami ovarici arteriae u.
ramus tubarius arteriae u.
tunica mucosa tubae u.
tunica muscularis tubae u.
tunica serosa tubae u.
venae u.
uterine
u. appendage
u. artery
u. canal
u. cavity
u. cervix
u. corpus
u. cuff
u. extremity
u. extremity of ovary
u. fundus
u. gland
u. hernia
u. horn
u. muscle
u. opening of uterine tube

u. orifice
u. ostium of uterine tube
u. part
u. part of uterine tube
u. relaxing factor (URF)
u. segment
u. sinus
u. sinusoid
u. tissue
u. tube fimbria
u. vein
u. venous plexus
u. vessel
u. wall
uterinus
plexus venosus u.
torus u.
uteroabdominal
uterocervical
uteroepichorial membrane
uteroovarian
uteroovaricum
ligamentum u.
uteroparietal suture
uteropelvic ligament
uteroplacental sinus
uterosacral
u. fold
u. ligament
uterotubal
uterovaginal
u. canal
u. nervous plexus
uterovaginalis
plexus nervosus u.
uteroventral
uterovesical
u. fold
u. junction (UVJ)
u. ligament
u. pouch
uterovesicalis
plica u.
uterque
auris u. (AU, au)
uterus, pl. **uteri**
u. acollis
adnexa u.
anatomical internal os of u.
anomalous u.
anterior lip of external os of u.
anterior surface of u.

arbor vitae u.
arcuate u.
u. arcuatus
arteriae helicinae u.
arteria ligamenti teretis u.
artery of round ligament of u.
u. bicameratus vetularum
bicornate u.
u. bicornate bicollis
u. bicornate unicollis
u. bicornis
bifid u.
u. bifidus
biforate u.
u. biforis
u. bilocularis
bipartite u.
u. bipartitus
body of u.
border of u.
broad ligament of u.
canalis cervicis u.
cavitas u.
cavity of u.
cavum u.
cervical gland of u.
cervical ligament of u.
cervix of u.
coiled artery of u.
cordiform u.
u. cordiformis
cornu u.
cornual portion of u.
corpus of u.
u. didelphys
double-mouthed u.
u. duplex
duplex u.
external os of u.
facies anterior u.
facies intestinalis u.
facies vesicalis u.
fornix u.
fundal portion of u.
fundus of u.
glandulae cervicales u.
heart-shaped u.
helicine artery of the u.
histological internal os of u.
horn of u.
incudiform u.
u. incudiformis

U

NOTES

uterus *(continued)*
    intestinal surface of u.
    isthmus of u.
    labia u.
    labium anterius ostii u.
    labium posterius ostii u.
    lateral angle of u.
    ligamenti teretis u.
    ligamentum latum u.
    ligamentum teres u.
    margo u.
    masculine u.
    u. masculinus
    membrane of cervix u.
    muscular coat of u.
    neck of u.
    one-horned u.
    opening of u.
    orifice of u.
    orificium externum u.
    orificium internum u.
    os u.
    ostium u.
    u. parvicollis
    plicae palmatae canalis cervicis u.
    posterior lip of external os of u.
    round ligament of u.
    rudimentary u.
    septate u.
    u. septus
    serosa of u.
    subseptate u.
    u. subseptus
    triangular u.
    u. triangularis
    tunica mucosa u.
    tunica muscularis u.
    tunica serosa u.
    unicorn u.
    u. unicornis
    vesical surface of u.
utile
    temps u.
utilization time
utricle
    macula of u.
    prostatic u.
    u. of vestibular labyrinth
utricular
    u. duct

    u. nerve
    u. recess of bony labyrinth
    u. recess of membranous labyrinth
    u. reflex
    u. spot
utricularis
    ductus u.
    nervus u.
utriculi (*pl. of* utriculus)
utriculoampullaris
    nervus u.
utriculoampullar nerve
utriculosaccular duct
utriculosaccularis
    ductus u.
    maculae u.
utriculus, pl. utriculi
    macula utriculi
utriculus prostaticus
utriform
uvaeformis
uvea
uveal
    u. part
    u. part of trabecular reticulum
    u. part of trabecular tissue of
      sclera
    u. tract
uvealis
    pars u.
uviform
UVJ
    uterovesical junction
uvomorulin
uvula, pl. uvuli
    bifid u.
    u. of bladder
    u. cerebelli
    Lieutaud u.
    muscle of u.
    u. palatina
    palatine u.
    u. of soft palate
    u. vermis
    u. vesicae
uvulae
    musculus azygos u.
uvularis
uvular muscle
uvuli (*pl. of* uvula)

VA

vertebral artery

vacuo

hydrocephalus ex v.

vacuolar

vacuolate

vacuolated

vacuolation

vacuole

autophagic v.

vacuolization

vacuome

vadum

vagal

v. ganglion

v. nerve trunk

v. part of accessory nerve

vagalis

pars v.

ramus communicans cum ramo
auriculari nervi v.

truncus v.

vagi

ganglion inferius nervi v.

ganglion superius nervi v.

nucleus dorsalis nervi v.

rami cardiaci cervicales inferiores
nervi v.

rami cardiaci cervicales superiores
nervi v.

rami cardiaci thoracici nervi v.

rami celiaci nervi v.

rami esophagei nervi v.

rami gastrici anteriores nervi v.

rami gastrici posteriores nervi v.

rami hepatici nervi v.

rami pharyngei nervi v.

rami renales nervi v.

ramus auricularis nervi v.

ramus communicans nervi
glossopharyngei cum ramo
auriculari nervi v.

ramus communicans plexus
tympanici cum ramo auriculari
nervi v.

ramus meningeus nervi v.

trigonum nervi v.

vagina, pl. **vaginae**

anterior part of fornix of v.

anterior wall of v.

arteria bulbi vaginae

azygos artery of v.

bipartite v.

body of v.

v. bulbi

bulbi vestibuli vaginae

bulbus vestibuli vaginae

carina urethralis vaginae

v. carotica

v. cellulosa

v. communis musculorum flexorum

v. communis tendinum musculorum
fibularium communis

v. communis tendinum musculorum
flexorum manus

v. externa nervi optici

vaginae fibrosa

vaginae fibrosae digitorum manus

vaginae fibrosae digitorum pedis

v. fibrosa tendinis

fornix vaginae

fossa navicularis vestibulae vaginae

fossa of vestibule of v.

fossa vestibuli vaginae

v. interna nervi optici

v. masculina

mucosa of v.

v. mucosa tendinis

mucous membrane of v.

muscular coat of v.

muscular layer of v.

v. musculi recti abdominis

v. musculorum obliqui superioris

v. musculorum peroneorum
communis

musculus sphincter vaginae

vaginae nervi optici

v. oculi

orificium vaginae

ostium vaginae

paries anterior vaginae

paries posterior vaginae

pars anterior fornicis vaginae

pars lateralis fornicis vaginae

pars lateralis fornix vaginae

pars posterior fornicis vaginae

pars posterior fornix vaginae

posterior fornix of v.

posterior wall of v.

v. processus styloidei

rudimentary v.

rugae of v.

rugal columns of v.

septate v.

v. septate

sphincter vaginae

spongy layer of v.

vaginae synoviales digitorum manus

vaginae synoviales digitorum pedis

v. synovialis

V

vagina *(continued)*

v. synovialis tendinis
v. synovialis trochleae
v. tendinis intertubercularis
v. tendinis musculi extensoris carpi ulnaris
v. tendinis musculi extensoris digiti minimi
v. tendinis musculi extensoris hallucis longi
v. tendinis musculi extensoris pollicis longi
v. tendinis musculi fibularis longi plantaris
v. tendinis musculi flexoris carpi radialis
v. tendinis musculi flexoris hallucis longi
v. tendinis musculi flexoris pollicis longi
v. tendinis musculi obliqui superioris
v. tendinis musculi peronei longi plantaris
v. tendinis musculi tibialis anterioris
v. tendinis musculi tibialis posterioris
vaginae tendinum carpales palmares
vaginae tendinum carpalium
vaginae tendinum carpalium dorsalium
vaginae tendinum digitorum pedis
v. tendinum musculi extensoris digitorum pedis longi
v. tendinum musculi flexoris digitorum pedis longi
v. tendinum musculorum abductoris longi et extensoris
v. tendinum musculorum abductoris longi et extensoris brevis pollicis
v. tendinum musculorum extensoris digitorum et extensoris
v. tendinum musculorum extensoris digitorum et extensoris indicis
v. tendinum musculorum extensorum carpi radiali
v. tendinum musculorum extensorum carpi radialium
v. tendinum musculorum fibularium communis
v. tendinum musculorum peroneorum communis
vaginae tendinum tarsales anteriores
vaginae tendinum tarsales fibulares
vaginae tendinum tarsales tibialis
tunica mucosa vaginae
tunica muscularis vaginae
tunica spongiosa vaginae

urethral carina of v.
vaginae vasorum
vestibule of v.
vestibulum vaginae

vaginal

v. artery
v. atresia
v. atrophy
v. cervix
v. columns
v. cornification test
v. cuff
v. fold
v. fornix
v. gland
v. hernia
v. introitus
v. ligament
v. lumen
v. mucification test
v. mucosa
v. muscle
v. nerve
v. opening
v. orifice
v. outlet
v. process
v. process of peritoneum
v. process of sphenoid bone
v. process of testis
v. ruga
v. synovial membrane
v. tissue
v. vault
v. venous plexus
v. wall

vaginales

nervi v.
ruga v.
rugae v.

vaginalis

arteria v.
paracentesis tunicae v.
plexus venosus v.
portio v.
saccus v.
tunica v.
vestige of processus v.
vestigium processus v.

vaginate
vaginoabdominal
vaginolabial

v. hernia

vaginoperineal
vaginoperitoneal
vaginovesical
vaginovulvar
vagoaccessorius
vagoglossopharyngeal

vagomimetic
vagotonia
vagotonic
vagotropic
vagovagal
vagus
    v. area
    dorsal motor nucleus of v.
    ganglion of trunk of v.
    inferior ganglion of v.
    v. nerve
    v. nerve root
    nervus v.
Valentin
    V. corpuscles
    V. ganglion
    V. nerve
valga
    coxa v.
    tibia v.
valgus
    cubitus v.
    pes v.
valla (*pl. of* vallum)
vallata
    papilla v.
vallatae
    papillae v.
vallate papilla
vallecula, pl. **valleculae**
    v. cerebelli
    epiglottic v.
    v. epiglottica
    v. sylvii
vallecular
vallis
vallum, pl. **valla**
    v. unguis
Valsalva
    V. antrum
    V. ligament
    V. maneuver
    V. muscle
    V. sinus
    teniae of V.
valva, pl. **valvae**
    v. aortae
    v. atrioventricularis dextra
    v. atrioventricularis sinistra
    v. ileocecalis
    v. mitralis
    v. tricuspidal

    v. tricuspidalis
    v. trunci pulmonalis
valval
valvar
valvate
valve
    abnormal cleavage of cardiac v.
    Amussat v.
    anal v.
    anterior cusp of left
      atrioventricular v.
    anterior cusp of mitral v.
    anterior cusp of right
      atrioventricular v.
    anterior cusp of tricuspid v.
    anterior urethral v.
    aortic v.
    atrioventricular v.
    auriculoventricular v.
    AV v.
    v. of Bauhin
    Bauhin v.
    Béraud v.
    Bianchi v.
    bicuspid aortic v.
    Bochdalek v.
    Braune v.
    cardiac v.
    caval v.
    colic v.
    commissural pulmonary v.
    congenital v.
    coronary v.
    v. of coronary sinus
    cusp of anterior mitral v.
    cusp of commissural mitral v.
    cusp of left aortic v.
    cusp of left atrioventricular v.
    cusp of posterior aortic v.
    cusp of posterior mitral v.
    cusp of pulmonic v.
    cusp of right aortic v.
    cusp of right atrioventricular v.
    eustachian v.
    v. of foramen ovale
    frenulum of ileocecal v.
    Gerlach v.
    Guérin v.
    Hasner v.
    heart v.
    Heister v.
    v. of Heister

V

NOTES

valve *(continued)*
- Hoboken v.
- Houston v.
- v. of Houston
- Huschke v.
- ileocecal v.
- ileocolic v.
- v. of inferior vena cava
- v. of Kerckring
- Kerckring v.
- Kohlrausch v.
- Krause v.
- left atrioventricular v.
- left coronary v.
- lunula of semilunar cusps of aortic/pulmonary v.
- Mercier v.
- mitral v.
- Morgagni v.
- nasal v.
- v. of navicular fossa
- nodule of semilunar v.
- O'Beirne v.
- v. of oval foramen
- parachute mitral v.
- peristaltic v.
- posterior cusp of left atrioventricular v.
- posterior cusp of mitral v.
- posterior cusp of right atrioventricular v.
- posterior cusp of tricuspid v.
- posterior urethral v.
- pulmonary v.
- v. of pulmonary trunk
- pulmonic v.
- pyloric v.
- quadricuspid pulmonary v.
- rectal v.
- right atrioventricular v.
- right coronary v.
- right septal v.
- Rosenmüller v.
- semilunar v.
- septal cusp of right atrioventricular v.
- septal cusp of tricuspid v.
- spiral v.
- sylvian v.
- Tarin v.
- Terrien v.
- thebesian v.
- tricuspid v.
- trunk v.
- Tulpius v.
- unicommissural aortic v.
- urethral v.
- v. of Varolius
- venous v.
- ventricular v.
- v. of vermiform appendix
- vesicoureteral v.
- Vieussens v.

valveless

valviform

valvula, pl. **valvulae**
- Amussat v.
- valvulae anales
- v. bicuspidalis
- valvulae conniventes
- v. foraminis ovalis
- v. fossae navicularis
- Gerlach v.
- v. ileocolica
- v. lymphatica
- v. processus vermiformis
- v. pylori
- v. semilunaris
- v. semilunaris anterior valvae trunci pulmonalis
- v. semilunaris dextra valvae aortae
- v. semilunaris dextra valvae trunci pulmonalis
- v. semilunaris posterior valvae aortae
- v. semilunaris sinistra valvae aortae
- v. semilunaris sinistra valvae trunci pulmonalis
- v. semilunaris tarini
- v. sinus coronarii
- v. spiralis
- v. tricuspidalis
- tuber valvulae
- v. venae cavae inferioris
- v. venosa
- v. vestibuli

valvular

valvule
- Foltz v.
- lymphatic v.

van
- v. der Waals force
- v. Helmont mirror
- v. Horne canal
- v.'s Slyke apparatus
- v.'s Slyke formula

vapor density

vaporthorax

Va/Q
- ventilation/perfusion ratio

vara
- coxa v.
- tibia v.

variant

variation

varicosum
- lymphangioma capillare v.

varix, pl. **varices**

varolii
>pons v.

Varolius
>V. sphincter
>valve of V.

varus
>cubitus v.
>metatarsus v.
>pes v.

vas, pl. **vasa**
>v. aberrans
>v. aberrans hepatis
>v. aberrans of Roth
>vasa afferentia
>v. anastomoticum
>vasa auris internae
>vasa brevia
>v. capillare
>vasa chylifera
>v. collaterale
>v. deferens
>vasa efferentia
>vasa lymphatica
>v. lymphaticum
>v. lymphaticum afferens
>v. lymphaticum efferens
>v. lymphaticum profundum
>v. lymphaticum superficiale
>vasa nervorum
>vasa previa
>v. prominens
>v. prominens ductus cochlearis
>vasa recta
>vasa recta renis
>v. rectum
>vasa sanguinea auris internae
>vasa sanguinea choroideae
>vasa sanguinea intrapulmonalia
>vasa sanguinea retinae
>v. sanguineum
>v. spirale
>vasa vasorum
>vasa vorticosa

vasal
vascula
vascular
>v. bed
>v. bud
>v. channel
>v. circle
>v. circle of optic nerve
>v. cone

>v. fold of cecum
>v. gland
>v. lacuna
>v. lamina of choroid
>v. layer
>v. layer of choroid coat of eye
>v. layer of eyeball
>v. layer of testis
>v. papilla
>v. pedicle
>v. plexus
>v. reflex
>v. ring
>v. sheath
>v. space of retroinguinal compartment
>v. spur
>v. stripe
>v. system
>v. tunic
>v. tunic of eye
>v. zone

vascularis
>corona v.
>nervus v.
>plica cecalis v.

vascularity
vascularization
vascularorum
>nervi v.

vasculature
vasculocardiac
vasculogenesis
vasculomotor
vasculosa
>Haller tunica v.
>meninx v.
>tela v.
>tunica v.
>zona v.

vasculosae
>stellulae v.

vasculosi
>coni v.

vasculosus
>circulus articularis v.
>plexus v.

vasculum
vasifaction
vasifactive
vasiform
vasis

NOTES

vasoactive
vasoconstriction
    active v.
    passive v.
vasoconstrictive
vasoconstrictor nerve
vasodepression
vasodepressor
vasodilatation
vasodilation
    active v.
    passive v.
vasodilative
vasodilator
vasofactive
vasoformation
vasoformative cell
vasoganglion
vasohypertonic
vasohypotonic
vasoinhibitor
vasoinhibitory
vasolabile
vasomotion
vasomotor nerve
vasopressor reflex
vasoreflex
vasorelaxation
vasorum
    lacuna v.
    vaginae v.
    vasa v.
vasosensory
vasostimulant
vasotonia
vasotonic
vasotrophic
vasotropic
vastoadductor fascia
vastoadductorium
    septum intermusculare v.
vastus
    v. intermedius
    v. intermedius muscle
    v. intermedius tendon
    v. lateralis
    v. lateralis muscle
    v. lateralis tendon
    v. medialis
    v. medialis muscle
    v. medialis obliquus musculus
    (VMO)
    v. medialis tendon
Vater
    ampulla of V.
    V. ampulla
    V. corpuscles
    V. fold

    papilla of V.
    V. papilla
Vater-Pacini corpuscles
vault
    cartilaginous v.
    cecal v.
    cranial v.
    v. of pharynx
    vaginal v.
VC
    vena cava
    ventral column
    vertebral canal
    visual cortex
    vocal cord
vegetality
vegetal pole
vegetation
vegetative
    v. life
    v. nervous system
    v. pole
vegetoanimal
veil
    v. cell
    Jackson v.
vein
    accessory cephalic v.
    accessory hemiazygos v.
    accessory saphenous v.
    accessory vertebral v.
    accompanying v.
    afferent v.
    anastomotic v.
    angular v.
    anonymous v.
    antebrachial v.
    anterior auricular v.
    anterior basal v.
    anterior basal branch of superior
      basal vein of right and left
      inferior pulmonary v.
    anterior cardiac v.
    anterior cardinal v.
    anterior cerebral v.
    anterior ciliary v.
    anterior circumflex humeral v.
    anterior facial v.
    anterior intercostal v.
    anterior interosseous v.
    anterior jugular v.
    anterior labial v.
    anterior pontomesencephalic v.
    anterior and posterior vestibular v.
    anterior scrotal v.
    anterior temporal diploic v.
    anterior tibial v.
    anterior vertebral v.
    apical v.

apical branch of right superior pulmonary v.
apicoposterior v.
apicoposterior branch of left superior pulmonary v.
appendicular v.
aqueous v.
arterial v.
ascending lumbar v.
auditory v.
auricular v.
axillary v.
azygos v.
basal v.
basilic v.
basivertebral v.
Baumgarten v.
brachial v.
brachiocephalic v.
Breschet v.
bronchial v.
Browning v.
buccal v.
bulb of jugular v.
v. of bulb of penis
v. of bulb of vestibule
Burow v.
canaliculus v.
capillary v.
cardiac v.
cardinal v.
carotid v.
caudate branch of left branch of portal v.
v. of caudate nucleus
central retinal v.
cephalic v.
cerebellar v.
v. of cerebellum
cerebral v.
v. of cerebrum
cervical v.
choroid v.
ciliary v.
cilioretinal v.
circumflex femoral v.
circumflex iliac v.
circumflex scapular v.
v. of cochlear aqueduct
v. of cochlear canaliculus
v. of cochlear window
colic v.

collateral v.
comitans v.
common anterior facial v.
common basal v.
common cardinal v.
common facial v.
common iliac v.
common modiolar v.
companion v.
condylar emissary v.
conjunctival v.
coronary v.
v. of corpus striatum
costoaxillary v.
cremasteric v.
cubital v.
cutaneous v.
cystic v.
deep cerebral v.
deep cervical v.
deep circumflex iliac v.
deep dorsal v.
deep epigastric v.
deep facial v.
deep femoral v.
deep lingual v.
deep middle cerebral v.
deep temporal v.
descending genicular v.
digital v.
diploic v.
dorsal callosal v.
dorsalis pedis v.
dorsal lingual v.
dorsal metacarpal v.
dorsal metatarsal v.
dorsal scapular v.
dorsispinal v.
duodenal v.
emissary v.
epigastric inferior v.
episcleral v.
esophageal v.
ethmoidal v.
external carotid v.
external iliac v.
external jugular v.
external mammary v.
external nasal v.
external palatine v.
external pterygoid v.
external pudendal v.

V

NOTES

vein (*continued*)

external pudic v.
external radial v.
extraocular part of central retinal
  artery and v.
facial v.
femoral v.
femoropopliteal v.
fetal intrahepatic v.
fibular v.
foramina of smallest v.
frontal v.
frontal diploic v.
v. of Galen
Galen v.
gastric v.
gastroepiploic v.
genicular v.
gluteal v.
gonadal v.
great cardiac v.
great cerebral v.
greater saphenous v.
great saphenous v.
groove for left brachiocephalic v.
groove for right brachiocephalic v.
groove for subclavian v.
v. of heart
hemiazygos v.
hemorrhoidal v.
hepatic v.
hepatic portal v.
highest intercostal v.
hypogastric v.
hypophyseoportal v.
ileal v.
ileocolic v.
iliac v.
ilioinguinal v.
iliolumbar v.
inferior alveolar v.
inferior anastomotic v.
inferior basal v.
inferior cardiac v.
inferior cerebellar v.
inferior cerebral v.
inferior choroid v.
inferior epigastric v.
v. of inferior eyelid
inferior gluteal v.
inferior hemiazygos v.
inferior hemorrhoidal v.
inferior interosseous v.
inferior labial v.
inferior laryngeal v.
inferior lateral genicular v.
inferior medial genicular v.
inferior mesenteric v.
inferior ophthalmic v.

inferior palpebral v.
inferior part of lingular vein of
  left superior pulmonary v.
inferior phrenic v.
inferior radicular v. (IRV)
inferior rectal v.
inferior thalamostriate v.
inferior thyroid v.
inferior ventricular v.
infralobar part of posterior vein of
  right superior pulmonary v.
infraorbital v.
infrasegmental v.
innominate cardiac v.
intercapital v.
intercapitular v.
intercostal v.
interlobar v.
intermediate antebrachial v.
intermediate basilic v.
intermediate cephalic v.
intermediate cubital v.
intermediate hepatic v.
internal auditory v.
internal cerebral v.
internal iliac v.
internal jugular v.
internal maxillary v.
internal pudendal v.
internal thoracic v.
interosseous v.
intersegmental part of pulmonary v.
interventricular v.
intervertebral v.
intestinal v.
intralobar part of the posterior
  vein of the right superior
  pulmonary v.
intrarenal v.
intrasegmental part of pulmonary v.
jejunal and ileal v.
jugular v.
v. of kidney
v. of knee
Krukenberg v.
v. of Labbé
Labbé v.
labial v.
labyrinthine v.
lacrimal v.
large saphenous v.
laryngeal v.
Latarjet v.
lateral atrial v.
lateral circumflex femoral v.
lateral direct v.
lateral mammary v.
lateral part of middle lobe vein of
  right superior pulmonary v.

lateral plantar v.
v. of lateral recess of fourth ventricle
lateral sacral v.
lateral thoracic v.
left brachiocephalic v.
left colic v.
left coronary v.
left gastric v.
left gastroepiploic v.
left gastroomental v.
left gonadal v.
left hepatic v.
left inferior pulmonary v.
left internal jugular v.
left marginal v.
left ovarian v.
left pericardiacophrenic v.
left phrenic v.
left and right brachiocephalic v.
left subclavian v.
left superior gluteal v.
left superior intercostal v.
left superior pulmonary v.
left suprarenal v.
left testicular v.
left umbilical v.
lesser ovarian v.
levoatriocardinal v.
lienal v.
lingual v.
lingular v.
long saphenous v.
long thoracic v.
v. of lower limb
lumbar v.
lumen of v.
lymph node of arch of azygos v.
main renal v.
mammary v.
Marshall oblique v.
masseteric v.
mastoid emissary v.
maxillary v.
Mayo v.
medial atrial v.
medial circumflex femoral v.
medial genicular v.
medial part of middle lobe vein of right superior pulmonary v.
medial plantar v.
median antebrachial v.

median basilic v.
median cephalic v.
median cubital v.
median sacral v.
mediastinal v.
medium v.
v. of medulla oblongata
meningeal v.
mesencephalic v.
mesenteric superior v.
metacarpal v.
middle cardiac v.
middle colic v.
middle genicular v.
middle hemorrhoidal v.
middle hepatic v.
middle lobe v.
middle lobe branch of right superior pulmonary v.
middle meningeal v.
middle rectal v.
middle temporal v.
middle thyroid v.
muscular v.
musculocutaneous v.
musculophrenic v.
nasal v.
nasofrontal v.
neck v.
obturator branch of pubic branch of inferior epigastric v.
occipital cerebral v.
occipital diploic v.
occipital emissary v.
omphalomesenteric v.
opening of pulmonary v.
opening of smallest cardiac v.
ophthalmic v.
orbital v.
ovarian v.
palatal v.
palatine v.
palmar digital v.
palmar metacarpal v.
palpebral v.
pampiniform v.
pancreatic v.
pancreaticoduodenal v.
paratonsillar v.
paraumbilical v.
paraventricular v.
parietal v.

V

NOTES

vein (*continued*)

parietal emissary v.
parotid v.
pectoral v.
perforating v.
pericallosal v.
pericardiac v.
pericardiacophrenic v.
pericardial v.
peroneal v.
petrosal v.
pharyngeal v.
phrenic v.
plantar digital v.
plantar metatarsal v.
v. of pons
pontine v.
popliteal v.
portal v.
postcardinal v.
postcentral v.
posterior anterior jugular v.
posterior auricular v.
posterior branch of right branch of
  portal v.
posterior branch of right superior
  pulmonary v.
posterior cardinal v.
posterior circumflex humeral v.
posterior facial v.
v. of posterior horn
posterior intercostal v.
posterior interosseous v.
posterior labial v.
posterior marginal v.
posterior parotid v.
posterior pericallosal v.
posterior scrotal v.
posterior temporal diploic v.
posterior tibial v.
precentral cerebellar v.
prefrontal v.
prepyloric v.
profunda femoris v.
v. of pterygoid canal
pubic branch of inferior
  epigastric v.
pudendal v.
pulmonary v.
pyloric v.
radial v.
ranine v.
rectal v.
rectosigmoid v.
renal v.
retinal v.
retromandibular v.
retroperitoneal v.
Retzius v.

right brachiocephalic v.
right branch of portal v.
right colic v.
right gastric v.
right gastroepiploic v.
right gastroomental v.
right gonadal v.
right hepatic v.
right inferior gluteal v.
right inferior pulmonary v.
right internal jugular v.
right ovarian v.
right pericardiacophrenic v.
right phrenic v.
right subclavian v.
right superior intercostal v.
right superior pulmonary v.
right suprarenal v.
right testicular v.
rolandic v.
Rolando v.
Rosenthal ascending v.
Ruysch v.
sacral v.
salvatella v.
Santorini v.
Santorini parietal v.
saphenous v.
Sappey v.
v. of scala tympani
v. of scala vestibuli
scleral v.
scrotal v.
v. of semicircular duct
v. of septum pellucidum
short gastric v.
short saphenous v.
sigmoid v.
small cardiac v.
smallest cardiac v.
small saphenous v.
Soemmering arterial v.
Soemmering external radial v.
spermatic v.
sphenoid emissary v.
spinal v.
v. of spinal cord
spiral v.
splenic v.
stellate v.
Stensen v.
sternocleidomastoid v.
stylomastoid v.
subcardinal v.
subclavian v.
subcostal v.
sublingual v.
sublobular v.
submental v.

subscapular v.
sulcus of umbilical v.
superficial cerebral v.
superficial circumflex iliac v.
superficial dorsal v.
superficial epigastric v.
superficial middle cerebral v.
superficial palmar v.
superficial perineal v.
superficial temporal v.
superficial vertebral v.
superior anastomotic v.
superior azygos v.
superior basal v.
superior branch of the right and
  left inferior pulmonary v.
superior bulb of internal jugular v.
superior cerebellar v.
superior cerebral v.
superior choroid v.
superior epigastric v.
v. of superior eyelid
superior gluteal v.
superior hemorrhoidal v.
superior intercostal v.
superior labial v.
superior laryngeal v.
superior lateral genicular v.
superior medial genicular v.
superior mesenteric v.
superior ophthalmic v.
superior palpebral v.
superior part of lingular vein of
  left superior pulmonary v.
superior phrenic v.
superior pulmonary v.
superior rectal v.
superior thalamostriate v.
superior thyroid v.
supracardinal v.
supraorbital v.
suprarenal v.
suprascapular v.
supratrochlear v.
supreme intercostal v.
sylvian v.
temporal v.
temporal diploic v.
v. of temporomandibular joint
temporomaxillary v.
terminal v.
testicular v.

thalamostriate v.
thebesian v.
v. of Thebesius
thoracic v.
thoracoacromial v.
thoracoepigastric v.
thymic v.
thyroid v.
tibial v.
trabecular v.
tracheal v.
transverse v.
transverse cervical v.
transverse facial v.
transverse part of left branch of
  portal v.
Trolard v.
tympanic v.
ulnar v.
umbilical part of left branch of
  portal v.
v. of uncus
v. of upper limb
uterine v.
vertebral v.
v. of vertebral column
vesalian v.
Vesalius v.
vesical v.
vestibular v.
v. of vestibular aqueduct
v. of vestibular bulb
vidian v.
Vieussens v.
vitelline v.
vortex v.
vorticose v.
veinlet
vela (*pl. of* velum)
velamen, pl. **velamina**
velamenta (*pl. of* velamentum)
velamentosa
  placenta v.
velamentous insertion
velamentum, pl. **velamenta**
  velamenta cerebri
velamina (*pl. of* velamen)
velar
veliform
veli palatini
Vella fistula
vellicate

NOTES

vellication
vellus olivae inferioris
velocity
    maximum v.
    nerve conduction v.
velopharyngeal
    v. portal
    v. sphincter
velopharynx
Velpeau
    V. canal
    V. fossa
velum, pl. **vela**
    anterior medullary v.
    aponeurosis of v.
    frenulum of superior medullary v.
    inferior medullary v.
    v. interpositum
    v. medullare inferius
    v. medullare superius
    palatine v.
    v. palatinum
    v. pendulum palati
    posterior medullary v.
    v. semilunare
    superior medullary v.
    v. tarini
    v. terminale
    v. transversum
    v. triangulare
vena, pl. **venae**
    v. advehens
    v. afferens hepatis
    agger valvae venae
    v. anastomotica inferior
    v. anastomotica superior
    v. angularis
    venae anteriores cerebri
    apertura mediana v.
    v. apicalis
    v. apicoposterior
    v. appendicularis
    v. aqueductus cochlea
    v. aqueductus vestibuli
    venae arcuatae renis
    v. arteriosa
    venae articulares tempo
    venae articulares
      temporomandibulares
    v. atrii lateralis
    v. atrii medialis
    v. auricularis anterior
    v. auricularis posterior
    v. axillaris
    v. azygos
    v. azygos major
    v. azygos minor inferior
    v. azygos minor superior
    v. basalis anterior

v. basalis communis
v. basalis inferior
v. basalis superior
v. basilica
v. basivertebralis
venae brachiales
venae brachiocephalicae
venae brachiocephalicae dextrae et
  sinistrae
venae bronchiales
v. bulbi penis
v. bulbi vestibuli
v. canaliculi cochlea
v. canalis pterygoidei
venae cardiacae anteriores
venae cardiacae minimae
v. cardiaca magna
v. cava (VC)
v. cava inferior
v. caval foramen
v. cava superior
venae cavernosae penis
venae cavernosae of spleen
v. centra
venae centrales hepatis
v. centralis glandulae suprarenalis
v. centralis retinae
v. cephalica accessoria
v. cephalica antebrachii
venae cerebelli inferiores
venae cerebelli superiores
v. cerebri anterior
venae cerebri inferiores
venae cerebri internae
v. cerebri magna
v. cerebri media profunda
v. cerebri media superficialis
venae cerebri profundae
venae cerebri superficiales
v. cervicalis profunda
venae choroideae oculi
v. choroidea inferior
v. choroidea superior
venae ciliares
venae ciliares anteriores
venae circumflexae femoris laterales
venae circumflexae femoris
  mediales
v. circumflexa humeri anterior
v. circumflexa humeri posterior
v. circumflexa iliaca profunda
v. circumflexa iliaca superficialis
v. circumflexa ilium profunda
v. circumflexa ilium superficialis
v. colica dextra
v. colica media
v. colica sinistra
v. colli profunda
venae columnae vertebralis

v. comitans
v. comitans of hypoglossal nerve
v. comitans nervi hypoglossi
venae conjunctivales
venae cordis
venae cordis anteriores
v. cordis magna
v. cordis media
venae cordis minimae
v. cordis parva
v. cornus posterioris
v. coronaria ventriculi
v. corporis callosi dorsalis
v. cutanea
v. cystica
venae digitales dorsales pedis
venae digitales palmares
venae digitales plantares
v. diploica
venae directae laterales
venae dorsales clitoridis
  superficiales
venae dorsales linguae
venae dorsales penis superficiales
v. dorsalis clitoridis profunda
v. dorsalis linguae
v. dorsalis penis profunda
venae ductuum semicircularium
v. emissaria
v. emissaria condylaris
venae emissariae
v. emissaria mastoidea
v. emissaria occipitalis
v. emissaria parietalis
venae encephali occipitales
venae epigastricae superiores
v. epigastrica inferior
v. epigastrica superficialis
venae episclerales
venae esophageae
venae ethmoidales
v. facialis
v. facialis anterior
v. facialis communis
v. facialis posterior
v. faciei profunda
v. femoralis
v. fenestrae cochlea
venae fibulares
venae frontales
v. gastrica dextra
venae gastricae breves

v. gastrica sinistra
v. gastroomentalis dextra
v. gastroomentalis sinistra
venae geniculares
venae genus
venae gluteae inferiores
venae gluteae superiores
v. gyri olfactorii
v. hemiazygos
v. hemiazygos accessoria
venae hemispherii cerebelli
  inferiores
venae hemispherii cerebelli
  superiores
venae hemorrhoidales inferiores
venae hemorrhoidales mediae
v. hemorrhoidalis superior
venae hepaticae
venae hepaticae dextrae
venae hepaticae intermediae
venae hepaticae mediae
venae hepaticae sinistrae
v. hypogastrica
v. ileocolica
v. iliaca communis
v. iliaca externa
v. iliaca interna
v. iliolumbalis
venae inferiores cerebri
v. innominata
venae insulares
venae intercapitales
venae intercapitulares
venae intercostales anteriores
venae intercostales posteriores
v. intercostalis superior dextra
v. intercostalis superior sinistra
v. intercostalis suprema
venae interlobares renis
venae interlobulares hepatis
interlobulares hepatis venae
venae interlobulares renis
v. intermedia antebrachii
v. intermedia basilica
v. intermedia cephalica
v. intermedia cubiti
v. intervertebralis
venae jejunales et ilei
v. jugularis anterior
v. jugularis externa
v. jugularis interna
venae labiales anteriores

V

NOTES

vena *(continued)*

venae labiales posteriores
v. labialis inferior
v. labialis superior
venae labyrinthi
v. lacrimalis
v. laryngea inferior
v. laryngea superior
v. lienalis
v. lingualis
v. lingularis
v. lobi medii
venae lumbales
v. lumbalis ascendens
v. magna cerebri
v. mammaria interna
v. maxillaris
v. mediana antebrachii
v. mediana basilica
v. mediana cephalica
v. mediana cubiti
venae mediastinales
venae medullae oblongatae
venae medullae spinalis
venae membri inferioris
venae membri superioris
venae meningeae
venae meningeae mediae
venae mesencephalicae
v. mesenterica inferior
v. mesenterica superior
venae metacarpeae dorsales
venae metacarpeae palmares
venae metatarseae dorsales
venae metatarseae plantares
v. modioli communis
venae musculophrenicae
venae nasales externae
v. nasofrontalis
venae nuclei caudati
v. obliqua atrii sinistri
v. obturatoria
venae obturatoriae
venae occipitales
v. ophthalmica inferior
v. ophthalmica superior
v. ovarica dextra
v. ovarica sinistra
v. palatina
v. palatina externa
venae palpebrales
venae palpebrales inferiores
venae palpebrales superiores
venae pancreaticae
venae pancreaticoduodenales
venae paraumbilicales
venae parietales
venae parotideae
venae pectorales

venae pedunculares
venae perforantes
venae pericardiacae
venae pericardiacophrenicae
venae peroneae
v. petrosa
venae pharyngeae
venae phrenicae inferiores
venae phrenicae superiores
v. phrenica inferior
venae pontis
v. pontomesencephalica anterior
v. poplitea
v. portae hepatis
v. portalis
venae posteriores ventriculi sinistri
v. posterior ventriculi sinistri
v. preauricularis
v. precentralis cerebelli
venae prefrontales
v. prepylorica
venae profundae clitoridis
venae profundae penis
v. profunda femoris
v. profunda linguae
v. profunda penis
venae pudendae externae
v. pudenda interna
venae pulmonales
v. pulmonalis inferior dextra
v. pulmonalis inferior sinistra
v. pulmonalis superior dextra
v. pulmonalis superior sinistra
venae radiales
venae rectae
venae rectales inferiores
venae rectales mediae
v. rectalis superior
venae renales
v. retromandibularis
venae retroperitoneales
v. revehens
venae sacrales laterales
v. sacralis mediana
v. saphena accessoria
v. saphena magna
v. saphena parva
v. scalae tympani
v. scalae vestibuli
v. scapularis dorsalis
venae sclerales
venae scrotales anteriores
venae scrotales posteriores
v. septi pellucidi anterior
v. septi pellucidi posterior
venae sigmoideae
v. spiralis modioli
v. splenica
venae stellatae

v. sternocleidomastoidea
venae striatae
v. stylomastoidea
v. subclavia
venae subcutaneae abdominis
v. sublingualis
v. submentalis
venae superiores cerebelli
venae superiores cerebri
v. supraorbitalis
v. suprarenalis dextra
v. suprarenalis sinistra
v. suprascapularis
venae supratrochleares
venae temporales profundae
venae temporales superficiales
v. temporalis media
v. terminalis
v. testicularis dextra
v. testicularis sinistra
venae thalamostriatae inferiores
v. thalamostriata superior
venae thoracicae internae
v. thoracica interna
v. thoracica lateralis
v. thoracoacromialis
v. thoracoepigastrica
venae thoracoepigastricae
venae thymicae
v. thyroidea ima
v. thyroidea inferior
v. thyroidea media
v. thyroidea superior
venae tibiales anteriores
venae tibiales posteriores
venae tracheales
venae transversae cervicis
venae transversae colli
v. transversa faciei
v. transversa scapulae
venae tympanicae
venae ulnares
v. umbilicalis
v. umbilicalis sinistra
venae uterinae
v. ventricularis inferior
v. ventriculi lateralis lateralis
v. ventriculi lateralis medialis
v. vermis inferior
v. vermis superior
v. vertebralis
v. vertebralis accessoria

v. vertebralis anterior
venae vesicales
venae vestibulares
venae vestibulares anterius et
  posterius
v. vitellina
venae vorticosae
venation
venectasia
veneris
  mons v.
venomotor
venopressor
venorespiratory reflex
venosa
  fossa v.
  valvula v.
venose
venosi
  fissura ligamenti v.
  fossa ductus v.
  ligamentum ductus v.
  sulci v.
venosinal
venosity
venostasis
venosum
  fissure for ligamentum v.
  foramen v.
  ligamentum v.
  ostium v.
  rete v.
venosus
  ductus v.
  fossa of ductus v.
  plexus v.
  v. sclerae sinus
  sinus v.
venous
  v. angle
  v. artery
  v. bifurcation
  v. capillary
  v. circle of mammary gland
  v. flow
  v. foramen
  v. groove
  v. heart
  v. hyperemia
  v. ligament
  v. plexus
  v. plexus of bladder

V

NOTES

venous *(continued)*
    v. plexus of canal of hypoglossal nerve
    v. plexus of foramen ovale
    v. plexus of hypoglossal canal
    v. pulse
    v. return
    v. segments of kidney
    v. segments of liver
    v. sheath
    v. sinuses
    v. sinus of sclera
    v. valve
vent
venter
    v. anterior musculi digastrici
    v. frontalis
    v. frontalis musculi occipitofrontalis
    v. inferior musculi omohyoidei
    v. occipitalis
    v. occipitalis musculi
    v. posterior musculi digastrici
    v. superior musculi omohyoidei
ventilate
ventilation
    alveolar v.
    maximum voluntary v. (MVV)
    v. meter
    pulmonary v.
    wasted v.
ventilation/perfusion ratio (Va/Q)
ventilatory compliance
ventrad
ventral
    v. anterior nucleus of thalamus
    v. aorta
    v. aspect
    v. border
    v. branch
    v. column (VC)
    v. column of spinal cord
    v. horn
    v. mesocardium
    v. mesogastrium
    v. nerve root
    v. nucleus of trapezoid body
    v. pancreas
    v. part of intertransversarii laterales lumborum muscle
    v. part of pons
    v. plate
    v. plate of neural tube
    v. posterior intermediate nucleus of thalamus
    v. posterior lateral nucleus of thalamus
    v. posterolateral nucleus of thalamus

    v. posteromedial nucleus of thalamus
    v. primary rami of cervical spinal nerve
    v. primary rami of lumbar spinal nerve
    v. primary rami of sacral spinal nerve
    v. primary rami of thoracic spinal nerve
    v. primary ramus of spinal nerve
    v. root of spinal nerve
    v. sacrococcygeal ligament
    v. sacrococcygeal muscle
    v. sacrococcygeus muscle
    v. sacroiliac ligament
    v. spinocerebellar tract
    v. spinothalamic tract
    v. splanchnic arteries
    v. surface
    v. surface of digit
    v. tegmental decussation
    v. thalamic peduncle
    v. tier thalamic nuclei
ventralis
    facies digitalis v.
    lamina v.
    musculus sacrococcygeus v.
    norma v.
    pedunculus thalami v.
    radix v.
    ramus v.
    sacrococcygeus v.
    sulcus v.
ventricle
    Arantius v.
    atrium of lateral v.
    bulb of lateral v.
    cerebral v.
    v. of cerebral hemisphere
    choroid plexus of fourth v.
    choroid plexus of lateral v.
    choroid plexus of third v.
    choroid tela of fourth v.
    choroid tela of third v.
    cornu of lateral v.
    v. of diencephalon
    Duncan v.
    fifth v.
    foramen of fourth v.
    fourth v.
    Galen v.
    v.'s of heart
    horn of lateral v.
    infundibulum of right v.
    laryngeal v.
    lateral aperture of fourth v.
    lateral recess of fourth v.
    lateral vein of lateral v.

left v.
medial vein of lateral v.
median aperture of fourth v.
median sulcus of fourth v.
medullary striae of fourth v.
Morgagni v.
posterior vein of left v.
v. of rhombencephalon
right v.
roof of fourth v.
sixth v.
sylvian v.
temporal horn of lateral v.
tenia of fourth v.
terminal v.
third v.
trabeculae carneae of right and
  left v.
trigone of lateral v.
vein of lateral recess of fourth v.
Verga v.
Vieussens v.
Wenzel v.
ventricular
v. afterload
v. artery
v. band of larynx
v. diverticulum
v. filling pressure
v. fluid
v. fold
v. ganglion
left v.
v. ligament
v. loop
v. plateau
v. preload
v. septal defect
v. septum
v. tachycardia (VT)
v. valve
v. wall
ventriculare
ligamentum v.
ventriculares
arteriae v.
ventricularis
musculus v.
plica v.
ventriculi (*pl. of* ventriculus)
ventriculoatrial
ventriculonector

ventriculoradial dysplasia
ventriculorum
septum membranaceum v.
septum musculare v.
ventriculosubarachnoid
ventriculus, pl. **ventriculi**
canalis ventriculi
v. cordis
v. cordis dexter/sinister
v. dexter
fibrae obliquae ventriculi
filtrum ventriculi
fundus ventriculi
incisura angularis ventriculi
incisura cardiaca ventriculi
v. laryngis
v. lateralis
Merkel filtrum ventriculi
paries anterior ventriculi
paries posterior ventriculi
pars cardiaca ventriculi
pars pylorica ventriculi
v. quartus
v. quintus
v. sinister
stratum circulare tunicae muscularis
  ventriculi
stratum longitudinale tunicae
  muscularis ventriculi
v. terminalis
v. tertius
trigonum ventriculi
tunica mucosa ventriculi
tunica muscularis ventriculi
tunica serosa ventriculi
vena coronaria ventriculi
ventriduct
ventriduction
ventrobasal nucleus
ventrodorsad
ventroinguinal
ventrolateral
ventromedial
v. nucleus
v. nucleus of hypothalamus
ventromedian
ventroposterolateral
Venturi
V. effect
V. meter
V. tube
venula, pl. **venulae**

V

NOTES

venula *(continued)*
v. macularis inferior
v. macularis superior
v. medialis retinae
v. nasalis retinae inferior
v. nasalis retinae superior
venulae rectae of kidney
venulae rectae renis
venulae stellatae
v. temporalis retinae inferior
v. temporalis retinae superior
venular
venule
high endothelial postcapillary v.
inferior macular v.
inferior nasal retinal v.
inferior temporal retinal v.
nasal v.
pericytic v.
postcapillary v.
stellate v.
superior macular v.
superior nasal retinal v.
superior temporal retinal v.
venulous
vera
conjugata v.
cutis v.
decidua v.
glottis v.
pelvis v.
placenta accreta v.
vertebra v.
verae
costae v.
Verga
accessory venous sinus of V.
V. lacrimal groove
V. ventricle
vergae
cavum v.
verge
anal v.
rectal v.
verheyenii
stellulae v.
Verheyen stars
vermes (*pl. of* vermis)
vermian fossa
vermicular
v. appendage
v. movement
vermiculation
vermicule
vermiform
v. appendage
appendix v.
v. appendix
v. process

vermiformis
appendix v.
folliculi lymphatici aggregati appendicis v.
mesenteriolum processus v.
noduli lymphoidei aggregati appendicis v.
ostium appendicis v.
processus v.
valvula processus v.
vermilion
v. border
lip v.
v. margin
v. surface of lip
v. transitional zone
vermis, pl. **vermes**
amis v.
folium v.
superior vein of v.
tuber v.
uvula v.
vermix
vernix caseosa
versicolor
membrana v.
vertebra, pl. **vertebrae**
accessory process of lumbar v.
anterior tubercle of cervical vertebrae
arcus vertebrae
basilar v.
block vertebrae
butterfly v.
carotid tubercle of v.
caudal vertebrae
v. C1–C7
central ramus of v.
centrum of v.
cervical v. (CV)
vertebrae cervicales
vertebrae coccygeae
coccygeal vertebrae
codfish vertebrae
corpus vertebrae
cranial v.
v. dentata
dorsal v.
dorsal rami of v.
endplate of vertebrae
false vertebrae
first cervical v.
hourglass vertebrae
inferior articular process of v.
inferior cervical facet of v.
inferior lumbar facet of v.
inferior thoracic facet of v.
lamina arcus vertebrae
vertebrae lumbales

lumbar vertebrae
v. magna
mammillary process of lumbar v.
neural arch of v.
odontoid v.
olisthetic v.
pedicle of arch of v.
pediculus arcus vertebrae
posterior tubercle of cervical
  vertebrae
processus spinosus vertebrae
processus transversus vertebrae
v. prominens
radix arcus vertebrae
sacral v.
vertebrae sacrales
second cervical v.
spinous process of v.
vertebrae spuriae
superior articular process of v.
superior costal facet of v.
superior lumbar facet of v.
superior thoracic facet of v.
tail vertebrae
vertebrae thoracicae
toothed v.
transitional v.
transverse costal facet of v.
transverse process of v.
true v.
uncinate process of cervical v.
uncinate process of first
  thoracic v.
v. vera
vertebral
v. arch
v. artery (VA)
body of v.
v. body
v. border of scapula
v. canal (VC)
v. column
v. disk
v. endplate
v. epidural space
v. foramen
v. formula
v. ganglion
v. groove
v. interspace
v. line of pleural reflection
v. nerve

v. notch
v. part
v. part of costal surface of the
  lung
v. part of diaphragm
v. prominence
v. pulp
v. region
v. rib
v. surface
v. vein
v. venous plexus
v. venous system
vertebral-basilar system
vertebrale
foramen v.
ganglion v.
vertebralis
arteria v.
canalis v.
chorda v.
columna v.
curvaturae secondariae columnae v.
curvatura primaria columnae v.
incisura v.
nervus v.
pars atlantica arteriae v.
pars cervicalis arteriae v.
pars intracranialis arteriae v.
pars transversaria arteriae v.
plexus periarterialis arteriae v.
plexus venosus v.
rami medullares laterales partis
  intracranialis arteriae v.
rami medullares mediales
  arteriae v.
rami musculares arteriae v.
ramus meningeus anterior
  arteriae v.
ramus meningeus partis
  intracranialis arteriae v.
regio v.
sulcus arteriae v.
theca v.
tubus v.
vena v.
venae columnae v.
vertebrarium
vertebrate
vertebroarteriale
foramen v.
vertebroarterial foramen

NOTES

vertebroarterialis
    foramen v.
vertebrochondral rib
vertebrocostal trigone
vertebrofemoral
vertebroiliac
vertebromediastinalis
    recessus v.
vertebromediastinal recess
vertebropelvic ligament
vertebrosacral
vertebrosternal rib
vertex, pl. **vertices**
    v. cordis
    v. of cornea
    v. corneae
    corneal v.
vertical
    v. aspect
    v. axis
    v. crest of internal acoustic meatus
    v. groove
    v. index
    v. muscle
    v. muscle of tongue
    v. plane
    v. plate
verticalis
    v. linguae
    norma v.
    sulcus v.
vertices (*pl. of* vertex)
verticillate
verticomental
verticosubmental projection
verum
    corpus luteum v.
verumontanum
vesalian
    v. bone
    v. vein
vesalianum
    os v.
Vesalius
    V. bone
    V. foramen
    V. vein
vesica, pl. **vesicae**
    apex vesicae
    v. biliaris
    collum vesicae
    corpus vesicae
    ectopia vesicae
    v. fellea
    ligamentum laterale vesicae
    musculus sphincter vesicae
    paracentesis vesicae
    v. prostatica
    sphincter vesicae

    stratum circulare musculi detrusoris
        vesicae
    trigonum vesicae
    v. urinaria
    uvula vesicae
vesical
    v. artery
    v. fascia
    v. gland
    v. nervous plexus
    v. orifice
    v. reflex
    v. retinaculum
    v. surface of uterus
    v. triangle
    v. vein
vesicales
    v. superiores
    venae v.
vesicalis
    v. inferior
    plexus venosus v.
vesicle
    acoustic v.
    acrosomal v.
    air v.
    allantoic v.
    amniocardiac v.
    auditory v.
    Baer v.
    blastodermic v.
    cerebral v.
    cervical v.
    chorionic v.
    encephalic v.
    excretory duct of seminal v.
    forebrain v.
    germinal v.
    graafian v.
    hindbrain v.
    lens v.
    lenticular v.
    Malpighi v.
    malpighian v.
    midbrain v.
    mucosa of seminal v.
    ocular v.
    optic v.
    otic v.
    pinocytotic v.
    prostatic v.
    pulmonary v.
    seminal v.
    spermatic v.
    synaptic v.
    telencephalic v.
    umbilical v.
vesicoabdominal
vesicocervical

vesicointestinal reflex
vesicoprostatic plexus
vesicopubic
vesicorectal
vesicosigmoid
vesicospinal
vesicoumbilical ligament
vesicoureteral valve
vesicourethral
    v. canal
    v. orifice
vesicouterina
    excavatio v.
    plica v.
vesicouterine
    v. ligament
    v. pouch
vesicouterinum
    cavum v.
vesicouterovaginal
vesicovaginal
    v. fascia
vesicovaginorectal
vesicovisceral
vesicula, pl. **vesiculae**
    v. fellis
    v. ophthalmica
    v. seminalis
    v. umbilicalis
vesicular
    v. appendage
    v. appendages of epoöphoron
    v. appendices of uterine tube
    v. mole
    v. ovarian follicle
    v. transport
    v. venous plexus
vesiculosa
    appendix v.
    glandula v.
vesiculosae
    appendices v.
    ductus excretorius glandulae v.
    tunica muscularis glandulae v.
vesiculosi
    stratum granulosum folliculi
        ovarici v.
vesiculosus
    folliculus ovaricus v.
Vesling line
vessel
    absorbent v.

accessory blood v.
afferent v.
anastomosing v.
anastomotic v.
anterior great v.
blood v.
capillary v.
cardinal v.
carotid v.
choroidal v.
choroid blood v.
chyle v.
ciliary v.
circumflex v.
collateral v.
deep epigastric v.
deep lymph v.
deep lymphatic v.
efferent v.
external iliac v.
gastroepiploic v.
gluteal v.
great v.
groove of middle meningeal v.
hyaloid v.
iliac v.
inferior v.
intercostal v.
v. of internal ear
internal spermatic v.
intraepithelial v.
intrapulmonary blood v.
lacteal v.
lymph v.
lymphatic v.
mammary v.
meningeal v.
nutrient v.
omphalomesenteric v.
opticociliary v.
peripheral v.
peroneal v.
posterior great v.
pudendal v.
pudic v.
pulmonary v.
retinal blood v.
sheath of v.
spermatic v.
submucosal v.
superficial blood v.
superficial lymph v.

V

NOTES

vessel *(continued)*
    surface v.
    transverse circumflex v.
    uterine v.
    vessel of v.
    vitelline v.
vestibula (*pl. of* vestibulum)
vestibular
    v. anus
    v. aqueduct
    v. area
    v. blind sac
    v. branch of labyrinthine artery
    v. canal
    v. canaliculus
    v. cecum of the cochlear duct
    v. crest
    v. fissure of cochlea
    v. fold
    v. fossa
    v. ganglion
    v. gland
    v. hair cells
    v. labium of limbus of spiral
      lamina
    v. labyrinth
    v. lamella of osseous spiral lamina
    v. ligament
    v. lip
    v. lip of limbus of spiral lamina
    v. lip of spiral limbus
    v. membrane
    v. nuclei
    v. organ
    v. part of vestibulocochlear nerve
    v. region
    v. root of vestibulocochlear nerve
    v. saccule
    v. surface of cochlear duct
    v. surface of tooth
    v. trough
    v. vein
    v. wall of cochlear duct
    v. window
vestibulare
    cecum v.
    ganglion v.
    labium limbi v.
    ligamentum v.
vestibulares
    nuclei v.
    rami v.
    venae v.
vestibularis
    v. ductus cochlearis
    labyrinthus v.
    membrana v.
    nervus v.
    pars inferior ganglii v.

    pars superior ganglii v.
    plica v.
    radix v.
vestibulate
vestibule
    aortic v.
    aqueduct of v.
    artery of bulb of v.
    buccal v.
    bulb of v.
    cochlear recess of v.
    crest of v.
    esophagogastric v.
    gastroesophageal v.
    v. of inner ear
    labial v.
    v. of larynx
    v. of mouth
    nasal v.
    v. of nose
    v. of omental bursa
    oral v.
    pyloric v.
    pyramid of v.
    Sibson aortic v.
    urogenital v.
    v. of vagina
    vein of bulb of v.
vestibuli
    apertura canaliculi v.
    apertura externa aqueductus v.
    aqueductus v.
    arteria v.
    arteria bulbi v.
    bulbi v.
    bulbus v.
    crista v.
    fenestra v.
    fossula fenestrae v.
    plica v.
    pyramis v.
    rima v.
    sacculus v.
    scala v.
    valvula v.
    vein of scala v.
    vena aqueductus v.
    vena bulbi v.
    vena scalae v.
vestibulocerebellum
vestibulocochlear
    v. artery
    v. nerve
    v. organ
vestibulocochleare
    organum v.
vestibulocochlearis
    arteria v.
    nervus v.

nuclei nervi v.
pars cochlearis nervi v.
pars vestibularis nervi v.
radix inferior nervi v.
radix superior nervi v.
ramus cochlearis arteriae v.
ramus vestibularis posterior
   arteriae v.
vestibuloequilibratory control
vestibulospinal
   v. reflex
   v. tract
vestibulospinalis
   tractus v.
vestibulourethral
vestibulum, pl. **vestibula**
   v. aortae
   v. bursae omentalis
   v. laryngis
   v. nasi
   v. oris
   v. pudendi
   v. vaginae
vestige
   v. of ductus deferens
   v. of processus vaginalis
   v. of vaginal process
vestigia (*pl. of* vestigium)
vestigial
   v. fold
   v. muscle
   v. organ
vestigialis
   ductus deferens v.
vestigium, pl. **vestigia**
   v. processus vaginalis
vetularum
   uterus bicameratus v.
viability
viable
viae
vibrans
   membrana v.
vibration
vibrissa, pl. **vibrissae**
vibrissal
vicarious hypertrophy
Vicq
   V. d'Azyr bundle
   V. d'Azyr centrum semiovale
   V. d'Azyr foramen

vidian
   v. artery
   v. canal
   v. nerve
   v. vein
Vieussens
   V. ansa
   V. anulus
   V. centrum
   V. foramen
   V. ganglion
   V. isthmus
   V. limbus
   loop of V.
   V. loop
   V. ring
   V. valve
   V. vein
   V. ventricle
VII
   segmentum bronchopulmonale basale
      mediale S VII
villi (*pl. of* villus)
villosa
   plica v.
villose
villosity
villous placenta
villus, pl. **villi**
   anchoring v.
   arachnoid v.
   chorionic v.
   duodenal v.
   finger-like v.
   floating v.
   free v.
   intestinal villi
   villi intestinales
   jejunal v.
   leaflike v.
   villi pericardiaci
   pericardial villi
   peritoneal villi
   villi peritoneales
   placental v.
   pleural villi
   villi pleurales
   primary v.
   secondary v.
   synovial villi
   villi synoviales
   tertiary v.

V

NOTES

vimentin
vinculum, pl. **vincula**
    v. breve
    v. breve digitorum manus
    v. breve of finger
    v. linguae
    vincula lingulae cerebelli
    long v.
    v. longum
    v. longum digitorum manus
    v. longum of finger
    v. preputii
    short v.
    vincula tendinea of digits of hand
      and foot
    vincula tendinum
    vincula tendinum digitorum manus
      et pedis
    vincula of tendon
Virchow
    V. angle
    V. cells
    V. corpuscles
Virchow-Hassall bodies
Virchow-Holder angle
Virchow-Robin space
vires (*pl. of* vis)
virga
virginale
    claustrum v.
virginal membrane
virgin generation
virile
    v. member
    membrum v.
virilia
virilis
    mamma v.
    urethra v.
virus-transformed cell
vis, pl. **vires**
    v. conservatrix
    v. a fronte
    v. a tergo
    v. vitae
    v. vitalis
viscance
viscera (*pl. of* viscus)
viscerad
visceral
    v. arches
    v. brain
    v. cavity
    v. cleft
    v. cranium
    v. inversion
    v. lamina
    v. layer
    v. layer of serous pericardium

    v. layer of tunica vaginalis of
      testis
    v. lymph node
    v. lymph node of abdomen
    v. mesoderm
    v. motor fiber
    v. motor neuron
    v. motor system
    v. muscle
    v. nervous system
    v. node
    v. pelvic fascia
    v. peritoneum
    v. plate
    v. pleura
    v. skeleton
    v. surface
    v. surface of liver
    v. surface of spleen
viscerale
    cranium v.
    peritoneum v.
viscerales
    lymphonodi abdominis v.
    nodi lymphoidei abdominis v.
    plexus v.
visceralis
    fascia pelvis v.
    lamina v.
    pars abdominalis plexus visceralis
      et ganglii v.
    pars pelvica plexus visceralis et
      ganglii v.
    pleura v.
    tunica serosa pleurae v.
visceralium
    pars craniocervicalis plexuum et
      gangliorum v.
    pars thoracica plexuum et
      ganglionorum v.
viscerimotor
viscerocardiac reflex
viscerocranium
    cartilaginous v.
    membranous v.
viscerogenic
viscerograph
visceroinhibitory
visceromegaly
visceromotor
visceroparietal
visceroperitoneal
visceropleural
viscerosensory reflex
visceroskeletal
visceroskeleton
viscerosomatic
viscerotome
viscerotrophic reflex

viscerotropic
viscerum
 situs inversus v.
viscosity
 absolute v.
 anomalous v.
 apparent v.
 coefficient of v.
 dynamic v.
 kinematic v.
 newtonian v.
 relative v.
viscus, pl. **viscera**
 abdominal viscera
 intraabdominal viscera
 intraperitoneal v.
 pelvic viscera
vision
 organ of v.
visual
 v. area
 v. cortex (VC)
 v. inattention
 v. receptor cells
visuoauditory
visus
 organum v.
vitae
 arbor v.
 vis v.
vital
 v. capacity
 v. center
 v. force
 v. knot
 v. node
 noeud v.
 v. tripod
vitalis
 turgor v.
 vis v.
vitalism
vitalistic
vitality
vitalize
vitamin D–binding protein
vitelliform
vitellina
 arteria v.
 membrana v.
 vena v.

vitelline
 v. artery
 v. circulation
 v. cord
 v. duct
 v. fistula
 v. membrane
 v. pole
 v. sac
 v. vein
 v. vessel
vitellinus
 pedunculus v.
vitellogenesis
vitellointestinal duct
vitellus
vitiligo, pl. **vitiligines**
vitrea
 camera v.
 lamina v.
 membrana v.
 substantia v.
 tunica v.
vitreoretinal
vitreous
 v. base
 v. body
 v. camera
 v. cavity
 v. cell
 central canal of the v.
 v. chamber
 v. chamber of eye
 v. fluid
 v. humor
 v. lamella
 v. membrane
 patellar fossa of v.
 persistent anterior hyperplastic
  primary v.
 persistent posterior hyperplastic
  primary v.
 primary v.
 secondary v.
 stroma of v.
 v. table
 tertiary v.
vitreum
 corpus v.
 stroma v.
vitreus
 humor v.

V

NOTES

viviparity
viviparous
viviperception
vivo
    ex v.
Vmax
VMO
    vastus medialis obliquus musculus
vocal
    v. cord (VC)
    v. fold
    v. ligament
    v. muscle
    v. process
    v. process of arytenoid cartilage
    v. shelf
vocale
    labium v.
    ligamentum v.
vocales
    chordae v.
vocalia
    labia v.
vocalis
    chorda v.
    glottis v.
    v. muscle
    musculus v.
    plica v.
    rima v.
Vogt angle
void
voiding
    v. flow rate
    v. internal urethral orifice
vola
volae
    superficialis v.
volar
    v. aspect
    v. carpal ligament
    v. interosseous artery
    v. interosseous nerve
    v. pad
    v. plate
    v. radiocarpal ligament
    v. region
    v. surface
volare
    ligamentum carpi v.
volaris
    arteria interossea v.
    musculus interosseus v.
Volkmann
    V. canals
    V. contracture
voltage-gated channel
volume
    cardiac v. (CV)

    closing v.
    corpuscular v.
    distribution v.
    expiratory reserve v.
    extracellular fluid v.
    forced expiratory v.
    inspiratory reserve v.
    minute v.
    residual v.
    respiratory minute v.
    resting tidal v.
    stroke v.
    tidal v.
    v. unit (VU)
voluntary
    v. area
    v. dehydration
    v. muscle
volute
vomer
    ala of v.
    v. bone
    v. cartilagineus
    cuneiform part of v.
    left v.
    wing of v.
vomeral
    v. groove
    v. sulcus
vomeralis
    sulcus v.
vomerine
    v. canal
    v. cartilage
    v. crest of choana
    v. groove
    v. ridge
vomeris
    ala v.
    crista choanalis v.
    pars cuneiformis v.
    sulcus v.
vomerobasilar canal
vomeronasal
    v. cartilage
    v. nerve
    v. organ
vomeronasale
    organum v.
vomeronasalis
    cartilago v.
vomerorostral canal
vomerorostralis
    canalis v.
vomerovaginal
    v. canal
    v. groove

vomerovaginalis
    canalis v.
    sulcus v.
vomiting
    dry v.
    v. reflex
vomiturition
Von Ebner gland
vortex, pl. **vortices**
    v. coccygeus
    v. cordis
    v. of heart
    v. lentis
    vortices pilorum
    v. vein
vorticosa
    vasa v.
vorticosae
    venae v.

vorticose vein
Voshell bursa
V-shaped area of esophagus
VT
    ventricular tachycardia
VU
    volume unit
vulva, pl. **vulvae**
    rima vulvae
vulval
vulvar slit
vulvocrural
vulvouterine
vulvovaginal
    v. anus
    v. gland

NOTES

V

Wachendorf membrane
waddingtonian homeostasis
waist
Waldeyer
- W. colon
- W. fascia
- W. fossae
- W. gland
- pelvic colon of W.
- W. sheath
- W. space
- W. sulcus
- W. throat ring
- W. tract
- W. zonal layer

wall
- abdominal w.
- anterior abdominal w.
- anterior rectus sheath w.
- w. of atrium
- bladder w.
- w. of body
- bowel w.
- cell w.
- chest w.
- w. of cranium
- deep anterior w.
- deep fascia of abdominal w.
- fibromuscular w.
- gallbladder w.
- inferior w.
- interior chest w.
- intermediate anterior w.
- jugular w.
- lateral pelvic w.
- medial w.
- membranous w.
- muscular w.
- musculofascial w.
- w. of nail
- nail w.
- nasal w.
- orbital w.
- parietal w.
- pelvic w.
- w. of pharynx
- posterior abdominal w.
- posterior oropharyngeal w.
- posterior rectus sheath w.
- pulpal w.
- splanchnic w.
- stomach w.
- superficial anterior w.
- thoracic w.
- tracheal w.

- urethral w.
- uterine w.
- vaginal w.
- ventricular w.

Walther
- W. canal
- W. duct
- W. ganglion
- W. plexus

wandering
- w. cell
- w. organ

Warburg apparatus
Warburg-Lipmann-Dickens-Horecker
- W.-L.-D.-H. shunt

Ward
- triangle of W.

warm-blooded
- w.-b. animal

Warthin-Finkeldey cells
Wasmann gland
wasserhelle cell
wasted ventilation
wasting
- salt w.

water
- bag of w.'s
- w. depletion
- w. diuresis
- intracellular w.
- total body w. (TBW)
- transcellular w.

water-clear cell of parathyroid
watershed
wave
- excitation w.
- w. number
- Traube-Hering w.

wax
- grave w.

web
- w. of finger
- w. of fingers/toes
- w. ligament
- w. space
- terminal w.
- thenar w.

webbed toes
webbing
Weber
- W. gland
- W. law
- W. organ
- W. paradox

W

Weber *(continued)*
    W. point
    W. triangle
Weber-Fechner law
Wedensky
    W. effect
    W. facilitation
    W. inhibition
wedge
    w. bone
    w. press
    w. pressure
wedge-and-groove
    w.-a.-g. joint
    w.-a.-g. suture
wedge-shaped
    w.-s. fasciculus
    w.-s. tubercle
Weibel-Palade bodies
Weigert law
weight
    birth w.
weightlessness
Weil
    W. basal layer
    W. basal zone
Weisbach angle
Weitbrecht
    W. cartilage
    W. cord
    W. fiber
    W. foramen
    W. ligament
weitbrechti
    apparatus ligamentosus w.
Welcker angle
Wenzel ventricle
Wepfer gland
Wernekinck commissure
Wernicke
    W. area
    W. area of brain
    W. center
    W. field
    W. radiation
    W. region
    W. zone
West
    zone 1–4 of W.
Westberg space
Wever-Bray phenomenon
Wharton
    W. duct
    W. jelly
Wheatstone bridge
white
    w. blood cell
    w. commissure
    w. corpuscle

    w. of eye
    w. fat
    w. fiber
    w. line
    w. line of anal canal
    w. line of Toldt
    w. matter
    w. muscle
    w. pulp
    w. pulp of spleen
    w. rami communicantes
    w. substance
    w. yolk
Whitnall
    W. ligament
    W. orbital tubercle
whorl
    coccygeal w.
    hair w.
whorled
wide plane
widow peak
width
    orbital w.
Wiener
    tract of Münzer and W.
Wilde cords
Wilhelmy balance
Wilkie artery
Willis
    antrum of W.
    arterial circle of W.
    artery of W.
    W. centrum nervosum
    circle of W.
    W. pancreas
    W. pouch
willisii
    accessorius w.
    chordae w.
Williston law
Wilson muscle
Winberger line
window
    aortic w.
    cochlear w.
    crest of round w.
    fossa of oval w.
    fossa of round w.
    little fossa of cochlear w.
    little fossa of oval vestibular w.
    little fossa of vestibular w.
    oval w.
    round w.
    vein of cochlear w.
    vestibular w.
windpipe
wing
    ashen w.

w. cell
w. of central lobule
w. of crista galli
gray w.
w. of ilium
Ingrassia w.
w. of nose
w. plate
w. of sacrum
sphenoidal w.
w. of vomer
Winslow
epiploic foramen of W.
foramen of W.
W. foramen
W. ligament
W. pancreas
W. stars
winslowii
stellulae w.
Wirsung
W. canal
duct of W.
W. pancreatic duct
wisdom tooth
withdrawal reflex
wolffian
w. body
w. cyst
w. duct
w. rest
w. ridge
w. tubules
Wolff law
Wölfler gland
Wolfring gland
womb
mouth of w.
neck of w.
Woodruff
nasopalatine plexus of W.
Woolner tip

work
unit of w.
worm
wormian bone
woven bone
wreath
ciliary w.
Wright respirometer
wrinkler muscle of eyebrow
Wrisberg
W. cartilage
W. ganglia
W. ligament
ligament of W.
nerve of W.
W. nerve
W. tubercle
wrist
anterior region of w.
w. bone
w. clonus
extensor retinaculum of w.
external collateral ligament of w.
flexor retinaculum of w.
internal collateral ligament of w.
w. joint
lateral ligament of w.
long radial extensor muscle of w.
medial ligament of w.
metacarpal of w.
posterior region of w.
radial collateral ligament of w.
radial eminence of w.
radial flexor muscle of w.
radiate ligament of w.
short radial extensor muscle of w.
triangular disk of w.
ulnar collateral ligament of w.
ulnar eminence of w.
ulnar extensor muscle of w.
ulnar flexor muscle of w.
wryneck

NOTES

W

xanthelasmoidea
xiphisternal joint
xiphisternum
xiphocostal angle
xiphoid
    x. angle
    x. cartilage
    x. process

xiphoideus
    processus x.
xiphoid-manubrial junction
xiphopagus
xiphosternalis
    symphysis x.
    synchondrosis x.
X zone

X

y-angle
yawn
yawning
Y cartilage
yellow
- y. body
- y. bone marrow
- y. cartilage
- y. fibers
- y. ligament
- y. spot
- y. yolk

yield stress
yin-yang
yoke
- alveolar y.

- y. bone
- y. of mandible
- y. of maxilla
- sphenoidal y.

yolk
- y. cells
- y. cleavage
- y. membrane
- y. sac
- y. stalk
- white y.
- yellow y.

Young modulus
Y-shaped
- Y-s. cartilage
- Y-s. ligament

Y

Z

    Z band
    Z disk
    Z filament
    Z line

Zaglas ligament

Zahn

    line of Z.
    Z. line
    Z. rib

Zeis gland

zero

    z. end-expiratory pressure
    z. gravity

Zimmermann

    Z. corpuscle
    Z. granule
    polkissen of Z.

Zinn

    annulus of Z.
    aponeurosis of Z.
    Z. aponeurosis
    Z. artery
    Z. corona
    Z. ligament
    Z. membrane
    Z. ring
    tendon of Z.
    Z. tendon
    Z. vascular circle
    Z. zone
    zonule of Z.
    Z. zonule

zinnii

    circulus z.

Z-line

zoetic

zoic

zona, pl. **zonae**

    z. arcuata
    z. ciliaris
    z. externa medullae renalis
    z. fasciculata
    z. glomerulosa
    z. hemorrhoidalis
    z. interna medullae renalis
    z. orbicularis
    z. orbicularis articulationis coxa
    z. pectinata
    z. pellucida
    z. perforata
    z. pupillaris
    z. radiata
    z. reticularis
    z. striata

    z. tecta
    z. transitionalis analis
    z. vasculosa

zonal

zonale

    stratum z.

zonary placenta

zonate

zone

    abdominal z.
    adherent z.
    anal transitional z.
    androgenic z.
    arcuate z.
    ciliary z.
    comfort z.
    cornuradicular z.
    entry z.
    ependymal z.
    fetal z.
    gingival z.
    Golgi z.
    grenz z.
    hemorrhoidal z.
    interpalpebral z.
    language z.
    Lissauer marginal z.
    Looser z.
    mantle z.
    Marchant z.
    marginal z.
    motor z.
    nucleolar z.
    Obersteiner-Redlich z.
    orbicular z.
    pectinate z.
    pellucid z.
    polar z.
    pupillary z.
    secondary X z.
    segmental z.
    Spitzka marginal z.
    subpapillary z.
    subplasmalemmal dense z.
    sudanophobic z.
    thymus-dependent z.
    trabecular z.
    transitional z.
    vascular z.
    vermilion transitional z.
    Weil basal z.
    Wernicke z.
    z. 1–4 of West
    X z.
    Zinn z.

Z

zonifugal
zonipetal
zonoskeleton
zonula, pl. **zonulae**
    z. adherens
    z. ciliaris
    z. occludens
zonular
    z. band
    z. fiber
    z. layer
    z. space
zonulares
    fibrae z.
zonularia
    spatia z.
zonule
    ciliary z.
    Zinn z.
    z. of Zinn
zoogonous
zoogony
zooid
zoosmosis
Zuckerkandl
    Z. body
    Z. convolution
    Z. fascia
    organ of Z.
zygal fissure
zygapophyseal
zygapophyseales
    articulationes z.
    juncturae z.
zygapophyses (*pl. of* zygapophysis)
zygapophysiales
    articulationes z.
    juncturae z.
zygapophysial joint
zygapophysis, pl. **zygapophyses**
    z. inferior
    z. superior
zygion
zygodactyly
zygoma
zygomati
    greater z.
zygomatic
    z. arch
    z. border of greater wing of
      sphenoid bone
    z. branch of facial nerve
    z. diameter
    z. fossa
    z. margin of greater wing of
      sphenoid bone
    z. muscle
    z. process

    z. process of frontal bone
    z. process of maxilla
    z. process of temporal bone
    z. reflex
    z. region
    z. suture line
zygomatica
    regio z.
zygomatici
    eminentia orbitalis ossis z.
    facies lateralis ossis z.
    processus frontalis ossis z.
    processus temporalis ossis z.
    rami z.
    ramus zygomaticofacialis nervi z.
    ramus zygomaticotemporalis
      nervi z.
    tuberculum marginale ossis z.
    tuberculum orbitale ossis z.
zygomatico
    ramus communicans nervi lacrimalis
      cum nervo z.
zygomaticoauricular index
zygomaticoauricularis
zygomaticofacial
    z. artery
    z. branch of zygomatic nerve
    z. foramen
zygomaticofaciale
    foramen z.
zygomaticofacialis
    ramus z.
zygomaticofrontal
    z. suture
    z. suture line
zygomaticofrontalis
    sutura z.
zygomaticomaxillaris
    sutura z.
zygomaticomaxillary
    z. suture
zygomaticoorbital
    z. artery
    z. foramen
zygomaticoorbitale
    foramen z.
zygomaticoorbitalis
    arteria z.
zygomaticosphenoid
zygomaticotemporal
    z. branch of zygomatic nerve
    z. foramen
    z. suture
zygomaticotemporale
    foramen z.
zygomaticotemporalis
    ramus z.
    sutura z.

zygomaticum
 os z.
 tuber z.
zygomaticus
 arcus z.
 z. major
 z. major muscle
 margo z.
 z. minor
 z. minor muscle
 musculus z.
 nervus z.
 processus z.

zygomaxillare
zygomaxillary point
zygon
zygonema
zygopodium
zygosis
zygosity
zygote
zygotene
zygotic
zymogenic cell

NOTES

Z

Appendix 1
Anatomical Illustrations

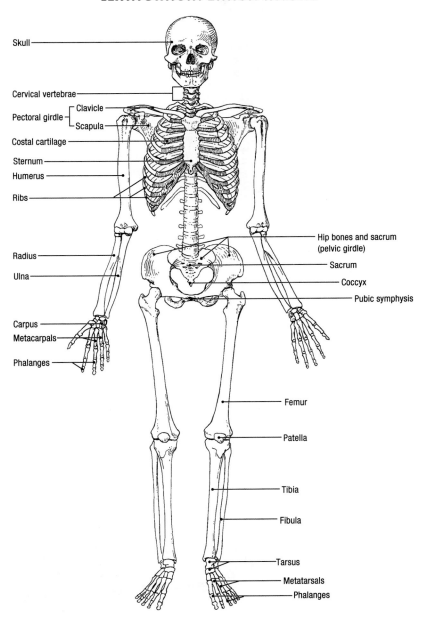

Skull

Cervical vertebrae

Clavicle
Pectoral girdle
Scapula

Costal cartilage

Sternum

Humerus

Ribs

Radius

Ulna

Carpus

Metacarpals

Phalanges

Hip bones and sacrum
(pelvic girdle)

Sacrum

Coccyx

Pubic symphysis

Femur

Patella

Tibia

Fibula

Tarsus

Metatarsals

Phalanges

Figure 1. Skeleton, adult, anterior view.

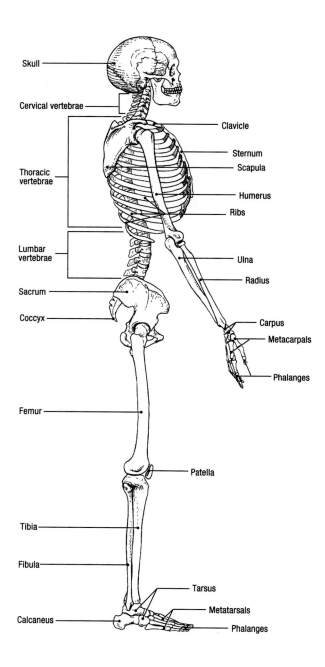

Figure 2. Skeleton, adult, lateral view.

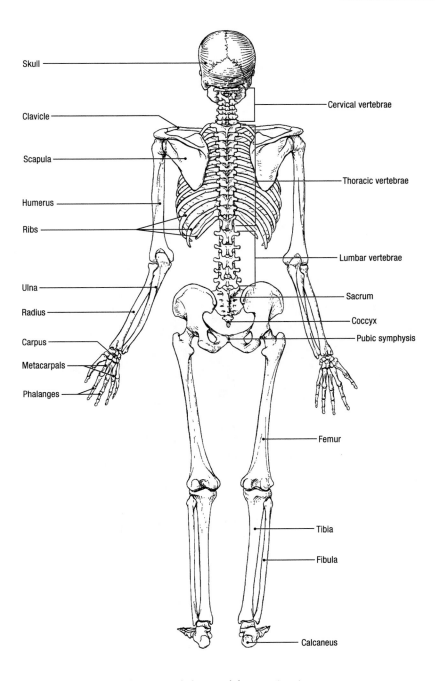

Figure 3. Skeleton, adult, posterior view.

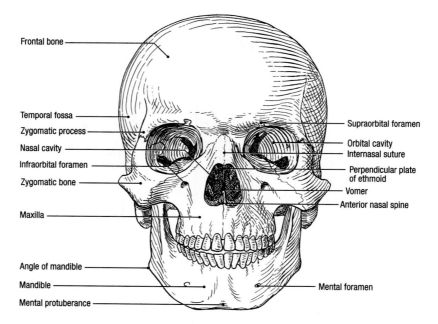

Figure 4. Frontal view of skull.

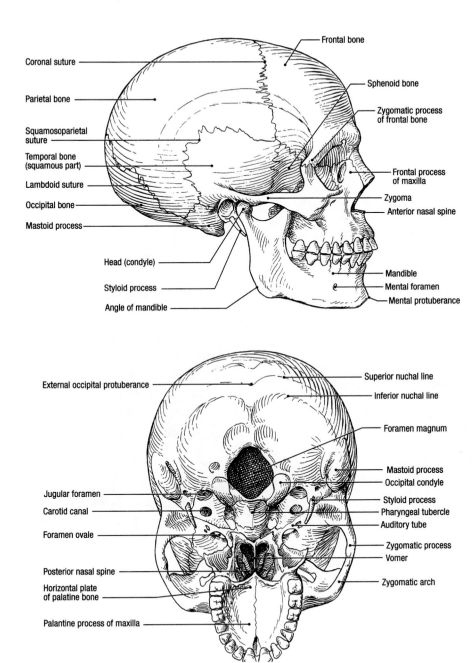

Figure 5. Lateral view (top) and inferior view (bottom).

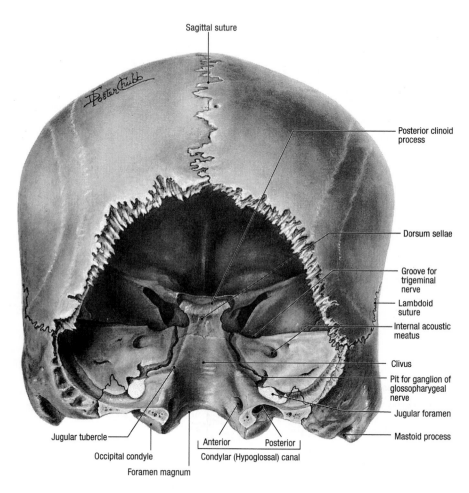

Figure 6. Skull. Bony features of posterior cranial fossa.

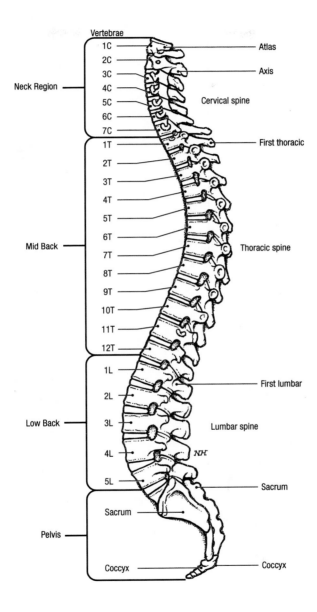

Figure 7. Vertebral column, lateral view.

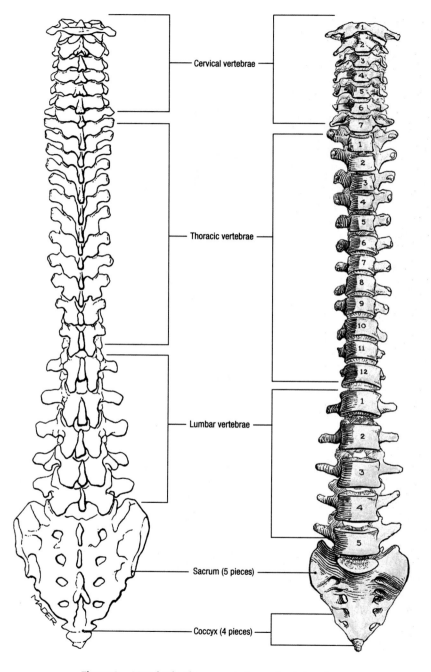

Figure 8. Vertebral column, posterior and anterior views.

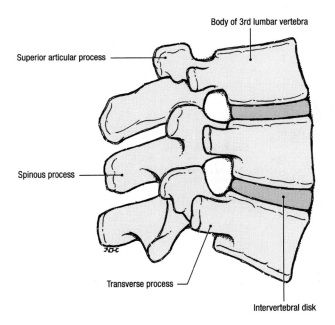

Figure 9. Vertebral processes in three lumbar vertebrae.

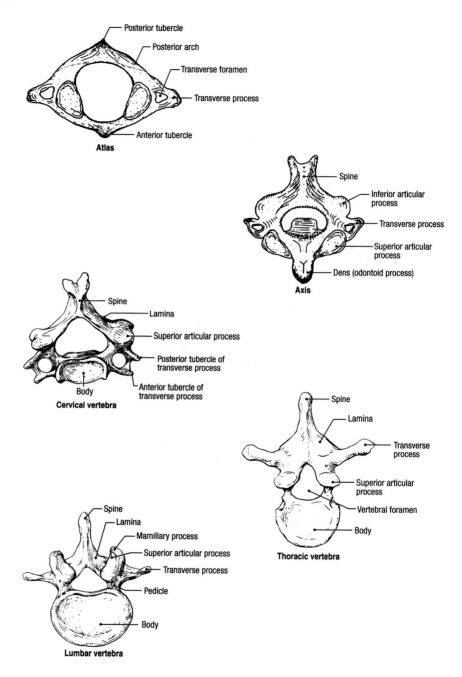

Figure 10. Typical atlas, axis, cervical, thoracic, and lumbar vertebrae.

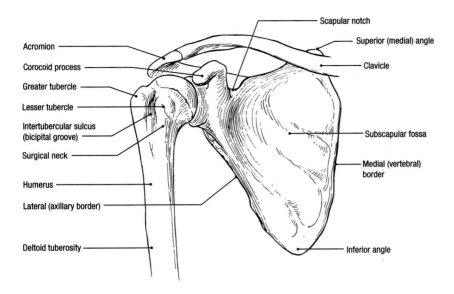

Scapular notch

Superior (medial) angle

Acromion

Corocoid process

Clavicle

Greater tubercle

Lesser tubercle

Intertubercular sulcus
(bicipital groove)

Subscapular fossa

Surgical neck

Medial (vertebral)
border

Humerus

Lateral (axillary border)

Deltoid tuberosity

Inferior angle

Figure 11.   Pectoral girdle and humerus, anterior view.

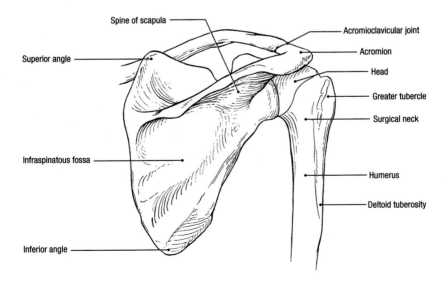

Spine of scapula

Acromioclavicular joint

Superior angle

Acromion

Head

Greater tubercle

Surgical neck

Infraspinatous fossa

Humerus

Deltoid tuberosity

Inferior angle

Figure 12.   Pectoral girdle and humerus, posterior view.

A11

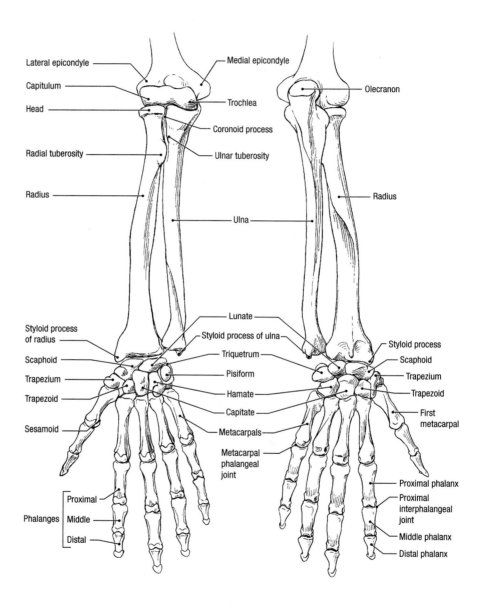

Figure 13. Bones of the forearm and hand. Anterior view (left) and posterior view (right).

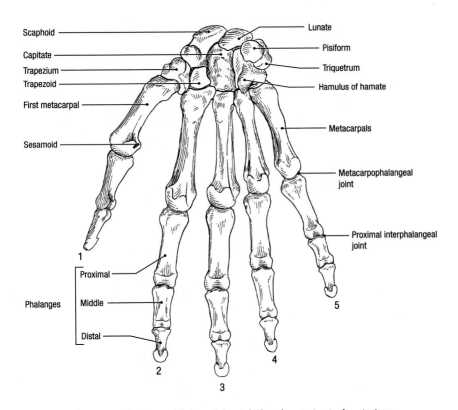

Scaphoid

Capitate

Trapezium

Trapezoid

First metacarpal

Sesamoid

Lunate

Pisiform

Triquetrum

Hamulus of hamate

Metacarpals

Metacarpophalangeal
joint

Proximal interphalangeal
joint

1

Phalanges

Proximal

Middle

Distal

2

3

4

5

Figure 14. Bones and joints of the right hand, anterior (palmar) view.

A13

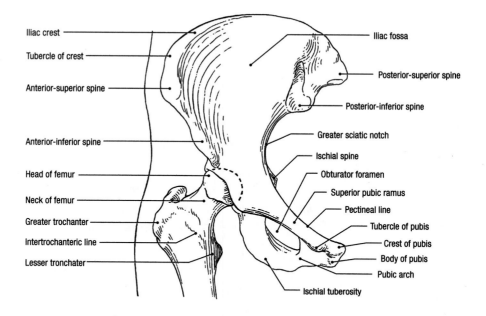

Iliac crest

Tubercle of crest

Anterior-superior spine

Anterior-inferior spine

Head of femur

Neck of femur

Greater trochanter

Intertrochanteric line

Lesser tronchater

Iliac fossa

Posterior-superior spine

Posterior-inferior spine

Greater sciatic notch

Ischial spine

Obturator foramen

Superior pubic ramus

Pectineal line

Tubercle of pubis

Crest of pubis

Body of pubis

Pubic arch

Ischial tuberosity

Figure 15. Bones of pelvis and hip, anterior view.

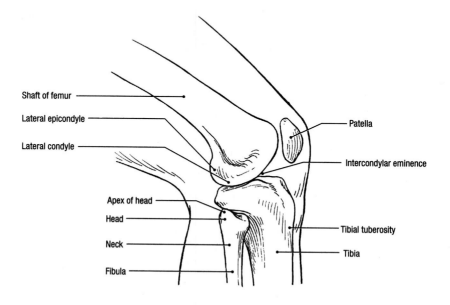

Shaft of femur

Lateral epicondyle

Lateral condyle

Apex of head

Head

Neck

Fibula

Patella

Intercondylar eminence

Tibial tuberosity

Tibia

Figure 16. Bones of knee region, lateral view.

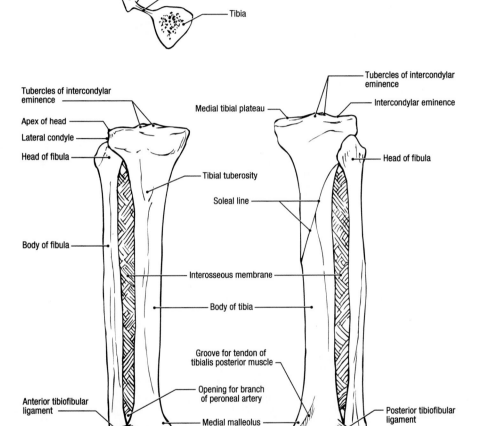

Figure 17. Bones of lower leg. Anterior view (left), posterior view (right), and cross-section (top).

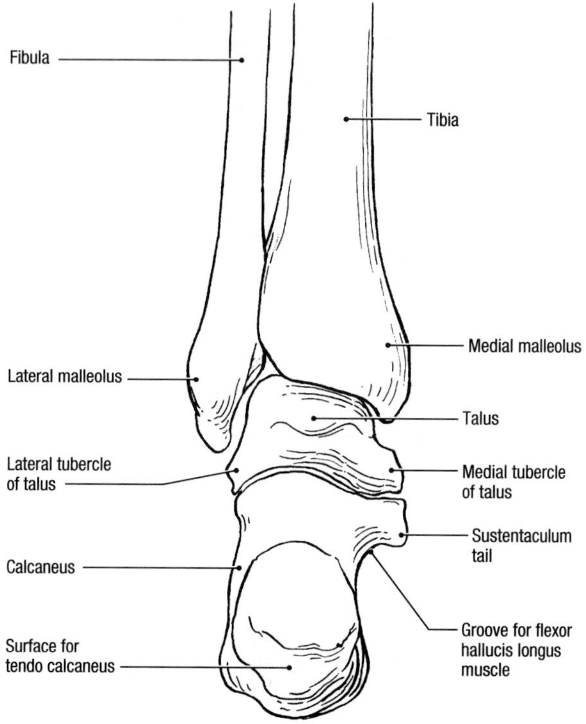

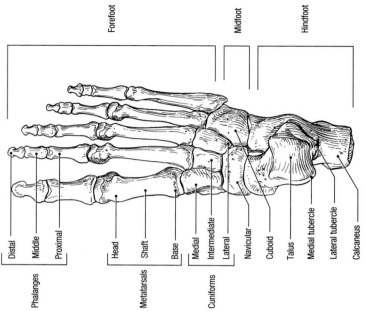

Figure 19. Bones of the ankle and foot, superior view.

Figure 18. Bones of the ankle and foot, dorsal view.

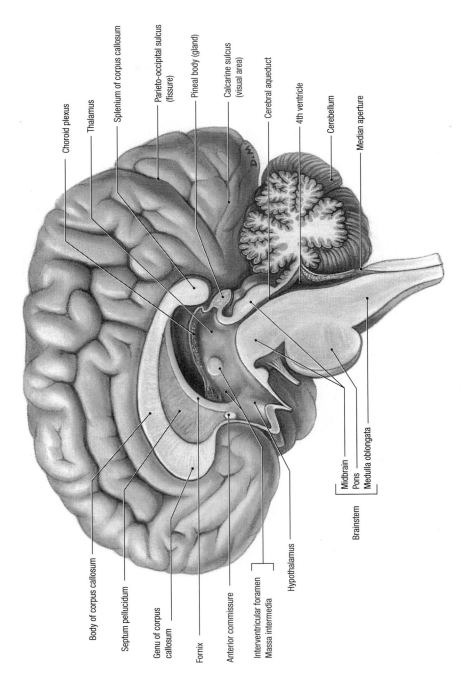

Choroid plexus

Thalamus

Splenium of corpus callosum

Parieto-occipital sulcus (fissure)

Pineal body (gland)

Calcarine sulcus (visual area)

Cerebral aqueduct

4th ventricle

Cerebellum

Median aperture

Body of corpus callosum

Septum pellucidum

Genu of corpus callosum

Fornix

Anterior commissure

Interventricular foramen
Massa intermedia

Hypothalamus

Brainstem
Midbrain
Pons
Medulla oblongata

Figure 20. Brain, median section.

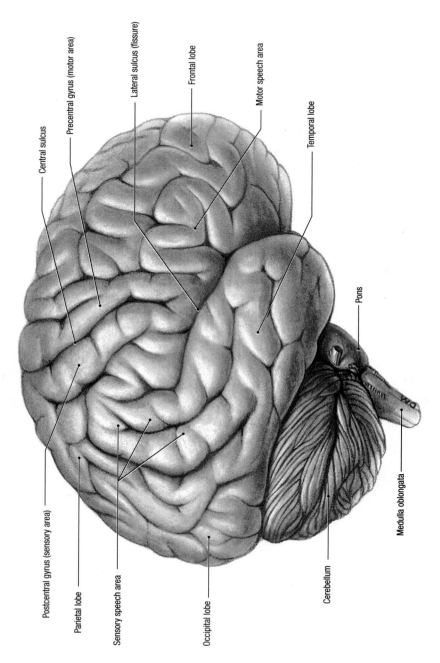

Central sulcus

Precentral gyrus (motor area)

Lateral sulcus (fissure)

Frontal lobe

Motor speech area

Temporal lobe

Pons

Postcentral gyrus (sensory area)

Parietal lobe

Sensory speech area

Occipital lobe

Cerebellum

Medulla oblongata

Figure 21. Brain, lateral view.

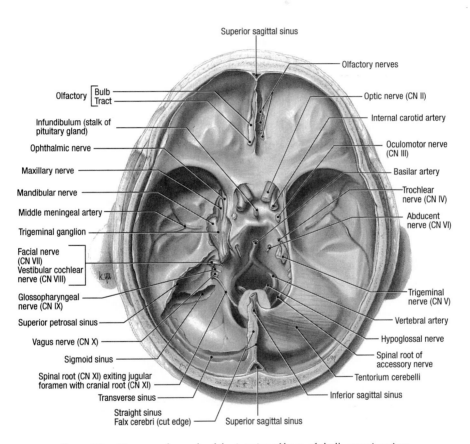

Figure 22. Nerves and vessels of the interior of base of skull, superior view.

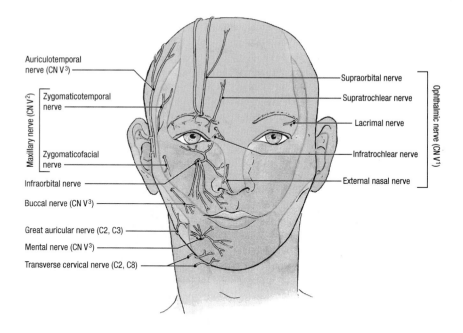

Figure 23. Sensory nerves of the face and scalp, anterior view.

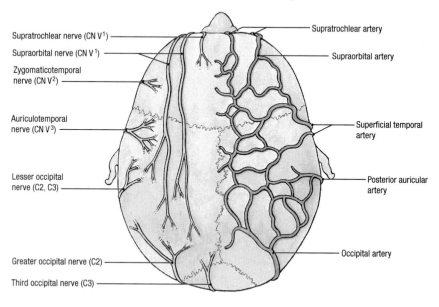

Figure 24. Sensory nerves and arteries of the face and scalp, superior view.

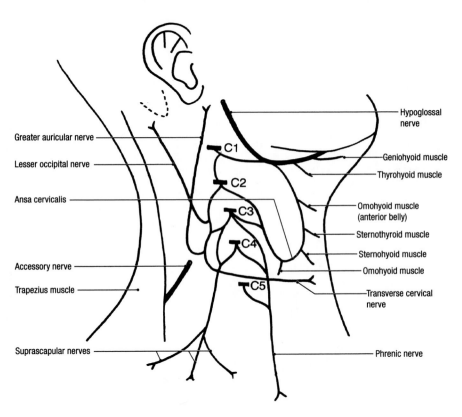

Figure 25. Cervical plexus.

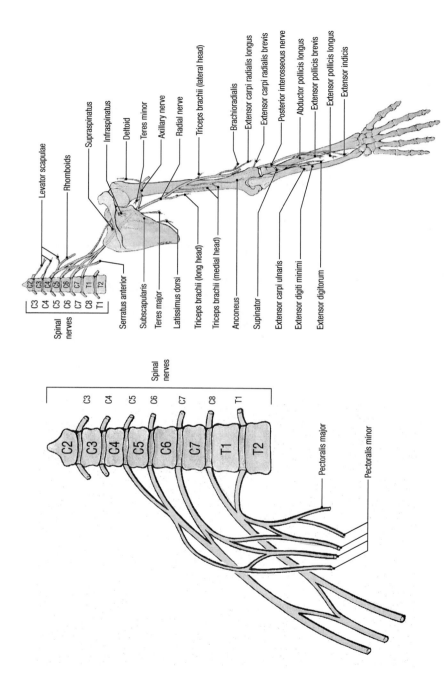

Figure 26. Nerves that innervate the muscles of the upper limb. Medial and lateral pectoral nerves (left) and radial nerve (right).

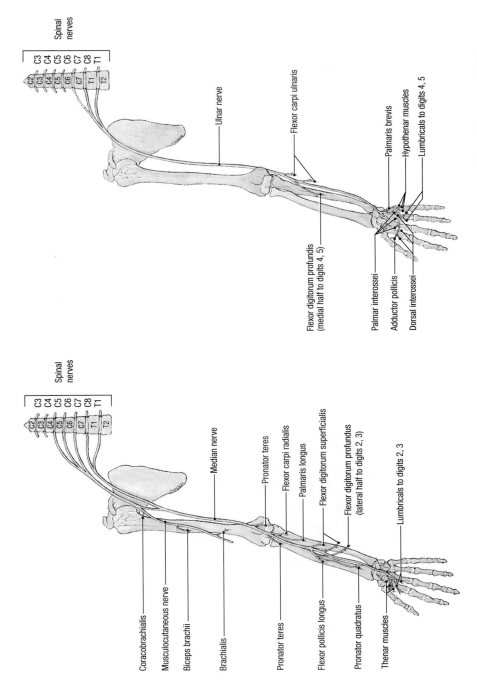

Figure 27. Nerves that innervate the muscles of the upper limb. Median and musculocutaneous nerves (left) and ulnar nerve (right).

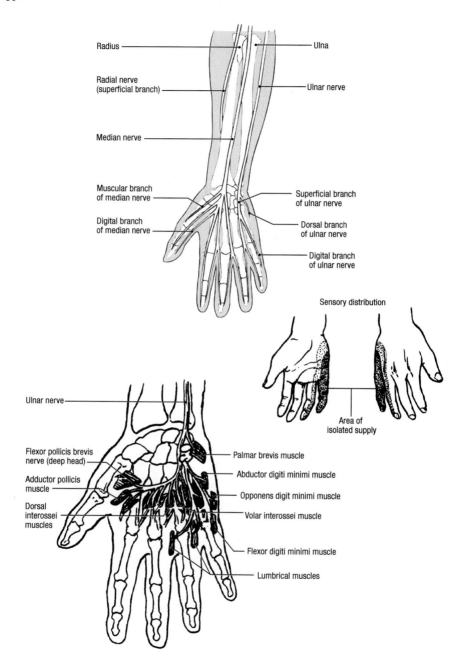

Figure 28. Nerves of the hand and sensory distribution.

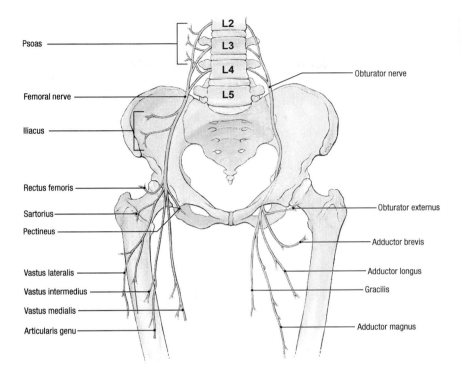

Psoas

Femoral nerve

Iliacus

Rectus femoris

Sartorius

Pectineus

Vastus lateralis

Vastus intermedius

Vastus medialis

Articularis genu

L2

L3

L4

L5

Obturator nerve

Obturator externus

Adductor brevis

Adductor longus

Gracilis

Adductor magnus

Figure 29. Femoral and obturator nerves.

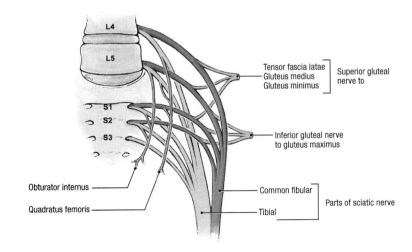

L4

L5

S1

S2

S3

Tensor fascia latae
Gluteus medius
Gluteus minimus

Superior gluteal nerve to

Inferior gluteal nerve to gluteus maximus

Obturator internus

Quadratus femoris

Common fibular

Tibial

Parts of sciatic nerve

Figure 30. Formation of the sciatic nerve in the pelvis.

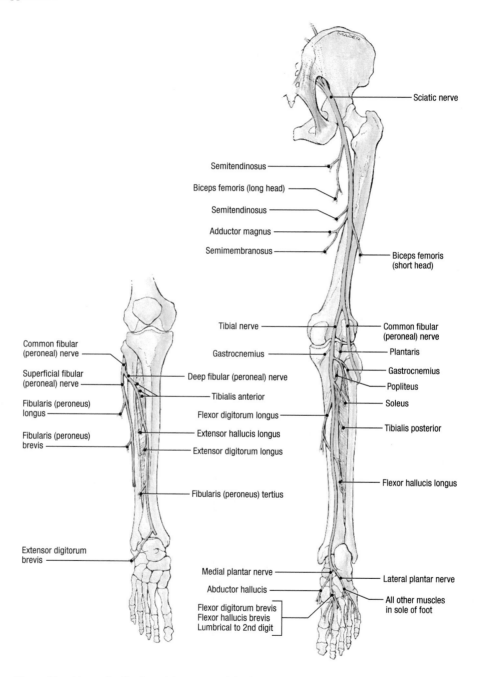

Figure 31. Motor distribution of the nerves of the lower limb. Common fibular (peroneal) nerve (left) and sciatic nerve (right).

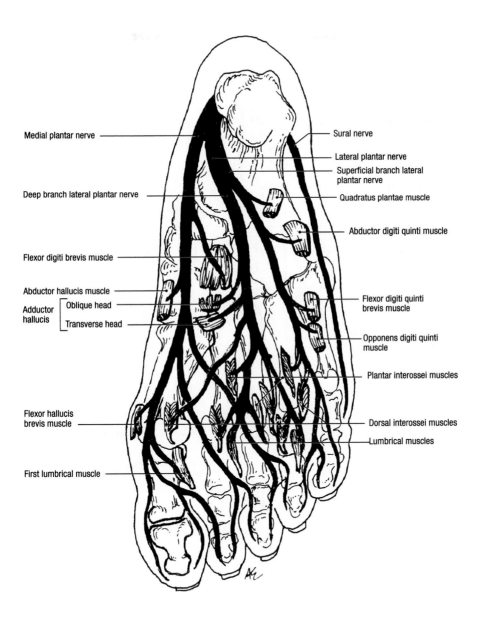

Medial plantar nerve

Deep branch lateral plantar nerve

Flexor digiti brevis muscle

Abductor hallucis muscle

Adductor hallucis — Oblique head

Transverse head

Flexor hallucis brevis muscle

First lumbrical muscle

Sural nerve

Lateral plantar nerve

Superficial branch lateral plantar nerve

Quadratus plantae muscle

Abductor digiti quinti muscle

Flexor digiti quinti brevis muscle

Opponens digiti quinti muscle

Plantar interossei muscles

Dorsal interossei muscles

Lumbrical muscles

Figure 32. Distribution of the tibial nerve in the foot.

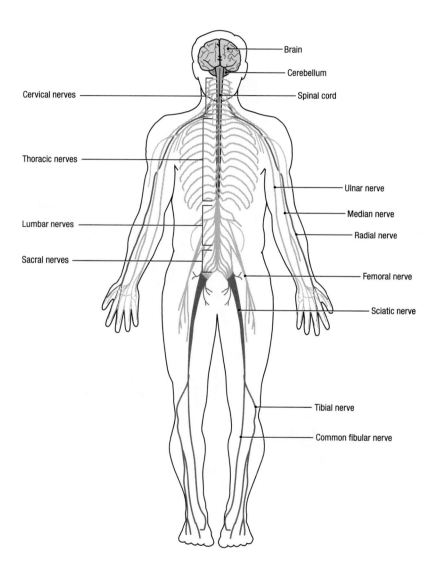

Figure 33. Anterior view of the major nerves of the body.

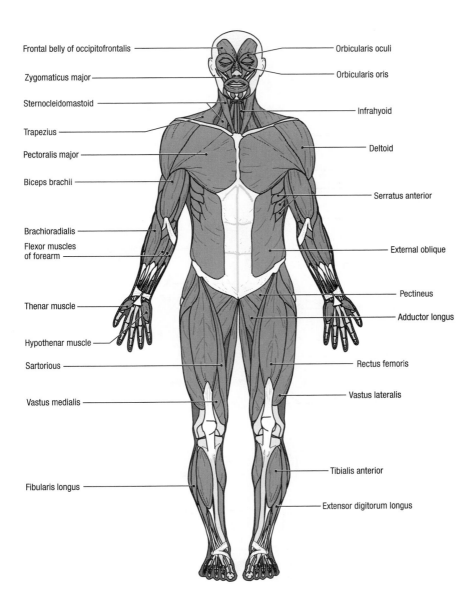

Frontal belly of occipitofrontalis

Zygomaticus major

Sternocleidomastoid

Trapezius

Pectoralis major

Biceps brachii

Brachioradialis

Flexor muscles of forearm

Thenar muscle

Hypothenar muscle

Sartorious

Vastus medialis

Fibularis longus

Orbicularis oculi

Orbicularis oris

Infrahyoid

Deltoid

Serratus anterior

External oblique

Pectineus

Adductor longus

Rectus femoris

Vastus lateralis

Tibialis anterior

Extensor digitorum longus

Figure 34.   Anterior view of the musculoskeletal system.

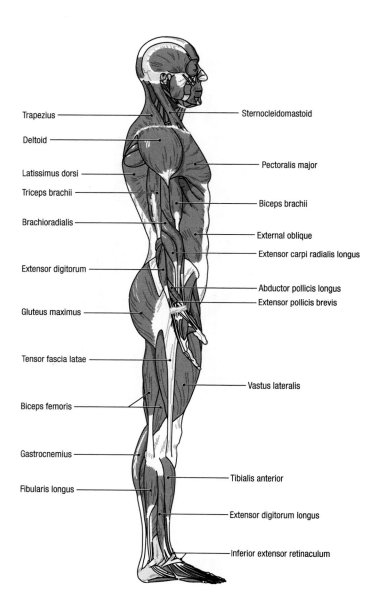

Trapezius

Deltoid

Latissimus dorsi

Triceps brachii

Brachioradialis

Extensor digitorum

Gluteus maximus

Tensor fascia latae

Biceps femoris

Gastrocnemius

Fibularis longus

Sternocleidomastoid

Pectoralis major

Biceps brachii

External oblique

Extensor carpi radialis longus

Abductor pollicis longus

Extensor pollicis brevis

Vastus lateralis

Tibialis anterior

Extensor digitorum longus

Inferior extensor retinaculum

Figure 35. Lateral view of the musculoskeletal system.

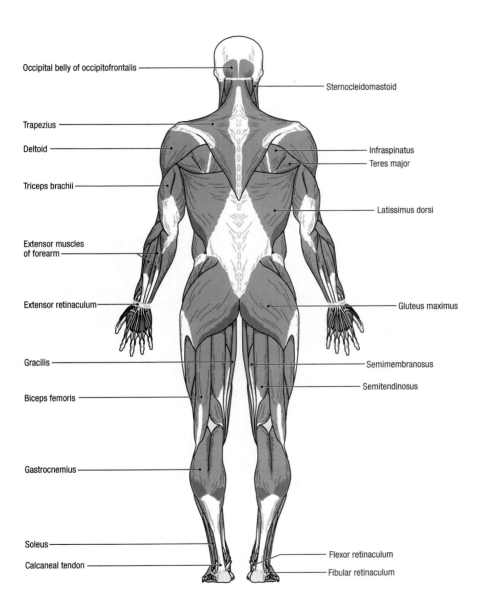

Occipital belly of occipitofrontalis

Sternocleidomastoid

Trapezius

Deltoid

Infraspinatus

Teres major

Triceps brachii

Latissimus dorsi

Extensor muscles
of forearm

Extensor retinaculum

Gluteus maximus

Gracilis

Semimembranosus

Semitendinosus

Biceps femoris

Gastrocnemius

Soleus

Flexor retinaculum

Calcaneal tendon

Fibular retinaculum

Figure 36. Posterior view of the musculoskeletal system.

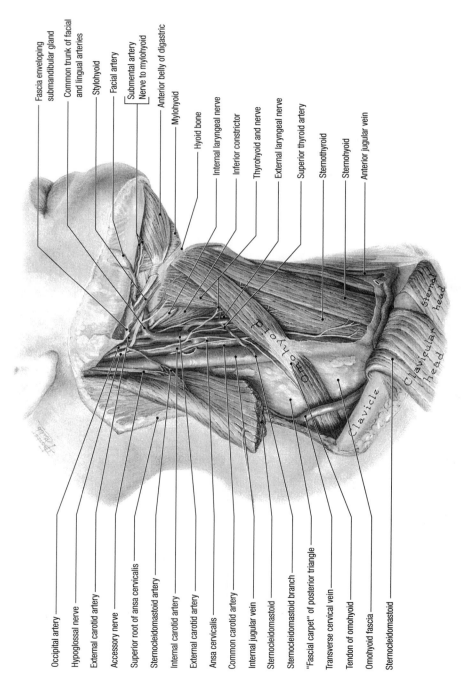

Figure 37. Deep muscles of the neck.

Fascia enveloping submandibular gland
Common trunk of facial and lingual arteries
Stylohyoid
Facial artery
Submental artery
Nerve to mylohyoid
Anterior belly of digastric
Mylohyoid
Hyoid bone
Internal laryngeal nerve
Inferior constrictor
Thyrohyoid and nerve
External laryngeal nerve
Superior thyroid artery
Sternothyroid
Sternohyoid
Anterior jugular vein

Sternal head
Clavicular head
Clavicle
Omohyoid

Occipital artery
Hypoglossal nerve
External carotid artery
Accessory nerve
Superior root of ansa cervicalis
Sternocleidomastoid artery
Internal carotid artery
External carotid artery
Ansa cervicalis
Common carotid artery
Internal jugular vein
Sternocleidomastoid
Sternocleidomastoid branch
"Fascial carpet" of posterior triangle
Transverse cervical vein
Tendon of omohyoid
Omohyoid fascia
Sternocleidomastoid

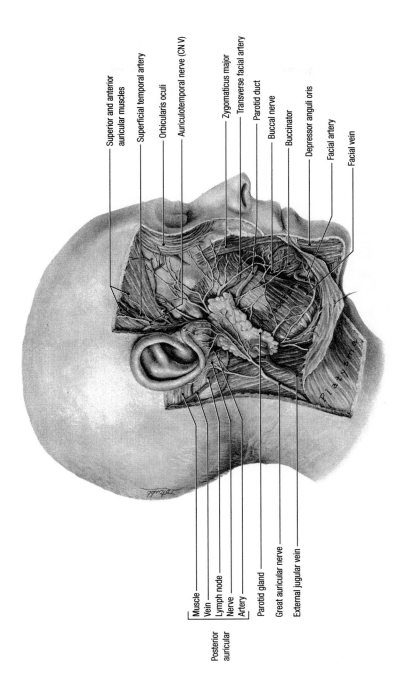

Muscle
Vein
Lymph node
Nerve
Artery

Posterior
auricular

Parotid gland
Great auricular nerve
External jugular vein

Superior and anterior auricular muscles
Superficial temporal artery
Orbicularis oculi
Auriculotemporal nerve (CN V)
Zygomaticus major
Transverse facial artery
Parotid duct
Buccal nerve
Buccinator
Depressor anguli oris
Facial artery
Facial vein

Figure 38. Relationships of the branches of the facial nerve and vessels to parotid gland and duct, lateral view.

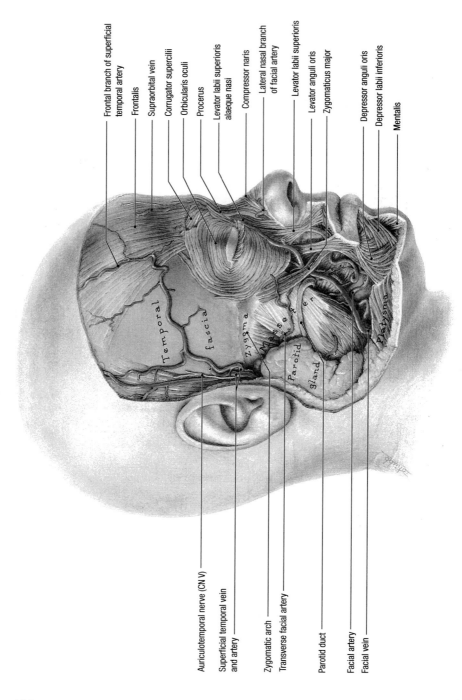

Figure 39. Muscles of facial expression and arteries and veins of face, lateral view.

Frontal branch of superficial temporal artery
Frontalis
Supraorbital vein
Corrugator supercilii
Orbicularis oculi
Procerus
Levator labii superioris alaeque nasi
Compressor naris
Lateral nasal branch of facial artery
Levator labii superioris
Levator anguli oris
Zygomaticus major
Depressor anguli oris
Depressor labii inferioris
Mentalis

Auriculotemporal nerve (CN V)
Superficial temporal vein and artery
Zygomatic arch
Transverse facial artery
Parotid duct
Facial artery
Facial vein

Temporal fascia
Zygoma
Masseter
Parotid gland
Platysma

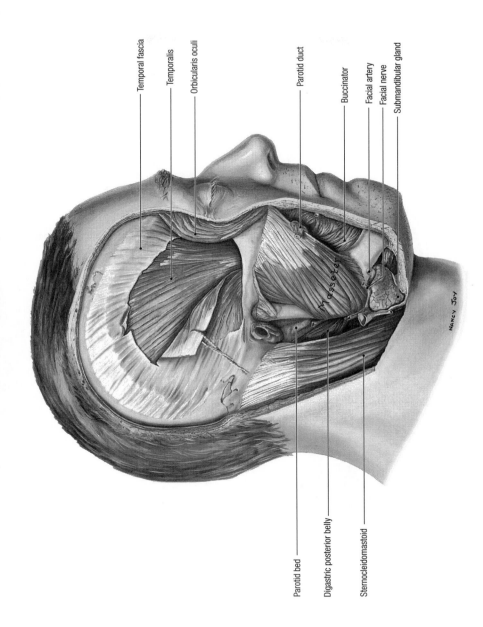

Figure 40. Great muscles of skull, lateral view.

Temporal fascia

Temporalis

Orbicularis oculi

Parotid duct

Buccinator

Facial artery

Facial nerve

Submandibular gland

Parotid bed

Digastric posterior belly

Sternocleidomastoid

Masseter

NANCY JOY

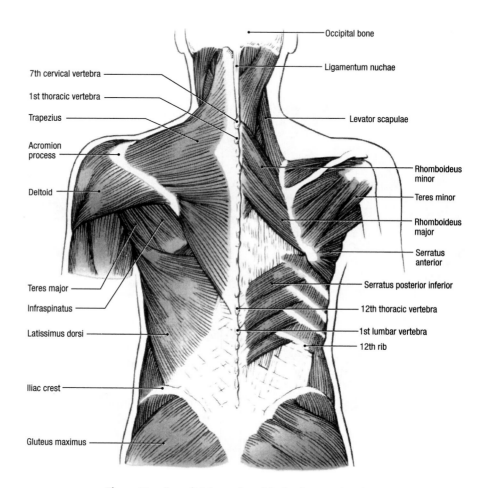

Figure 41. Superficial muscles of the back, posterior view.

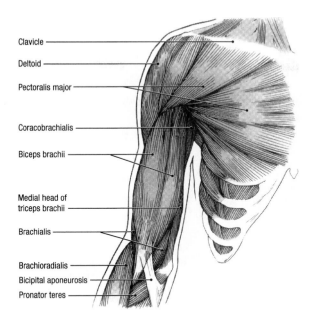

Clavicle

Deltoid

Pectoralis major

Coracobrachialis

Biceps brachii

Medial head of
triceps brachii

Brachialis

Brachioradialis

Bicipital aponeurosis

Pronator teres

Figure 42. Superficial muscles of the shoulder and chest, anterior view.

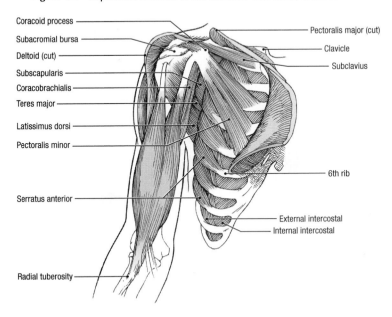

Coracoid process

Subacromial bursa

Deltoid (cut)

Subscapularis

Coracobrachialis

Teres major

Latissimus dorsi

Pectoralis minor

Serratus anterior

Radial tuberosity

Pectoralis major (cut)

Clavicle

Subclavius

6th rib

External intercostal
Internal intercostal

Figure 43. Deep muscles of the shoulder and chest, anterior view.

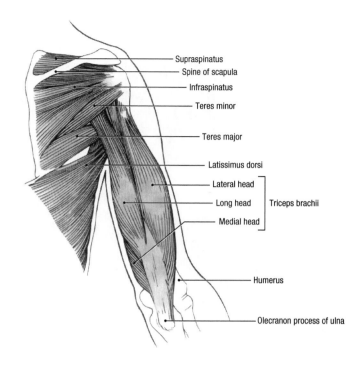

Figure 44. Muscles of the arm, posterior view.

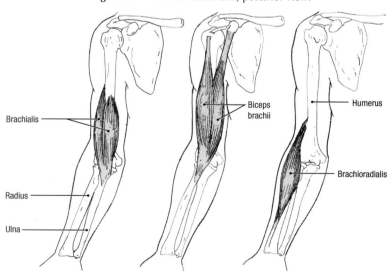

Figure 45. Muscles of the arm, posterior view.

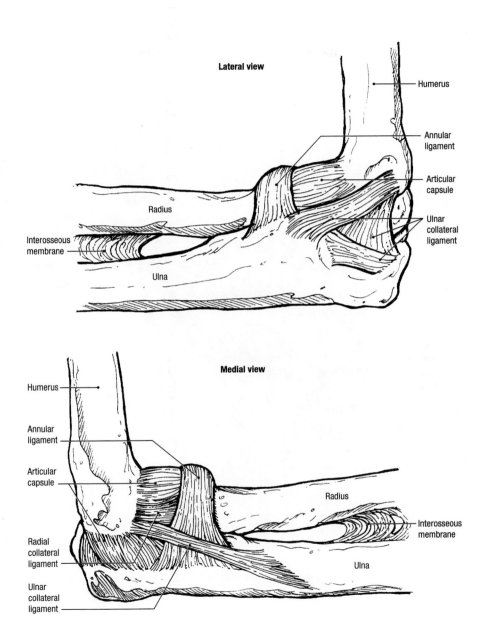

Figure 46.   Ligaments of the elbow. Lateral view (top) and medial view (bottom).

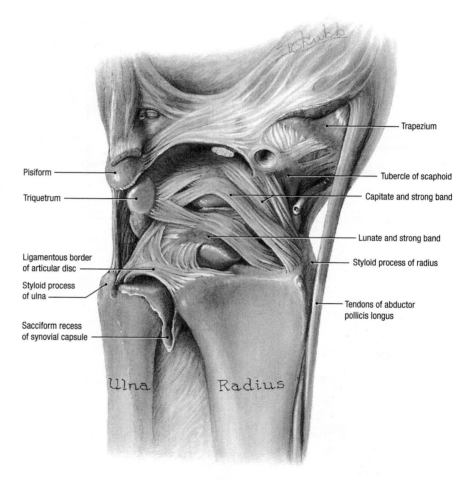

Pisiform

Triquetrum

Ligamentous border
of articular disc

Styloid process
of ulna

Sacciform recess
of synovial capsule

Trapezium

Tubercle of scaphoid

Capitate and strong band

Lunate and strong band

Styloid process of radius

Tendons of abductor
pollicis longus

Ulna   Radius

Figure 47.   Ligaments of the distal radioulnar, radiocarpal, and intercarpal joints.

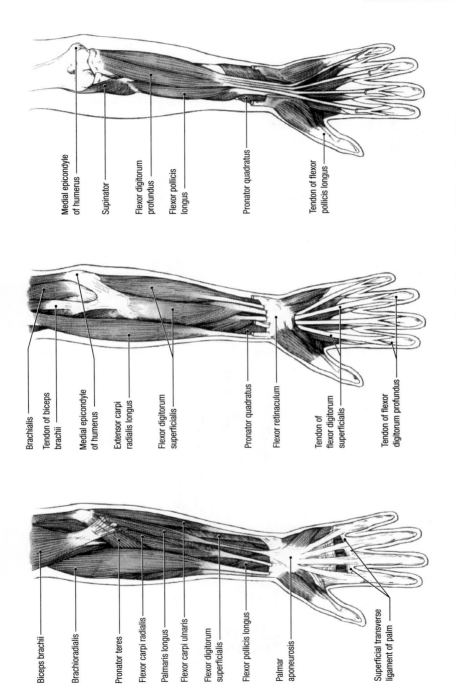

Figure 48. Muscles of the wrist and hand, anterior view. Superficial (left), mid-level (middle), and deep (right).

A41

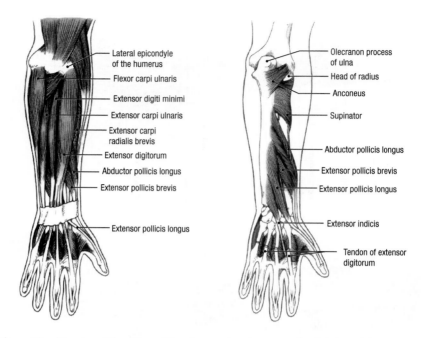

Figure 49. Muscles of the wrist and hand, posterior view. Superficial (left) and deep (right).

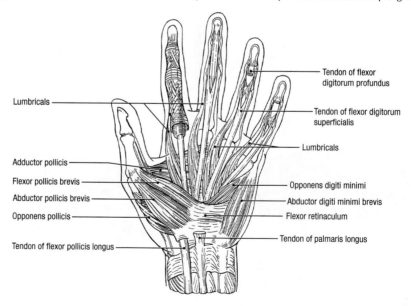

Figure 50. Muscles of the hand, anterior (palmar) view.

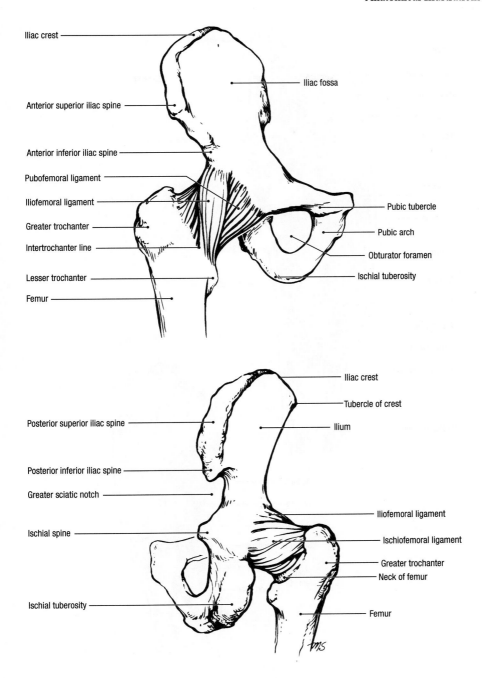

Figure 51. Ligaments of the hip. Anterior (top) and posterior (bottom) views.

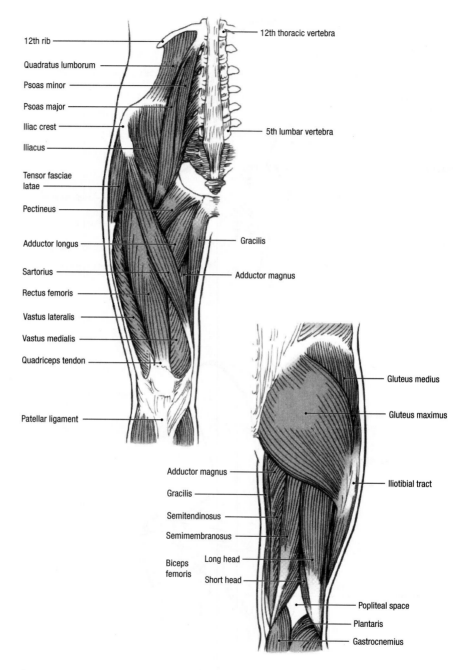

Figure 52. Superficial muscles of the hip and thigh. Anterior (left) and posterior (right) views.

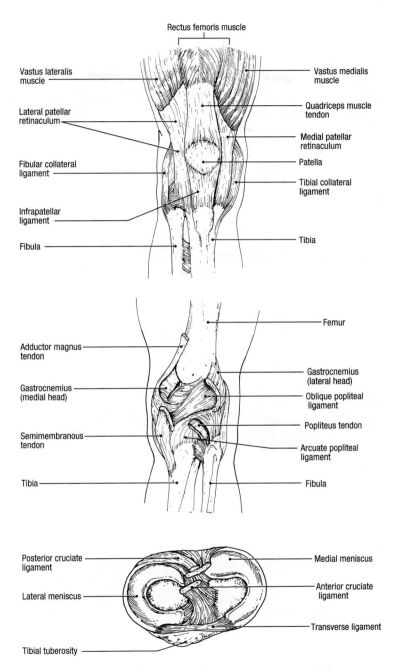

Figure 53. Ligaments of the knee. Anterior view (top), posterior view (middle), and superior view (bottom).

A45

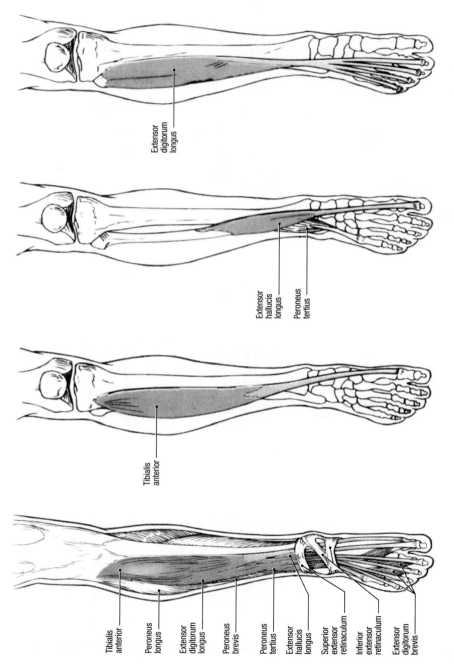

Figure 54. Muscles of the lower leg, anterior compartment, anterior view.

Extensor digitorum longus

Extensor hallucis longus

Peroneus tertius

Tibialis anterior

Tibialis anterior

Peroneus longus

Extensor digitorum longus

Peroneus brevis

Peroneus tertius

Extensor hallucis longus

Superior extensor retinaculum

Inferior extensor retinaculum

Extensor digitorum brevis

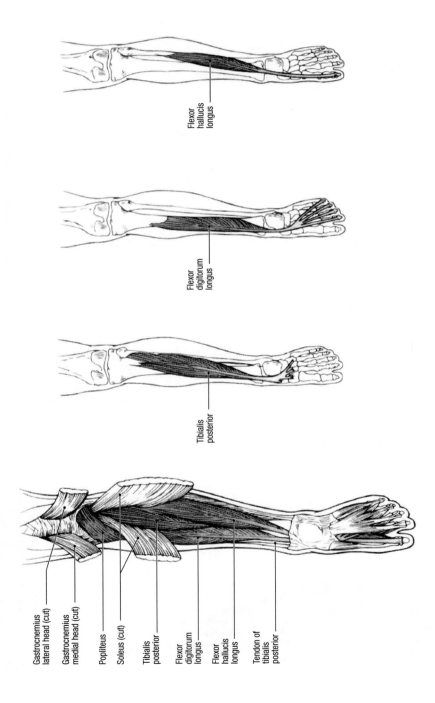

Figure 55. Muscles of the lower leg, deep compartment, posterior view.

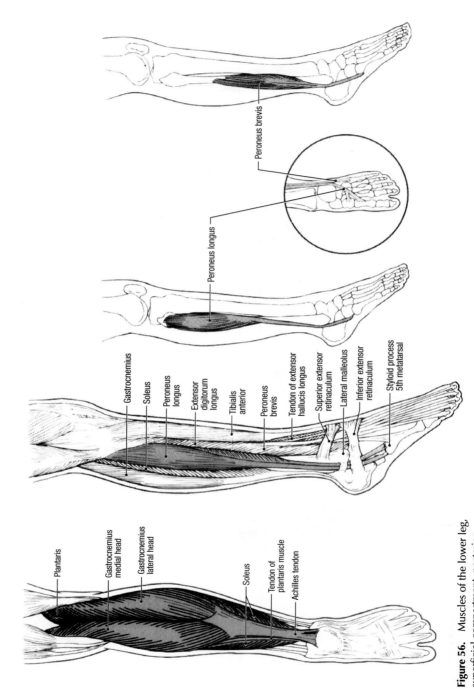

Peroneus brevis

Figure 57. Muscles of the lower leg, lateral compartment, lateral view.

Peroneus longus

Gastrocnemius
Soleus
Peroneus longus
Extensor digitorum longus
Tibialis anterior
Peroneus brevis
Tendon of extensor hallucis longus
Superior extensor retinaculum
Lateral malleolus
Inferior extensor retinaculum
Styloid process 5th metatarsal

Plantaris
Gastrocnemius medial head
Gastrocnemius lateral head
Soleus
Tendon of plantaris muscle
Achilles tendon

Figure 56. Muscles of the lower leg, superficial compartment, posterior view.

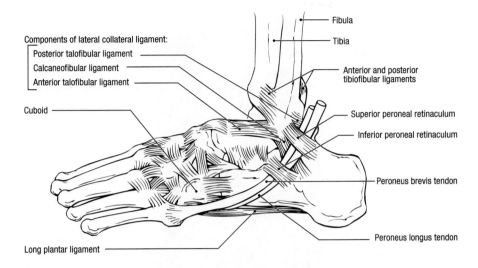

Components of lateral collateral ligament:
Posterior talofibular ligament
Calcaneofibular ligament
Anterior talofibular ligament

Cuboid

Long plantar ligament

Fibula

Tibia

Anterior and posterior
tibiofibular ligaments

Superior peroneal retinaculum

Inferior peroneal retinaculum

Peroneus brevis tendon

Peroneus longus tendon

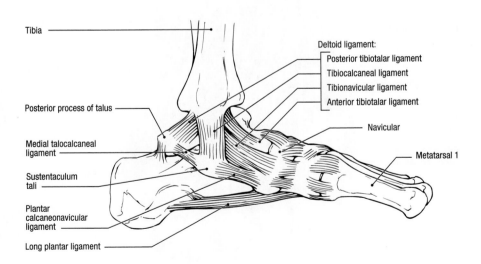

Tibia

Posterior process of talus

Medial talocalcaneal
ligament

Sustentaculum
tali

Plantar
calcaneonavicular
ligament

Long plantar ligament

Deltoid ligament:
Posterior tibiotalar ligament
Tibiocalcaneal ligament
Tibionavicular ligament
Anterior tibiotalar ligament

Navicular

Metatarsal 1

Figure 58. Ligaments of the foot. Lateral view (top) and medial view (bottom).

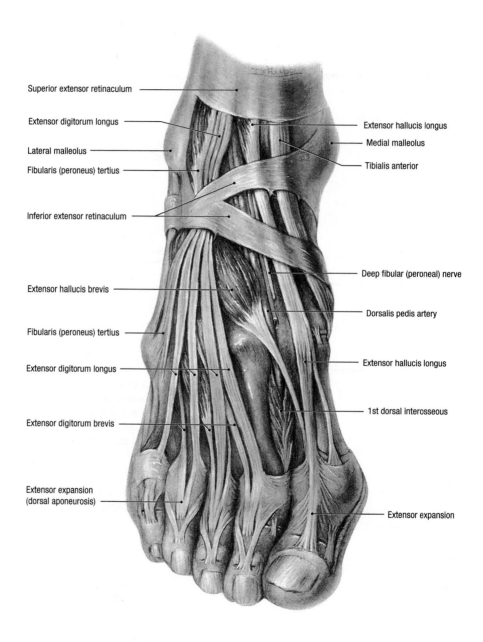

Superior extensor retinaculum

Extensor digitorum longus

Lateral malleolus

Fibularis (peroneus) tertius

Inferior extensor retinaculum

Extensor hallucis brevis

Fibularis (peroneus) tertius

Extensor digitorum longus

Extensor digitorum brevis

Extensor expansion
(dorsal aponeurosis)

Extensor hallucis longus

Medial malleolus

Tibialis anterior

Deep fibular (peroneal) nerve

Dorsalis pedis artery

Extensor hallucis longus

1st dorsal interosseous

Extensor expansion

Figure 59. Dorsum of foot, superior view.

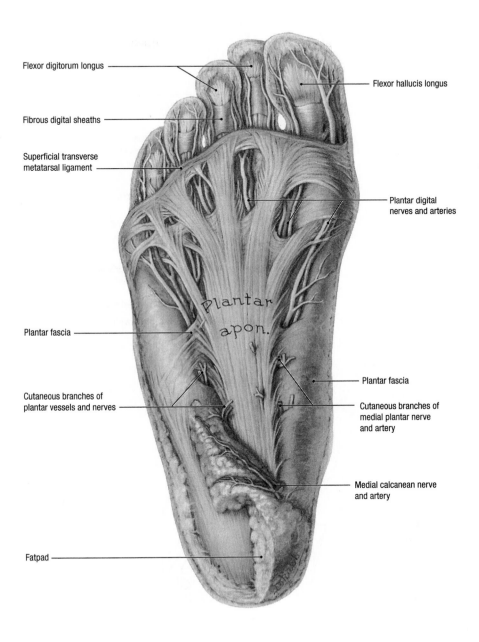

Flexor digitorum longus

Flexor hallucis longus

Fibrous digital sheaths

Superficial transverse metatarsal ligament

Plantar digital nerves and arteries

Plantar fascia

Plantar fascia

Cutaneous branches of plantar vessels and nerves

Cutaneous branches of medial plantar nerve and artery

Medial calcanean nerve and artery

Fatpad

Figure 60. Superficial dissection of the sole of foot.

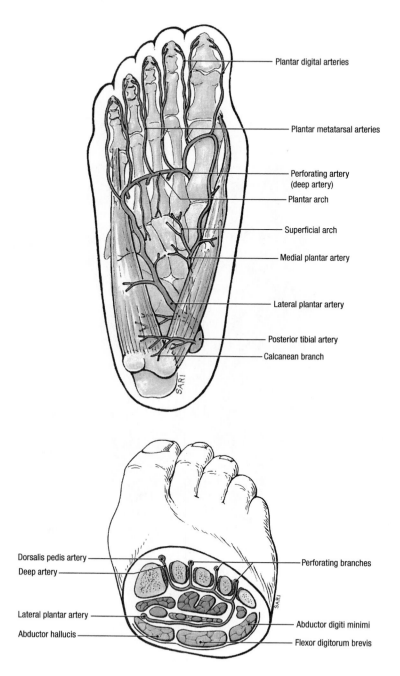

Figure 61. Blood supply of foot. Plantar view (top) and transverse section (bottom).

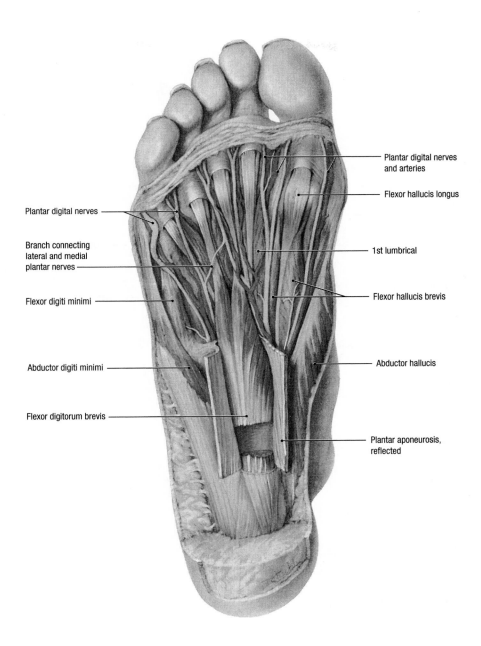

Plantar digital nerves and arteries

Flexor hallucis longus

Plantar digital nerves

Branch connecting lateral and medial plantar nerves

1st lumbrical

Flexor digiti minimi

Flexor hallucis brevis

Abductor digiti minimi

Abductor hallucis

Flexor digitorum brevis

Plantar aponeurosis, reflected

Figure 62. First layer of muscles of sole of foot, digital nerves, and arteries.

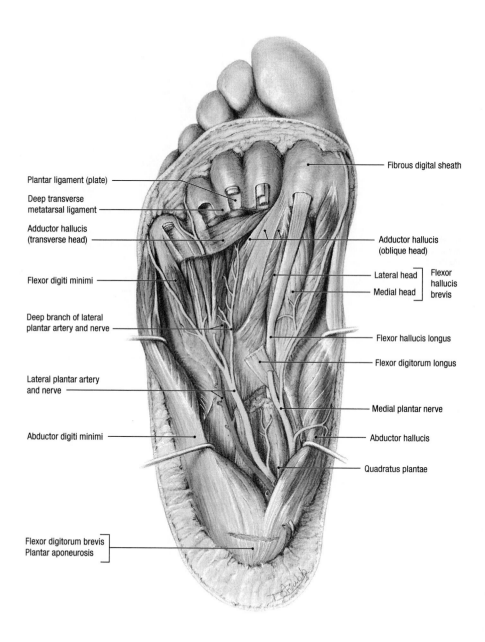

Plantar ligament (plate)

Deep transverse
metatarsal ligament

Adductor hallucis
(transverse head)

Flexor digiti minimi

Deep branch of lateral
plantar artery and nerve

Lateral plantar artery
and nerve

Abductor digiti minimi

Flexor digitorum brevis
Plantar aponeurosis

Fibrous digital sheath

Adductor hallucis
(oblique head)

Lateral head ⎤ Flexor
⎥ hallucis
Medial head ⎦ brevis

Flexor hallucis longus

Flexor digitorum longus

Medial plantar nerve

Abductor hallucis

Quadratus plantae

Figure 63.   Third layer of muscles of sole of foot.

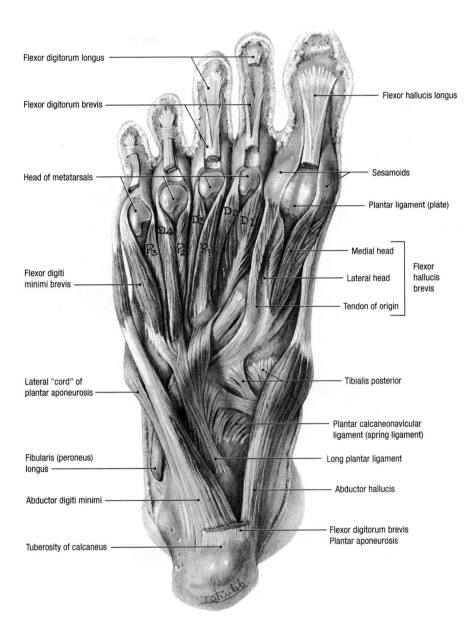

Flexor digitorum longus

Flexor digitorum brevis

Head of metatarsals

Flexor digiti
minimi brevis

Lateral "cord" of
plantar aponeurosis

Fibularis (peroneus)
longus

Abductor digiti minimi

Tuberosity of calcaneus

Flexor hallucis longus

Sesamoids

Plantar ligament (plate)

Medial head

Lateral head

Tendon of origin

Flexor
hallucis
brevis

Tibialis posterior

Plantar calcaneonavicular
ligament (spring ligament)

Long plantar ligament

Abductor hallucis

Flexor digitorum brevis
Plantar aponeurosis

Figure 64. Fourth layer of muscles of sole of foot.

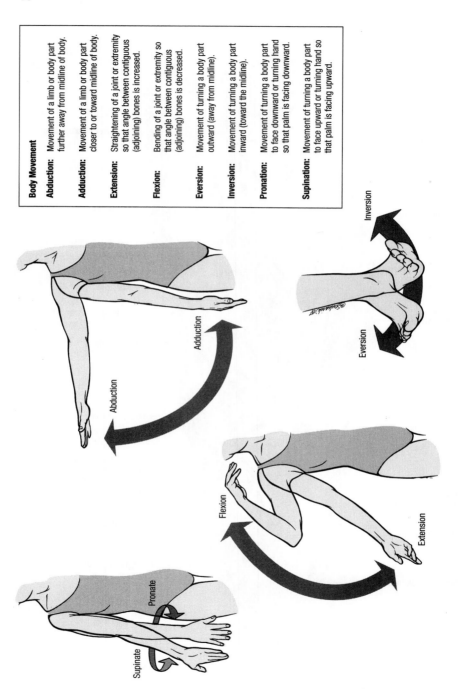

Body Movement

Abduction: Movement of a limb or body part further away from midline of body.

Adduction: Movement of a limb or body part closer to or toward midline of body.

Extension: Straightening of a joint or extremity so that angle between contiguous (adjoining) bones is increased.

Flexion: Bending of a joint or extremity so that angle between contiguous (adjoining) bones is decreased.

Eversion: Movement of turning a body part outward (away from midline).

Inversion: Movement of turning a body part inward (toward the midline).

Pronation: Movement of turning a body part to face downward or turning hand so that palm is facing downward.

Supination: Movement of turning a body part to face upward or turning hand so that palm is facing upward.

Figure 65. Terms of movement.

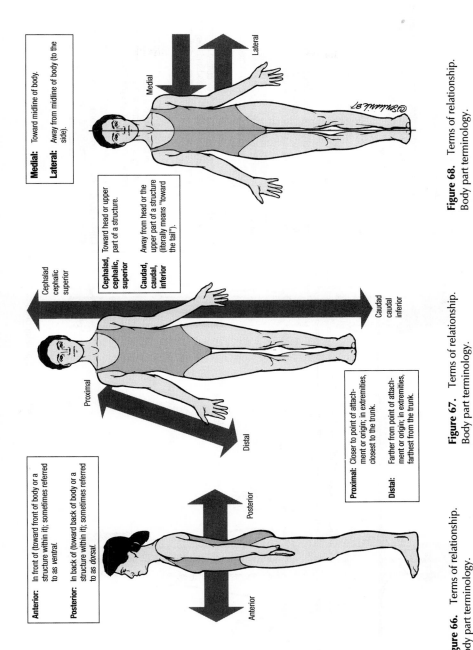

Anterior: In front of (toward front of body or a structure within it); sometimes referred to as *ventral*.

Posterior: In back of (toward back of body or a structure within it); sometimes referred to as *dorsal*.

Proximal: Closer to point of attachment or origin; in extremities, closest to the trunk.

Distal: Farther from point of attachment or origin; in extremities, farthest from the trunk.

Cephalad, cephalic, superior: Toward head or upper part of a structure.

Caudad, caudal, inferior: Away from head or the upper part of a structure (literally means "toward the tail").

Medial: Toward midline of body.

Lateral: Away from midline of body (to the side).

Figure 66. Terms of relationship. Body part terminology.

Figure 67. Terms of relationship. Body part terminology.

Figure 68. Terms of relationship. Body part terminology.

A57

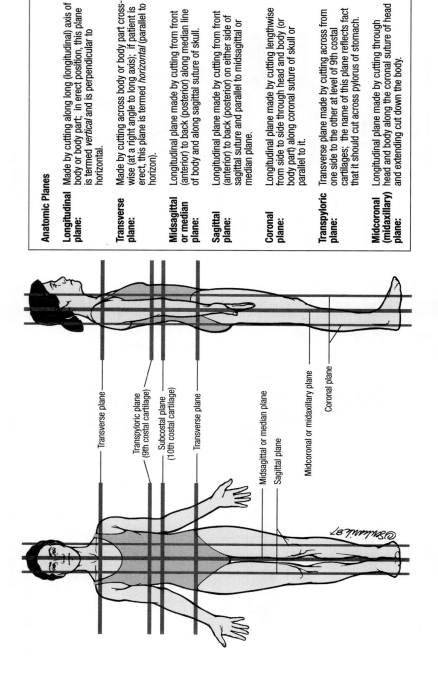

Figure 69. Terms of relationship. Anatomic planes.

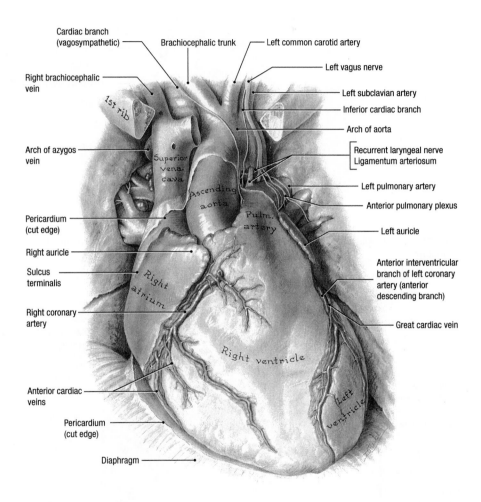

Cardiac branch (vagosympathetic)

Brachiocephalic trunk

Left common carotid artery

Left vagus nerve

Right brachiocephalic vein

1st rib

Left subclavian artery

Inferior cardiac branch

Arch of aorta

Arch of azygos vein

Superior vena cava

Recurrent laryngeal nerve
Ligamentum arteriosum

Ascending aorta

Left pulmonary artery

Anterior pulmonary plexus

Pericardium (cut edge)

Pulm. artery

Left auricle

Right auricle

Right atrium

Sulcus terminalis

Anterior interventricular branch of left coronary artery (anterior descending branch)

Right coronary artery

Great cardiac vein

Right ventricle

Anterior cardiac veins

Left ventricle

Pericardium (cut edge)

Diaphragm

Figure 70. Sternocostal (anterior) surface of the heart and great vessels in situ.

A59

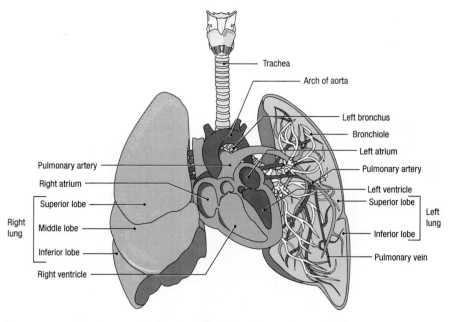

Trachea

Arch of aorta

Left bronchus

Bronchiole

Left atrium

Pulmonary artery

Pulmonary artery

Left ventricle

Right atrium

Superior lobe

Superior lobe

Left lung

Right lung

Middle lobe

Inferior lobe

Inferior lobe

Pulmonary vein

Right ventricle

Figure 71. Cardiopulmonary system shown with cutaway of heart and left lung revealing internal anatomy.

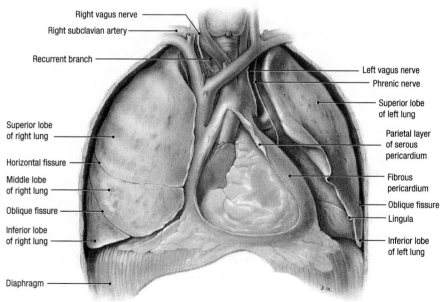

Right vagus nerve

Right subclavian artery

Recurrent branch

Left vagus nerve

Phrenic nerve

Superior lobe
of left lung

Superior lobe
of right lung

Parietal layer
of serous
pericardium

Horizontal fissure

Fibrous
pericardium

Middle lobe
of right lung

Oblique fissure

Oblique fissure

Lingula

Inferior lobe
of right lung

Inferior lobe
of left lung

Diaphragm

Figure 72. Thoracic contents in situ, anterior view.

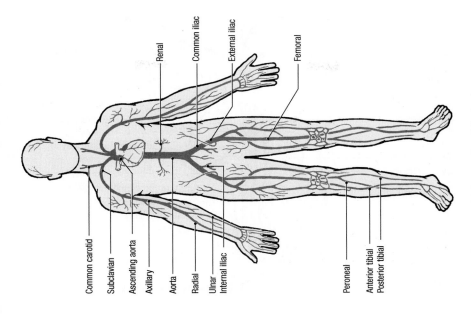

Common carotid

Subclavian

Ascending aorta

Axillary

Aorta

Radial

Ulnar

Internal iliac

Renal

Common iliac

External iliac

Femoral

Peroneal

Anterior tibial

Posterior tibial

Figure 74. Major arteries of the body.

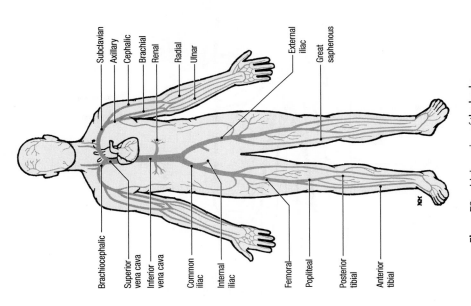

Brachiocephalic

Superior vena cava

Inferior vena cava

Common iliac

Internal iliac

Femoral

Popliteal

Posterior tibial

Anterior tibial

Subclavian

Axillary

Cephalic

Brachial

Renal

Radial

Ulnar

External iliac

Great saphenous

Figure 73. Major veins of the body.

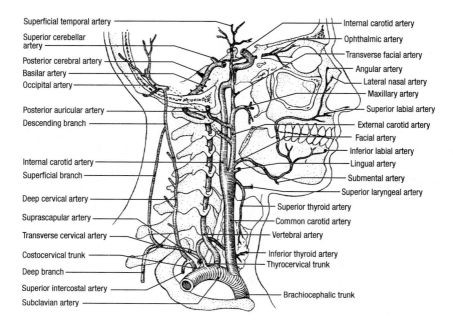

Figure 75. Subclavian and carotid arteries and their branches.

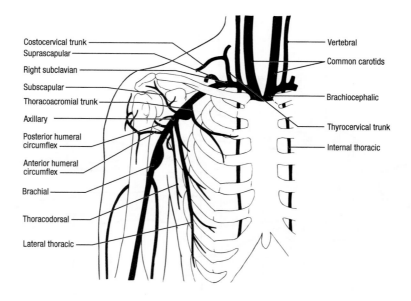

Figure 76. Blood supply to the shoulder.

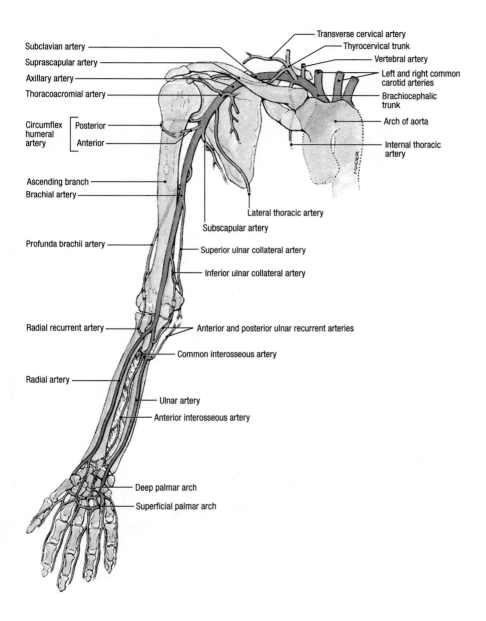

Transverse cervical artery
Thyrocervical trunk
Vertebral artery
Left and right common carotid arteries
Brachiocephalic trunk
Arch of aorta
Internal thoracic artery

Subclavian artery
Suprascapular artery
Axillary artery
Thoracoacromial artery

Circumflex humeral artery — Posterior
Anterior

Ascending branch
Brachial artery

Profunda brachii artery

Radial recurrent artery

Radial artery

Lateral thoracic artery
Subscapular artery

Superior ulnar collateral artery
Inferior ulnar collateral artery

Anterior and posterior ulnar recurrent arteries
Common interosseous artery

Ulnar artery
Anterior interosseous artery

Deep palmar arch
Superficial palmar arch

Figure 77.   Arteries of the upper limb, anterior view.

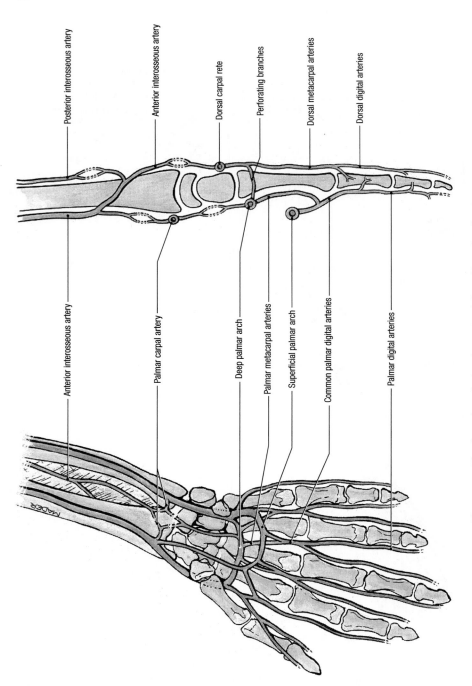

Figure 78. Blood supply to the hand, anterior (left) and lateral (right) views.

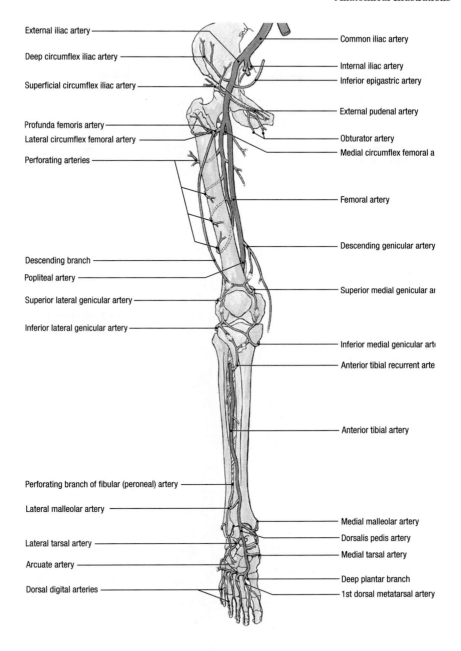

External iliac artery

Deep circumflex iliac artery

Superficial circumflex iliac artery

Profunda femoris artery

Lateral circumflex femoral artery

Perforating arteries

Descending branch

Popliteal artery

Superior lateral genicular artery

Inferior lateral genicular artery

Perforating branch of fibular (peroneal) artery

Lateral malleolar artery

Lateral tarsal artery

Arcuate artery

Dorsal digital arteries

Common iliac artery

Internal iliac artery

Inferior epigastric artery

External pudenal artery

Obturator artery

Medial circumflex femoral a

Femoral artery

Descending genicular artery

Superior medial genicular a

Inferior medial genicular art

Anterior tibial recurrent arte

Anterior tibial artery

Medial malleolar artery

Dorsalis pedis artery

Medial tarsal artery

Deep plantar branch

1st dorsal metatarsal artery

Figure 79. Arteries of the lower limb, anterior view.

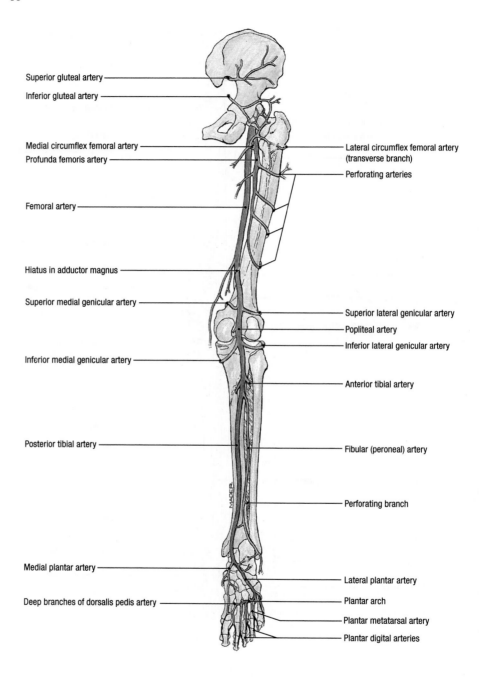

Superior gluteal artery

Inferior gluteal artery

Medial circumflex femoral artery

Profunda femoris artery

Femoral artery

Hiatus in adductor magnus

Superior medial genicular artery

Inferior medial genicular artery

Posterior tibial artery

Medial plantar artery

Deep branches of dorsalis pedis artery

Lateral circumflex femoral artery (transverse branch)

Perforating arteries

Superior lateral genicular artery

Popliteal artery

Inferior lateral genicular artery

Anterior tibial artery

Fibular (peroneal) artery

Perforating branch

Lateral plantar artery

Plantar arch

Plantar metatarsal artery

Plantar digital arteries

Figure 80. Arteries of the lower limb, posterior view.

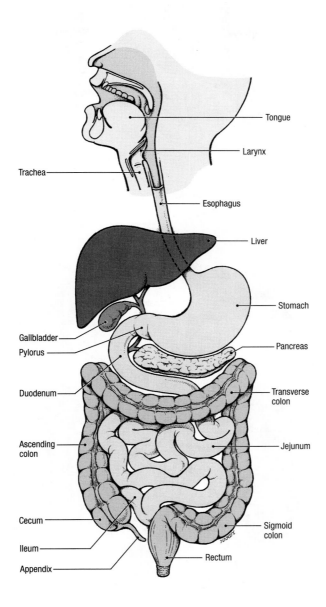

Figure 81. Digestive organs and associated structures.

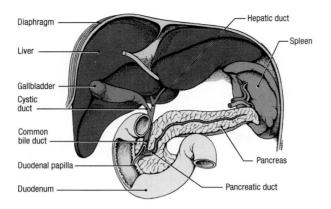

Figure 82. Gallbladder, liver, and biliary system.

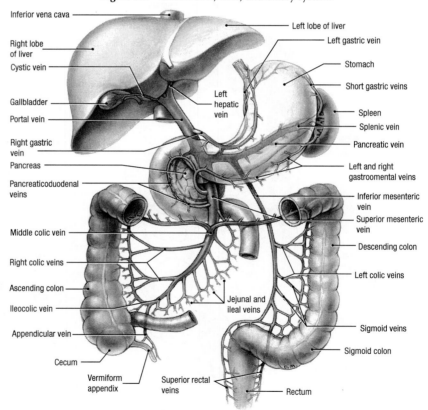

Figure 83. Portal venous system, anterior view.

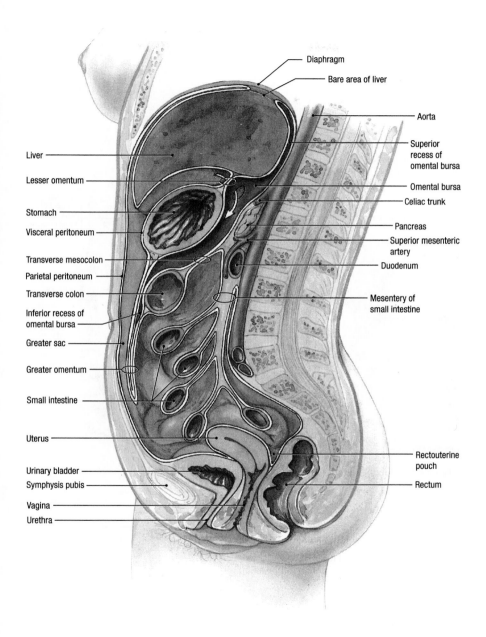

Diaphragm

Bare area of liver

Aorta

Superior recess of omental bursa

Omental bursa

Celiac trunk

Pancreas

Superior mesenteric artery

Duodenum

Mesentery of small intestine

Rectouterine pouch

Rectum

Liver

Lesser omentum

Stomach

Visceral peritoneum

Transverse mesocolon

Parietal peritoneum

Transverse colon

Inferior recess of omental bursa

Greater sac

Greater omentum

Small intestine

Uterus

Urinary bladder

Symphysis pubis

Vagina

Urethra

Figure 84. Peritoneal cavity, median section.

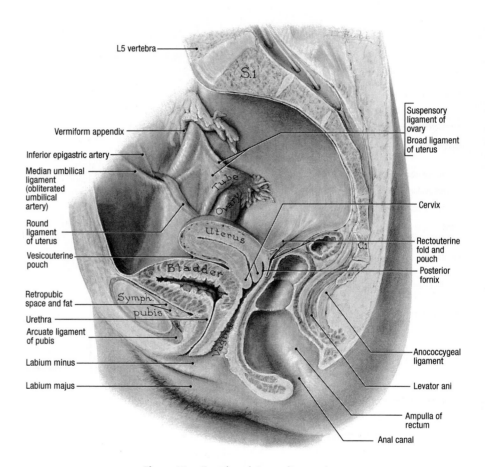

L5 vertebra
S.1

Suspensory ligament of ovary

Broad ligament of uterus

Vermiform appendix

Inferior epigastric artery

Median umbilical ligament (obliterated umbilical artery)

Round ligament of uterus

Vesicouterine pouch

Retropubic space and fat

Urethra

Arcuate ligament of pubis

Labium minus

Labium majus

Tube
Ovary
Uterus
Bladder
Symph. pubis
Vagina
C.1

Cervix

Rectouterine fold and pouch

Posterior fornix

Anococcygeal ligament

Levator ani

Ampulla of rectum

Anal canal

Figure 85. Female pelvis, median section.

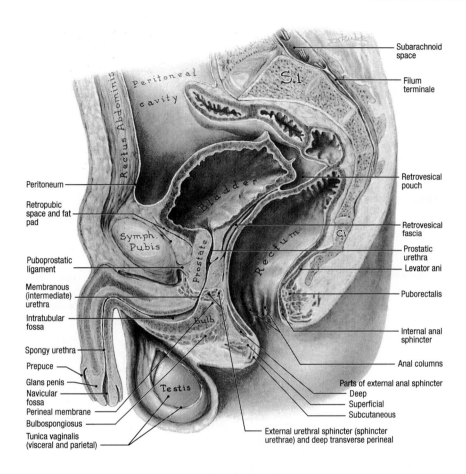

Subarachnoid space

Filum terminale

Retrovesical pouch

Peritoneum

Retropubic space and fat pad

Puboprostatic ligament

Membranous (intermediate) urethra

Intratubular fossa

Spongy urethra

Prepuce

Glans penis

Navicular fossa

Perineal membrane

Bulbospongiosus

Tunica vaginalis (visceral and parietal)

Retrovesical fascia

Prostatic urethra

Levator ani

Puborectalis

Internal anal sphincter

Anal columns

Parts of external anal sphincter
Deep
Superficial
Subcutaneous

External urethral sphincter (sphincter urethrae) and deep transverse perineal

Figure 86. Male pelvis, median section.

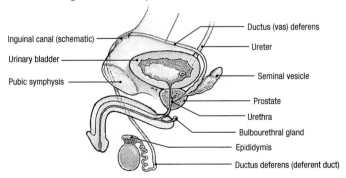

Ductus (vas) deferens

Inguinal canal (schematic)

Ureter

Urinary bladder

Seminal vesicle

Pubic symphysis

Prostate

Urethra

Bulbourethral gland

Epididymis

Ductus deferens (deferent duct)

Figure 87. Overview of the male urogenital system, median section.

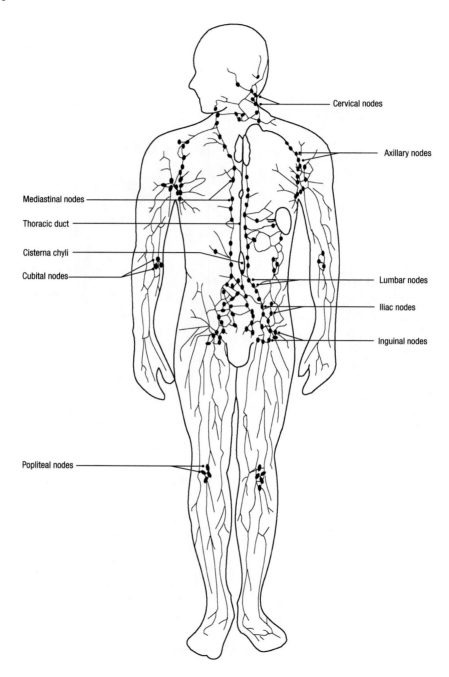

Figure 88. Overview of the major lymphatic channels and groups.

Table of Muscles

1. Muscles of the Shoulder

Muscle	Origin	Insertion	Nerve	Action
Deltoid	Lateral third of clavicle, acromion, and spine of scapula	Deltoid tuberosity of humerus	Axillary, C5, C6	Abducts, adducts, flexes, extends, and rotates arm medially
Infraspinatus	Infraspinous fossa of scapula	Middle facet of greater tubercle of humerus	Suprascapular, C5, C6	Rotates arm laterally; helps to hold humeral head in glenoid cavity of scapula
Latissimus dorsi	Spines of T7-T12 thoracolumbar fascia, iliac crest, ribs 9–12	Floor of bicipital groove of humerus	Thoracodorsal, C6, C7, C8	Extends, adducts, and medially rotates humerus; raises body towards arms during climbing
Pectoralis major	Clavicular head: anterior surface of medial half of clavicle	Lateral lip of intertubercular groove of humerus	Lateral and medial pectoral nerves; clavicular head C5 and C6, sternocostal head C7, C8, and T1	Abducts, medially rotates humerus; draws scapula anteriorly and inferiorly
Pectoralis minor	3rd to 5th ribs near the costal cartilages	Medial border and superior surface of coracoid process of scapula	Medial pectoral nerve C8 and T1	Stabilizes scapula against thoracic wall
Subscapularis	Subscapular fossa	Lesser tubercle of humerus	Upper and lower subscapular, C5, C6, C7	Medially rotates arm and adducts it; helps to hold humeral head in glenoid cavity

(*continued*)

Muscle	Origin	Insertion	Nerve	Action
Supraspinatus	Supraspinous fossa of scapula	Superior facet of greater tubercle of humerus	Suprascapular, C4, C5, C6.	Initiates and assists deltoid in abduction of arm and acts with rotator cuff muscles
Teres minor	Superior portion of lateral border of scapula	Inferior facet of greater tubercle of humerus	Axillary, C5, C6	Laterally rotates arm; helps to hold humeral head in glenoid cavity of scapula
Teres major	Dorsal surface of inferior angle of scapula	Medial lip of intertubercular groove of humerus	Lower subscapular, C6, C7	Adducts and rotates arm medially

2. Muscles of the Arm

Muscle	Origin	Insertion	Nerve	Action
Anconeus	Lateral epicondyle of humerus	Lateral surface of olecranon and superior part of posterior surface of ulna	Radial, C7, C8, and T1	Assists triceps in extending forearm; abducts ulna during pronation
Biceps brachii	Long head, supraglenoid tubercle; short head, coracoid process	Radial tuberosity of radius	Musculocutaneous, C5, C6	Flexes arm and forearm, supinates forearm
Brachialis	Distal anterior surface of humerus	Coronoid process of ulna and ulnar tuberosity	Musculocutaneous, C5, C6	Flexes forearm in all positions
Coracobrachialis	Tip of coracoid process of scapula	Middle third of medial surface of humerus	Musculocutaneous, C5, C6, C7	Flexes and adducts arm

(continued)

Muscle	Origin	Insertion	Nerve	Action
Triceps	Long head, infraglenoid tubercle; lateral head, superior to radial groove of humerus; medial head to radial groove	Posterior surface of olecranon process of ulna	Radial, C6, C7, C8	Chief extensor of forearm at elbow

3. Muscles of the Anterior Forearm

Muscle	Origin	Insertion	Nerve	Action
Flexor carpi radialis	Medial epicondyle of humerus	Bases of second and third metacarpals	Median, C6, C7	Flexes forearm; flexes and abducts hand
Flexor digitorum profundus	Anteromedial surface of ulna, interosseous membrane	Base of distal phalanges of medial four fingers	Ulnar and median, C8, T1	Flexes distal interphalangeal joints and hand
Flexor carpi ulnaris	Humeral head medial epicondyle of humerus; ulnar head; olecranon and posterior border of ulna	Pisiform bone, hook of hamate, and base of fifth metacarpal	Ulnar, C7, C8	Flexes and adducts hand, flexes forearm
Flexor digitorum superficialis	Medial epicondyle, coronoid process of ulna, superior anterior border of radius	Middle phalanges of finger	Median, C7, C8, T1	Flexes proximal interphalangeal joints; flexes hand at wrist
Flexor pollicis longus	Anterior surface of radius, interosseous membrane, and coracoid process	Base of distal phalanx of thumb	Median, C8, T1	Flexes thumb
Palmaris longus	Medial epicondyle of humerus	Distal half of flexor retinaculum, palmar aponeurosis	Median, C7, C8	Flexes hand at wrist and forearm

(continued)

Muscle	Origin	Insertion	Nerve	Action
Pronator quadratus	Anterior surface of distal ulna	Anterior surface of distal radius	Anterior interosseous nerve from median, C8, and T1	Pronates forearm; helps hold radius and ulna together
Pronator teres	Medial epicondyle of humerus and coracoid process of ulna	Middle of lateral surface of radius	Median, C6, C7	Pronates and flexes arm at elbow

4. Muscles of the Posterior Forearm

Muscle	Origin	Insertion	Nerve	Action
Abductor pollicis longus	Interosseous membrane, middle third of posterior surface of radius and ulna	Lateral surface of base of first metacarpal	Radial, deep branch, posterior interosseous nerve C7, C8	Abducts thumb and hand
Articularis cubiti	Distal portion of posterior aspect of shaft of humerus	Posterior fibrous capsule of elbow joint	Radial, C7, C8	Retracts posterior joint capsule during extension of elbow
Brachioradialis	Lateral supracondylar ridge of humerus	Base of radial styloid process	Radial, C5, C6, C7	Flexes forearm
Extensor digitorum	Lateral epicondyle of humerus	Extensor expansion, base of middle and digital phalanges	Radial, deep branch, posterior interosseous nerve C7, C8	Extends fingers and hand at wrist
Extensor digiti minimi	Common extensor tendon and interosseous membrane	Extensor expansion, base of middle and distal phalanges	Radial, deep branch, posterior interosseous nerve C7, C8	Extends little finger
Extensor carpi radialis brevis	Lateral epicondyle of humerus	Posterior base of third metacarpal	Radial, deep branch, C7, C8	Extends fingers and abducts hands at wrist

(continued)

Muscle	Origin	Insertion	Nerve	Action
Extensor carpi radialis longus	Lateral supra-condylar ridge of humerus	Dorsum of base of second meta-carpal	Radial, C6, C7	Extends and abducts hand at wrist and joint
Extensor carpi ulnaris	Lateral epicon-dyle and poste-rior surface of ulna	Base of fifth metacarpal	Radial, deep branch, poste-rior interosseous nerve C7, C8	Extends and adducts hand at wrist joint
Extensor pollicis longus	Interosseous membrane, middle third of posterior sur-face of ulna	Base of distal phalanx of thumb	Radial, deep branch, poste-rior interosseous nerve C7, C8	Extends distal phalanx of thumb and abducts hand
Extensor pollicis brevis	Interosseous membrane and posterior sur-face of middle third of radius	Base of proxi-mal phalanx of thumb	Radial, deep branch, poste-rior interosseous nerve C7, C8	Extends proxi-mal phalanx of thumb and abducts hand
Extensor indicis	Posterior sur-face of ulna and interosseous membrane	Extensor expan-sion of index finger	Radial, deep branch, poste-rior interosseous nerve C7	Extends index finger, helps extend hand
Supinator	Lateral epicon-dyle of hume-rus, radial collateral and annular liga-ments, crest of ulna	Lateral side of upper part of radius	Radial, deep branch, C5, C6	Supinates forearm

5. Muscles of the Hand

Muscle	Origin	Insertion	Nerve	Action
Abductor digiti minimi	Pisiform, piso-hamate liga-ment, flexor retinaculum	Medial side of base of proxi-mal phalanx of little finger	Ulnar, deep branch, C8, T1	Abducts little finger
Abductor pollicis brevis	Flexor retinacu-lum, scaphoid, and trapezium	Lateral side of base of proxi-mal phalanx of thumb	Median, recur-rent branch, C8, T1	Abducts thumb

(continued)

Appendix 2

Muscle	Origin	Insertion	Nerve	Action
Adductor pollicis	Capitate and bases of second and third metacarpals, (oblique head); palmar surface of third metacarpal (transverse head)	Medial side of base of proximal phalanx of the thumb	Ulnar, deep branch, C8, T1	Adducts thumb towards middle digit
Dorsal interossei (4)	Adjacent sides of metacarpal bones	Extensor expansions and bases of phalanges of digits 2–4	Ulnar, deep branch, C8, T1	Abducts fingers; flexes metacarpophalangeal joints; extends interphalangeal joints
Flexor digiti minimi brevis	Flexor retinaculum and hook of hamate	Medial side of base of proximal phalanx of little finger	Ulnar, deep branch, C8, T1	Flexes proximal phalanx of little finger
Flexor pollicis brevis	Flexor retinaculum and trapezium	Base of proximal phalanx of thumb	Median, recurrent branch, C8, T1	Flexes thumb
Lumbricals (4)	1–2 lateral, 3–4 medial side of tendons of flexor digitorum profundus	Lateral side of extensor expansion	Median (two lateral), ulnar (two medial)	Flexes metacarpophalangeal joints, extends interphalangeal joints
Opponens digiti minimi	Flexor retinaculum and hook of hamate	Medial side of fifth metacarpal	Ulnar, deep branch, C8, T1	Opposes little finger with thumb
Opponents pollicis	Flexor retinaculum and tubercles of scaphoid and trapezium	Lateral side of first metacarpal	Median, recurrent branch, C8, T1	Opposes thumb to other digits
Palmar interossei (3)	Palmar surfaces of 2nd, 4th, and 5th metacarpals (unipennate muscles)	Bases of proximal phalanges in same sides as their origins; extensor expansion	Ulnar, deep branch, C8, T1	Adducts fingers; flexes metacarpophalangeal joints; extends interphalangeal joints

(continued)

Muscle	Origin	Insertion	Nerve	Action
Palmaris brevis	Ulnar side of flexor retinaculum, palmar aponeurosis	Skin of ulnar side of hand	Ulnar, superficial, T1	Wrinkles skin on palmar side of hand

6. Anterior Muscles of the Thigh

Muscle	Origin	Insertion	Nerve	Action
Iliacus	Iliac crest, iliac fossa; ala of sacrum	Lesser trochanter, psoas major tendon	Femoral, L2, L4	Flexes and rotates thigh medially with psoas major
Sartorius	Anterior-superior iliac spine, superior part of notch inferior to it	Upper medial side of tibia	Femoral, L2, L3	Flexes, abducts, rotates thigh at hip laterally; flexes, rotates leg at knee joint
Rectus femoris	Anterior-inferior iliac spine; ilium rim of acetabulum	Base of patella; tibial tuberosity	Femoral, L2, L3, L4	Extends leg at knee joint; stabilizes hip; helps iliopsoas flex thigh
Vastus medialis	Intertrochanteric line; linea aspera; medial intermuscular septum	Medial side of patella; tibial tuberosity	Femoral, L2, L3, L4	Extends leg at knee joint
Vastus lateralis	Intertrochanteric line; greater trochanter; linea aspera; gluteal tuberosity; lateral intermuscular septum	Lateral side of patella; tibial tuberosity	Femoral, L2, L3, L4	Extends leg at knee joint
Vastus intermedius	Upper shaft of femur; lower lateral intermuscular septum	Base of patella; by patellar ligament to tibial tuberosity	Femoral	Extends leg at knee joint

(*continued*)

Muscle	Origin	Insertion	Nerve	Action
Abductor longus	Body of pubis below its crest	Middle third of linea aspera of femur	Obturator, branch of anterior division, L2, L3, L4	Adducts, flexes, rotates thigh laterally
Abductor brevis/minimus	Inferior pubic ramus	Pectineal line; uppermost linea aspera of femur	Obturator, L2, L3, L4	Adducts, flexes, rotates thigh laterally
Adductor magnus	Ischiopubic ramus; ischial tuberosity	Linea aspera; medial supracondylar line; adductor tubercle	Obturator, L2, L3, L4; sciatic L4	Adducts, flexes, rotates, and extends thigh
Pectineus	Pectineal line of pubis	Pectineal line of femur	Femoral, L3, L4; obturator	Adducts and flexes thigh; helps with rotation
Gracilis	Body and inferior ramus of pubis	Superior part of medial surface of tibia	Obturator, L2, L3	Adducts and flexes thigh; flexes and rotates leg medially
Obturator externus	Margin of obturator foramen and obturator membrane	Trochanteric fossa of femur	Obturator, L3, L4	Rotates thigh laterally

7. Medial Muscles of the Thigh

Muscle	Origin	Insertion	Nerve	Action
Abductor brevis	Body and inferior ramus of pubis	Pectineal line and proximal part of linea aspera of femur	Obturator, L2, L3, L4	Adducts thigh; aids in flexion
Adductor longus	Body of pubis inferior to pubic crest	Middle third of linea aspera of femur	Obturator L2, L3, L4	Adducts thigh
Adductor magnus	Ischiopubic ramus; ischial tuberosity	Linea aspera; medial supracondylar line; adductor tubercle	Obturator and sciatic	Adducts, flexes, and extends thigh

(continued)

Muscle	Origin	Insertion	Nerve	Action
Gracilis	Body of inferior pubic of ramus	Superior part of medial surface of tibia	Obturator, L2, L3	Adducts thigh; flexes leg; helps rotate thigh medially
Obturator externus	Margin of obturator foramen and obturator membrane	Trochanteric fossa of femur	Obturator, L3, L4	Laterally rotates thigh; steadies head of femur in acetabulum
Pectineus	Pectineal line of pubis	Pectineal line of femur	Femoral L2 and L3; branch of obturator	Adducts and flexes thigh; aids in medial rotation of thigh

8. Muscles of the Gluteal Region

Muscle	Origin	Insertion	Nerve	Action
Coccygeus (ischiococcygeus)	Ischial spine	Inferior end of spine	Branches of S4 and S5 nerves	Forms small part of pelvic diaphragm that supports pelvic viscera and flexes coccyx
Gluteus maximus	Ilium; sacrum; coccyx; sacrotuberous ligament	Gluteal tuberosity; iliotibial tract	Inferior gluteal, L5, S1, S2	Extends and rotates thigh laterally
Gluteus medius	Ilium between iliac crest and anterior and posterior gluteal lines	Greater trochanter	Superior gluteal, L5, S1	Abducts and rotates thigh medially; helps keep pelvis level
Gluteus minimus	Ilium between anterior and posterior gluteal lines	Greater trochanter	Superior gluteal, L5, S1	Abducts and rotates thigh medially; helps keep pelvis level
Inferior gemellus	Ischial tuberosity	Obturator internus tendon	Nerve to quadratus femoris	Rotates thigh laterally *(continued)*

Muscle	Origin	Insertion	Nerve	Action
Obturator internus	Ischiopubic rami; obturator membrane	Greater trochanter	Nerve to obturator internus, L5, S1	Abducts and rotates thigh laterally
Piriformis	Pelvic surface of sacrum; sacrotuberous ligament	Superior border of greater trochanter	Sacral, S1, S2	Rotates thigh medially
Quadratus femoris	Ischial tuberosity	Intertrochanteric crest	Nerve to quadratus femoral, L5, S1	Rotates thigh laterally

9. Posterior Muscles of the Thigh *

Muscle	Origin	Insertion	Nerve	Action
Biceps femoris	Long head from ischial tuberosity; short head from linea aspera and upper supracondylar line	Lateral side of head of fibula; tendon split here by fibular collateral ligament of knee	Tibial (long head), common peroneal; division of sciatic nerve, L5, S1, S2	Flexes leg medially; extends thigh
Semimembranosus	Ischial tuberosity	Medial condyle of tibia	Tibial portion of sciatic, L5, S1, S2	Extends thigh; rotates leg medially; helps raise trunk of body against gravity
Semitendinosus	Ischial tuberosity	Medial surface of superior part of tibia	Tibial division of sciatic nerve, L5, S1, S2	Extends thigh; flexes leg; rotates knee medially when flexed

* These three muscles collectively are called hamstrings.

10. Muscles of the Anterior and Lateral Leg

Muscle	Origin	Insertion	Nerve	Action
Anterior				
Articularis genus	Distal portion of anterior aspect of shaft of femur	Synovial membrane of suprapatellar bursa of knee joint	Femoral, L2-L4	Retracts synovial membrane during extension of the knee
Tibialis anterior	Lateral tibial condyle; interosseous membrane	First cuneiform; first metatarsal	Deep peroneal, L4, L5	Dorsiflexes and inverts foot
Extensor hallucis longus	Middle half of anterior surface of fibula; interosseous membrane	Base of distal phalanx of great toe	Deep peroneal, L5, S1	Extends great toe; dorsiflexes and inverts foot
Extensor digitorum longus	Lateral tibial condyle; upper two-thirds of fibula	Bases of middle and distal phalanges	Deep peroneal, L5, S1	Extends great toe and dorsiflexes ankle
Peroneus tertius	Distal one-third of fibula; interosseous membrane	Base of fifth metatarsal	Deep peroneal	Dorsiflexes and inverts foot
Lateral				
Peroneus brevis	Lower lateral side of fibula; interosseous membrane	Base of fifth metatarsal	Superficial peroneal	Everts and plantar flexes foot
Peroneus longus	Lateral tibial condyle; head and upper lateral side of fibula	Base of first metatarsal; medial cuneiform	Superficial peroneal	Everts and plantar flexes foot

11. Posterior Muscles of the Leg

Muscle	Origin	Insertion	Nerve	Action
Superficial group				
Gastrocnemius	Lateral (head) and medial (head) femoral condyle	Posterior aspect of calcaneus via tendo calcaneus	Tibial, S2, S2	Flexes knee; plantar flexes ankle when knee extended
Soleus	Upper fibular head; soleal line on tibia	Posterior aspect of calcaneus via tendo calcaneus	Tibial, S1, S2	Plantar flexes foot and ankle
Plantaris	Lower lateral supracondylar line and oblique popliteal ligament	Posterior surface of calcaneus	Tibial, S1, S2	Assist gastrocnemius in plantar flexing ankle and flexing knee
Deep group				
Popliteus	Lateral surface of lateral condyle of femur and lateral meniscus	Posterior surface of tibia, superior to soleal line	Tibial, L4, L5, S1	Weakly flexes knee and unlocks knee
Flexor hallucis longus	Inferior two-thirds of posterior surface of fibula and interior part of interosseous membrane	Base of distal phalanx of great toe	Tibial, S2, S3	Flexes great toe at all joints; weakly plantar flexes ankle; supports medial longitudinal arches of foot
Flexor digitorum longus	Medial portion of posterior surface of tibia inferior to soleal line	Base of distal phalanges of lateral four digits	Tibial, S2, S3	Flexes lateral four digits; plantar flexes ankle; supports arches of foot
Tibialis posterior	Interosseous membrane; posterior surface of tibia inferior to soleal line; posterior surface of fibula	Tuberosity of navicular cuneiform and cuboid; base of 2–4 metatarsals	Tibial, L4, L5	Plantar flexes ankle; inverts foot

12. Muscles of the Foot

Muscle	Origin	Insertion	Nerve	Action
Dorsum of foot				
Extensor digitorum brevis	Dorsal surface of calcaneus	Lateral side of long extensor tendons with slips to proximal phalanges 2–4 toes	Deep peroneal, L5, S1	Assist in extending middle three toes
Extensor hallucis brevis	Dorsal surface of calcaneus	Base of proximal phalanx of great toe	Deep peroneal, L5, S1	Extends great toe
Sole of foot				
Abductor digit minimi	Medial and lateral tubercles of calcaneus, plantar aponeurosis and intermuscular septa	Lateral side of base of proximal phalanx of fifth digit	Lateral plantar, S2, S3	Abducts and flexes fifth digit
Abductor hallucis	Medial tubercle of calcaneus; flexor retinaculum and plantar aponeurosis	Medial side of base of proximal phalanx of first digit	Medial plantar, S2, S3	Abducts and flexes great toe
Adductor hallucis; oblique head	Base of metatarsals 2–4	Proximal phalanx of great toe	Deep branch of lateral plantar, S2, S3	Adducts great toe; assists in maintaining transverse arch
Adductor hallucis; transverse head	Capsule of lateral four metatarsophalangeal joints	Tendon of head attached to lateral sides of base of proximal phalanx of first digit	Deep branch of plantar, S2, S3	Adducts great toe; assists in maintaining transverse arch
Dorsal interossei (4)	Adjacent shafts of metatarsals	Proximal phalanges of second toe medial and lateral sides; third and fourth toes lateral sides	Lateral plantar, S2, S3	Abducts toes; flexes and extends proximal and distal phalanges

(continued)

Muscle	Origin	Insertion	Nerve	Action
Flexor digitorum brevis	Medial tubercle of calcaneus, plantar aponeurosis and intermuscular septa	Middle phalanges of lateral four toes	Medial plantar, S2, S3	Flexes middle phalanges of lateral four toes
Flexor digiti minimi brevis	Base of fifth metatarsal	Proximal phalanx of fifth toe	Lateral plantar, S2, S3	Flexes fifth toe
Flexor hallucis brevis	Cuboid; third cuneiform	Proximal phalanx of great toe	Medial plantar, S2, S3	Flexes great toe
Lumbricals (4)	Tendons of flexor digitorum longus	Proximal phalanges; extensor expansion	First by medial plantar nerve; lateral three by lateral plantar nerve, S2, S3	Flexes metatarsophalangeal joints and extends interphalangeal joints
Plantar interossei (3)	Medial sides of metatarsals 3–5	Medial side of base of proximal phalanges 3–5	Lateral plantar, S2, S3	Adducts toes; flexes proximal and extends distal phalanges
Quadratus plantae	Medial and lateral side of calcaneus	Tendons of flexor digitorum longus	Lateral plantar, S2, S3	Assists in flexing toes

13. Muscles of the Thoracic Wall

Muscle	Origin	Insertion	Nerve	Action
External intercostals	Lower border of ribs	Upper border of rib below	Intercostal	Elevates rib in inspiration
Internal intercostals	Lower border of ribs	Upper border of rib below	Intercostal	Depresses ribs; interchondral part elevates ribs
Transverse thoracic	Posterior surface of lower sternum and xiphoid	Inner surface of costal cartilages 2–6	Intercostal	Depresses ribs

(continued)

Muscle	Origin	Insertion	Nerve	Action
Subcostals	Inner surface of lower ribs near their angles	Upper borders of ribs 2 or 3 below	Intercostal	Elevates ribs
Levator costarum	Tips of transverse processes of C7 and T7-T11 vertebrae	Subjacent ribs between tubercle and angle	Dorsal primary rami of C8-T11	Elevates ribs; assists with lateral bending

14. Muscles of the Anterior Abdominal Wall

Muscle	Origin	Insertion	Nerve	Action
External oblique	External surface of lower eight ribs, 5–12	Anterior half of iliac crest; anterior-superior iliac spine; pubic tubercle; linea alba	Intercostal, T7-T11; subcostal T12	Compresses abdomen; flexes trunk; assists in forced expiration
Internal oblique	Lateral two-thirds of inguinal ligament; iliac crest; thoraco-lumbar fascia	Lower four costal cartilages; lineal alba; pubic crest; pectineal line	Intercostal T7-T11; subcostal T12; iliohypogastric and ilioinguinal L1	Compresses abdomen; flexes trunk; assists in forced expiration
Transverse abdominis	Lateral one-third of inguinal ligament; iliac crest; thoraco-lumbar fascia; lower six costal cartilages	Linea alba; pubic crest; pectineal line	Intercostal T7-T12; subcostal T12; iliohypogastric and ilioinguinal L1	Compresses abdomen; depresses ribs
Rectus abdominis	Pubic crest and pubic symphysis	Xiphoid process and costal cartilages 5–7	Intercostal T7-T12	Depresses ribs; flexes trunk
Pyramidalis	Pubic body	Linea alba	Subcostal T12	Tenses linea alba

15. Muscles of the Posterior Abdominal Wall

Muscle	Origin	Insertion	Nerve	Action
Quadratus lumborum	Medial half of inferior border of twelfth rib and tips of lumbar transverse processes	Iliolumbar ligament and internal tip of iliac crest	Ventral branches of T12, L1-L4	Extends, laterally flexes vertebral column; flexes twelfth rib during inspiration
Psoas major	Sides of T12-L5 vertebra and disc; transverse processes	Lesser trochanter of femur	Anterior rami of L1, L2, and L3	Flexes and rotates thigh laterally at hip; flexes lumbar vertebral anteriorly and laterally
Psoas minor	Sides of T12-L1 vertebra and intervertebral disc	Pectineal line, iliopectineal eminence via iliopectineal arch	Anterior rami of L1, L2	Works conjointly with psoas major to flex thigh at hip; stabilizes joint

16. Superficial Muscles of the Back

Muscle	Origin	Insertion	Nerve	Action
Erector spinae	Arises by a broad tendon from posterior part of iliac crest, posterior surface of sacrum, sacral and inferior lumbar spinous processes and supraspinous ligament	Iliocostal—lumborum, thoracis and cervicis; Longissimus-thoracis, cervicis and capitis; spinalis: thoracis, cervicis and capitis	Posterior rami of spinal nerves	Extend vertebral column and head
Interspinales	Superior surface of spinous processes of cervical and lumbar vertebrae	Inferior surface of spinous process of vertebrae superior to vertebrae of origin	Posterior rami of spinal nerves	Assist in extension and rotation of vertebral column

(continued)

Muscle	Origin	Insertion	Nerve	Action
Intertrans-versarii	Transverse process of cervical and lumbar vertebrae	Transverse process of adjacent vertebrae	Posterior and anterior rami of spinal nerves	Assists in lateral bending of vertebral column; stabilizes vertebral column
Latissimus dorsi	Spines of T5-T12	Floor of bicipital groove	Thoracodorsal	Adducts, extends, and rotates arm medially
Levator scapulae	Transverse process of C1-C4	Superior part of medial border of scapula	Dorsal scapular C5; cervical C3-C4	Elevates scapula
Rhomboid major	Spines of T2-T5	Medial border of scapula	Dorsal scapular, C4, C5	Adducts scapula
Rhomboid minor	Spines of C7-T1	Root of spine of scapula	Dorsal scapular, C4, C5	Adducts scapula
Serratus anterior	External surface of lateral parts of 1–8 ribs	Anterior surface of medial border of scapula	Long thoracic nerve, C5, C6, C7	Protracts and rotates scapula
Serratus posterior-superior	Ligamentus nuchae, supraspinal ligament and spines of C7-T3	Superior borders of 2–4 ribs	Intercostal, 2–5	Elevates ribs
Serratus posterior-inferior	Spinal processes of T11 to L2 vertebrae	Inferior border of 8–12 ribs near their angles	Anterior thoracic spinal 9–12	Depresses ribs
Transverso-spinal	Transverse process of C4-T12 vertebrae; multifidus arises from sacrum, ilium, transverse process of T1-T3 and articular process of C4-C7	Thoracis, cervicis, and capitis	Spinal, posterior rami	Extends and rotates vertebral column; stabilizes vertebrae during movement

(*continued*)

A89

Muscle	Origin	Insertion	Nerve	Action
Trapezius	External occipital protuberance, superior nuchal line, ligamentum nuchae, spines of C7–T12	Lateral third of clavicle, acromion, and spine of scapula	Spinal accessory, C3-C4	Adducts, rotates, elevates, and depresses scapula

17. Suboccipital Muscles

Muscle	Origin	Insertion	Nerve	Action
Obliquus capitis inferior	Spine of axis C2	Transverse process of atlas, C1	Suboccipital	Extends and laterally rotates head
Obliquus capitis superior	Transverse process of atlas, C1	Occipital bone above inferior nuchal line	Suboccipital	Extends, rotates, and laterally flexes head
Rectus capitis posterior major	Spine of axis, C2	Lateral portion of inferior nuchal line	Suboccipital	Extends, rotates, and laterally flexes head
Rectus capitis posterior minor	Posterior tubercle of atlas, C1	Occipital bone below inferior nuchal line	Suboccipital	Extends, rotates, and laterally flexes head

18. Muscles of the Neck

Muscle	Origin	Insertion	Nerve	Action
Cervical muscles				
Platysma	Superficial fascia over upper part of deltoid and pectoralis major	Mandible; skin and muscles over the mandible and angle of mouth	Facial, CN VII	Depresses lower jaw and lip and angle of mouth; wrinkles skin of neck

(continued)

Muscle	Origin	Insertion	Nerve	Action
Sternocleido-mastoid	Lateral surface of mastoid process of temporal bone and lateral half of superior nuchal line	Mastoid process and lateral one-half of superior nuchal line	Spinal accessory, C2-C3	Tilts head; laterally flexes and rotates face to opposite side, raises thorax

Suprahyoid muscles

Muscle	Origin	Insertion	Nerve	Action
Digastric	Anterior belly from digastric fossa of mandible posterior belly from mastoid notch	Intermediate tendon attached to body of hyoid	Anterior belly: mylohyoid nerve, branch of alveolar nerve; Posterior belly: facial nerve CN 7	Depresses mandible and elevates hyoid and tongue
Mylohyoid	Mylohyoid line of mandible	Median raphe and body of hyoid bone	Mylohyoid and trigeminal, CN 3	Elevates hyoid and tongue; depresses mandible
Styloid	Styloid process	Body of hyoid	Facial, CN 7	Elevates and retracts hyoid
Geniohyoid	Genial tubercle of mandible	Body of hyoid	C1 via the hypoglossal nerve	Pulls hyoid bone antero-superiorly, shortens floor of mouth; widens pharynx

Infrahyoid muscles

Muscle	Origin	Insertion	Nerve	Action
Omohyoid	Inferior belly from medial lip of suprascapular notch and supra-scapular liga-ment; superior belly from intermediate tendon	Inferior belly to intermediate tendon; superior belly to body of hyoid	Ansa cervicalis, C1-C3	Depresses and retracts hyoid and larynx
Sternohyoid	Manubrium sterni and medial end of clavicle	Body of hyoid	Ansa cervicalis, C1-C3	Depresses hyoid and larynx

(*continued*)

Muscle	Origin	Insertion	Nerve	Action
Sternothyroid	Manubrium sterni; first costal cartilage	Oblique line of thyroid cartilage	Ansa cervicalis, C1-C3	Depresses thyroid cartilage and larynx
Thyrohyoid	Oblique line of thyroid cartilage	Body and greater horn of hyoid	C1 via hypoglossal nerve	Depresses and retracts hyoid and larynx

19. Prevertebral Muscles

Muscle	Origin	Insertion	Nerve	Action
Lateral vertebral				
Anterior scalene	Transverse process of C4-C6 vertebrae	First rib	Cervical, C4, C5, C6	Elevates first rib; laterally flexes and rotates neck
Middle scalene	Posterior tubercles of transverse processes of C4-C5 vertebrae	Superior surface of first rib, posterior groove for subclavian artery	Cervical spine, anterior rami	Elevates first rib during forced inspiration; flexes neck laterally
Posterior scalene	Posterior tubercles of transverse processes of C4-C5 vertebrae	External border of second rib	Anterior rami of cervical spine, C7, C8	Elevates second rib during forced inspiration; flexes neck laterally
Anterior vertebral				
Longus capitis	Basilar part of occipital bone	Anterior tubercles of C3-C6 transverse processes	Spinal, anterior rami, C1-C3	Flexes
Longus colli	Anterior tubercle of C2 vertebra; bodies of C1-C3 and transverse processes of C3-C6 vertebrae	Bodies of C5-T3 vertebrae, transverse processes of C3-C5 vertebrae	Spinal, anterior rami, C2-C6	Flexes and rotates head to opposite side

(*continued*)

Muscle	Origin	Insertion	Nerve	Action
Rectus capitis anterior	Anterior surface of lateral mass of atlas, C1	Base of skull, anterior to occipital condyle	C1 and C2	Flexes head
Rectus capitis lateralis	Transverse processes of C1	Jugular process of occipital bone	C1 and C2	Flexes head; helps stabilize head
Rectus capitis posterior	Spinous processes of C2 vertebra	Middle of inferior nuchal line of occipital bone	Suboccipital	Extends head
Splenic capitis, et cervicis	Inferior half of ligamentum nuchae, spinous processes of C7-T3 of T4 vertebrae	Splenius capitis: superolaterally to mastoid process of temporal bone, lateral third of superior nuchal line of occipital bone Splenius cervicis: posterior tubercles of transverse C1-C3 or C4 vertebrae		

20. Muscles of Facial Expression

Muscle	Origin	Insertion	Nerve	Action
Auricularis anterior, posterior, and superior	Epicranial aponeurosis and mastoid part of temporal bone	Auricle (external ear)	Facial	Protraction, retraction, and elevation of external ear
Buccinator	Mandible; pterygo-mandibular raphe; alveolar processes	Angle of mouth	Facial, CN 7	Presses cheek against molar teeth to aid in chewing

(*continued*)

Appendix 2

Muscle	Origin	Insertion	Nerve	Action
Corrugator supercilii	Medial supra-orbital margin	Skin of medial eyebrow	Facial, CN 7	Draws eyebrow medially and inferiorly producing vertical wrinkles above nose
Depressor anguli oris	Oblique line of mandible	Angle of mouth	Facial, CN 7	Depresses angle of mouth
Depressor labii inferioris	Mandible below mental foramen	Orbicularis oris and skin of lower lip	Facial, CN 7	Depresses lower lip
Depressor septi	Incisor fossa of maxilla	Mobile part of nasal septum	Facial	Helps dilate nostril during inspiration; depresses nasal septum
Levator angulioris	Canine fossa of maxilla	Angle of mouth	Facial, CN 7	Elevates angle of mouth medially
Levator labii superioris	Maxilla above infraorbital foramen	Skin of upper lip and alar cartilage of nose	Facial, CN 7	Elevates upper lip, dilates nose
Levator labii superioris alaeque nasi	Frontal process of maxilla	Skin of upper lip	Facial	Elevates ala of nose and upper lip
Mentalis	Incisor fossa of mandible	Skin of chin	Facial, CN 7	Elevates and protrudes lower lip
Nasalis	Maxilla lateral to incisor fossa	Nasal cartilages	Facial, CN 7	Draws ala (side) of nose toward nasal septum
Occipito-frontalis	Superior nuchal line; upper orbital margin	Epicranial aponeurosis	Facial, CN 7	Elevates eyebrows; wrinkles forehead

(*continued*)

Muscle	Origin	Insertion	Nerve	Action
Orbicularis oculi	Medial orbital margin; medial palpebral ligament; lacrimal bone	Skin and rim of orbit; tarsal plate; lateral palpebral raphe	Facial, CN 7	Closes and/or squints eyelids
Procerus	Nasal bone and cartilage	Skin between eyebrows	Facial, CN 7	Depresses medial end of eyebrow; produces wrinkles over nose
Zygomaticus major	Zygomatic arch	Angle of mouth	Facial, CN 7	Draws angle of mouth backwards and upwards
Zygomaticus minor	Zygomatic arch	Angle of mouth	Facial, CN 7	Elevates upper lip

21. Muscles of Mastication

Muscle	Origin	Insertion	Nerve	Action
Lateral pterygoid	Superior head from infratemporal surface of sphenoid; inferior head from lateral surface of lateral pterygoid plate	Neck of mandible; articular disk and capsule of temporomandibular joint	Trigeminal, CN V3	Protracts and depresses mandible; produces side-to-side movements of mandible
Masseter	Lower border and medial surface of zygomatic arch	Lateral surface of coronoid process, ramus and angle of mandible	Trigeminal, CN V3	Elevates mandible
Medial pterygoid	Tuber of maxillary; medial surface of lateral pterygoid plate; pyramidal process of palatine bone	Medial surface of angle and ramus of mandible	Trigeminal, CN V3	Protracts, protrudes, and elevates mandible, closes jaw; produces grinding motion

22. Muscles of Eye Movement

Muscle	Origin	Insertion	Nerve	Action
Ciliary	Scleral spur	Meridional, radial, and circular fibers are intrinsic to ciliary body	Parasympathetic fibers of oculomotor nerve and ciliary ganglion	Relieve tension on lens of eye, allowing it to become more convex for near vision
Inferior rectus	Common tendinous ring	Sclera just behind cornea	Oculomotor, CN 3	Depresses, adducts, and rotates eyeball medially
Lateral rectus	Common tendinous ring	Sclera just behind cornea	Abducens, CN 6	Abducts eyeball
Medial rectus	Common tendinous ring	Sclera just behind cornea	Oculomotor, CN 3	Adducts eyeball
Superior rectus	Common tendinous ring	Sclera just behind cornea	Oculomotor, CN 3	Elevates, adducts, and rotates eye medially
Inferior oblique	Floor of orbit lateral to lacrimal groove	Sclera beneath lateral rectus	Oculomotor, CN 3	Rotates eyeball upward and laterally, elevates adducted eye
Superior oblique	Body of sphenoid bone above optic canal	Sclera beneath superior rectus	Trochlear, CN 4	Levator palpebrae superioris
Abducts, depresses, and medially rotates eyeball	Lesser wing of sphenoid above and anterior to optic canal	Tarsal plate and skin of upper eyelid	Oculomotor, CN 3	Elevates upper eyelid

23. Muscles of the Palate

Muscle	Origin	Insertion	Nerve	Action
Levator veli palatini	Petrous part of temporal bone; cartilage of auditory tube	Aponeurosis of soft palate	Vagus via pharyngeal plexus, CN 10, 11	Elevates soft palate

(continued)

Muscle	Origin	Insertion	Nerve	Action
Musculus uvulae	Posterior nasal spine of palatine bone; palatine aponeurosis	Mucous membrane of uvula	Vagus nerve via pharyngeal plexus, CN 10, 11	Elevates uvula
Palato-pharyngeus	Hard palate and palatine aponeurosis	Lateral wall of pharynx	Cranial part of accessory nerve CN 11 through pharyngeal branch of vagus nerve CN 10 via pharyngeal plexus	Tenses soft palate; moves walls of pharynx superiorly, anteriorly, and medially during swallowing
Tensor veli palatini	Scaphoid fossa; spine of sphenoid; cartilage of auditory tube	Palatine aponeurosis	Mandibular branch nerve, CN V3, via otic ganglion	Tenses soft palate and opens auditory tube during swallowing and yawning

24. Muscles of the Tongue

Muscle	Origin	Insertion	Nerve	Action
Genioglossus	Superior part of mental spine of mandible	Dorsum of tongue and body of hyoid bone	Hypoglossal, CN 12	Depresses tongue; pulls tongue anteriorly for protrusion
Hyoglossus	Body of greater horn of hyoid bone	Side and inferior aspect of tongue	Hypoglossal, CN 12	Depresses and retracts tongue
Inferior long muscle of tongue	Root of tongue and body of hyoid bone	Apex of tongue	Hypoglossal, CN 12	Curls tip of tongue inferiorly and shortens tongue
Palatoglossus	Palatine aponeurosis	Side of tongue	Cranial part of accessory nerve CN 12 through pharyngeal branch of vagus nerve CN 10 via pharyngeal plexus	Elevates posterior tongue and draws soft palate onto tongue

(continued)

Muscle	Origin	Insertion	Nerve	Action
Superior long muscle of tongue	Submucous fibrous layer and median fibrous septum	Margins of tongue and mucous membrane	Hypoglossal, CN 12	Curls tip and sides of tongue superiorly and shortens tongue
Transverse muscle of tongue	Median fibrous septum	Fibrous tissue at margins of tongue	Hypoglossal, CN 12	Narrows and elongates tongue; aids in protrusion of tongue
Vertical muscle of tongue	Superior surface of borders of tongue	Inferior surface of borders of tongue	Hypoglossal, CN 12	Flattens tongue; aids in protrusion of tongue

25. Muscles of the Pharynx

Muscle	Origin	Insertion	Nerve	Action
Circular muscles				
Crico-pharyngeus	Posterolateral cricoid cartilage on one side	Posterolateral cricoid cartilage of other side	Vagus, CN 10	Serves as upper esophageal sphincter
Geniohyoid	Inferior mental spine of mandible	Body of hyoid bone	C1 via hypoglossal	Pulls hyoid bone superiorly; shortens floor of mouth; widens pharynx
Inferior constrictor	Arch of cricoid and oblique line of thyroid cartilage	Median raphe of pharynx	Vagus via pharyngeal plexus; recurrent and external laryngeal, CN 10, 11	Constricts lower pharynx
Middle constrictor	Greater and lesser horns of hyoid; stylohyoid ligament	Median raphe	Vagus via pharyngeal plexus, CN 10, 11	Constricts lower pharynx

(continued)

Muscle	Origin	Insertion	Nerve	Action
Superior constrictor	Medial pterygoid plate; pterygoid hamulus; pterygomandibular raphe; mylohyoid line of mandible; side of tongue	Median raphe and pharyngeal tubercle of skull	Vagus via pharyngeal plexus	Constricts upper pharynx

Longitudinal muscles

Muscle	Origin	Insertion	Nerve	Action
Palato-pharyngeus	Hard palate; aponeurosis of soft palate	Thyroid cartilage and muscles of the pharynx	Vagus via pharyngeal plexus, CN 10, 11	Elevates pharynx and closes nasopharynx
Salpingo-pharyngeus	Cartilage of auditory tube	Muscles of the pharynx	Vagus via pharyngeal plexus	Elevates nasopharynx; opens auditory tube
Stylo-pharyngeus	Styloid process	Thyroid cartilage and muscles of the pharynx	Glossopharyngeal, CN 9	Elevates pharynx and larynx

26. Muscles of the Larynx

Muscle	Origin	Insertion	Nerve	Action
Aryepiglottic	Apex of arytenoid cartilage	Side of epiglottic cartilage	Recurrent laryngeal	Adducts
Cricothyroid	Arch of cricoid cartilage	Inferior horn and lower lamina of thyroid cartilage	External laryngeal	Tenses and stretches vocal fold
Lateral crico-arytenoid	Arch of cricoid cartilage	Muscular process of arytenoid cartilage	Recurrent laryngeal, CN 10	Adducts
Oblique arytenoid	Muscular process of arytenoid cartilage	Apex of opposite arytenoid	Recurrent laryngeal, CN 10	Closes inter-cartilaginous portion of rima glottidis

(continued)

A99

Muscle	Origin	Insertion	Nerve	Action
Posterior crico-arytenoid	Posterior surface of lamina of cricoid cartilage	Muscular process of arytenoid cartilage	Recurrent laryngeal, CN 10	Abducts
Thyro-arytenoid	Inner surface of thyroid lamina	Anterolateral surface of arytenoid cartilage	Recurrent laryngeal, CN 10	Adducts; relaxes vocal fold
Thyroepiglottic	Anteromedial surface of lamina of thyroid cartilage	Lateral margin of epiglottic cartilage	Recurrent laryngeal, CN 10	Adducts
Transverse arytenoid	Posterior surface of arytenoid cartilage	Opposite arytenoid cartilage	Recurrent laryngeal, CN 10	Adducts
Vocalis	Anteromedial surface of lamina of thyroid cartilage	Vocal process	Recurrent laryngeal, CN 10	Relaxes posterior vocal ligaments; maintains tension of anterior part of ligament

27. Summary of Autonomic Ganglia of the Head and Neck

Ganglion	Location	Parasympathetic Fibers	Sympathetic Fibers	Chief Distribution
Ciliary	Behind eyeball between optic nerve and lateral rectus muscle	Inferior division oculomotor nerve via short ciliary nerves	Internal carotid artery, long ciliary nerve	Ciliary muscle and sphincter pupillae (parasympathetic); dilator pupillae and tarsal muscle (sympathetic)
Pterygopalatine	In pterygopalatine fossa below maxillary nerve, lateral to the sphenopalatine foramen and anterior pterygoid canal	Facial nerve, greater petrosal nerve, and pterygoid nerve	Internal carotid plexus	Nasal, palatine, and lacrimal glands via maxillary, zygomatic, and lacrimal nerves

(continued)

Ganglion	Location	Parasympathetic Fibers	Sympathetic Fibers	Chief Distribution
Submandibular	Lateral surface of hypoglossus muscle, deep to the mylohyoid muscle, suspended from the lingual nerve	Facial nerve, chorda tympani and lingual nerve	Plexus on facial aretery	Submandibular and sublingual glands
Otic	Below foramen ovale	Glossopharyngeal nerve, its tympanic branch, lesser petrosal nerve	Plexus on middle meningeal	Parotid gland

28. Muscles of the Ears

Muscle	Origin	Insertion	Nerve	Action
Stapedius	Internal walls of pyramidal eminence of posterior wall of tympanic cavity	Neck of the stapes	Facial, cranial nerve V2	Dampens vibrations of stapes reflexively in response to loud nose
Tensor tympani	Canal for tensor tympani of petrous part of temporal bone and cartilage of pharyngotympanic (auditory) tube	Handle of malleus	Branch of mandibular nerve, cranial nerve V3 via otic ganglion	Tenses tympanic membrane to dampen excessive vibration

29. Cranial Nerves

Nerve	Cranial Exit	Cell Bodiess	Components	Chief Function
I: Olfactory	Cribriform plate	Nasal mucosa	SVA	Smell
II: Optic	Optic canal	Ganglion cells of retina	SSA	Vision

(*continued*)

Nerve	Cranial Exit	Cell Bodiess	Components	Chief Function
III: Oculomotor	Superior orbital fissure	Nucleus CN III (midbrain)	GSE	Eye movements (superior, inferior, and medial recti, inferior oblique, and levator palpebrae superioris muscles)
		Edinger-Westphal nucleus (midbrain)	GVE	Constriction of pupil (sphincter pupillae muscle) and accommodation (ciliary muscle)
IV: Trochlear	Superior orbital fissure	Nucleus CN IV (midbrain)	GSE	Eye movements (superior oblique muscle)
V: Trigeminal	Superior orbital fissure; foramen rotundum and foramen ovale	Motor nucleus CN V (pons)	SVE	Muscles of mastication, (mylohyoid, anterior belly of digastric, tensor veli palatini, and tensor tympani muscles)
		Trigeminal ganglion	GSA	Sensation in head (skin and mucous membranes of face and head)
VI: Abducens	Superior orbital fissure	Nucleus CN VI (pons)	GSE	Eye movement (lateral rectus muscle)
VII: Facial	Stylomastoid foramen	Motor nucleus CN VII (pons)	SVE	Muscle of facial expression (posterior belly of digastric stylohyoid and stapedius muscles)

(continued)

Nerve	Cranial Exit	Cell Bodiess	Components	Chief Function
		Salivatory nucleus (pons)	GVE	Lacrimal and salivary secretion
		Geniculate ganglion	SVA	Taste from anterior two-thirds of tongue and palate
		Geniculate ganglion	GVA	Sensation from palate
		Geniculate ganglion	GSA	Sensation from external acoustic means
VIII: Vestibulo-cochlear	Does not leave skull	Vestibular ganglion	SSA	Equilibrium, hearing
IX: Glosso-pharyngeal	Jugular foramen	Nucleus ambiguus (medulla)	SVE	Elevation of pharynx (stylopharyngeus muscle)
		Dorsal nucleus (medulla)	GVE	Secretion of saliva (parotid gland)
		Inferior ganglion	GVA	Sensation in carotid sinus and body, tongue, and pharynx
		Inferior ganglion	SVA	Taste from posterior one-third of tongue
		Inferior ganglion	GSA	Sensation in external and middle ear
X: Vagus	Jugular foramen	Nucleus ambiguus	SVE	Muscles of movements of pharynx, larynx, and palate

(*continued*)

Nerve	Cranial Exit	Cell Bodiess	Components	Chief Function
		Dorsal nucleus (medulla)	GVE	Involuntary muscle and gland control in thoracic and abdominal viscerae
		Inferior ganglion	GVA	Sensation in pharynx, larynx, and other viscerae
		Inferior ganglion	SVA	Taste from root of tongue and epiglottis
		Superior ganglion	GSA	Sensation in external ear and external acoustic meatus
XI: Accessory	Jugular foramen	Spinal cord (foramen)	SVE	Movement of head and shoulder (sternocleidomastoid and trapezius muscles)
XII: Hypoglossal	Hypoglossal canal	Nucleus CN XII (medulla)	GSE	Muscles of movements of tongue

From Chung KW. Gross anatomy, 4th ed. Baltimore: Lippincott Williams & Wilkins, 2000.

Table of Ligaments and Tendons

Shoulder/Upper Arm

Latin name	English name	Articulation
La. acromioclaviculare	acromioclavicular l.	Connects acromion to clavicle; strengthens articular capsule
La. annulare radii	annular l. of radius	Connects head of radius in radial notch
La. collaterale ulnare	collateral ulnar l.	Connects medial epicondyle to humerus and coronoid process of ulna and olecranon
La. conoideum	conoid l.	Connects coracoid process of scapula to clavicle
La. coracoacromiale	coracoacromial l.	Connects coracoid process to acromion
La. coracoclaviculare	coracoclavicular l.	Connects coracoid process of scapula to clavicle
La. coracohumerale	coracohumeral l.	Connects coracoid process of scapula to humerus
La. costoclaviculare	costoclavicular l.	Connects 1st costal cartilage to clavicle
La. glenohumeralia	glenohumeral ligs.	Connects articular capsule of humerus to glenoid cavity and anatomical neck of humerus
La. interclaviculare	interclavicular l.	Connects clavicle to opposite clavicle
La. orbiculare	annular l. of radius	Ligament that encircles and holds the head of the radius in the radial notch of the ulna
La. sternoclaviculare anterius	anterior sternoclavicular l.	Fibrous band that reinforces the sternoclavicular joint anteriorly

(continued)

Abbreviations used: l., ligament; La. ligamenta; ligs., ligaments.

Shoulder/Upper Arm

Latin name	English name	Articulation
La. sternoclavicular posterius	posterior sternoclavicular l.	Fibrous band that reinforces the sternoclavicular joint posteriorly
La. suspensorium axillae	suspensory l.	Connects between the clavipectoral fascia downward to the axillary fascia
La. transversum humeri	transverse humeral l.	Connects obliquely from the greater to the lesser tuberosity of the humerus
La. transversum scapulae inferius	inferior transverse l.	Connects scapula to glenoid cavity; creates foramen of scapula for vessels/nerves
La. transversum scapulae superius	superior transverse l.	Connects coracoid process to scapular notch of scapula
La. trapezoideum	trapezoid l.	Connects coracoid process to clavicle

Hand/Forearm

Latin name	English name	Articulation
La. anulare radii	annular l. of radius	Connects radius to ulna
La. carpi radiatum	radiate l. of wrist	Multiple fibrous bands on palmar surface of metacarpal joint
La. carpi transversum	transverse carpal l.	Continuous with antebrachial fascia
La. carpi volare	transverse carpal l.	Reinforcing fibers in antebrachial fascia, palmar surface of wrist
La. carpometacarpalia dorsalia	dorsal carpometacarpal ligs.	Join carpal bones to bases of metacarpals
La. carpometacarpalia palmaria	palmar carpometacarpal metacarpals	Join carpal bones to metacarpals ligs.
La. collateralia articulationum interphalangealium manus	collateral ligs. of interphalangeal articulations	Fibrous bands on each side of interphalangeal joints of fingers

(continued)

Hand/Forearm

Latin name	English name	Articulation
La. collateralia articulationum metacarpophalangealium	collateral ligs. of metacarpophalangeal articulations	Fibrous bands on sides of each metacarpophalangeal joint
La. collaterale carpi radiale	radial carpal collateral l.	Connects styloid process of radius to scaphoid
La. collaterale carpi ulnare	ulnar carpal collateral l.	Connects styloid process of ulna to triquetral and pisiform bones
La. collaterale radiale	collateral radial l.	Connects lateral epicondyle of humerus to annular l. of radius
La. intercarpalia dorsalia interossea	dorsal intercarpal ligs.	Connect carpal bones together
La. intercarpalia interossea	interosseous intercarpal ligs.	Connect various carpal bones
La. intercarpalia palmaria	palmar intercarpal ligs.	Connect various carpal bones
La. metacarpalia dorsalia	dorsal metacarpal ligs.	Interconnects bases of metacarpal bones
La. metacarpalia interossea	interosseous metacarpal ligs.	Interconnects bases of metacarpal bones
La. metacarpalia palmaria	palmar metacarpal ligs.	Interconnects bases of metacarpals
La. metacarpeum transversum profundum	deep transverse metacarpal l.	Interconnects heads of metacarpals
La. metacarpale transversum superficiale	superficial transverse metacarpal l.	Between longitudinal bands of palmar aponeurosis.
La. natatorium	superficial transverse metacarpal l.	Thickening of the deep fascia in most distal part of the base of the triangular palmar

(continued)

Hand/Forearm

Latin name	English name	Articulation
La. palmaria	palmar l.	Connects anterior aspect of each metacarpophalangeal and interphalangeal joint of the hand
La. palmaria articulationum	palmar ligs. of interphalangealium interphalangeal	Interphalangeal articulations of hand between collateral articulations
La. palmaria articulationis metacarpophalangeae	palmar l. of metacarpal joint	Connects metacarpophalangeal joints to the collateral ligs.
La. pisohamatum	pisohamate l.	Connects pisiform bone to hook of hamate bone
La. pisometacarpeum	pisometacarpal l.	Connects pisiform bone to bases of metacarpals
La. quadratum	quadrate l.	Connects radial notch of ulna to neck of radius
La. radiocarpale dorsale	dorsal radiocarpal l.	Connects radius to carpal bones
La. radiocarpale palmare	palmar radiocarpal l.	Connect radius to lunate, triquetral, capitate, and hamate bones
La. ulnocarpale palmare	palmar ulnocarpal l.	Connects styloid process of ulna to carpal bones

Head/Neck

Latin name	English name	Articulation
La. annulare stapedis	annular l. of stapes	Connects stapes to fenestra vestibuli
La. annularia	annular l. of trachea	Connects adjacent tracheal cartilages
La. articulare anterius	anterior l. of auricle	Connects zygomatic process to helix
La. articulare posterius	anterior l. of auricle	Connects mastoid process to conchal eminence

(continued)

Head/Neck

Latin name	English name	Articulation
La. articulare superius	superior l. of auricle	Connects osseous external acoustic meatus to helix
La. ceratocricoideum	ceratocricoid l.	One of three ligs. reinforcing the cricothyroid articulation capsule
La. corniculopharyngeal	cricopharyngeal l.	Connects corniculate cartilage and cricoid cartilage
La. cricoarytenoideum posterius	cricoarytenoid l.	Connects arytenoid cartilage to lamina of cricoid cartilage
La. cricopharyngeum	cricopharyngeal l.	Connects tip of corniculate cartilage and lamina of cricoid cartilage
La. cricotracheale	cricotracheal l.	Connects cricoid cartilage with first ring of trachea
La. hyaloideo-capsulare	hyalocapsular l.	Connects vitreous body to the posterior surface of the lens of the eye
La. hyoepiglotticum	hyoepiglottic l.	Connects epiglottis to the upper border of the hyoid bone
La. hyothyroideum laterale	lateral thyroid l.	Connects superior horn of thyroid cartilage to tip of greater horn of hyoid cartilage
La. hyothyroideum medium	median thyroid l.	Central portion of the thyroid membrane.
La. incudis posterius	posterior l. of incus	Ligamentus band extending from short crus of incus
La. incudis superius	superior l. of incus	Connects body of incus with root at tympanic recess
La. intracapsularia	intrascapular l.	Ligaments located within and separate from the articular capsule of synovial joint

continued)

Head/Neck

Latin name	English name	Articulation
La. jugale	cricopharyngeal l.	Connects tip at corniculate cartilage and the lamina at the cricoid cartilage and pharyngeal mucosa
La. laterale articulationis temporomandibularis	lateral l. of temporo-mandibular joint	Capsular ligament that passes down and backward across lateral surface of temporomandibular joint
La. mallei anterius	anterior l. of malleus	Connects base of anterior process to spine of sphenoid
La. mallei laterale	lateral l. of malleus	Connects the posterior half of the tympanic notch to neck of malleus
La. mallei superius	superior l. of malleus	Connects from the head of the malleus to epitympanic recess
La. mediale articulationis temporomandibularis	medial l. of temporo-mandibular joint	Strengthens the medial part of the articular capsule
La. ossiculorum auditus	l. of auditory ossicles	Connects the ear bone with each other and with walls of the tympanic cavity
La. palpebrale externum	lateral palpebral l.	Connects tarsal plates to orbital eminence of zygomatic bone
La. palpebrale laterale	lateral palpebral l.	Connects tarsal plate to orbital eminence of zygomatic bone
La. palpebral mediale	medial palpebral l.	Connects medial ends of tarsal plates to maxilla at medial orbital margin
La. sphenomandibulare	sphenomandibular l.	Connects from the spine to sphenoid bone to lingula of mandible
La. spirale cochleae	spiral l. of cochlear duct	Forms outer wall of cochlear duct to which the basal lamina attaches

(*continued*)

Head/Neck

Latin name	English name	Articulation
La. spirale ductus cochlearis	spiral l. of cochlear duct	Forms outer wall of cochlear duct to which the basal lamina attaches
La. stylohyoideum	stylohyoid l.	Connects from the tip of the styloid process to the lesser cornu of the hyoid bone
La. stylomandibulare	stylomandibular l.	Connects from the tip of the styloid process to the temporal bone
La. suspensorium bulb	suspensory l.	Connects between the lateral and medial orbital margins
La. suspensorium glandulae thyroideae	suspensory l.	Connects from the sheath of the thyroid gland to the thyroid and cricoid cartilages
La. tarsale externum	lateral palpebral l.	Connects the tarsal plates to the orbital eminence of the zygomatic bone
La. tarsale internum	medial palpebral l.	Connects between the medial ends of the tarsal plates to the maxilla at the medial orbital margin
La. temporomandibular	lateral temporomandibular l.	Capsular l. that passes obliquely down and backward across the lateral surface of the temporomandibular joint
La. thyroepiglotticum	thyroepiglottic l.	Connects the petiole of the epiglottis to the interior of the thyroid cartilage
La. thyrohyoideum laterale	lateral thyroid l.	Connects the superior horn of the thyroid cartilage to the tip of the greater horn of the hyoid cartilage
La. thyrohyoideum medium	median thyrohyoid l.	Central thickened portion of the thyroid membrane
La. trachealia	annular l.	Connects adjacent tracheal cartilages

(continued)

Head/Neck

Latin name	English name	Articulation
La. ventriculare	vestibular l.	Inferior border of the quadrangular membrane that underlies the ventricular fold of the larynx
La. vestibulare	vestibular l.	Inferior border of the quadrangular membrane that underlies the ventricular fold of the larynx
La. vocale	vocal l.	Connects on either side from the thyroid cartilages to the vocal process of the arytenoid cartilages

Thorax/upper abdomen

Latin name	English name	Articulation
La. colli costae	costotransverse l.	Connects neck of rib to corresponding transverse process
La. costotransversarium anterius	superior costotransverse l.	Connects transverse rib to next highest vertebra
La. costotransversarium laterale	lateral costotransverse l.	Connects tip of transverse process to neck of rib
La. costotransversarium posterius	lateral costotransverse l.	Connects tip of process to neck of ribs
La. costotransversarium superius	superior costotransverse l.	Connects neck of ribs to transverse process of next higher vertebrae
La. costoxiphoideum	costoxiphoid l.	Connects xiphoid process to 7th and often 6th cartilages
La. pulmonale	pulmonary l.	Two-layered fold formed as the pleura of the mediastinum is reflected onto the lung inferior to the root of the lung
La. sternocostale intraarticulare	intraarticular sternocostal l.	Connects between a costal cartilage and the sternum within the articular capsule

(continued)

Thorax/upper abdomen

Latin name	English name	Articulation
La. sternocostalia radiata	radiate sternocostal l.	Fibers of the articular capsule that radiate from the costal cartilages to the anterior surface of the sternum
La. sternopericardiaca	sternopericardial l.	Connects from the pericardium to the sternum
La. suspensoria mammaria	suspensory l. of breast	Connects from the fibrous stroma of the mammary gland to the overlying skin
La. tuberculi costae	lateral costotransverse l.	Connects the tip of the transverse process to the posterior surface of the neck and the rib
La. vena cava sinistrae	left vena caval l.	Connects from the left brachiocephalic vein to the oblique vein of the left atrium

Spine

Latin name	English name	Articulation
La. alaria	alar l.	Connects axis to occiput; limits rotation of head
La. apicis dentis axis	apical dental l.	Connects axis to occiput
La. atlantooccipitale laterale	lateral atlantooccipital l.	Connects occiput to atlas
La. capitis costae intraarticulare	interarticular l. of head of rib	Connects crest of rib to intervertebral disk
La. capitis costae radiatum	radiate l. of head of rib	Connects head of rib to adjacent vertebrae/disks
La. caudale integumenti communis	caudal retinaculum	Forms coccygeal foveola
La. costotransversarium	costotransverse l.	Connects neck of rib to transverse process of corresponding vertebra

(continued)

Spine

Latin name	English name	Articulation
La. costotransversarium laterale	lateral costotransverse l.	Connects transverse process of vertebra to corresponding rib
La. costotransversarium superius	superior costotransverse l.	Connects neck of rib to transverse process of vertebra above
La. cruciforme atlantis	cruciform l. of atlas	Connects transverse l. of atlas to longitudinal fascicles
La. flava	yellow ligs.	Joins laminae of 2 adjacent vertebrae
La. iliofemorale	iliofemoral l.	Connects anterior/inferior iliac spine and intertrochanteric femur
La. iliolumbale	iliolumbar l.	Connects L4-L5 to iliac crest
La. interspinalia	interspinal ligs.	Interconnects spinous processes
La. intertransversaria	intertransverse ligs.	Interconnects vertebral transverse processes
La. longitudinale anterius	ant. longitudinal l.	Extends from occiput/atlas to sacrum
La. longitudinale posterius	post. longitudinal l.	Extends from occiput to coccyx
La. lumbocostale	lumbocostal l.	Connects 12th rib to transverse processes of L1-L2
La. nuchae	radiate l.	Connects head of each rib to bodies of the two vertebrae with which it articulates
La. sacrococcygeum anterius	anterior sacrococcygeal l.	Connects sacrum to coccyx
La. sacrococcygeum laterale	lateral sacrococcygeal l.	Connects 1st coccygeal vertebra to sacrum; completes foramen of S-5

(continued)

Spine

Latin name	English name	Articulation
La. sacrococcygeum posterius	deep posterior sacrococcygeal l.	Terminal portion of posterior longitudinal l.; unites S-5 and profundum coccyx
La. sacrococcygeum posterius	superficial posterior sacrococcygeal l.	Connects sacral hiatus to coccyx superficiale
La. sacroiliaca anteriora	anterior sacroiliac ligs.	Connects sacrum to ilium
La. sacroiliaca interossea	interosseous sacroiliac ligs.	Numerous bundles connecting tuberosities of sacrum to those of ilium
La. sacroiliaca posteriora	posterior sacroiliac ligs.	Connects ilium and iliac spines to sacrum
La. sacrospinalum	sacrospinal l.	Connects ischium to lateral margins of sacrum
La. sacrotuberale	sacrotuberal l.	Connects ischial tuberosity to sacrum and coccyx and iliac spine
La. supraspinale	supraspinal l.	Interconnects tips of spinous processes of vertebrae
La. transversum atlantis	transverse l. of atlas	Horizontal portion of cruciform l. of atlas

Abdominal/Pelvic

Latin name	English name	Articulation
La. arcuatum laterale	lateral arcuate l.	Connects first lumbar vertebrae and 12th rib to diaphragm
La. arcuatum mediale	medial arcuate l.	Connects body of first lumbar vertebra to transverse process
La. arcuatum medianum	median arcuate l.	Connects crura of diaphragm that arches over aorta
La. arcuatum pubis	inferior pubic l.	Arches across pubic symphysis

(continued)

A115

Abdominal/Pelvic

Latin name	English name	Articulation
La. cardinale	cardinale l.	Connects uterine, cervix and vault of lateral fornix of vagina
La. coronarium hepatis	coronary l. of liver	Connects peritoneal reflections to diaphragm at margins of bare area of liver
La. duodenorenale	duodenorenal l.	Connects termination of hepatoduodenal to front of right kidney
La. falciforme	falciform process of sacro-tuberous l.	Passes from ischial tuberosity to ilium, sacrum, and coccyx
La. falciforme hepatis	falciform l. of liver	Connects liver to diaphragm and anterior abdominal wall
La. fundiforme clitoris	fundiform l. of clitoris	Connects linea alba with fascia of the clitoris
La. fundiforme penis	fundiform l. of penis	Connects linea alba with fascia of penis
La. gastrophrenicum	gastrocolic l.	Connects stomach with transverse colon
La. gastrophrenicum	gastrophrenic l.	Connects greater curvature of stomach with inferior surface of diaphragm
La. gastrosplenicum	gastrosplenic l.	Connects greater curvature of stomach with ilium of spleen
La. genitoinguinale	genitoinguinal l.	In a fetus, a fold of mesorchium containing gubernaculum testis
La. hepatocolicum	hepatocolic l.	Connects hepatoduodenal l. to transverse colon
La. hepatoesophageum	hepatoesophageal l.	Connects between the liver and the part of the esophagus
La. hepatogastricum	hepatogastric l.	Connects liver to lesser curvature of stomach

(continued)

Abdominal/Pelvic

Latin name	English name	Articulation
La. hepatorenale	hepatorenal l.	A prolongation of the coronary ligs. downward over the right kidney
La. ischiocapsulare	ischiofemoral l.	Connects from the ischium upward and laterally over the femoral neck
La. lacunare	lacunar l.	Connects from medial end of the inguinal ligament to the pectineal line
La. laterale vesicae	lateral bladder l.	Passes from one side of the bladder to blend with the pelvic fascia
La. pectineale	pectineal l.	A strong fibrous band that passes laterally from the lacunar l. along the pectineal line of the pubis
La. phrenicocolicum	phrenicocolic l.	Connects from the left flexure of the colon to the diaphragm
La. phrenicolienal	phrenosplenic l.	Connects between the diaphragm and the spleen
La. phrenicosplenicum	phrenicosplenic l.	Connects between the diaphragm and the spleen
La. pubicum inferius	inferior pubic l.	Arches across the inferior aspect of the pubic symphysis
La. pubicum superius	superior pubic l.	Passes transversely above the pubic symphysis
La. pubofemorale	pubofemoral l.	Connects from the superior ramus of the pubis to the intertrochanteric femur
La. puboprostaticum	puboprostatic l.	Anchors the prostate and neck of the bladder to the pubis on each side

(*continued*)

Abdominal/Pelvic

Latin name	English name	Articulation
La. puboprostaticum mediale	puboprostatic l.	Anchors the prostate and neck of the bladder to the pubis on each side
La. pubovesicale	pubovesical l. (female)	Fascial thickening comparable with puboprostatic l.
La. pubovesicale	pubovesical l. (male)	Connects between the lower part of the pubic symphysis and the prostate and bladder
La. sacrodurale	sacrodural l.	Connects between the midline of the inferior part of the dorsal sac to the posterior longitudinal l. of the sacrum.
La. sacroiliacum posterius	posterior sacroiliac l.	Connects from the ilium to the sacrum posterior to the sacroiliac joint
La. sacrospinale	sacrospinal l.	Connects between the ischial spine and the sacrum and coccyx
La. serosum	serous l.	Connects certain viscera to the abdominal wall or to each other
La. splenorenale	splenorenal l.	Extends from the anterior aspect of the left kidney to the splenic hilum
La. suspensorium clitoris	suspensory l.	Connects from the pubic symphysis to the deep fascia of the clitoris
La. suspensorium ovarii	suspensory l.	Extends upward from the upper pole of the ovary
La. suspensorius penis	suspensory l.	Connects from the pubic symphysis to the deep fascia of the penis
La. teres hepatis	round l.	Connects from umbilicus to the liver where it continues to the origins of the left portal vein

(continued)

Abdominal/Pelvic

Latin name	English name	Articulation
La. teres uteri	round l.	Attached to uterus on either side of the front and below the opening of the uterine tube and connects to the labium majus
La. transversale cervicis	cardinal l.	Connects the uterine cervix and the vault of the lateral fornix of the vagina
La. transversum pelvis	transverse perineal l.	Thickened anterior border of the perineal membrane
La. triangulare dextrum hepatis	right triangular l.	Connects from the right lobe of the liver to the diaphragm
La. triangulare sinistrum hepatis	left triangular l.	Connects from the left lobe of the liver to the diaphragm

Hip/Thigh

Latin name	English name	Articulation
La. capitis femoris	l. of head of femur	Connects femur, acetabular notch, and transverse l. of acetabulum
La. inguinale	inguinal l.	Connects ilium to pubis
La. ischiofemorale	ischiofemoral l.	Connects ischium to femur
La. transversum acetabuli	transverse l. of acetabulum	Connects acetabular lip of hip joint to acetabular notch

Knee/Calf

Latin name	English name	Articulation
La. capitis fibulae anterius	anterior l. of head of fibula	Connects head of fibula to lateral condyle of tibia
La. capitis fibulae posterius	posterior l. of head of fibula	Connects head of fibula to lateral condyle of tibia
La. collaterale fibulare	collateral fibular l.	Connects lateral epicondyle of femur to head of fibula

(continued)

Knee/Calf

Latin name	English name	Articulation
La. collateral tibiale	collateral tibial l.	Connects medial epicondyle of femur to medial meniscus and tibia
La. cruciatum anterius genus	anterior cruciate l. of knee.	Connects lateral condyle of femur to condylar eminence of tibia
La. cruciata genus	cruciate ligs. of knee	Bundles in knee joint between condyles of femur
La. cruciatum posterius genus	posterior cruciate l. of knee	Connects medial condyle of femur to intercondylar area of tibia
La. menisci lateralis	posterior meniscofemoral l.	Connects between the medial condyle of the femur to the posterior crus of the lateral meniscus
La. meniscofemorale anterius	anterior meniscofemoral l.	Connects lateral meniscus to posterior cruciate l.
La. meniscofemorale posterius	posterior meniscofemoral l.	Connects lateral meniscus to medial condyle of femur
La. patellae	patellar l.	Connects patella to tibial tuberosity
La. popliteum arcuatum	arcuate popliteal l.	Connects fibula to articular capsule
La. popliteum obliquum	oblique popliteal l.	Connects medial condyle of tibia to lateral epicondyle of femur
La. teres femoris	head of femur l.	Connects from the fovea in the head of the femur to the borders of the acetabular notch
La. tibiofibulare anterius	anterior tibiofibular l.	Connects tibia to fibula
La. tibiofibulare posterius	posterior tibiofibular l.	Connects tibia to distal fibula

(continued)

Knee/Calf

Latin name	English name	Articulation
La. tibionaviculare	medial l.	Connects from medial malleolus of the tibial downward to the tarsal bones
La. transversum genus	transverse l. of knee	Connects lateral meniscus to medial meniscus

Foot and Ankle

Latin name	English name	Articulation
La. bifurcatum	bifurcate l.	Dorsum of foot; comprises calcaneonavicular and calcaneocuboid ligs.
La. calcaneocuboideum	calcaneocuboid l.	Connects calcaneus to cuboid
La. calcaneocuboideum plantare	plantar calcaneocuboid l. short plantar l.	Connects calcaneus to cuboid
La. calcaneofibulare	calcaneofibular l.	Connects fibula to calcaneus
La. calcaneonaviculare	calcaneonavicular l.	Connects calcaneus to navicular bone
La. calcaneonaviculare dorsale	dorsal calcaneonavicular l.	Connects calcaneus to navicular bone
La. calcaneonaviculare plantare	plantar calcaneonavicular l.	Connects sustentaculum tali to navicular; supports talus
La. calcaneotibiale	calcaneotibial l.	Connect medial malleolus to sustentaculum tali of calcaneus
La. collateralia articulationum	collateral ligs. of metatarsophalangeal articulations	Fibrous bands on sides of each metatarsophalangeal joint
La. cruciatum cruris	inferior extensor of foot	Joins malleolus to dorsum of foot
La. cuboideonaviculare dorsale	dorsal cuboideonavicular l.	Connects cuboid and navicular bones

(*continued*)

Foot and Ankle

Latin name	English name	Articulation
La. cuboideonaviculare plantare	plantar cuboideonavicular l.	Connects cuboid and navicular bones
La. cuneocuboideum dorsale	dorsal cuneocuboid l.	Connects cuboid and lateral cuneiform bones
La. cuneocuboideum interosseum	interosseus cuneocuboid l.	Connects cuboid and lateral cuneiform bones
La. cuneocuboideum plantare	plantar cuneocuboid l.	Connects cuboid and lateral cuneiform bones
La. cuneometatarsalia interossea	interosseous cuneometatarsal ligs.	Connects cuneiform and metatarsal bones
La. cuneonavicularia dorsalia	dorsal cuneonavicular ligs.	Connects navicular and cuneiform bones
La. cuneonavicularia plantaria	plantar cuneonavicular ligs.	Connects navicular to cuneiform bones
La. intercuneiformia dorsalia	dorsal intercuneiform ligs. cuneiform bones	Connects dorsal surfaces of
La. intercuneiformia interossea	interosseous intercuneiform ligs.	Connects adjacent cuneiform bones
La. intercuneiformia plantaria	plantar intercuneiform ligs.	Joins plantar surfaces of cuneiform bones
La. laterale articulationis talocruralis	lateral l. of ankle joint	Lateral side of ankle joint
La. mediale articulationis talocruralis	medial l. of ankle	Connects medial malleolus of tibia to tarsal bones
La. meniscofemoralia	meniscofemoral ligs.	Connects from posterior part of lateral meniscus to the lateral surface of the medial meniscus
La. metatarsale transversum profundum	deep transverse metatarsal l.	Joins heads of metatarsals
La. metatarsale transversum superficiale	superficial transverse metatarsal l.	Lies on sole of foot beneath heads of metatarsals

(continued)

Foot and Ankle

Latin name	English name	Articulation
La. metatarsalia dorsalia	dorsal metatarsal ligs.	Interconnects bases of metatarsal bones
La. metatarsalia interossea	interosseous metatarsal ligs.	Interconnects bases of metatarsal bones
La. metatarsalia plantaria	plantar metatarsal ligs.	Plantar surface of metatarsal bones
La. plantaria articulationum interphalangealium pedis	plantar ligs. of interphalan-geal articulations	Interphalangeal articulations of foot between collateral ligs.
La. plantaria articulationum metatarsophalangeal	plantar ligs. of metatarso-phalangeal articulations	Plantar surface of meta-tarsophalangeal articulations between collateral ligs.
La. plantare longum	long plantar l.	Connects calcaneus to bases of metatarsal bones
La. talocalcaneare laterale	lateral talocalcaneal l.	Connects talus to calcaneus
La. talocalcaneare mediale	medial talocalcaneal l.	Connects tubercle of talus to sustentaculum tali of calcaneus
La. talocalcaneum	talocalcaneal l.	Connects talus and the calcaneus
La. talocalcaneum interosseum	interosseous talocalcaneal l.	Connects calcaneus to talus
La. talofibulare anterius	anterior talofibular l.	Connects lateral malleolus of fibula to posterior process of talus.
La. talonaviculare	talonavicular l.	Connects neck of talus to navicular bone
La. talotibiale	medial tibiotalar l.	Connects downward from the medial malleolus of the tibia of the tarsal bones.
La. tarsi	ligs. of tarsus	Connects bones of tarsus

(continued)

Foot and Ankle

Latin name	English name	Articulation
La. tarsi dorsalia	dorsal ligs. of tarsus	Collectively, bifurcate, dorsal cuboideonavicular, cuneocuboid, cuneonavicular, intercuneiform, and talonavicular ligaments
La. tarsi interossea	interosseous ligs. of tarsus	Collectively, interosseous, cuneocuboid, intercuneiform, and talocalcaneal ligs.
La. tarsi plantaria	plantar ligs. of tarsus	Inferior ligs. of foot (long plantar, plantar calcaneocuboid, calcaneonavicular, cuneonavicular, cuboideonavicular, intercuneiform, cuneocuboid)
La. tarsometatarsalia dorsalia	dorsal tarsometatarsal ligs.	Connects bases of metatarsals to dorsal cuboid and cunei-form bones
La. tarsometatarsalia plantaria	plantar tarsometatarsal ligs.	Connects metatarsal bones to cuboid and cuneiform bones
La. transversum cruris	superior extensor retinaculum of foot	Connects tibia to fibula; holds extensor tendons in place
Tendo calcaneus	achilles tendon calcaneal tendon	Connects triceps surae muscle to tuberosity of calcaneus

Table of Nerves

1. Nerves of the Head and Neck Region

Nerve	Origin	Course	Innervation
Abducent	Pons	Intradural on clivus; traverses cavernous sinus and superior orbital fissure to enter orbit	Lateral rectus
Ansa cervicalis	Hypoglossal	Descends on external surface of carotid sheath	Omohyoid, sternohyoid, and sternothyroid
Deep petrosal	Internal carotid plexus	Traverses cartilages of foramen lacerum, joins greater petrosal nerve at entrance of pterygoid canal	Lacrimal gland, mucosa of nasal cavity, palate, and upper pharynx
Great auricular	Cervical plexus	Ascends over sternocleidomastoid; anterior and parallel to external jugular	Skin of auricle, adjacent scalp, and over angle of jaw
Greater petrosal	Genu of facial nerve	Exits facial canal via hiatus for greater petrosal nerve	Pterygoid ganglion for innervation of lacrimal, nasal, palatine, and upper pharyngeal mucous glands
Glossopharyngeal	Rostral end of medulla	Exits cranium via jugular foramen, passes between superior and middle constrictors of pharynx to tonsillar fossa, enters posterior third of tongue	Somatic to stylopharyngeus; visceral to parotid gland; sensory of posterior tongue, pharynx, tympanic cavity, auditory tube, carotid body, and sinus
Hypoglossal	Between pyramid and olive of myencephalon	Hypoglossal canal, medial to angle of mandible, between mylohyoid and hypoglossus to muscles of tongue	Intrinsic and extrinsic muscles of tongue

(continued)

A125

Nerve	Origin	Course	Innervation
Intermediate	Facial nerve	Acoustic meatus to distal end of facial nerve	Pterygopalatine and submandibular ganglia via greater petrosal nerve, chorda tympani; tongue and palate
Lesser occipital	Cervical plexus	Parallel to antero-superior border of sternocleidomastoid	Skin of posterior surface of auricle and adjacent scalp
Lesser petrosal	Tympanic plexus	Tympanic cavity to middle cranial fossa; sphenopetrosal fissure or foramen ovale	Otic ganglion for secretomotor innervation of parotid gland
Long thoracic	Anterior rami	Distally on external surface of serratus anterior	Serratus anterior
Nerve to mylohyoid	Inferior alveolar nerve	Inferior alveolar nerve of mandibular foramen to groove on medial aspect of ramus of mandible	Mylohyoid and anterior belly of digastric muscle
Nerve to tensor tympani	Otic ganglion	Cartilaginous portion of pharyngotympanic tube to hemicranial of tensor tympani	Tensor tympani
Nerve to tensor veli palatini	Anterior mandibular nerve	Branch of nerve to medial pterygoid	Tensor veli palatini
Olfactory	Olfactory cells in olfactory epithelium of roof of nasal cavity	Foramen of cribriform plate to ethmoid, to olfactory bulbs	Olfactory mucosa; sense of smell
Phrenic	Cervical plexus	Superior thoracic aperture between mediastinal pleura and pericardium	Diaphragm; pericardial sac, mediastinal pleura, diaphragmatic peritoneum
Posterior inferior nasal	Greater palatine	Greater palatine canal through plate of palatine bone	Mucosa of inferior concha and walls of inferior and middle meatuses

Nerve	Origin	Course	Innervation
Subclavian	Brachial plexus	Posterior to clavicle, anterior to brachial plexus, and subclavian artery	Subclavius; sterno-clavicular joint
Supraclavicular, lateral, intermediate, and medial	Cervical plexus	Center or posterior border of sterno-cleidomastoid; fan out as they descend into lower neck, upper thorax, and shoulder	Skin of lower anterolateral neck, uppermost thorax and shoulder
Supraorbital	Frontal nerve	Supraorbital foramen, breaks up into small branches	Mucous membrane of frontal sinus, conjunctivae, and skin of forehead
Suprascapular	Brachial plexus	Posterior triangle of neck; under superior transverse scapular ligament	Supraspinatus, infraspinatus muscles; superior and posterior glenohumeral joint
Supratrochlear	Facial nerve	Supraorbital nerve, divides into two or more branches	Skin in middle of forehead to hairline
Transverse cervical	Cervical plexus	Posterior border of sternocleidomastoid muscle, runs anteriorly across muscle	Skin overlying anterior triangle of neck
Trochlear	Dorsolateral aspect of mesocephalon below inferior colliculus	Passes around brain-stem to enter dura in edge of tentorium close to posterior clinoid process; runs in lateral wall of cavernous sinus, entering orbit via superior orbital fissures	Superior oblique muscle
Upper scapular	Brachial plexus	Posteriorly enters subscapularis	Superior portion of subscapularis

2. Nerves of the Facial Region

Nerve	Origin	Course	Innervation
Auriculotemporal	Mandibular nerve	Passes between neck of mandible and external acoustic meatus to accompany superficial temporal artery	Skin anterior to auricle, posterior temporal region, tragus, helix of auricle, exterior acoustic meatus, upper tympanic membrane
Buccal	Mandibular nerve	Infratemporal fossa, passes anteriorly to reach cheek	Skin and mucosa of cheek, buccal gingiva
Chorda tympani	Facial nerve	Traverses tympanic cavity, passes between incus and malleus; exits temporal bone via petrotympanic fissure; enters infratemporal fossa, merges with lingual nerve	Submandibular and sublingual glands; taste sensation from anterior tongue
Deep temporal	Mandibular nerve	Temporal fossa to temporalis muscle	Temporalis; periosteum of temporal fossa
External nasal	Anterior ethmoidal nerve	Runs in nasal cavity and emerges on face between nasal bone and lateral nasal cartilage	Skin on dorsum of nose including tip of nose
Facial	Posterior border of pons	Runs through internal acoustic meatus and facial canal of petrous part of temporal bone, exiting via stylomastoid foramen; intraparotid plexus	Stapedius, posterior belly of digastric, stylohyoid facial and scalp muscles; skin of external acoustic meatus
Greater palatine	Branch of pterygopalatine ganglion (maxillary nerve)	Passes inferiorly through greater palatine canal and foramen	Palatine glands; mucosa of hard palate

(*continued*)

Nerve	Origin	Course	Innervation
Inferior alveolar	Terminal branch of posterior mandibular nerve	Lateral and medial pterygoid muscles of infratemporal fossa to enter mandibular canal of mandible	Lower teeth, periodontium, periosteum and gingiva of lower jaw
Infraorbital	Terminal branch of maxillary nerve	Runs in floor of orbit and emerges at infraorbital foramen	Skin of cheek, lower lid, lateral side of nose and inferior septum and upper lip, upper premolar incisors and canine teeth; mucosa of maxillary sinus and upper lip
Lesser palatine	Pterygopalatine ganglion (maxillary nerve)	Passes inferior through palatine canal and lesser palatine foramen	Glands of soft palate; mucosa of soft palate
Lingual	Terminal branch of posterior mandibular nerve	Joins chorda tympani, passes anteroinferiorly between lateral and medial pterygoid muscles, oral cavity	Submandibular ganglion and submandibular and sublingual salivary glands
Mandibular	Trigeminal ganglion	Foramen ovale to infratemporal fossa, divides into anterior and posterior trunks, ramifying into smaller branches, bifurcating into lingual and inferior alveolar nerve	Muscles of mastication, mylohyoid, anterior belly of digastric, tensor tympanic, tensor veli palatini; skin overlying mandible, teeth, gingiva, tongue, and temporomandibular joint
Masseteric	Mandibular nerve	Passes laterally through mandibular notch	Masseter; temporomandibular joint

(continued)

Nerve	Origin	Course	Innervation
Maxillary	Trigeminal nerve	Anteriorly through foramen rotundum, to pterygopalatine fossa, sends roots to pterygoid ganglion (maxillary nerve); continues anteriorly through infraorbital fissures as infraorbital nerve	Pterygopalatine ganglion, lacrimal gland, mucosal glands of nasal cavity, palate, and upper pharynx; skin overlying maxillary mucosa of postero-inferior nasal cavity, maxillary sinus, upper half of mouth, (teeth, gingiva and mucosa of palate, vestibule, and cheek)
Mental	Terminal branch of inferior alveolar nerve	Mandibular canal at mental foramen	Skin of chin; skin and mucosa of lower lip
Nasopalatine	Pterygopalatine ganglion (maxillary nerve)	Exits pterygopalatine fossa via spheno-palatine foramen; runs anteroinferiorly across nasal septum, to incisive foramen to palate	Mucosal glands of nasal septum; mucosa of nasal septum, anterior-most hard palate
Nerve to lateral and medial pterygoid	Anterior mandibular nerve	Arises in infratem-poral fossa, inferior to foramen ovale	Lateral and medial pterygoid muscles
Nerve to pterygoid canal	Formed by merger of greater and deep petrosal nerves	Traverses pterygoid canal, to pterygoid ganglion in ptery-goid fossa	Pterygopalatine ganglion
Nerve to stapedius	Facial nerve	Arises as facial nerve, descends posterior to muscle in facial canal	Stapedius
Pharyngeal	Pterygopalatine ganglion	Passes posteriorly through palatovagi-nal canal	Supplies mucosa of nasopharynx posterior to the pharyngotympanic tubes

(continued)

Nerve	Origin	Course	Innervation
Superior alveolar	Maxillary nerve	Posteriorly emerges from pterygomaxillary fissure into infratemporal fossa to posterior aspect of maxilla; Middle and anterior: arises from infraorbital nerve of maxillary sinus, descends walls of sinus	Mucosa of maxillary sinus, maxillary teeth and gingiva
Trigeminal	Lateral surface of pons by two roots; motor and sensory	Crosses medial part of crest of petrous part of temporal bone, trigeminal cave of dural mater lateral to body of sphenoid and cavernous sinus; motor root passes ganglion to become part of mandibular nerve	Motor: somatic; muscles of mastication, mylohyoid, anterior belly of digastric, tensor tympanic, tensor veli palatini; Sensory: dura of anterior and middle cranial fossa, skin of face, teeth, gingiva, mucosa of nasal cavity, paranasal sinuses, and mouth
Zygomatic	Maxillary nerve	Arises in floor of orbit, divides into two temporal nerves, traverses foramina of same; communicating branch joins lacrimal nerve	Skin over zygomatic arch, anterior temporal region; conveys secretory postsynaptic parasympathetic fibers from pterygopalatine ganglion to lacrimal gland

3. Nerves of the Eye Region

Nerve	Origin	Course	Innervation
Anterior ethmoid	Nasociliary nerve	Arises in orbit, passes via anterior ethmoidal foramen, cranial cavity via cribriform plate of ethmoid to nasal cavity	Dural of anterior cranial fossa; mucous membranes of sphenoidal sinus, ethmoid cells and upper nasal cavity

(continued)

Nerve	Origin	Course	Innervation
Ciliary, long and short	Nasociliary nerve; short ciliary ganglion	Passes to posterior aspect of eyeball	Cornea, conjunctiva; ciliary body and iris
Frontal	Ophthalmic nerve	Crosses orbit on superior aspect of levator palpebrae superioris; divides into supraorbital and supratrochlear branches	Skin of forehead, scalp, eyelid, and nose; conjunctiva of upper lid and mucosa of frontal sinus
Infratrochlear	Nasociliary nerve	Follows medial wall of orbit to upper eyelid	Skin, conjunctiva, lining of upper eyelid
Lacrimal	Ophthalmic nerve	Palpebral fascia of upper eyelid near lateral angle of eye	Small area of skin and conjunctiva of lateral part of upper eyelid
Nasociliary	Ophthalmic nerve	Arises in superior orbital fissure, anteromedially across retrobulbar orbit, providing sensory root to ciliary ganglion, terminates as infratrochlear nerve	Ciliary ganglion (short) coveys postsynaptic sympathetic and parasympathetic to ciliary body and iris; tactile sensation for eyeball; mucous membrane of ethmoid cells, anterosuperior nasal cavity; skin of dorsum and apex of nose
Oculomotor	Interpeduncular fossa of mesencephalon	Dura of posterior clinoid process, lateral wall of cavernous sinus, enters orbit through superior orbital fissure and divides into superior and inferior branches	All extraocular muscles except superior oblique and lateral rectus; presynaptic parasympathetic fibers to ciliary ganglions for ciliary body and sphincter pupillae

(continued)

Nerve	Origin	Course	Innervation
Ophthalmic	Trigeminal ganglion	Anteriorly in lateral wall of cavernous sinus to enter orbit through superior orbital fissure, branching into frontal, nasociliary, and lacrimal nerve	General sensation from eyeball; mucous membrane of ethmoid cells, frontal sinus, dura of anterior cranial fossa, falx cerebri, and tentorium cerebelli, antero-superior nasal cavity; skin of forehead, upper lid and dorsum and apex of nose
Optic	Ganglion cells of retina	Exits orbit visa optic canals; fibers from nasal half of retina crosses to contralateral side at chiasm; passes via optic tracts to geniculate bodies, superior colliculus and pretectum	Vision from retina
Posterior ethmoidal	Nasociliary	Leaves orbit via posterior ethmoid foramen	Supplies ethmoid and sphenoid paranasal sinuses

4. Nerves of the Ear Region

Nerve	Origin	Course	Innervation
Cochlear	Division of vestibulocochlear nerve	Traverses internal acoustic meatus, enters modiolus with spiral ganglia and peripheral processes in spiral lamina	Spiral organ for hearing
Posterior auricular	As first extracranial branch of facial nerve	Passes posteriorly to ear, sending branch to occipital region	Posterior auricular muscle and intrinsic auricular muscles, occipital belly of occipitofrontalis

(continued)

Nerve	Origin	Course	Innervation
Tympanic	As first extracranial branch of glosso-pharyngeal nerve from inferior petrosal glossopharyngeal ganglion	Passes into tympanic canaliculus, enters tympanic cavity, ramifies on promon-tory of labyrinthine wall a tympanic plexus	Otic ganglion for secretomotor inner-vation of parotid gland; mucosa of tympanic cavity, mastoid cells, and pharyngotympanic tube
Vestibular	As a division of the vestibulocochlear nerve	Traverses internal acoustic meatus to vestibular ganglion at fundus; branches pass to vestibule of bony labyrinth	Cristae of ampullae of semicircular ducts, maculae of saccule and utricle (for sense of equilibrium)
Vestibulocochlear	Groove between pons and myence-phalon	Traverses internal acoustic meatus, dividing into co-chlear and vestibular nerve	Spiral organ for hearing and cristae of ampullae of semicircular ducts, maculae of saccule and utricle (for sense of equili-brium)

5. Nerves of the Thoracic Region

Nerve	Origin	Course	Innervation
Abdominopelvic splanchnic	Lower thoracic and lumbar segments of sympathetic trunk	Passes medially and inferiorly to prever-tebral ganglion of paraaortic plexus	Abdominopelvic blood vessels and viscera
Cardiac plexus	Cervical and cardiac branches of vagus nerve and cardiopul-monary splanchnic nerve from sympa-thetic trunk	From arch of aorta, posterior surface of heart, extends along coronary arteries and to SA node	SA nodal tissue, coronary arteries; parasympathetic fibers slow rate, reduce force of heartbeat, constrict arteries; sympathetic fibers have opposite effect
Cardiopulmonary splanchnic	Cervical and upper thoracic ganglia of sympathetic trunk	Descends anterome-dially to cardiac, pulmonary and esophageal plexuses	Conveys postsynap-tic sympathetic fibers to nerve plexuses of thoracic viscera

(continued)

Nerve	Origin	Course	Innervation
Cervical splanchnic	Cervical ganglia of sympathetic trunk	Passes medially and inferior to cardiac and pulmonary plexuses	Conducting tissue (SA and AV nodes) and coronary arteries
Esophageal plexus	Vagus nerve; greater splanchnic nerve	Tracheal bifurcation, vagus and sympathetic nerve from plexus around esophagus	Vagal and sympathetic fibers to smooth-muscles and glands of inferior two-thirds of esophagus
Greater splanchnic	Thoracic sympathetic ganglion	Highest abdomino-pelvic splanchnic nerve; anteromedially passes on bodies of thoracic vertebrae, through diaphragm to celiac trunk	Celiac ganglia, innervation of celiac arteries
Intercostal	Anterior rami of T1-T11 nerve	Intercostal spaces between internal innermost layers of intercostal muscles	Intercostal muscles, muscles of antero-lateral abdominal wall; skin overlying pleura/peritoneum deep to muscles
Lateral pectoral	Brachial plexus	Clavipectoral fascia to deep surface of pectoral muscles	Pectoralis major, medial pectoralis nerve that innervates pectoralis minor
Least splanchnic	Lowest thoracic ganglion of sympathetic trunk	Diaphragm with sympathetic trunk, ends in renal plexus	Renal arteries and derivatives
Lesser splanchnic	10th and 11th thoracic ganglia of sympathetic trunk	Descends antero-medially to perforate diaphragm to reach aorticorenal ganglion	Prevertebral ganglia; visceral afferents from upper GI tract
Lumbar splanchnic	Lumbar ganglia of sympathetic trunk	Passes antero-medially on bodies of lumbar vertebrae to prevertebral ganglia of paraaortic plexus	Lower abdominal wall and pelvic viscera; visceral afferents from upper GI tract

(*continued*)

A135

Nerve	Origin	Course	Innervation
Medial pectoral	Medial cord of brachial plexus	Passes between axillary artery and vein, enters deep surface of pectoralis minor	Pectoralis minor and part of pectoralis major
Pulmonary plexus	Vagus nerve, cardio-pulmonary splanch-nic nerve from sympathetic trunk	Forms on primary bronchi, extends along root of lung and brachial sub-divisions	Parasympathetic fibers constrict bronchioles; sym-pathetic fibers dilate them
Recurrent laryngeal	Vagus nerve	Subclavian on right; left runs around aortic arch, ascends in tracheoesophageal groove	Intrinsic muscles of larynx (except cricothyroid); Sensory: inferior to level of vocal cords
Subcostal	Anterior ramus of T12 spinal nerve	Inferior border of 12th rib in same manner as intercostal nerve	Muscles of antero-lateral abdominal wall; lateral cutane-ous branch supplies skin to anterior iliac crest
Superior laryngeal	Vagus nerve	Descends in para-pharyngeal space; lateral to thyroid cartilage, divides into internal and external laryngeal nerve; inferior pierces thyrohyoid membrane; external runs inferomedially to gap between cricoid and thyroid cartilages	Cricothyroid muscle (external laryn-geal); supraglottic
Thoracic splanchnic	Thoracic ganglia of sympathetic trunk	Anteromedially on thoracic vertebrae as lower cardiopul-monary splanchnic nerve to thoracic plexus (cardiac, pul-monary, esophageal); upper abdomino-pelvic splanchnic nerve to prevertebral ganglia of paraaortic plexus	Thoracic: 1st–5th splanchnic nerve (heart, lungs, esoph-agus); 6th–12th (greater, lesser, least splanchnic nerve); presynaptic sympa-thetic fibers to pre-vertebral ganglia

(continued)

Nerve	Origin	Course	Innervation
Thoracoabdominal	Lower intercostal nerve	Costal margin between 2nd and 3rd layers of abdominal muscle	Anterolateral abdominal muscles; overlying skin, underlying perito-neum, periphery of diaphragm
Thoracodorsal	Posterior cord of brachial plexus	Between upper and lower subscapular nerve, runs infero-laterally along posterior axillary wall to latissimus dorsi	Latissimus dorsi
Vagus	Via 8–10 rootlets from medulla of brainstem	Superior medias-tinum posterior to sternoclavicular joint and brachiocephalic vein; gives rise to recurrent laryngeal nerve; continues into abdomen	Voluntary muscle of larynx and upper esophagus; invol-untary muscle/glands of tracheo-bronchial tree and heart via pulmonary and cardiac plexuses; Sensory: pharynx, larynx, reflex afferens from same areas as above

6. Nerves of the Back and Spinal Regions

Nerve	Origin	Course	Innervation
Accessory	Cranial root: medulla; spinal root: cervical spinal cord	Spinal root ascends into cranial cavity via foramen mag-num; exits via jugular foramen; traverses posterior triangle of neck	Sternocleido-mastoid and trapezius
Dorsal scapular	Anterior ramus of C5 with contribution from C4	Scalenus medius, descends deep to levator scapulae, enters deep surface of rhomboids	Rhomboids; occa-sionally supplies levator scapulae

(*continued*)

Nerve	Origin	Course	Innervation
Greater occipital	Medial branch of posterior ramus of spinal nerve	Deep muscles of neck and trapezius to ascend posterior scalp to vertex	Multifidus cervicis, semispinalis capitis; posterior scalp
Suboccipital	Posterior ramus of C1 spinal nerve	Between occipital bone and atlas, inferior to transverse part of vertebral artery, into suboccipital triangle; communicates with occipital nerve	Suboccipital muscles (rectus capitis, major and minor, obliquus capitis inferior and superior)

7. Nerves of the Shoulder and Arm Region

Nerve	Origin	Course	Innervation
Anterior interosseous	Median nerve	Inferiorly on interosseous membrane	Flexor digitorum profundus, flexor pollicis longus, pronator quadrates
Axillary	Terminal branch of posterior cord of brachial plexus	Posterior aspect of arm; posterior circumflex humeral artery; winds around surgical neck of humerus; gives rise to brachial cutaneous nerve	Teres minor and deltoid; shoulder joint and skin over inferior part of deltoid
Deep branch of radial nerve	Radial nerve distal to elbow	Neck of radius in supinator; posterior compartment of forearm, becomes posterior interosseous nerve	Extensor carpi radialis brevis and supinator
Deep branch of ulnar nerve	Ulnar nerve at wrist, passes between pisiform and hamate	Deep between muscles of hypothenar eminence, across palm with deep palmar arch	Hypothenar muscles, lumbricales of digits 4–5, all interossei, adductor pollicis and deep head of flexor pollicis brevis

(continued)

Nerve	Origin	Course	Innervation
Lateral cutaneous nerve of forearm	Musculocutaneous nerve	Descends along lateral border of forearm to wrist	Skin of lateral aspect of forearm
Lower subscapular	Posterior cord of brachial plexus	Passes inferolaterally to subscapular artery and vein, to subscapularis and teres major	Inferior portion of subscapularis and teres major
Medial cutaneous nerve of arm	Medial cord of brachial plexus	Runs along medial side of axillary vein; communicates with intercostobrachial nerve	Skin on medial side of arm
Medial cutaneous of forearm	Medial cord of brachial plexus	Runs between axillary artery and vein	Skin over medial side of forearm
Musculocutaneous	Lateral cord of brachial plexus	Deep surface of coracobrachialis, descends between biceps brachii and brachialis	Flexor muscles of arm; lateral antebrachial cutaneous nerve
Palmar cutaneous branch of ulnar nerve	Arises from ulnar nerve near middle of forearm	Ulnar artery, perforates deep fascia in the distal third of forearm	Skin at base of medial palm, overlying medial carpals
Posterior cutaneous nerve of forearm	Arises in arm from radial nerve	Perforates lateral head of triceps, descends along lateral side of arm and posterior aspect of forearm to wrist	Skin of distal posterior arm, posterior aspect of forearm
Posterior interosseous	Terminal branch of deep branch of radial nerve	Between superficial and deep layers of posterior forearm; between extensor pollicis longus and interosseous membrane	Extensor carpi ulnaris, extensors of digits, abductor pollicis longus

(*continued*)

Nerve	Origin	Course	Innervation
Radial	Terminal branch of posterior cord of brachial plexus	Descends posterior to axillary artery; radial groove with deep brachial artery; passes between long and medial head of triceps; bifurcates in cubital fossa into superficial and deep radial nerve	Triceps brachii, anconeus, brachio-radialis, extensor carpi radialis longus muscle; skin on posterior aspect of arm and forearm via posterior cutaneous nerve of arm and forearm
Superficial branch of ulnar nerve	Arises from ulnar nerve at wrist, passes between pisiform and hamate bones	Palmaris brevis, divides into two common palmar digital nerve	Palmaris brevis; skin of the palmar and distal dorsal aspects of digit 5 and medial side of digit 4, proximal portion of palm
Ulnar	Terminal branch of medial cord of brachial plexus	Runs down medial aspect of arm; does not branch in the brachium	Majority of intrinsic muscles of hand; deep head of flexor pollicis brevis; medial lumbricales for digits 4 and 5; skin of palmar and distal dorsal aspects of medial 1–1/2 digits and adjacent palm

8. Nerves of the Hand

Nerve	Origin	Course	Innervation
Common palmar digital	Median and superficial branch of ulnar nerve	Runs distally between long flexor tendons of palm, bifurcating in distal palm	Proper palmar digital nerve; skin and joints of palmar and dorsal aspect of fingers
Dorsal branch of ulnar nerve	Ulnar nerve about 5 cm proximal to flexor retinaculum	Passes distally deep to flexor carpi ulnaris, dorsally to perforate deep fascia, medial side of dorsum of hand, dividing into 2 or 3 dorsal digital nerve	Skin of medial aspect of dorsum of hand, proximal portions of little and medial half of ring finger; adjacent sides of proximal portion of ring and middle fingers

(*continued*)

Nerve	Origin	Course	Innervation
Lateral branch of median nerve	Median nerve as it enters palm of hand	Runs laterally to palmar thumb and radial side of index finger	First lumbrical; skin of palmar and distal dorsal aspects of thumb, radial half of index finger
Medial branch of median nerve	Median nerve as it enters palm	Runs medially to adjacent sides of index, middle, and ring fingers	Second lumbrical; skin of palmar and distal dorsal aspects of adjacent sides of index, middle, and ring fingers
Median	Arises by two roots; one from lateral cord of brachial plexus; one from medial cord; root joint lateral to axillary artery	Medial side of brachial artery; cubital fossa, btween heads of pronator teres, intermediate and deep layers of anterior forearm; becomes superficial proximal to wrist; passes deep to flexor retinaculum	Some flexor muscles in forearm; some thenar muscles, lateral lumbricals; skin of palmar and distal dorsal aspects of lateral digits and palm
Palmar cutaneous branch of ulnar nerve	Arises from ulnar nerve, near middle of forearm	Passes between tendons of palmaris longus and flexor carpi radialis; runs superficial to flexor retinaculum	Skin of central palm
Recurrent branch of median nerve	Median nerve distal to flexor retinaculum	Distal border of flexor retinaculum, enters the muscles	Abductor pollicis brevis; opponens pollicis, superficial head of flexor pollicis brevis
Superficial branch of radial nerve	Radial nerve	Anterior to pronator teres, to brachioradialis; deep fascia at wrist, passes onto dorsum of hand	Skin of lateral half of dorsum of hand and thumb; proximal portions of digits 2, 3, and lateral half of 4

9. Nerves of the Abdomen and Pelvic Regions

Nerve	Origin	Course	Innervation
Cavernous nerve	Parasympathetic fibers of prostatic nerve plexus	Perforates perineal membrane to reach erectile bodies of penis	Helicine artery of cavernous bodies; simulation produces engorgement of arterial pressure
Clunial (superior, middle, inferior)	Superior: posterior rami of L1, L2, and L3; Middle: posterior rami of S1, S2, and S3; Inferior: posterior cutaneous nerve of thigh	Superior nerve cross iliac crest: middle nerve, exit through posterior sacral foramina, entering gluteal region; inferior nerve curve around inferior border of gluteus maximus	Skin of buttock as far as greater trochanter
Coccygeal	Conus medullaris of spinal cord	Anterior and posterior rami joint adjacent rami of S4 and S5; anterior rami form coccygeal plexus, gives rise to anococcygeal nerve	Skin over coccyx
Genitofemoral	Lumbar plexus	Descends on anterior surface of psoas major, divides into genital and femoral branches	Femoral: Skin over femoral triangle; genital branch supplies scrotum or labia minora; genital branch to cremaster muscle
Hypogastric	Superior hypogastric plexus into pelvis	Sacrum with hypogastric sheath, merges with pelvic splanchnic nerve in inferior hypogastric plexus	Pelvic viscera; intraperitoneal pelvic viscera (fundus, body of uterus)
Iliohypogastric	Lumbar plexus	Iliac crest; traverses abdominal muscle; external oblique aponeurosis to reach inguinal and pubic regions	Internal oblique and transverse abdominal muscles; superolateral quadrant of buttocks; skin over iliac crest and hypogastric region

(continued)

Nerve	Origin	Course	Innervation
Ilioinguinal	Lumbar plexus	Passes between 2nd and 3rd layers of abdominal muscle; inguinal canal, divides into femoral and scrotal or labial branches	Lower part of internal oblique, transverse abdominal muscles; skin over femoral triangle; mons pubis, adjacent skin of labia majora or scrotum
Inferior gluteal	Sacral plexus	Pelvis through greater sciatic foramen inferior to piriformis, divides into several branches	Gluteus maximus
Inferior rectal	Pudendal nerve	Pudendal canal, medially through ischioanal fat pad to anal canal	External anal sphincter; perianal skin
Lateral cutaneous nerve	Lumbar plexus	Deep to inguinal ligament, medial to anterior superior iliac	Skin on anterior and lateral aspects of thigh
Nerve to obturator internus	Sacral plexus	Gluteal region, greater sciatic foramen, inferior to piriformis; descends posterior to ischial spine; lesser sciatic foramen, to obturator internus	Superior gemellus and obturator internus
Quadratus	Sacral plexus	Leaves pelvis through greater sciatic foramen deep to sciatic nerve	Inferior gemellus and quadratus femoris
Obturator	Lumbar plexus	Enters thigh through obturator foramen, divides into anterior and posterior branches	Adductor longus, adductor brevis, gracilis, and pectineus; obturator externus, adductor magnus; skin of medial thigh above knee

(*continued*)

Nerve	Origin	Course	Innervation
Pelvic splanchnic	Sacral plexus	Runs anteriorly and inferiorly to merge with inferior hypogastric plexus	Motor: parasympathetic fibers for pelvic viscera, descending and sigmoid colon Sensory: uterus, upper vagina, floor of bladder, rectum and upper anal canal, prostate
Perineal	Terminal branch of pudendal nerve	Pudendal nerve from pudendal canal to superficial perineum dividing into superficial cutaneous and deep motor branch	Urogenital triangle; skin of posterior urogenital triangle; posterior aspect of scrotum
Posterior cutaneous nerve	Sacral plexus	Leaves pelvis through greater sciatic foramen inferior to piriformis, deep to gluteus maximus	Skin of buttock; skin over posterior and lateral aspects of thigh, calf, lateral perineum
Posterior labial	Perineal nerve	Pudendal canal and ramifies in subcutaneous tissue	Skin of posterior portion of labium majus
Pudendal	Sacral plexus	Enters gluteal region through greater sciatic foramen inferior to piriformis; descends to sacrospinous ligament; perineum through lesser sciatic foramen	Most motor and sensory innervation to the perineum
Sciatic	Sacral plexus	Enters gluteal region through greater sciatic foramen inferior to piriformis; descends along posterior aspect of thigh, divides proximal to knee into tibial and common fibular peroneal nerve	Hamstrings; provides articular branches to hip and knee joints

(continued)

Nerve	Origin	Course	Innervation
Superior gluteal	Sacral plexus	Leaves pelvis through greater sciatic foramen, superior to piriformis and rubs between gluteus medius and minimus	Gluteus medius, gluteus minimus, tensor fascia latae

10. Nerves of the Legs and Feet

Nerve	Origin	Course	Innervation
Anterior femoral cutaneous	Femoral nerve	Arises in femoral triangle, pierces fascia lata of thigh along path of sartorius muscle	Skin on medial and anterior aspect of thigh
Calcaneal branches	Tibial and sacral nerve	Passes from distal part of posterior aspect of leg to skin on heel	Skin of heel
Common fibular	Terminal branch of sciatic nerve	Begins at apex of popliteal fossa; follows medial border of biceps femoris muscle, to posterior aspect of head of fibula; bifurcates into superficial and deep fibular nerve	Skin on lateral part of posterior aspect of leg; knee joint via articular branch; short head of biceps femoris
Common plantar digital	Medial and lateral plantar nerve	Runs anteriorly in sole of foot between flexor tendons; bifurcates in distal sole	Proper plantar digital nerve; skin of plantar and distal dorsal aspect of toes
Deep fibular	Common fibular nerve	Arises between fibularis longus and neck of fibula; extensor digitorum; interosseous retinaculum; distal end of tibia and enters dorsum of foot	Muscles of anterior compartment of leg, dorsum of foot; skin of first interdigital cleft; sends articular branches to the joints it crosses

(*continued*)

Nerve	Origin	Course	Innervation
Femoral	Lumbar plexus	Passes deep to midpoint of inguinal ligament; lateral to femoral vessels, divides into muscular and cutaneous branches	Anterior thigh muscles; hip and knee joints; skin on anteromedial side of thigh and leg
Lateral plantar	Terminal branch of tibial nerve	Passes laterally in foot between quadratus plantae, flexor digitorum brevis muscles, divides into superficial and deep branches	Quadratus plantae, abductor digiti minimi, flexor digiti minimi brevis; plantar and dorsal interosseous, lateral three lumbricales, abductor hallucis; skin on sole later to a line splitting 4th digit
Medial cutaneous nerve of leg	Saphenous nerve	Descends medial side of leg with greater saphenous vein	Skin of anteromedial side of leg and medial side of foot
Medial dorsal cutaneous nerve	Superficial fibular	Descends across ankle anteriorly running into medial aspect of dorsum of foot	Most of skin of dorsum of foot, proximal portion of toes, except for web between great and 2nd toes
Medial plantar	Terminal branch of the tibial nerve	Passes distally in foot between abductor hallucis and flexor digitorum brevis; divides into muscular and cutaneous branches	Abductor hallucis, flexor digitorum brevis, flexor hallucis brevis and first lumbrical; skin of medial side of sole of foot and sides of first three digits
Saphenous	Femoral nerve	Descends with femoral vessels through femoral triangle and adductor canal, descends with great saphenous vein	Skin on medial side of leg and foot

(*continued*)

Nerve	Origin	Course	Innervation
Superficial fibular	Common fibular nerve	Arises between fibularis longus and neck of fibula, descends in lateral compartment of leg; deep fascia at distal third of leg to become cutaneous and send branches to foot and digits	Fibularis; skin on distal third of anterior surface of leg and dorsum of foot and all digits except lateral side of 5th and adjoining sides of 1st and 2nd digits
Sural	Arises between heads of gastrocnemius and becomes superficial at the middle of the leg; descends with small saphenous vein and passes posterior to the lateral malleolus to lateral side of foot	Descends between heads of gastrocnemius, becomes superficial at middle of leg; descends with small saphenous vein, passes posteriorly to the lateral malleolus to the lateral side of foot	Skin on posterior and lateral aspects of leg and lateral side of foot
Tibial	Sciatic nerve	Forms as sciatic, bifurcates at apex of popliteal fossa; descends through same, lies on popliteus; runs inferiorly on tibialis posterior with posterior tibial vessels; terminates beneath floor of retinaculum, dividing into medial and lateral plantar nerve	Motor: Muscles of posterior compartment of thigh; popliteal fossa, posterior compartment of leg, sole of foot: Sensory: knee joint; skin of leg, sole of foot

Appendix 5
Table of Arteries

1. Arteries of the Brain

Artery	Origin	Course	Distribution
Anterior cerebral	Terminal branch of internal carotid artery	Passes anteriorly, loops around genus of corpus callosum and passes posteriorly in interhemispheric fossa	A1 segment: thalamus and corpus striatum A2 segment: cortex of medial aspects of frontal and parietal lobes
Anterior communicating	Anterior cerebral artery	Connects anterior cerebral arteries in prechiasmatic cistern to complete cerebral arterial circle	Anteromedial central perforating artery
Anterior inferior cerebellar	Lower part of basilar artery	Runs posterolaterally in and out of internal acoustic meatus	Inferior aspect of lateral lobes of cerebellum, inferolateral pons, choroid plexus in cerebellopontine angle; gives rise to labyrinthine artery
Basilar	Formed by intercranial union of vertebral artery	Ascends clivus in pontine cistern; terminates by bifurcating into posterior cerebral artery	Anterior inferior cerebellar, labyrinthine, pontine, mesencephalic, and superior cerebellar artery
Middle cerebral	Larger terminal branch of the internal carotid artery	Runs in lateral cerebral sulcus, then posterosuperiorly on the insula	Insula and most of lateral surface of cerebral hemispheres
Posterior cerebral	Terminal branch of basilar artery	Passes laterally, winding around cerebral peduncle to reach the terminal cerebral surface	Inferior aspect of temporal lobe and occipital lobe of cerebrum

(continued)

Artery	Origin	Course	Distribution
Posterior inferior cerebellar	Intracranial portion of vertebral artery	Passes posteriorly around side of medulla to reach inferior aspect of cerebellum	Medial portion of inferior aspect of cerebellum, postero-lateral medulla oblongata, and choroid plexus of fourth ventricle
Superior cerebellar	Terminal part of basilar artery	Curves around cerebral peduncle	Superior aspect of cerebellum, colli-culi; most cerebellar nuclei; pons; pineal body; superior medullary velum; choroid plexus of third ventricle

2. Arteries of the Head and Neck

Artery	Origin	Course	Distribution
Ascending cervical	Terminal branch of thyrocervical trunk	Ascends on preverte-bral fascia	Anterior and prevertebral muscles
Common carotid artery	Left: Second branch of arch of aorta Right: terminal branch of brachio-cephalic artery	Ascends to deep sternoclavicular joint in carotid sheath under cover of sternocleidomas-toid to level of hyoid bone	Terminal branches: internal and external carotid artery
Deep cervical	Costocervical trunk	Passes posteriorly between transverse process of C7 and neck of first rib and ascends between semispinalis cervicis and capitis to C2 level	Deep posterior muscles of neck; descending branch of occipital artery; branches of vertebral artery
Dorsal scapular	Subclavian artery	Passes laterally through brachial plexus then deep to levator scapulae; joins dorsal scapular nerve, running along vertebral border of scapula, deep to rhomboid muscles	Trapezius, rhomboids, latissi-mus dorsi, and around shoulder

(*continued*)

A149

Artery	Origin	Course	Distribution
External carotid	Common carotid artery at superior border of thyroid cartilage	Ascends slightly and then inclines posteriorly and laterally, passing between mastoid process and mandible; enters substance of parotid gland, bifurcating into terminal branches deep to neck of mandible	Anterior branches: superior thyroid, facial, and lingual arteries Posterior branches: occipital and posterior auricular artery Medial branch: ascending pharyngeal Terminal branches: maxillary and superficial temporal artery
Inferior thyroid	Terminal branch (with ascending branch of cervical artery) of thyrocervical trunk	Ascends anterior to anterior scalene, medially passing between vertebral vessels and carotid sheath, descends on longus coli to lower border of thyroid gland	Branches: inferior laryngeal artery; pharyngeal, tracheal, esophageal, and inferior and ascending glandular branches (lateral to parathyroid gland); main visceral artery of neck
Internal carotid	Common carotid artery at superior border of thyroid cartilage	Ascends vertically in neck to enter carotid canal, horizontally runs anteromedially through cavernous sinus, turns under anterior clinoid process, bifurcates into anterior and middle cerebral artery	Walls of cavernous sinus, pituitary gland, trigeminal ganglion; provides blood supply to orbit, eyeball, upper nasal cavity, and brain
Occipital	External carotid artery	Passes medially to posterior belly of digastric and mastoid processes	Scalp of back of head
Posterior auricular	External carotid arteries	Passes superiorly, deep to parotid, along styloid process between mastoid process and ear	Scalp posterior to auricle

(continued)

Artery	Origin	Course	Distribution
Posterior communicating	Anastomosis between internal carotid and posterior cerebral artery	Passes superiorly to oculomotor nerve	Optic tract, cerebral peduncle, internal capsule, and thalamus
Sphenopalatine	Third part of maxillary artery	Passes medially via sphenopalatine foramen, dividing into septal and posterior lateral nasal arteries	Mucosa of postero-inferior half of nasal cavity, ethmoid cells and maxillary and sphenoidal para-nasal sinuses
Superficial cervical	Thyrocervical trunk	Passes laterally between sternoclei-domastoid and anterior scalene; across brachial plexus and posterior triangle of neck; bifurcates with accessory nerve on deep aspect of trapezius	Anterior scalene, sternocleidomastoid plexus, muscles of posterior triangle of neck, and trapezius
Superficial temporal	Small terminal branch of external carotid artery	Ascends anterior to ear to temporal region and ends in scalp	Facial muscles and skin of frontal and temporal regions
Superior thyroid	First branch from anterior aspect of external carotid artery	Passes deep to infra-hyoid muscles to the superior pole of the thyroid gland; anastomosis with inferior thyroid artery	Superior laryngeal artery; infrahyoid; sternocleidomastoid; cricothyroid; anterior, posterior, and lateral glandular branches
Suprascapular	Thyrocervical trunk	Passes inferiorly over anterior scalene muscle and phrenic nerve	Supraspinatus and infraspinatus muscles
Supratrochlear	Thyrocervical trunk	Passes from supra-trochlear notch to medial forehead and anterior scalp	Skin and muscles of medial part of forehead and scalp

(continued)

Artery	Origin	Course	Distribution
Thyrocervical trunk	Anterior aspect of first part of subclavian	Ascends as short, wide trunk near the medial border of the anterior scalene and posterior to carotid sheath	Branches from trunk; transverse cervical and suprascapular; terminal branches ascending cervical and inferior thyroid arteries
Thyroid ima	Brachiocephalic arch of aorta	Ascends on anterior aspect of trachea to thyroid gland	Medial aspect of both lobes of thyroid

3. Arteries of the Eyes

Artery	Origin	Course	Distribution
Anterior ciliary	Muscular (rectus) branches of ophthalmic artery	Pierces sclera at attachment of rectus muscle, forms network in iris and ciliary body	Iris and ciliary body
Central artery of retina	Ophthalmic artery	Runs in dural sheath of optic nerve, pierces nerve near eyeball; ramifying from center of optic disc to retinal arterioles	Optic retina; branches; macular, nasal, and temporal retinal arterioles
Infraorbital	Third part of maxillary artery	Passes along infraorbital groove and foramen to face	Inferior rectus and oblique muscles, inferior eyelid, lacrimal sac, maxillary sinus, maxillary incisor, canine teeth, anterior cheek
Lacrimal	Ophthalmic artery	Passes along superior border of lateral rectus muscle	Superior border of lateral rectus muscle; lacrimal gland, conjunctivae, and eyelids
Long posterior ciliaries	Ophthalmic artery	Pierces sclerae to supply ciliary body and iris	Ciliary body and iris

(continued)

Artery	Origin	Course	Distribution
Ophthalmic	Internal carotid artery	Traverses optic foramen to reach orbital cavity	Optic foramen, orbital cavity
Short posterior ciliaries	Ophthalmic artery	Pierces sclerae at optic nerve; supplies choroid, cones, and rods of optic retina	Periphery of optic nerve
Supraorbital	Terminal branch of the ophthalmic artery	Passes superiorly and posteriorly from supraorbital foramen to forehead and scalp	Supplies muscles and skin of most of forehead and anterior scalp

4. Arteries of the Nose

Artery	Origin	Course	Distribution
Anterior ethmoid	Ophthalmic artery	Passes through anterior ethmoid foramen to anterior cranial fossa and into nasal cavity, sending branches to skin of nose	Anterior and middle ethmoid cells, dura matter of anterior cranial fossa, anterosuperior nasal cavity, skin on dorsum of nose
Dorsal nasal	Ophthalmic artery	Courses along dorsal aspect of nose and supplies its surface	Dorsal aspect of nose and skin surface
Lateral nasal	Facial artery	Passes to ala of nose	Skin on ala and dorsum of nose
Posterior ethmoidal	Ophthalmic artery	Passes through posterior ethmoidal foramen to posterior ethmoidal cells	Posterior ethmoidal foramen, posterior ethmoidal cells
Posterior lateral nasal	Sphenopalatine artery	Split over conchae and meatuses; anastomoses with nasal branches of ethmoid and greater palatine artery	Lateral walls of posterior interior nasal cavity, ethmoid cells, maxillary, and sphenoid sinuses

(*continued*)

Artery	Origin	Course	Distribution
Posterior septal	Sphenopalatine artery	Crosses inferior surface to reach nasal septum, courses anteroinferiorly on vomer to incisive canals	Nasal septum, greater palatine artery, septal branch of superior tibial artery
Sphenopalatine	Third part of maxillary artery	Passes medially through sphenopalatine foramen, dividing immediately into septal and posterior lateral nasal	Mucosa of postero-inferior half of nasal cavity, ethmoid cells, maxillary and sphenoid paranasal sinuses
Supraorbital	Terminal branch of ophthalmic artery	Passes superiorly and posterior from supraorbital foramen to forehead and scalp	Supplies muscles and skin of most of forehead and anterior scalp

5. Arteries of the Face and Mouth

Artery	Origin	Course	Distribution
Angular	Terminal branch of the facial artery	Passes to medial angle (canthus) of eye	Superior part of cheek, lower eyelid
Artery of pterygoid canal	Third part of maxillary artery; greater palatine	Passes posteriorly through pterygoid canal	Mucosa of uppermost pharynx; auditory tube and tympanic cavity
Ascending palatine	Facial artery	Ascends along side and crosses over superior border of superior constrictor of pharynx to reach soft palate and tonsillar fossa	Lateral wall of pharynx, tonsils, auditory tube, and soft palate
Descending palatine	Third part of maxillary artery	Arises in pterygopalatine fossa; descends in palatine canal	Greater and lesser palatine artery

(continued)

Artery	Origin	Course	Distribution
Ascending pharyngeal	Medial aspect of external carotid artery	Ascends between internal carotid artery and pharynx to cranial base through jugular foramen; branches through jugular foramen and hypoglossal canal	Pharyngeal wall, palatine tonsil, soft palate, and dura of posterior cranial fossa
Anterior superior alveolar	Infraorbital artery	Arises within infraorbital canal and ascends through anterior alveolar canals	Mucosa of maxillary sinus; maxillary superior incisor and canine teeth
Inferior alveolar	First part of maxillary artery	Descends posterior maxillary nerve between medial pterygoid and ramus of mandible to enter mandibular canal via mandibular foramen	Mylohyoid, dental, mental branches; muscles of floor or mouth, mandible and lower teeth, soft tissue of chin
Posterior superior alveolar	Third part of maxillary artery	Exits from pterygopalatine fossa via pterygomaxillary fissure; penetrates infratemporal surface of maxillary; alveolar canals	Mucosa of maxillary sinuses, maxillary molar, premolar teeth and adjacent gingiva
Facial	External carotid artery	Ascends deep to submandibular gland, winds around inferior border or mandible and enters face, ascending obliquely across cheek and side of nose to medial angle of eye	Branches: ascending palatine, tonsillary, glandular, submental, inferior, and superior labial and lateral nasal Terminal branch: angular artery
Buccal	Maxillary artery	Runs anterolaterally with buccal nerve, emerging from underneath anterior border of ramus of mandible	Buccinator muscle, overlying skin, underlying oral mucosa; facial and infraorbital artery

(*continued*)

Artery	Origin	Course	Distribution
Deep lingual	Third part of lingual artery	Anterior border of hypoglossus, passes anteriorly flanking frenulum	Genioglossus, inferior longitudinal muscle and mucosa of underside and tip of tongue
Deep temporal (anterior and posterior)	Second part of maxillary artery	Ascends between temporalis and bone of temporal fossa	Temporalis muscle, periosteum, and bone
Inferior labial	Facial artery near angle of mouth	Runs medially in lower lip	Lower lip and chin
Lesser palatine	Descending palatine	Descend inferoposteriorly through lesser palatine foramen	Soft palate
Lingual	External carotid artery	Loops over greater horn of hyoid, passes medial to hypoglossus and ascends to run along side of tongue	Suprahyoid branch; dorsal and deep lingual artery and sublingual artery
Masseteric	Second part of maxillary artery	Passes posterior to temporalis tendon, accompanying masseteric nerve through mandibular notch	Masseter and temporomandibular joint; facial and transverse facial artery
Maxillary	Terminal branch of external carotid	Passes posterior and medial (1st part) to neck of mandible, deep to head of lateral pterygoid (2nd part), into pterygopalatine fossa (3rd part)	1st Part: deep auricular, anterior tympanic, middle meningeal, accessory meningeal, inferior alveolar 2nd Part: deep temporal, pterygoid (branches), masseteric, buccal 3rd Part: posterior superior alveolar, descending palatine, artery of pterygoid canal, pharyngeal, sphenopalatine, infraorbital

(continued)

Artery	Origin	Course	Distribution
Mental (branch)	Terminal branch of inferior alveolar artery	Emerges from mental foramen and passes to chin	Facial muscles and skin of chin
Middle meningeal	First part of maxillary artery	Ascends vertically through foramen spinosum into middle cranial fossa; lateral walls of the cranial dura matter	Branches: ganglionic branches, petrosal branches superior tympanic artery; temporal branches, anastomotic branch to lacrimal artery; most blood is distributed to periosteum, bone, and red bone marrow
Mylohyoid	Inferior alveolar	Pierces sphenomandibular ligament to run anteroinferiorly with nerve in groove on medial aspect of ramus of mandible	Floor of mouth; submental artery
Sublingual	Trigeminal branch of lingual artery	Runs on genioglossus muscle superior to mylohyoid	Muscles and mucous membrane of floor of mouth and anterior lingual gingiva
Submental	Facial artery, distal to submandibular gland in submental bridge	Courses along inferior aspect of mylohyoid, adjacent to its attachment to the mandibular, to the mandibular symphysis	Mylohyoid, anterior belly of digastric, submental lymph nodes; inferior labial and mental artery; lower lip
Superior labial	Facial artery near angle of mouth	Runs medially in upper lip	Upper lip and ala (side) of septum and nose
Supraorbital	Terminal branch of the ophthalmic artery	Passes superiorly and posteriorly from supraorbital foramen to forehead and scalp	Supplies muscles of skin and most forehead and anterior scalp
Transverse facial	Superficial temporal artery within parotid gland	Crosses face superficial to masseter and inferior zygomatic arch	Parotid gland and duct; muscles and skin of face

6. Arteries of the Larynx

Artery	Origin	Course	Distribution
Superior laryngeal	Superior thyroid	Runs deep to thyrohyoid membrane with internal laryngeal nerve	Larynx

7. Arteries of the Ears

Artery	Origin	Course	Distribution
Deep auricular	First part of maxillary artery	Ascends in parotid gland posterior to temporomandibular joint; wall of external acoustic meatus	Temporomandibular joint; skin of external acoustic meatus; tympanic membrane
Labyrinthine	Basilar or via a common trunk with anterior inferior cerebellar	Exits cranial cavity via internal acoustic meatus; enters bony labyrinth	Membranous labyrinth
Stylomastoid	Posterior auricular	Enters stylomastoid foramen, ascends facial canal and supplies facial nerve	Branches: posterior tympanic artery (to tympanic membrane); mastoid (to mastoid cells), and stapedial (to stapedius, stapes, and secondary tympanic membrane) branches

8. Arteries of the Shoulder and Arm

Artery	Origin	Course	Distribution
Axillary	Continuation of subclavian artery	Runs inferolaterally through axillary fossa, changing to brachial artery when it crosses the inferior border of the teres major; medial (1st); posterior (2nd); lateral (3rd) to pectoralis minor	Superior thoracic; thoracoacromial and lateral thoracic arteries; subclavian and anterior and posterior circumflex humeral arteries

(continued)

Artery	Origin	Course	Distribution
Brachial	Continuation of axillary artery past inferior border of teres minor	Courses in medial intermuscular septum with median nerve; ends by bifurcating into radial and ulnar arteries in cubital fossa	Main artery of arm; superior and inferior ulnar collateral
Circumflex humeral anterior and posterior	Third part of axillary artery; opposite origin of subscapular artery	Forms a circle around neck of humerus, larger posterior circumflex humeral artery passes through quadrangular space with axillary nerve	Shoulder joint, muscles of the proximal arm; deltoid, teres major and minor; long and lateral heads of triceps
Circumflex scapular	Terminal branch of subscapular artery	Curves around axillary border of scapula and enters infraspinous fossa	Subscapular and infraspinatus muscles; collateral anastomosis of shoulder around scapula
Common interosseous	Ulnar artery, distal to bifurcation of brachial artery	Passes deep to bifurcate into terminal branches after a very short course	Anterior and posterior interosseous arteries
Deep artery of arm	Brachial artery	Accompanies radial nerve through radial groove in humerus; terminal branches take part in anastomosis around elbow joint	Branches: Deltoid, muscular (to head of triceps) and nutrient (to humerus) branches Terminal branches: middle and radial collateral arteries
Interosseous anterior and posterior	Common interosseous artery	Pass to anterior and posterior sides of interosseous membrane	Anterior and posterior compartments of forearm and distal forearm; anastomosis around the elbow
Middle collateral	Deep artery of arm	Descends to anastomose with recurrent interosseous artery	Part of collateral pathway around elbow; lateral and medial heads of triceps

(continued)

Artery	Origin	Course	Distribution
Radial	Smaller terminal division of brachial artery in cubital fossa	Inferolaterally under cover of brachioradialis, distally lies lateral to flexor carpi radialis tendon; crosses snuff box	Muscles of lateral portion of both anterior and posterior compartments of forearm, lateral aspect of wrist, skin of dorsum of hand
Radial collateral	Terminal branch of deep artery of arm	Perforates lateral intermuscular septum with radial nerve runs between brachialis and brachioradialis to anastomose with radial recurrent, anterior to lateral epicondyle of humerus	Forms part of cubital anastomosis; upper brachialis and brachioradialis; anterolateral aspect of elbow
Radial recurrent	Lateral side of radial artery distal to its origin	Ascends on supinator, passes between brachioradialis and brachialis to anastomose with radial collateral, interior to lateral epicondyle of humerus	Forms part of the cubital anastomosis; supinator, lower brachialis and brachioradialis, anterolateral aspect of elbow joint
Subclavian	Left: aortic arch Right: brachiocephalic trunk	Passes posterior to sternoclavicular joint; arches over cervical pleura anterior to apex of lung; crosses first rib posterior to anterior scalene, becoming axillary artery at rib's outer edge	Branches: vertebral, internal thoracic, thyrocervical; dorsal scapular and costocervical on left side; medial (1st); posterior (2nd); and lateral (3rd) to scalenus anterior muscle
Subscapular	Third part of axillary artery	Largest (but short 4-cm) branch of axillary artery; descends along lateral border of subscapularis and axillary border of scapula to bifurcate at the level of the inferior angle	Circumflex scapular and thoracodorsal arteries; muscles of both side of the scapula, latissimus dorsi and posterior chest wall

(*continued*)

Artery	Origin	Course	Distribution
Thoracoacromial	Second part of axillary artery deep to pectoralis minor	Curls around superomedial border of pectoralis minor; clavipectoral fascia and divides into four branches	Branches: acromial, clavicular, pectoral and deltoid
Thoracodorsal	Subscapular artery	Continues course of subscapular artery; thoracodorsal nerve to latissimus dorsi	Latissimus dorsi
Ulnar	Large terminal branch of brachial artery in cubital fossa	Passes inferomedially, directly inferiorly, deep to pronator teres, palmaris longus, flexor digitorum superficialis; medial side of retinaculum at wrist, continues as superficial palmar arch	Medial part of anterior compartment of forearm, wrist, and hand; superficial structures of central palm, palmar, and distal aspects of fingers
Ulnar collateral, superior and inferior	Superior ulnar collateral artery; brachial artery	Ulnar collateral with ulnar nerve to posterior aspect of elbow	Anastomoses distally with anterior and posterior ulnar recurrent arteries
Ulnar recurrent (anterior and posterior)	Ulnar artery, distal to elbow joint	Passes superiorly and posterior ulnar collateral artery passes posteriorly	Anterior and posterior ulnar collateral arteries

9. Arteries of the Hand

Artery	Origin	Course	Distribution
Carpal branches, dorsal and palmar	Radial and ulnar arteries at level of wrist	Anastomoses with corresponding branches of counterpart artery	Provide collateral circulation at wrist
Common palmar digital	Superficial palmar arch	Passes distally anterior to lumbricals to bifurcate proximal to webbings between digits	Palmar metacarpal arteries from deep palmar arch; proper palmar digital arteries

(continued)

A161

Artery	Origin	Course	Distribution
Deep palmar arch	Direct continuation of radial artery completed on medial side by deep branch of ulnar artery	Curves medially, deep to long flexor tendons in contact with bases of metacarpals	Palmar metacarpal arteries
Dorsal carpal arch	Radial and ulnar arteries	Arches within fascia on dorsum of hand	Dorsal metacarpal arteries
Dorsal digit	Dorsal metacarpal arteries	Run distally on the posterolateral aspects of the proximal one and a half phalanges	Dorsal aspects of proximal one and a half phalanges of fingers
Dorsal metacarpal	Dorsal carpal arch	Runs on 2nd–4th dorsal interossei	Bifurcates into dorsal distal arteries; supplies skin, muscle and bone of dorsum of hand and fingers to center of phalanx
Palmar metacarpal	Deep palmar arch from radial arteries	Runs distally on plane between adductor pollicis and interosseous muscle	Anastomose distally with common palmar arteries.
Princeps pollicis	Radial artery as it turns into palm	Descends on palmar aspect of first metacarpal and divides at the base of proximal phalanx into two branches that run along sides of thumb	Thumb
Proper palmar digits	Common palmar digital arteries	Run along of digits 2–5; base of middle phalanx; gives rise to dorsal branch which replaces dorsal digit arteries	All palmar and distal part including nail beds of dorsal aspect of fingers
Radial indicis	Radial artery	Passes along lateral side of index finger to its distal end	All of palmar and distal part including nail beds of dorsal aspect of index finger

(*continued*)

Artery	Origin	Course	Distribution
Superficial palmar arch	Direct continuation of ulnar artery; arch is completed on lateral side by superficial branch of radial artery	Curves laterally deep to palmar aponeurosis and superficial to long flexor tendons; curve of arch lies across palm at level of distal border of extended thumb	Three common palmar digital arteries

10. Arteries of the Heart and Thoracic Region

Artery	Origin	Course	Distribution
Anterior intercostals	Internal thoracic and musculophrenic arteries	Passes between internal and innermost intercostal muscles	Intercostal muscles overlying skin and underlying parietal pleura
Anterior interventricular	Left coronary artery	Passes along anterior interventricular groove to apex of heart	Walls of right and left ventricles including most of interventricular septum and contained atrioventricular bundle and branches
Arch of aorta	Continuation of ascending aorta	Arches posteriorly on left side of trachea and esophagus and superior to root of left lung	Brachiocephalic, left common carotid, left subclavian
Ascending aorta	Aortic orifice of left ventricle	Ascends approximately 5 cm to level of sternal angle where it becomes arch of aorta	Right and left coronary arteries
Atrioventricular (AV) nodal branch	Right coronary artery near origin of posterior interventricular artery	Runs anteriorly in uppermost part of interventricular septum to AV node	AV node
Brachiocephalic	First and largest branch of arch of aorta	Ascends posterolaterally to the right, running anterior to right trachea; deep to the sternoclavicular joint, bifurcating into terminal branches	Right common carotid and right subclavian arteries

(continued)

A163

Artery	Origin	Course	Distribution
Circumflex (branch)	Left coronary artery	Passes to left atrio-ventricular groove to posterior surface of heart	Primarily left atrium and left ventricle; atrial and marginal branches
Internal thoracic	Inferior surface of subclavian artery	Descends antero-medially, posterior to sternal end of clavicle and costal cartilages; lateral to sternum; divides at level of 6th costal cartilage into superior epigastric and musculophrenic arteries	Sternum and skin anterior to it; medial aspect of breast
Lateral thoracic	Second part of axillary artery	Descends along axillary border of pectoralis minor and follows it onto thoracic wall	Lateral chest wall, serratus anterior, and breasts
Left coronary	Left coronary sinus	Runs in AV groove and gives off anterior interventricular and circumflex branches	Most of left atrium and ventricle, interventricular septum and AV bundles; may supply AV node
Left marginal	Circumflex branch	Follows left border of heart	Left ventricle
Musculophrenic	Terminal branch of internal thoracic artery	Arising in 6th intercostal space, descends inferolaterally, paralleling costal margin	Branches: anterior intercostal arteries of 7th–9th intercostal spaces; upper abdominal muscles and pericardium
Pericardiacophrenic	Internal thoracic artery	Descends parallel to phrenic nerve between mediastinal parietal pleura and pericardium	Mediastinal parietal pleura and pericardium; anastomoses with phrenic and musculophrenic arteries

(continued)

Artery	Origin	Course	Distribution
Posterior intercostal	Posterior aspect of thoracic aorta	Passes laterally, then anteriorly parallel to ribs	Lateral and anterior cutaneous branches
Posterior intercostals	Superior intercostal artery, thoracic aorta	Passes between internal and innermost intercostal muscles	Intercostal muscles and overlying skin, parietal pleura
Posterior interventricular	Right coronary artery	Runs from posterior interventricular groove to apex of heart	Right and left ventricles and interventricular septum
Right coronary	Right aortic sinus	Follows coronary AV groove between the atria and ventricles	Right atrium, SA and AV nodes, posterior part of interventricular septum
Right marginal	Right coronary artery	Passes to inferior margin of heart and apex	Right ventricle and apex of heart
Sinoatrial (SA) nodal	Right coronary artery near its origin; circumflex branch of left coronary	Winds around right (60%) or left (40%) side of ascending aorta and ascending SA node	Left atrium and SA node
Superior epigastric	Femoral artery	Runs in superficial fascia toward umbilicus	Subcutaneous tissue and skin over suprapubic region
Superior phrenic	Anterior aspects of thoracic aorta	Arises at aortic hiatus and passes to superior aspect of diaphragm	Diaphragm and diaphragmatic parts of pericardium and parietal pleura
Superior thoracic	Only branch of first part of axillary artery	Runs anteromedially along superior border of pectoralis minor and then passes between it and pectoralis major to thoracic wall	Helps supply 1st and 2nd intercostal spaces and superior part of serratus anterior

(*continued*)

Artery	Origin	Course	Distribution
Thoracic aorta	Continuation of arch of aorta	Descends in posterior mediastinum to left of vertebral column	Posterior intercostal arteries, subcostal, some phrenic arteries, and visceral branches (tracheal and esophageal)

11. Arteries of the Lungs

Artery	Origin	Course	Distribution
Bronchial (1–2 branches)	Anterior aspect of 1st part of thoracic aorta, posterior intercostal artery	Runs on the posterior aspect of the primary bronchial and follow the tracheobronchial tree	Bronchial and peribronchial tissue, visceral pleura
Left pulmonary	Pulmonary trunk	Joins left bronchus and pulmonary veins to form root of left lung; descends in lung	Supplies the left lung Branches: superior and inferior lobar arteries; segmental arteries
Lingular, inferior and superior	Superior lobar artery of left lung	Descends anteriorly to lingula	Lingular division; superior, inferior and bronchopulmonary segments of left lung
Right pulmonary	Pulmonary trunk	Passes beneath arch of aorta to join right branches and pulmonary veins to form root of right lung; descends in lung	Supplies the right lung Branches: superior, middle, and inferior lobar arteries; segmental arteries
Segmental artery of lung	Lobar arteries	Arises within lung as tertiary branches of the right and left pulmonary arteries	Each segmental artery serves a bronchopulmonary segment of the lung

12. Arteries of the Spine and Back

Artery	Origin	Course	Distribution
Anterior spinal	Merging of intra-cranial branches, one from each vertebral artery; inferiorly by bifurcation of ante-rior segmental medullary arteries	Forms a continuous anastomotic chain that descends to the length of the spinal cord in the entrance to the anterior median fissure	Supplies anterior portion of spinal cord by means of sulcal branches; extending into ante-rior median fissure, pial plexus; ramifies over the surface of the cord
Posterior spinal	Superiorly from intercranial branch of vertebral artery; continued inferiorly bifurcating of poste-rior segmental medullary arteries	Forms continuous anastomotic chain that descends to the length of the spinal cord in the postero-lateral sulcus, adja-cent to dorsal roots of spinal nerves	Posterolateral aspect of spinal cord via pial plexus and peripheral branches
Radicular, anterior and posterior	Spinal branches of segmental arteries (vertebral, posterior intercostal, lumbar, and sacral arteries)	Courses along ante-rior and posterior roots of spinal nerves	Anterior and poste-rior roots of spinal nerves and dural sheaths and arachnoid
Segmental medullary, anterior and posterior	Spinal branches of segmental arteries (vertebral, posterior intercostal, lumbar, and sacral arteries)	Course along ante-rior and posterior roots of spinal nerves medially to anastomose with longitudinal anterior and posterior spinal arteries	Dorsal and ventral roots of selected spinal nerves; spinal cord; major anterior segmental medul-lary artery; lower thoracic, upper lumbar level on left side
Vertebral	First part of subcla-vian artery	Ascends via trans-verse foramina of vertebrae C6-C2, passes laterally to transverse C1; runs horizontal and medial to enter fore-arm magnum; forms basilar artery	Radicular and segmental medul-lary arteries menin-geal; anterior and posterior spinal, posterior and infe-rior cerebellar; medial and lateral medullary

13. Arteries of the Abdomen and Pelvic Regions

Artery	Origin	Course	Distribution
Abdominal aorta	Continuation of thoracic aorta	Runs on anterior aspect of bodies of lumbar vertebrae	Visceral branches: celiac, superior and inferior mesenteric, renal, middle suprarenal, gonadal; Parietal branches: lumbar, median sacral
Anterior division of internal iliac	Internal iliac	Passes anteriorly along lateral wall of lesser pelvis in hypogastric sheath, divides into visceral and parietal branches	Parietal branch: obturator artery; Visceral branches: umbilical artery, inferior vesicle, uterine, vaginal middle rectal, and pudendal
Appendicular	Ileocolic artery	Passes between layers of meso-appendix	Vermiform appendix
Celiac	Abdominal aorta, distal to aortic hiatus of diaphragm	Runs a short course (1.25 cm); giving rise to left gastric, bifurcating into splenic and common hepatic arteries	Supplies inferior-most esophagus, stomach, duodenum, liver and biliary apparatus, and pancreas
Common hepatic	Terminal branch of celiac artery trunk	Passes to right along superior border of pancreas anterior to portal vein	Terminal branches: hepatic artery proper, gastro-duodenal artery
Common iliac, left and right	Terminal branches of abdominal aorta	Anterior to L4 vertebral body, diverging as they descend to terminate at L5/S1 level, anterior to sacroiliac joints	Terminal branches: external and internal iliac arteries
Cystic	Right hepatic artery	Arises within hepatoduodenal ligament	Gallbladder and cystic duct
Deep circumflex iliac	External iliac artery	Runs on deep aspect of anterior abdominal wall, parallel to inguinal ligament	Iliacus muscle and inferior part of anterolateral abdominal wall

(continued)

Artery	Origin	Course	Distribution
Dorsal pancreatic	Splenic artery	Descends posterior to pancreas, dividing into right and left branches	Middle portion of pancreas
Esophageal	Anterior aspect of thoracic aorta	Runs anteriorly to esophagus	Esophagus
Gastroduodenal	Hepatic artery	Descends retroperitoneally, posterior to gastroduodenal junction	Stomach, pancreas, first part of duodenum, distal part of bile duct
Greater pancreatic	Splenic artery	Penetrates left portion of pancreas splitting into right and left branches, paralleling pancreatic duct	Anastomoses with other pancreatic branches; supplies mostly the tail of the pancreas and contained duct
Hepatic artery proper	Celiac trunk	Passes retroperitoneally to reach hepatoduodenal ligament, passes between its layers to porta hepatis; bifurcates into right and left hepatic arteries	Branches: right gastric, supraduodenal; right and left hepatic arteries; supplies liver and gallbladder, (stomach, pancreas, duodenum)
Ileocolic	Terminal branch of superior mesenteric artery	Runs along root of mesentery and divides into ileal and colic branches	Ileum, cecum, and ascending colon
Iliolumbar	Posterior division of internal iliac	Ascends anterior to sacroiliac joint and posterior to common iliac vessels and psoas major	Psoas major, iliacus, quadratus lumborum muscles, cauda equina in vertebral canal
Inferior epigastric	External iliac artery	Runs superiorly and enters rectus sheath; runs deep to rectus abdominus	Rectus abdominus and medial part of anterolateral abdominal wall

(continued)

Artery	Origin	Course	Distribution
Inferior gluteal	Anterior division of internal iliac	Exits pelvis near gluteal region via greater sciatic foramen inferior to piriformis, descends on medial side of sciatic nerve; anastomoses with superior gluteal artery; cruciate anastomosis of thigh, involving 1st perforating artery of deep femoral, medial and lateral circumflex femoral arteries	Pelvic: diaphragm (coccygeus and levator ani) piriformis, quadratus femoris, uppermost hamstrings, gluteus maximus, and sciatic nerve
Inferior mesenteric	Abdominal aorta	Descends retroperitoneally to left of abdominal aorta	Gastrointestinal tract
Inferior pancreaticoduodenal (anterior and posterior)	Superior mesenteric artery	Ascends retroperitoneally on head of pancreas	Distal portion of duodenum and inferior head of uncinate process of pancreas
Inferior phrenic	First branches of abdominal aorta	Ascending crus to underside of domes; medial branches anastomoses with each other, pericardiacophrenic arteries; lateral branches approach thoracic wall, anastomose with posterior intercostal and musculophrenic arteries	Diaphragm, inferior vena cava, esophagus, suprarenal glands
Inferior suprarenal	Renal	Ascends vertically to gland	Posterior and inferior aspect of suprarenal gland
Internal iliac	Common iliac	Passes over pelvic brim to reach pelvic cavity	Main blood supply to pelvic organs, gluteal muscles and perineum

(*continued*)

Artery	Origin	Course	Distribution
Intestinal (n = 15–18)	Superior mesenteric artery	Passes between two layers of mesentery	Jejunum and ileum
Lateral sacral, superior and inferior	Posterior division of internal iliac	Runs on anteromedial aspect of piriformis to send branches into pelvic sacral foramen	Piriformis, structures in sacral canal, erector spina and overlying skin
Left colic	Inferior mesenteric artery	Passes retroperitoneally toward left to descending colon	Descending colon
Left gastric	Celiac trunk	Ascends retroperitoneally to esophageal hiatus, passes between layers of hepatogastric ligament	Distal portion of esophagus and lesser curvature of stomach
Left gastroomental	Splenic artery in hilum of spleen	Passes between layers of gastrosplenic ligament to greater curvature of stomach	Left portion of greater curvature of stomach
Lumbar	Abdominal aorta	Runs horizontal course posteriorly around sides of lumbar vertebrae and laterally on posterior abdominal wall	Branches: dorsal to deep muscles of back and overlying skin; spinal to vertebrae, vertebral canal, roots, and spinal cord
Marginal artery of colon	Formed by anastomoses between right, middle, left colic and sigmoid arteries	Anastomotic channel parallels the colon at its mesenteric border	Anterior and posterior aspect of colon
Medium sacral	Posterior aspect of abdominal aorta	Descends in median line over L4 and L5 vertebrae and the sacrum and coccyx	Lower lumbar vertebrae, sacrum and coccyx
Middle colic	Superior mesenteric artery	Ascends retroperitoneally and passes between layers of transverse mesocolon	Transverse colon

(*continued*)

Artery	Origin	Course	Distribution
Middle rectal	Anterior division of internal iliac	Descend in pelvis to lower part of rectum	Seminal vesicles and lower part of rectum
Middle suprarenal	Abdominal aorta	Arises at level of superior mesenteric artery short course over crura of diaphragm	Suprarenal glands; suprarenal branches of inferior phrenic and renal arteries
Obturator	Anterior division of internal iliac	Runs anterolaterally on lateral pelvic wall to exit pelvis via obturator canal	Pelvic muscles, nutrient artery to ilium, head of femur, muscles of medial compartment of thigh
Ovarian	Abdominal aorta, inferior of renal arteries	Runs inferolaterally on psoas major, passes medially to cross pelvic brim and descend in suspensory ligament of ovary	Tubal, ureteric, and ovarian branches
Perineal	Internal pudendal artery	Leaves pudendal canal and enters superficial perineal space	Superficial perineal muscles and scrotum or labia
Posterior division of internal iliac	Internal iliac	Passes posteriorly and gives rise to parietal branches	Pelvic wall and gluteal region
Posterior gastric	Splenic artery	Ascends retroperitoneally in posterior wall of omental bursa, passes to gastric fundus via gastrophrenic fold	Posterior wall of stomach
Prostatic branches	Inferior vesicle artery	Descends on posterolateral aspect of prostate	Prostate

(continued)

Artery	Origin	Course	Distribution
Renal, left and right	Posterolateral aspect of abdominal aorta	Runs horizontally and laterally across crura of diaphragm and psoas major, posterior to renal vein; bifurcating into anterior and posterior divisions into segmental arteries near renal hiatus	Blood to kidneys; inferior suprarenal, capsular branches, anterior division giving rise to superior, anterior superior, anterior inferior, inferior segmental arteries; posterior division becomes posterior segmental artery
Retroduodenal	Gastroduodenal artery	Arises and runs posteriorly to first part of duodenum	First part of duodenum, bile duct, head of pancreas
Right colic	Superior mesenteric artery	Passes retroperitoneally to reach ascending colon	Ascending colon
Right gastric	Hepatic artery	Runs between layers of hepatogastric ligament	Right portion of lesser curvature of stomach
Right gastroomental	Gastroduodenal artery	Passes between layers of greater omentum to greater curvature of stomach	Right portion of greater curvature of stomach
Segmental arteries of kidney (superior, anterior, anterior inferior, and posterior)	Anterior and posterior divisions from renal arteries	Arises at hilum, through perirenal fat of renal sinus, around to renal pelvis to renal segment	Renal segment
Segmented arteries of liver, (right anterior, right posterior, left medial, and left lateral	Left and right branches of hepatic artery proper	Arises within liver; right and left branches course horizontally	Liver
Short gastric (n = 4–5)	Splenic aretery in hilum of spleen	Passes between layers of gastrosplenic ligament to fundus of stomach	Fundus of stomach
Sigmoid (n = 3–4)	Inferior mesenteric artery	Passes retroperitoneally to descending colon	Descending and sigmoid colon

(continued)

Artery	Origin	Course	Distribution
Splenic	Celiac trunk	Runs retroperitoneally along superior border of pancreas, passes between layers of splenorenal ligament to hilum of spleen	Body of pancreas, spleen, greater curvature of stomach
Subcostal	Thoracic aorta	Courses along inferior border of 12th rib	Muscles of anterolateral abdominal wall
Superficial circumflex iliac	Femoral artery	Runs in superficial fascia along inguinal ligament	Subcutaneous tissue, skin over inferior part of anterolateral abdominal wall
Superior epigastric	Internal thoracic artery	Descends in rectus sheath deep to rectus abdominis	Rectus abdominus and superior part of anterolateral abdominal wall
Superior gluteal	Posterior division of internal iliac	Enters gluteal region through greater sciatic foramen superior to piriformis, into superficial and deep branches; anastomoses with inferior gluteal and medial circumflex femoral arteries	Piriformis muscle; gluteus maximus; gluteus medius and minimus muscles; tensor of fascia lata
Superior mesenteric	Abdominal aorta	Runs in root of mesentery to ileocecal junction	Part of gastrointestinal tract derived from midgut
Superior pancreaticoduodenal, anterior and posterior	Gastroduodenal artery	Descends on head of pancreas	Proximal portion of duodenum and head of pancreas
Superior rectal	Terminal branch inferior mesenteric artery	Crosses left common iliac vessels and descends into the pelvis between the layers of sigmoid colon	Upper part of rectum; middle and inferior rectal arteries

(*continued*)

Artery	Origin	Course	Distribution
Superior suprarenal	Inferior phrenic	Short, multiple branches arising from the trunks of the inferior phrenic arteries ascend diaphragmatic crura, along superomedial aspect of gland	Superior part of suprarenal glands
Superior vesicle	Proximal part of umbilical	Usually multiple, passes to superior aspect of urinary bladder	Superior aspect of urinary bladder, pelvic portion of ureter
Supraduodenal	Gastroduodenal, hepatic, right gastric, or retroduodenal arteries	Passes superior to 1st part of duodenum	Upper and proximal portion of superior part of duodenum
Umbilical	Anterior division of internal iliac	Obliterates becoming medial umbilical ligament after running a short pelvic course; gives rise to superior vesicle arteries	Superior aspect of urinary bladder; artery to ductus deferens (males)

14. Arteries of the Perineum

Artery	Origin	Course	Distribution
Artery of bulb of penis or vestibule of vagina	Internal pudendal artery	Pierces perineal membrane to bulb of penis or vestibule of vagina	Bulb of penis or vestibule and bulbourethral gland (male) and greater vestibular gland (female)
Artery to ductus deferens	Inferior or superior vesicle	Runs retroperitoneally to ductus deferens	Ductus deferens
Cremasteric	Inferior epigastric	Accompanies spermatic cord through inguinal canal to scrotal sac	Cremaster muscle, covering of cord in males, round ligament in females

(continued)

Artery	Origin	Course	Distribution
Deep artery of penis or clitoris	Terminal branch of internal pudendal artery	Pierces perineal membrane to reach erectile bodies of clitoris or penis	Helicine arteries uncoils to engorge erectile sinuses with arterial blood
Dorsal artery of penis or clitoris	Terminal branch of internal pudendal artery	Pierces perineal membrane and passes through suspensory ligament of penis or clitoris	Skin of penis and erectile tissue of penis or clitoris
External pudendal, superficial and deep branches	Femoral artery	Pass medially across the thigh to scrotum or labial majora	Skin of mons pubis and anterior labia or root of penis and anterior scrotum
Inferior rectal	Internal pudendal artery	Leaves pudendal canal, crosses ischio-anal fossa to anal canal	Distal portion of anal canal, inferior to pectinate line
Inferior vesicle	Anterior division of internal iliac	Passes retroperitoneally to inferior aspect of male urinary bladder	Inferior aspect of urinary bladder; ductus deferens, seminal vesicles and prostate
Internal pudendal	Anterior division of internal iliac	Leaves pelvis through greater sciatic foramen to ischial spine, enters perineum via lesser sciatic foramen, into pudendal canal, to urogenital triangle	Main artery to perineum, muscles, and skin of anal and urogenital triangles; erectile bodies
Posterior scrotal or labial	Terminal branches of perineal artery	Runs in superficial fascia of posterior scrotum or labium majus	Skin of scrotum or labium majus
Testicular	Abdominal aorta, inferior to renal arteries	Descends inferomedially across psoas muscle; passes through inguinal canal, part of spermatic cord, reaches testis in scrotum	Abdominal part provides arterial blood to ureters, iliac lymph nodes; inguinal scrotal part supplies cremaster and covering of cords

(continued)

Artery	Origin	Course	Distribution
Uterine	Anterior division of internal iliac	Runs medially in base of broad ligament superior to cardinal ligament, crossing superior to ureter, to sides of uterus	Uterus, ligaments of uterus, uterine tube, and vagina
Vaginal	Uterine artery	Arises lateral to ureter and descends inferior to lateral aspect of vagina	Vagina: branches to inferior part of urinary bladder and termination of ureter

15. Arteries of the Leg and Thigh

Artery	Origin	Course	Distribution
Anterior tibial	Terminal branch of popliteal artery	Passes between tibia and fibula to superior part of interosseous membrane, between tibialis anterior and extensor digitorum longus	Anterior compartment of leg
Deep artery of thigh	Femoral artery	Passes inferiorly on medial intermuscular septum, deep to adductor longus	Adductor magnus muscle, posterior and lateral anterior compartments of thigh
Descending genicular	Femoral artery	Vastus medialis, anterior tendon of adductor magnus, superior medial genicular artery	Saphenous nerve, medial skin of leg, muscular branches of vastus medialis and adductor magnus
Femoral	Continuation of external iliac artery, distal to inguinal ligament	Descends via femoral triangle, adductor canal, and changes to popliteal at adductor hiatus	Anterior and anteromedial surface of thigh
Fibular (peroneal)	Posterior tibial	Descends in posterior compartment adjacent to posterior intermuscular septum	Posterior compartment of leg; perforating branches supply lateral compartment of leg

(continued)

Artery	Origin	Course	Distribution
Genicular (superior lateral and medial; inferior lateral and medial and middle)	Popliteal	Arises in knee joint around patella, femoral and tibial condyles; middle genicular pierces oblique popliteal ligament in posterior joint capsule	Descending genicular, descending branch of lateral circumflex femoral, circumflex fibular, recurrent tibial arteries
Lateral circumflex femoral	Deep artery of thigh; femoral artery	Passes laterally to sartorius and rectus femoris and divides into three branches	Ascending: anterior gluteal region; Transverse: around femur; Descending: knee and joins genicular anastomosis
Medial circumflex femoral	Deep artery of thigh	Passes medially and posteriorly between pectineus and iliopsoas, gluteal region and divides into two branches	Supplies most blood to head and neck of femur; transverse branch joins with cruciate anastomosis of thigh; ascending branch joints inferior gluteal artery
Popliteal	Continuation of femoral artery at adductor hiatus in adductor magnus	Passes via popliteal fossa to leg; ends at lower border of popliteus muscle, dividing into anterior and posterior tibial arteries	Superior, middle, inferior genicular arteries to both lateral and medial aspects of knee
Posterior tibial	Popliteal	Passes via posterior compartment of leg; terminates distal to flexor retinaculum, dividing into medial and lateral plantar arteries	Posterior and lateral compartments of leg; circumflex fibular branch joins anastomosis around knee; nutrient artery to tibia
Sural, left and right	Popliteal	Large branch arises at level of femoral condyles and passes directly into head of gastrocnemius; branches to soleus	Medial and lateral heads of gastrocnemius, plantaris, and soleus muscles

16. Arteries of the Foot

Artery	Origin	Course	Distribution
Arcuate of foot	Continuation of dorsalis pedis	Passes laterally, dorsally to bases of metatarsals	2nd, 3rd, and 4th dorsal metatarsal arteries
Common plantar digital arteries	Terminal portions of plantar metatarsal	Short segments distal to transverse head of adductor hallucis proximal to webs between toes	Terminal branches: plantar distal arteries proper
Deep plantar arch	Continuation of lateral plantar artery	Anteromedially between 3rd and 4th layers of muscles of sole of foot; deep plantar artery between 1st and 2nd metatarsal base	Plantar metatarsal arteries
Dorsal digital	Dorsal metatarsal arteries	Distally on the posterolateral aspect of proximal one and a half phalanges	Dorsal aspect of proximal one and a half phalanges of toes
Dorsal metatarsal	Termination of dorsalis pedis (1st); arcuate artery (2nd, 3rd, and 4th)	Distally on the superficial aspect of corresponding dorsal interosseous muscles	Dorsal digital arteries
Dorsal pedis	Continuation of anterior tibial artery distal to inferior extensor retinaculum	Descends anteromedially to first interosseous space, divides into plantar and arcuate arteries	Muscles of dorsum of foot; first dorsal interosseus muscle as deep plantar artery; contributes to plantar arch
Lateral plantar	Terminal branch of posterior tibial artery	Courses anterolaterally between 1st and 2nd muscle layers of sole of foot to base of 5th metatarsal; anterolaterally between 3rd and 4th layers as deep plantar arch	Muscles of 1st and 2nd layers; skin and subcutaneous tissue of lateral sole; lateral tarsal, arcuate arteries

(continued)

Appendix 5

Artery	Origin	Course	Distribution
Medial plantar	Terminal branch of posterior tibial artery	Medial to calcaneus, distally along medial side of foot between 1st and 2nd layers of plantar muscle	Flexor hallucis brevis, abductor hallucis; skin and subcutaneous tissue of medial sole
Plantar metatarsal	Junction between lateral plantar and dorsalis pedis arteries (1st); deep plantar arch (2nd)	Extend distally between metatarsal bones on plantar aspect of interosseous muscles	Perforating branch, common plantar digital arteries

Anatomy Words (English-Latin)

Arteries

accessory artery	arteria obturatoria accessoria
acetabular branch	arteria acetabuli
anterior conjunctival artery	arteria conjunctivalis anterior
anterior choroidal artery	arteria choroidea anterior
anterior superior alveolar arteries	arteria arteriae alveolares superiores anteriores
anterior cecal artery	arteria cecalis anterior
anterior communicating artery	arteria communicans anterior
anterior medial malleolar artery	arteria malleolaris anterior medialis
anterior cerebral artery	arteria cerebri anterior
anterior circumflex humeral artery	arteria circumflexa humeri anterior
anterior ethmoidal artery	arteria ethmoidalis anterior
anterior labial branches of deep external	N/A
anterior interosseous artery	arteria interossea anterior
anterior ciliary arteries	arteriae ciliares anteriores
anterior inferior cerebellar artery	arteria inferior anterior cerebelli
anterior lateral malleolar artery	arteria malleolaris anterior lateralis
anterior meningeal branch (of anterior ethmoidal artery)	arteria meningea anterior
anterior interosseous artery	arteria interossea volaris
anterior perforating arteries	arteriae perforantes anteriores
anterior parietal artery	arteria parietalis anterior
anterior tibial recurrent artery	arteria recurrens tibialis anterior
anterior spinal artery	arteria spinalis anterior
anterior basal segmental artery	arteria segmentalis basalis anterior
anterior tympanic artery	arteria tympanica anterior
anterior tibial artery	arteria tibialis anterior
anterior temporal branch	arteria temporalis anterior
anterior vestibular artery	arteria vestibularis anterior
anterolateral central arteries	arteria arteriae centrales anterolaterales
anteromedial central arteries	arteria arteriae centrales anteromediales
aorta	arteria aorta
appendicular artery	arteria appendicularis
arcuate arteries of kidney	arteria arteriae arcuatae renis
arcuate artery of foot	arteria arcuata pedis
arteries of brain	arteriae encephali
arteries of lower limb	arteriae membri inferioris
arteries of upper limb	arteriae membri superioris
artery to ductus deferens	arteria deferentialis
artery to sciatic nerve	arteria comitans nervi ischiadici

artery to tail of pancreas	arteria caudae pancreatis
artery of bulb of vestibule	arteria bulbi vestibuli
artery of bulb of vagina	arteria bulbi vaginae
artery of bulb of penis	arteria bulbi penis
artery to ductus deferens	arteria ductus deferentis
artery of pterygoid canal	arteria canalis pterygoidei
artery of round ligament of uterus	arteria ligamenti teretis uteri
artery of tuber cinereum	arteria tuberis cinerei
artery of caudate lobe	arteria lobi caudati
artery of precentral sulcus	arteria sulci precentralis
artery of postcentral sulcus	arteria sulci postcentralis
artery of central sulcus	arteria sulci centralis
ascending palatine artery	arteria palatina ascendens
ascending artery	arteria intermesenterica
ascending pharyngeal artery	arteria pharyngea ascendens
ascending cervical artery	arteria cervicalis ascendens
atrial anastomotic branch of circumflex	N/A
atrial arteries	arteria arteriae atriales
axillary artery	arteria axillaris
basilar artery	arteria basilaris
brachial artery	arteria brachialis
branch to angular gyrus	arteria angularis
branch of left coronary artery	arteria anastomotica auricularis magna
branch to angular gyrus	arteria gyri angularis
buccal artery	arteria buccalis
calcarine branch of medial occipital artery	arteria calcarina
callosomarginal artery	arteria callosomarginalis
caroticotympanic arteries	arteria arteriae caroticotympanicae
celiac arterial trunk	arteria celiaca
central retinal artery	arteria retinae centralis, arteria centralis retinae
cervicovaginal artery	arteria cervicovaginalis
circumflex scapular artery	arteria circumflexa scapulae
collicular artery	arteria collicularis
common iliac artery	arteria iliaca communis
common plantar digital artery	arteria digitalis plantaris communis
common interosseous artery	arteria interossea communis
common carotid artery	arteria carotis communis
common hepatic artery	arteria hepatica communis
common cochlear artery	arteria cochlearis communis
cortical radiate arteries	arteriae corticales radiatae, arteria interlobulares (renis)

cremaster artery	arteria cremasterica
cystic artery	arteria cystica
deep plantar branch of dorsalis pedis artery	arteria plantaris profundus
deep lingual artery	arteria ranina
deep artery of penis	arteria profunda penis
deep auricular artery	arteria auricularis profunda
deep plantar artery	arteria plantaris profunda arteriae
deep temporal artery	arteria temporalis profunda
deep artery of clitoris	arteria profunda clitoridis
deep circumflex iliac artery	arteria circumflexa iliaca profunda
deep artery of thigh	arteria profunda femoris
deep cervical artery	arteria cervicalis profunda
deep lingual artery	arteria profunda linguae
descending palatine artery	arteria palatina descendens
descending genicular artery	arteria descendens genus
dorsal nasal artery	arteria nasi externa
dorsal scapular artery	arteria scapularis dorsalis, arteria dorsalis scapulae
dorsal pancreatic artery	arteria pancreatica dorsalis
dorsal metacarpal artery	arteria metacarpalis dorsalis
dorsal scapular artery	arteria scapularis descendens
dorsal digital artery	arteria digitalis dorsalis
dorsal artery of penis	arteria dorsalis penis
dorsal metatarsal artery	arteria metatarsalis dorsalis
dorsal nasal artery	arteria dorsalis nasi
dorsal artery of clitoris	arteria dorsalis clitoridis
dorsalis pedis artery	arteria dorsalis pedis
episcleral artery	arteria episcleralis
external carotid artery	arteria carotis externa
external pudendal arteries (superficial and deep)	arteriae pudendae externae
external iliac artery	arteria iliaca externa
facial artery	arteria facialis
femoral artery	arteria femoralis
fibular artery	arteria fibularis
first and second posterior intercostal arteries	arteriae intercostales posteriores I et II
gastroduodenal artery	arteria gastroduodenalis
gastroomental arteries	arteriae gastro-omentales
great segmental medullary artery	arteria radicularis magna
greater palatine artery	arteria palatina major
greater pancreatic artery	arteria pancreatica magna

helicine arteries of the uterus	arteriae helicinae uteri
helicine arteries of penis	arteriae helicinae penis
hepatic artery proper	arteria hepatica propria
humeral nutrient arteries	arteriae nutriciae humeri
hyaloid artery	arteria hyaloidea
ileal arteries	arteriae ileales
ileocolic artery	arteria ileocolica
iliolumbar artery	arteria iliolumbalis
inferior alveolar artery	arteria alveolaris inferior
inferior epigastric artery	arteria epigastrica inferior
inferior gluteal artery	arteria glutea inferior, arteria ischiadica
inferior hypophysial artery	arteria hypophysialis inferior
inferior labial branch of facial artery	arteria labialis inferior
inferior laryngeal artery	arteria laryngea inferior
inferior lateral genicular artery	arteria inferior lateralis genus
inferior lingular artery	arteria lingularis inferior
inferior medial genicular artery	arteria genus inferior medialis, arteria inferior medialis genus
inferior mesenteric artery	arteria mesenterica inferior
inferior pancreatic artery	arteria pancreatica inferior
inferior pancreaticoduodenal artery	arteria pancreaticoduodenalis inferior
inferior phrenic artery	arteria phrenica inferior
inferior rectal artery	arteria rectalis inferior
inferior suprarenal artery	arteria suprarenalis inferior
inferior thyroid artery	arteria thyroidea inferior
inferior tympanic artery	arteria tympanica inferior
inferior ulnar collateral artery	arteria collateralis ulnaris inferior
inferior vesical artery	arteria vesicalis inferior
infraorbital artery	arteria infraorbitalis
insular arteries	arteriae insulares
interlobar arteries of kidney	arteriae interlobares renis
interlobular arteries of liver	arteria interlobulares (hepatis)
internal carotid artery	arteria carotis interna
internal iliac artery	arteria iliaca interna
internal pudendal artery	arteria pudenda interna
internal thoracic artery	arteria thoracica interna
intrarenal arteries	arteriae intrarenales
jejunal arteries	arteriae jejunales
labyrinthine artery	arteria labyrinthi, arteria auditiva interna
lacrimal artery	arteria lacrimalis
lateral basal segmental artery	arteria segmentalis basalis lateralis
lateral circumflex femoral artery	arteria circumflexa femoris lateralis
lateral frontobasal artery	arteria frontobasalis lateralis

lateral malleolar branch	arteriae malleolares posteriores laterales
lateral and medial palpebral arteries	arteriae palpebrales (laterales et mediales)
lateral occipital artery	arteria occipitalis lateralis
lateral plantar artery	arteria plantaris lateralis
lateral sacral arteries	arteriae sacrales laterales
lateral tarsal artery	arteria tarsea lateralis
lateral thoracic artery	arteria thoracica lateralis
left colic artery	arteria colica sinistra
left coronary artery	arteria coronaria sinistra
left gastric artery	arteria gastrica sinistra
left gastroomental artery	arteria gastroomentalis sinistra, arteria gastroepiploica sinistra
left pulmonary artery	arteria pulmonalis sinistra
lesser palatine artery	arteria palatina minor
lingual artery	arteria lingualis
long posterior ciliary arteries	arteria arteriae ciliares posteriores longae
lowest lumbar arteries	arteriae lumbales imae
lumbar arteries	arteriae lumbales
mammillary arteries	arteriae mammillares
marginal artery of colon	arteria marginalis colon, arteriae jejunales
masseteric artery	arteria masseterica
maxillary artery	arteria maxillaris
medial striate artery	arteria striata medialis distalis
medial frontobasal artery	arteria frontobasalis medialis
medial malleolar branches (of posterior tibial artery)	arteriae malleolares posteriores mediales
medial circumflex femoral artery	arteria circumflexa femoris medialis
medial plantar artery	arteria plantaris medialis
medial occipital artery	arteria occipitalis medialis
medial striate artery	arteria recurrens
medial basal segmental artery	arteria segmentalis basalis medialis
medial tarsal arteries	arteria tarsea medialis
median artery	arteria comitans nervi mediani
median commissural artery	arteria commissuralis mediana
median sacral artery	arteria sacralis mediana
median artery	arteria mediana
median callosal artery	arteria callosa mediana
mental branch (of inferior alveolar artery)	arteria mentalis
metatarsal artery	arteria metatarsalis

middle cerebral artery	arteria cerebri media, arteria temporalis intermedia
middle colic artery	arteria colica media
middle genicular artery	arteria genus media, arteria media genus
middle meningeal artery	arteria meningea media
middle suprarenal artery	arteria suprarenalis media
middle rectal artery	arteria rectalis media
middle temporal artery	arteria temporalis media
middle temporal branch of insular part of middle collateral artery	arteria collateralis media
muscular arteries (of ophthalmic artery)	arteriae musculares
musculophrenic artery	arteria musculophrenica
nutrient artery	arteria nutricia
nutrient artery of femur	arteria nutriciae femoris
nutrient artery of radius	arteria radii nutricia
nutrient artery of ulna	arteria nutricia ulnae
obturator artery	arteria obturatoria
occipital artery	arteria occipitalis
ophthalmic artery	arteria ophthalmica
ovarian artery	arteria ovarica
palmar metacarpal artery	arteria metacarpalis palmaris
paracentral branches (of pericallosal artery)	arteria paracentralis
parietal arteries (lateral and medial)	arteriae parietales (laterales et mediales)
parieto-occipital branches (of anterior cerebral artery)	arteriae parieto-occipitales
perforating radiate arteries (of kidney)	arteriae perforantes radiatae (renis)
perforating arteries of penis	arteriae perforantes penis
perforating arteries (of deep femoral artery)	arteriae perforantes arteriae, profundae femoris
pericallosal artery	arteria pericallosa
pericardiacophrenic artery	arteria pericardiacophrenica, arteria comes nervi phrenici
perineal artery	arteria perinealis
plantar metatarsal artery	arteria metatarsalis plantaris
polar frontal artery	arteria polaris frontalis
polar temporal artery	arteria polaris temporalis
pontine arteries	arteriae pontis
popliteal artery	arteria poplitea
posterior auricular artery	arteria auricularis posterior
posterior cecal artery	arteria cecalis posterior
posterior cerebral artery	arteria cerebri posterior
posterior choroidal artery	arteria choroidea posterior

posterior circumflex humeral artery	arteria circumflexa humeri posterior
posterior communicating artery	arteria communicans posterior
posterior conjunctival artery	arteria conjunctivalis posterior
posterior ethmoidal artery	arteria ethmoidalis posterior
posterior gastric artery	arteria gastrica posterior
posterior inferior cerebellar artery	arteria inferior posterior cerebelli
posterior intercostal arteries 3–11	arteriae intercostales posteriores III–XI
posterior interosseous artery	arteria interossea posterior
posterior lateral nasal arteries	arteriae nasales posteriores laterales
posterior meningeal artery	arteria meningea posterior
posterior parietal artery	arteria parietalis posterior
posterior septal branch of nose	arteria nasalis posterior septi
posterior spinal artery	arteria spinalis posterior
posterior superior alveolar artery	arteria alveolaris superior posterior
posterior temporal branch of middle cerebral artery	arteria temporalis posterior
posterior tibial recurrent artery	arteria recurrens tibialis posterior
posterior tibial artery	arteria tibialis posterior
posterior tympanic artery	arteria tympanica posterior
posterolateral central arteries	arteria arteriae centrales posterolaterales
posteromedial central arteries	arteria arteriae centrales posteromediales
precuneal branches (of anterior cerebral artery)	arteria precunealis
prepancreatic artery	arteria prepancreatica
princeps pollicis artery	arteria princeps pollicis
profunda brachii artery	arteria profunda brachii
proper cochlear artery	arteria cochlearis propria
proper digital arteries	arteria digitalis palmaris propria
proper palmar digital arteries	arteriae digitales palmares propriae
proper plantar digital artery	arteria digitalis plantaris propriae
proximal medial striate arteries	arteria centralis brevis
pterygomeningeal artery	arteria pterygomeningealis
pudendal artery	arteriae labiales anteriores
pulmonary trunk	arteria pulmonalis
radial recurrent artery	arteria recurrens radialis
radial artery	arteria radialis
radial collateral artery	arteria collateralis radialis
radialis indicis artery indicis radialis	arteria radialis indicis, arteria volaris
radicular arteries (anterior and posterior)	arteriae radiculares (anterior et posterior)
recurrent interosseous artery	arteria interossea recurrens
renal artery	arteria renalis

retroduodenal artery	arteria retroduodenalis
right colic artery	arteria colica dextra
right coronary artery	arteria coronaria dextra
right flexural artery	arteria flexurae dextrae
right gastric artery	arteria gastrica dextra
right gastroomental artery	arteria gastroepiploica dextra, arteria gastroomentalis dextra
right pulmonary artery	arteria pulmonalis dextra
segmental arteries of liver	arteriae segmenti hepaticae
segmental arteries of kidney	arteriae renis
segmental medullary arteries	arteriae medullares segmentales
short circumferential arteries	arteriae circumferentiales brevis
short gastric arteries	arteriae gastricae breves
short posterior ciliary arteries	arteria ciliares posterior brevis
sigmoid arteries	arteriae sigmoideae
sphenopalatine artery	arteria sphenopalatina
splenic artery	arteria splenica, arteria lienalis
stylomastoid artery	arteria stylomastoidea
subclavian artery	arteria subclavia
subcostal artery	arteria subcostalis
sublingual artery	arteria sublingualis
submental artery	arteria submentalis
subscapular artery	arteria subscapularis
superficial brachial artery	arteria brachialis superficialis
superficial cervical artery	arteria cervicalis superficialis
superficial circumflex iliac artery	arteria circumflexa iliaca superficialis
superficial epigastric artery	arteria epigastrica superficialis, arteria epigastrica superior
superior cerebellar artery	arteria superior cerebelli
superior gluteal artery	arteria glutea superior
superior hypophysial artery	arteria hypophysialis superior
superior labial branch of facial artery	arteria labialis superior
superior laryngeal artery	arteria laryngea superior
superior lateral genicular artery	arteria superior lateralis genus
superior lingular artery	arteria lingularis superior
superior medial genicular artery	arteria superior medialis genus
superior mesenteric artery	arteria mesenterica superior
superior pancreaticoduodenal artery (anterior and posterior)	arteria pancreaticoduodenalis
superior phrenic artery	arteria phrenica superior
superior rectal artery	arteria rectalis superior
superior suprarenal arteries	arteriae suprarenales superiores
superficial temporal artery	arteria temporalis superficialis superior

superior thoracic artery	arteria thoracica superior
superior thyroid artery	arteria thyroidea superior
superior tympanic artery	arteria tympanica superior
superior ulnar collateral artery	arteria collateralis ulnaris superior
superior vesical artery	arteria vesicalis superior
suprachiasmatic artery	arteria suprachiasmatica
supraduodenal artery	arteria supraduodenalis
supraoptic artery	arteria supraoptic artery
supraorbital artery	arteria supraorbitalis
suprascapular artery	arteria suprascapularis
supratrochlear artery	arteria frontalis, arteria supratrochlearis
supreme intercostal artery	arteria intercostalis suprema
sural arteries	arteriae surales
testicular artery	arteria spermatica interna, arteria testicularis
thoracoacromial artery	arteria thoracoacromialis
thoracodorsal artery	arteria thoracodorsalis
thyroid ima artery	arteria thyroidea ima
tibial nutrient artery	arteria nutricia tibiae
transverse cervical artery	arteria transversa colli
transverse facial artery	arteria transversa faciei
ulnar recurrent artery	arteria recurrens ulnaris
ulnar artery	arteria ulnaris
umbilical artery	arteria umbilicalis
uncal artery	arteria uncalis
urethral artery	arteria urethralis
uterine artery	arteria uterina
vaginal artery	arteria vaginalis
ventricular arteries	arteriae ventriculares
vertebral artery	arteria vertebralis
vestibulocochlear artery	arteria vestibulocochlearis
vitelline artery	arteria vitellina
zygomatico-orbital artery	arteria zygomatico-orbitalis

Bones

acetabulum	os acetabuli
acromial bone	os acromiale
ankle bone	talus
atlas bone (neck)	atlas
axis bone (neck)	axis
basilar bone	os basilare
Bertin bone	N/A
blade bone	scapula

breast bone	sternum
Breschet bone	os suprasternale
calcaneus (foot)	calcaneus; os calcis; os tarsi fibulare
capitate bone (wrist)	os capitatum; os carpale distale tertium
carpal bone (wrist)	os carpi; os carpalia
central bone (wrist)	os centrale
clavicle (shoulder)	clavicula
coccyx (lower back)	os coccygis
compact bone	substantia compacta
concha, inferior nasal (skull)	concha nasalis inferior
cortical bone	substantia corticalis
cranial bone	os cranii; os cranialis
cuboid bone (foot)	N/A
os cuboideum cuneiform bone, lateral (foot)	os cuneiforme laterale
cuneiform bone, medial (foot)	os cuneiforme mediale
cuneiform bone, intermediate (foot)	os cuneiforme intermedium
digit, bone of	os digitorum
digit of hand	os digitorum manus
elbow bone (olecranon process of ulna)	cubitus
epipteric bone (Flower bone)	N/A
ethmoid bone	os ethmoidale
facial bone	os facialis; os faciei
femur (thigh)	femur
fibula (leg)	fibula
flat bone	os planum
foot, bone of	os pedis
frontal bone (skull)	os frontale
greater multangular bone	trapezium
hamate bone (wrist)	os hamatum; os carpale distale quartum
hand, bones of carpals, metacarpals phalanges	ossa manus
heel bone	calcaneus
hip bone (pelvis and hip)	os coxae
hooked bone—hamate bone	N/A
humerus (arm)	humerus
hyoid bone (neck)	N/A
os hyoideum iliac bone; ilium (pelvis)	os ilii; os ilium
incisive bone	os incisivum
incus (ear)	incus
inferior limb, bones of os coxae, pelvis, tibia, fibula, tarsus metatarsus, digits of foot	ossa membri inferioris

innominate bone (hip bone)	os coxae
intermaxillary bone	os incisivum
interparietal bone	os interparietale
irregular bone	os irregulare
ischial bone; ischium (pelvis)	os ischii
jaw bone (mandible)	mandibula
Krause bone (small bone)	N/A
lacrimal bone (skull)	os lacrimale
lamellar bone	N/A
lesser multangular bone (trapezoid)	os trapezoideum
lone bone (pipe bone)	os longum
lunate bone (wrist)	os lunatum
malar bone	os zygomaticum
malleus (ear)	malleus
mandible (lower jaw)	mandibula
mastoid bone of occipital bone	margo mastoideus squamae occipitalis
maxilla (skull, upper jaw)	maxilla
metacarpal bone, third or middle	N/A
os metacarpale tertium metacarpal bone (hand)	os metacarpalia
metatarsal bone (foot)	os metatarsalia
nasal bone	os nasale
navicular bone of foot	os naviculare; os centrale tarsi
navicular bone of hand	os scaphoideum
nonlamellar bone (woven bone)	N/A
occipital bone (skull)	os occipitale
palatine bone (skull)	os palatinum
parietal bone (skull)	os parietale
patella (knee)	patella
pelvis	os pelvicus
penis bone	os penis
perichondral bone (periosteal bone)	N/A
peroneal bone (fibula)	N/A
petrosal bone	N/A
phalanx, proximal	os phalanx proximalis
phalanx, middle	os phalanx media
phalanx, distal	os phalanx distalis
ping-pong bone	N/A
pisiform bone (wrist)	os pisiforme
pneumatic bone	os pneumaticum
preinterparietal bone	N/A
premaxillary bone	os incisivum
pubic bone (pelvis)	os pubis; mons pubis

pyramidal bone (triquetral)	os triquetrum
radius (forearm)	radius
rib, ribs	os costae; os costale; ossa costalis
Riolan bone	N/A
sacrum (lower back)	os sacrum
scaphoid bone (wrist)	os scaphoideum
scapula	scapula
septal bones (interalveolar septum)	N/A
sesamoid bones of foot	os sesamoidea pedis
sesamoid bone of hand	os sesamoidea manus
shin bone (tibia)	N/A
short bone	os breve
sieve bone	N/A
skull, bone of	os cranii
sphenoid bone (base of skull)	os sphenoidale
splint bone (fibula)	N/A
spongy bone	substantia spongiosa
stapes (ear)	stapes
sternum (chest)	sternum
suprainterparietal bone	N/A
suprasternal bone	N/A
os suprasternalia sutural bone	os suturalia; os fonticulorum
tail bone (coccyx)	os coccygis
talus (ankle)	talus; os tarsi tibiale
temporal bone (skull)	os temporale
thigh bone (femur)	N/A
trabecular bone	substantia spongiosa
trapezium bone (wrist)	os trapezium; os carpale distale primum
trapezoid bone (wrist)	os trapezoideum; os carpale distale secundum
triangular bone	os trigonum
triquetral bone (wrist)	os triquetrum
tympanic bone	N/A
ulna (forearm)	ulna
vertebrae, thoracic (dorsal)	vertebrae thoracicae
vertebrae, lumbar	vertebrae lumbales
vertebrae, cervical	vertebrae cervicales
vertebrae, sacral	vertebrae sacrales
vertebrae, coccygeal	vertebrae coccygeae
Vesalius' bone; vesalian bone	os vesalianum pedis
vomer (skull)	vomer
wedge bone (intermediate cuneiform)	N/A
zygomatic bone, yoke bone	os zygomaticum

Ligaments

acromioclavicular	ligamentum acromioclaviculare
anterior cruciate	ligamentum cruciatum anterius
anterior ligament of fibular head	ligamentum capitis fibulae anterius
anterior ligament of Helmholtz	N/A
anterior ligament of malleus	ligamentum mallei anterius
anterior longitudinal	ligamentum longitudinal anterius
anterior meniscofemoral	ligamentum meniscofemorale anterius
anterior sacrococcygeal	ligamentum sacrococcygeum anterius
anterior sacroiliac	ligamenta sacroiliaca anteriora
anterior sternoclavicular	ligamentum sternoclaviculare
anterior talofibular	ligamentum talofibulare anterius
anterior tibiofibular	ligamentum tibiofibulare anterius
anular	ligamentum anulare
anular ligament of radius	ligamentum anulare radii
anular ligament of stapes	ligamentum anulare stapedis
anular ligament of trachea	ligamentum anularia trachealia
apical ligament of dens	ligamentum apicis dentis
arcuate popliteal	ligamentum popliteum arcuatum
auditory ligament of ossicles	ligamenta ossiculorum auditus
auricle ligament	ligamenta auricularia
Bardinet ligament	N/A
Barkow ligament	N/A
Bellini ligament	N/A
Berry ligament	N/A
Bertin ligament	N/A
Bichat ligament	N/A
bifurcate	ligamentum bifurcatum
Bigelow ligament	N/A
Botallo ligament	N/A
Bourgery ligament	N/A
broad ligament of uterus	ligamentum latum uteri
Brodie ligament	N/A
Burns ligament	N/A
calcaneocuboid	ligamentum calcaneocuboideum
calcaneofibular	ligamentum calcaneofibulare
calcaneonavicular	ligamentum calcaneonaviculare
Caldani ligament	N/A
Campbell ligament	N/A
Camper ligament	N/A
capsular	ligamentum capsulare
cardinal	ligamentum cardinale
carpometacarpal	ligamentum carpometacarpalia

ceratocricoid	ligamentum ceratocricoideum
Civinini ligament	N/A
Clado ligament	N/A
collateral	ligamentum collaterale
conoid	ligamentum conoideum
Cooper ligament	N/A
coracoacromial	ligamentum coracoacromiale
coracoclavicular	ligamentum coracoclaviculare
coracohumeral	ligamentum coracohumerale
coronary ligament of liver	ligamentum coronarium hepatis
costoclavicular	ligamentum costoclaviculare
costotransverse	ligamentum costotransversarium
costoxiphoid	ligamentum costoxiphoideum
Cowper ligament	N/A
cricoarytenoid	ligamentum cricoarytenoideum posterius
cricopharyngeal	ligamentum cricopharyngeum
cricotracheal	ligamentum cricotracheale
cruciate ligament of atlas	ligamentum cruciatum atlas
cruciate ligament of knee	ligamenta cruciata genus
Cruveilhier ligament	N/A
cuboideonavicular	ligamenta cuboideonavicular
cuneocuboid	ligamentum cuneocuboideum
cuneocuboid interosseous	ligamentum cuneocuboideum interosseum
cuneometatarsal interosseous	ligamenta cuneometatarsalia interossea
deep posterior sacrococcygeal	ligamentum sacrococcygeum posterius profundum
deep transverse metatarsal	ligamentum metatarsale transversum profundum
deltoid	N/A
Denonvilliers ligament	N/A
denticulate	ligamentum denticulatum
Denuce ligament	N/A
dorsal carpometacarpal	ligamenta carpometacarpalia dorsalia
dorsal cuboideonavicular	ligamentum cuboideonaviculare dorsale
dorsal cuneocuboid	ligamentum cuneocuboideum dorsale
dorsal cuneonavicular	ligamenta cuneonavicularia dorsalia
dorsal metacarpal	ligamenta metacarpalia dorsalia
dorsal metatarsal	ligamenta metatarsalia dorsalia
dorsal radiocarpal	ligamenta radiocarpale dorsale
dorsal tarsal	ligamenta tarsi dorsalia
dorsal tarsometatarsal	ligamenta tarsometatarsalia dorsalia
duodenorenal	ligamentum duodenorenale

epididymis	ligamenta epididymidis
extracapsular	ligamenta extracapsularia
falciform ligament of liver	ligamentum falciforme hepatis
Ferrein ligament	N/A
fibular collateral	ligamentum collaterale fibulare
Flood ligament	N/A
fundiform ligament of clitoris	ligamentum fundiforme clitoridis
fundiform ligament of penis	ligamenta fundiforme penis
gastrocolic	ligamentum gastrocolicum
gastrophrenic	ligamentum gastrophrenicum
gastrosplenic	ligamentum gastrosplenicum
genitoinguinal	ligamentum genitoinguinale
Gerdy ligament	N/A
Gillette suspensory ligament	N/A
Gimbernat ligament	N/A
glenohumeral	ligamenta glenohumeralia
Gunz ligament	N/A
head of femur ligament	ligamentum capitis femoris
Helmholtz axis ligament	N/A
Hensing ligament	N/A
hepatocolic	ligamentum hepatocolicum
hepatoduodenal	ligamentum hepatoduodenale
hepatoesophageal	ligamentum hepatoesophageum
hepatogastric	ligamentum hepatogastricum
hepatorenal	ligamentum hepatorenale
Hesselbach ligament	N/A
Hey ligament	N/A
Hueck ligament	N/A
Humphry ligament	N/A
Hunter ligament	N/A
hyalocapsular	ligamentum hyaloideocapsular
hyoepiglottic	ligamentum hyoepiglotticum
hypsiloid	N/A
iliofemoral	ligamentum iliofemorale
iliolumbar	ligamentum iliolumbale
iliopectineal	N/A
inferior ligament of epididymis	ligamentum epididymidis inferius
inferior pubic	ligamentum pubicum inferius
inferior transverse scapular	ligamentum transversum scapulae inferius
inguinal	ligamentum inguinale
intercarpal	ligamenta intercarpalia
interclavicular	ligamentum interclaviculare

intercuneiform	ligamenta intercuneiformia
interfoveolar	ligamentum interfoveolare
interosseous metacarpal	ligamenta metacarpalia interossea
interosseous sacroiliac	ligamenta sacroiliaca interossea
interspinous	ligamentum interspinale
intertransverse	ligamentum intertransversarium
intraarticular ligament of head of rib	ligamentum capitis costae intraarticulare
intraarticular sternocostal	ligamentum sternocostale intraarticulare
intrascapular	ligamenta intracapsularia
ischiofemoral	ligamentum ischiofemorale
jugal	N/A
Krause ligament	N/A
laciniate	N/A
lacunar	ligamentum lacunare
Lannelongue ligament	N/A
lateral ligament of ankle	ligamentum collaterale laterale
lateral arcuate	ligamentum arcuatum laterale
lateral ligament of bladder	ligamentum laterale vesicae
lateral costotransverse	ligamentum costotransversarium
lateral ligament of malleus	ligamentum mallei laterale
lateral palpebral	ligamentum palpebrale laterale
lateral sacrococcygeal	ligamentum sacrococcygeum laterale
lateral talocalcaneal	ligamentum talocalcaneum laterale
lateral ligament of temporomandibular joint	ligamentum laterale articulationis temporomandibularis
lateral thyroid	ligamentum thyrohyoideum laterale
Lauth ligament	N/A
left triangular ligament of liver	ligamentum triangulare sinistrum hepatis
left vena cava ligament	ligamentum venae cavae sinistrae
Lisfranc ligament	N/A
Lockwood ligament	N/A
longitudinal	ligamenta longitudinalia
long plantar	ligamentum plantare longum
lumbocostal	ligamentum lumbocostale
Luschka ligament	N/A
Mackenrodt ligament	N/A
Mauchart ligament	N/A
Meckel ligament	N/A
medial ligament of ankle joint	ligamentum collaterale mediale
medial arcuate	ligamentum arcuatum mediale
medial palpebral	ligamentum palpebral

medial talocalcaneal	ligamentum talocalcaneum mediale
medial temporomandibular joint	ligamentum mediale articulationis temporomandibularis
median arcuate	ligamentum arcuatum medianum
median thyroid	ligamentum thyrohyoideum medianum
median umbilical	ligamentum umbilicale medianum
meniscofemoral	ligamenta meniscofemoralia
metatarsal interosseous	ligamenta metatarsalia interossea
nuchal	N/A
oblique popliteal	ligamentum popliteum obliquum
ovarian ligament	ligamentum ovarii proprium
palmar	ligamenta palmaria
palmar carpometacarpal	ligamenta carpometacarpalia palmaria
palmar ligament of interphalangeal joint of hand	ligamenta palmaria articulationis interphalangeae
palmar metacarpal	ligamenta metacarpalia palmaria
palmar ligament of metacarpo-phalangeal joint	ligamenta palmaria articulationis metacarpophalangeae
palmar radiocarpal	ligamentum radiocarpale palmare
palmar ulnocarpal	ligamentum ulnocarpale palmare
patellar	ligamentum patellae
pectinate ligament of iris	N/A
pectineal	ligamentum pectineale
phrenicocolic	ligamentum phrenicocolicum
phrenicosplenic	ligamentum phrenicosplenicum
pisohamate	ligamentum pisohamatum
pisometacarpal	ligamentum pisometacarpeum
plantar	ligamentum plantaria
plantar calcaneocuboid	ligamentum calcaneocuboideum
plantar calcaneonavicular	ligamentum calcaneonaviculare
plantar cuboideonavicular	ligamenta cuboideonavicularia plantaria
plantar cuneonavicular	ligamenta cuneonavicularia plantaria
plantar ligament of interphalangeal joint of foot	ligamenta plantaria articulationis interphalangeae pedis
plantar ligament of metatarsophalangeal joint	ligamenta plantaria articulationis metatarsophalangeae
plantar metatarsal	ligamenta metatarsalia plantaria
plantar tarsal	ligamenta tarsi plantaria
plantar tarsometatarsal	ligamenta tarsometatarsalia plantaria
posterior cruciate	ligamentum cruciatum posterius
posterior ligament of fibular head	ligamentum capitis fibulare posterius
posterior ligament of incus	ligamentum incudis posterius

posterior longitudinal	ligamentum longitudinale posterius
posterior meniscofemoral	ligamentum meniscofemorale posterius
posterior sacroiliac	ligamenta sacroiliaca posteriora
posterior sternoclavicular	ligamentum sternoclaviculare posterius
posterior talocalcaneal	ligamentum talocalcaneum posterius
posterior talofibular	ligamentum talofibulare posterius
posterior tibiofibular	ligamentum tibiofibulare posterius
Poupart ligament	N/A
proper ligament of ovary	N/A
pterygospinous	ligamentum pterygospinale
pubofemoral	ligamentum pubofemorale
puboprostatic	ligamentum puboprostaticum
pubovesical ligament of female	ligamentum pubovesicale
pubovesical ligament of male	ligamentum puboprostaticum
pulmonary	ligamentum pulmonale
quadrate	ligamentum quadratum
radial collateral ligament of elbow	ligamentum collaterale radiale articulationis cubiti
radial collateral ligament of wrist	ligamentum collaterale carpi radiale articulationis radiocarpalis
radiate carpal	ligamentum carpi radiatum
radiate ligament of head of rib	ligamentum capitis costae radiatum
radiate sternocostal	ligamenta sternocostalia radiata
reflected inguinal	ligamentum reflexum
Retzius ligament	N/A
rhomboid	N/A
right triangular ligament of liver	ligamentum triangulare dextrum hepatis
ring	N/A
round ligament of liver	ligamentum teres hepatis
round ligament of uterus	ligamentum teres uteri
sacrodural	ligamentum sacrodurale
sacrospinous	ligamentum sacrospinale
sacrotuberous	ligamentum sacrotuberale
serous	ligamentum serosum
Simonart ligament	N/A
Soemmerring ligament	N/A
sphenomandibular	ligamentum sphenomandibulare
spiral ligament of cochlear duct	ligamentum spirale ductus cochlearis
splenorenal	ligamentum splenorenale
Stanley cervical	N/A
sternoclavicular	ligamenta sternoclavicularia
sternopericardial	ligamenta sternopericardiaca
stylohyoid	ligamentum stylohyoideum

stylomandibular	ligamentum stylomandibulare
superficial posterior sacrococcygeal	ligamentum sacrococcygeum posterius superficiale
superficial transverse metacarpal	ligamentum metacarpale transversum superficiale
superficial transverse metatarsal	ligamentum metatarsal transversum superficiale
superior costotransverse	ligamentum costotransversarium superius
superior epididymis	ligamentum epididymidis superius
superior ligament of incus	ligamentum incudis superius
superior ligament of malleus	ligamentum mallei superius
superior pubic	ligamentum pubicum superius
superior transverse scapular	ligamentum transversum scapulae superius
supraspinous	ligamentum supraspinale
suspensory ligament of axilla	ligamentum suspensorium axillae
suspensory ligament of breast	ligamenta suspensoria mammaria
suspensory ligament of clitoris	ligamentum suspensorium clitoridis
suspensory ligament of eye	ligamentum suspensorium bulbi
suspensory ligament of ovary	ligamentum suspensorium ovarii
suspensory ligament of penis	ligamentum suspensorium penis
suspensory ligament of thyroid	ligamentum suspensorium glandulae thyroideae
talocalcaneal interosseous	ligamentum talocalcaneare interosseum
talonavicular	ligamentum talonaviculare
tarsal	ligamenta tarsi
tarsal interosseous	ligamenta tarsi interossea
tarsometatarsal	ligamenta tarsometatarsalia
Teutleben ligament	N/A
Thompson ligament	N/A
thyroepiglottic	ligamentum thyroepiglotticum
tibial collateral	ligamentum collaterale tibiale
transverse acetabular	ligamentum transversum acetabuli
transverse ligament of atlas	ligamentum transversum atlas
transverse carpal	N/A
transverse cervical	N/A
transverse crural	N/A
transverse genicular	N/A
transverse humeral	ligamentum transversum humeri
transverse ligament of knee	ligamentum transversum genus
transverse metacarpal	N/A
transverse metatarsal	N/A

transverse perineal	ligamentum transversum perinei
transverse tibiofibular	N/A
trapezoid	ligamentum trapezoideum
triangular	N/A
ulnar collateral ligament of elbow	ligamentum collaterale ulnar articulationis cubiti
ulnar collateral ligament of wrist	ligamentum collaterale carpi ulnare articulationis radiocarpalis
uterovesical	N/A
Valsalva ligament	N/A
venous	N/A
ventral sacrococcygeal	N/A
ventral sacroiliac	N/A
vesicoumbilical	N/A
vesicouterine	N/A
vestibular	ligamentum vestibulare
vocal	ligamentum vocale
volar carpal	N/A
Weitbrecht ligament	N/A
Winslow ligament	N/A
Wrisberg ligament	N/A
Zaglas ligament	N/A
Zinn ligament	N/A

Muscles

abdominal muscle	musculi abdominis
abdominal external oblique muscle	musculus obliquus externus abdominis
abdominal internal oblique muscle	musculus obliquus internus abdominis
abductor digiti minimi muscle of foot	musculus abductor digiti minimi pedis
abductor digiti minimi muscle of hand	musculus abductor digiti minimi manus
abductor hallucis muscle	musculus abductor hallucis
abductor pollicis brevis muscle	musculus abductor pollicis brevis
abductor pollicis longus muscle	musculus abductor pollicis longus
adductor muscle, short	musculus adductor brevis
adductor muscle of great toe	musculus adductor hallucis
adductor muscle, long	musculus adductor longus
adductor muscle, great	musculus adductor magnus
adductor minimus muscle	musculus adductor minimus
adductor pollicis muscle	musculus adductor pollicis
accessory flexor muscle of foot	musculus flexor accessorius
Aeby cutaneomucous muscle	N/A
agonistic muscle	N/A

Albinus muscle	N/A
anconeus muscle	musculus anconeus
antagonistic muscle	N/A
anterior auricular muscle	musculus auricularis anterior
anterior cervical intertransverse muscle	musculi intertransversarii anteriores cervicis
anterior rectus muscle of head	musculus rectus capitis anterior
anterior scalene muscle	musculus scalenus anterior
anterior serratus muscle	musculus serratus anterior
anterior tibial muscle	musculus tibialis anterior
antigravity muscle	N/A
antitragus muscle	musculus antitragicus
arrector pili muscle	musculi arrectores pilorum
articular muscle	musculus articularis
articular muscle of elbow	musculus articularis cubiti
articular muscle of knee	musculus articularis genus
articular muscle of knee	articularis genu musculus
aryepiglottic muscle	musculus aryepiglotticus
auditory ossicles, muscles of	musculi ossiculorum auditus
axillary arch muscle	pectorodorsalis musculus
back muscles	musculi dorsi
Bell muscle	N/A
biceps brachii muscle	musculus biceps brachii
biceps femoris muscle	musculus biceps femoris
bipennate muscle	musculus bipennatus
Bochdalek muscle	musculus triticeoglossus
Bovero muscle	musculus cutaneomucosus
Bowman muscle	musculus ciliaris
brachial muscle	musculus brachialis
brachiocephalic muscle	musculus brachiocephalicus
brachioradial muscle	musculus brachioradialis
Braune muscle	musculus puborectalis
broadest muscle of the back	musculus latissimus dorsi
bronchoesophageal muscle	musculus bronchoesophageus
Brücke muscle	N/A
buccinator muscle	musculus buccinator
buccopharyngeal muscle	musculi pars buccopharyngea
bulbocavernous muscle	musculus bulbospongiosus
canine muscle	musculus levator anguli oris
cardiac muscle	N/A
Casser perforated muscle	musculus coracobrachialis
ceratocricoid muscle	musculus ceratocricoideus

cervical iliocostal muscle	musculus iliocostalis cervicis
cervical interspinal muscles	musculi interspinalis cervicis, musculus longissimus cervicis
cervical muscles	musculi colli
cervical rotator muscles	musculi rotatores cervicis
Chassaignac axillary muscle	N/A
chin muscle	musculus mentalis
chondroglossus muscle	musculus chondroglossus
chondropharyngeal muscles	musculi pars chondropharyngea
ciliary muscle	muscularis ciliaris
coccygeal (coccygeus) muscle(s)	musculus coccygeus, musculi coccygei
Coiter muscle	musculus corrugator supercilii
compressor muscle of naris	N/A
congenerous muscles	N/A
constrictor muscle of pharynx, inferior	musculus constrictor pharyngis inferior
constrictor muscle of pharynx, middle	musculus constrictor pharyngis medius
constrictor muscle of pharynx, superior	musculus constrictor pharyngis superior
coracobrachial muscle	musculus coracobrachialis
corrugator muscle	corrugator supercilii musculus
corrugator cutis muscle of anus	musculus corrugator cutis ani
corrugator supercilii muscle	musculus corrugator supercilii
Crampton muscle	N/A
cremaster muscle	musculus cremaster
cricoarytenoid muscle, lateral	musculus cricoarytenoideus lateralis
cricoarytenoid muscle, posterior	musculus cricoarytenoideus posterior
cricopharyngeal muscle	musculus cricopharyngeus
cricothyroid muscle	musculus cricothyroideus
cruciate muscle	musculus cruciatus
cutaneomucous muscle	musculus cutaneomucosus
cutaneous muscle	musculus cutaneus
dartos muscle of scrotum	musculus tunica dartos
deep muscles of back	musculi dorsi
deep flexor muscles of fingers	musculus flexor digitorum profundus
deep transverse perineal muscle	musculus transversus perinei profundus
deltoid muscle	musculus deltoideus
depressor muscle of angle of mouth	musculus depressor anguli oris
depressor muscle of epiglottis	musculus thyroepiglottic
depressor muscle of eyebrow	musculus depressor supercilii
depressor muscle of lower lip	musculus depressor labii inferioris
depressor muscle of septum of nose	musculus depressor septi
depressor superciliary muscle	musculus depressor supercilii
detrusor muscle of urinary bladder	musculus detrusor urinae
diaphragm, diaphragmatic muscle	diaphragma

digastric muscle	musculus digastricus
dilator muscle	musculus dilator
dilator muscle of ileocecal sphincter	musculus dilator pylori ilealis
dilator muscle of pupil	musculus dilator pupillae
dilator muscle of pylorus	musculus dilator pylori gastroduodenalis
dorsal interosseous muscles of foot	musculi interossei dorsalis pedis
dorsal interosseous muscles of hand	musculi interossei dorsalis manus
dorsal muscles	musculi dorsi
dorsal sacrococcygeal muscle	musculus sacrococcygeus dorsalis
Dupré muscle	musculus articularis genu
Duverney muscle	musculus orbicularis oculi
elevator muscle of anus	musculus levator ani
elevator muscle of prostate	musculus levator prostatae
elevator muscles of rib	musculi levatores costarum
elevator muscle of scapula	musculus levator scapulae
elevator muscle of soft palate	musculus levator veli palatini
elevator muscle of upper eyelid	musculus levator palpebrae superioris
elevator muscle of upper lip	musculus levator labii superioris
elevator muscle of upper lip and wing of nose	musculus levator labii superioris alaeque nasi
emergency muscles	N/A
epicranial muscle	musculus epicranius
epimeric muscle	N/A
epitrochleoanconeus muscle	musculus epitrochleoanconaeus
erector muscles of hairs	musculi arrectores pilorum
erector muscle of penis	musculus ischiocavernosus
erector muscle of spine	musculus erector spine
eustachian muscle	musculus tensor tympani
expressions, muscles of	musculi faciales
extensor muscle of fingers	musculus extensor digitorum
extensor muscle of great toe, long	musculus extensor hallucis longus
extensor muscle of great toe, short	musculus extensor hallucis brevis
extensor muscle of hand, short	musculus extensor digitorum brevis manus
extensor muscle of index finger	musculus extensor indicis
extensor muscle of little finger	musculus extensor digiti minimi
extensor muscle of thumb, long	musculus extensor pollicis longus
extensor muscle of thumb, short	musculus extensor pollicis brevis
extensor muscle of toes, long	musculus extensor digitorum longus
extensor muscle of toes, short	musculus extensor digitorum brevis
extensor muscle of wrist, radial, long	musculus extensor carpi radialis longus
extensor muscle of wrist, radial, short	musculus extensor carpi radialis brevis
extensor muscle of wrist, ulnar	musculus extensor carpi ulnaris

external intercostal muscles	musculi intercostales externi
external oblique muscle	musculus obliquus externus abdominis
external obturator muscle	musculus obturator externus
external pterygoid muscle	lateral pterygoid musculus
external sphincter muscle of anus	N/A
extraocular muscles	musculi bulbi
extrinsic muscles	N/A
facial and masticatory muscles	musculi faciales et masticatores
facial expression, muscles of	musculi faciales
facial muscles	musculi faciales
fast muscle	N/A
fauces (the throat), muscles of	musculi palati et faucium
femoral muscle	musculus vastus intermedius
fibular muscle, long	musculus peroneus longus
fibular muscle, short	musculus peroneus brevis
fibular muscle, third	musculus peroneus tertius
fixation muscles, fixator muscles	N/A
fixator muscle of base of stapes	musculus fixator baseos stapedis
flexor muscle of fingers, deep	musculus flexor digitorum profundus
flexor muscle of fingers, superficial	musculus flexor digitorum superficialis
flexor muscle of great toe, long	musculus flexor hallucis longus
flexor muscle of great toe, short	musculus flexor hallucis brevis
flexor muscle of little finger, short	musculus flexor digiti minimi brevis manus
flexor muscle of little toe, short	musculus flexor digiti minimi brevis pedis
flexor muscle of thumb, short	musculus flexor pollicis brevis
flexor muscle of thumb, long	musculus flexor pollicis longus
flexor muscle of toes, short	musculus flexor digitorum brevis
flexor muscle of toes, long	musculus flexor digitorum longus
flexor muscle of wrist, radial	musculus flexor carpi radialis
flexor muscle of wrist, ulnar	musculus flexor carpi ulnaris
fusiform muscle	musculus fusiformis
Gantzer muscle	N/A
gastrocnemius muscle	musculus gastrocnemius
Gavard muscle	N/A
gemellus muscle, inferior	musculus gemellus inferior
gemellus muscle, superior	musculus gemellus superior
genioglossal muscle	musculus genioglossus
geniohyoid muscle	musculus geniohyoideus
glossopalatine muscle	musculus palatoglossus
glossopharyngeal muscle	pars glossopharyngea musculi

gluteal muscle, greatest	musculus gluteus maximus
gluteal muscle, least	musculus gluteus minimus
gluteal muscle, middle	musculus gluteus medius
gracilis muscle	musculus gracilis
great adductor muscle	musculus adductor magnus
greater pectoral muscle	musculus pectoralis major
greater posterior rectus muscle of head	musculus rectus capitis posterior major
greater psoas muscle	musculus psoas major
greater rhomboid muscle	musculus rhomboideus major
greater zygomatic muscle	musculus zygomaticus major
Guthrie muscle	sphincter urethrae
hamstring muscles	N/A
head, muscles of	musculi capitis
Hilton muscle	musculus aryepiglotticus
Horner muscle	musculus orbicularis oculi
Houston muscle	compressor venae dorsalis penis
hyoglossal muscle	musculus hyoglossus
iliac muscle	musculus iliacus
iliacus minor muscle	musculus iliacus minor
iliococcygeal muscle	musculus iliococcygeus
iliocostal muscle	musculus iliocostalis
iliocostal muscle of neck	musculus iliocostalis cervicis
iliocostal muscle of loins	musculus iliocostalis lumborum
iliocostal muscle of thorax	musculus iliocostalis thoracis
iliopsoas muscle	musculus iliopsoas
incisive muscles of inferior lip	musculi incisivi labii inferioris
incisive muscles of lower lip	musculi incisivi labii inferioris
incisive muscles of superior lip	musculi incisivi labii superioris
incisive muscles of upper lip	musculi incisivi labii superioris
index extensor muscle	musculus extensor indicis
inferior constrictor muscle of pharynx	musculus constrictor pharyngis inferior
inferior gemellus muscle	musculus gemellus inferior
inferior longitudinal muscle of tongue	musculus longitudinalis inferior
inferior oblique muscle	musculus obliquus inferior
inferior oblique muscle of head	musculus obliquus capitis inferior
inferior posterior serratus muscle	serratus posterior inferior musculus
inferior rectus muscle	musculus rectus inferior
inferior tarsal muscle	musculus tarsalis inferior
infrahyoid muscles	musculi infrahyoidei
infraspinous muscle	musculus infraspinatus
innermost intercostal muscle	musculus intercostalis intimus
inspiratory muscles	N/A

intercostal muscles	musculi intercostales
interfoveolar muscle	ligamentum interfoveolare
intermediate great muscle	musculus vastus intermedius
internal intercostal muscle	musculus intercostalis internus
internal oblique muscle	musculus obliquus internus abdominis
internal obturator muscle	musculus obturator internus
internal pterygoid muscle	musculus pterygoideus medialis
interosseous muscles	musculi interossei
interspinal muscles	musculi interspinales
intertransverse muscles	musculi intertransversarii
intra-auricular muscles	N/A
intraocular muscles	N/A
intrinsic muscles	N/A
intrinsic muscles of foot	N/A
involuntary muscles	N/A
iridic muscles	N/A
ischiocavernous muscle	musculus ischiocavernosus
Jarjavay muscle	N/A
Jung pyramidal auricular muscle	N/A
Klein cutaneomucous muscle	N/A
Kohlrausch muscle	N/A
Koyter muscle	musculus corrugator supercilii
Krause cutaneomucous muscle	musculus cutaneomucosus
Lindström muscle	N/A
Langer axillary arch muscle	N/A
larynx, muscles of	musculi laryngis
lateral cricoarytenoid muscle	musculus cricoarytenoideus lateralis
lateral great muscle	musculus vastus lateralis
lateral lumbar intertransversarii muscles	musculi intertransversarii laterales lumborum
lateral pterygoid muscle	musculus pterygoideus lateralis
lateral rectus muscle	musculus rectus lateralis
lateral rectus muscle of the head	musculus rectus capitis lateralis
lateral vastus muscle	musculus vastus lateralis
latissimus dorsi muscle	musculus latissimus dorsi
lesser rhomboid muscle	musculus rhomboid minor
lesser zygomatic muscle	musculus zygomaticus minor
levator ani muscle	musculus levator ani
levator muscles	N/A
lingual muscles	musculi linguae
long abductor muscle of thumb	musculus abductor pollicis longus
long adductor muscle	musculus adductor longus
long extensor muscle of great toe	musculus extensor hallucis longus

long extensor muscle of thumb	musculus extensor pollicis longus
long extensor muscle of toes	musculus extensor digitorum longus
long fibular muscle	musculus peroneus longus
long flexor muscle of great toe	musculus flexor hallucis longus
long flexor muscle of thumb	musculus flexor pollicis longus
long flexor muscle of toes	musculus flexor digitorum longus
long muscle of head	musculus longus capitis
longissimus muscle	musculus longissimus
longissimus muscle of back	musculus longissimus thoracis
longissimus muscle of head	musculus longissimus capitis
longissimus muscle of neck	musculus longissimus cervicis
longitudinal muscle of tongue, inferior	musculus longitudinalis inferior linguae
longitudinal muscle of tongue, superior	musculus longitudinalis superior linguae
long muscle of head	musculus longus capitis
long muscle of neck	musculus longus colli
long palmar muscle	musculus palmaris longus
long peroneal muscle	musculus peroneus longus
long radial extensor muscle of wrist	musculus extensor carpi radialis longus
lumbar interspinal muscles	musculus interspinalis lumborum
lumbar quadrate muscle	musculus quadratus lumborum
lumbar rotator muscles	musculi rotatores lumborum
lumbrical muscles of foot	musculi lumbricales pedis
lumbrical muscles of hand	musculi lumbricales manus
Marcacci muscle	N/A
masseter muscle	musculus masseter
medial great muscle	musculus vastus medialis
medial lumbar intertransverse muscles	musculi intertransversarii mediales lumborum
medial pterygoid muscle	musculus pterygoideus mediales
medial rectus muscle	musculus rectus medialis
medial vastus muscle	musculus vastus medialis
mentalis muscle	musculus mentalis
Merkel muscle	musculus ceratocricoideus
mesothenar muscle	musculus adductor pollicis
middle constrictor muscle of pharynx	musculus constrictor pharyngis medius
middle scalene muscle	musculus scalenus medius
mucocutaneous muscle	N/A
Mueller muscle	musculus orbitalis
multifidus muscles	musculi multifidi
multipennate muscle	musculus multipennatus
mylohyoid muscle	musculus mylohyoideus
nasal muscle	musculus nasalis
neck, muscles of	musculi colli

nonstriated muscle	N/A
notch of helix, muscles of	musculus incisurae helicis
oblique arytenoid muscle	musculus arytenoideus obliquus
oblique auricular muscle	musculus obliquus auriculae
oblique muscle of abdomen, external	musculus obliquus externus abdominis
oblique muscle of abdomen, internal	musculus obliquus internus abdominis
oblique muscle of head, inferior	musculus obliquus capitis inferior
oblique muscle of head, superior	musculus obliquus capitis superior
obturator muscle, external	musculus obturator externus
obturator muscle, internal	musculus obturator internus
occipitofrontal muscle	musculus occipitofrontalis
Ochsner muscles	N/A
ocular muscles	musculi bulbi
Oddi muscle	N/A
Oehl muscles	N/A
omohyoid muscle	musculus omohyoideus
opposing muscle of little finger	musculus opponens digiti minimi
opposing muscle of thumb	musculus opponens pollicis
orbicular muscle	musculus orbicularis
orbicular muscle of eye	musculus orbicularis oculi
orbicular muscle of mouth	musculus orbicularis oris
orbital muscle	musculus orbitalis
organic muscle	N/A
palate and fauces, muscles	musculi palati et faucium
palatine muscles	musculi palati
palatoglossus muscle	musculus palatoglossus
palatopharyngeal muscle	musculus palatopharyngeus
palmar interosseous muscle	musculus interosseus palmaris
palmar muscle, short	musculus palmaris brevis
palmar muscle, long	musculus palmaris longus
papillary muscle	musculus papillaris
pectinate muscles	musculi pectinati
pectineal muscle	musculus pectineus
pectoral muscle, greater	musculus pectoralis major
pectoral muscle, smaller	musculus pectoralis minor
pectorodorsalis muscle	N/A
penniform muscle	musculus unipennatus
perineal muscles	musculi perinei
peroneal muscle, long	musculus fibularis longus
peroneal muscle, short	musculus fibularis brevis
peroneal muscle, third	musculus fibularis tertius
pharyngopalatine muscle	musculus palatopharyngeus
Phillip muscle	N/A

piriform muscle	musculus piriformis
plantar interosseous muscle	musculus interosseus plantaris
plantar muscle	musculus plantaris
plantar quadrate muscle	musculus quadratus plantae
platysma muscle	musculus platysma
pleuroesophageal muscle	musculus pleuroesophageus
popliteal muscle	musculus popliteus
posterior auricular muscle	musculus retrahens aurem
posterior cervical intertransverse muscles	musculi intertransversarii posteriores cervicis
posterior cricoarytenoid muscle	musculus cricoarytenoideus posterior
posterior scalene muscle	musculus scalenus posterior
posterior tibial muscle	musculus tibialis posterior
Pozzi muscle	musculus extensor digitorum brevis manus
procerus muscle	musculus procerus
pronator muscle, quadrate	musculus pronator quadratus
pronator muscle, round	musculus pronator teres
psoas muscle, greater	musculus psoas major
psoas muscle, smaller	musculus psoas minor
pterygoid muscle	musculus pterygoideus
pubococcygeal muscle	musculus pubococcygeus
puboprostatic muscle	musculus puboprostaticus
puborectal muscle	musculus puborectalis
pubovaginal muscle	musculus pubovaginalis
pubovesical muscle	musculus pubovesicalis
pyloric sphincter muscle	musculus sphincter pyloricus
pyramidal auricular muscle	musculus pyramidalis auriculae
quadrate (four-sided) muscle	musculus quadratus
quadrate muscle of loin	musculus quadratus lumborum
quadrate muscle of lower lip	musculus depressor labii inferioris
quadrate muscle of sole	musculus quadratus plantae
quadrate muscle of thigh	musculus quadratus femoris
quadrate muscle of upper lip	musculus quadratus labii superioris
radial flexor muscle of wrist	musculus flexor carpi radialis
rectococcygeus muscle	musculus rectococcygeus
rectourethral muscle	musculus rectourethralis
rectouterine muscle	musculus rectouterinus
rectovesical muscle	musculus rectovesicalis
rectus abdominis muscle	musculus rectus abdominis
rectus muscle of head, anterior	musculus rectus capitis anterior
rectus muscle of head, lateral	musculus rectus capitis lateralis
rectus muscle of head, greater posterior	musculus rectus capitis posterior major

rectus muscle of head, smaller posterior	musculus rectus capitis posterior minor
rectus femoris muscle	musculus rectus femoris
red muscle	N/A
Reisseisen muscles	N/A
rhomboid muscle, greater	musculus rhomboideus major
rhomboid muscle, lesser	musculus rhomboideus minor
ribbon muscles	musculi infrahyoidei
Rider muscles	N/A
Riolan muscle	musculus cremaster
risorius muscle	musculus risorius
rotator muscles	musculi rotatores
rotator muscles of neck	musculi rotatores cervicis
rotator muscles of back	musculi rotatores lumborum
rotator muscles of thorax	musculi rotatores thoracis
Rouget muscle	N/A
round pronator muscle	musculus pronator teres
Ruysch muscle	N/A
sacrococcygeal muscle	musculus sacrococcygeus
salpingopharyngeal muscle	musculus salpingopharyngeus
Santorini muscle	musculus risorius
sartorius muscle	musculus sartorius
scalene muscle, anterior	musculus scalenus anterior
scalene muscle, middle	musculus scalenus medius
scalene muscle, posterior	musculus scalenus posterior
scalene muscle, smallest	musculus scalenus minimus
scalp muscle	N/A
Sebileau muscle	N/A
second tibial muscle	musculus tibialis secundus
semimembranous muscle	musculus semimembranosus
semispinal muscle	musculus semispinalis
semispinal muscle of head	musculus semispinalis capitis
semispinal muscle of neck	musculus semispinalis cervicis
semispinal muscle of thorax	musculus semispinalis thoracis
semitendinous muscle	musculus semitendinosus
serratus anterior muscle	musculus serratus anterior
serratus posterior inferior muscle	musculus serratus posterior inferior
serratus posterior superior muscle	musculus serratus posterior superior
shawl muscle	N/A
short adductor muscle	musculus adductor brevis
short extensor muscle of great toe	musculus extensor hallucis brevis
short extensor muscle of thumb	musculus extensor pollicis brevis
short extensor muscle of toes	musculus extensor digitorum brevis
short fibular muscle	musculus peroneus brevis

short flexor muscle of great toe	musculus flexor hallucis brevis
short flexor muscle of little finger	musculus flexor digiti minimi brevis
short flexor muscle of little toe	musculus flexor digiti minimi brevis
short flexor muscle of thumb	musculus flexor pollicis brevis
short flexor muscle of toes	musculus flexor digitorum brevis
short palmar muscle	musculus palmaris brevis
short peroneal muscle	musculus peroneus brevis
short radial extensor muscle of wrist	musculus extensor carpi radialis brevis
Sibson muscle	musculus scalenus minimus
skeletal muscles	musculi skeleti
slow muscle	N/A
smaller muscle of helix	musculus helicis minor
smaller pectoral muscle	musculus pectoralis minor
smaller posterior rectus muscle of head	musculus rectus capitis posterior minor
smaller psoas muscle	musculus psoas minor
smallest scalene muscle	musculus scalenus minimus
smooth muscle	N/A
Soemmerring muscle	N/A
soleus muscle	musculus soleus
somatic muscles	musculi skeleti
sphincter muscle of anus	musculus sphincter ani
sphincter muscle of bile duct	musculus sphincter ductus choledochi
sphincter muscle of hepatopancreatic ampulla	musculus sphincter ampulla hepatopancreaticae
sphincter muscle of pupil	musculus sphincter pupillae
sphincter muscle of pylorus	musculus sphincter pyloricus
sphincter muscle of urethra	musculus sphincter urethrae
sphincter muscle of urinary bladder	musculus sphincter vesicae urinariae
spinal muscle	musculus spinalis
spinal muscle of head	musculus spinalis capitis
spinal muscle of neck	musculus spinalis cervicis
spinal muscle of throat	musculus spinalis thoracis
spindle-shaped muscle	musculus fisiform
splenius muscle of head	musculus splenius capitis
splenius muscle of neck	musculus splenius cervicis
stapedius muscle	musculus stapedius
sternal muscle	musculus sternalis
sternochondroscapular muscle	musculus sternochondroscapularis
sternoclavicular muscle	musculus sternoclavicularis
sternocleidomastoid muscle	musculus sternocleidomastoideus
sternocostal muscle	musculus transversus thoracis
sternohyoid muscle	musculus sternohyoideus
sternomastoid muscle	N/A

sternothyroid muscle	musculus sternothyroideus
strap muscles	N/A
striated muscle	N/A
styloauricular muscle	musculus styloauricularis
styloglossus muscle	musculus styloglossus
stylohyoid muscle	musculus stylohyoideus
stylopharyngeal muscle	musculus stylopharyngeus
subanconeus muscle	musculus articularis cubiti
subclavian muscle	musculus subclavius
subcostal muscle	musculus subcostalis
subcrural muscle	musculus articularis genu
suboccipital muscles	musculi suboccipitales
subquadricipital muscle	musculus articularis genu
subscapular muscle	musculus subscapularis
subvertebral muscles	musculi hypaxial
superficial back muscles	N/A
superficial flexor muscle of fingers	musculus flexor digitorum superficialis
superficial lingual muscle (of tongue)	N/A
superficial transverse perineal muscle	musculus transversus perinei superficialis
superior auricular muscle	musculus auricularis superior
superior constrictor muscle of pharynx	musculus constrictor pharyngis superior
superior gemellus muscle	musculus gemellus superior
superior longitudinal muscle of tongue	musculus longitudinalis superior
superior oblique muscle	musculus obliquus superior
superior oblique muscle of head	musculus obliquus capitis superior
superior posterior serratus muscle	musculus serratus posterior superior
superior rectus muscle	musculus rectus superior
superior tarsal muscle	musculus tarsalis superior
supinator muscle	musculus supinator
supraclavicular muscle	musculus supraclavicularis
suprahyoid muscles	musculi suprahyoidei
supraspinalis muscle	musculus supraspinalis
supraspinous muscle	musculus supraspinatus
suspensory muscle of duodenum	musculus suspensorius duodeni
synergic or synergistic muscle	N/A
Tailor muscle	N/A
temporal muscle	musculus temporalis
temporoparietal muscle	musculus temporoparietalis
tensor muscle of fascia lata	musculus tensor fasciae latae
tensor muscle of soft palate	musculus tensor veli palati
tensor tarsi muscle	musculus orbicularis oculi
tensor muscle of tympanic membrane	musculus tensor tympani

teres major muscle	musculus teres major
teres minor muscle	musculus teres minor
Theile muscle	N/A
third peroneal muscle	musculus peroneus tertius
thoracic interspinal muscle	musculus thoracic interspinalis
thoracic intertransverse muscles	musculi intertransversarii thoracis
thoracic longissimus muscle	musculus longissimus thoracis
thoracic rotator muscles	musculi rotatores thoracis
thorax, muscles of	musculi thoracis
thyroarytenoid muscle	musculus thyroarytenoideus
thyroepiglottic muscle	musculus thyroepiglotticus
thyrohyoid muscle	musculus thyrohyoideus
tibial muscle, anterior	musculus tibialis anterior
tibial muscle, posterior	musculus tibialis posterior
Tod muscle	N/A
tongue, muscles of	musculi linguae
Toynbee muscle	musculus tensor tympani
tracheal muscle	musculus trachealis
tracheloclavicular muscle	musculus trachecloclavicularis
trachelomastoid muscle	musculus longissimus capitis
tragicus muscle	N/A
transverse arytenoid muscle	musculus arytenoideus transversus
transverse muscle of abdomen	musculus transversus abdominis
transverse muscle of auricle	musculus transversus auriculae
transverse muscle of chin	musculus transversus menti
transverse muscle of nape	musculus transversus nuchae
transverse muscle of neck	musculus transversus nuchae
transverse muscle of thorax	musculus transversus thoracis
transverse muscle of tongue	musculus transversus linguae
transversospinal muscle	musculus transversospinalis
trapezius muscle	musculus trapezius
Treitz muscle	N/A
triangular muscle	musculus triangularis
triceps muscle of arm	musculus triceps brachii
triceps muscle of hip	musculus triceps coxae
triceps muscle of calf	musculus triceps surae
trigonal muscle	N/A
true (deep) muscles of back	musculi dorsi
two-bellied muscle	musculus digastricus
ulnar extensor of wrist	musculus extensor carpi ulnaris
ulnar flexor muscle of wrist	musculus flexor carpi ulnaris
unipennate muscle	musculus unipennatus
unstriated muscle, unstriped muscle	N/A

urogenital diaphragm, muscles of	musculi diaphragmatis urogenitalis
uvula, muscles of	musculus uvulae
Valsalva muscle	N/A
vastus intermedius muscle	musculus vastus intermedius
vastus lateralis muscle	musculus vastus lateralis
vastus medialis muscle	musculus vastus medialis
ventral sacrococcygeal muscle	musculus sacrococcygeus ventralis
vertical muscle of tongue	musculus verticalis linguae
visceral muscle	N/A
vocal muscle	musculus vocalis
voluntary muscle	N/A
white muscle	N/A
Wilson muscle	musculus sphincter urethrae
wrinkler muscle of eyebrow	musculus corrugator supercilii
yoked muscles	N/A
zygomatic muscle, greater	musculus zygomaticus major
zygomatic muscle, lesser	musculus zygomaticus minor

Nerves

abdominopelvic splanchnic nerve	N/A
abducent nerve	nervus abducens
accelerator nerve	N/A
accessory nerve	nervus accessorius
accessory nerve, vagal	ramus internus nervi accessorii
accessory phrenic nerve	nervi phrenici accessorii
acoustic nerve	N/A
afferent nerve (centripetal; esodic)	N/A
alveolar nerve, inferior	nervus alveolaris inferior
alveolar nerve, superior	nervi alveolares superiores
ampullar nerve, anterior	nervus ampullaris anterior
ampullar nerve, inferior	nervus ampullaris inferior
ampullar nerve, lateral	nervus ampullaris lateralis
ampullar nerve, superior	nervus ampullaris superior
anal nerve, inferior	nervi rectales inferiores
Andersch nerve (tympanic)	N/A
anococcygeal nerve	nervi anococcygei
anterior ampullar nerve	nervus ampullaris anterior
anterior antebrachial nerve (anterior interosseous nerve)	N/A
anterior auricular nerve	nervi auriculares anteriores
anterior crural nerve (femoral nerve)	N/A
anterior cutaneous nerves of abdomen (thoracoabdominal nerves)	N/A

anterior ethmoidal nerve	nervus ethmoidalis anterior
anterior femoral cutaneous nerve	rami cutanei anteriores nervi femoralis
anterior interosseous nerve	nervus interosseous anterior
anterior labial nerves	nervi labiales anteriores
anterior scrotal nerves	nervi scrotales anteriores
anterior supraclavicular nerve (medial supraclavicular nerve)	N/A
anterior tibial nerve (deep peroneal nerve)	N/A
aortic nerve (Cyon nerve, depressor nerve of Ludwig, Ludwig nerve)	N/A
Arnold nerve	ramus auricularis nervi vagi
articular nerve	nervus articularis
auditory nerve (cochlear)	nervus vestibulocochlearis
augmentor nerve (cervical splanchnic nerve)	N/A
auricular nerve, anterior	nervi auriculares anteriores
auricular nerve, great	nervus auricularis magnus
auricular nerve, internal	ramus posterior nervi auricularis magni
auricular nerve, posterior	nervus auricularis posterior
auricular nerve of vagus nerve	ramus auricularis nervi vagi
auriculotemporal nerve	nervus auriculotemporalis
autonomic nerve	nervus autonomicus
axillary nerve	nervus axillaris
baroreceptor nerve (pressoreceptor nerve)	N/A
Bell long thoracic nerve	nervus thoracicus longus
Bock nerve	ramus pharyngeus ganglii pterygopalatini
buccal nerve	nervus buccalis buccinator nerve
cardiac nerve, cervical, inferior	nervus cardiacus cervicalis inferior
cardiac nerve, cervical, middle	nervus cardiacus cervicalis medius
cardiac nerve, cervical, superior	nervus cardiacus cervicalis superior
cardiac nerve, inferior	nervus cardiacus cervicalis inferior
cardiac nerve, middle	nervus cardiacus cervicalis medius
cardiac nerve, superior	nervus cardiacus cervicalis superior
cardiac nerve, supreme	rami cardiaci cervicales superiores nervi vagi
cardiac nerve, thoracic	rami cardiaci thoracici
cardiopulmonary splanchnic nerve	N/A
caroticotympanic nerve	nervus caroticotympanicus
carotid sinus nerve (Hering sinus)	ramus sinus carotici
cavernous nerves of clitoris	nervi cavernosi clitoridis

cavernous nerves of penis	nervi cavernosi penis
celiac nerve	rami coeliaci nervi vagi
centrifugal nerve (efferent)	N/A
centripetal nerve (afferent)	N/A
cerebral nerve	nervi craniales
cervical nerve	nervi cervicales
cervical splanchnic nerve (augmentor nerve)	N/A
chorda tympani nerve	chorda tympani
ciliary nerve	N/A
nervi ciliares circumflex nerve (axillary)	nervus axillaris
cluneal nerve	rami clunium
coccygeal nerve	nervus coccygeus
cochlear nerve	nervus cochlearis
common fibular nerve (common peroneal)	N/A
common palmar digital nerve	nervi digitales palmares communes
common peroneal nerve	nervus peroneus communis
common plantar digital nerve	nervi digitales plantares communes
cranial nerves	nervi craniales
cranial nerve, first (olfactory)	nervi olfactorii
cranial nerve, second (optic)	nervus opticus
cranial nerve, third (oculomotor)	nervus oculomotorius
cranial nerve, fourth (trochlear)	nervus trochlearis
cranial nerve, fifth (trigeminal)	nervus trigeminus
cranial nerve, sixth (abducens)	nervus abducens
cranial nerve, seventh (facial)	nervus facialis
cranial nerve, eighth (acoustic)	nervus vestibulocochlearis
cranial nerve, ninth (glossopharyngeal)	nervus glossopharyngeus
cranial nerve, tenth (vagal)	nervus vagus
cranial nerve, eleventh (accessory)	nervus accessorius
cranial nerve, twelfth (hypoglossal)	nervus hypoglossus
crural interosseous nerve	nervus interosseus cruris
cubital nerve (ulnar nerve)	nervus ulnaris
cutaneous nerves	nervus cutaneus
cutaneous femoral nerve, lateral	nervus cutaneus femoris lateralis
cutaneous nerve, femoral	nervus cutaneus femoralis
cutaneous nerve of arm, lateral, inferior	nervus cutaneus brachii lateralis inferior
cutaneous nerve of calf, medial	nervus cutaneus surae medial
cutaneous nerve of foot, dorsal, lateral	nervus cutaneus dorsalis lateralis
cutaneous nerve of forearm, medial	nervus cutaneus antebrachii medialis
cutaneous nerve of neck, anterior	nervus transversus colli
cutaneous nerve of thigh, posterior	nervus cutaneus femoralis posterior

Cyon nerve (aortic)	N/A
dead nerve (nonvital dental pulp)	N/A
deep fibular nerve (deep peroneal)	N/A
deep peroneal nerve	nervus peroneus profundus
deep petrosal nerve	nervus petrosus profundus
deep temporal nerve	nervi temporales profundi
dental nerve, inferior	nervus alveolaris inferior
depressor nerve of Ludwig (aortic)	N/A
diaphragmatic nerve	nervus phrenicus
digastric nerve	ramus digastricus nervi facialis
digital nerves, dorsal, radial	nervi digitales dorsales nervi radialis
dorsal nerve of clitoris	nervus dorsalis clitoridis
dorsal digital nerves	N/A
dorsal digital nerves of foot	nervi digitales dorsales pedis
dorsal digital nerves of hand	N/A
dorsal interosseous nerve (posterior interosseous)	N/A
dorsal lateral cutaneous nerve (lateral dorsal cutaneous)	N/A
dorsal medial cutaneous nerve (medial dorsal cutaneous nerve)	N/A
dorsal nerve of penis	nervus dorsalis penis
dorsal nerve of scapula (dorsal scapular)	N/A
dorsal nerves of toes	N/A
dorsal scapular nerve	nervus dorsalis scapulae
efferent nerve (centrifugal nerve)	N/A
encephalic nerves	nervi craniales
esodic nerve (afferent nerve)	N/A
ethmoidal nerve, anterior	nervus ethmoidalis anterior
excitor nerve	N/A
excitoreflex nerve	N/A
exodic nerve	N/A
external acoustic meatus, nerve of	nervus meatus acustici externi
external carotid nerves	nervi carotici externi
external respiratory nerve of Bell (long thoracis nerve)	N/A
external saphenous nerve (sural nerve)	N/A
external spermatic nerve (genital branch of genitofemoral nerve)	N/A
facial nerve	nervus facialis
femoral cutaneous nerve	nervus cutaneus femoralis (lateralis, medialis)
femoral nerve	nervus femoralis

fibular nerve, superficial	nervus fibularis superficialis
fourth lumbar nerve (furcal nerve)	nervus furcalis
frontal nerve	nervus frontalis
fusimotor nerves	N/A
Galen nerve	N/A
gangliated nerve	N/A
gastric nerves	truncus vagalis anterior; truncus vagalis posterior
genitocrural nerve (genitofemoral nerve)	N/A
genitofemoral nerve	nervus genitofemoralis
gluteal nerve, inferior	nervus gluteus inferior
great auricular nerve	nervus auricularis magnus
greater occipital nerve	nervus occipitalis major
greater palatine nerve	nervus palatinus major
greater petrosal nerve	N/A
greater splanchnic nerve	nervus splanchnicus major
greater superficial petrosal nerve	nervus petrosus major
great sciatic nerve	N/A
gustatory nerves	N/A
hemorrhoidal nerves	N/A
hemorrhoidal nerves, inferior	nervi rectales inferiores
Hering sinus nerve (carotid sinus)	N/A
hypogastric nerve	nervus hypogastricus
hypoglossal nerve	nervus hypoglossus
iliohypogastric nerve	nervus iliohypogastricus
ilioinguinal nerve	nervus ilioinguinalis
inferior alveolar nerve	nervus alveolaris inferior
inferior cervical cardiac nerve	nervus cardiacus cervicalis inferior
inferior cluneal nerves	nervi clunium inferiores
inferior dental nerve (inferior alveolar)	N/A
inferior gluteal nerve	nervus gluteus inferior
inferior laryngeal nerve	nervus laryngeus inferior
inferior lateral brachial cutaneous nerve	nervus cutaneus brachii lateralis inferior
inferior maxillary nerve	N/A
inferior rectal nerves	nervi rectales inferiores
inferior vesical nerves	N/A
infraoccipital nerve	nervus suboccipitalis
infraorbital nerve	nervus infraorbitalis
infratrochlear nerve	nervus infratrochlearis
inferior laryngeal nerve	nervus laryngeus inferior
inhibitory nerve	N/A
intercarotid nerve	N/A

intercostal nerve	nervus intercostalis
intercostobrachial nerve	nervus intercostobrachiales
intercostohumeral nerves	N/A
intermediary nerve	nervus intermedius
intermediate dorsal cutaneous nerve	nervus cutaneus dorsalis intermedius
Jacobson nerve (tympanic)	nervus tympanicus
jugular nerve	nervus jugularis
lacrimal nerve	nervus lacrimalis
Lancisi, nerves of	N/A
Langley nerves (pilomotor nerves)	N/A
Latarjet nerve (superior hypogastric plexus)	N/A
laryngeal nerve, recurrent	nervus laryngealis recurrens
lateral ampullar nerve	nervus ampullaris lateralis
lateral antebrachial cutaneus	nervus cutaneous antebrachii lateralis
lateral antebrachial cutaneous nerve	nervus cutaneus antebrachii lateralis
lateral anterior thoracic nerve	nervus pectoralis lateralis
lateral cutaneous nerve of calf	N/A
lateral cutaneous nerve of forearm	N/A
lateral cutaneous nerve of thigh	N/A
lateral dorsal cutaneous nerve	nervus cutaneus dorsalis lateralis
lateral femoral cutaneous nerve	nervus cutaneus femoris lateralis
lateral pectoral nerve	nervus pectoralis lateralis
lateral plantar nerve	N/A
lateral popliteal nerve	N/A
lateral supraclavicular nerve	nervus supraclavicularis lateralis
lateral sural cutaneous nerve	nervus cutaneus surae lateralis
lesser internal cutaneous nerve	N/A
lesser occipital nerve	nervus occipitalis minor
lesser palatine nerves	nervi palatini minores
lesser petrosal nerve	N/A
lesser splanchnic nerve	nervus splanchnicus minor
lesser superficial petrosal nerve	nervus petrosus minor
lingual nerve	nervus lingualis
long buccal nerve	N/A
long ciliary nerve	nervus ciliaris longus
longitudinal nerves of Lancisi	N/A
long saphenous nerve	N/A
long subscapular nerve	N/A
long thoracic nerve	nervus thoracicus longus
lower lateral cutaneous nerve of arm	N/A
lower splanchnic nerve	nervus splanchnicus imus
Ludwig nerve (aortic nerve)	N/A

lumbar nerves	nervi lumbales
lumbar splanchnic nerves	nervi splanchnici lumbales
lumboinguinal nerve	ramus femoralis nervi genitofemoralis
Luschka, nerve of	N/A
masseteric nerve	nervus massetericus
masticator nerve	N/A
maxillary nerve	nervus maxillaris
medial antebrachial cutaneous nerve	nervus cutaneus antebrachii medialis
medial anterior thoracic nerve	N/A
medial cutaneous nerve of arm	N/A
medial cutaneous nerve of forearm	N/A
medial cutaneous nerve of leg	N/A
medial dorsal cutaneous nerve	nervus cutaneus dorsalis medialis
medial pectoral nerve	nervus pectoralis medialis
medial plantar nerve	nervus plantaris medialis
medial popliteal nerve	N/A
medial supraclavicular nerve	nervus supraclavicularis medialis
medial sural cutaneous nerve	nervus cutaneus surae medialis
median nerve	nervus medianus
meningeal nerve	ramus meningeus medius nervi maxillaris
mental nerve	nervus mentalis
middle cervical cardiac nerve	nervus cardiacus cervicalis medius
middle cluneal nerves	nervi clunium medii
middle meningeal nerve	N/A
middle supraclavicular nerve	N/A
mixed nerve	nervus mixtus
motor nerve	nervus motorius
motor nerve of tongue	nervus hypoglossus
musculocutaneous nerve	nervus musculocutaneus
musculospiral nerve (radial)	nervus radialis
myelinated nerve	N/A
mylohyoid nerve	nervus mylohyoideus
nasal nerve	N/A
nasociliary nerve	nervus nasociliaris
nasopalatine nerve	nervus nasopalatinus
nervus plantaris lateralis mandibular nerve	nervus mandibularis
nervus trigeminus trigeminal nerve	nervus trigeminus
obturator nerve	nervus obturatorius
oculomotor nerve	nervus oculomotorius
olfactory nerves	nervi olfactorii
ophthalmic nerve	nervus ophthalmicus

optic nerve (second cranial nerve)	nervus opticus
orbital nerve	N/A
palatine nerve, anterior	nervus palatinus major
parasympathetic nerve	N/A
parotid nerves	rami parotidei nervi auriculotemporalis
pathetic nerve	N/A
pectoral nerve, lateral	nervus pectoralis lateralis
pelvic splanchnic nerves	nervi pelvici splanchnici
parasympathetic nerve	N/A
perforating cutaneous nerve	N/A
perineal nerves	nervi perineales
peroneal nerve, common	nervus fibularis communis
phrenic nerve	nervus phrenicus
pneumogastric nerve	nervus vagus
popliteal nerve, external	nervus fibularis communis
popliteal nerve, internal	nervus tibialis
popliteal nerve, lateral	nervus fibularis communis
popliteal nerve, medial	nervus tibialis
posterior ampullar nerve	nervus ampullaris posterior
posterior antebrachial nerve	N/A
posterior antebrachial cutaneous nerve	nervus cutaneus antebrachii posterior
posterior auricular nerve	nervus auricularis posterior
posterior brachial cutaneous nerve	nervus cutaneus brachii posterior
posterior cutaneous nerve of arm	N/A
posterior cutaneous nerve of forearm	N/A
posterior cutaneous nerve of thigh	N/A
posterior ethmoidal nerve	nervus ethmoidalis posterior
posterior femoral cutaneous nerve	nervus cutaneus femoris posterior
posterior interosseous nerve	nervus interosseus posterior
posterior labial nerves	nervi labiales posteriores
posterior scapular nerve	N/A
posterior scrotal nerves	nervi scrotales posteriores
posterior supraclavicular nerve	N/A
posterior thoracic nerve	N/A
presacral nerve	plexus hypogastricus superior
pressor nerve	N/A
pressoreceptor nerve	N/A
proper palmar digital nerves	nervi digitales palmares proprii
proper plantar digital nerves	nervi digitales plantares proprii
pterygoid nerve	nervus pterygoideus
pterygoid canal, nerve of	nervus canalis pterygoidei
pterygopalatine nerves	nervi pterygopalatini
pudendal nerve	nervus pudendus

pudic nerve (pudendal)	nervus pudendus
quadrate muscle of thigh, nerve of	nervus musculi quadrati femoris
radial nerve	nervus radialis
recurrent laryngeal nerve	nervus laryngeus recurrens
recurrent nerve	nervus laryngealis recurrens
recurrent meningeal nerve	nervus meningeus recurrens
recurrent ophthalmic nerve	ramus tentorii nervi ophthalmici
saccular nerve	nervus saccularis
sacral nerve	nervus sacralis
sacral splanchnic nerves	nervi splanchnici sacrales
saphenous nerve	nervus saphenus
sartorius, nerve to	N/A
Scarp nerve	nervus nasopalatinus
sciatic nerve	nervus sacralis
secretomotor nerve	N/A
secretory nerve	N/A
sensory nerve	nervus sensorius
short ciliary nerve	nervus ciliaris brevis
short saphenous nerve	nervus saphenus brevis
sinus nerve	ramus sinus carotici nervi glossopharyngei
sinus nerve of Hering	N/A
sinuvertebral nerves	N/A
small deep petrosal nerve	N/A
smallest splanchnic nerve	N/A
small sciatic nerve	N/A
smell, nerve of (olfactory nerve)	nervi olfactorii
somatic nerve	N/A
space nerve	N/A
spinal accessory nerve	nervus accessorius
spinal nerve	nervus spinalis
splanchnic nerve	nervus splanchnicus
stapedius muscle, nerve to	nervus stapedius
statoacoustic nerve	N/A
subclavian nerve	nervus subclavius
subcostal nerve	nervus subcostalis
sublingual nerve	nervus sublingualis
suboccipital nerve	nervus suboccipitalis
subscapular nerves	nervi subscapulares
sudomotor nerves	N/A
superficial cervical nerve	nervus cervicalis superficialis
superficial fibular nerve	nervus fibularis superficialis
superficial peroneus nerve	nervus peroneus superficialis

superior alveolar nerves	nervi alveolares superiores
superior cervical cardiac nerve	nervus cardiacus cervicalis
superior cluneal nerves	nervi clunium superiores
superior dental nerves	N/A
superior gluteal nerve	nervus gluteus superior
superior laryngeal nerve	nervus laryngeus superior
superior lateral brachial cutaneous nerve	nervus cutaneus brachii lateralis superior
superior maxillary nerve	N/A
supraorbital nerve	nervus supraorbitalis
suprascapular nerve	nervus suprascapularis
supratrochlear nerve	nervus supratrochlearis
sural nerve	nervus suralis
sympathetic nerve	N/A
temporomandibular nerve	N/A
tensor tympani muscle, nerve of	nervus tensoris tympani
tensor veli palatini muscle, nerve of	nervus tensoris veli palatini
tentorial nerve	ramus tentorii nervi ophthalmici
terminal nerves	nervi terminales
third occipital nerve	nervus occipitalis tertius
thoracic cardiac nerves	nervi cardiaci thoracici
thoracic nerve	nervus thoracis
thoracic splanchnic nerve	nervus splanchnicus thoracici
thoracoabdominal nerves	nervi thoracoabdominales
thoracodorsal nerve	nervus thoracodorsalis
thyrohyoid muscle, nerve to	ramus thyrohyoideus ansae certibial
nerve	snervus tibialis
tibial communicating nerve	N/A
tibial nerve	nervus tibialis
Tiedemann nerve	N/A
tonsillar nerves	rami tonsillares nervi glossopharyngei
transverse nerve of neck	nervus transversus coli
trifacial nerve (trigeminal nerve)	N/A
trochlear nerve	nervus trochlearis
vicalis	
tympanic nerve	nervus tympanicus
tympanic membrane, nerve of	N/A
ulnar nerve	nervus ulnaris
unmyelinated nerve	N/A
upper lateral cutaneous nerve of arm	N/A
upper subscapular nerve	N/A
upper thoracic splanchnic nerves	N/A
utricular nerve	nervus utricularis
utriculoampullar nerve	nervus utriculoampullaris

vaginal nerves	nervi vaginales
vagus nerve	nervus vagus
Valentin nerve	N/A
vascular nerve	nervus vascularis
vasoconstrictor nerve	N/A
vasodilator nerve	N/A
vasomotor nerve	N/A
vasosensory nerve	N/A
vertebral nerve	nervus vertebralis
vestibular nerve	nervus vestibularis
vestibulocochlear nerve	nervus vestibulocochlearis
vidian nerve	nervus canalis pterygoidei
visceral nerve	nervus autonomicus
volar interosseous nerve	N/A
Willis, nerve of	nervus accessorius
Wrisberg nerve	nervus intermedius
zygomatic nerve	nervus zygomaticus
zygomaticotemporal nerves	ramus zygomaticotemporalis nervi

Veins

accessory cephalic	vena cephalica accessoria
accessory hemiazygos	vena hemiazygos accessoria
accessory saphenous	vena saphena accessoria
accessory vertebral	vena vertebralis accessoria
accompanying	vena comitans
angular	vena angularis
anterior auricular preauricularis	vena auricularis anterior/vena
anterior basal	vena basalis anterior
anterior cardiac	vena cardiacae anteriores
anterior cerebral	vena anteriores cerebri
anterior ciliary	vena ciliares anteriores
anterior circumflex	vena circumflexa humeri anterior
anterior facial	vena facial anterior
anterior intercostal	vena intercostales anteriores
anterior jugular	vena jugularis anterior
anterior labial	vena labiales anteriores
anterior pontomesencephalic	vena pontomesencephalica anterior
anterior (posterior)vestibular	vena vestibulares anterius, posterius
anterior scrotal	vena scrotales anteriores
anterior vein of septum pellucidum	vena anterior septi pellucidi
anterior tibial	vena tibiales anteriores
anterior vertebral	vena vertebralis anterior

apical	vena apicalis
apicoposterior	vena apicoposterior
appendicular	vena appendicularis
arterial	vena arteriosa
ascending lumbar	vena lumbalis ascendens
axillary	vena axillaris
azygos	vena azygos
basal	vena basalis
basilic	vena basilica
basivertebral	vena basivertebrales
brachial	vena brachiales
bronchial	vena bronchiales
bulb of penis vein	vena bulbi penis
bulb of vestibuli	vena bulbi vestibuli
caudate nucleus	vena nuclei caudati
cavernous vein of penis	vena cavernosae penis
central vein of liver	vena centrales hepatis
central retinal	vena centralis retinae
central vein of suprarenal gland	vena centralis glandulae suprarenalis
cephalic	vena cephalica
cephalic vein of forearm	vena cephalica antebrachii
cerebellar	vena cerebelli
cochlear canaliculus	vena canaliculi cochleae
cochlear window	vena fenestrae cochleae
common basal	vena basalis communis
common facial	vena facialis communis
common iliac	vena iliaca communis
common modiolar	vena spiralis modioli
condylar emissary	vena emissaria condylaris
conjunctival	vena conjunctivales
cystic	vena cystica
deep cerebral	vena profundae cerebri
deep cervical	vena colli profunda
deep circumflex iliac	vena circumflexa iliaca profunda
deep vein of clitoris	vena profundae clitoridis
deep dorsal vein of clitoris	vena dorsalis clitoridis profunda
deep dorsal vein of penis	vena dorsalis penis profunda
deep facial	vena faciei profunda
deep femoral	vena profunda femoris
deep lingual	vena profunda linguae
deep middle cerebral	vena media profunda cerebri
deep vein of penis	vena profundae penis
deep temporal	vena temporales profundae

diploic	vena diploica
direct lateral	vena directae laterales
dorsal digit vein of foot	vena digitales dorsales pedis
dorsal lingual	vena dorsales linguale
dorsal metacarpal	vena metacarpeae dorsales
dorsal metatarsal	vena metatarseae
dorsal scapular	vena scapularis dorsalis
emissary	vena emissaria
episcleral	vena episclerales
ethmoid	vena ethmoidales
external iliac	vena iliaca externa
external jugular	vena jugularis externa
external nasal	vena nasales externa
external palatine	vena palatine externa
external pudendal	vena pudendae externae
facial	vena facialis
femoral	vena femoralis
fibular	vena fibulares
genicular	vena geniculares
great cardiac	vena cardiaca magna
great cerebral vein of Galen	vena magna cerebri
heart	vena cordis
hemiazygos	vena hemiazygos
hepatic	vena hepaticae
hepatic portal	vena portae hepatis
ileocolic	vena ileocolica
iliolumbar	vena iliolumbalis
inferior anastomotic	vena anastomotica inferior
inferior basal	vena basalis inferior
inferior vein of cerebellar hemisphere	vena inferiores cerebelli
inferior cerebral	vena inferiores cerebri
inferior choroid	vena choroidea inferior
inferior epigastric	vena epigastrica inferior
inferior gluteal	vena gluteae inferiores
inferior labial	vena labialis inferior
inferior laryngeal	vena laryngea inferior
inferior mesenteric	vena mesenterica inferior
inferior ophthalmic	vena ophthalmica inferior
inferior palpebral	vena palpebrales inferiores
inferior phrenic	vena phrenica inferior
inferior rectal	vena rectales inferiores
inferior thalamostriate	vena thalamostriatae inferiores
inferior thyroid	vena thyroideae inferior

inferior ventricular	vena ventricularis inferior
inferior vein of vermis	vena inferior vermis
intercapitular	vena intercapitulares
interlobar veins of kidney	vena interlobares renis
interlobular veins of kidney	vena interlobulares renis
interlobular veins of liver	vena interlobulares hepatis
intermediate basilic	vena intermedia basilica
intermediate cephalic	vena intermedia cephalica
intermediate hepatic	vena hepaticae intermediae
internal cerebral	vena internae cerebri
internal iliac	vena iliaca interna
internal jugular	vena jugularis interna
internal pudendal	vena pudenda interna
internal thoracic	vena sacrales laterales
lateral thoracic	vena thoracica lateralis
left colic	vena colica sinistra
left gastric	vena gastrica sinistra
left hepatic	vena hepaticae sinistrae
left inferior pulmonary	vena pulmonalis inferior sinistra
left ovarian	vena ovarica sinistra
left and right brachiocephalic	vena brachiocephalicae (dextrae et sinistra)
left superior intercostal	vena intercostalis superior sinistra
left superior pulmonary	vena pulmonalis superior sinistra
left suprarenal	vena testicularis sinistra
left umbilical	vena umbilicalis
lingual	vena lingularis
long thoracic	vena membri inferioris
lumbar	vena lumbales
mastoid emissary	vena emissaria mastoidea
maxillary	vena maxillaris
medial vein of lateral ventricle	vena medialis ventriculi lateralis
median antebrachial	vena mediana antebrachii
median cubital	vena intermedia cubiti
median sacral	vena sacralis mediana
mediastinal	vena mediastinales
medulla oblongata	vena medullae oblongatae
meningeal	vena meningeae
mesencephalic	vena mesencephalicae
middle cardiac	vena cordis media
middle colic	vena colica media
middle lobe (pulmonary)	vena lobi medii
middle meningeal	vena meningeae

middle rectus	vena rectales mediae
middle temporal	vena temporalis media
middle thyroid	vena thyroidea media
musculophrenic	vena musculophrenicae
nasofrontal	vena nasofrontalis
oblique vein of left atrium	vena obliqua atrii sinistri
obturator	vena obturatoria
occipital	vena occipitalis
occipital cerebral	vena encephali occipitales
occipital emissary	vena emissaria occipitalis
olfactory gyrus	vena gyri olfactorii
palmar digital	vena digitales palmares
palmar metacarpal	vena metacarpeae palmares
palpebral	vena palpebrales
pancreatic	vena pancreaticae
pancreaticoduodenal	vena pancreaticoduodenales
paraumbilical	vena paraumbilicales
parietal	vena parietales
parietal emissary	vena emissarium parietales
parotid	vena parotideae
pectoral	vena pectorales
peduncular	vena pedunculares
perforating	vena perforantes
pericardial	vena pericardiacae
pharyngeal	vena pharyngeae
plantar digital	vena digitales plantares
plantar metatarsal	vena metatarseae plantares
pontine	vena pontis
popliteal	vena poplitea
posterior auricular	vena auricularis posterior
posterior circumflex humeral	vena circumflexa humeri posterior
posterior horn	vena cornus posterioris
posterior intercostal	vena intercostales posteriores
posterior labial	vena labiales posteriores
posterior scrotal	vena scrotales posteriores
posterior vein of septum pellucidum	vena posterior septi pellucidi
posterior vein of left ventricle	vena posterior ventriculi sinistri
posterior tibial	vena tibiales posteriores
precentral cerebelli	vena precentralis cerebelli
prefrontal	vena prefrontales
prepyloric	vena prepylorica
profunda femoris	vena profunda femoris
pterygoid canal vein	vena canalis pterygoidei